COMPUTING METHODS IN APPLIED SCIENCES AND ENGINEERING, VI

Sixth International Symposium on
Computing Methods in Applied Sciences and Engineering

Organized by

Institut National de
Recherche en Informatique et en Automatique (INRIA)

Sponsored by

ADI, GAMNI, IFIP TC7 - WG 7.2

Scientific Secretaries

A. DERVIEUX
J. ERHEL

Secretaries

TH. BRICHETEAU
S. GOSSET

COMPUTING METHODS IN APPLIED SCIENCES AND ENGINEERING, VI

Proceedings of the Sixth International Symposium on Computing Methods in Applied Sciences and Engineering

Versailles, France, December 12-16, 1983

Edited by

R. GLOWINSKI

J.-L. LIONS

Institut National de Recherche en Informatique et en Automatique
Paris, France

INRIA Institut National de Recherche en Informatique et en Automatique

1984

NORTH-HOLLAND
AMSTERDAM · NEW YORK · OXFORD

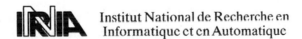

ISBN: 0 444 87597 2

Published by:
ELSEVIER SCIENCE PUBLISHERS B.V.
P.O. Box 1991
1000 BZ Amsterdam
The Netherlands

Sole distributors for the U.S.A. and Canada:
ELSEVIER SCIENCE PUBLISHING COMPANY, INC.
52 Vanderbilt Avenue
New York, N.Y. 10017
U.S.A.

PRINTED IN THE NETHERLANDS

PREFACE

Numerical simulation is a most important field of Science and Engineering. Therefore, every other year, since 1973, the International Symposia on Computing Methods in Applied Sciences and Engineering, organized by INRIA, have gathered the best specialists in the world. Although the sixth symposium has been the continuation of the previous ones, the organizers have tried this time to emphasize on the very topics which are at the interface of numerical mathematics and computer sciences, such as parallel computing. It is clear that vector machines will completely modify the landscape of numerical simulation and will make possible the simulation of phenomena related to turbulence, combustion, physics and three-dimension mechanics, etc... However efficient numerical methods will still have to be used and theoretically analyzed.
The program of the 1983 December Symposium has reflected the remarks mentioned above. The following topics have been discussed within the scope of the numerical simulation:

- Structural Mechanics
- Fluid Mechanics
- Nonlinear Analysis
- Numerical Algebra and Software
- Semi-Conductors
- Reservoir engineering
- Particle and Spectral Methods
- Multigrid Methods
- Parallel Computing.

A full session has been devoted to the numerical treatment of the Euler equations for compressible inviscid fluids.

This book contains the text of the lectures which have been presented during the symposium.
The organizers wish to express their thanks to the three organisations which accepted to sponsor this meeting - Agence de l'Informatique ADI, Groupement pour l'Avancement des Méthodes Numériques de l'Ingénieur GAMNI,

and International Federation for Information Processing IFIP TC7-WG7.2 - to the sessions chairpersons who directed very interesting discussions, to the lecturers and to the 350 scientists and engineers from 22 different countries who attended with interest the sessions of the Symposium.

Sincere gratitude is also expressed both to the INRIA Public Relations Department the help of which greatly contributed to the success of the Symposium and to the Scientific Secretaries, Miss J. ERHEL and Mr. A. DERVIEUX.

R. GLOWINSKI

J.L. LIONS

PREFACE

La simulation numérique joue un rôle fondamental dans les sciences pures et appliquées. C'est la raison pour laquelle, tous les deux ans, depuis 1973, le Colloque International sur les Méthodes de Calcul Scientifique et Technique qu'organise l'INRIA, rassemble les meilleurs spécialistes du monde. Ce sixième colloque est donc le prolongement des précédents, mais les organisateurs ont essayé de mettre en exergue les thèmes qui se trouvent à la frontière des mathématiques numériques et de l'informatique, comme par exemple le calcul parallèle. Il est clair que les ordinateurs vectoriels vont complètement modifier le domaine de la simulation numérique et rendre possible l'étude des phénomènes liés à la turbulence, à la combustion, à la physique et à la mécanique tri-dimensionnelle, etc... Cependant malgré l'existence de ces super-calculateurs, il faudra toujours utiliser des méthodes numériques très performantes et en faire l'analyse théorique.
Le programme du Colloque de Décembre 1983 reflète fidèlement les remarques formulées ci-dessus. Les thèmes suivants ont été étudiés du point de vue de la simulation numérique:

- Mécanique des Structures
- Mécanique des Fluides
- Analyse Non Linéaire
- Algèbre Numérique et Logiciels
- Semi-Conducteurs
- Problèmes Multiphasiques en Gisement de Pétrole
- Méthodes Particulaires et Spectrales
- Méthodes Multigrilles
- Calcul Parallèle.

De plus une session entière a été consacrée au traitement numérique des équations d'Euler pour les fluides parfaits compressibles.

Ce livre contient le texte des Conférences qui ont été présentées durant le Colloque organisé par l'INRIA.

Les organisateurs remercient vivement les trois organismes qui ont bien voulu accorder leur patronage à cette manifestation - Agence de l'Informatique ADI, Groupement pour l'Avancement des Méthodes Numériques de l'Ingénieur GAMNI et International Federation for Information Processing IFIP TC7-WG7.2 - les présidents de séances qui ont dirigé des discussions très intéressantes, les conférenciers et les 350 scientifiques et ingénieurs de 22 pays différents qui ont suivi avec assiduité l'ensemble des sessions.

Leurs remerciements s'adressent également au Service des Relations Extérieures de l'INRIA dont l'aide apportée contribua au succès de ce Colloque, et à Mademoiselle J. ERHEL et Monsieur A. DERVIEUX, Secrétaires Scientifiques.

R. GLOWINSKI J.L. LIONS

TABLE OF CONTENTS

SEMI-CONDUCTORS/
SEMI-CONDUCTEURS

NUMERICAL ALGEBRA AND SOFTWARE

ALGEBRE ET LOGICIEL NUMERIQUES

Computing Methods in Applied Sciences and Engineering, VI
R. Glowinski and J.-L. Lions (Editors)
Elsevier Science Publishers B.V. (North-Holland)

The Use of Pre-Conditioning over Irregular Regions

Gene H. Golub*
Computer Science Department
Stanford University
Stanford, California 94305
USA

David Mayers
Oxford University Computing Laboratory
19 Parks Road
Oxford OX1 3PL
England

Abstract

Some ideas and techniques for solving elliptic pde.'s over irregular regions are discussed. The basic idea is to break up the domain into subdomains and then to use the pre-conditioned conjugate gradient method for obtaining the solution over the entire domain. The solution of Poisson's equaton over a T-shaped region is described in some detail and a numerical example is given.

1. Introduction.

In the last several years, there has been great interest in solving a variety of problems using *domain decomposition* or *substructuring* (cf. [1], [2], [3], [4], [9]). The basic idea is to break up a domain into subdomains and then to solve for an approximate solution on each subdomain. There is a need then to "paste" the approximate solutions together to get the solution to the original problem.

It turns out that a very effective numerical procedure for obtaining the solution is the pre-conditioned conjugate gradient method (cf. [3]). It provides a particularly good method because advantage is taken of the eigenvalue structure of certain matrices.

It is obvious that the ideas of domain decomposition are quite applicable to parallel processing. We will not delve into that subject at this time.

We will concentrate on solving Poisson's equation on a T-shaped region. The general ideas expressed here will be applicable to a much wider class of problems.

2. The Pre-Conditioned Conjugate Gradient Method.

Our discussion in this section closely follows that given in [3]. Consider the system of equations

$$A\mathbf{x} = \mathbf{b} \tag{2.1}$$

where A is a symmetric, positive definite matrix. We re-write (2.1) as

$$M\mathbf{x} = N\mathbf{x} + \mathbf{b}$$

where M is symmetric and positive definite. We assume that given a vector $\mathbf{d}$ it is "easy" to solve the system

$$M\mathbf{z} = \mathbf{d}. \tag{2.2}$$

* The work of this author was in part supported by NSF and the DOE.

The pre-conditioned conjugate gradient (CG) method proceeds as follows.

Algorithm.

Let $\mathbf{x}^{(0)}$ be a given vector and arbitrarily define $\mathbf{p}^{(-1)}$. For $k = 0, 1, \ldots$

(1) Solve $M\mathbf{z}^{(k)} = \mathbf{b} - A\mathbf{x}^{(k)} \equiv \mathbf{r}^{(k)}$.

(2) Compute

$$b_0 = 0\,,$$

$$b_k = \frac{\mathbf{z}^{(k)^T} M\,\mathbf{z}^{(k)}}{\mathbf{z}^{(k-1)^T} M\,\mathbf{z}^{(k-1)}} \qquad k \geq 1$$

$$\mathbf{p}^{(k)} = \mathbf{z}^{(k)} + b_k\,\mathbf{p}^{(k-1)}\,.$$

(3) Compute

$$a_k = \frac{\mathbf{z}^{(k)^T} M\,\mathbf{z}^{(k)}}{\mathbf{p}^{(k)^T} A\,\mathbf{p}^{(k)}}$$

$$\mathbf{x}^{(k+1)} = \mathbf{x}^{(k)} + a_k\,\mathbf{p}^{(k)}\,.$$

Note that in the computation of the numerator of a_k and b_k one need not compute $M\,\mathbf{z}^{(k)}$ since $M\,\mathbf{z}^{(k)} = \mathbf{r}^{(k)}$. Furthermore, the matrix A need not be known explicitly since we need compute the vector $\mathbf{q}^{(k)} = A\,\mathbf{p}^{(k)}$. Finally, we note that the residual $\mathbf{r}^{(k+1)}$ can be computed by the relationship

$$\mathbf{r}^{(k+1)} = \mathbf{r}^{(k)} - a_k\,A\,\mathbf{p}^{(k)}\,.$$

The convergence properties of the conjugate gradient method are fairly well understood. Let

$$K = M^{-1}A$$

and

$$\kappa = \frac{\lambda_{\max}(M^{-1}A)}{\lambda_{\min}(M^{-1}A)}$$

where $\lambda_{\max}(M^{-1}A)$ is the largest eigenvalue of the argument and $\lambda_{\min}(M^{-1}A)$ is the smallest.

The following is known about the CG method.

(A) $$\frac{(\mathbf{x}^{(\ell)} - \mathbf{x})^T A\,(\mathbf{x}^{(\ell)} - \mathbf{x})}{(\mathbf{x}^{((0)} - \mathbf{x})^T A\,(\mathbf{x}^{(0)} - \mathbf{x})} \leq 4\left(\frac{\sqrt{\kappa}-1}{\sqrt{\kappa}+1}\right)^{2\ell}.$$

(B) If $K = M^{-1}A$ has p distinct eigenvalues then the CG method converges in at most p iterations.

Thus if we wish to solve the system

$$(A + \mathbf{u}\mathbf{u}^T)\,\mathbf{x} = \mathbf{b} \qquad (\mathbf{u} \neq \mathbf{0})$$

with $M = A$ and $N = -\mathbf{u}\,\mathbf{u}^T$, the CG method will converge in at most 2 iterations since $M^{-1}N$ is a matrix of rank one and hence has at most two distinct eigenvalues.

Our aim will be to devise matrices M so that κ is as small as possible and that the matrix N be of low rank.

In many cases the matrix A can be written as

$$A = \begin{pmatrix} M_1 & F \\ F^T & M_2 \end{pmatrix}$$

where the systems

$$M_1 \mathbf{z}_1 = \mathbf{d}_1 \qquad \text{and} \qquad M_1 \mathbf{z}_2 = \mathbf{d}_2$$

are easy to solve and for such matrices, it is convenient to choose

$$M = \begin{pmatrix} M_1 & 0 \\ 0 & M_2 \end{pmatrix}.$$

Suppose we are given $\mathbf{x}_1^{(0)}$ and we compute $\mathbf{x}_2^{(0)}$ so that

$$M_2 \mathbf{x}_2^{(0)} = \mathbf{r}_2^{(0)}.$$

A short calculation shows that $\mathbf{z}_1^{(1)} = \mathbf{0}$ and hence $\mathbf{a}_1 = 1$. Generalizing a result of Reid [7], it is shown in [3], then for $j = 0, 1, \ldots$

$$a_j = 1, \quad \mathbf{z}_1^{(2j+1)} = \mathbf{0}, \quad \mathbf{z}_2^{(2j)} = \mathbf{0}. \tag{2.3}$$

Using (2.3), we are able to eliminate roughly half the number of operations (cf. [7]).

3. Some Preliminary Results.

In many applications (cf. [5], [6]), matrices of the following structure arise quite naturally:

$$A = \begin{pmatrix} A_1 & & & & B_1 \\ & \cdot & & \bigcirc & B_2 \\ & & \cdot & & \vdots \\ & \bigcirc & & \cdot & \\ & & & A_r & B_r \\ B_1 & \cdot & \cdot & B_r & Q \end{pmatrix}.$$

The matrix A is symmetric and positive definite so that each A_i $(n_i \times n_i)$ is also positive definite and so is the $p \times p$ matrix Q.

Suppose we wish to solve the system

$$A\mathbf{x} = \mathbf{b} \tag{3.1}$$

where

$$\mathbf{x} = \begin{pmatrix} \mathbf{x}_1 \\ \mathbf{x}_2 \\ \vdots \\ \mathbf{x}_r \\ \xi \end{pmatrix} \qquad \text{and} \qquad \mathbf{b} = \begin{pmatrix} \mathbf{b}_1 \\ \mathbf{b}_2 \\ \vdots \\ \mathbf{b}_r \\ \mathbf{c} \end{pmatrix}.$$

One possibility is to apply the CG method to (3.1) with

$$M = \begin{pmatrix} A_1 & & & \\ & \cdot & & \bigcirc \\ & & \cdot & \\ & \bigcirc & & A_r & \\ & & & & Q \end{pmatrix}.$$

Note the matrix $M^{-1}N$ is 2-cyclic and as indicated in Section 2 various simplifications can be made in the procedure and about half the computations can be eliminated.

Let us examine the eigenvalues of $M^{-1}N$. Consider the matrix equation

$$M^{-1}N\mathbf{u} = \lambda\mathbf{u}.$$

Then if $\mathbf{u} = \begin{pmatrix} \mathbf{u}_1 \\ \vdots \\ \mathbf{u}_r \\ \mathbf{v} \end{pmatrix}$, a short manipulation shows

$$\sum_{i=1}^{r} B_i^T A_i^{-1} B_i \mathbf{v} = \lambda^2 Q \mathbf{v}.$$

Instead of solving (3.1) directly, we could eliminate $\mathbf{x}_1, \mathbf{x}_2, \ldots, \mathbf{x}_r$ and solve for ξ. This leads to the equation

$$\left(Q - \sum_{i=1}^{r} B_i^T A_i^{-1} B_i\right)\xi = \mathbf{c} - \sum_{i=1}^{r} B_i^T A_i^{-1} \mathbf{b}_i. \tag{3.3}$$

We can apply the CG method to (3.3) and if Q is a $p \times p$ matrix we will obtain the solution in at most p iterations. The matrix

$$G \equiv Q - \sum_{i=1}^{r} B_i^T A_i^{-1} B_i$$

is the Schur complement of Q in A and hence G is symmetric and positive definite.

Associated with (3.3), we can choose a pre-conditioner $\tilde{M}$. There are various choices of $\tilde{M}$. One possibility is to choose $\tilde{M} = I$. Perhaps a more natural pre-conditioner is $\tilde{M} = Q$. Note if

$$\tilde{M}\xi = \tilde{N}\xi + \tilde{\mathbf{b}},$$

then the convergence properties of the algorithm are determined by the eigenvalues of $\tilde{M}^{-1}\tilde{N}$. Let

$$\tilde{M}^{-1}\tilde{N}\mathbf{w} = \gamma\mathbf{w}.$$

Then if $\tilde{M} = Q$, we have

$$\sum_{i=1}^{r} B_i^T A_i^{-1} B_i \mathbf{w} = \gamma Q \mathbf{w}. \tag{3.4}$$

Hence the eigenvalues of (3.2) are the squares of the eigenvalues of (3.4). The rate of convergence of the two procedures are essentially the same because we are able to eliminate half of the numerical operations. In Section 4, we shall indicate how we can further improve the convergence properties by choosing various pre-conditioners related to the differential equations.

In order to solve (3.3) with $\tilde{M} = Q$, we need to solve the system

$$Q\mathbf{z}^{(k)} = \mathbf{c} + \sum_{i=1}^{r} B_i^T A_i^{-1}(B_i\eta - \mathbf{b}_i)$$

where η is an approximation to the solution. Note that one need not compute the matrix $B_i^T A_i^{-1} B_i$ but instead one solves the system

$$A_i \varphi_i = B_i \eta - \mathbf{b}_i .$$

If A_i has a simple structure this can be solved without much difficulty.

4. The Model Problem.

We shall explain the method by applying it to the problem studied in [3]. We wish to solve

$$\text{(4.1)} \qquad \begin{cases} -\Delta u = f & (x,y) \in T \\ u = g & (x,y) \in \partial T \end{cases}$$

where T is the T-shaped domain shown in Figure 1. We use a uniform square grid of size h, where $1/2h$ is an integer N, and $0 < \ell < N$. Using the standard 5-point finite difference approximation we obtain the system of equations in matrix form

$$\text{(4.2)} \qquad \begin{pmatrix} P & J & O \\ J^T & R & K^T \\ O & K & Q \end{pmatrix} \begin{pmatrix} \mathbf{u} \\ \mathbf{w} \\ \mathbf{v} \end{pmatrix} = \begin{pmatrix} \mathbf{a} \\ \mathbf{b} \\ \mathbf{c} \end{pmatrix} .$$

The vector $\mathbf{u}$ comprises the unknown values of the solution at the interior points of the lower square, $\mathbf{v}$ comprises the unknowns in the interior of the upper square, and $\mathbf{w}$ comprises the unknown values at grid points on the dividing line. Thus P, Q, and R are square matrices of orders $(2N-1)^2$, $(2\ell-1)^2$, and $(2\ell-1)$ respectively.

There are several possible pre-conditioners one can use for solving (4.2). The obvious choices are

$$M^{(1)} = \begin{pmatrix} P & J & O \\ J^T & R & O \\ O & O & Q \end{pmatrix}$$

and

$$M^{(2)} = \begin{pmatrix} P & O & O \\ O & R & O \\ O & O & Q \end{pmatrix} .$$

The pre-conditioner $M^{(1)}$ requires that for each iteration two fast Poisson solvers are used whereas for pre-conditioner $M^{(2)}$ two fast Poisson solvers are used and a system of equations involving the matrix Q must be solved. Which pre-conditioner is to be preferred? Using a theorem of Varga [8, pg. 90], we see that the spectral radius is smaller when $M^{(1)}$ is used as the pre-conditioner and hence $M^{(1)}$ is to be preferred in many situations.

Theorem (Varga):
Let $A = M^{(1)} - N^{(1)} = M^{(2)} - N^{(2)}$ be two regular splittings of A where $A^{-1} > 0$. If $N^{(2)} \geq N^{(1)} \geq 0$, equality excluded, then

$$1 \geq \rho({M^{(2)}}^{-1} N^{(2)}) \geq \rho({M^{(1)}}^{-1} N^{(1)}) > 0 .$$

Here $\rho({M^{(i)}}^{-1} N^{(i)})$ denotes the spectral radius of the argument. Even though it appears that $M^{(1)}$ is to be preferred over $M^{(2)}$, we will consider below a pre-conditioner related to $M^{(2)}$!

We obtain the **capacitance matrix** by eliminating $\mathbf{u}$ and $\mathbf{v}$ from these equations, giving

$$\text{(4.3)} \qquad (R - J^T P^{-1} J - K^T Q^{-1} K)\mathbf{w} = \mathbf{b} - J^T P^{-1} \mathbf{a} - K^T Q^{-1} \mathbf{c}$$

which we shall write

$$C\mathbf{w} = \mathbf{d}. \tag{4.4}$$

As indicated in Section 3, it follows at once that C is symmetric and positive definite.

We now wish to solve (4.4) by a pre-conditioned conjugate gradient method. If we can find a suitable pre-conditioning matrix this iterative method converges very rapidly, and it will certainly terminate in at most $(2\ell - 1)$ iterations, this being the order of the matrix C.

We now consider the computation of $C\mathbf{p}^{(k)}$. Returning to the original equation (4.2) it will be seen that the matrices P and Q correspond to the solution of the discrete form of Poisson's equation on the lower and upper squares respectively, given values on the boundaries of these squares. The matrices J and K consist of unit matrices, augmented by blocks of zeros. Hence to compute $C\mathbf{p}^{(k)}$ we need to solve the discrete Laplace equation in the two squares, given zero boundary conditions at all boundary points, except at the points on the dividing line between the two squares, where the given boundary values are the elements of the vector $\mathbf{p}^{(k)}$. The required vector $C\mathbf{p}^{(k)}$ then consists of the residuals of the 5-point formula evaluated at each of these dividing points. The two discrete Laplace problems are both Dirichlet problems on a rectangle, and can be solved very efficiently by a fast Poisson Solver, using some form of Fast Fourier Transform.

For the most rapid rate of convergence we wish the matrix M to be close to C, and in particular for the eigenvalues of $M^{-1}C$ to be clustered as closely as possible. If we examine the elements of C in some particular cases we find that it is quite close to a Toeplitz form, with the element C_{ij} being mainly a function of $|i - j|$ only, and with the largest element on the diagonal, the elements decreasing quite rapidly as $|i - j|$ increases.

This suggests that the elements of C do not depend very much on the shape and size of the two rectangles in Figure 1, and that we could find a useful approximation to C by letting the boundaries of the two squares move away to infinity. We must then find solutions of Laplace's equation in the two half planes. In this approximation the elements C_{ij} are in fact dependent only on $|i - j|$, and the elements in each row are obtained by solving the discrete Laplace equation in a half plane, the solution being required to vanish at infinity, and also at all points on the axis except at the origin, at which it is equal to 1. We thus wish to solve

$$u_{r,s+1} + u_{r,s-1} + u_{r+1,s} + u_{r-1,s} - 4u_{rs} = 0, \quad s > 0$$
$$u_{rs} \to 0 \quad \text{as} \quad r \to \pm\infty, \quad \text{and as} \quad s \to \infty$$
$$u_{r,0} = 0 \quad (r \neq 0)$$
$$u_{0,0} = 1.$$

Defining the generating function

$$\phi_s(t) = \sum_{r=-\infty}^{\infty} t^r u_{rs}$$

we thus obtain

$$\phi_{s+1} + \phi_{s-1} + \left(t + \frac{1}{t} - 4\right)\phi_s = 0, \qquad (s > 0)$$
$$\phi_0 = 1.$$

The general solution of the recurrence relation for ϕ_s is

$$\phi_s = A_1\lambda_1^s + A_2\lambda_2^s$$

where A_1 and A_2 are arbitrary constants, and λ_1 and λ_2 are the roots of

$$\lambda^2 + \left(t + \frac{1}{t} - 4\right)\lambda + 1 = 0 .$$

The condition at infinity eliminates one of these two terms, and the condition $\phi_0 = 1$ determines the remaining arbitrary constant, giving the solution

$$\phi_s(t) = \left[2 - \frac{1}{2}\left(t + \frac{1}{t}\right) - \left(\left\{2 - \frac{1}{2}\left(t + \frac{1}{t}\right)\right\}^2 - 1\right)^{1/2}\right]^s .$$

The solution in the other half plane is of course determined by symmetry, and the residuals at the grid points on the axis are given by

$$\begin{aligned}\rho_r &= u_{r-1,0} + u_{r+1,0} + u_{r,1} + u_{r,-1} - 4u_{r,0} \\ &= u_{r-1,0} + u_{r+1,0} + 2u_{r,1} - 4u_{r,0}\end{aligned}$$

for which the generating function is

$$\begin{aligned}\psi(t) &= \left[t + \frac{1}{t} - 4\right]\phi_0 + 2\phi_1 \\ &= -2\left\{\left[2 - \frac{1}{2}\left(t + \frac{1}{t}\right)\right]^2 - 1\right\}^{1/2} .\end{aligned} \tag{4.5}$$

To determine ρ_r we now expand $\psi(t)$ in positive and negative powers of t; ρ_r is then the coefficient of t^r. The simplest way to do this is to write $t = e^{i\theta}$, $t + \frac{1}{t} = 2\cos\theta$, and then expand $\psi(t)$ as a Fourier cosine series in $\cos\theta$. This gives

$$\begin{aligned}\psi &= -2\{[2 - \cos\theta]^2 - 1\}^{1/2} \\ &= \frac{1}{2}\rho_0 + 2\Sigma\, \rho_r \cos r\theta .\end{aligned}$$

It is then convenient to make the substitution $\theta = 2\alpha$ and obtain the final result

$$\begin{aligned}\rho_r &= \frac{1}{2\pi}\int_{-\pi}^{\pi} -2\cos r\theta [(2 - \cos\theta)^2 - 1]^{1/2}\, d\theta \\ &= -\frac{4}{\pi}\int_0^{\pi} \cos 2r\alpha \sin\alpha [1 + \sin^2\alpha]^{1/2}\, d\alpha .\end{aligned} \tag{4.6}$$

These integrals are easily evaluated numerically for small values of r. For moderate and large values of r, it is sufficient to integrate twice by parts and obtain the asymptotic approximation

$$\rho_r = \frac{2}{\pi r^2} + O\left(\frac{1}{r^4}\right).$$

A possible pre-conditioning matrix M is then obtained by writing

$$M_{ij} = \rho_{|i-j|} .$$

This matrix has in fact been found to give good convergence, but a minor modification improves it still further. Dryja [4] used as pre-conditioning matrix

$$M = K^{1/2}$$

where K is the matrix

$$K = \begin{pmatrix} 2 & -1 & & \\ -1 & 2 & -1 & \\ & -1 & 2 & -1 \\ & & & \cdots \end{pmatrix}.$$

In our notation this corresponds to a different generating function

$$\psi_D = \left[-t + 2 - \frac{1}{t}\right]^{1/2}.$$

This suggests replacing $2 - t - \frac{1}{t}$ by K in the generating function. A multiplying scale factor is unimportant for our purpose, so for comparison we use instead of Dryja's form, $M = -(4K)^{1/2}$, and from(4.5) we suggest trying also

$$-2\left\{\left(1 + \frac{1}{2}K\right)^2 - 1\right\}^{1/2} = -\{4K + K^2\}^{1/2}.$$

The eigenvalues and eigenvectors of the matrix K are well known, so it is a simple matter to write down the elements of these matrices giving three possible pre-conditioners:

$$M^{(1)}_{ij} = \rho_{|i-j|}$$

where ρ_r is given by (4.6)

$$M^{(2)}_{ij} = -\frac{8}{n}\sum_{k=1}^{n-1} \sin\left(\frac{ik\pi}{n}\right) \sin\left(\frac{k\pi}{2n}\right) \sin\left(\frac{jk\pi}{n}\right)$$

$$M^{(3)}_{ij} = -\frac{8}{n}\sum_{k=1}^{n-1} \sin\left(\frac{ik\pi}{n}\right) \sin\left(\frac{k\pi}{2n}\right) \left[1 + \sin^2\left(\frac{k\pi}{2n}\right)\right] \sin\left(\frac{jk\pi}{n}\right)$$

where n is the order of the matrix, in our case $(2\ell - 1)$. Here $M^{(2)}$ is Dryja's pre-conditioner, and $M^{(3)}$ is our modification. Owing to the particular form of $M^{(2)}$ and $M^{(3)}$, the solution of a system of equations $M\mathbf{z} = \mathbf{r}$ is particularly simple in these two cases, and can be done by a simple application of the Fast Fourier Transform.

For the model problem of Figure 1 we can now evaluate the eigenvalues of the matrix $M^{-1}C$ for each of the three pre-conditioning matrices M. We have used $\ell = \frac{1}{2}N$, with $N = 8$ and 16, giving two sets of matrices, of orders 7 and 15 respectively, for which the eigenvalues are given in Table 1. It will be seen that in each case the eigenvalues are clustered quite closely about 1, the clustering being most marked for the matrix $M^{(3)}$.

Table 2 gives some results which illustrate the rate of convergence of the method. The domain is as in Figure 1, and the function f and the boundary conditions g are chosen to give a smooth solution as in [3]. The entries in Table 2 are the values of $(\mathbf{z}^k, M\mathbf{z}^k)$, which give an indication of how fast z^k, and hence r^k, is tending to zero. The initial approximation was constructed by simply taking $\mathbf{w}^{(0)} = \mathbf{0}$, corresponding to making $\mathbf{u} = \mathbf{0}$ at these grid points.

Acknowledgment. Much of the work of the first author was performed while a guest at the Oxford University Computing Lab. He is pleased to thank Professor Leslie Fox for his generous hospitality.

The authors wish to thank Gérard Meurant and Petter Bjøstad for their helpful comments.

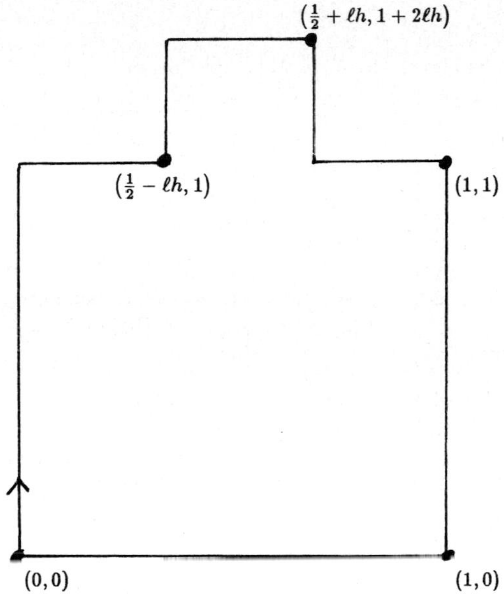

Figure 1

Table 1

Eigenvalues of $M^{-1}C$, for three matrices M

dimension 7

$M^{(1)}$	$M^{(2)}$	$M^{(3)}$
1.29279	1.40048	1.00000
1.04310	1.36048	1.00000
1.00953	1.29815	0.99999
1.00135	1.21928	0.99968
0.99771	1.13432	0.99736
0.98384	1.04073	0.96727
0.98265	0.93631	0.91185

dimension 15

$M^{(1)}$	$M^{(2)}$	$M^{(3)}$
1.66158	1.41079	1.00000
1.15918	1.40058	1.00000
1.05216	1.38385	1.00000
1.01339	1.36098	1.00000
1.01081	1.33257	1.00000
1.00885	1.29930	1.00000
1.00617	1.26220	1.00000
1.00090	1.22217	1.00000
0.99844	1.18079	1.00000
0.99125	1.13894	0.99995
0.99092	1.09911	0.99971
0.98361	1.06133	0.99731
0.98262	1.02975	0.98958
0.97753	0.96949	0.93837
0.97735	0.89807	0.88376

Table 2

Values of $(\mathbf{z}^{(k)}, M\mathbf{z}^{(k)})$, for three pre-conditioning matrices M

dimension 7

k	$M^{(1)}$	$M^{(2)}$	$M^{(3)}$
1	0.4, -1	0.9, -1	0.3, -2
2	0.2, -4	0.7, -3	0.5, -6
3	0.4, -8	0.4, -5	0.2, -11
4	0.3, -12	0.3, -7	
5		0.6, -10	
6		0.2, -12	

dimension 15

k	$M^{(1)}$	$M^{(2)}$	$M^{(3)}$
1	0.3	0.1	0.7, -2
2	0.5, -3	0.1, -2	0.3, -5
3	0.2, -5	0.1, -4	0.5, -9
4	0.3, -9	0.1, -6	
5		0.1, -8	
6		0.1, -10	

References

[1] Bjørstad, P. and O. Widlund, "Solving elliptic problems on regions partitioned into substructures," in *Elliptic Problem Solvers*, G. Birkhoff and A. Schoenstadt, eds., Academic Press, New York, 1984.

[2] Buzbee, B. L., Dorr, F. W., George, J. A. and Golub, G. H., "The direct solution of the discrete Poisson equation on irregular region," *SIAM J. Numer. Anal.* **8** (1971), pp. 722–736.

[3] Concus, P., Golub, G. and O'Leary, D., "A generalized conjugate gradient method for the numerical solution of elliptic partial differential equations," in *Sparse Matrix Computation*, J. Bunch and D. Rose, eds., Academic Press, New York, 1976, pp. 309–322.

[4] Dryja, M., "A capacitance matrix method for Dirichlet problems on polygonal region," *Num. Math.*, (1982), pp. 51–64.

[5] Golub, G., H., Luk, F. and Pagano, M., "A large sparse least squares problem in photogrammetry," *Proceedings of the 12th Annual Symposium on the Interface*, University of Waterloo, May, 1979.

[6] Golub, G. H. and Plemmons, R., "Sparse least squares problems," *Proceedings of the IRIA Fourth International Symposium on Computing Methods in Applied Science and Engineering*, Versailles, France, Dec. 1979.

[7] Reid, J., "The use of conjugate gradients for systems of linear equations possessing 'Property A'," *SIAM J. Numer. Anal.* **9** (1972), pp. 325–332.

[8] Varga, Richard S., *Matrix Iterative Analysis*, Prentice-Hall, Englewood Cliffs, New Jersy, 1962.

[9] Dingh, Q. V., Glowinski, R. and Periaux, J., "On the solution of elliptic problems by domain decomposition methods Applications," *manuscript*.

Computing Methods in Applied Sciences and Engineering, VI
R. Glowinski and J.-L. Lions (Editors)
Elsevier Science Publishers B.V. (North-Holland)

MATRIX COMPUTATIONAL PROCESSES IN SUBSPACES

Yu.A.Kuznetsov

Department of Computational Mathematics
USSR Academy of Sciences
Moscow, USSR

In recent years was formed and now is intensely developed the new direction in solution of the finite dimensional problems. This direction is called the computational processes in subspaces. The present paper shows some results concerning applications of these methods to solution of systems of linear algebraic equations with large sparse matrices arising in finite difference and finite element discretizations of partial differential equations.

1. PARTIAL SOLUTIONS

Consider the system of linear algebraic equations

$$Au = f \tag{1}$$

with real $N \times N$-matrix A and with vector $f \in \operatorname{im} A = AE_N$, where E_N is the space of N-dimensional real vectors with conventional scalar product $(\, , \,)$ and with the norm $\| \cdot \| = (\, , \,)^{1/2}$. Let $U_A \subseteq \operatorname{im} A$ be some subspace of E_N, and R be some $M \times N$-matrix with rank $R < N$. Here M and N are positive integers with $M \times N$.

<u>Definition</u>. For given A, R, $f \in U_A$ and some solution u of (1) the vector $v = Ru$ is called the partial solution of this system.

If $\det A \neq 0$, then partial solution is equal to

$$v = RA^{-1}f. \tag{2}$$

The problems of finding these solutions often arise in solving partial differential equations by finite difference of finite element methods. The corresponding examples will be given later. In this section we shall present the algorithms for finding the partial solutions for systems, arising in approximations of the Poisson equation in rectangle.

Consider the problem

$$\begin{aligned} -\Delta u + cu &= f, \quad x = (x_1, x_2) \in G, \\ \alpha \frac{\partial u}{\partial \nu} + \beta u &= g, \quad x \in \partial G, \end{aligned} \tag{3}$$

where G is a rectangle with sides parallel to coordinate axes and with boundary ∂G, and where f and g are given functions, $c = \text{const} \geqslant 0$, α and β are some non-negative functions equal to constant at each side of G and satisfying the condition $\alpha + \beta > 0$, ν is the outward normal to ∂G. Let us construct in G the rectangular mesh and approximate (3) by a finite difference method. This will bring us to the system (1) with matrix

$$A = A_1 \otimes I_2 + I_1 \otimes A_2 \tag{4}$$

of order $N = N_1 \times N_2$, where A_1 and A_2 are three-diagonal Jacobi matrices of orders N_1 and N_2, respectively.

Remark. Here and further the symbol I stands for unit matrices. Moreover, it will be assumed throughout the paper that A is non-singular, and $N_i = O(N^{1/2})$ when $N \to \infty$, $i = 1, 2$.

Introduce the diagonal matrices P_1 and P_2 with diagonal elements equal to 0 or 1 and with rank $P_i = O(N^{1/2})$, $i = 1,2$. Then the following lemma is valid [3] (the same result is established in [14], [19] for particular case).

Lemma 1. Let $U_A = \text{im } P_1$ and $R = P_2$. Then partial solution $v = P_2 A^{-1} f$ of system (1), (4) is computed by separation of variables method with $O(N)$ arithmetic operations and $O(N^{1/2})$ computer memory.

Now consider a more simple problem. Let at sides of G parallel to Ox_1 axis $\alpha\beta = 0$, the mesh on the Ox_2 axis is uniform, and the non-zero elements of P_1 and P_2 correspond only to two mesh lines parallel to Ox_2 (the lines may coincide). The problem is to find the partial solutions $P_2 A^{-1} f$ of system (1) with right sides $f \in \text{im } P_1$. Then from [14], [19] we have the following

Lemma 2. Solution of the formulated problem by separation of variables method requires

$$O(N) + r \times O(N^{1/2} \ln N)$$

arithmetic operations with $O(N^{1/2})$ computer memory.

This result can easily be extended to the case of the non-zero elements of P_1 and P_2 corresponding to several parallel lines when the number of these lines remains bounded as $N \to \infty$. Fast algorithms for partial solutions in the case of orthogonal mesh lines were considered in [4], [30].

2. FAST DIRECT METHODS

Consider the problem of solving (1),(4) with right side $f \in E_N$ when the mesh is piecewise uniform in both directions, i.e. for each variable x_i, $i = 1,2$, the number of nodes at which the mesh size is altered remains bounded at $N \to \infty$. More precisely, we assume that the mesh size at the variable x_i is altered $m_i = O(1)$, $i = 1,2$, times. This means that the domain G can be divided into $(m_1 + 1) \times (m_2 + 1)$ rectangles $G_{s,t}$, $s = 0,\ldots,m_1$, $t = 0,\ldots,m_2$, as shown at Fig. 1, and in each rectangle the mesh remains uniform

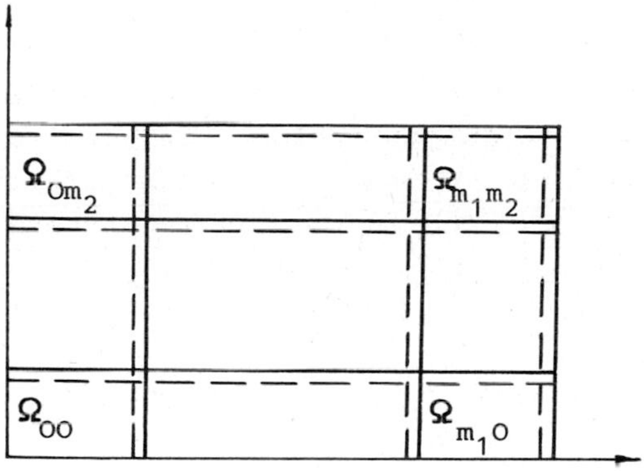

Fig. 1

Following [19], let us describe the algorithm for solution of this problem. The algorithm consists of three steps.

<u>Step 1</u>. For $s = 0,1,\ldots,m_1$ and $t = 0,1,\ldots,m_2$ we place the rectangles Ω_{st} in which the mesh sizes are equal to $h_{1,s}$ and $h_{2,t}$ at the directions x_1 and x_2, respectively, into rectangles $\tilde{\Omega}_{st}$ with sides of length $h_{1,s} \times 2^{k_s}$ and $h_{2,t} \times 2^{l_t}$, respectively, where k_s and l_t are minimal positive integers for which these embeddings are possible (see Fig. 2).

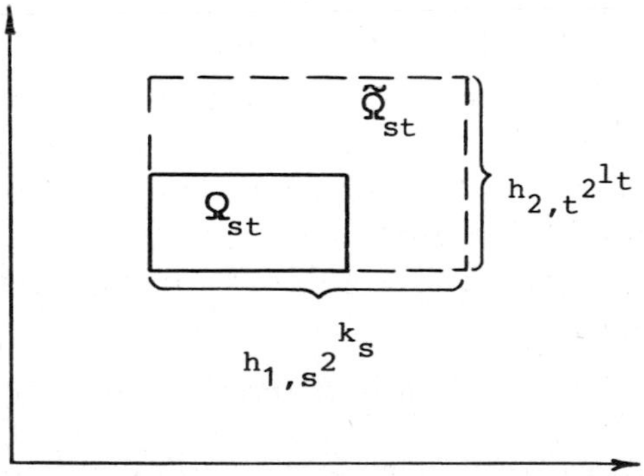

Fig. 2

Then in domains $\tilde{\Omega}_{st}$ we solve the systems corresponding to finite difference approximations of Poisson equation with homogeneous Dirichlet conditions at the boundary and with the mesh sizes $h_{1,s}$ and $h_{2,t}$ in corresponding directions. The right sides f_{st} for these systems are equal to f in the nodes belonging to Ω_{st} and are equal to zero **in the** remaining nodes. When all these systems are solved, the vector $\tilde{u}$ is formed with its components equal to computed values at the internal nodes of Ω_{st} and equal to zero at all the nodes belonging to the boundaries of Ω_{st}. Denote $u = \tilde{u} + w$. Then for w we have the system

$$Aw = F \tag{5}$$

with the right side $F = f - A\tilde{u} \in P_1 E_N$, where the diagonal elements of P_1 are non-zero for the nodes belonging to the sides of rectangle G ($\alpha \neq 0$), to the boundaries of rectangles Ω_{st} inside G, and to adjacent lines (dotted at Fig. 1).

If the FACR algorithm [29] is used for solution of equations in Ω_{st} (now the faster algorithms are available), then the vectors $\tilde{u}$ and F can be computed by $O(N \ln \ln N)$ operations.

Step 2. This step consists in partial solution of (5) with matrix $P_1 = P_2$. Due to the lemma 1 this would require $O(N)$ arithmetic operations and $O(N^{1/2})$ of computer memory.

Step 3. After all the components of w corresponding to the nodes at the boundaries of Ω_{st} and the adjacent lines dotted at Fig. 1 are computed, the remaining components of w are found step by step in the domains Ω_{st} using the algorithm of Step 1, by $O(N \ln \ln N)$ arithmetic operations.

This algorithm can be summarized as follows.

Theorem 1. The system (1),(4) can be solved by the proposed direct method for $O(N \ln \ln N)$ arithmetic operations.

The method was published in a slightly different formulation in [19].

3. GENERALIZED CONJUGATE GRADIENTS METHOD IN SUBSPACE

Let some non-singular matrix B be given, and $C = B - A$. Then the subspace

$$U_A = \text{im } [(I - AB^{-1})\ A] \subseteq \text{im } C \cap \text{im } A \tag{6}$$

is invariant with respect to matrix $S = AB^{-1} = I - CB^{-1}$, so that the vectors $\xi^k = Au^k - f$ of any iterative method

$$\xi^0 \in U_A,\ u^k = u^0 - B^{-1} \sum_{i=1}^{k} \gamma_{k,i}\, S^{i-1} (Au^0 - f), \tag{7}$$

where $\gamma_{k,i}$ are some real numbers, belong to U_A.

If the iterative method (7) converges, i.e. for every initial vector $u^0 \in E_N$, for which $\xi^0 \in U_A$, the sequence $\{u^k\}$ in (7) converges to some solution u' of the system (1), then the partial solution Cu' of this system can be found from the iterative process

$$Cu^k = Cu^0 - CB^{-1} \sum_{i=1}^{k} \gamma_{k,i} (I - CB^{-1})^{i-1} \xi^0.$$

Note that the condition $f \in U_A$ is replaced here by the condition $\xi^0 \in U_A$, which corresponds to transformation of the original system (1) to the system $Aw = F$ with vectors $w = u - u^0$ and $F = Au^0 - f \in U_A$. Then the iterative process for Cw has the form

$$Cw^k = - CB^{-1} \sum_{i=1}^{k} \gamma_{k,i} (I - CB^{-1})^{i-1} F$$

Assume that for given matrix B and subspace U_A the symmetric positive definite in U_A matrix D is known, for which S is D-symmetric and D-positive definite operator in the subspace U_A, i.e. for any $\xi, \eta \in U_A$

$$(S\xi, \eta)_D = (DS\xi, \eta) = (\xi, DS\eta) = (\xi, S\eta)_D$$

and for any nonzero $\xi \in U_A$ $(S\xi, \xi)_D > 0$. Then the partial solution Cu' of system (1) can be found [21] (see e.g. [24], [23]) by generalized conjugate gradients method in subspace U_A:

$$\xi^0 = CB^{-1} (A\varphi - f) \in U_A,$$

$$Cg_k = \begin{cases} CB^{-1} \xi^0, & k=1, \\ CB^{-1} \xi^{k-1} - \alpha_k Cg_{k-1}, & k>1, \end{cases} \qquad \alpha_k = \frac{((I-CB^{-1})\xi^{k-1}, Ag_k)_D}{\|Ag_k\|_D^2} = -\frac{\|\xi^{k-1}\|_{DS}^2}{\|\xi^{k-2}\|_{DS}^2},$$

$$Ag_k = \begin{cases} (I-CB^{-1})\, \xi^0 & k=1, \\ (I-CB^{-1})\, \xi^0 - \alpha_k Ag_{k-1}, & k>1, \end{cases} \tag{8}$$

$$Cu^k = Cu^{k-1} - \beta_k Cg_k,$$

$$\xi^k = \xi^{k-1} - \beta_k Ag_k, \qquad \beta_k = \frac{(\xi^{k-1}, Ag_k)_D}{\|Ag_k\|_D^2} = \frac{\|\xi^{k-1}\|_{DS}^2}{\|Ag_k\|_D^2},$$

$$k = 1, 2, \ldots, \hat{k}.$$

Here φ is an arbitrary vector from E_N and $\hat{k}$ is the number of steps of the method, sufficient either for computation of the exact vector Cu', or to obtain the inequality

$$\|Au^{\hat{k}} - f\|_D \leq \varepsilon \|Au^0 - f\|_D . \tag{9}$$

with $\varepsilon < 1$. In the last case we will write $\hat{k} = k_\varepsilon$. It is clear that for $A = A^T > 0$, $B = I$, $D = A^{-1}$ and $U_A = E_N$ the method (8) is the classical conjugate gradients method [10].

Theorem 2. Let for given $\xi^0 \in U_A$ the value $p = p(\xi^0)$ denote the maximal integer for which the vectors $\xi^0, S\xi^0, \ldots, S^{p-1}\xi^0$ are linearly independent. Then for every $k \leq p$ the vectors ξ^{k-1}, Ag_k are nonzero, belong to U_A and satisfy the orthogonality conditions

$$(Ag_k, Ag_l)_D = \delta_{k,l} \|Ag_k\|_D^2 , \quad l = \overline{1,k},$$

$$(\xi^k, Ag_l)_D = 0 , \quad l = \overline{1,k}, \tag{10}$$

$$(S\xi^k, \xi^l)_D = 0 , \quad l = 0,\overline{k-1},$$

where $\delta_{k,l}$ is the Kronecker symbol.

Corollary. For all $k \leq p$ the inequalities $\alpha_k < 0$, $\beta_k > 0$ are valid.

Using the results of [6], [10], [27] it can be shown [17], [24] that the method (8) can be implemented with three-term formulae

$$Cu^k = Cu^{k-1} - \frac{1}{q_k} [CB^{-1}\xi^{k-1} - e_{k-1}(Cu^{k-1} - Cu^{k-2})],$$

$$\xi^k = \xi^{k-1} - \frac{1}{q_k} [(I - CB^{-1})\xi^{k-1} - e_{k-1}(\xi^{k-1} - \xi^{k-2})], \tag{11}$$

where $e_0 = 0$ and

$$q_k = \frac{\|S\xi^{k-1}\|_D^2}{\|\xi^{k-1}\|_{DS}} - e_{k-1} , \quad e_k = q_k \frac{\|\xi^k\|_{DS}^2}{\|\xi^{k-1}\|_{DS}^2} . \tag{12}$$

Lemma 3. Under given assumptions the subspace U_A is the linear span of the D-orthonormal system of eigenvectors of matrix S, corresponding to its positive eigenvalues $\lambda_1 \leq \lambda_2 \leq \ldots \leq \lambda_m$ not equal to one, where $m = \dim U_A$.

Note, that $\lambda = 1$ can be one of the eigenvalues of S, however it is excluded from the set λ_i, $i = 1,\dots,m$ due to conditions on S and U_A.

The following theorem (see e.g. [11], [23]) is valid.

Theorem 3. Let all the eigenvalues λ_i, $i = 1,\dots,m$, of the matrix S from the last lemma belong to the interval $[a,b]$ with positive a and b, $a < b$. Then for any $k \leqslant p$ the estimate

$$\| \xi^k \|_D \leqslant \frac{1}{T_k\left(\frac{b+a}{b-a}\right)} \| \xi^0 \|_D , \tag{13}$$

is valid for the method (8), where $T_k(x)$ is the Chebyshev polynomial of the first kind and of order k, and

$$T_k\left(\frac{b+a}{b-a}\right) = \frac{1 + q^{2k}}{2q^k} ,$$

with $q = (1 - d^{1/2}) / (1 + d^{1/2})$, $d = a/b$.

Corollary. In order to reduce the D-norm of the initial residual vector, or, equivalently, the A^TDA-norm of the initial error vector $\psi^0 = u^0 - u'$ by $1/\varepsilon$ times ($\varepsilon < 1$), it is sufficient to make k_ε steps of method (8), where k_ε is the minimal integer satisfying the inequality

$$k_\varepsilon \geqslant \ln\left[\frac{1}{\varepsilon} + \sqrt{\frac{1}{\varepsilon^2} - 1}\right] \Big/ \ln\left[\frac{1 + d^{1/2}}{1 - d^{1/2}}\right]. \tag{14}$$

It follows from (14) that for $d, \varepsilon << 1$ the k_ε is given by the formula

$$k_\varepsilon \simeq \frac{d^{-1/2}}{2} \ln \frac{2}{\varepsilon} . \tag{15}$$

Note, that if for some $k \geqslant 1$ the vectors Cu^k and ξ^k are known, then the vector u^k satisfies the system

$$Bu^k = Cu^k + f + \xi^k. \tag{16}$$

Moreover, if $\rho(B^{-1}C) \leqslant 1$, then instead of approximate solution $u^{\hat{k}}$ of system (1) one can look for the vector $v^{\hat{k}}$ being the solution of the system

$$Bv^{\hat{k}} = Cu^{\hat{k}} + f , \tag{17}$$

for which the following inequality takes place,

$$\| Av^{\hat{k}} - f \|_D < \| Au^{\hat{k}} - f \|_D . \tag{18}$$

4. FICTITIOUS COMPONENTS METHOD

Let the system of equations

$$A_1 u_1 = f_1 \tag{19}$$

be qiven, where A_1 is symmetric positive semi-definite $N_1 \times N_1$-matrix with $N_1 < N$ and $f_1 \in \operatorname{im} A_1$. This system can be transformed into the system (1), if we introduce the $N \times N$-matrix

$$A = \begin{bmatrix} A_1 & 0 \\ 0 & A_2 \end{bmatrix} , \tag{20}$$

with A_2 being some symmetric positive semi-definite $N_2 \times N_2$-matrix where $N_2 = N - N_1$, and vector $f = \begin{bmatrix} f_1 \\ 0 \end{bmatrix} \in E_N$. The resulting system (1) is then solved by a generalized conjugate gradients method (11), (12) with some fixed symmetric positive definite $N \times N$-matrix B and with matrix $D = A^+$, yielding for u^k the formula

$$u^k = u^{k-1} - \frac{1}{q^k} \left[B^{-1} \xi^{k-1} - e_{k-1} (u^{k-1} - u^{k-2}) \right] , \tag{21}$$

with $e_0 = 0$ and

$$q_k = \frac{\| B^{-1} \xi^{k-1} \|_A^2}{\| \xi^{k-1} \|_{B^{-1}}^2} - e_{k-1}, \quad e_k = q_k \frac{\| \xi^k \|_{B^{-1}}^2}{\| \xi^{k-1} \|_{B^{-1}}^2} . \tag{22}$$

This method of transformation of the system (19) was proposed in [22], where it was called the fictitious components method.

According to theorem 3, for the method (21), (22) in the subspace $U_A = \operatorname{im} \left[(1 - AB^{-1}) A \right]$ the estimate (13) takes place, in which

$$a = \min_{\substack{\lambda \in \sigma(S) \\ \lambda \neq 0,\ \lambda \neq 1}} \lambda , \quad b = \max_{\substack{\lambda \in \sigma(S) \\ \lambda \neq 1}} \lambda . \tag{23}$$

Here $\sigma(S)$ is the set of all eigenvalues of S. Since $T_k\left(\frac{b+a}{b-a}\right)$ is monotonely increasing function of $d = a/b$, we obtain the problem of finding the matrix A_2 by given matrices A_1 and B, for which the ratio a/b reaches its maximal value. The solution of this problem is given by the following theorem (see [18], [22]).

Theorem 4. The maximum value of the ratio a/b is reached with $A_2 = 0$.

Therefore in the class of transformations (20) of the system (19) to (1) the optimal choice of the matrix in the fictitious components method is given by

$$A = \begin{bmatrix} A_1 & 0 \\ 0 & 0 \end{bmatrix} \tag{24}$$

Consider an application of this variant of the fictitious components method for solution of finite element approximation to the problem (3) in the case of G being some polygonal plane domain with boundary ∂G, $\beta = \beta(x_1, x_2)$ being non-negative and not totally vanishing on ∂G, $\alpha > 0$ on ∂G and $c \equiv 0$ in G.

Let us include G into a rectangular domain $\tilde{G}$ with sides parallel to coordinate axes, and introduce in $\tilde{G}$ the rectangular mesh with standard triangulation on it. Assume that ∂G belongs to the union of all sides of triangles, as shown at Fig. 3. The problem (3) is then approximated in G by finite element method with piecewise-linear base functions, yielding the system (19), and in the same way the problem

$$\begin{aligned} -\Delta v &= \tilde{f} \text{ in } \tilde{G}, \\ \frac{dv}{d\tilde{\nu}} + v &= 0 \text{ on } \partial\tilde{G} \end{aligned} \tag{25}$$

is approximated, with $\tilde{\nu}$ being the outward normal to $\partial\tilde{G}$. This approximation results in the system

$$Bv = \tilde{f} \tag{26}$$

with symmetric positive definite $N \times N$-matrix B. From [1], [2] we now obtain the following

Lemma 4. For the matrices A and B the ratio a/b is bounded from below by a positive constant independent of N.

Corollary. For the same problem we have $k_\varepsilon = O(\ln 1/\varepsilon)$.

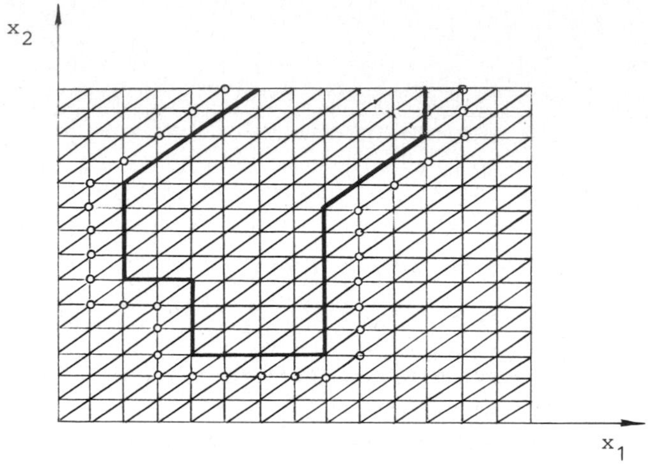

Fig. 3

Let us discuss in brief the realization of the method (21), (22) as a process in the subspace U_A from (6). Matrix B can be represented in block form

$$B = \begin{bmatrix} B_{11} & B_{12} \\ B_{21} & B_{22} \end{bmatrix}, \tag{27}$$

where B_{ij} are the blocks of order $N_i \times N_j$, $i,j = 1, 2$. Define the matrix

$$\tilde{C} = \begin{bmatrix} B_{11}-A_1 & B_{12} \\ 0 & 0 \end{bmatrix}, \tag{28}$$

It is easy to see that $U_A = \operatorname{im} \tilde{C} \subseteq RE_N$, where R is the diagonal matrix with diagonal elements equal to one for the nodes belonging to ∂G and equal to zero otherwise.

From this and from the equality $\tilde{C}B^{-1}\xi = CB^{-1}\xi$, which is true for any $\xi \in \operatorname{im} A$, the formulas (21), (22) take the form

$$\begin{aligned} \tilde{C}u^k &= \tilde{C}u^{k-1} - \frac{1}{q_k}\left[\tilde{C}B^{-1}\xi^{k-1} - e_{k-1}(\tilde{C}u^{k-1} - \tilde{C}u^{k-2})\right], \\ \xi^k &= \tilde{\xi}^{k-1} - \frac{1}{q_k}\left[(I - \tilde{C}B^{-1})\xi^{k-1} - e_{k-1}(\xi^{k-1} - \xi^{k-2})\right], \end{aligned} \tag{29}$$

with $\ _0 = 0$ and

$$q_k = \frac{(RB^{-1}\xi^{k-1}, (I-\tilde{C}B^{-1})\xi^{k-1})}{(RB^{-1}\cdot\xi^{k-1}, \xi^{k-1})} \ _{k-1}, \quad _k = q_k \frac{(RB^{-1}\xi^k, \xi^k)}{(RB^{-1}\xi^{k-1}, \xi^{k-1})}. \tag{30}$$

Implementation of (29), (30) consists, as well as in general case, of three steps. The first step, called the entrance into subspace, consists in finding for given $\varphi \in E_N$ the vectors

$$\tilde{C}u^0 = \tilde{C}\varphi - \tilde{C}B^{-1}(A\varphi - f), \quad \xi^0 = (I - \tilde{C}B^{-1})(A\varphi - f),$$

This can be done, according to the results of section 2, by $O(N \ln\ln N)$ arithmetic operations.

The second step (iterations in the subspace) consists of $\hat{k}$ steps by the formulae (29), (30). The main procedure of this step consists in finding the vector $\tilde{C}B^{-1}\xi^{k-1}$ by given vector $\xi^{k-1} \in$ $\in U_A$. In order to do this one has to find the partial solution $P_2B^{-1}\xi^{k-1}$ of the system $B\varphi^k = f$, where the diagonal elements of P_2 are equal to one for the nodes belonging to ∂G and for the nodes marked at Fig. 3 by circles. According to lemma 1 this problem can be solved by $O(N)$ arithmetic operations with $O(N^{1/2})$ computer memory. All other computations at each step of the process require just $O(N^{1/2})$ operations.

The third step, called the exit from the subspace, consists in finding ($(C-\tilde{C})u^k - (C-\tilde{C})u^{k-1}$ for all $k \geq 1$) the vector $u^{\hat{k}}$ being the solution of

$$Bu^k = \tilde{C}u^k + (C-\tilde{C})u^0 + \xi^k + f. \tag{31}$$

This is done by $O(N \ln\ln N)$ arithmetic operations using the ideas of section 2.

Therefore we have the following

<u>Theorem 5</u>. The system (19) can be solved by the method (29), (30) with the accuracy $\varepsilon < 1$ (i.e. when for given $\varphi \in E_N$ the inequality (9) is satisfied) by

$$O(N \ln\ln N) + O(N \ln 1/\varepsilon)$$

arithmetic operations.

Other variants of the fictious components method and in particular those concerned with construction of the matrix

$$A = \begin{bmatrix} A_1 & A_{12} \\ O & A_2 \end{bmatrix}, \tag{32}$$

were treated in [12],[25]. These methods are closely related to the matrix capacitance method [5], [7], [26].

5. DOMAIN DECOMPOSITION METHODS

Let G be the union of $r \geqslant 1$ non-intersecting rectangles G_i with sides parallel to the coordinate axes, $\overline{G} = \bigcup_{i=1}^{r} \overline{G}_i$, $G_i \cap G_j = \emptyset$, $i \neq j$, and let the non-empty intersections of $\overline{G}_i$ be the lines parallel to one coordinate axis only, say to Ox_2.

Consider the problem

$$\begin{aligned} -\mathrm{div} Q \,\mathrm{grad}\, u &= f \quad \text{in } G, \\ \alpha \frac{\partial u}{\partial \nu} + \beta u &= g \quad \text{on } \partial G, \end{aligned} \tag{33}$$

with f, g, α, β and ν satisfying the same conditions as before and Q being positive and constant at each rectangle G_i. Now we assume that at the sides of G_i, parallel to x_1, either $\alpha \equiv 0$, or $\beta \equiv 0$. In the domain G we construct the rectangular mesh, uniform at direction Ox_2 with $h = N^{-1/2}$ (it is assumed that we can do it), and approximate the problem (33) by finite difference or finite element method with piecewise linear base functions. This will bring us at the system (1) with symmetric positive definite matrix A.

Let us write down the matrix A in block form, dividing all the unknowns and corresponding equations into $(r+1)$ groups. The group with index $i \leqslant r$ would contain all the unknowns corresponding to the nodes belonging to G_i and to its boundaries (if $\alpha \neq 0$) parallel to Ox_1, and the remaining group with index $r+1$ would contain all other unknowns. Then the matrix A can be written in the form

$$A = \begin{bmatrix} A_{11} & A_{12} \\ A_{21} & A_{22} \end{bmatrix} \tag{34}$$

where $A_{11} = A^{(1)} \oplus A^{(2)} \oplus \dots \oplus A^{(r)}$. The system with matrix A can be solved (since the matrix DS for $D = A^{-1}$ is self-adjoint and positive definite operator in U_A, see [13]) by generalized conjugate gradients method (21),(22) in subspace U_A with matrix

$$B = \begin{bmatrix} A_{11} & A_{12} \\ O & A_{22} \end{bmatrix} \tag{35}$$

It is easy to see [9] that this method is equivalent to the generalized conjugate gradients method

$$w^k = w^{k-1} - \frac{1}{g_k}\left[\tilde{B}^{-1}(\tilde{A}w^{k-1} - \tilde{f}) - e_{k-1}(w^{k-1} - w^{k-2})\right], \tag{36}$$

with $e_0 = 0$ and

$$q_k = \frac{\|\tilde{B}^{-1}\tilde{\xi}^{k-1}\|^2_{\tilde{A}}}{\|\tilde{\xi}^{k-1}\|^2_{\tilde{B}^{-1}}} - e_{k-1}, \quad e_k = q_k \frac{\|\tilde{\xi}^k\|^2_{\tilde{B}^{-1}}}{\|\tilde{\xi}^{k-1}\|^2_{\tilde{B}^{-1}}}, \tag{37}$$

applied to the system $\tilde{A}w = f$. Here $\tilde{A} = A_{22} - A_{21}A_{11}^{-1}A_{12}$, $\tilde{B} = A_{22}$, $\tilde{f} = F_2 - A_{21}A_{11}^{-1}f_1$ and $\tilde{\xi}^k = \tilde{A}w^k - \tilde{f}$. The main computatuonal procedure of (36), (37) consists in finding for given $w^{k-1} \in E_{\tilde{N}}$, $\tilde{N} = O(N^{1/2})$ the vector $A_{21}A_{11}^{-1}A_{12}w^{k-1}$. According to lemma 2 this would require $O(N)$ arithmetic operations at the first step of the method and $O(N^{1/2}\ln N)$ operations at other steps.

If we now assume that the mesh at the direction Ox_1 is piecewise uniform and that $\tilde{f}$ is found using the algorithm of section 2, then we obtain [14], [15] the following

Theorem 6. The $k_\varepsilon = O(N^{-1/4}\ln 1/\varepsilon)$ steps of the method (36), (37) can be implemented by

$$O(N) + O(N^{3/4}\ln N \ln 1/\varepsilon) \tag{38}$$

arithmetic operations and with $O(N^{1/2})$ computer memory, if $\tilde{f}$ is already computed.

6. DOMAIN DECOMPOSITION METHOD FOR EIGENVALUE PROBLEMS

Let us consider the following eigenvalue problem for Laplace operator

$$\begin{aligned} -\Delta u &= \lambda u \quad \text{in } G, \\ u &= 0 \quad \text{on } \partial G, \end{aligned} \tag{39}$$

where the domain G is shown at Fig. 4, i.e. $\bar{G}$ is a union of two rectangles $\overline{G}_1$ and $\overline{G}_2$ with $G_1 \cap G_2 = 0$. Using the simpliest discretization of (39) by finite difference method with uniform rectangular mesh (we assume that it is possible) we obtain the algebraic eigenvalue problem

$$Au = \lambda u \tag{40}$$

with symmetric positive definite block M-matrix

$$A = \begin{bmatrix} A_{11} & A_{12} \\ A_{21} & A_{22} \end{bmatrix} \tag{41}$$

of order N . Here A_{11} corresponds to internal mesh nodes of G_1 and A_{22} corresponds to mesh nodes in $G_2 \cup (\overline{G}_1 \cap \overline{G}_2)$.

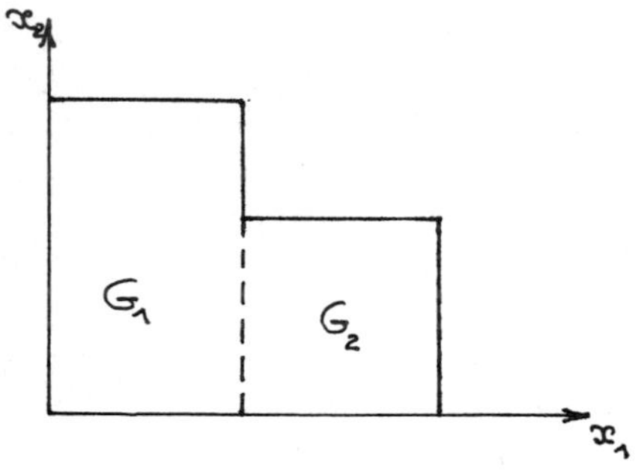

Fig. 4

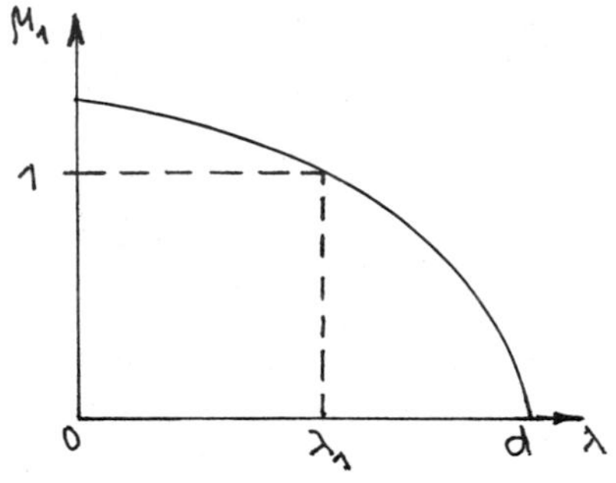

Fig. 5

Let us introduce the regular splitting

$$A = B - C \tag{42}$$

of matrix A in which $B = A_{11} \oplus A_{22}$ and consider the new eigenvalue problem

$$B_\lambda v = \mu Cv \tag{43}$$

with

$$B_\lambda = B - \lambda I. \tag{44}$$

It can be shown [32] that the minimal eigenvalue $\lambda_1 = 1/\rho(A^{-1})$ of matrix A is the unique solution to the nonlinear equation

$$\mu_1(\lambda) = 1 , \tag{45}$$

where

$$\mu_1(\lambda) = 1/\rho(B^{-1} C). \tag{46}$$

Using the theory of M-matrices [31] it is easy to show that $\mu_1(\lambda)$ is strictly monotone positive function of λ on the interval (0, d), where $d = 1/\rho(B^{-1})$ can be computed (in our case) by explicit formula, and $\lambda_1 \in (0, d)$ (see Fig. 5).

We shall apply for solution of the equation (45), i.e. for approximate computation of λ_1, the following two-stage iterative process, which consists of external and internal iterations.

As an external iterations we shall use, for example, the bisection procedure with $a_0 = 0$ and $b_0 = d$. Then for all $k \geqslant 1$ we first compute the value

$$\lambda^{(k)} = \frac{a_{k-1} + b_{k-1}}{2}$$

and set $a_k = \lambda^{(k)}$, $b_k = b_{k-1}$, if $\rho(B^{-1}_{\lambda^{(k)}} C) < 1$, and $a_k = a_{k-1}$, $b^{(k)} = \lambda^{(k)}$, if $\rho(B^{-1}_{(k)} C) \geqslant 1$. It is clear, that in this manner we have

$$|\lambda^{(k)} - \lambda_1| \leqslant \frac{d}{2^k} \tag{47}$$

As for internal iterations (computation of $\rho(B^{-1} C)$) we will use the simpliest iterative process in fixed subspace $U_A = \text{im } C$

$$v^0 \in \text{im } C,\ v^s = CB_\lambda^{-1} v^{s-1} \in \text{im } C,\ s = 1,2,\ldots,$$
$$\rho(B^{-1}C) = \lim_{s \to \infty} \frac{\| v^s \|}{\| v^{s-1} \|}. \tag{48}$$

According to lemma 2 of section 1 the implementation of this process requires O (N) arithmetic operations at the first step and $O(N^{1/2} \ln N)$ operations for each other. Moreover, the realization of (48) requires only $O(N^{1/2})$ of computer memory.

CONCLUSION

In the last two sections we have considered only the simpliest variants of iterative methods in subspaces. Other versions of the domain decomposition method (or the subdomains iterations method) were studied in [15], [16], [20], [25], [28]. The new and important direction in this field is connected with the use of operators, which are equivalent by energy [8], in the two-step iterative methods in subspaces (see [7] [15], [25]).

REFERENCES

1. Astrachantsev, G.P., Iterative methods of solution of variational-difference schemes for two-dimensional elliptic equations of the second order, Ph.D.Thesis, LOMI AN SSSR (Leningrad, 1972). /Russian/
2. Astrachantsev, G.P., Method of fictive domains for an elliptic equation of second order with natural boundary conditions, Zh. Vych. Matem. Matem. Fiz., 18 (1978), 118-125. /Russian/
3. Banegas, A., Fast Poisson solvers for problems with sparsity, Math. Comput., 1978, v. 3, No. 142, 441-446.
4. Bachvalov, N.S., Orechov, M.Yu., On fast methods for solution of Poisson equation, Zh. Vych. Matem. Matem. Fiz., 22 (1982), 1386-1392. /Russian/
5. Buzbee, B.L., Dorr, J.A., George, J.A., Golub, G.H.,The direct solution of the discrete Poisson equation on irregular regions, SIAM J. Numer. Anal., 1971, v. 8, 722-736.
6. Concus,P., Golub, G.H., O'Leary, D.P. A generalized conjugate gradient method for the numerical solution of elliptic partial differential equations, in: Sparse Matrix computations, N.-Y.: Acad. Press, 1976, 309-332.
7. Drya, M., Matrix capacitance algorithm for variational-difference schemes, in: Variational-difference methods in mathematical physics (Novosibirsk, 1981), 63-73. /Russian/
8. Dyakonov, E.G., Construction of iterative methods using spectrum-equivalent operators, Zh. Vych. Matem. Matem Fiz., 6 (1966), 12-34. /Russian/
9. Dyakonov, E.G., Direct and iterative methods using matrix bordering, in: Numerical methods in mathematical physics (Novosibirsk, 1979), 45-68. /Russian/
10. Hestenes, M.R., Stiefel, E., Method of conjugate gradients for solving linear systems, J.res. Nat. Bur. Stand., 1952, v. 49, 409-436.
11. Il'in, V.P., Kuznetsov, Ju.A., Iterative methods in numerical solution of differential equations, in: Lect. Notes in Math., 1979, No. 704, pp. 23-36.
12. Kaporin, I.E., Nikolaev, E.S., Method of fictive unknowns for solution of finite-difference elliptic boundary value problems in irregular domains, Different. Uravnenija, 16 (1980), 1211-1225. /Russian/
13. Kuznetsov, Yu.A., On the theory of iterative processes, DAN SSSR, 184 (1969), 274-277. /Russian/
14. Kuznetsov, Yu.A., Block-relaxation methods in subspace, their optimization and applications, in: Variational-difference methods in mathematical physics (Novosibirsk, 1978), 178-212. /Russian/
15. Kuznetsov, Yu.A., Block-relaxation methods in subspace for two-dimensional elliptic equations, in: Numerical methods in mathematical physics (Novosibirsk, 1979), 20-44. /Russian/
16. Kuznetsov, Yu.A., Block-relaxation method in subspace of solution of two- and three-dimensional diffusion problems in multizone domains, in: Solution of systems of variational-difference equations (Novosibirsk, 1979), 24-59. /Russian/
17. Kuznetsov, Yu.A., Conjugate gradients method, its generalizations and application, in: Computational processes and systems (Nauka, Moscow, 1983), 267-300. /Russian/
18. Kuznetsov, Yu.A., Matsokin, A.M., On optimization of the method of fictive components, in: Numerical methods of linear algebra (Novosibirsk, 1977), 79-86. /Russian/

19. Kuznetsov, Yu.A., Matsokin, A.M., On partial solution of systems of linear algebraic equations, in: Numerical methods of linear algebra (Novosibirsk, 1978), 62-89. /Russian/
20. Lebedev, V.I., Agoshkov, V.I., Generalized Shwartz algorithm with variable parameters, Preprint VINITI No. 19 (Seminar "Problems of computational mathematics", Moscow, 1981). /Russian/
21. Marchuk, G.I., Kuznetsov, Yu.A., To the theory of optimal iterative processes, DAN SSSR, 181 (1969), 1331-1334. /Russian/
22. Marchuk, G.I., Kuznetsov, Yu.A., Some problems of iterative methods, in: Numerical methods of linear algebra (Novosibirsk, 1972), 4-20. /Russian/
23. Marchuk, G.I., Kuznetsov, Yu.A., Iterative methods and quadratic functionals (Novosibirsk, 1972). /Russian/ (See also: Marchuk, G.I., Methods of Numerical Mathematics (Nauka, Novosibirsk, 1975), 4-143).
24. Marchuk, G.I., Kuznetsov, Yu.A., Theory and application of generalized conjugate gradients method, Preprint No. 72 (Computing Center, Novosibirsk, 1980). /Russian/
25. Matsokin, A.M., Method of fictive components and modified difference analog of Shwartz method, in: Numerical methods of linear algebra (Novosibirsk, 1980), 66-77. /Russian/
26. Proskurowski, W., Widlund, O., On the numerical solution of Helmholtz equation by the capacitance matrix method, Math. Comput., 1976, v. 30, 433-468.
27. Reid, J.K., On the method of conjugate gradients for the solution of large sparse systems of linear equations, in: Large Sparse Sets Linear Equat., London-New York: Acad. Press., 1971, 231-252.
28. Smelov, V.V., Iteration over the domains in problems with transport equations, Zh. Vych. Matem. Matem. Fiz., 21 (1981), 1488-1504. /Russian/
29. Swarztrauber, P.N., The method of cyclic reduction, Fourier analysis and the FACRalgorithms for the discret solution of Poisson equation on rectangule, SIAM Rev., 1976, v. 19, 490-501.
30. Tyrtyshnikov, E.E., On algorithms of discrete Fourier transform, in: Numerical methods of algebra (Moscow, 1981), 10-26. /Russian/
31. Varga, R.S. Matrix iterative analysis. Englewood Cliffs, N.J. Prentice-Hall, 1962, XIV, 322 pp.
32. Kuznetsov, Yu.A. On the symmetrization of approximate problems in transport theory, in : Computational methods in Math. Physics, geophysics and optimal control (Novosibirsk Nauka, 1978), 125-137, /Russian/.

Computing Methods in Applied Sciences and Engineering, VI
R. Glowinski and J.-L. Lions (Editors)
Elsevier Science Publishers B.V. (North-Holland)

ITERATIVE METHODS FOR ELLIPTIC PROBLEMS ON REGIONS PARTITIONED INTO SUBSTRUCTURES AND THE BIHARMONIC DIRICHLET PROBLEM

Olof B. Widlund

Courant Institute of Mathematical Sciences
New York University
251 Mercer Street, New York, N.Y. 10012
U.S.A.

A finite element problem can often naturally be partitioned into subproblems which correspond to subregions into which the region has been partitioned or from which it was originally assembled. A class of methods are discussed in which these subproblems are solved by direct methods while the interaction between the subregions is handled by a conjugate gradient algorithm. The mathematical framework for this work is provided by regularity theory for elliptic finite element equations and block-Gaussian elimination. The same tools can be used to provide a reinterpretation of Bjørstad's work on very fast methods for the biharmonic Dirichlet problem on rectangles and a variant of an algorithm introduced by Glowinski and Pironneau for mixed finite element approximations of the same elliptic problem on general regions. The relationship between substructuring and finite element-capacitance matrix techniques is also briefly explored.

INTRODUCTION

In this paper, we will discuss some special iterative methods for elliptic finite element problems defined on regions regarded as unions of subregions. We are interested, in particular, in the design of algorithms for which the interaction between the subproblems, i.e. the discrete elliptic problems on the subregions, is computed with the aid of a conjugate gradient method, while the subproblems themselves are solved by a direct method. This approach differs from the standard one used in industry where direct methods are employed throughout.

The partition of elliptic problems into subproblems is a very natural idea that is widely used in practice and which has been discussed in the engineering literature for at least twenty years; see e.g. Prezemieniecki [18,19]. Thus the modeling of an entire mechanical system can profitably be organized by discretizing the partial differential equations by finite elements on subregions. These tasks and the factorization of the resulting stiffness matrices for the subproblems can be assigned to different engineering groups and computer systems or processors with coordination required only at the interfaces between the substructures to assure matching of finite element triangulations and solutions. These ideas are particularly attractive for parallel computing and in the case when some of the substructures are identical or previously analyzed as in the case when a simulation is repeated after the redesign or damage of one or a few of the substructures. The final phase, in which the interaction is accounted for, does not lend itself equally well to parallel computing and it also represents a significant fraction of the total computational work even on a sequential computer.

Three alternative methods are discussed in this paper and numerical results for a model problem are presented. These algorithms are analyzed in detail in a forthcoming paper by Bjørstad and Widlund [4], in which a complete theory for conforming Lagrangian finite element approximations of second order elliptic problems in the plane are given. The development of this theory requires the use of elliptic regularity theory for problems on Lipschitz regions; see Grisvard [15], and some apparently new finite element results. We regard the extension of these results to more general elliptic problems as a significant open problem.

Earlier work on algorithms of this kind is reported in Concus, Golub and O'Leary [6], who gave numerical results for Poisson's problem on T-shaped regions. For an analysis of this method see Bjørstad and Widlund [4]. Two interesting new preconditioners for the same special problem are introduced in the contribution to these proceedings by Golub and Mayers [12]. In their numerical experiments a comparison is made between these algorithms and the one characterized as second best in our previous paper; cf. Bjørstad and Widlund [3]. Methods inspired by Schwarz' alternating method and by control theory are considered in Glowinski, Periaux and Dihn [10]. For a discussion of one of these, see Bjørstad and Widlund [4]. There has also been extensive activity in the Soviet Union in this area. We are not yet well acquainted with this work and are grateful to Yu. Kuznetsov and Maximilian Dryja for drawing our attention to it. After our first series of numerical experiments, we received a preprint of one of Dryja's papers [8], from which we first learned of what we currently regard as the best algorithm. In that paper Dryja analyzes the case of Laplace's equation on L-shaped regions.

The techniques developed for the substructuring problems, which combine block Gaussian elimination and regularity theorems for elliptic problems, are also useful in examining certain iterative methods for the biharmonic problem. Even on rectangular regions, this problem is of considerable difficulty, since separation of variables cannot be used directly. By introducing an auxiliary variable, known as vorticity in fluid dynamics, the problem can be reformulated as a second order elliptic system. Many iterative methods have been developed to solve this system in which a different, simpler boundary value problem is solved in each step; see Bjørstad [1,2] and Glowinski and Pironneau [11] and the many references given in those papers. In this paper, we introduce an alternative to an optimal method given by Glowinski and Pironneau [11] and give a reinterpretation of Bjørstad's work.

While for substructured problems we seek to solve a large linear system by an iterative method which involves smaller problems, there are occasions when we are willing to solve a larger linear system repeatedly, to obtain the solution of a smaller system. Such algorithms are known as capacitance matrix methods; see O'Leary and Widlund [17], Proskurowski and Widlund [20,21] for a discussion and references to the literature. In our last section, we explore the relationship between substructuring and finite element-capacitance matrix methods.

SUBSTRUCTURED FINITE ELEMENT PROBLEMS AND BLOCK FORM OF THE STIFFNESS MATRICES

To simplify the discussion, we confine ourselves to problems defined on the union Ω of Ω_1, Ω_2 and Γ_3 and to the Dirichlet case. Here Ω_1 and Ω_2 are plane, bounded, nonintersecting regions and Γ_3 the intersection of their closures. The boundaries of Ω_1 and Ω_2 are $\Gamma_1 \cup \Gamma_3$ and $\Gamma_2 \cup \Gamma_3$, respectively, and the boundary of Ω is $\Gamma_1 \cup \Gamma_2$. We assume that these subregions are curvilinear polygons, i.e. in particular they are Lipschitz regions; see Grisvard [13]. A linear, second order, positive definite, selfadjoint elliptic operator is defined on Ω. Its symmetric bilinear form is denoted by $a_\Omega(u,v)$; see Ciarlet [5] or Strang and Fix [22]. A simple example is given by the Laplace operator with homogeneous Dirichlet condition for which

$$a_\Omega(u,v) = \int_\Omega \nabla u \cdot \nabla v \, dx \ , \quad u,v \in H_0^1(\Omega) \ .$$

Here $H_0^1(\Omega)$ is the subspace of elements, with zero boundary values, of the Sobolev space of square integrable functions with square integrable first derivatives. Triangulations of Ω_1 and Ω_2 are introduced in such a way that the nodes on Γ_3 coincide and Γ_3 follows element boundaries. We assume that each degree of freedom of the finite element subspaces is associated with a node and a basis function in the finite element space and that the support of any basis function coincides with the triangles to which its node belongs. An element of the stiffness matrix has the form $a_\Omega(\phi_i,\phi_j)$, where ϕ_i and ϕ_j are basis functions, and it therefore vanishes unless the two nodes belong to a common triangle.

It is easy to see from the definition of the bilinear form that

$$a_{\Omega_1}(u,v) + a_{\Omega_2}(u,v) = a_\Omega(u,v) \tag{2.1}$$

and that therefore the stiffness matrix of a problem on Ω can be constructed from those of Ω_1 and Ω_2. The same relation holds for any pair of nonoverlapping subregions. This fact is frequently used in practice to construct stiffness matrices from the stiffness matrices of the individual triangles; see Strang and Fix [22].

We write the stiffness matrices corresponding to Ω_1 and Ω_2 respectively as

$$K^{(1)} = \begin{pmatrix} K_{11} & K_{13} \\ K_{13}^T & K_{33}^{(1)} \end{pmatrix} \quad \text{and} \quad K^{(2)} = \begin{pmatrix} K_{22} & K_{23} \\ K_{23}^T & K_{33}^{(2)} \end{pmatrix} , \tag{2.2}$$

where K_{11}, represents couplings between the pairs of nodes in Ω_1, K_{13} couplings between the pairs belonging to Ω_1 and Γ_3 respectively and $K_{33}^{(1)}$ couplings between nodes on Γ_3, etc.

Written in variational form a Dirichlet problem on Ω_1 has the form

$$\int_{\Omega_1} \nabla u_h \cdot \nabla v_h \, dx = \int_{\Omega_1} f \, v_h \, dx \ , \quad \forall v_h \in S^h(\Omega_1) \cap H_0^1(\Omega_1) \ ,$$

$$u_h \in S^h(\Omega_1) \ , \qquad \gamma_0 u_h = g_D \ ,$$

where $S^h(\Omega_1) \subset H^1(\Omega)$ is a finite element subspace and $\gamma_0 u_h$ is the trace, i.e. the restriction of u_h to the boundary. In matrix form this problem can be written as

$$\begin{pmatrix} K_{11} & K_{13} \\ 0 & I \end{pmatrix} \begin{pmatrix} x_1 \\ x_3 \end{pmatrix} = \begin{pmatrix} b_1 \\ b_3 \end{pmatrix} \tag{2.3}$$

where x_1 and x_3 are the vectors of nodal values corresponding to the open set Ω_1 and Γ_3 respectively. The components of b_1 are constructed from the right hand side f and the Dirichlet values on Γ_1 and b_3 is the vector of Dirichlet values on Γ_3.

If on the other hand Neumann data g_N are given on Γ_3, while we still have Dirichlet data g_D on Γ_1, the problem is of the form,

$$\int_{\Omega_1} \nabla u_h \cdot \nabla v_h \, dx = \int_{\Omega_1} f \, v_h \, dx + \int_{\Gamma_3} g_N v_h \, ds \;, \quad \forall \, v_h \in S^h \cap H_0^1(\Omega_1, \Gamma_1) \;,$$
$$u_h \in S^h \;, \qquad \gamma_0^{\Gamma_1} u_h = g_D .$$

Here $\gamma_0^{\Gamma_1} u_h$ is the trace on Γ_1 and $H_0^1(\Omega_1,\Gamma_1)$ the subspace of $H^1(\Omega_1)$ with vanishing trace on Γ_1. In this case x_3 is a vector of unknowns and the linear system is

$$\begin{pmatrix} K_{11} & K_{13} \\ K_{13}^T & K_{33}^{(1)} \end{pmatrix} \begin{pmatrix} x_1 \\ x_3 \end{pmatrix} = \begin{pmatrix} b_1 \\ b_3 \end{pmatrix} . \tag{2.4}$$

We note that the vector b_1 vanishes if f and g_D are zero in which case the vector b_3 represents the Neumann data on Γ_3.

The stiffness matrix of the entire problem is of the form,

$$K = \begin{pmatrix} K_{11} & 0 & K_{13} \\ 0 & K_{22} & K_{23} \\ K_{13}^T & K_{23}^T & K_{33} \end{pmatrix}$$

where, by (2.1) and a simple computation,

$$K_{33} = K_{33}^{(1)} + K_{33}^{(2)} . \tag{2.5}$$

We note that the degrees of freedom have been partitioned into three sets of which the third, the separator set, corresponds to the nodes of Γ_3. From the point of view of graph theory, the undirected graph of K becomes disconnected into two components if the nodes of the separator set and their incident edges are removed. If conforming finite elements are used, then it also follows from the assumptions on $a_\Omega(u,v)$ that K is positive definite, symmetric and as a consequence so are K_{11}, K_{22}, and K_{33}.

We consider the linear system of algebraic equations of the form

$$Kx = \begin{pmatrix} K_{11} & 0 & K_{13} \\ 0 & K_{22} & K_{23} \\ K_{13}^T & K_{23}^T & K_{33} \end{pmatrix} \begin{pmatrix} x_1 \\ x_2 \\ x_3 \end{pmatrix} = \begin{pmatrix} b_1 \\ b_2 \\ b_3 \end{pmatrix} . \tag{2.6}$$

By using block-Gaussian elimination, we can reduce this system to the positive definite, symmetric system,

$$\begin{aligned} Sx_3 &= (K_{33} - K_{13}^T K_{11}^{-1} K_{13} - K_{23}^T K_{22}^{-1} K_{23})x_3 \\ &= b_3 - K_{13}^T K_{11}^{-1} b_1 - K_{23}^T K_{22}^{-1} b_2 = \tilde{b}_3 . \end{aligned} \tag{2.7}$$

It is common practice to complete the process by solving (2.7) by a direct method.

The right hand side $\tilde{b}_3$ can be obtained at the expense of solving the two subproblems on Ω_1 and Ω_2, multiplying the resulting vectors by the sparse matrices K_{13}^T and K_{23}^T respectively and by subtracting the resulting vectors from b_3. From now on, we will consider only the case when b_1 and b_2 are zero. Such a reduction can of course easily be accomplished.

The matrix S, is a so-called Schur complement; see Cottle [7], and can be expensive to compute and store. However, we notice that Sy can be computed for a given vector y at the expense of solving the two subproblems with the sparse right hand sides $K_{13}y$ and $K_{23}y$, respectively, and certain sparse matrix and vector operations. In the next section, we will develop iterative methods which only require S in terms of such matrix-vector products.

The cost of computing Sy depends primarily on the efficiency of the solvers for the subproblems. It should also be noted that if a Gaussian elimination method is used, advantage can be taken of the sparsity of the vectors $K_{13}y$ and $K_{23}y$. Thus when the lower triangular systems of equations are solved, the computation can begin with the first equation which has a nonzero right hand side. Similarly, the solution of the upper triangular systems can be stopped as soon as all the components of $K_{11}^{-1} K_{13} y$ and $K_{22}^{-1} K_{23}y$, necessary for computing $K_{13}^T(K_{11}^{-1} K_{13}y)$ and $K_{23}^T(K_{22}^{-1} K_{23}y)$, have been found. This can effectively reduce the size of the triangular systems necessary to carry out the iteration steps. It is thus particularly advantageous if all the variables at nodes adjacent to Γ_3 are ordered late. It should be noted, however, that such a constraint may be hard to impose on existing software or that it may lead to an increase in the time required to factor K_{11} and K_{22} into their triangular factors.

We will need the Schur complements with respect to the matrices $K^{(1)}$ and $K^{(2)}$ defined in (2.2). They are,

$$S^{(1)} = K_{33}^{(1)} - K_{13}^T K_{11}^{-1} K_{13} \quad \text{and} \quad S^{(2)} = K_{33}^{(2)} - K_{23}^T K_{22}^{-1} K_{23} , \tag{2.8}$$

Using (2.7) and (2.8), we find that

$$S = S^{(1)} + S^{(2)} . \tag{2.9}$$

The mappings S and $S^{(1)}$ play an important role in what follows. The vector $S^{(1)}y$ can be computed by solving a Dirichlet problem (2.3) and then applying the matrix of (2.4), which corresponds to a Neumann case, to the solution vector. This can be seen by a straightforward computation:

$$\begin{pmatrix} K_{11} & K_{13} \\ K_{13}^T & K_{33}^{(1)} \end{pmatrix} \begin{pmatrix} K_{11} & K_{13} \\ 0 & I \end{pmatrix}^{-1} \begin{pmatrix} 0 \\ y \end{pmatrix} = \begin{pmatrix} 0 \\ S^{(1)}y \end{pmatrix} .$$

From our previous discussion we see that this is a Dirichlet-Neumann map which takes the Dirichlet data y on Γ_3 into the discrete Neumann data $S^{(1)}y$ on the same set. For a discussion of the continuous case, see next section.

CONJUGATE GRADIENT ALGORITHMS FOR SUBSTRUCTURED PROBLEMS AND AN INFORMAL THEORY.

The general theory of conjugate methods is quite well known and it will therefore be discussed only very briefly; see Concus, Golub and O'Leary [6], Hestenes [14] or Luenberger [16].

Let $Ax = b$ be a linear system of algebraic equations with a positive definite, symmetric matrix A. Let $x^{(0)}$ be an initial guess and $r^{(0)} = b - Ax^{(0)}$ the initial residual. The k-th iterate in the standard conjugate gradient method, $x^{(k)}$, can then be characterized as the minimizing element for the variational problem

$$\min(\tfrac{1}{2})y^T Ay - y^T b$$

where $y-x^{(0)}$ varies in the linear space spanned by $r^{(0)}, Ar^{(0)},\ldots,A^{k-1}r^{(0)}$. By expanding in eigenvectors of A, it can be established that

$$(x^{(k)}-x)^T A(x^{(k)}-x)/(x^{(0)}-x)^T A(x^{(0)}-x)$$

is bounded from above by,

$$\min_{p\in P_{k-1}} \max_{\lambda\in\sigma(A)} (1-\lambda p(\lambda))^2 , \qquad (3.1)$$

See Luenberger [16]. Here x is the exact solution, P_{k-1} the space of all polynomial of degree k-1 and $\sigma(A)$ the spectrum of A. This bound can be used to establish that the convergence is rapid if A is well conditioned and that the rate of convergence can be bounded uniformly for entire families of operators if all the eigenvalues fall in a fixed interval.

Preconditioned conjugate gradient methods have been studied extensively in recent years. The idea goes back to the mid-fifties; see Hestenes [14]. Let A_0 be another positive definite, symmetric operator for which it is feasible to solve auxiliary systems of the form $A_0y = c$ repeatedly for different right hand sides. In one of the versions of the method, the original problem $Ax = b$ is transformed into $AA_0^{-1}y = b$. The iterate $y^{(k)}$ is sought as the sum of an initial guess $y^{(0)}$ and a linear combination of $r^{(0)}, AA_0^{-1}r^{(0)},\ldots,(AA_0^{-1})^{k-1}r^0$. If an appropriate inner product is used, a convenient recursion formula results, see e.g. Proskurowski and Widlund [21]. Each step of this algorithm requires the solution of an auxiliary linear system. It is important to note that the estimate (3.1) still holds, but that now the eigenvalues of AA_0^{-1}, i.e. those of the symmetric generalized eigenvalue problem $A\phi = \lambda A_0\phi$, are of relevance rather than those of A. It is also worth noting that the estimate (3.1) can be used to show particularly rapid convergence if the eigenvalues are clustered.

For the problem at hand, we first consider the solution of equation (2.7) without preconditioning. From (2.9), we see that $Sy = S^{(1)}y + S^{(2)}y$. It is therefore at least plausible that S will be ill conditioned if $S^{(1)}$ and $S^{(2)}$ are. As shown in section 2, $S^{(1)}$ represents a Dirichlet-Neumann map and therefore involves a loss of a derivative in $L_2(\Gamma_3)$. Such a map will have a spectral condition number proportional to the number of nodes on Γ_3. In order to clarify this point, we consider the continuous case, leaving the details on the finite element case to the forthcoming paper by Bjørstad and Widlund [4].

Thus consider two harmonic functions u_1 and u_2 defined on Ω_1 and Ω_2 respectively. These functions vanish on Γ_1 and Γ_2 and have the same trace on Γ_3. They can therefore be combined to form $u \in H_0^1(\Omega)$. This function satisfies

$$a_\Omega(u,v) = f(v) \ , \quad \forall\, v \in H_0^1(\Omega) \ , \tag{3.2}$$

where the linear functional f has its support on Γ_3. It is easy to show that

$$a_\Omega(u,v) = \int_{\Omega_1} \nabla u_1 \cdot \nabla v dx + \int_{\Omega_2} \nabla u_2 \cdot \nabla v \, dx = \int_{\Gamma_3} \frac{\partial u}{\partial \nu} v \, ds - \int_{\Gamma_3} \frac{\partial u}{\partial \nu} v \, ds = \int_{\Gamma_3} [\frac{\partial u}{\partial \nu}] v \, ds \ ,$$

where ν is the normal outward with respect to Ω_1. We can therefore rewrite equation (3.2) as

$$[\frac{\partial u}{\partial \nu}] = f \ ,$$

where $[\frac{\partial u}{\partial \nu}]$ corresponds to Sy. Following Lions and Magenes [15] we see that for $u \in H_0^1(\Omega)$,

$$\gamma_0^{\Gamma_3} u \in H_{00}^{1/2}(\Gamma_3) = \{u \in H_0^{1/2}(\Gamma_3) \ ; \ \|\rho^{-1/2} u\|_{L^2(\Gamma_3)} < \infty\} \ ,$$

where ρ is the distance of a point on Γ_3 to its end points. It then follows, from a standard variational argument, that $\frac{\partial u}{\partial \nu}$, $\frac{\partial u}{\partial \nu}$ and $[\frac{\partial u}{\partial \nu}]$ belong to the dual space of $H_{00}^{1/2}(\Gamma_3)$; i.e. a derivative is lost in comparison with $\gamma_0^{\Gamma_3} u$.

In view of what we have just learned, it is natural to try to find a preconditioner which also involves the loss of a derivative in $L_2(\Gamma_3)$. A natural choice would be a tangential derivative but that is not a symmetric operator. Instead we can use the square root of the negative of a discretization of the Laplacian on Γ_3. Such a method is practical for problems in the plane and has been tested; see section 4. We denote this operator as J.

An even better method, involves the solution of a system,

$$\begin{pmatrix} K_{11} & 0 & K_{13} \\ 0 & K_{22} & K_{23} \\ K_{13}^T & 0 & K_{33}^{(1)} \end{pmatrix} \begin{pmatrix} x_1 \\ x_2 \\ x_3 \end{pmatrix} = \begin{pmatrix} 0 \\ 0 \\ y \end{pmatrix} .$$

This solution can be accomplished by solving equation (2.4) with the right hand side $(0,y)^T$ and then the discrete Dirichlet problem on Ω_2, using x_3 as data. The relevant mapping is now $SS^{(1)^{-1}}$ since

$$\begin{pmatrix} K_{11} & 0 & K_{13} \\ 0 & K_{22} & K_{23} \\ K_{13}^T & K_{23}^T & K_{33} \end{pmatrix} \begin{pmatrix} K_{11} & 0 & K_{13} \\ 0 & K_{22} & K_{23} \\ K_{13}^T & 0 & K_{33}^{(1)} \end{pmatrix}^{-1} \begin{pmatrix} 0 \\ 0 \\ y \end{pmatrix} = \begin{pmatrix} 0 \\ 0 \\ SS^{(1)^{-1}} y \end{pmatrix} .$$

It can be shown that this operator is uniformly well conditioned; see Bjørstad and Widlund [4].

We note that we need not construct any auxiliary operator when this preconditioner is used. It is also interesting to note that if a rectangular region is cut in half, and treated fully symmetrically, then $S = 2S^{(1)}$ and the conjugate gradient iteration converges in one step.

NUMERICAL EXPERIMENTS WITH A SUBSTRUCTURED PROBLEM.

We will only give a brief report on a few of the many experiments which we have carried out for the five point approximation of Poisson's equation on regions which are unions of two rectangles. The choice of such regions greatly simplifies the experiments and makes it feasible to conduct many experiments with very many degrees of freedom since fast Poisson solvers can be used to solve the subproblems. We note that this simple finite difference approximation can be viewed as a conforming finite element approximation on a mesh of right triangles using piecewise linear functions; see Strang and Fix [22].

In the experiments considered here, we consider the union of two rectangles with corners at the points (0,0), (1,0), (1,1/2), (0,1/2) and (1/8,1/2), (5/8,1/2), (5/8,1), (1/8,1) respectively. The relevant parameters are the number of mesh points q on Γ_3, the interval between (1/8,1/2) and (5/8,1/2), and the total number of degrees of freedom. We have used data which are consistent with an exact solution $u(x,y) = x^2+y^2 - xe^x\cos y$, We have found no real difference between the performance of our method for this and other cases.

We first show, in Table I, how the number of iterations grows as a function of q using the operator $SS^{(1)-1}$. We stopped the iterations at the level of the truncation error. The initial guess was the zero function. We note that the overall number of degrees of freedom increases quadratically with q and equals 48641 for q = 127.

TABLE I

q	Number of Iterations	Maximum Error on Ω
3	2	3.66×10^{-4}
7	3	9.59×10^{-5}
15	3	2.45×10^{-5}
31	4	6.09×10^{-6}
63	4	1.49×10^{-6}
127	5	3.02×10^{-7}

In Table II, we compare this preconditioner, $S^{(1)}$ with J, which is the square root of the negative of the discrete Laplacian on Γ_3, for the case when q = 127.

TABLE II

No. of Iterations	Preconditioner: $S^{(1)}$	Preconditioner: J
0	3.79×10^{-1}	3.79×10^{-1}
1	1.25×10^{-2}	3.22×10^{-2}
2	7.48×10^{-4}	4.01×10^{-3}
3	2.56×10^{-5}	5.26×10^{-4}
4	4.42×10^{-7}	8.74×10^{-5}
5	3.02×10^{-7}	1.05×10^{-5}
6	3.02×10^{-7}	1.33×10^{-6}
7	3.02×10^{-7}	3.08×10^{-7}
8	3.02×10^{-7}	3.03×10^{-7}

Ten to twelve iterations were required to decrease the maximum error by a factor of 10 if no preconditioner was used.

We also give some information on the spectrum of $SS^{(1)^{-1}}$ for $q = 63$. The smallest eigenvalues were found to be $\lambda_1 = 1.713$, $\lambda_2 = 1.777$, ..., $\lambda_5 = 1.992$ with all eigenvalues less than or equal to 2.000. The corresponding spectrum for SJ^{-1} shows a somewhat less pronounced cluster with $\lambda_1 = 1.733$, $\lambda_2 = 1.804$, ..., $\lambda_5 = 2.008$, ..., $\lambda_{63} = 2.828$.

DIRICHLET'S PROBLEM FOR THE BIHARMONIC EQUATION.

The biharmonic Dirichlet problem has the form,

$$\begin{aligned} \Delta^2\psi &= f \quad \text{in } \Omega, \\ \gamma_0\psi &= g_0 \quad , \\ \gamma_1\psi &= g_1 \quad , \end{aligned} \tag{5.1}$$

where $\gamma_0\psi$ and $\gamma_1\psi$ are traces; i.e. the restrictions of ψ and $\partial\psi/\partial\nu$ to the boundary. For a discussion of the existence of traces, a definition of the Sobolev spaces used in this section and a general introduction to elliptic problems; see Lions and Magenes [15]. We assume that the region Ω is plane, bounded and simply connected and denote the boundary by Γ.

In our discussion, we will use the following regularity result; cf. Glowinski and Pironneau [11]:

Let Γ be sufficiently smooth, and assume that $f \in L^2(\Omega)$, $g_0 \in H^{3/2}(\Gamma)$ and $g_1 \in H^{1/2}(\Gamma)$. Then the solution of (5.1) satisfies $\psi \in H^2(\bar{\Omega})$ and $\gamma_0(\Delta\psi) \in H^{-1/2}(\Gamma)$.

It is easy to show that we only need to consider the case where $f = 0$ and $g_0 = 0$ in what follows. Following Glowinski and Pironneau [11], we attempt to solve equation (5.1) by finding λ, the trace of the vorticity $\omega = -\Delta\psi$, for which $\gamma_1\psi = g_1$. The relation between $\gamma_0\omega$ and $\gamma_1\psi$ is given by the system,

$$\begin{aligned} -\Delta\omega &= 0 \quad \text{in } \Omega, \\ \gamma_0\omega &= \lambda \quad , \\ -\Delta\psi &= \omega \quad \text{in } \Omega, \\ \gamma_0\psi &= 0 \quad , \end{aligned} \tag{5.2}$$

which can be shown to define a selfadjoint, $H^{-1/2}(\Gamma)$-elliptic operator A which maps $H^{-1/2}(\Gamma)$ onto $H^{1/2}(\Gamma)$. At the expense of solving two Dirichlet problems for the simplest second order elliptic problem, we can thus compute $A\lambda$ for a given λ and can then use the conjugate gradient method to find λ and the solution of problem (5.1).

There have been many attempts to use related methods for finite difference approximations of (5.1); see Bjørstad [1,2] and Glowinski and Pironneau [11] for a discussion and references. The idea is systematically developed by Glowinski and Pironneau in the framework of mixed finite elements. These have proven to be the most successful class of methods for solving problem (5.1). We review some of their work stressing the similarity with the work discussed in sections 2 and 3.

The mixed finite element methods are defined by,

$$\int_\Omega \nabla\omega_h\cdot\nabla v_h\,dx = 0\ , \qquad \forall\, v_h \in S_0^h\ ,\quad \omega_h \in S^h\ ,$$

$$\int_\Omega \nabla\psi_h\cdot\nabla v_h\,dx = \int_\Omega \omega_h v_h\,dx,\quad \forall\, v_h \in S_0^h\ ,\quad \psi_h \in S_0^h\ ,$$

and

$$\int_\Omega \nabla\psi_h\cdot\nabla\mu_h\,dx = \int_\Omega \omega_h\mu_h\,dx + \int_\Gamma g_1\mu_h\,ds\ ,\ \forall\mu_h \in M_h\ .$$

Here $S^h \subset H^1(\Omega)$ is the finite element subspace, $S_0^h = S^h \cap H_0^1(\Omega)$ and M_h the space spanned by the basis functions associated with the modes on Γ. M_h is thus a subspace of S^h which is complementary to S_0^h. With ψ_1 denoting the vector of nodal values of ψ_h in Ω, ω_1 the vector of the negative of the nodal values of ω_h in Ω and ω_2 the vector values of $-\omega_h$ on Γ, we obtain a linear system of equations with a symmetric, indefinite coefficient matrix,

$$\begin{pmatrix} 0 & K_{11} & K_{12} \\ K_{11} & M_{11} & M_{12} \\ K_{12}^T & M_{12}^T & M_{22} \end{pmatrix} \begin{pmatrix} \psi_1 \\ \omega_1 \\ \omega_2 \end{pmatrix} = \begin{pmatrix} 0 \\ 0 \\ b_2 \end{pmatrix}\ . \tag{5.3}$$

Here K_{11} is the stiffness submatrix which represents the couplings between pairs of nodes in Ω while K_{12} represents couplings between Ω and Γ. The matrices M_{11}, M_{12} and M_{22} are mass submatrices with elements $\int_\Omega \phi_i\phi_j\,dx$. The matrix K_{11} and the mass matrix are positive definite, symmetric. Diagonal subsystems associated with the vectors ψ_1 and ω_1 can be solved at the expense of two discrete problems for the Laplacian. As in the continuous case, we can reduce the problem (5.3) to a linear system for the trace of the vorticity. Using block Gaussian elimination, we can regard the resulting matrix, A_h , as a Schur complement. The symmetry of A_h follows by inspection. The fact that it is positive definite can be shown by using a result given in Cottle [7]. He uses Sylvester's theorem to relate the inertia of a block matrix with that of a principal minor and the corresponding Schur complement.

The matrix A_h is an approximation of the operator A. Glowinski and Pironneau used techniques from mathematical analysis to prove that in the case of a convex region, the condition number of A_h grows linearly with the number of nodes on Γ. This is not surprising since A maps $H^{-1/2}(\Gamma)$ into $H^{1/2}(\Gamma)$, i.e. a derivative is gained. They therefore suggested a preconditioner, based on the square root of a discretization of the operator $-(d^2/ds^2)+1$. This provides an optimal method in the sense that the rate of convergence becomes independent of the number of degrees of freedom of the discrete model. This preconditioner is of course basically the same as the inverse of the operator J described in section 3.

In view of the success of the preconditioner $S^{(1)}$ for the substructured problems, we have initiated numerical experiments which will use the Dirichlet-Neumann map as a preconditioner for A_h. Given a vector y, $S^{(1)}y$ can be computed by solving the same discrete Laplace problem which is required for the computation of $A_h y$. Since the continuous as well as the discrete Dirichlet-Neumann map has a one dimensional null space of constants in this case, the operator $S^{(1)}$ will be modified by adding cee^T, where $e^T = (1,1,\ldots,1)$ and $c > 0$.

Fast algorithms for the biharmonic problem on rectangles were considered in the 1980 Stanford dissertation of Peter Bjørstad [1]; see also Bjørstad [2]. Although the region is quite special such algorithms are nevertheless very useful in a number of applications. The discrete model is the standard 13-point difference approximation of problem (5.1) on uniform meshes, but to simplify the notations we will instead describe Bjørstad's algorithm for the continuous case.

The task is thus the solution of equation (5.1) with a rectangular region Ω. The boundary $\Gamma = \Gamma_H \cup \Gamma_V$, where Γ_H is the union of the two horizontal sides. Because of the simple geometry there are alternatives to equation (5.2) when we search for problems which are easy to solve. Thus we can solve,

$$\begin{aligned} \Delta^2\psi &= 0 , \\ \gamma_0\psi &= 0 \quad \text{on} \quad \Gamma_H \cup \Gamma_V , \\ \gamma_1\psi &= 0 \quad \text{on} \quad \Gamma_H , \\ \gamma_0\omega &= \lambda \quad \text{on} \quad \Gamma_V , \end{aligned} \tag{5.4}$$

by separation of the variables. We do this by expanding in Fourier series in the horizontal direction and solving the resulting boundary value problems for a set of fourth order ordinary differential equations. We note that the data for equation (5.4) is quite special, but that we can always reduce our problem to this form by solving a preliminary separable problem. In the discrete case the appropriate algorithmic tools are the fast Fourier transform and the Choleski algorithm for five diagonal linear systems of equations.

Equation (5.4) can be used to define a mapping from $\gamma_0\omega$ to $\gamma_1\psi$ on Γ_V. To find a preconditioner, we consider another separable problem for which we can use a Fourier expansion in the vertical direction.

$$\begin{aligned} \Delta^2\tilde{\psi} &= 0 , \\ \gamma_0\tilde{\psi} &= 0 \quad \text{on} \quad \Gamma_H \cup \Gamma_V , \\ \gamma_0\tilde{\omega} &= 0 \quad \text{on} \quad \Gamma_H , \\ \gamma_1\tilde{\psi} &= \mu \quad \text{on} \quad \Gamma_V . \end{aligned} \tag{5.5}$$

By evaluating $\gamma_0\tilde{\omega}$ on Γ_V for the solution of (5.5), we define the preconditioner which has proven to be highly successful in Bjørstad's work. A tight uniform bound on the spectrum of the preconditioned operator is rigorously established in his work. In computational practice, the error decreases by a factor of about 10 in each iteration step. We also note that advantage can be taken of the sparsity of data of the problem and the fact that in the iterations, the solution is only required on and close to Γ_V.

FINITE ELEMENT-CAPACITANCE MATRIX METHODS

There are occasions when a linear system of equations can be imbedded in a larger system which is easier to solve. The larger problem can for example be a discrete elliptic problem on a region for which a fast Poisson solver can be used. These techniques are also of interest if a series of problems, e.g. the same elliptic equation on different regions, can be imbedded in the same larger problem. In such a case, possibly extensive preprocessing of the larger problem might pay off if a sufficient number of smaller problems are to be solved. There are a number of different versions of these so called capacitance matrix methods. Here we will only consider iterative variants; see Dryja [9], O'Leary and Widlund [17], and Proskurowski and Widlund [20,21]. We also limit our discussion to self adjoint second order elliptic problems defined on a region which is imbedded in a larger region.

We can then adopt the same notations as in section 2, and regard Γ_3 as the boundary of Ω_1, which is the region of interest. The complement of $\Omega_1 \cup \Gamma_3$ with respect to the larger region is denoted by Ω_2. The vectors x_1, x_2 and x_3 are associated with the nodal values in Ω_1, Ω_2 and Γ_3 respectively. To simplify our arguments, we assume that all of the problems have symmetric, positive definite coefficient matrices.

For a Neumann problem on Ω_1, we then obtain,

$$\begin{pmatrix} K_{11} & K_{13} \\ K_{13}^T & K_{33}^{(1)} \end{pmatrix} \begin{pmatrix} x_1 \\ x_3 \end{pmatrix} = \begin{pmatrix} b_1 \\ b_3 \end{pmatrix} \tag{6.1}$$

while the larger problem has the form

$$\begin{pmatrix} K_{11} & 0 & K_{13} \\ 0 & K_{22} & K_{23} \\ K_{13}^T & K_{23}^T & K_{33} \end{pmatrix} \begin{pmatrix} x_1 \\ x_2 \\ x_3 \end{pmatrix} = \begin{pmatrix} \tilde{b}_1 \\ \tilde{b}_2 \\ \tilde{b}_3 \end{pmatrix} \tag{6.2}$$

The value of $\tilde{b}_2$ is irrelevant and we can also, at the expense of one solution of equation (6.2), reduce the right hand side of (6.1) to the case when $b_1 = 0$. Since we can solve (6.2) in each iteration step, we can use the operator S as a preconditioner. Given $\tilde{b}_3 = y$ and with $\tilde{b}_1$ and $\tilde{b}_2$ zero, we can compute the solution of equation (6.2) and substitute x_1 and x_3 into equation (6.1). We then obtain the residual $b_3 - S^{(1)}S^{-1}y$. We know from section 3 that this operator has good spectral properties.

A Dirichlet problem on Ω_1 can be reduced to a Neumann problem on Ω_2 by the following device. The finite element Dirichlet problem has the form,

$$\begin{pmatrix} K_{11} & K_{13} \\ 0 & I \end{pmatrix} \begin{pmatrix} x_1 \\ x_3 \end{pmatrix} = \begin{pmatrix} b_1 \\ b_3 \end{pmatrix}$$

The same solution can also be obtained from,

$$\begin{pmatrix} K_{11} & 0 & K_{13} \\ 0 & K_{22} & K_{23} \\ 0 & K_{23}^T & K_{33}^{(2)} \end{pmatrix} \begin{pmatrix} x_1 \\ x_2 \\ x_3 \end{pmatrix} = \begin{pmatrix} b_1 \\ K_{23} \quad b_3 \\ K_{33}^{(2)} \quad b_3 \end{pmatrix} \tag{6.3}$$

which has the unique solution $(x_1, 0, b_3)^T$. Equation (6.3) can now be solved by the same technique as the previous problem on Ω_1 since it essentially is a Neumann problem on Ω_2. The resulting iteration matrix is $S^{(2)}S^{-1}$.

It is also of interest to allow more general coefficient matrices in equation (6.2). These issues will be discussed in a forthcoming paper, see Widlund [23].

REFERENCES

[1] Bjørstad, P. E., Numerical Solution of the Biharmonic Equation, Ph.D. Thesis, Stanford Univ. (1980).

[2] Bjørstad, P. E., SIAM J. Numerical Analysis 20 (1983) 59-77.

[3] Bjørstad, P. E. and Widlund, O. B., Solving elliptic problems on regions partitioned into substructures, in: Birkhoff, G. and Schoenstadt, A. (eds.), Elliptic Problem Solvers (Academic Press, New York, 1984).

[4] Bjørstad, P. E. and Widlund, O. B., Extended version of reference 3, in preparation.

[5] Ciarlet, P. G., The finite element method for elliptic problems (North Holland, Amsterdam, 1978).

[6] Concus, P., Golub, G., and O'Leary, D., Proc. Symp. Sparse Matrix Computations, Argonne National Lab. (Academic Press, 1975).

[7] Cottle, R., Lin. Alg. Appl. 8 (1974) 189-211.

[8] Dryja, M., Numer. Math. 39 (1982) 51-69.

[9] Dryja, M., A finite element - capacitance matrix method for elliptic problems, to appear.

[10] Glowinski, R., Periaux, J. and Dihn, Q.V., Domain decomposition methods for nonlinear problems in fluid dynamics, INRIA Report No. 147 (July 1982).

[11] Glowinski, R. and Pironneau, O., SIAM Review 21 (1979) 167-212.

[12] Golub, G. H. and Mayers, D., These Proceedings.

[13] Grisvard, P., Boundary value problems in nonsmooth domains, Dept. of Math., Univ. of Maryland, Lecture Notes 19 (1980).

[14] Hestenes, M., Proc. Symp. Appl. Math. VI (McGraw-Hill, New York, 1956) 83-102.

[15] Lions, J. L. and Magenes, E., Nonhomogeneous boundary value problems and applications, I (Springer, New York, 1972).

[16] Luenberger, D. G., Introduction to linear and nonlinear programming, (Addison-Wesley, 1973).

[17] O'Leary, D. and Widlund, O., Math. Comp. 33 (1979) 849-879.

[18] Prezemieniecki, J. S., Am. Inst. Aero. Astro. 7 (1963) 138-147.

[19] Prezemieniecki, J. S., Theory of matrix structural analysis (McGraw Hill, New York, 1968).

[20] Proskurowski, W. and Widlund, O., Math. Comp. 30 (1976) 433-468.

[21] Proskurowski, W. and Widlund, O., SIAM J. Sci. Stat. Comput. 1 (1980) 410-468.

[22] Strang, G. and Fix, G. J., An analysis of the finite element method, (Prentice-Hall, 1973).

[23] Widlund, O., On the convergence of finite element - capacitance matrix methods, in preparation.

ACKNOWLEDGEMENTS

This work was supported by the National Science Foundation under Contract NSF-MCS-8203236 and by the U. S. Department of Energy under contract DE-AC02-76-ER03077-V at the Courant Mathematics and Computing Laboratory.

Computing Methods in Applied Sciences and Engineering, VI
R. Glowinski and J.-L. Lions (Editors)
Elsevier Science Publishers B.V. (North-Holland)

ITERATIVE METHODS FOR DISCRETE ELLIPTIC EQUATIONS[1]

Seymour V. Parter
Department of Computer Sciences
University of Wisconsin-Madison
Madison, Wisconsin 53706

Michael Steuerwalt
University of California
Los Alamos National Laboratory
Los Alamos, New Mexico 87545

We describe a basic theory for the estimation of the rates of convergence of iterative methods for the solution of the systems of linear equations which arise in the numerical solution of elliptic boundary value problems. This theory is then applied to finite-element equations solved via certain block or point iterative methods. There is a special emphasis on the "point" SOR iterative method.

INTRODUCTION

Consider the system of linear algebraic equations

(1) $$AU = F$$

which arises from the discretization of a boundary-value problem for an elliptic partial equation

(1.1) $$Lu = f \quad \text{in} \quad \Omega, \qquad \beta u = 0 \quad \text{on} \quad \partial\Omega \; .$$

A direct iterative method for the solution of (1) is provided by a "splitting" $A = M - N$, where the matrix M has an inverse and it is not too difficult to solve problems of the general form $MX = Y$. Then, after choosing a guess U^0 one uses the splitting to construct successive iterates $\{U^k\}$ by the formula

(1.2) $$MU^{k+1} = NU^k + F \; .$$

This iterative scheme leads one to study the eigenvalue problem

(1.3) $$\lambda MU = NU \; .$$

It is well-known that this scheme is convergent iff

(1.4) $$\rho := \max|\lambda| < 1 \; .$$

Moreover, smaller ρ implies faster convergence (see [17]).

This report is concerned with a method for determining the asymptotic behavior of $\rho = \rho(n)$, $n \to \infty$ where n is the order of the matrix A. This theory is of interest both for mathematical reasons and for practical reasons. At this time let us concentrate on the practical value of the theory.

A typical result obtained from this theory is of the form

(1.5) $$\rho \approx 1 - \Lambda_0\left(\frac{1}{n}\right)^p \; ,$$

where the exponent p is known exactly and the coefficient Λ_0 is given as $\Lambda_0 = \mathrm{Re}\,\Lambda_{min}$ where Λ_{min} is the "minimal" eigenvalue of an eigenvalue problem

(1.6) $$Lu = \Lambda Q u \quad \text{in} \quad \Omega, \qquad \beta u = 0 \quad \text{on} \quad \partial\Omega .$$

In the formula (1.6), the operator Q is a differential operator of lower (than L) degree. Hence, p, the "order" (in $1/n$)) of the rate of convergence is well-determined. But, the coefficient Λ_0 is - in general - given implicitly as an eigenvalue of a problem which is just as hard as (if not harder) than the original problem (1.1).

Of course, a knowledge of the order p is very useful in comparing competitive schemes. Still, one can only make a complete comparison of the efficiency of these schemes if one knows the coefficients Λ_0. Nevertheless, even without exact knowledge of Λ_0, the theory is still useful.

(A) There are cases - model problems - in which one can compute the eigenvalues. For these model problems a precise comparison is possible. And, as is frequently the case, one can hope the qualitative nature of this comparison extends to more general problems.

(B) In many cases the operator Q is a zero'th order operator i.e. there is a function $q > 0$ and $Qu = qu$. In these circumstances there are many cases in which Λ_0 is monotone decreasing in q. That is, if

(1.7) $$q_1 \geq q_2, \ \forall x \in \Omega; \quad \text{then} \quad \Lambda_0(q_2) \geq \Lambda_0(q_1) .$$

Thus, qualitative information about Q provides qualitative information about Λ_0. This is the case for second order elliptic operators with nice boundary conditions (e.g. Dirichlet Conditions) and for all self-adjoint problems.

(C) There are cases where one is considering a "scale" of schemes, e.g., the "k-line schemes", and the parameter of the scale (say k) appears explicitly in the formula for $\Lambda_0(k)$. In these cases one can make useful comparisons of the schemes.

(D) Finally, there are cases where one is dealing with a continuous family of iterative schemes - say depending on a parameter ω (e.g., SOR type schemes). In such a case it is desirable to have qualitative information about the dependency of $\rho(\omega)$ on ω. This qualitative information may then be used to formulate adaptive schemes to optimize the choice of ω.

Such eigenvalue problems have been studied intensively - see [5-9,12,15,17-19]. While we cannot give a complete discussion of the earlier results, it is useful to recall some of this history and comment on some of the current activity along these lines.

One of the most useful ideas is the concept of schemes which satisfy "Property A" or "Block Property A" - see [1,19]. This theory enables one to connect the eigenvalues of the block Jacobi method with the associated block SOR methods. Thus, when this important condition holds, one can estimate SOR methods in terms of the Jacobi methods. More importantly, there is a simple algorithm for an adaptive method for determining the optimal ω (see [7]).

Garabedian [6] considered the case where L is a second order operator. Let

(1.8) $$\varepsilon^k = U - U^k, \quad \text{then} \quad M(\varepsilon^{k+1} - \varepsilon^k) = -A\varepsilon^k .$$

Formally taking $\Delta t = \alpha h$, α = constant, Garabedian recognized - in the point SOR case - that (1.8) is a formal difference approximation to a time dependent equation of the form

$$b \frac{\partial \varepsilon}{\partial t} = L\varepsilon + b_1 \frac{\partial^2 \varepsilon}{\partial x \partial t} + b_2 \frac{\partial^2 \varepsilon}{\partial y \partial t} . \tag{1.9}$$

Thus, the separation of variables transformation

$$\varepsilon(x,y,t) = \varepsilon_0(x,y)e^{-\lambda t}$$

leads to the elliptic eigenvalue problem

$$L\varepsilon = \lambda Q\varepsilon \tag{1.10a}$$

where

$$Q = b - b_1 \frac{\partial}{\partial x} - b_2 \frac{\partial}{\partial y} . \tag{1.10b}$$

Garabedian then argues that the slowest rate of decay of $\varepsilon(x,y,t)$ yields the rate of convergence. This heuristic approach leads one to a "formula" for the asymptotic behavior of ρ as $n \to \infty$. While Garabedian never completed the details of a rigorous proof; in all the cases he studied, the results obtained by this method are correct. Garabedian then went on to use this "result" to obtain a formula for the optimal choice of ω.

Some years ago, Parter [8,9], developed a theory which was limited to self adjoint L, symmetric A and N. More recently Parter and Steuerwalt [12-14] have extended that theory to include non-self-adjoint L, non-self-adjoint splittings, parabolic problems and SOR methods. The treatment of SOR raises some interesting mathematical questions which we mention in Section 3.

In the last few years a group of active researchers at the Université Libre de Bruxelles under the guidance of R. Beauwens (see [2-4]) have been studying these problems in depth. They - and Parter and Steuerwalt [14] have discussed appropriate generalizations of the method of Garabedian. However, it seems that the Belgian school has not discussed many of the rigorous details.

In Section 2 we describe the general theory. In Section 3 we discuss the application to finite-element methods. In this section we describe the theoretical results of [13] for k-line block Jacobi, Gauss-Seidel and SOR and the results of [14] point SOR.

The results for finite-difference equations are contained in [12].

A GENERAL APPROACH

For simplicity we develop the basic ideas within the framework of the simplest finite-element approach to the boundary-value-problem (1.1), (1.2).

Consider the following "elliptic boundary value problem". Let Ω be a smooth domain in $\mathbf{R}^m$ and let $H_m(\Omega)$ be the usual space of functions with generalized L_2 derivatives of order m. Let

$$\tilde{H}_m \subset H_m(\Omega)$$

be a subspace and let $B(u,v)$ be a bilinear form which is continuous and coercive over $\tilde{H}_m$. We seek a function $u \in \tilde{H}_m$ which satisfies

(2.1) $$B(u,\phi) = F(\phi), \quad \forall\phi \in \tilde{H}_m$$

where $F(\phi)$ is a continuous linear functional defined on $\tilde{H}_m$.

Let $\{S_h\}$, $0 < h \leq h_0$, be a family of finite-dimensional subspaces of $\tilde{H}_m$. Let $\{\phi_1^h,\phi_2^h,\ldots,\phi_n^h\}$ be a basis for S_h. The discrete problem (1) arises from problem (2.1) restricted to S_h. That is, find $U^h \in S_h$ such that

(2.2) $$B(U^h,\phi) = F(\phi) \quad \forall\phi \in S_h .$$

Setting

(2.3) $$a_{ij} = B(\phi_j^h,\phi_i^h), \quad f_j = F(\phi_j^h), \quad U^h = \sum u_j\phi_j^h$$

the problem (2.2) takes the form (1) with

(2.4) $$U = (u_1,u_2,\ldots,u_n)^T, \quad F = (f_1,f_2,\ldots,f_n)^T .$$

Following the discussion in Section 1 we imagine a splitting (1.3) of the $n \times n$ matrix A. Our theory starts with the following heuristic approach:

Assumption A.1: Suppose there is an exponent $p > 0$ and an operator Q defined on $H_m(\Omega)$ such that $h^pN \sim Q$ in the following weak sense: for all sufficiently smooth $\phi,\psi \in \tilde{H}_m$ let Φ^h, Ψ^h be their projection into S_h and let $\hat{\Phi}$ and $\hat{\Psi}$ be the corresponding vectors of coefficients. Then, we assume that

(2.5) $$h^p\hat{\Phi}^*N\hat{\Psi} \to \iint \bar{\phi}[Q\psi]dx, \quad h \to 0 .$$

Remark: Given a splitting, the discovery of p and Q is part of the "art" of this method. We will have more to say about this subject later.

Having made this assumption, let $\lambda \neq 0$ be an eigenvalue of (1.5) with eigenvector U. Then $\lambda MU = NU$ and $\lambda(M - N)U = (1 - \lambda)NU$. Thus

$$AU = \frac{(1 - \lambda)}{\lambda h^p}(h^pN)U .$$

That is, for every $\phi \in S_h$ we have

(2.6) $$B(U^h,\phi) = \mu[\iint \bar{\phi}[NU^h]dx + E(\phi,U^h)] ,$$

where

$$\mu = \frac{1 - \lambda}{\lambda h^p}, \quad U^h = \sum u_j\phi_j^h$$

and the "error" $E(\phi,u)$ is "small" if ϕ,U^h are smooth.

From this starting point we see that we need some further technical assumptions to complete the theory.

Assumption A.2: Consider the eigenvalue problem: Find Λ and $v \in \tilde{H}_m$ such that

(2.9) $$B(v,\phi) = \Lambda \iint \bar{\phi}[Qv]dx, \quad \forall\phi \in \tilde{H}_m .$$

We assume there is a minimal eigenvalue $\Lambda_m = \Lambda_0 + iT$. By minimal we mean that for any eigenvalue λ, it is true that

(2.8a) $$0 < \Lambda_0 \leq \operatorname{Re} \lambda \ ,$$

and, if $\operatorname{Re} \lambda = \Lambda_0$ then

(2.8b) $$|\lambda| \geq |\Lambda_m| \ .$$

<u>Assumption A.3</u>: The eigenvalues of the family of discrete problems (2.6) approximate the eigenvalues of the limit problem (2.9) and vice-versa.

<u>Remark</u>: It is usually not difficult to prove A.3. In many cases it follows from the standard theories of spectral approximation (see [8]). However, it is an essential point. In dealing with this condition one raises many technical questions about "spectral approximations" and estimates which are used to establish (A.1).

Finally, we require some information about the eigenvalues λ of (1.5) which satisfy $|\lambda| = \rho$.

<u>Assumption A.4</u>: Either

(2.9a) $\rho < 1$ and ρ itself is an eigenvalue,

<u>or</u> there is a constant $C_0 > 0$ and for every h, there is an eigenvalue λ with

(2.9b) $$|\lambda| = \rho \quad \text{and} \quad \left|\frac{1-\lambda}{h^p}\right| \leq C_0 \ .$$

<u>Remarks</u>: Both (2.9a) and (2.9b) have been used. In particular, (2.9b) was used in [13] to give a new convergence theorem - as well as an estimation of rates of convergence - for certain non-self-adjoint, nonpositive type problems.

The basic results are:

<u>Theorem 1</u>: Let (A.1), (A.2) and (A.3) hold. Then

$$\rho \geq 1 - \Lambda_0 h^p + o(h^p) \ .$$

If (A.4) also holds, then

$$\rho \approx 1 - \Lambda_0 h^p \ .$$

The proof is relatively straightforward. See [13].

EXAMPLES: THEORY

This approach has been used to estimate the rates of convergence for many block iterative schemes for elliptic difference equations - particularly by Parter [9,10]; Parter and Steuerwalt [12] and the Belgian school [2-4]. In this section we describe the application to certain finite-element equations.

Consider the Dirichlet problem for a second order elliptic operator defined on Ω, the unit square. That is

(3.1) $$Lu = f, \quad (x,y) \in \Omega; \quad u(x,y) = 0 \quad \text{for} \quad (x,y) \in \partial\Omega$$

where

$$Lu := -[(au_x)_x + (bu_x)y + (bu_y)_x + (cu_y)_y] + d_1u_x + d_2u_y + d_0u$$

and L is uniformly elliptic with $d_0 \geq 0$.

Let $\Delta x, \Delta y$ be chosen and consider the finite-element subspace S_h of Tensor products of hermite cubic splines. That is, on each rectangle

$$x_k \leq x \leq x_{k+1}, \quad y_j \leq y \leq y_{j+1}$$

the elements of S_h are cubic polynomials in (x,y) which are completely determined by the 4-vectors

$$U_{kj} = (u_{kj}, h(u_x)_{kj}, h(u_y)_{kj}, h^2(u_{xy})_{kj}) \ . \tag{3.2}$$

Moreover, we choose the basis of S_h to be that basis for which the $\{U_{kj}\}$ are the correct interpolation conditions. We use these subspaces in a regular finite element approach to the approximate solution of the boundary-value-problem (3.1). In this way the equations (2.1) lead to a system of equations (1.).

Because the basic unknowns are described by the 4-vectors U_{jk} we choose to write the matrix A in a double subscript notation. That is: a typical term of AU has the form

$$\sum a^0_{jk,\sigma\mu} u_{\sigma\mu} + \sum a^1_{jk,\sigma\mu}(hu_x)_{\sigma\mu} + \sum \tilde{a}^1_{jk,\sigma\mu}(hu_y)_{\sigma\mu} + \sum a^2_{jk,\sigma\mu}(h^2u_{xy})_{\sigma\mu} \ . \tag{3.3}$$

The matrix A is sparse, but nevertheless, in the general case each row of A contains 36 non-zero entries. This complexity in A is frequently reflected in N as well. Hence, it appears it is not too easy to study the bilinear form V^*NU and verify (A.1). However, in this case - and the discerning reader will appreciate that similar simplifications occurs in all "nodal" finite-element spaces - the basic nature of S_h yield estimates which enable one to ignore certain terms in V^*NU. Specifically we have:

Theorem 3.1: Let

$$h = \sqrt{\Delta x \Delta y} \ .$$

There is a constant $K > 0$ such that, for every $u,v \in S_h$ and every $\phi \in C^1(\bar{\Omega})$ we have

$$h^2 \sum_{i,j} |v_{ij}|^2 \leq K\|v\|^2_{L_2} \ , \tag{3.5a}$$

$$h^2 \sum_{i,j} u_{ij}v_{ij}\phi_{ij} = \iint uv\phi\,dxdy + \delta(u,v,\phi) \ , \tag{3.5b}$$

$$h^2 \sum_{i,j} [|v_{i,j} - v_{i+1,j}|^2 + |v_{ij} - v_{i,j+1}|^2] \leq Kh^2\|\nabla v\|^2_{L_2} \ , \tag{3.5c}$$

$$h^2 \sum_{i,j} \sum_{|\alpha|=1}^{6} (\Delta x)^{\alpha_1}(\Delta y)^{\alpha_2}|(D^\alpha u)_{ij}|^2 \leq Kh^2\|\nabla u\|^2_{L_2} \ , \tag{3.5d}$$

$$h^2 \sum_{i,j} \{ \sum_{|\alpha|+|\beta|\geq 1} (\Delta x)^{\alpha_1+\beta_1}(\Delta y)^{\alpha_2+\beta_2}|(D^\alpha u)_{ij}(D^\beta v)_{ij}|\} \leq K\eta(u,v,h) \tag{3.5e}$$

where

(3.6a) $$|\delta(u,v,\phi)| \leq K[1 + \|\nabla\phi\|_\infty]\eta(u + v, u - v, h)$$

with

(3.6b) $$\eta(u,v,h) = h[\|u\|_{L_2}\|\nabla v\|_{L_2} + \|v\|_{L_2}\|\nabla u\|_{L_2} + h\|\nabla u\|_{L_2}\|\nabla u\|_{L_2}] .$$

Proof: See Section 7 of [13].

Having established these estimates we now turn to a class of splittings which includes the block Jacobi methods and the point Gauss-Seidel methods but not the SOR methods.

We suppose our splitting satisfies

Property S:

(i) If $a_{ij} = 0$, then $n_{ij} = m_{ij} = 0$.
(ii) If $a_{ij} \neq 0$ and $n_{ij} \neq 0$, then $n_{ij} = -a_{ij}$.

For such schemes we consider the bilinear form $V^*(h^2N)U$.

As we attempt to find Q and verify Assumption (A.1) our task is simplified by the estimates of Theorem 3.1. Using (3.5c) we see that we need only consider the terms of the form

$$h^2 \sum_{i,j,\sigma,\mu} n_{ij,\sigma\mu}\bar{v}_{ij}u_{\sigma\mu} .$$

Then, using (3.5c) we may rewrite this expression as

(3.7) $$h^2 \sum_{i,j} \bar{v}_{ij}u_{ij}\left(\sum_{\sigma,\mu} n_{ij,\sigma\mu}\right) + o(1) .$$

Moreover (3.5b) suggests two basic facts:

(3.8) $$h^2 \sum_{i,j} \bar{v}_{ij}u_{ij}\left(\sum_{\sigma,u} n_{ij,\sigma u}\right) \sim \iint q(x,y)\bar{v}(x,y)u(x,y)dxdy$$

and, in order to determine the function q we need only a rough evaluation of the integrals which constitute the $\{n_{ij,\sigma\mu}\}$.

These arguments have been carried out in complete detail in [12]. If

$$\sum_{\sigma,\mu} n_{ij,\sigma\mu} = \bar{q}_{ij}(h)$$

converges to a function $q(x,y)$, then A.1 is satisfied where the phrase sufficiently smooth merely means $\phi,\psi \in \overset{\circ}{H}_1(\Omega)$. If $q(x,y) > 0$ then A.2 is satisfied. Because we only require $\phi,\psi \in H_1$, it is an easy matter to obtain A.3 from the standard theory of Spectral Approximation [8]. Finally, the verification A.4 rests on the fact that these block schemes satisfy block property A. Even so, in the non-self-adjoint case, it is not easy. The main results are

Theorem 3.2: Consider the k-line (horizontal block Jacobi scheme) applied to (3.3a). Let Γ_0 be the minimal eigenvalue of the elliptic eigenvalue problem

(3.9) $$L\phi = \lambda c(x,y)\phi \text{ in } \Omega, \quad \phi = 0 \text{ in } \partial\Omega .$$

Then, the radius of convergence of this iterative scheme is given by

(3.10a) $$\rho_J(k) \approx 1 - \frac{5}{12} k\Gamma_0(\Delta y)^2 .$$

Since these schemes satisfy block property A the theory of Young [18] gives estimates

$$\rho_{GS}(k) \approx 1 - \frac{5}{6} k\Gamma_0(\Delta_y)^2, \quad \rho(k,\omega_b) \approx 1 - 2\left(\frac{5}{6} k\Gamma_0\right)^{1/2}\Delta y .$$

Theorem 3.3: Consider the "point" Gauss-Seidel iterative scheme. Let $r := \Delta y/\Delta x$ and let

$$q(x,y) := \frac{156}{175} [ra(x,y) + \frac{1}{r} c(x,y)] .$$

Then

$$\rho \geqq 1 - \Lambda_0 h^2 + o(h^2)$$

where Λ_0 is the minimal eigenvalue of the problem

$$Lu = \lambda qu \quad \text{in} \quad \Omega, \qquad u = 0 \quad \text{on} \quad \partial\Omega .$$

The "point" SOR splitting does not satisfy property S. Moreover, this scheme does not satisfy property A. To illustrate the application of the theory in this case we consider the simplest case, i.e.

(3.11) $$Lu = -(u_{xx} + u_{yy}) .$$

We write the matrix A as $A = D - W - W^*$ where D is the diagonal of A and W is strictly lower triangular. With $\omega = 2 - Ch$, $C > 0$, the SOR iteration is described by

$$\frac{1}{\omega} (D - \omega W)U^{k+1} = \frac{1}{\omega} [\omega W^* + (1 - \omega)D]U^k + F .$$

Then

(3.12) $$hN = h[W^* - \frac{1}{2} D] + \frac{Ch^2}{4} D + O(h^3)D .$$

Using the estimates of Theorem 3.1 we find that

(3.13)
$$V^*(hN)U = h^2 \sum_{i,j} C \frac{156}{175} u_{ij}\bar{v}_{ij} + h^2 \sum_{i,j} \bar{v}_{ij}\left[\frac{210}{175} \frac{u_{i,j+1} - u_{i,j}}{h}\right]$$
$$+ h^2 \sum_{i,j} \bar{v}_{ij}\left[\frac{102}{175} \frac{u_{i+1,j} - u_{i,j}}{h}\right] + \frac{h^2}{10} \sum_{i,j} [(\bar{v}_y)_{ij}u_{ij} - (u_y)_{ij}\bar{v}_{ij}] + \varepsilon(u,v)$$

where $\varepsilon(u,v)$ is $o([\|u\|_{H_1} + \|v\|_{H_1}]^2)$. Thus (A.1) holds with

(3.14) $$Qu = C \frac{156}{175} u + \frac{102}{175} u_x + u_y .$$

Unfortunately, in this case the phrase "sufficiently smooth" would seem to require more than $\phi,\psi \in \overset{\circ}{H}_1(\Omega)$. This fact also makes the proof of A.3 more difficult. Nevertheless it is true that A.3 holds. As for (A.4), we do not know. Nevertheless we may apply Theorem I and obtain an inequality. Then, assuming that equality actually holds we use the Garabedian substitution

$$u = v\exp\{-\frac{\Lambda}{2}(\frac{102}{175}x + y)\}$$

to obtain an "exact" formula for $\rho(\omega)$ provided that $\omega = 2 - Ch > 0$. For each ω, let $\tilde{\rho}(\omega)$ denote the value so obtained. Then $\tilde{\rho}(\omega)$ is a lower bound (and is probably equal to) the true value of $\rho(\omega)$. Using this formula we obtain a candidate $\tilde{\omega}_b$ for the optimal ω. This approach yields

(3.15a) $$\tilde{\omega}_b = 2 - c_b h, \quad c_b = \frac{\sqrt{2}\,\pi}{156}[41029]^{1/2}$$

(3.15b) $$\rho(\tilde{\omega}_b) > 1 - \frac{2\sqrt{2}\,\pi(175)}{[4109]^{1/2}} h = \tilde{\rho}(\tilde{\omega}_b) .$$

It is both more interesting and more convenient to take $c = (2 - \omega)/h$ as the independent variable and consider the function

(3.16) $$\Lambda_0(c) = \frac{1 - \tilde{\rho}(\omega)}{h} .$$

The results are qualitatively like the results obtained by Young in the case of (block) property A. We have

Case 1: $0 < c < c_b$. Then

(3.17) $$\Lambda_0(c) = \frac{2\sqrt{2}\,\pi(175)}{[41029]^{1/2}}\frac{c}{c_b} .$$

Notice, (3.17) asserts that in the range $\tilde{\omega}_b \leq \omega \leq 2$, the quantity $\tilde{\rho}(\omega)$ is asymptotically linear in ω.

Case 2: $c_b < c < 2/h$. Then

(3.18) $$\Lambda_0(c) = \frac{2\sqrt{2}\,\pi(175)}{[41029]^{1/2}}\frac{c}{c_b} - \frac{1}{[41029]}[(156c)^2 - 2\pi^2(41029)]^{1/2} .$$

Observe that

(3.19) $$\frac{d}{dc}\Lambda_0(c)\Big|_{c=c_b+} = -\infty .$$

Thus, just as in the case when property A applies, it is better to over-estimate $\tilde{\omega}_b$ than to underestimate $\tilde{\omega}_b$. Moreover, the function is linear for $\omega > \tilde{\omega}_b$ and decreases sharply for $\omega < \tilde{\omega}_b$. Finally, we emphasize that $\tilde{\rho}(\omega)$ has a unique minimum. Hence, if we could believe that $\rho(\omega) = \tilde{\rho}(\omega)$ we could use these facts to develop an adaptive approach for the determination of $\tilde{\omega}_b$.

FOOTNOTES

[1]Supported by the U. S. Department of Energy under Contract W-7405 Eng-36, and by the Air Force Office of Scientific Research under Contract AFOSR-82-0275.

REFERENCES

[1] Arms, R. J., Gates, L. D. and Zondek, B., A method of block iteration, J. Soc. Ind. Appl. Math. 4 (1956) 220-229.

[2] Bardiaux, M., Etude sur l'estimation des taux de convergence des méthodes itératives de factorisation du type FP(k) en présence de relaxation, Rapport Technique, Université Libre de Bruxelles (1978).

(3) Beauwens, R., Garabedian estimates for single relaxation, Technical Report, Université Libre de Bruxelles (1977).

[4] Beauwens, R. and Surlezanoska, S., Sur les taux de convergence asymptotiques des méthodes itératives de factorisation, Rapport Technique, Université Libre de Bruxelles, Bruxelles (1976).

[5] Frankel, S. P., Convergence rates of iterative treatments of partial differential equations, M. T. A. C. 4 (1950) 65-76.

[6] Garabedian, P. R., Estimation of the relaxation factor for small mesh size, M. T. A. C. 10 (1956) 183-185.

[7] Hageman, L. A. and Young, D. M., Applied Iterative Methods (Academic Press, New York, 1981).

[8] Osborn, J. E., Spectral approximation for compact operators, Math. Comp. 29 (1975) 712-725.

[9] Parter, S. V., Multi-line iterative methods for elliptic difference equations and fundamental frequencies, Numer. Math. 3 (1961) 305-319.

[10] ________________, On estimating the "rates of convergence" of iterative methods for elliptic difference equations, Trans. Amer. Math. Soc. 114 (1965) 320-354.

[11] ________________, On the eigenvalues of second order elliptic difference operators, SIAM J. Numer. Anal. 19 (1982) 518-530.

[12] Parter, S. V. and Steuerwalt, M., Block iterative methods for elliptic and parabolic difference equations, SIAM J. Numer. Anal. 19 (1982) 1173-1195.

[13] ________________, Block iterative methods for elliptic finite element equations, SIAM J. Numer. Anal., to appear.

[14] ________________, On point SOR for elliptic finite-element equations, in preparation.

[15] Shortly, G. H. and Weller, R., The numerical solution of LaPlaces equation, J. Appl. Physics 9 (1938) 334-348.

[16] Varga, R. S., Factorization and normalized iterative methods, in: Langer, R. E. (ed.), Boundary Problems in Differential Equations (University of Wisconsin Press, Madison, 1960).

[17] Varga, R. S., Matrix Iterative Analysis, (Prentice-Hall, Englewood Cliffs, New Jersey, 1962).

[18] Young, D. M., Iterative methods for solving partial difference equations of elliptic type, Trans. Amer. Math. Soc. 76 (1954) 92-111.

[19] ________________, Iterative solution of large linear systems (Academic Press, New York, 1971).

Computing Methods in Applied Sciences and Engineering, VI
R. Glowinski and J.-L. Lions (Editors)
Elsevier Science Publishers B.V. (North-Holland)

THE SOLUTION OF NEARLY SYMMETRIC SPARSE LINEAR EQUATIONS

Iain S. Duff
Computer Science and Systems Division
Atomic Energy Research Establishment,
Harwell, Oxfordshire,
England

There exist very satisfactory iterative and direct methods for the solution of sparse sets of positive-definite symmetric equations. We examine the extension of these methods to more general systems and look at techniques which are particularly efficacious when the system is nearly symmetric. We define what we mean by this term and consider direct methods employing special pivoting strategies and some based on frontal approaches. The iterative methods discussed include preconditioned Lanczos and conjugate gradients and techniques based on Chebychev acceleration.

INTRODUCTION

Over the last few years there has been a considerable amount of work both in algorithms and software for the solution of sparse symmetric positive definite linear equations. In the direct solution of such equations, approaches based on minimum degree orderings and variable-band or profile methods are the most popular and have been refined so that their implementation is singularly efficient (see for example, George and Liu, 1981). Few open problems exist in this area and the availability of good software is generally very satisfactory. The situation regarding the use of iterative methods to solve such systems is almost as satisfactory, only the availability of good software falls short of ideal. The most popular and powerful iterative methods are based on variants of point or block incomplete Cholesky conjugate gradient (ICCG), where the conjugate gradient algorithm is preconditioned by a partial factorization of the coefficient matrix.

Unfortunately these highly refined and efficient techniques do not trivially extend to the more general case of symmetric indefinite and unsymmetric systems. There are efficient extensions to the indefinite symmetric case, for example the use of 2x2 pivoting in direct methods (Duff and Reid, 1983) and the use of Lanczos recurrences rather than the related conjugate gradient iteration for iterative methods (for example, SYMMLQ of Paige and Saunders, 1975). We will not, however, concern ourselves here with these minor (but non-trivial) extensions.

In this paper, we will consider the solution of sparse sets of linear equations

$$A\underline{x} = \underline{b} \qquad (1.1)$$

where the matrix A is either symmetric in structure only or is completely unsymmetric. The methods we discuss will, however, only be efficient if A is nearly symmetric. We define what is meant by this in section 2 of this paper.

Fortunately, the class of nearly symmetric systems includes many practically occurring problems so there is some justification for designing algorithms and software specifically for this class. We give some examples of where such systems arise in section 2. In section 3 we discuss the solution of the system (1.1) by direct methods and, in section 4, the solution using iterative methods is considered. The terms iterative and direct are somewhat fuzzy in this context since many approaches include elements of both. We include comments on this intersection in both sections 3 and 4.

We have intentionally avoided reference to techniques based on multigrids. We do this because multigrid approaches are being considered in some detail by several other contributors to this volume rather than because we do not believe them to be useful. A good description of multigrid methods can be found in the Proceedings edited by Hackbusch and Trottenberg (1982). Multigrid methods have already shown themselves competitive on symmetric systems (for example, Behie and Forsyth, 1983) and they can be used to solve non-self-adjoint problems (see, for example, the above conference volume).

NEARLY SYMMETRIC SYSTEMS

In this section we clarify the term "nearly symmetric" and indicate some problems giving rise to such systems. Unfortunately, the term is not independent of the method of solution and so the classification cannot be used indiscriminately on a given problem.

For direct methods, a system is "nearly" symmetric if most off-diagonal entries occur in pairs, that is $a_{ij} \neq 0$ if and only if $a_{ji} \neq 0$ with only a few exceptions. We quantify this in section 3 where we look at the performance of a direct method designed for such a system. With direct methods it is also important to maintain the near symmetry during the elimination process. To this end, diagonal pivoting is to be preferred but will be potentially unstable unless some form of diagonal dominance is present. Again we quantify the diagonal dominance requirements in section 3.

For iterative methods, numerical symmetry is much more important since then three term recurrence relations are sufficient to maintain conjugacy and orthogonality in conjugate direction methods and the fact that the spectrum is real can be utilized by other approaches. Indeed Faber and Manteuffel (1982) have recently shown that for an s-term conjugate gradient method to exist (s a fixed integer) the matrix A must either be Hermitian or of the form $A=(dI+B)\,e^{i\theta}$, with B skew-Hermitian. Perhaps even more important is how close the matrix is to being positive definite since the property that the eigenvalues lie in a half-plane not containing the origin is necessary for many iterative techniques, notably methods based on Chebychev acceleration (Manteuffel, 1977).

A common source of near symmetric matrices is obtained from discretizations of the convection-diffusion equation

$$\epsilon\, \nabla^2 u + \nabla u = f \qquad (2.1)$$

where the use of either central or upwind differencing for the convection term will yield a matrix symmetric in structure but unsymmetric in value. The asymmetry will become more pronounced as $\epsilon \to 0$. If upwind differencing is used the matrix will normally be diagonally dominant (depending on the boundary conditions) but this would not be the case when central differences are in use on highly convective equations ($\epsilon \ll 1$). We use simplified systems derived from (2.1) in section 3.

A more complicated example arises when solving coupled differential equations arising in fluid-flow calculations. For example, the Navier-Stokes equations when using Boussinesq approximation for bouyancy driven flow can be written

$$\underline{u}.\nabla\underline{u} = -\nabla p - \nabla^2\underline{u} + \gamma.\underline{g}.T$$
$$\nabla.\underline{u} = 0$$
$$\underline{u}.\nabla T - \kappa\nabla^2 T = 0$$

whence after a finite-element discretization the algebraic equations can be written

$$\begin{bmatrix} N & C & F \\ C^T & 0 & 0 \\ G & 0 & K \end{bmatrix} \begin{bmatrix} u \\ p \\ T \end{bmatrix} = 0$$

which are clearly not positive definite and are not even symmetric since $F \neq G^T$.

Other examples which have nonsymmetric structure but are "nearly symmetric" arise in oil reservoir modelling where complications such as multiply-connected wells are present. These give rise to "off-band" terms which are unsymmetric since the flow is assumed to go one way... i.e. in at an injection well and out at a production well.

DIRECT METHODS

The main reason why direct methods are so efficient when the system is positive definite is that pivoting for sparsity can be completely decoupled from the numerical factorization and indeed diagonal pivoting in any order is numerically stable. Additionally, since the choice of pivots is restricted to diagonal entries, algorithms for their selection and the subsequent adjustments to data structures are enormously simplified. Of a more minor nature, in the sparse case, are the savings in storage and arithmetic operations because of symmetry in the numerical factorization.

When the coefficient matrix is more general then diagonal pivoting may be unstable and, in some cases, impossible. If the system is symmetric indefinite, a result of Bunch and Parlett (1971) shows that it is always possible to choose block 2x2 pivots which ensure stability while maintaining symmetry, and this approach can be extended to the sparse case (Duff and Reid, 1982a, 1983). One can, of course, use techniques for general unsymmetric systems but these can be relatively expensive (see, for example, Duff, 1982).

Indeed, pivoting strategies which are well-established for general systems may be very inefficient on symmetric or near-symmetric systems particularly those with regular structures, for example from discretizations of partial differential equations. After collaboration with Zlatev and Wasniewski of Copenhagen and earlier experiments by Zlatev (1980), alternative pivoting strategies were included in the Harwell Subroutine library code MA28 (Duff, 1977), which uses the pivotal strategy of Markowitz (1957), subject to a potential pivot a_{ij}, say, satisfying a threshold condition

$$|a_{ij}| \geq u.\max_k |a_{kj}| \tag{3.1}$$

where u, the threshold parameter, has some preset value in the range (0,1]. The main additions were to allow an arbitrary restriction on the number of rows in which the pivot search can be conducted (a value of 3 is recommended by Zlatev, 1980) and to choose the best numerical pivot from those with equal sparsity claims. The effect of introducing these changes can be quite dramatic as is witnessed in the results of Table 3.1. The limitation of the search reduces the time

Pivoting option	Time (Seconds on IBM 3081K)	Non-zeros in factors	Total storage	Scaled residual
Threshold with Markowitz	51.0	91487	263520	10^{-12}
Restricted search	25.9	90797	259021	10^{-12}
Stricter numerical control	11.6	68108	195177	10^{-15}

Table 3.1 The effect of changes in the pivot strategy when using a general unsymmetric code. Problem is five-diagonal of order 2000 with semi-bandwidth 42.

to select the pivots, in turn reducing the overall factorization time, which is evident if one compares rows 1 and 2 in the table. The main problem with an exhaustive Markowitz search is that the regular structure means that many rows have the same number of non-zeros present in them and often a long search is performed to establish that the current best Markowitz count cannot be bettered. The limitation on the search does not noticeably affect either the storage for the factors or the stability. The code will still show no preference for diagonal entries but, if the numerically best pivot is also selected, then diagonal dominance will cause the diagonal to be favoured. For our example in Table 3.1 this has two advantages. First, diagonal pivots will, although having the same Markowitz cost, give lower fill-in and the greater maintenance of stability means that more entries are suitable for subsequent pivoting because excessive growth has not occurred. The effect is clear from line 3 of the Table where not only is the decomposition more accurate but the reduction in fill-in through using diagonal pivots halves the overall factorization time.

The same effect gives rise to a counter-intuitive performance of threshold pivoting codes when solving systems of the form used in

Table 3.1. Normally, we would expect that increasing the threshold parameter would improve stability at the cost of sparsity but the results, shown in Table 3.2, of the earlier version of MA28 on a discretization of Poisson's equation in polar coordinates indicate that the opposite can occur. This is again caused by forcing more pivots to be chosen from the diagonal, since, as the threshold parameter, u, increases, more off-diagonal pivots are unsuitable because of the diagonal dominance of the system.

Value of threshold parameter, u	.1	.4	.7	.9
Non-zeros in factors	8677	7239	6657	6643
Growth in size of intermediate entries	1.3×10^9	3.6×10^4	2.9×10^3	3.1×10^3

Table 3.2 The effect of changing the threshold parameter when running MA28 on diagonally-dominant systems. Poisson's equation in polar coordinates on a 21x21 grid.

The major drawback to using direct methods based on Gaussian elimination, particularly on equations from the discretization of partial differential equations, lies in the storage required for the matrix factors. For some purposes, however, only factors of an approximate inverse are required, for example when they are being used as a preconditioning for an iterative method as in the popular ICCG method where an incomplete Cholesky decomposition is used to precondition the conjugate gradient method. One might expect that the more accurate a factorization, the better preconditioner one would have until, in the extreme case, a direct method is obtained. Thus a whole spectrum of semi-direct methods can be devised. Some of these we will discuss in section 4 but here we will emphasise an approach where the stress is on the direct factorization rather than the iterative technique being accelerated.

A long-established technique for improving the solution of linear equations is that of iterative refinement which is a particular form of the more general defect correction (for example, Stetter, 1978). Here, an approximate solution $\underline{x}^{(k)}$ is improved through the correction

$$\underline{\delta x}^{(k)} = B\ \underline{r}^{(k)}$$

where the residual $\underline{r}^{(k)} = \underline{b}-A\underline{x}^{(k)}$, and B is an approximation to the inverse of A. This approximation, B, need not be very accurate for convergence of the iterative process. All that is required is that the spectral radius of I-BA is less than 1.0. Of course, for poor approximations convergence may be very slow. We can easily extend the incomplete factorization idea embodied in ICCG to quite general systems by performing Gaussian elimination on the system and ignoring fill-ins on the basis of their position or size. For general systems we prefer the latter and choose to ignore entries a_{ij} (dropping them completely from the data structure) if

$$|a_{ij}| < TOL * \max_k |a_{ik}|$$

where TOL is some preset non-negative drop tolerance. Drop tolerances were included in the Harwell Subroutine Library code for sparse symmetric positive-definite systems, MA31 (Munksgaard, 1980) and have been recommended for use in unsymmetric codes by Zlatev et al (1982). In collaboration with Zlatev and Wasniewski, the use of drop tolerances with iterative refinement has been incorporated in the Harwell MA28 package and we show in Table 3.3 the results of running the code with a number of drop tolerances on the structure

Value of drop tolerance	0.	10^{-4}	10^{-2}	.05
Storage for "factors"	48055	42701	38911	28325
Time for factorization	2.184	1.719	1.410	.526
Number of iterations	1	2	3	8
Time for solution	.066	.094	.121	.235

Table 3.3 The effect of using drop tolerances with iterative refinement in MA28 on five-diagonal systems. Accuracy of 10^{-3} requested. Times are in seconds on an IBM 3081K. Order 1000, semi-bandwidth 100.

used in Table 3.1. At higher values of the drop tolerance (say 0.1), the iteration fails to converge. Since lowering the drop tolerance decreases the number of iterations at the cost of more storage and more work per iteration, a balance must be sought and the optimal value for the drop tolerance will depend on the relative importance of storage and time.

Another way of reducing the number of iterations is to use a more powerful iterative method. An example of this is the Chebychev iteration of Manteuffel which only requires that the eigenvalues of A (in our case of the preconditioned system) lie in the positive half-plane. Duff and Zlatev are currently performing experiments in combining a preconditioning using drop tolerances with Manteuffel's method to provide a more general algorithm than that of van der Vorst (1981). Initial results look encouraging.

Approaches using incomplete factorizations can be considered hybrid direct-iterative methods. Another form of hybrid can be obtained by solving part of the system by a direct method and the remainder by an iterative one. Li et al (1982) give one example of this approach but, to my belief, such a hybrid method has not been used successfully on unsymmetric problems.

So far in this section, we have concentrated on how a general solution scheme for unsymmetric systems might be used efficiently on symmetric or near-symmetric systems. Other efforts attempt to extend the highly successful symmetric methods to more general cases. Both of the popular codes SPARSPAK (George et al, 1980) and YSMP (Eisenstat et al, 1982) offer entries for diagonal pivoting on structurally symmetric systems and YSMP has a version for diagonal pivoting on unsymmetric systems corresponding to a sparsity ordering on the pattern of the upper triangle of A reflected in the diagonal. None of these codes uses any numerical control. A more robust approach is provided by NSPFAC (a derivative of NSPIV, Sherman, 1978) and MA37 (Duff and Reid, 1982b) where an initial ordering is perturbed during the factorization to maintain stability.

Order	534	1224	183	216
Non-zeros	3474	9613	1069	876
Measure of asymmetry	.26	.39	.58	1.0
Non-zeros in factors				
MA37	9714	91136	2775	7058
MA28	9297	63313	1271	3677
Total storage for subsequent solution				
MA37	16271	116471	5172	10607
MA28	21787	133971	3641	8651

Table 3.4 The performance of MA37 on problems with varying degrees of asymmetry. All problems were made diagonally dominant.

In the case of NSPFAC, the initial ordering is prescribed, for MA37 an ordering produced by the minimum degree ordering on the pattern of $A+A^T$ is used. One might expect that the latter ordering would do badly if the matrix is far from symmetric and we illustrate this in Table 3.4 where the index of asymmetry has been calculated as

$$\frac{\text{Number of pairs such that } a_{ij}=0,\ a_{ji}\neq 0}{\text{Total number of off-diagonal entries}}$$

and we compare the ordering with that obtained when off-diagonal pivoting is allowed. Another problem with techniques which perturb a diagonal-pivoting strategy is that, even if the original matrix is nearly symmetric, the forced selection of off-diagonal pivots for stability reasons will cause increasing asymmetry in successive reduced matrices and so make later diagonal pivots inappropriate. We illustrate this with the runs in Table 3.5 where we have used

Scaling on diagonal	4	1	.01	.001
Non-zero in factors	8048	8344	13640	14660
Total storage for factors	13134	13440	18118	19226
Multiplications	52838	56264	140460	170380

Table 3.5 Effect of non-diagonal dominance on the performance of MA37. On model problem on 20×20 grid with off-diagonals -1 and diagonal entries e.

five-diagonal matrices with the diagonal entries a_{ii} equal to e and off-diagonal entries -1. As e is reduced the loss of diagonal dominance causes MA37 to perform increasingly badly.

In Tables 3.4 and 3.5 we have decoupled asymmetry from non-diagonal dominance. If one combines the effects of both, the performance of

MA37 degrades even further. For example, on a chemical engineering problem of order 205 with 572 non-zeros the MA37 code yields factors with 11581 non-zeros as against the 5719 forecast if diagonal dominance were present as against 619 if a general off-diagonal pivoting scheme were used.

In conclusion, these extensions of symmetric methods can be made robust and are powerful and efficient on "near-symmetric" matrices but can perform very poorly on more general systems. An approach of Pagallo and Maulino (1983) attempts to overcome the defect in starting with an ordering based on some symmetric derivative of A but, to my knowledge, there is no efficient implementation of this and the harder problem of maintaining stability while preserving the good sparsity properties of the original ordering has not been tackled.

A common source of problems of symmetric structure but unsymmetric values arises in finite-element discretizations where the geometry dictates symmetry but the individual element matrices, and hence the assembled matrix, are unsymmetric. In this case, we can use a frontal scheme where the pivot selection and elimination are performed within full frontal matrices. Our experience with a code implementing this approach (MA32, Duff, 1981) has been very encouraging particularly on the CRAY-1 where vectorization of the inner-loops can give them a runtime performance of over 100 million floating point operations per second (Duff, 1983). Three-dimensional problems with front-widths of nearly 700 are currently being solved on the CRAY-1 at Harwell using this code (Jackson, private communication).

ITERATIVE METHODS

In this section on iterative methods, it is our intention to give a flavour of the methods for solving nearly symmetric systems. In no sense is our treatment intended to be exhaustive. Readers interested in a detailed survey should consult Eisenstat (1982).

Many efficient iterative methods for general symmetric systems are based on the conjugate-gradient algorithm (Hestenes and Steifel, 1952) or the closely allied method due to Lanczos (1950,1952). The relationship is very neatly detailed in the thesis of Simon (1982) inter alia. With a suitable starting approximation these methods generate "best" approximations to the solution which lie in Krylov subspaces of increasing dimension. Thus, at the kth stage, the approximation lies in

$$\underline{x}^{(0)}+\mathrm{Sp}(\underline{r}^{(0)},A\underline{r}^{(0)},A^2\underline{r}^{(0)},A^3\ \underline{r}^{(0)},\ldots\ldots,A^{k-1}\underline{r}^{(0)})$$

where $\underline{r}^{(0)}$ is the residual $(\underline{b}-A\underline{x}^{(0)})$ corresponding to the initial estimate of the solution $\underline{x}^{(0)}$. The difference between the various algorithms depends on the definition of "best". One chooses an approximation which minimizes the error or residual in some norm and the mathematical differences depend on which is minimized and what norm is used. Of course, there may be additional numerical differences between the various formulations.

While the convergence of all these methods is guaranteed if the system is positive definite and many methods require only symmetry, the rate of convergence is governed by the spectrum of the iteration matrix, both on its range and distribution. Recently methods have been suggested which precondition the matrix yielding a matrix whose spectrum is better suited for fast convergence (for example, Concus et al, 1976). Normally, one would expect to obtain a good approximation to the solution in far fewer than n iterations where n is the order of the system. For example, Jackson and Robinson (1982) have generalized the conjugate-gradient algorithm into a scheme for solving (1.1) as follows:

Given $\underline{x}^{(0)}$

$$\underline{r}_o = \underline{b} - A\underline{x}_o$$

$$\underline{p}_o = P\underline{r}_o$$

i=0,1...

$$\left.\begin{aligned}
\alpha_i &= (A\underline{p}_i, \underline{r}_i)_K / (A\underline{p}_i, A\underline{p}_i)_K \\
\underline{x}_{i+1} &= \underline{x}_i + \alpha_i \underline{p}_i \\
\underline{r}_{i+1} &= \underline{r}_i - \alpha_i A\underline{p}_i \\
\beta_t &= -(AP\underline{r}_{i+1}, A\underline{q}_t)_K / (A\underline{p}_i, A\underline{q}_t)_K \qquad t=1,\ldots,s \\
\underline{p}_{i+1} &= P\underline{r}_{i+1} + \sum_{t=1}^{s} \beta_t \underline{q}_t
\end{aligned}\right| \qquad (4.1)$$

where the $\underline{q}_t$ are s previously generated search directions chosen from $\underline{p}_j$ (j<i+1), A includes its preconditioning and may be unsymmetric, P is an auxiliary matrix which we use later to define a biconjugate gradient algorithm, and the notation $(\underline{x},\underline{y})_K$ denotes the K norm $\underline{x}^T K \underline{y}$ for any positive definite K. ($K=A^{-1}$, $P=I$ for the usual conjugate-gradient algorithm).

The crucial thing which makes this class of methods so successful for symmetric systems is the self-adjoint nature of the operator with respect to the K-norm. This enables one to generate a recurrence relation for bases of the successive Krylov spaces which avoids any explicit reorthogonalization against all previously generated basis vectors. Thus, s in (4.1) has the value 1,$\underline{q}_i = \underline{p}_i$ and the storage is a very modest multiple of n. The work per iteration is usually dominated by the matrix vector multiplication and the solution of the preconditioned equations (which are designed to be simple) since the only other work is again a small multiple of n.

When the matrix is not symmetric (even though its structure may be), it may not be possible to use an s-term recurrence, for a fixed integer s. Faber and Manteuffel (1982) examine in some detail the conditions which are necessary and sufficient for this. Thus, when using Lanczos recurrences, an upper Hessenberg rather than tridiagonal system is generated and must be solved either implicitly or explicitly, and, when using conjugate methods, the new search direction must be orthogonalized, at each step, against all previous search directions. The extra work and storage necessary makes these methods unattractive on any reasonably sized problems.

There are basically two main approaches for overcoming this problem. The first is to, in some sense, symmetrize the system while the other is to truncate one of the above techniques so that work and storage requirements are manageable. We discuss each in turn. Naturally in all cases a good preconditioning is desirable if not essential. We will discuss these also.

A long-established technique for solving the system (1.1) when techniques for solving only symmetric systems are available is to solve instead the normal equations

$$A^TA\underline{x} = A^T\underline{b} \tag{4.2}$$

where, particularly in the sparse case, it is inadvisable to form A^TA explicitly. There are really two main problems with this approach. Because we must multiply by both A and A^T at each iteration, the amount of work per iteration is almost doubled. The more significant problem lies in the poorer eigenvalue distribution of (4.2) compared with (1.1). Clearly the singular values are squared but, when A is far from symmetric, the effect can be much worse since the convergence depends on the spread of eigenvalues rather than the singular values and that spread can be more than squared when the normal equations are formed. Even with preconditioning, the formation of the normal equations (implicitly or explicitly) is not very satisfactory.

Recently, Chan and Elman (private communication) have been investigating the use of the Krylov (or perhaps twin Krylov) sequence(s)

$$\underline{r}^{(0)}, S\underline{r}^{(0)}, (A^TA)\underline{r}^{(0)}, (A^TA)S\underline{r}^{(0)}, (A^TA)^2\underline{r}^{(0)}, \ldots \tag{4.3}$$

where the matrix S is equal to the square root of A^TA, and they have shown that a Lanczos method based on this dual sequence has potentially very good convergence properties. Yip (Boeing Computer Services), Simon (Stony Brook, now BCS) and Saunders (Stanford) have all experimented with this method but using S=A or A^T. They have found that the results are poor even when the matrix is very near to symmetric. The problem occurs because, when A is not symmetric, A or A^T may be a very poor approximation to $(A^TA)^{1/2}$. No cheap way of obtaining the square root of a sparse matrix is known, but we are currently investigating better approximations to the square root than A or A^T (or $(A+A^T)/2$) and their effect when used as S in the sequence (4.3).

Another technique for extending conjugate gradients to unsymmetric systems is to use the augmented system

$$\begin{pmatrix} \alpha I & A \\ A^T & 0 \end{pmatrix} \begin{pmatrix} \underline{y} \\ \underline{x} \end{pmatrix} = \begin{pmatrix} \underline{b} \\ \underline{0} \end{pmatrix} \tag{4.4}$$

where the coefficient matrix is now symmetric. Although it is not positive definite, conjugate gradients can be applied but again there are problems both in conditioning and the fact that we are minimizing errors in the whole system (4.4) rather than (1.1) at each stage. If $\alpha=1$, then we obtain, using the bidiagonalization algorithm of Golub and Kahan (1965), the LSQR algorithm of Paige and Saunders (1982). If $\alpha=0$ and the auxiliary matrix P of (4.1) is of the form

$$\begin{bmatrix} O & I \\ I & O \end{bmatrix}$$

then one obtains the biconjugate gradient algorithm. Jackson and Robinson (1982) give an example on which a biconjugate algorithm becomes unstable due to an explosion of components in the eigenspace corresponding to the largest eigenvalues.

Of course, for some unsymmetric systems it is possible to split the matrix, viz.

$$A = M-N$$

so that M $(=(A+A^T)/2)$ is positive definite and then use the generalized conjugate-gradient algorithm of Concus and Golub (1975). This method will work well unless the norm of N is large compared with the norm of M, but does require the solution of the system with M as a coefficient matrix at each step. Much work is still being performed on how to implement this but it looks only really attractive for mildly convective systems or equivalent.

If one does not try to blend or encapsulate the system to make it symmetric, then, in order to create a computationally feasible method, the amount of reorthogonalization and past history retained by the unsymmetric versions must be curtailed by truncation (for example, Vinsome, 1976, Axelsson, 1980, and Saad, 1981a). The two main parameters are the extent of the truncation (that is, how many past search directions are kept, the value of s in (4.1)) and which vectors are kept. Unfortunately, the number of previous search directions needed will depend on the spectrum of the iteration matrix since we will at least want the invariant subspace corresponding to a cluster of largest eigenvalues to be well represented. Jackson and Robinson (1982) illustrate this and also show that it is much better to remove search directions most orthogonal to the current search direction rather than on a first-in first out basis.

Obtaining good preconditionings for unsymmetric systems is as much an art as a science and the theory is difficult and undeveloped. As in the symmetric case, however, a good preconditioning can make all the difference between the acceptability or otherwise of an iterative scheme. We now give some examples of preconditionings which have been found useful in practice.

Most preconditionings are based on a partial LU factorization of A. By "partial" we mean that not all the fill-in is kept; usually either small entries are ignored (as in the drop tolerances of the last section) or, more commonly, fill-ins are allowed in only pre-specified positions in the matrix in analogy with the class of ICCG methods for symmetric systems. Unless the structure is very regular, the so-called ILU(0) method is usually employed, where no fill-ins at all are allowed and the partial factors have the same sparsity pattern as the original matrix. The ILU(0) preconditioning, however, is not itself well-defined since the pivot order may have a significant effect on the performance of the preconditioning. This is illustrated from some results by Duff and Meurant in Table 4.1. Here we see the effect of

Problem	1	2	3
Ordering			
Pagewise	16	21	32
Cuthill McKee	16	21	32
Nested dissection	20	91	37
D4	26	106	52
Minimum degree	28	104	49

Table 4.1 Effect of orderings on the number of iterations required by ICCG(0) for a set of problems on a 20×20 grid. The first two orderings can be proven always to have the same performance.

different orderings for ICCG(0) on model problems. It appears that local orderings (that is, orderings where neighbouring grid points are, as far as possible, numbered consecutively) give preconditionings which require far fewer iterations of the conjugate-gradient method. Duff and Meurant plan to investigate this phenomenon further. Chan et al (1981) have observed and analyzed significant reductions in the number of iterations when preconditioning with an incomplete factorization using an alternating direction ordering.

On more regular structures, for example five or nine diagonal matrices resulting from finite-difference discretizations, it is much easier to prescribe positions in which fill-in during partial factorization is allowed. A whole class of methods ILU(k) are used where k denotes the number of extra diagonals of non-zeros permitted. Thus the pattern of ILU(2) on a five-diagonal system is shown in Figure 4.1 where the dashed lines indicate where fill-in has been allowed. As might be expected, the more diagonals allowed (it makes sense to work by diagonals since, in general, entries on the same diagonal have a similar order of magnitude and decrease on successively inner diagonals) the better the preconditioning and the fewer iterations required. However, as k increases both the work to generate and to use ILU(k) increases and it is the experience of many workers (for example, Meijerink and van der Vorst, 1978) that the best overall compromise is around ILU(3) although, for operational reasons (and storage), ILU(0) is sometimes preferred.

If one writes the original matrix in the form

$$(a+b+c+d+e)$$

by diagonals as shown in Figure 4.1, then ILU(0) only modifies the diagonal according to the elimination steps

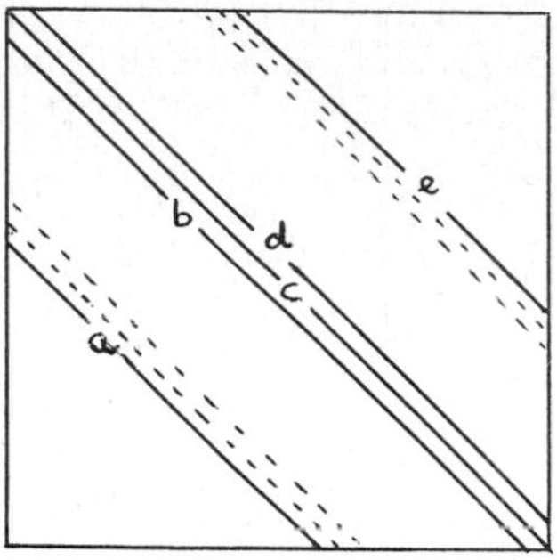

Figure 4.1 ILU(2) pattern on 5 diagonal matrix

$$\gamma_1 = c_1$$
$$\gamma_2 = c_2 - b_1 \gamma_1^{-1} d_1$$
$$\vdots$$
$$\gamma_j = c_j - b_{j-1}\gamma_{j-1}^{-1}d_{j-1} - a_{j-m}\gamma_{j-m}^{-1}e_{j-m}$$
$$\vdots$$

A minor addition to this preconditioning modifies the diagonal γ so that the error matrix E, defined as

$$E = A - \hat{L}\hat{D}\hat{U}$$

where $\hat{L}\hat{D}\hat{U}$ are the partial factors of A, has zero row sum. The effect of this on ILU(0), now called MILU(0) for modified ILU, is to add two terms to the formula for γ_j viz.

$$\gamma_j = c_j - b_{j-1}\gamma_{j-1}^{-1}d_{j-1} - a_{j-m}\gamma_{j-m}^{-1}e_{j-m} - a_{j-m}\gamma_{j-m}^{-1}d_{j-m}$$
$$- b_{j-1}\gamma_{j-1}^{-1}e_{j-1}$$

which, in the ICCG algorithm, corresponds to the method of Dupont Kendall and Rachford (DKR). This produces a noticeable improvement as we can see from the results in Table 4.2 of runs performed by Meurant.

Test problem	1		2		3	
Preconditioning	#its	work/N	#its	work/N	#its	work
None	109	1199	-		195	2145
Diagonal scaling	109	1199	137	1507	189	2079
ICCG(0)	33	495	47	705	58	870
MICCG(0)	23	345	32	480	35	525
Block	15	285	22	418	26	494
Modified block	11	209	17	323	19	361

Table 4.2 Results from paper of Concus et al (1982) illustrating benefits of block preconditioning. In all cases N, the order of the system, was equal to 2500.

More recently block preconditioning has been used with great success. In the five-diagonal case, the matrix can be written in the form

$$\begin{matrix} B_1 & C_1 & & & \\ A_1 & B_2 & C_2 & & \\ & A_2 & B_3 & C_3 & \\ & & \ddots & \ddots & \ddots \end{matrix}$$

so that when block elimination is performed, the diagonal blocks are modified according to

$$\hat{B}_1 = B_1$$

$$\hat{B}_2 = \hat{B}_1 - A_1 \hat{B}_1^{-1} C_1$$

$$\vdots$$

A variety of techniques can be used for approximating the $\hat{B}_j^{-1}$, for example, another partial factorization giving rise to the nested factorization technique of Appleyard et al (1981) or some of the strategies of Concus et al (1982) in the symmetric case. Results from the latter of runs on test problems are shown in Table 4.2 and illustrate the benefits of block-preconditioning methods.

We have already discussed the Chebychev semi-iterative method of Manteuffel. Although this can only be employed when the spectrum of A, or rather the preconditioned system, lies in the positive half plane, Saad (1981b) has extended the method to use two polynomials on disjoint intervals which can be either side of the imaginary axis and Smolarski (1981) has extended the method by enclosing the spectrum in a polygon rather than an ellipse. Elman (1982) has proposed a hybrid conjugate gradient-Chebychev method where the former step generates the parameters for the latter. Little experimental work has, however, been performed on these extensions.

Of course, the original work of Lanczos (1950,1952) gave algorithms for general systems and methods based on Lanczos biorthogonalization have been studied recently (Saad, 1982), although, to our knowledge, there has not been many results on the effects of preconditionings with such algorithms.

As was the case with the frontal scheme with direct methods, one can run most of the abovementioned iterative techniques on an element problem without first assembling the matrix. This is easy since usually only the product of the matrix with a vector is required which can be done by summing the contributions from the elements. However, preconditioning then presents a problem. A start on this has been made by Nour-Omid and Parlett (1982) but their results have not been very encouraging since the savings in the number of iterations are balanced by the extra work in calculating and using the element-preconditioning matrices.

As we said in the introduction, the situation regarding software implementing iterative schemes for unsymmetric matrices is unsatisfactory. Apart from the codes of Paige and Saunders

(1975,1982) most are experimental. A future release of the ITPACK package from Texas (Kincaid and Young, 1983) is planned which will include codes for unsymmetric matrices based on work by Jea and Young (1983) among others.

ACKNOWLEDGEMENTS

I am indebted to my colleagues Peter Jackson, Ian Jones and John Reid for reading an early draft of this paper and making several constructive suggestions and to Gene Golub and Michael Saunders for their helpful comments.

REFERENCES

Appleyard, J.R., Cheshire, I.M. and Pollard, R.K. (1981). Special techniques for fully implicit simulators. Proc. Eur. Sym. on Enhanced Oil Recovery, Bournemouth, Sept. 1981, pp.395-408.

Axelsson, O. (1980). Conjugate gradient type methods for unsymmetric and inconsistent systems of linear equations. Lin. Alg. and its Applics. 29, pp.1-16.

Behie, A. and Forsyth, P. Jnr. (1983). Comparison of fast iterative methods for symmetric systems. IMA J. Numer. Anal. 3, pp.41-63.

Bunch, J.R. and Parlett, B.N. (1971). Direct methods for solving symmetric indefinite systems of linear equations. SIAM J. Numer. Anal. 8, pp.639-655.

Chan, T.F., Jackson, K.R. and Zhu, B. (1981). Alternating-direction incomplete factorizations. Report 208, Department of Computer Science, Yale University.

Concus, P. and Golub, G.H. (1975). A generalized conjugate gradient method for nonsymmetric systems of linear equations. Report STAN-CS-75-535. Computer Science Department, Stanford University.

Concus, P., Golub, G.H. and Meurant, G. (1982). Block preconditioning for the conjugate gradient method. Report LBL-14856. Physics, Computer Science and Mathematics Division. Lawrence Berkeley Laboratory.

Concus, P., Golub, G.H. and O'Leary, D.P. (1976). A generalized conjugate gradient method for the numerical solution of elliptic partial differential equations. Report STAN-CS-76-533. Computer Science Department. Stanford University.

Duff, I.S. (1977). MA28 - a set of Fortran subroutines for sparse unsymmetric linear equations. AERE Report R 8730, HMSO, London.

Duff, I.S. (1981). MA32 - a package for solving sparse unsymmetric systems using the frontal method. AERE Report R 10079, HMSO, London.

Duff, I.S. (1982). Direct methods for solving sparse systems of linear equations. AERE Report CSS 131. SIAM J. Sci. Stat. Comput. To appear.

Duff, I.S. (1983). The solution of sparse linear equations on the

CRAY-1. AERE Report CSS 125 (revised). To appear in "High Speed Computation" edited by J.S. Kowalik, Springer Verlag.

Duff, I.S. and Reid, J.K. (1982a). MA27 - a set of Fortran subroutines for solving sparse symmetric sets of linear equations. AERE Report R 10533, HMSO, London.

Duff, I.S. and Reid, J.K. (1982b). The multifrontal solution of unsymmetric sets of linear equations. AERE Report CSS 133. Presented at Sparse Matrix Symposium 1982, Fairfield Glade, Tn. October 24-27, 1982.

Duff, I.S. and Reid, J.K. (1983). The multifrontal solution of indefinite sparse symmetric linear systems. ACM Trans. Math. Softw. 9, pp.302-325.

Eisenstat, S.C., Gursky, M.C., Schultz, M.H. and Sherman, A.H. (1982). Yale Sparse Matrix Package 1: The symmetric codes. Int. J. Numer. Meth. Engng. 18, pp.1145-1151.

Eisenstat, S.C. (1982). Iterative methods for solving large sparse linear systems. Invited paper at Sparse Matrix Symposium 1982, Fairfield Glade, Tn. October 24-27, 1982.

Elman, H.C. (1982). Iterative methods for large, sparse, nonsymmetric systems of linear equations. Report 229. Department of Computer Science, Yale University.

Faber, V. and Manteuffel, T. (1982). Conjugate gradient methods. SIAM J. Numer. Anal. To appear.

George, A. and Liu, J.W.-H. (1981). Computer Solution of Large Sparse Positive Definite Systems. Prentice Hall.

George, A., Liu, J. and Ng, E. (1980). User Guide for SPARSPAK: Waterloo Sparse Linear Equations Package. Report CS-78-30 (Revised Jan. 1980). Department of Computer Science, University of Waterloo.

Golub, G. and Kahan, W. (1965). Calculating the singular values and pseudo-inverse of a matrix. J. SIAM Numer. Anal. 2, pp.205-224.

Hackbusch, W. and Trottenberg, U. (Eds) (1982). Multigrid Methods. Proceedings of Conference held at Koln-Porz, November 23-27, 1981. Lecture Notes in Mathematics 960. Springer-Verlag.

Hestenes, M.R. and Stiefel, E. (1952). Methods of conjugate gradients for solving linear systems. J. Res. Nat. Bur. Standards 49, pp.409-436.

Jackson, C.P. and Robinson, P.C. (1982). A numerical study of various algorithms related to the preconditioned conjugate gradient method. AERE Report TP 950. Submitted for publication.

Jea, K.C. and Young, D.M. (1983). On the simplification of generalized conjugate-gradient methods for nonsymmetrizable linear systems. Lin. Alg. and its Applics. 52/53, pp.399-417.

Kincaid, D.R. and Young, D.M. Jnr. (1983). The ITPACK Project: Past, Present and Future. Report CNA-180, Center for Numerical

Analysis, Austin, Texas.

Lanczos, C. (1950). An iteration method for the solution of the eigenvalue problem of linear differential and integral operators. J. Res. Nat. Bur. Standards 45, pp.255-281.

Lanczos, C. (1952). Solution of systems of linear equations by minimized iteration. J. Res. Nat. Bur. Standards 49, pp.33-53.

Li, M.R., Nour-Omid, B. and Parlett, B.N. (1982). A fast solver free of fill-in for finite element problems. SIAM J. Numer. Anal. 19, pp.1233-1242.

Manteuffel, T.A. (1977). The Tchebychev iteration for nonsymmetric linear systems. Numer. Math. 28, pp.307-327.

Markowitz, H.M. (1957). The elimination form of the invese and its application to linear programming. Management Science 3, pp.255-269.

Mejerink, J.A. and van der Vorst, H.A. (1978). Guide lines for the usage of incomplete decompositions in solving sets of linear equations as they occur in practical problems. J. Comput. Phys. 44, pp.134-155.

Munksgaard, N. (1980). Solving sparse symmetric sets of linear equations by preconditioned conjugate gradients. ACM Trans. Math. Softw. 6, pp.206-219.

Nour-Omid, B. and Parlett, B.N. (1982). Element preconditioning. Report PAM-103, Center for Pure and Applied Mathematics, Univ. of California, Berkeley.

Pagallo, G. and Maulino, C. (1983). A bipartite quotient graph model for unsymmetric matrices. Proceedings of InterAmerican Workshop on Numerical Methods, Caracas, June 1982. Springer-Verlag. To appear.

Paige, C.C. and Saunders, M.A. (1975). Solution of sparse indefinite systems of linear equations. SIAM J. Numer. Anal. 12, pp.617-629.

Paige, C.C. and Saunders, M.A. (1982). LSQR: An algorithm for sparse linear equations and sparse least squares. ACM Trans. Math. Softw. 8, pp.43-71 and pp.195-209.

Saad, Y. (1981a). Krylov subspace methods for solving large unsymmetric linear systems. Math. Comp. 37, pp.105-126.

Saad, Y. (1981b). Iterative solution of indefinite symmetric systems by methods using orthogonal polynomials over two disjoint intervals. Report 212. Department of Computer Science, Yale University.

Saad, Y. (1982). The Lanczos biorthogonalization algorithm and other oblique projection methods for solving large unsymmetric systems. SIAM J. Numer. Anal. 19, pp.485-506.

Sherman, A.H. (1978). Algorithms for sparse Gaussian elimination with partial pivoting. ACM Trans. Math. Softw. 4, pp.330-338 and pp.391-398.

Simon, H.D. (1982). The Lanczos algorithm for solving symmetric linear systems. Report PAM-74. Center for Pure and Applied Mathematics. University of California, Berkeley.

Smolarski, D.C. (1981). Optimum semi-iterative methods for the solution of any linear algebraic system with a square matrix. Report UIUCDCS-R-81-1077, Department of Computer Science, University of Illinois.

Stetter, H.J. (1978). The defect correction principle and discretization methods. Numer. Math. 29, pp.425-443.

van der Vorst, H.A. (1981). Iterative solution methods for certain sparse linear systems with a nonsymmetric matrix arising from pde problems. J. Comput. Phys. 44, pp.1-19.

Vinsome, P.K.W. (1976). Orthomin, an iterative method for solving sparse sets of simultaneous linear equations. Paper number SPE 5729. 4th SPE Symposium on Numerical Simulation of Reservoir Performance.

Zlatev, Z. (1980). On some pivotal strategies in Gaussian elimination by sparse techniques. SIAM J. Numer. Anal. 17, pp.18-30.

Zlatev, Z., Wasniewski, J. and Schaumburg, K. (1982). Comparison of two algorithms for solving large linear systems. SIAM J. Sci. Stat. Comput. 3, pp.486-501.

Computing Methods in Applied Sciences and Engineering, VI
R. Glowinski and J.-L. Lions (Editors)
Elsevier Science Publishers B.V. (North-Holland)

TOEPLITZ MATRICES AND THEIR APPLICATIONS

V.V.Voevodin, E.E.Tyrtyshnikov

Department of Computational Mathematics
USSR Academy of Sciences
Moscow
USSR

In many applications one has to solve the problems of linear algebra with matrices whose elements, though not equal to zero, are nevertheless determined by a small number of independent parameters. The special structure of such matrices can be used for construction of efficient numerical methods.

The paper deals with matrices of Toeplitz type. Some results concerning algebraic computations for these matrices will be presented here; the proofs of them can be found in [1] .

Let us assume that the elements of matrices belong to some ring Ω with unit I . We also assume that if for matrix A and some vector y with elements from Ω the equations $Ax=y$ and $z'A=y'$ are uniquely solvable, then A is invertible. Further conditions on the ring will be specified later.

The matrix A_n of order $n+1$ whose elements a_{ij} depend on the difference $i-j$ only is called Toeplitz; this matrix will be denoted as

$$A_n = \begin{bmatrix} a_0 & a_{-1} \cdots & a_{-n} \\ a_1 & a_0 \cdots & a_{-n+1} \\ \cdots & & \\ a_n & a_{n-1} \cdots & a_0 \end{bmatrix}$$

Toeplitz matrix C_n is called circulant if for all $i\neq 0$ $c_i = c_{n-i+1}$.

Toeplitz matrices arise in problems of applied electrodynamics, acoustics, optics, automatic control, image processing, differential and integral equations, probability theory and statistics. We will give two examples of such problems which are typical for the whole class.

If for given observations of a stationary random process y_t at times $t=0,1,\ldots,n$ one has to predict the value $\tilde{y}_{n+1}=x_0y_0+\ldots+x_ny_n$ of the random value y_{n+1} , then the quadratic risk $M(\tilde{y}_{n+1}-y_{n+1})^2$ will be minimal in case of the coefficients x_i satisfying the system

$$A_n(x_0,\ldots,x_n)'=(a_n,\ldots,a_0)',$$

with Toeplitz matrix A_n with elements $a_{ij}=My_iy_j=a_{i-j}$.

Historically, these and more complicated statistical problems had been the first applications in which the efficient algorithms for solution of Toeplitz systems were required [2].

Consider the linear integral equations with two properties [3] :

1). The kernel is invariant to transformations from the set Φ ;

2). The domain of integration S is obtained from some its part by successive transformations with $R \in \Phi$, i.e. $S = S_0 \cup \ldots \cup S_n$ and $S_{i+1} = RS_i$ for all $i < n$.

Assume that the integral over S is equal to the sum of the integrals over S_i , which are approximated in the same manner using the nodes $x_{ij} \in S_i$, $j = 0, 1, \ldots, m$, and $Rx_{ij} = x_{i+1,j}$, for all j and $i < n$. Then for suitable numeration of the nodes the integral equation of the first kind will be equivalent to the system of linear algebraic equations $A_n D x = b$ with block-Toeplitz matrix A_n in which a_i are the blocks of order $m+1$, and with diagonal matrix D containing the coefficients of the quadrature formula.

If there exists one more transformation $Q \in \Phi$ with $Qx_{ij} = x_{i,j+1}$ for all i and $j < m$, then each block a_i is the Toeplitz matrix of order $m+1$.

If $Rx_{nj} = x_{0j}$ for all j , then A_n is block circulant matrix. If $Qx_{in} = x_{i0}$ for all i , then each block a_i is circulant.

The typical example on the plane is the system of intervals which could be obtained from a given interval either by multiple shifts with some fixed vector, or by rotations with the angle $2\pi/(n+1)$. If the kernal depends on the distance between the points only, then the first situation leads to the block-Toeplitz matrix, and the second one - to the block-circulant matrix. In the first case as an additional transformation Q one can take the shift in direction parallel to given intervals, and to obtain the block-Toeplitz matrix with Toeplitz blocks.

Let now S be the cylinder with axis l , and R be the rotation with axis l by the angle $2\pi/(n+1)$. Consider the parts of the cylinder, S_i , $i = 0, \ldots, n$, for which $RS_i = S_{i+1}$, and represent each S_i as the sum of domains S_{ij} , $j = 0, 1, \ldots, m$, for which $QS_{ij} = S_{i,j+1}$, with Q being the shift by the vector parallel to l . To approximate the integral over S_{ij} choose the points x_{ijk} , $k = 0, 1, \ldots, q$, for which

$$Rx_{ijk} = x_{i+1,jk}, \quad i < n, \quad Rx_{njk} = x_{0jk};$$

$$Qx_{ijk} = x_{i,j+1,k}, \quad j < m.$$

Under suitable numeration of the nodes x_{ijk} we obtain the "three-level" matrix, which in this case is the block-circulant matrix with block-Toeplitz blocks.

These integral equations are widely used in problems of electrodynamics, in particular, for study of diffraction for the cases when the wave length is comparable with the size of scattering body.

The product of Toeplitz matrices is, in general, not Toeplitz. At the same time, the inverse of Toeplitz matrix can be represented as a sum of products of Toeplitz matrices. Therefore the multiplicative matrix operations generate the set of matrices, which can be represented as a sum of products of Toeplitz matrices. In some applications this class of matrices arise quite naturally [4], while in other cases their use is caused by the necessity to change the formulation of the problem, for example, when one has to find the regularized solution satisfying the equation $(A^*A + \alpha I)x_\alpha = A^* b$.

Therefore when considering the Toeplitz matrices, one has to study the more broad class of matrices. Moreover, in some applications the "multilevel" block-Toeplitz matrices and their products can arise.

A large number of types of special matrices (including the Toeplitz ones) can be described as follows [5].

1. Let the elements $\alpha_{ij}^{(q)}$, $\beta_{ij}^{(q)}$, $\gamma^{(q)} \in \Omega$, $q = 1, \ldots, Q$, $i, j = 1, \ldots, n$, be given. Consider the equalities

$$\sum_{ij=1}^{n} \alpha_{ij}^{(q)} z_{ij} \beta_{ij}^{(q)} = \gamma^{(q)}, \quad q=1,\dots,Q.$$

We say that the block matrix $A=\{A_{ij}\}_{ij=1}^{n}$ is of the block type Γ if these equalities take place for $z_{ij}=A_{ij}$.

Consider the following important examples. Let $A=\{A_{ij}\}$, $A_{ij}\in\Omega$ be the Toeplitz matrix of order n. Then the equalities will take the form $A_{i+1,j+1} - A_{ij}=0$, $ij=1,\dots,n-1$. The type of this matrix will be denoted by T_n. If A is circulant matrix, then we have the additional equalities $A_{i1}-A_{i-1,n}=0$, $i=2,3,\dots,n$. This type is denoted as C_n. The corresponding notations can also be used for matrices, which are symmetric, triangular, three-diagonal, sparse with certain structure, etc. We will use mainly the notations D_n, T_n, C_n, G_n corresponding to diagonal, Toeplitz, circulant and general type matrices of block order n.

2. Let the quadratic matrix A be divided into $n_1\times n_1$ quadratic blocks $A_{i_1 j_1}$. Assume that we have the splitting-up of A into blocks $A_{i_1 j_1,\dots,i_{\kappa-1} j_{\kappa-1}}$, $\kappa>1$, and let each block be divided into $n_\kappa\times n_\kappa$ quadratic blocks $A_{i_1 j_1,\dots,i_\kappa j_\kappa}$. Then the set of blocks $A_{i_1 j_1,\dots,i_\kappa j_\kappa}$ is called the κ-th level of splitting-up of the matrix A, and the number n_κ is called the order of κ-th level. We say that A has p levels, if $A_{i_1 j_1,\dots,i_{p+1} j_{p+1}}$ are the elements from Ω and $n_{p+1}=1$.

3. The matrix A is of composite type $\alpha_1\dots\alpha_p$ if it is the p-level one, and the blocks of level κ satisfy for $1\le\kappa\le p$ the relations 1 which determine the type α_κ. This operation is associative.

4. The lemma of level transposition [5]. Let the types α_i, $i=1,\dots,p$ and their compositions

$$\Gamma_1=\alpha_1\dots\alpha_\kappa\alpha_{\kappa+1}\dots\alpha_p, \quad \Gamma_2=\alpha_1\dots\alpha_{\kappa+1}\alpha_\kappa\dots\alpha_p$$

be given. There exists the permutation matrix P determined by the orders of types α_i, such that for any matrix A of the type Γ_1 the matrix $B=PAP'$ has the type Γ_2, and for any matrix B of the type Γ_2 the matrix $A=P'BP$ has the type Γ_1.

5. Any multilevel matrix containing the splittings-up of the types D, T, C and G of any number and in any order, can be transformed into similar matrix of the type $D\dots DC\dots CT\dots TG\dots G$ by transpositions of rows and columns [3].

Note that if A is of the type $D\Gamma$, then it is block-diagonal with blocks of the type Γ at the diagonal. In this case solution of all problems of linear algebra with matrix $D\Gamma$ is reduced to solution of several analogous problems with matrix of type Γ. The number of these problems coinsides with the block order of D. The unification of the levels of the type Γ at the end of composition of types can be regarded as a transition to another ring Ω.

The lemma of level transposition has two meanings. On one side, it is a tool for solution of a number of problems. On the other side, it is very useful for investigation of the algebraic properties of the classes of matrices being the sums of products of Toeplitz matrices, and for construction of efficient methods for these matrices from given methods for block-Toeplitz matrices.

Algorithmically the problem of transposing the two levels is equivalent to transposing the rectangular matrix with order equal to the orders of the types of these levels. The storage requirement for $m\times n$ matrix is equal to mn words, however it is difficult to use the same storage area for transposed matrix when labeling of words and utilization of additional memory independent of m, n is prohibited. Nevertheless this problem can be resolved with $O(mn)$ operations required with additional memory of $min(m,n)$ [6].

6. Let the ring Ω contain not less than n different elements. By transposing the corresponding rows and columns the levels of types $G_{n_1},\dots,G_{n_p}$ can be united into one level of the type G_n with $n=n_1\dots n_p$ if and only if each pair of the set $n_1,\dots,n_p$ is mutually simple [1]. All the levels of the types D or G can be united into one level.

If Ω is the field of complex numbers or the ring of matrices, then matrix of the type $G_{n_1}\dots G_{n_p}$ is unitarily similar to diagonal matrix. The corresponding transformation can be done using the discrete Fourier transformation algorithms of orders $n_1,\dots,n_p$ [5]. Let $\gamma^{(q)}=0$ in the equalities 1 for all q, and $\alpha_{ii}^{(q)}\beta_{ii}^{(q)}=0$ for all i, q.

7. The set of matrices of the same type and order is closed in respect to addition of matrices operation and in respect to left and right side multiplication by scalars from Ω in the case of these multipliers being commutative with all the coefficients $\alpha_{ij}^{(q)}$, $\beta_{ij}^{(q)}$, which determine this type.

The product of the matrices of the same type is not, in general, the matrix of this type. In important cases, however, the structure of inverse to the matrix, which is obtained from matrices of the type Γ by operations of addition and multiplication, can be described by the known structure of inverses to the matrices of type Γ.

8. Let Γ and Σ be the types of matrices such that for any l the inverse to the matrix of the type ΓG_l can be expressed as the sum k of paired products of the matrices of type ΣG_l. Then if the invertible matrix is equal to the sum of matrix of the type Γ and the sum m of paired products of matrices of the type Γ, then the inverse matrix can be represented as a sum of paired products of matrices of the types Σ with the number of terms equal to $k(m+1)$ [1].

9. Let under conditions of 8 the invertible matrix be represented as a sum of the matrix of type Γ and the sum of m products of matrices of the type Γ with the number of multipliers equal to r. Then the inverse matrix can be represented as a sum of paired products of matrices of the type Γ with the number of addends equal to $k(m+1)3^s$, where $s\le\lceil \log_2 r\rceil$ ($\lceil\dots\rceil$ stands for the nearest larger integer).

The type T_n corresponds to situation described in 8, 9.

10. Let A_n be the Toeplitz matrix. Consider the values α_i, β_i, γ_i, $\delta_i\in\Omega$, satisfying the equations

$$A_n(\alpha_0,\dots,\alpha_n)'=(\varphi_0,\dots,\varphi_n)',$$

$$A_n(\beta_0,\dots,\beta_n)'=(I,\dots,I)',$$

$$(\gamma_0,\dots,\gamma_n)A_n=(I,\dots,I),$$

$$(\delta_0,\dots,\delta_n)A_n=(\psi_0,\dots,\psi_n),$$

where $\varphi_0=\psi_0=0$, $\varphi_i=a_1+\dots+a_i$, $\psi_i=a_{-1}+\dots+a_{-i}$, and consider the Toeplitz matrices

$$Q_l=\begin{bmatrix} I & & 0\\ -I & \ddots & \\ 0 & -I & I\end{bmatrix},\qquad Q_u=\begin{bmatrix} I & -I & & 0\\ & \ddots & \ddots & \\ 0 & & I & -I\end{bmatrix}$$

$$\alpha = \begin{bmatrix} \alpha_n & \cdots & \alpha_0 \\ & \ddots & \vdots \\ 0 & & \alpha_n \end{bmatrix}, \qquad \beta = \begin{bmatrix} \beta_n & \cdots & \beta_0 \\ & \ddots & \vdots \\ 0 & & \beta_n \end{bmatrix},$$

$$\gamma = \begin{bmatrix} \gamma_n & & 0 \\ \vdots & \ddots & \\ \gamma_0 & \cdots & \gamma_n \end{bmatrix}, \qquad \delta = \begin{bmatrix} \delta_n & & 0 \\ \vdots & \ddots & \\ \delta_0 & \cdots & \delta_n \end{bmatrix}.$$

If the equations for α_i, β_i, γ_i, δ_i are solvable, then the matrix A_n is non-degenerate, and A_n^{-1} has the following form [1, 7] :

$$A_n^{-1} = Q_u(\beta a_0 \gamma' + \alpha \gamma' + \beta \delta) Q_l .$$

11. Let A_n be the Toeplitz matrix, and let $\bar{\alpha}_i$, $\bar{\delta}_i \in \Omega$ satisfy the equations

$$A_n (\bar{\alpha}_0, \ldots, \bar{\alpha}_n)' = (\psi_n, \ldots, \psi_0)',$$

$$(\bar{\delta}_0, \ldots, \bar{\delta}_n) A_n = (\varphi_n, \ldots, \varphi_0).$$

If the equations for $\bar{\alpha}_i$, $\bar{\delta}_i$, β_i, γ_i are solvable, then the matrix A_n is non-degenerate, and A_n^{-1} has the following form [1, 7] :

$$A_n^{-1} = Q_l(\bar{\beta} a_0 \bar{\gamma} + \bar{\alpha} \bar{\gamma} + \bar{\beta} \bar{\delta}) Q_u .$$

Here

$$\bar{\alpha} = \begin{bmatrix} \bar{\alpha}_0 & & 0 \\ \vdots & \ddots & \\ \bar{\alpha}_n & \cdots & \bar{\alpha}_0 \end{bmatrix}, \qquad \bar{\beta} = \begin{bmatrix} \beta_0 & & 0 \\ \vdots & \ddots & \\ \beta_n & \cdots & \beta_0 \end{bmatrix},$$

$$\bar{\gamma} = \begin{bmatrix} \gamma_0 & \cdots & \gamma_n \\ & \ddots & \vdots \\ 0 & & \gamma_n \end{bmatrix}, \qquad \bar{\delta} = \begin{bmatrix} \bar{\delta}_0 & \cdots & \bar{\delta}_n \\ & \ddots & \vdots \\ 0 & & \bar{\delta}_0 \end{bmatrix}.$$

The representations 10, 11 for A_n^{-1} can be easily transformed to the form which contains the sums of two products of triangular Toeplitz matrices, and in 10 the left and right multipliers are the right and left triangular matrices, respectively, while in 11 they are the left and right triangular matrices. These representations of the inverse to Toeplitz matrix are the most general ones. They are completely determined by solution of some systems with matrices A_n and A'_n under special choice of the right sides when these systems are solvable. If some additional constraints are imposed on A_n, then for representation of A_n^{-1} as solutions of such systems one can take some rows and columns of A_n^{-1} [9].

Let us call the representation of the matrix in the form of paired products of Toeplitz matrices the tt -representation or the representation of the type tt. If in such representation all left multipliers are left (right) triangular, and all right - right (left) triangular matrices, then it is called lu (ul) - representation. If the matrix is represented as a sum of Toeplitz matrix and the expression of the type tt, lu or ul, then this is the representation of the type $t+tt$, $t+lu$ or $t+ul$.

The left triangular Toeplitz matrix with first column $x=(x_0,\dots,x_n)'$ will be denoted by $L(x)=L(x_0,\dots,x_n)$, and the right triangular Toeplitz matrix with first row $y=(y_0,\dots,y_n)$ will be denoted by $U(y)=U(y_0,\dots,y_n)$.

The minimal possible numbers of addends in the representation of A in the form tt, lu or ul are denoted by $r_{tt}(A)$, $r_{lu}(A)$, $r_{ul}(A)$ and called the Toeplitz ranks of A. The number $r_{lu}(A)$ is called the left, and $r_{ul}(A)$ - the right Toeplitz rank of A. If Ω is the field of complex numbers, then the following propositions take place [10]:

- Left (right) Toeplitz rank of quadratic matrix of order n with elements a_{ij} coincides with an ordinary rank of the matrix of the same order, which is constructed from the original matrix as follows: first (last) row and column are taken as they are, and for all $i,j>1(i,j<n)$ the corresponding element is replaced by the difference $a_{ij}-a_{i-1,j-1}(a_{ij}-a_{i+1,j+1})$.
- The matrix is represented as a sum of paired products of left and right (right and left) triangular Toeplitz matrices S_iT_i if and only if the matrix constructed when considering the left (right) Toeplitz rank can be decomposed into the sum of matrices s_it_i, where s_i is the first (last) column of S_i, and t_i is the first (last) row of T_i.
- For any matrix the left and right Toeplitz ranks are not greater than its order and their difference do not exceed 2.
- For any non-degenerate matrix the left (right) Toeplitz rank is equal to the right (left) Toeplitz rank of the inverse matrix.

12. If the matrix is represented in the form tt with the number of addends m, then it can be represented in the form $t+lu$ and $t+ul$ with the number of addends equal to $2m+1$. If the matrix is represented in the form lu or ul with k addends, then it can be represented in the form $t+tt$ with $1+[(k+1)/2]$ addends. In all cases the following inequality holds [1]:

$$r_{tt}(A)\leq \min\left(\left[\frac{r_{lu}(A)+1}{2}\right],\left[\frac{r_{ul}(A)+1}{2}\right]\right)+1.$$

13. The inverse to the matrix determined by $t+tt$ - representation with $m+1$ addends can always be represented in the form $t+tt$ with $m+2$ addends [1].

The representation of the inverse is constructed with solution of the systems with this matrix and specially chosen right sides. The number of these systems is equal to $2(m+1)$. Existence of these solutions implies the invertibility of the matrix [1, 7]. This representation of the inverse is the most general one. However, from computational point of view the more convenient are some other representations.

14. Let A be the matrix of the type $t+lu$, and

$$A = T + G_1 R_1 + \dots + G_m R_m,$$

where

$$T = \{t_{i-j}\}_{ij=0}^{n}, \quad G_l = L(0, g_1^{(l)}, \dots, g_n^{(l)}), \quad R_l = U(0, r_1^{(l)}, \dots, r_n^{(l)}), \; l = 1, \dots, m.$$

Consider the values x_{il} , y_{i0} , z_{li} , $w_{0i} \in \Omega$, satisfying the equations

$$A \begin{bmatrix} x_{00} \dots x_{0m} \\ \dots \\ x_{n0} \dots x_{nm} \end{bmatrix} = \begin{bmatrix} I & 0 & \dots & 0 \\ 0 & g_1^{(1)} & \dots & g_1^{(m)} \\ \dots \\ 0 & g_n^{(1)} & \dots & g_n^{(m)} \end{bmatrix},$$

$$A(y_{00}, \dots, y_{n0})' = (0 \dots 0I)',$$

$$\begin{bmatrix} z_{00} \dots z_{0n} \\ \dots \\ z_{m0} \dots z_{mn} \end{bmatrix} A = \begin{bmatrix} 1 & 0 & \dots & 0 \\ 0 & r_1^{(1)} & \dots & r_n^{(1)} \\ \dots \\ 0 & r_1^{(m)} & \dots & r_n^{(m)} \end{bmatrix},$$

$$(w_{00}, \dots, w_{0n})A = (0 \dots 0I).$$

and quadratic matrix

$$\mathcal{X} = \begin{bmatrix} I & 0 & \dots & 0 \\ 0 & r_1^{(1)} & \dots & r_n^{(1)} \\ \dots \\ 0 & r_1^{(m)} & \dots & r_n^{(m)} \end{bmatrix} \begin{bmatrix} x_{00} \dots x_{0m} \\ \dots \\ x_{n0} \dots x_{nm} \end{bmatrix} - \begin{bmatrix} 0 & 0 & \dots & 0 \\ 0 & I & & \\ \vdots & & \ddots & \\ 0 & & & I \end{bmatrix}$$

If the equations for x_{il} , y_{i0} , z_{li} , w_{0i} are solvable and the element y_{n0} is invertible, then there exists the matrix $\mathcal{X}^{-1}$, the matrix A is invertible and A^{-1} has the following form [1, 5]

$$A^{-1} = \sum_{j=0}^{m} \sum_{i=0}^{m} L(x_{0i}, \dots, x_{ni}) \{\mathcal{X}^{-1}\}_{ij} U(z_{j0}, \dots, z_{jn}) -$$

$$- L(0, y_{00}, \dots, y_{n-1,0}) y_{n0}^{-1} U(0, w_{00}, \dots, w_{0,n-1}).$$

By reordering the addends this lu -representation can be transformed into lu -representation with $m+2$ addends.

In [4] for the case of Ω being the ring of matrices another formula of the type 14 was obtained,

$$A^{-1} = L(x_{00},\dots,x_{n0})x_{00}^{-1}U(z_{00},\dots,z_{0n}) -$$

$$- L(0,y_{00},\dots,y_{n-1,0})y_{n0}^{-1}U(0,w_{00},\dots,w_{0,n-1}) +$$

$$+ \sum_{j=1}^{m}\sum_{i=1}^{m} L(0,x'_{1i},\dots,x'_{ni})\{W^{-1}\}_{ij}\,U(0,z'_{j1},\dots,z'_{jn}),$$

where

$$W = \begin{bmatrix} r_1^{(1)} \dots r_n^{(1)} \\ \dots \\ r_1^{(m)} \dots r_n^{(m)} \end{bmatrix} \begin{bmatrix} x'_{11} \dots x'_{1m} \\ \\ x'_{n1} \dots x'_{nm} \end{bmatrix} - \begin{bmatrix} I & & 0 \\ & \ddots & \\ 0 & & I \end{bmatrix}$$

and x'_{ij} , y'_{ij} satisfy the equations

$$\begin{bmatrix} a_{11} \dots a_{1n} \\ \dots \\ a_{n1} \dots a_{nn} \end{bmatrix} \begin{bmatrix} x'_{11} \dots x'_{1m} \\ \dots \\ x'_{n1} \dots x'_{nm} \end{bmatrix} = \begin{bmatrix} g_1^{(1)} \dots g_1^{(m)} \\ \dots \\ g_n^{(1)} \dots g_n^{(m)} \end{bmatrix}$$

$$\begin{bmatrix} z'_{11} \dots z'_{1n} \\ \dots \\ z'_{m1} \dots z'_{mn} \end{bmatrix} \begin{bmatrix} a_{11} \dots a_{1n} \\ \dots \\ a_{n1} \dots a_{nn} \end{bmatrix} = \begin{bmatrix} r_1^{(1)} \dots r_n^{(1)} \\ \dots \\ r_1^{(m)} \dots r_n^{(m)} \end{bmatrix}.$$

This representation is more restrictive for it requires the invertibility of x_{00} .

Let Ω be the ring of matrices of order r .

15. Let matrix A with elements from Ω be of the form 14 and has non-degenerate leading submatrices. Then each set x_{il} , y_{i0} and z_{li} , w_{0i} in the representation 14 of A^{-1} can be computed with $(2.5mn^2 + 2n^2 + O(m^2n))r^3$ multiplications by one algorithm, or with $(3.5mn^2 + 3.5n^2 + O(mn))r^3$ multiplications by another algorithm. The estimates for additions and subtractions are the same [5].

The analogous algorithms were proposed in [4]. In [5] the algorithms were obtained as a corollary of the well-known algorithms for solution of Toeplitz systems obtained with the lemma on level transpositions.

16. For Toeplitz system $Ax = b$ consider the regularized system $(A^*A + \alpha I)u_\alpha = A^*b$ and the system

$$\begin{bmatrix} \alpha I & A^* \\ A & -I \end{bmatrix} \begin{bmatrix} z_\alpha \\ w_\alpha \end{bmatrix} = \begin{bmatrix} A^*b \\ 0 \end{bmatrix}$$

Both systems are uniquely solvable for $\alpha > 0$, and $z_\alpha = u_\alpha$ [5].

The matrix of second system is of the type G_2T , so that by lemma on level transpositions it can be transformed into similar matrix of the type TG_2 by transpositions of rows and columns, i.e. into block-Toeplitz matrix with blocks of the second order. The solutions of the original and the transformed systems coinside up to the known transposition of components. Therefore for solution of the regularized system one can use the algorithms for solution of block-Toeplitz system with blocks of the second order.

The possibility of efficient solution of regularized systems makes the necessity for stability analysis in these algorithms less urgent. The representation 14 can also be used to organize the refinement procedure.

17. The multiplication of block-Toeplitz matrix of block order n with blocks of order r by some vector can be implemented with $(6n\log_2 n)r^2$ complex multiplications and $(12n\log_2 n)r^2$ complex additions and subtractions [1].

18. If the representation 14 of A^{-1} is found, then the solution of the system with matrix A for some right side, or the refinement of the solution, can be done with $O(r^2 mn\log_2 n)$ operations required.

19. If together with the conditions of 14 the following equalities take place,

$$T=T^*,\quad R_i^*=G_i\,diag\{\sigma_i,\dots,\sigma_i\},$$

where $\sigma_i^2=I$, $\sigma_i\in\Omega$, then x_{il}, y_{io}, z_{li}, w_{oi} are linked by the equations [1, 5]

$$\begin{bmatrix} z_{00}\dots z_{0n}\\ \dots\\ z_{m0}\dots z_{mn}\end{bmatrix}=\begin{bmatrix} I & & \\ & \sigma_1\ddots & \\ & & \sigma_m\end{bmatrix}\begin{bmatrix} x_{00}\dots x_{0m}\\ \dots\\ x_{n0}\dots x_{nm}\end{bmatrix}^*,$$

$$(w_{00},\dots,w_{0n})=(y_{00},\dots,y_{n0})'^*.$$

If

$$T=T^*,\quad R_i^*=diag\{\sigma_i,\dots,\sigma_i\},$$

then these values satisfy the equations

$$\begin{bmatrix} z_{00}\dots z_{0n}\\ \dots\\ z_{m0}\dots z_{mn}\end{bmatrix}=\begin{bmatrix} x_{00}\dots x_{0m}\\ \\ x_{n0}\dots x_{nm}\end{bmatrix}^*\begin{bmatrix} I & & \\ & \sigma_1\ddots & \\ & & \sigma_m\end{bmatrix}^*,$$

$$(w_{00},\dots,w_{0n})=(y_{00},\dots,y_{n0})'^*.$$

In applications these dependences between the parameters which determine the matrix A are often encounted. In all these cases the statement 19 enables to reduce by a factor of two the computational expences in determining the values which form A^{-1}.

When constructing the algorithms one can also take into account other properties of the original matrices, such as special types of the ring Ω, band form of a matrix, etc. These investigations are important for practice. However, as it became clear from [11, 12], such investigations do not give an answer to the question on what is the real computational complexity of these problems. It was shown in [11,12] how one can solve the Toeplitz system of order n by $O(n\log_2^2 n)$ arithmetic operations. This result shows that the algorithms of solution of Toeplitz systems, which are used in practice, are not optimal by order. In the same manner not optimal are the algorithms of 15.

Unfortunately, there are a lot of unclear points in the new algorithms. In particular, it is not known whether or not they will be competitive with classical algorithms for not very large n, say for

$n < 1000$.

Let us give some results concerning the new or the so-called asymptotically fast algorithms.

20. Let A be the non-degenerate matrix of order n and Toeplitz rank α. The solution of the system with matrix A can be found by $O(\alpha^2 n \log_2^2 n + \alpha^3 n)$ arithmetic operations [13].

The corresponding method was obtained by certain modification of the algorithm described in [12], which required $O(\alpha^4 n \log_2^2 n)$ operations.

21. Let A be the Toeplitz matrix of order n and with elements a_{ij} satisfying the conditions

$$a_{ij} = 0, \quad l \leqslant i-j, \quad j-i \leqslant m$$

for some non-negative l, m. The solution of the system with matrix A can be found by $O(n \log_2 (p+q) + \min(p,q) \log_2^2 \min(p,q))$ operations [1].

We note also that the algorithm of [11] requires $O(n \log_2 n + (p+q) \log_2^2 (p+q))$ operations.

REFERENCES

1. Voevodin V.V., Tyrtyshnikov E.E. Computations with Toeplitz matrices. - In: Computational processes and systems. Moscow, Nauka, 1983, 124-266 (Russian).
2. Levinson N. The Wiener RMS (Root-Mean-Square). Error Criterion in filter design and prediction. - J. Math. and Phys., 1947, 25, 261-278.
3. Voevodin V.V. On solution of some systems with block-Toeplitz matrices. - In: Numerical analysis for FORTRAN. Moscow University Publ., 1967, v. 17, 43-47 (Russian).
4. Friedlander B., Kailath T., Morf M., Ljung L. New inversion formulas for matrices classified in terms of their distance from Toeplitz matrices. - Linear Alg. and Appl., 1979, v. 27, 31-60.
5. Voevodin V.V., Tyrtyshnikov E.E. Numerical methods for systems with Toeplitz matrices. JVM and MF, 1981, v. 21, No. 3, 531-544 (Russian).
6. Tretiakov A.A., Tyrtyshnikov E.E. Transposing of large rectangular matrices. - In: Numerical methods of algebra. Moscow University Publ., 1981, 58-68 (Russian).
7. Tyrtyshnikov E.E. On sums of products of Toeplitz matrices. - In: Computational methods of linear algebra (Proc. of the National Conference). Novosibirsk, 1980, 107-119 (Russian).
8. Sachnovich A.L. On some method for inversion of Toeplitz matrices. - In: Math. Investigations, Kishinev, 1973, No. 4 (30), 180-186 (Russian).
9. Gochberg I.S., Heinig G. Inversion of finite Toeplitz matrices composed from elements of non-commutative algebra. - Rev. Roumaine Math. Pures Appl., 1974, v. 19, No. 5, 623-665 (Russian).
10. Kailath T., Kung S.Y., Morf M. Displacement ranks of matrices and linear equations. - J. Math. Anal. Appl., 1979, v. 68, No.2, 395-407.
11. Brent R.P., Gustavson F.G., Yun D.Y.Y. Fast computation of Pade approximants and the solution of Toeplitz systems of equations. - Research Report RC 8173, IBM Research Center, Yorktown Heights, NY, Jan. 1980.
12. Bitmead R.R., Anderson B.D.O. Asymptotically fast solution of Toeplitz and related systems of linear equations. Linear Alg. and Appl., 1980, v. 34, 103-116.

13. Tyrtyshnikov E.E. On complexity of some algebraic problems. - In: Methods and algorithms of numerical analysis. Moscow University Publ., 1982, 98-107 (Russian).

Computing Methods in Applied Sciences and Engineering, VI
R. Glowinski and J.-L. Lions (Editors)
Elsevier Science Publishers B.V. (North-Holland)

THE LOOK AHEAD LANCZOS ALGORITHM FOR LARGE UNSYMMETRIC EIGENPROBLEMS

B.N. Parlett, D.R. Taylor, Z-S Liu

Mathematics Department
University of California
Berkeley, California 94720

(with support from Office of Naval Research Contract N00014-76-C-0013)

The two-sided Lanczos algorithm is an attractive way to compute a few eigenelements of large unsymmetric linear operators. Unfortunately it can suffer from serious breakdowns which occur when an associated moment matrix does not permit triangular factorization. We modify the algorithm slightly so that it corresponds to using a 2×2 "pivot" whenever it offers advantages over the standard 1×1 pivot. This device reduces the incidence of breakdown.

The two techniques are compared on problems that model the onset of instability in a hot plasma.

1. THE LANCZOS ALGORITHM AND ITS BREAKDOWN

The most popular way to obtain all the eigenvalues of a nonsymmetric $n \times n$ matrix B is to use the QR algorithm that is readily available in most computing centers. As the order n increases above 100 the QR algorithm becomes less and less attractive, especially if only a few of the eigenvalues are wanted. It is important to think of B as an operator and not as a conventional array. In our main application $B = (F - \sigma M)^{-1} M$ where M is symmetric positive definite and F is unsymmetric. Both are banded. Any product $y = Bx$ is realized by a subprogram which solves $(F - \sigma M)y = Mx$ using precomputed factors of $F - \sigma M$.

That is where the Lanczos algorithm comes into the picture. It does not alter B at all but simply uses vectors Bx. It constructs a tridiagonal matrix J gradually by adding a row and column at each step. After several steps some of the eigenvalues θ_i of J will be close to some eigenvalues λ_k of B and by the nth step, if nothing goes wrong, $\theta_i = \lambda_i$, $i = 1,\ldots,n$. This description is correct in the context of exact arithmetic. Unfortunately, things can go wrong, even in the absence of rounding errors.

In order to discuss these troubles we must give some details about the algorithm. For the relation between these troubles and orthogonal polynomials see [2]. Let J_k be the $k \times k$ tridiagonal produced at step k of the algorithm. There are infinitely many tridiagonal matrices similar to B and J_n is one of them. Thus for some matrix $Q_n \equiv (q_1,\ldots,q_n)$ we have

$$Q_n^{-1} B Q_n = J_n . \qquad (1)$$

It simplifies the exposition considerably to introduce a redundant symbol and write P_n^* instead of Q_n^{-1}. The superscript * indicates conjugate transpose.

Let $P_n \equiv (p_1, \ldots, p_n)$ and replace (1) by three separate relations:

$$P_n^* Q_n = I \tag{2}$$

$$BQ_n = Q_n J_n \tag{3}$$

$$P_n^* B = J_n P_n^* \quad . \tag{4}$$

We mention in passing that when $B^* = B = A$ then we can arrange that $P_n = Q_n$. The difficulty we are going to describe cannot occur when $P_n = Q_n$.

By equating elements on each side of $BQ_n = Q_n J_n$ and $P_n^* B = J_n P_n^*$ in the natural increasing order we shall see that B, p_1 and q_1 essentially determine all the other elements of P_n, Q_n and J_n.

Let us see how this works. Usually one chooses $p_1 = q_1$ and $\|q_1\|^2 = q_1^* q_1 = 1$, but all that is necessary is that $p_1^* q_1 = 1$. On writing

$$J_n \equiv \begin{bmatrix} \alpha_1 & \gamma_2 & & 0 \\ \beta_2 & \alpha_2 & \gamma_3 & \\ & \beta_3 & \cdot & \cdot \\ 0 & & \cdot & \cdot \end{bmatrix}$$

we find, on equating the (1,1) elements in $P_n^* B Q_n = J_n$, that

$$p_1^* B q_1 = \alpha_1 \quad ,$$

and, equating first columns in (3) and (4), that

$$Bq_1 = q_1 \alpha_1 + q_2 \beta_2 \quad , \qquad p_1^* B = \alpha_1 p_1^* + \gamma_2 p_2^* \quad .$$

Hence q_2 and p_2^* are, respectively, multiples of "residual" vectors

$$r_2 \equiv Bq_1 - q_1 \alpha_1 \quad , \qquad s_2^* \equiv p_1^* B_1 - \alpha_1 p_1^* \quad .$$

Furthermore, since $p_2^* q_2 = 1$ by (2), we have

$$s_2^* r_2 = \gamma_2 p_2^* q_2 \beta_2 = \gamma_2 \beta_2 \equiv \omega_2 \quad .$$

If $\omega_2 \neq 0$ and β_2 is given any nonzero value then γ_2, q_2, and p_2^* are all determined uniquely. A good choice is $\beta_2 = \sqrt{|\omega_2|}$.

The general pattern emerges at the next step, on equating the second columns on each side of $BQ_n = Q_n J_n$ and $P_n^* B = J_n P_n^*$,

$$p_2^* B q_2 = \alpha_2 \quad ,$$

$$Bq_2 = q_1 \gamma_2 + q_2 \alpha_2 + q_3 \beta_3 \quad , \qquad p_2^* B = \beta_2 p_1^* + \alpha_2 p_2^* + \gamma_3 p_3^* \quad .$$

At this point we can compute the "residual" vectors

$$r_3 \equiv Bq_2 - q_1\gamma_2 - q_2\alpha_2 \quad , \quad s_3^* \equiv p_2^*B - \beta_2 p_1^* - \alpha_2 p_2^*$$

and

$$\omega_3 \equiv s_3^* r_3 = \gamma_3\beta_3 \quad .$$

If $\omega_3 \neq 0$ and β_3 is given any nonzero value then γ_3, q_3, and p_3^* are all determined uniquely. And so it goes on until some ω_j vanishes.

This is the Lanczos algorithm. It must terminate at the nth step with $\omega_{n+1} = 0$ but it may stop sooner.

Premature termination at, say, step j $(< n)$ can occur in two ways:

(I) either $r_{j+1} = 0$ or $s_{j+1}^* = 0^*$, or both;

or

(II) $r_{j+1} \neq 0$, $s_{j+1}^* \neq 0^*$, but $\omega_{j+1} = 0$.

In the 1950's, when the Lanczos algorithm was regarded as a way to compute J_n, Case I was regarded as a mild nuisance. If $r_{j+1} = 0$ then *any* nonzero vector orthogonal to $p_1,\ldots,p_j$ can be chosen as q_{j+1}. Similarly $s_{j+1} = 0$ gives ample choice for p_{j+1}.

Today, regarding Lanczos as a way to find a few eigenvalues of large B, it seems better to stop if Case I occurs in the knowledge that *every* eigenvalue of J_j is an eigenvalue of B. If more eigenvalues are wanted then it is best to start the Lanczos algorithm afresh with new, carefully chosen, starting vectors q_1 and p_1.

The real trouble, which cannot occur when $B = B^* = A$, is Case II. Wilkinson calls this a *complete* breakdown [7, p.389]. There seems to be no choice but to start again, but no one has been able to suggest a practical way to choose the new q_1 and p_1^* so as to avoid another wasted run of the algorithm. That is why the Lanczos method has not been used much when $B^* \neq B$. In this article we propose a modification of the algorithm which greatly reduces the occurrence of Case II. The price paid for this convenience is that J is not quite tridiagonal. There is a small bump (or bulge) in the tridiagonal form to mark each occurrence (or near occurrence) of Case II.

The Lanczos algorithm can be regarded as the two-sided Gram-Schmidt process applied to the columns of the special matrices

$$K = K_n \equiv (q_1, Bq_1, B^2q_1, \ldots, B^{n-1}q_1)$$

and

$$\tilde{K} = \tilde{K}_n = (p_1, B^*p_1, \ldots, (B^*)^{n-1}p_1) \quad .$$

We will not establish this result, see [11, p.8], but content ourselves with stating the key observation, namely

$$\mathrm{Span}(q_1, q_2, \ldots, q_j, Bq_j) = \mathrm{Span}(q_1, \ldots, q_j, B^j q_1) \quad .$$

The K matrices are called Krylov matrices and the pleasant fact is that most of the work required for general Gram-Schmidt disappears in this case because

$$p_i^* B q_j = 0 \quad \text{for } i < j-1 \ .$$

Thus the general formula

$$\mu q_{j+1} = B^j q_1 - \sum_{i=1}^{j} q_i (p_i^* B^j q_1/\omega_i)$$

collapses to

$$\beta_{j+1} q_{j+1} = Bq_j - q_j \alpha_j - q_{j-1}\gamma_j \quad .$$

Similarly

$$\gamma_{j+1} p_{j+1}^* = p_j^* B - \alpha_j p_j^* - \beta_j p_{j-1}^* \quad .$$

2. TRIANGULAR FACTORIZATION OF MOMENT MATRICES

Consider again the matrices K and $\tilde{K}^*$ defined in the previous section. Note that the (i,j) element of $\tilde{K}^*K$ is $(p_1^* B^{i-1})(B^{j-1}q_1)$, so

$$\tilde{K}^* K = M = M(p_1, q_1) \ ; \qquad m_{i+1,j+1} = p_1^* B^{i+j} q_1 \quad .$$

In order to use this observation we need some basic facts about the Lanczos algorithm (see [4], [5] or [7]). If it does not break down it produces matrices P and Q such that

$$q_{j+1} = \chi_j(B) q_1 / \Big(\prod_{i=2}^{j+1} \beta_i \Big) \ ,$$

$$p_{j+1}^* = p_1^* \chi_j(B) / \Big(\prod_{i=2}^{j+1} \gamma_i \Big) \ ,$$

where

$$\chi_{j+1}(t) = (t - \alpha_j)\chi_j(t) - \omega_j \chi_{j-1}(t) \quad , \quad (\chi_0(t) = 1, \ \chi_{-1}(t) = 0) \ ,$$

and χ_j is the characteristic polynomial of the tridiagonal matrix J_j. In other words, for each j, q_j is a linear combination of the first j columns of K while p_j is the *same* linear combination of the columns of $\tilde{K}$, up to scaling. This can be expressed compactly in matrix notation as

$$Q = KL^{-*}\Pi^{-1} \ , \qquad P = \tilde{K}L^{-*}\tilde{\Pi}^{-1} \ . \tag{5}$$

Here

$$\Pi = \mathrm{diag}(1, \beta_2, \beta_2\beta_3, \dots) \ ,$$

$$\tilde{\Pi} = \mathrm{diag}(1, \gamma_2, \gamma_2\gamma_3, \dots) \ ,$$

and L is the unit lower triangular matrix such that $L^{-*} \equiv (L^{-1})^*$ has the coefficients of χ_j above the diagonal in the jth column.

Using (5) we can rewrite $I = P^*Q$ as

$$I = P^*Q = (\tilde{\Pi}^{-1} L^{-1} \tilde{K}^*)(KL^{-*}\Pi^{-1})$$

i.e.

$$M = \tilde{K}^* K = L \Omega L^* \tag{6}$$

where

$$\Omega = \tilde{\Pi}\Pi = \mathrm{diag}(1, \omega_2, \omega_2\omega_3, \ldots, \omega_2 \ldots \omega_n)$$

This result is not new (see Householder [2]) but it is worth emphasizing that the product $\omega_2 \ldots \omega_i$ is the ith pivot which arises in performing Gaussian elimination on the moment matrix M associated with B, q_1 and p_1.

When $B \neq B^*$ the moment matrix M need not be positive definite and so triangular factorization need not be stable, even when M is nonsingular.

The best known remedy for instability is to use some form of row or column interchange whenever ω_i is too small. The standard "partial pivoting" strategy is not available to us because a whole column of M is not known in the middle of the Lanczos process. An alternate remedy is to enlarge the notion of a "pivot" to include 2×2, or even larger submatrices. This idea is discussed in Parlett and Bunch [6]. It is the basis of our method. In triangular factorization the use of a 2×2 pivot at step j means annihilating columns j and $j+1$ simultaneously (below row $j+1$).

In the context of the Lanczos process our remedy for a small value of ω_j requires us to compute q_{j+1} at the same time as q_j, and p_{j+1} at the same time as p_j. The J matrix will bulge out of tridiagonal form temporarily. The disturbance is local. The formulas for these "Lanczos" vectors are somewhat different from the standard ones. We call our modification the "look-ahead Lanczos algorithm" because it computes at the current step some quantities which are not usually needed until the next step in the standard Lanczos process. Nevertheless, no matrix-vector products are ever wasted.

3. THE NEXT PIVOT

The triangular factorization $M = L\Omega L^*$ is not developed explicitly. Yet in order to look ahead and decide whether to do a single step or a double step, our algorithm needs to determine the top 2×2 submatrix of what would be the reduced matrix in Gaussian elimination, as indicated in the snapshot below.

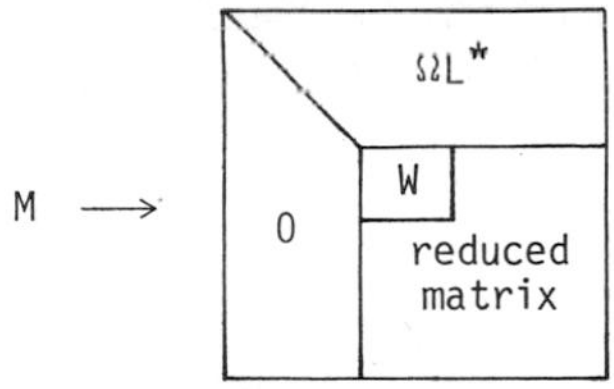

It turns out that after $(j-1)$ steps the top 2×2 is a multiple of

$$W_j = \begin{pmatrix} \omega_j & \theta_j \\ \theta_j & \omega_{j+1} \end{pmatrix}$$

where

$$\theta_j := s_j^* \tilde{r}_{j+1} = \tilde{s}_{j+1}^* r_j, \quad \tilde{r}_{j+1} := B r_j - q_{j-1}\omega_{j-1}, \quad \tilde{s}_{j+1}^* := s_j^* B - \omega_{j-1} p_{j-1}^*.$$

Details of this and other aspects of the algorithm are given in [10] and [11].

If both $\omega_j = 0$ and W_j is singular then more drastic measures are needed to salvage the algorithm. However, in many cases a tiny ω_j is accompanied by a decent value of θ_j and our modification allows the algorithm to proceed. There are several reasonable ways to choose q_j and q_{j+1} as linear combinations of r_j and $\tilde{r}_{j+1}$.

The extra work of computing θ_j at each step confers an unanticipated benefit. There is no need to wait for a near breakdown in order to take a double step, i.e. to use a 2×2 "pivot" in factoring M. The stability of the algorithm is measured by certain angles, namely, $\angle(p_i, q_i)$, $i = 1,2,\ldots$. The smaller the better. In the symmetric case the angles are all zero. By looking ahead we may always choose the pivot to keep $\angle(p_j, q_j)$ and $\angle(p_{j+1}, q_{j+1})$ as small as we can.

In order to keep J in upper Hessenberg form $(J(i,k) = 0$ if $i > k+1)$ we always choose q_j to be a multiple of r_j. Then either p_j is a multiple of s_j (a normal Lanczos step) or p_j is a multiple of s_{j+1} (a double step). In the double step both q_{j+1} and p_{j+1} are then determined by biorthonormality (i.e. $p_i^\star q_k = \delta_{ik}$). This gives two angles $\phi_+ = \angle(p_j, q_j)$ and $\phi_- = \angle(p_{j+1}, q_{j+1})$. Moreover, both angles may be computed *before* forming these vectors. Our criterion is

If $\angle(r_j, s_j) < \max\{\phi_+, \phi_-\}$ then

[take a single step

else

[take a double step

The numbers ω_j, θ_j, ω_{j+1} together with norms of the residual vectors are required in making this decision.

It is worth mentioning that when it is decided to use ω_j as a 1×1 pivot (a normal Lanczos step), the residual vectors $\tilde{r}_{j+1}$ and $\tilde{s}_{j+1}^\star$ are not wasted but are readily modified to give r_{j+1} and $s_{j+1}^\star$.

There is not enough space to describe the algorithm in more detail but one practical feature should be mentioned. It has proved convenient to normalize each q_i and each p_i to have Euclidean norm one. As a result, $P_j^\star Q_j = \Psi_j = \mathrm{diag}(\psi_1, \ldots, \psi_n)$. Very small values of the ψ_i indicate a failure of the algorithm. Approximate eigenvalues λ must be computed from a generalized eigenproblem

$$(J_j - \lambda \Psi_j)u = 0 \quad .$$

This causes no inconvenience.

4. ERROR ESTIMATES

By the end of step j the Look-Ahead Lanczos Algorithm has produced matrices $Q_j := (q_1, \ldots, q_j)$, $P_j =: (p_1, \ldots, p_j)$, J_j, and $\Psi_j = \mathrm{diag}(\psi_1, \ldots, \psi_j)$. Moreover, $q_i^\star q_i = p_i^\star p_i = 1$, $\psi_i = \cos\angle(p_i, q_i) = p_i^\star q_i$, $i = 1, \ldots, j$. For simplicity we will assume that the last step was a normal single Lanczos step so that J_j is tridiagonal in its last row and column. Because of our change in normalization, the equations governing the algorithm in exact arithmetic are

$$BQ_j - Q_j\Psi_j^{-1}J_j = q_{j+1}e_j^*\beta_{j+1}/\psi_{j+1} = q_{j+1}(0,\ldots,0,1)\beta_{j+1}/\psi_{j+1} \quad ,$$

$$P_j^*B - J_j\Psi_j^{-1}P_j^* = e_jp_{j+1}^*\gamma_{j+1}/\psi_{j+1} \quad ,$$

$$P_j^*Q_j = \Psi_j \quad .$$

Approximate eigenelements of B are derived from exact eigenelements of the pair (J_j,Ψ_j). Let

$$J_ju = \Psi_ju\theta \quad , \quad v^*J_j = \theta v^*\Psi_j \quad , \quad \|u\| = \|v\| = 1 \quad , \quad v^*u > 0 \quad .$$

Multiplying the governing equations by u and v^* and simplifying

$$B(Q_ju) - (Q_ju)\theta = q_{j+1}u(j)\beta_{j+1}/\psi_{j+1} \quad ,$$

$$(v^*P_j^*)B - \theta(v^*P_j^*) = p_{j+1}^*\overline{v(j)}\gamma_{j+1}/\psi_{j+1} \quad .$$

The corresponding approximate eigenelement of B is

$$(\theta, Q_ju, v^*P_j^*) \quad .$$

It is important to observe that the column and row residual norms associated with this element, namely, $\|B(Q_ju) - (Q_ju)\theta\|$ and $\|(v^*P_j^*)B - \theta(v^*P_j^*)\|$, can be evaluated without forming Q_ju and P_jv by $|u(j)\beta_{j+1}/\psi_{j+1}|$ and $|v(j)\beta_{j+1}/\psi_{j+1}|$. We use the Euclidean vector norm here. When $j = 30$ and $n \geq 1000$ this feature is very attractive.

For normal operators B, the residual norms are actual error bounds for θ. There is an eigenvalue λ of B satisfying

$$|\lambda - \theta| \leq \|B(Q_ju) - (Q_ju)\theta\| / \|Q_ju\|$$

In the normal case $P_j = Q_j$ and $\|Q_ju\| = 1$.

In the general case it is possible to have tiny residual norms but no λ close to θ. However this can only happen when the closest eigenelement of B is very ill-conditioned. Fortunately there is a useful interpretation of the residual norms whatever the nature of B. They bound the change which must be made to B in order to make our approximations exact. The following theorem is proved in [3].

THEOREM. Let B and unit vectors x, y^* with $y^*x \neq 0$ be given. For any scalar γ define $\mathcal{E} := \{E : (B-E)x = x\gamma,\ y^*(B-E) = \gamma y^*\}$. Then

$$\min_{E\in\mathcal{E}} \|E\| = \max\{\|Bx - x\gamma\|, \|y^*B - \gamma y^*\|\}$$

Here $\|E\|$ is the spectral norm, $\max_{w\neq 0} \|Ew\|/\|w\|$. Moreover, in [3] the result is generalized to sets of approximate eigenvectors and to the Frobenius norm.

This result justifies us in computing Qu and Pv when the residual norms are small enough for the application. Indeed, if Qu and Pv are not accurate eigenvectors then this misfortune will usually be signalled by the easily computed quantity

$$\text{cond}(\theta) = 1/v^*u = \text{secant} \angle(u,v) \geq 1 .$$

Recall that $\|v\| = \|u\| = 1$, $v^*u > 0$.

Our algorithm accepts eigenelements when the residual norms drop below a user defined threshold. The norms and cond(θ) are returned as output information along with $\|Q_j u\|$ and $\|P_j v\|$. More details are given in [3].

In exact arithmetic all approximations become exact when $j = n$ and so there is no doubt of formal convergence. What is interesting is that well isolated and extremal eigenvalues of B are discovered for quite small values of j. The same phenomenon occurs in the symmetric case where it is predicted by theory (see [12], Chapter 12).

We have neglected the influence of roundoff error because, at present, we use full reorthogonalization of q_{j+1} against all previous p_i, and p_{j+1} against all previous q_i. Consequently our output is close to the predictions of exact arithmetic.

5. INCURABLE BREAKDOWN

In section 3 we mentioned that it is possible for both the 1×1 pivot ω_j and the 2×2 pivot W_j to be singular. In principle one could then consider the leading principal 3×3 submatrix of the reduced moment matrix as a pivot. If it too were singular one could consider the 4×4 and so on. If *all* such principal submatrices are singular then we say that the breakdown at step j is incurable. In his thesis Taylor proves the following surprising consequence (Theorem 4.2) of this ultimate disaster.

> THEOREM (Taylor). Let B have distinct eigenvalues and let J_j be the block tridiagonal matrix produced by the Look Ahead Lanczos algorithm at step j. If incurable breakdown occurs at step j+1 then each eigenvalue of J_j is an eigenvalue of B.

The proof is given in [11].

6. NUMERICAL RESULTS

Here are a few results computed in single precision on a VAX 780 under the Berkeley UNIX system. This does not permit us to tackle large problems. More results will be presented at the symposium.

Example 1:

$$B = \begin{bmatrix} 0 & 0 & 0 & 0 & 0 & 1 \\ 1 & 0 & 0 & 0 & 0 & 0 \\ 0 & 1 & 0 & 0 & 0 & 0 \\ 0 & 0 & 1 & 0 & 0 & 0 \\ 0 & 0 & 0 & 1 & 0 & 0 \\ 0 & 0 & 0 & 0 & 1 & 0 \end{bmatrix}, \qquad r_1 = s_1 = \begin{bmatrix} 1 \\ 2 \\ 3 \\ 4 \\ 5 \\ 6 \end{bmatrix}$$

The eigenvalues of B are the sixth roots of unity. Our algorithm produced a J matrix with eigenvalues equal to those of B to working accuracy.

To give snapshots of the progress of the algorithms we exhibit

ℓ the order of J , $\rho(J)$ the spectral radius of J ,

$\phi_1 = \cos \angle(r,s)$, $\phi_2 = \min\{\cos \phi_+, \cos \phi_-\}$ (see Section 3) .

If a double step occurs for a particular value of ℓ then ϕ_1 and ϕ_2 are not defined at $\ell+1$ and such places are indicated by dashes. Recall that tiny values of ϕ_1 signal disaster for Lanczos.

TABLE 1

	Regular Lanczos			Look-Ahead Lanczos		
ℓ	ϕ_1	ϕ_2	$\rho(J)$	ϕ_1	ϕ_2	$\rho(J)$
1	1.000	.1277	.8351	1.000	.1277	.8351
2	.1281	.0077	1.503	.1281	.0077	1.503
3	.0072	10^{-6}	1.012	.0072	10^{-6}	1.012
4	10^{-6}	.0488	10^{+6}	10^{-6}	.0488	---
5	10^{-7}	10^{-7}	---	---	---	4.781
6	---	---	---	.0068	.0068	---
7	---	---	---	---	---	1.000

Example 2:

B was a far from normal real 12×12 matrix (Example II in [10]) with eigenvalues scattered in the complex plane and a real cluster (1 at 99, 2 at 98). Ordinary Lanczos did not break down and with full reorthogonalization both algorithms give accurate approximations to all eigenvalues by step 12. One conjugate pair converged to 5 decimals at step 8.

Without reorthogonalization there was a difference. At the end of step 12 our algorithm had all the eigenvalues correct to 6 decimals except the clustered ones which were good to 3. However, simple Lanczos at this step had one spurious eigenvalue near 0 and one eigenvalue in the cluster was not represented at all. Roundoff error allows the algorithm to continue but approximations to the clustered eigenvalues did not improve.

We have begun to apply our program to problems arising in plasma physics. The eigenvalues come in $\pm$ pairs and are, for the most part, on or near the imaginary axis. The simple Lanczos version has not broken down yet. As in the examples above its performance is mildly inferior to the look ahead program; eigenvalues sometimes stabilize a few steps later and sometimes have errors in the last two decimal places.

The look ahead algorithm was run for 20 or 30 steps without reorthogonalization, using an inverted form of the operator. The speed of the Lanczos algorithm in approximating certain eigenvalues is indicated in the table below.

TABLE 2: Lanczos step at which eigenvalues stabilized to working precision.

	n = 34	n = 94
Smallest conjugate pair	9, 15, 21	7
2nd smallest conjugate pair	11	8
3rd smallest conjugate pair	17	17

In order to proceed to larger cases we need to use double precision ($\varepsilon = 10^{-17}$) to invert the operator and we need some form of reorthogonalization to avoid wasted effort.

REFERENCES

[1] C.Davis and W.Kahan, "The rotation of eigenvectors by a perturbation - III," SIAM J. on Num. Anal. **7** (1970), 1-46.

[2] W.Gragg, Notes from a "Kentucky Workshop" on the moment problem and indefinite metrics.

[3] W.Kahan, B.N.Parlett, and E.Jiang, "Residual bounds on approximate eigensystems of nonnormal matrices," SIAM J. on Num. Anal. **19** (1982), 470-483.

[4] A.S.Householder, The Theory of Matrices in Numerical Analysis, (Blaisdell Publishing Co., 1964).

[5] C.Lanczos, "An iteration method for the solution of the eigenvalue problem of linear differential and integral operators," Jour. Res. Natl. Bur. Stds. **45** (1950), 255-282; see pages 266-270.

[6] B.N.Parlett and J.R.Bunch, "Direct methods for solving symmetric indefinite systems of linear equations," SIAM J. on Num. Anal. **8** (1971), 639-655.

[7] J.H.Wilkinson, The Algebraic Eigenvalue Problem, (Oxford University Press, 1965).

[8] Y.Saad, "Variations on Arnoldi's method for computing eigenelements of large unsymmetric matrices," Lin. Alg. Appl. **34** (1980), 269-295.

[9] Y.Saad, "The Lanczos biorthogonalization algorithm and other oblique projection methods for solving large unsymmetric eigenproblems," SIAM Jour. Num. Anal. **19** (1982), 485-509.

[10] B.N.Parlett and D.R.Taylor, "A Look Ahead Lanczos Algorithm for Unsymmetric Matrices," Technical report PAM-43 of the Center for Pure and Applied Mathematics, University of California, Berkeley (1981).

[11] D.R.Taylor, "Analysis of the Look Ahead Lanczos Algorithm," Technical report PAM-108 of the Center for Pure and Applied Mathematics, University of California, Berkeley (1982).

[12] B.N.Parlett, The Symmetric Eigenvalue Problem, (Prentice Hall, Inc., 1980).

Computing Methods in Applied Sciences and Engineering, VI
R. Glowinski and J.-L. Lions (Editors)
Elsevier Science Publishers B.V. (North-Holland)

On the constructive study of the wellposedness of the inverse Sturm-Liouville problem

Akira MIZUTANI

Department of Mathematics
Faculty of Science
Gakushuin University
Mejiro, Tokyo, 171 Japan

We consider the inverse Sturm-Liouville problem. It is well known that "spectral characteristics" determine uniquely the Sturm-Liouville problem, i.e., the potential function q(x) and the boundary conditions. In this paper, we prove that small perturbations in spectral characteristics lead to small perturbations in q(x) when boundary conditions are fixed.

§1. Introduction and Summary

Let q(x) be a real continuous function in the interval $0 \le x \le \pi$ and let h and H be real numbers. We shall consider a differential equation

$$(1.1) \qquad y'' + (\lambda - q(x))y = 0, \quad 0<x<\pi,$$

with boundary conditions

$$(1.2) \qquad y'(0) - hy(0) = 0, \quad y'(\pi) + Hy(\pi) = 0.$$

Let us denote this Sturm-Liouville problem, the eigenvalue problem (1.1)-(1.2), by E(q(x);h,H) and its eigenvalues by $\{\lambda_n\}_{n=0}^{\infty}(-\infty<\lambda_0<\lambda_1<\lambda_2<\cdots)$. We put

$$(1.3) \qquad \rho_n = \int_0^\pi \phi(x,\lambda_n)^2 dx,$$

where $\phi(x,\lambda)$ is the solution of (1.1) which satisfies the initial conditions

$$(1.4) \qquad y(0) = 1, \quad y'(0) = h.$$

We call the numbers $\{\lambda_n,\rho_n\}_{n=0}^{\infty}$ the spectral characteristics of the eigenvalue problem E(q(x);h,H).

The objective of the so-called inverse Sturm-Liouville problem is to find q(x) from the given knowledge of eigenvalues and so on. V.A.Marchenko[9] showed that the spectral characteristics determine the Sturm-Liouville problem, and I.M. Gel'fand and B.M.Levitan[2] gave an algorithm for the construction of q(x),h and H.

In this paper, we shall consider the question whether the inverse Sturm-Liouville problem is well posed. We assume that a real function $q(x)\in C^1[0,\pi]$ and real numbers h and H are known and that the spectral characteristics $\{\lambda_n,\rho_n\}_{n=0}^{\infty}$ of E(q(x);h,H) are also known. Suppose that we are given different spectral characteristics $\{\mu_n,\sigma_n\}_{n=0}^{\infty}$ of E(p(x);h,H) with unknown $p(x)\in C^1[0,\pi]$. Then it will be shown that the difference $|p(x)-q(x)|$ is small if $\{\mu_n,\sigma_n\}_{n=0}^{\infty}$ are close to $\{\lambda_n,\rho_n\}_{n=0}^{\infty}$ in a certain sense. More precisely, we have

<u>Theorem 1.</u>

If $A \equiv \sum_{n=0}^{\infty}\{\sqrt{\lambda_n}|\sigma_n-\rho_n| + |\mu_n-\lambda_n|\}$ is sufficiently small, then we have

(1.5) $\max_{0\le x\le\pi}|p(x)-q(x)| \le C_1\cdot A,$

where $C_1>0$ is a constant depending only on $q(x)$, h and H.

Now we assume further that $q'(x)$, the derivative of $q(x)$, is absolutely continuous. Then the following asymptotic formulae for the spectral characteristics hold:

(1.6) $\sqrt{\lambda_n} = n + \frac{a_0}{n} + O(\frac{1}{n^3}),$ (1.7) $\rho_n = \frac{\pi}{2} + \frac{b_0}{n^2} + O(\frac{1}{n^3}),$

where a_0 and b_0 are constants.

As a generalization of Theorem 1, we have

Theorem 2.

Suppose that the given numbers $\{\mu_n,\sigma_n\}_{n=0}^{\infty}$ satisfy the following asymptotic formulae:

(1.8) $\sqrt{\mu_n} = n + \frac{a_0'}{n} + O(\frac{1}{n^3}),$ (1.9) $\sigma_n = \frac{\pi}{2} + \frac{b_0'}{n^2} + O(\frac{1}{n^3}).$

If $B \equiv \sum_{n=1}^{\infty}\{\sqrt{\lambda_n}|\sigma_n-\rho_n-\frac{b_0'-b_0}{n^2}| + |\mu_n-\lambda_n-2(a_0'-a_0)|\} +$

$+ |\mu_0-\lambda_0| + |\sigma_0-\rho_0| + |a_0'-a_0| + |b_0'-b_0|$

is sufficiently small, then we have

(1.10) $\max_{0\le x\le\pi}|p(x)-q(x)| \le C_2\cdot B,$

where $C_2>0$ is a constant depending only on $q(x)$, h and H.

Here we refer to some other papers. As was described above, Gel'fand-Levitan showed that the spectral characteristics determine $q(x)$. There seems no result on the wellposedness due to perturbations of spectral characteristics. While B.M.Levitan and M.G.Gasymov[7] showed that two sets of eigenvalues $\{\lambda_n\}$ and $\{\lambda_n^*\}$, which correspond to the eigenvalue problems $E(q(x);h,H)$ and $E(q(x);h,J)$ with $H\neq J$, also determine $q(x)$. For such a inverse problem, H.Hochstadt[4] proved that small perturbations in a finite set of eigenvalues $\{\lambda_n\}$ (but with another set of eigenvalues $\{\lambda_n^*\}$ invariant) lead to small perturbations in $q(x)$.

Recently the inverse problems of Sturm-Liouville type are looked at again by T. Suzuki and R.Murayama. Especially Suzuki obtained many results mainly on the inverse problems for parabolic equations. We refer to Suzuki[11]~[14] and Murayama[10].

This paper is composed of four sections. In §2, we shall take some preliminaries. In §§3~5, we shall prove Theorems 1 and 2. In §Appendix, as an application, we shall state a ner algorithm to solve the inverse Sturm-Liouville problem. The algorithm is a slight modification of the well-known Gel'fand-Levitan's.

Acknowledgements

I wish to express my sincere gratitude to Professor Hiroshi Fujita for his unceasing encouragement and for the interest he has shown in my work. I also wish to express my sincere thanks to Dr. Takashi Suzuki who gave me valuable advice and provided me with many informations on the inverse problems.

§2.Preliminaries

Let us consider the eigenvalue problem $E(q(x);h,H)$:

(2.1) $y'' + (\lambda-q(x))y = 0, \quad 0<x<\pi,$

(2.2) $y'(0) - hy(0) = 0, \quad y'(\pi) + Hy(\pi) = 0,$

where $q(x)$ is a real continuous function in the interval $0\le x\le\pi$, h and H are real constants and λ is a real parameter.

We denote the eigenvalues of $E(q(x);h,H)$ by $\{\lambda_n\}_{n=0}^{\infty}$ $(\lambda_0<\lambda_1<\lambda_2<\cdots)$. As is known, the eigenvalues of the problem under consideration are bounded below. So we may assume without loss of generality that the smallest eigenvalue $\lambda_0\geq 1$. Otherwise it is sufficient to consider a differential equation

$$y'' + (\lambda-(q(x)-\lambda_0+1))y = 0,$$

in place of (2.1). We put

$$\rho_n = \int_0^\pi \phi(x,\lambda_n)^2 dx, \quad n=0,1,2,\cdots,$$

where $\phi(x,\lambda)$ is the solution of (2.1) which satisfies the initial conditions

(2.3) $\quad y(0) = 1, \quad y'(0) = h.$

It is well known that the function $\phi(x,\lambda)$ satisfies an integral equation

(2.4) $$\phi(x,\lambda) = \cos\sqrt{\lambda}x + h\cdot\frac{\sin\sqrt{\lambda}x}{\sqrt{\lambda}} + \frac{1}{\sqrt{\lambda}}\int_0^x \sin\sqrt{\lambda}(x-s)\cdot q(s)\phi(s,\lambda)ds.$$

From (2.4), we easily obtain

Lemma 2.1.[1]

There exists a constant $M>0$ such that

$$|\phi(x,\lambda)| + |\phi'(x,\lambda)|/\sqrt{\lambda} + \sqrt{\lambda}|\dot{\phi}(x,\lambda)| + |\dot{\phi}'(x,\lambda)| < M$$

holds for every $\lambda\geq 1$ and $0\leq x\leq \pi$.

When we make the repeated use of (2.4), we obtain

Lemma 2.2.[2]

If $q(x)\in A^2[0,\pi]$, then we have

(2.5) $$\phi(x,\lambda) = (1+\frac{C_2(x)}{4\xi^2})\cos\xi x + \frac{S_1(x)}{2\xi}\sin\xi x + O(\frac{1}{\xi^3}),$$

(2.6) $$\phi'(x,\lambda) = (-\xi+\frac{S_1^*(x)}{4\xi})\sin\xi x + \frac{C_0^*(x)}{2}\cos\xi x + O(\frac{1}{\xi^2}).$$

Where

$$\xi = \sqrt{\lambda}, \quad Q(x) = \int_0^x q(s)ds, \quad S_1(x) = C_0^*(x) = 2h + Q(x),$$

$$C_2(x) = q(x)-q(0)-2hQ(x)-\frac{1}{2}Q(x)^2$$

and $$S_1^*(x) = q(x)+q(0)+2hQ(x)+\frac{1}{2}Q(x)^2.$$

Because of the asymptotic expansions for the functions $\phi(x,\lambda)$ and $\phi'(x,\lambda)$ in Lemma 2.2, we obtain asymptotic formulae for the spectral characteristics.

Lemma 2.3.

If $q(x)\in A^2[0,\pi]$, then we have

(2.7) $$\sqrt{\lambda_n} = n + \frac{a_0}{n} + O(\frac{1}{n^3}),$$ (2.8) $$\rho_n = \frac{\pi}{2} + \frac{b_0}{n^2} + O(\frac{1}{n^3}),$$

where $$Q = \int_0^\pi q(t)dt, \quad a_0 = (h+H+\tfrac{1}{2}Q)/\pi \text{ and } b_0 = \{h+H+\tfrac{1}{2}Q+\pi(h^2-\tfrac{1}{2}q(0))\}/2.$$

---

[1] $' = \partial/\partial x, \quad \dot{} = \partial/\partial\lambda.$

[2] By $A^m[0,\pi]$ $(m=1,2,3,\cdots)$, we denote the set of all real-valued functions defined in $0\leq x\leq\pi$ which have m-1 continuous derivatives and whose m-1th derivatives are absolutely continuous in $[0,\pi]$.

Now suppose that we are given a real function $p(x)\in C^1[0,\pi]$ and a real number j besides $q(x)\in C^1[0,\pi]$ and h. Then we can derive a "deformation formula".

Lemma 2.4(Suzuki[11],cf.[2]).

(i) There exists a unique solution $K=K(x,s)\in C^2(\overline{D})$ which satisfies a differential equation

(2.9) $K_{xx} - p(x)K = K_{ss} - q(s)K, \quad \text{in } D,$

with boundary conditions

(2.10) $K(x,x) = j - h + \frac{1}{2}\int_0^x (p(t)-q(t))dt, \quad 0\le x\le\pi,$

(2.11) $K_s(x,0) - hK(x,0) = 0, \quad 0<x<\pi,$

where $D=\{(x,s);0<s<x<\pi\}$.

(ii) We put

(2.12) $\psi(x,\lambda) = \phi(x,\lambda) + \int_0^x K(x,s)\phi(s,\lambda)ds,$

where $\phi(x,\lambda)$ is the solution of (2.1)-(2.3).
Then $\psi(x,\lambda)$ satisfies

(2.13) $\psi''(x,\lambda) + (\lambda-p(x))\psi(x,\lambda) = 0,$

(2.14) $\psi(0,\lambda) = 1, \quad \psi'(0,\lambda) = j.$

In place of (2.10), we consider the boundary condition

(2.10)' $K(x,x) = f(x), \quad 0\le x\le\pi.$

We denote the solution of (2.9),(2.10)' and (2.11) by $K(x,s;p,q,f,h)$. Then we have

Lemma 2.5(Suzuki[11]).
There is a monotonously increasing continuous function $\gamma:[0,\infty)\times[0,\infty)\times[0,\infty)\to(0,\infty)$ such that

$$||K(\cdot,\cdot;p,q,f,h)||_{C^2(\overline{D})}\le \gamma(||p||_{C^1[0,\pi]},||q||_{C^1[0,\pi]},L)\cdot||f||_{C^2[0,\pi]}$$

holds for $h\le L$.

§3. Derivation of the linear integral equation

In the present and the posterior sections, we shall study the wellposedness of the inverse Sturm-Liouville problem.
Let us consider the eigenvalue problem $E(q(x);h,H)$:

(3.1) $y'' + (\lambda-q(x))y = 0, \quad 0<x<\pi,$

(3.2) $y'(0) - hy(0) = 0, \quad y'(\pi) + Hy(\pi) = 0,$

where $q(x)\in C^1[0,\pi]$, h and H are real constants and λ is a real parameter.
We denote the spectral characteristics of $E(q(x);h,H)$ by $\{\lambda_n,\rho_n\}_{n=0}^{\infty}$ and the solution of (3.1) which satisfies the initial conditions

(3.3) $y(0) = 1, \quad y'(0) = h,$

by $\phi(x,\lambda)$. Similarly we consider the second eigenvalue problem $E(p(x);h,H)$:

(3.4) $y'' + (\lambda-p(x))y = 0; \quad 0<x<\pi,$

(3.2) $y'(0) - hy(0) = 0, \quad y'(\pi) + Hy(\pi) = 0,$

where $p(x)\in C^1[0,\pi]$. We denote its spectral characteristics by $\{\mu_n,\sigma_n\}_{n=0}^{\infty}$ and the solution of (3.4) with (3.3) by $\psi(x,\lambda)$.

Let K(x,s) be the solution of a differential equation

(3.5) $K_{xx} - p(x)K = K_{ss} - q(s)K, \quad 0<s<x<\pi,$

with boundary conditions

(3.6) $K(x,x) = \frac{1}{2}\int_0^x (p(t)-q(t))dt, \quad 0\leq x\leq\pi,$

(3.7) $K_s(x,0) - hK(x,0) = 0, \quad 0<x<\pi.$

Then it follows from Lemma 2.4 that

(3.8) $\psi(x,\lambda) = \phi(x,\lambda) + \int_0^x K(x,s)\phi(s,\lambda)ds.$

Henceforth we assume that $q(x), h, H, \{\lambda_n,\rho_n\}_{n=0}^{\infty}$ and $\{\mu_n,\sigma_n\}_{n=0}^{\infty}$ are known but that $p(x)$ is unknown. To begin with, we shall proceed to the derivation of the linear integral equation for K(x,s).

Theorem 3.1.
If we put

(3.9) $F(x,s) = \sum_{n=0}^{\infty}\left[\frac{\phi(x,\mu_n)\phi(s,\mu_n)}{\sigma_n} - \frac{\phi(x,\lambda_n)\phi(s,\lambda_n)}{\rho_n}\right],$

then we have

(3.10) $K(x,s) + \int_0^x K(x,t)F(t,s)dt + F(x,s) = 0, \quad \text{for } 0\leq s\leq x\leq\pi.$

Proof.
We define the spectral functions $\rho(\lambda)$ and $\sigma(\lambda)$ by

(3.11) $\rho(\lambda) = \sum_{\lambda_n\leq\lambda}\rho_n^{-1}, \qquad \sigma(\lambda) = \sum_{\mu_n\leq\lambda}\sigma_n^{-1},$

and we put $\chi(\lambda)=\sigma(\lambda)-\rho(\lambda)$. Then we have

(3.12) $\int_{-\infty}^{\infty}[\int_a^x\psi(u,\lambda)du][\int_b^s\phi(t,\lambda)dt]d\sigma(\lambda) = 0, \quad \text{for } 0\leq b\leq s<a\leq x\leq\pi.$

Let us prove (3.12). Let $\Gamma(x,s)$ $(0\leq s\leq x\leq\pi)$ be the resolvent kernel of K(x,s), i.e.,

$$\Gamma(x,s) = \sum_{n=1}^{\infty}(-1)^{n-1}K^{(n)}(x,s),$$

where $K^{(1)}(x,s)=K(x,s)$ and $K^{(n)}(x,s)=\int_s^x K(x,t)K^{(n-1)}(t,s)dt$ for $n=2,3,\cdots$.

Then we have

(3.13) $\phi(t,\lambda) = \psi(t,\lambda) - \int_0^t\Gamma(t,u)\psi(u,\lambda)du.$

We obtain from (3.13) that

$$\int_{-\infty}^{\infty}[\int_a^x\psi(u,\lambda)du][\int_b^s\phi(t,\lambda)dt]d\sigma(\lambda)$$
$$= \int_{-\infty}^{\infty}[\int_a^x\psi(u,\lambda)du][\int_0^b(-\int_b^s\Gamma(t,u)dt)\psi(u,\lambda)du]d\sigma(\lambda) +$$
$$+ \int_{-\infty}^{\infty}[\int_a^x\psi(u,\lambda)du][\int_b^s(1-\int_b^s\Gamma(t,u)dt)\psi(u,\lambda)du]d\sigma(\lambda) = 0.$$

The last equality follows from the Parseval equality and $[a,x]\cap[0,s]=\phi$. Thus (3.12) is proved.
We often use the following easily verified identity later.

Lemma 3.1.

If we put

$$(3.14)\qquad H(x,t;a) = \int_a^x K(u,t)du \text{ for } 0\le t\le a,\ \int_t^x K(u,t)du \text{ for } a\le t\le x,$$

then we have

$$(3.15)\qquad \int_0^x H(x,t;a)\theta(t)dt = \int_a^x[\int_0^u K(u,t)\theta(t)dt]du,$$

for every continuous function $\theta(t)$.

Now let us prove that

$$(3.16)\qquad \int_a^x du\int_b^s dv\{F(u,v)+K(u,v)+\int_0^u K(u,t)F(t,v)dt\} = 0, \quad \text{for } 0\le b\le s<a\le x\le\pi.$$

If (3.16) is true, then we obtain (3.10) by differentiating (3.16) with respect to x and s. We calculate the left side of (3.16) as

$$(3.17)\qquad \int_{-\infty}^{\infty}[\int_a^x\phi(u,\lambda)du][\int_b^s\phi(v,\lambda)dv]d\chi(\lambda) + \int_a^x(\int_b^s K(u,v)dv)du + \\ + \int_a^x du\int_b^s dv\int_0^u K(u,t)F(t,v)dt.$$

The third term of (3.17) becomes

$$\begin{aligned} &\int_0^x H(x,t;a)(\int_b^s F(t,v)dv)dt \quad \text{(by Lemma 3.1)}\\ &= \int_{-\infty}^{\infty}[\int_0^x H(x,t;a)\phi(t,\lambda)dt][\int_b^s\phi(v,\lambda)dv]d\chi(\lambda)\\ &= \int_{-\infty}^{\infty}[\int_0^x H(x,t;a)\phi(t,\lambda)dt][\int_b^s\phi(v,\lambda)dv]d\sigma(\lambda) - \int_b^s H(x,t;a)dt.\end{aligned}$$

Therefore the left side of (3.16) becomes

$$\begin{aligned} &\int_{-\infty}^{\infty}[\int_a^x\phi(u,\lambda)du][\int_b^s\phi(v,\lambda)dv]d\chi(\lambda) + \int_a^x(\int_b^s K(u,v)dv)du +\\ &+ \int_{-\infty}^{\infty}[\int_0^x H(x,t;a)\phi(t,\lambda)dt][\int_b^s\phi(v,\lambda)dv]d\sigma(\lambda) - \int_b^s(\int_a^x K(u,t)du)dt\\ &= \int_{-\infty}^{\infty}[\int_a^x\phi(u,\lambda)du+\int_0^x H(x,t;a)\phi(t,\lambda)dt][\int_b^s\phi(v,\lambda)dv]d\sigma(\lambda)\\ &\qquad\qquad \text{(by the Parseval equality)}\\ &= \int_{-\infty}^{\infty}[\int_a^x\{\phi(u,\lambda)+\int_0^u K(u,t)\phi(t,\lambda)dt\}du][\int_b^s\phi(v,\lambda)dv]d\sigma(\lambda) \quad \text{(by (3.14))}\\ &= \int_{-\infty}^{\infty}[\int_a^x\psi(u,\lambda)du][\int_b^s\phi(v,\lambda)dv]d\sigma(\lambda) = 0 \quad \text{(by (3.12))}.\end{aligned}$$

Thus Theorem 3.1 is proved.

§4. Proof of Theorem 1

In §3, we showed that $K(x,s)$ satisfies the linear integral equation (3.10). Let us consider to solve (3.10). We start from $F(s,t)$ and construct the iterated kernels $F^{(n)}(s,t;x)(n=1,2,3,\cdots)$ as follows:

$$(4.1)\qquad F^{(1)}(s,t;x)=F(s,t), \quad F^{(n+1)}(s,t;x)=\int_0^x F(s,u)F^{(n)}(u,t;x)du \text{ for } n=1,2,\cdots.$$

We put

$$(4.2)\qquad S(s,t;x) = \sum_{n=1}^{\infty}(-1)^n F^{(n)}(s,t;x),$$

and assume that

$$(4.3)\qquad \int_0^{\pi}\int_0^{\pi}|F(s,t)|^2 dsdt < 1.$$

Then it is easily seen that the right side of (4.2) converges absolutely and uniformly and that

(4.4) $$K(x,s) = S(x,s;x), \quad \text{for } 0\le s\le x\le \pi.$$

While it follows from (3.6) that

(4.5) $$\frac{1}{2}(q(x)-p(x)) = -\frac{d}{dx}K(x,x).$$

Lemma 4.1.[3]
Assume that the function F(x,s) defined by (3.9) has a continuous derivative and that the condition (4.3) is satisfied. Then we have

$$\frac{1}{2}(q(x)-p(x)) = \frac{d}{dx}F(x,x) - K(x,x)^2 + 2\int_0^x F_x(x,u)K(x,u)du.$$

Proof.
We obtain in view of (4.1)~(4.5) that

(4.6) $$\frac{1}{2}(q(x)-p(x)) = \frac{d}{dx}F(x,x) + \sum_{n=1}^{\infty}(-1)^n\frac{d}{dx}F^{(n+1)}(x,x;x)$$

$$= \frac{d}{dx}F(x,x) + \sum_{n=1}^{\infty}(-1)^n\{2\int_0^x F_x(x,u)F^{(n)}(x,u;x)du+\sum_{k=1}^{n}F^{(k)}(x,x;x)F^{(n+1-k)}\}$$

$$= \frac{d}{dx}F(x,x) + 2\int_0^x F_x(x,u)K(x,u)du - K(x,x)^2,$$

which proves Lemma 4.1.

Now we shall prove Theorem 1.
Recall that

(3.9) $$F(x,s) = \sum_{n=0}^{\infty}\left[\frac{\phi(x,\mu_n)\phi(s,\mu_n)}{\sigma_n} - \frac{\phi(x,\lambda_n)\phi(s,\lambda_n)}{\rho_n}\right].$$

Differentiating formally the right side of (3.9) with respect to x, yields

$$\sum_{n=0}^{\infty}\left[\frac{\phi'(x,\mu_n)\phi(s,\mu_n)}{\sigma_n} - \frac{\phi'(x,\lambda_n)\phi(s,\lambda_n)}{\rho_n}\right]$$

$$= \sum_{n=0}^{\infty}\Big[\left(\frac{1}{\sigma_n} - \frac{1}{\rho_n}\right)\phi'(x,\lambda_n)\phi(s,\lambda_n) +$$

$$+ \frac{1}{\sigma_n}\int_{\lambda_n}^{\mu_n}(\dot\phi'(x,\lambda)\phi(s,\lambda)+\phi'(x,\lambda)\dot\phi(s,\lambda))d\lambda\Big].$$

If $A\equiv\sum_{n=0}^{\infty}[\sqrt{\lambda_n}|\sigma_n-\rho_n|+|\mu_n-\lambda_n|]$ is finite, then we find from Lemma 2.1 that F(x,s) has a continuous derivative and that

(4.7) $$|F_x(x,s)| \le C'\cdot\sum_{n=0}^{\infty}[\sqrt{\lambda_n}|\sigma_n-\rho_n|+|\mu_n-\lambda_n|] \equiv C'A.$$

We also obtain

(4.8) $$\left|\frac{d}{dx}F(x,x)\right| \le 2C'A.$$

In a similar way, we obtain from (3.9) that

$$|F(x,s)| \le C''\cdot\sum_{n=0}^{\infty}[|\sigma_n-\rho_n|+|\sqrt{\mu_n}-\sqrt{\lambda_n}|] \equiv C''\cdot A'.$$

Where C' and C" are constants depending only on q(x),h and H.

[3] For Lemma 4.1 to hold, the assumption (4.3) is superfluous.

If $\pi C''A'$ is small, e.g., $\pi C''A'<\frac{1}{2}$, then we have

(4.9) $|K(x,s)| = |\sum_{n=1}^{\infty}(-1)^n F^{(n)}(x,s;x)| \le \sum_{n=1}^{\infty}\frac{1}{\pi}(\pi C''A')^n \le 2C''A'$.

Consequently from Lemma 4.1 and (4.7)~(4.9), we obtain the desired estimate

$$|p(x)-q(x)| \le C_1\cdot\sum_{n=0}^{\infty}[\sqrt{\lambda_n}|\sigma_n-\rho_n|+|\mu_n-\lambda_n|],$$

if the right side is sufficiently small.

Q.E.D.

§5. Outline of the proof of Theorem 2

Since by assumption $q(x)\in A^2[0,\pi]$, the spectral characteristics $\{\lambda_n,\rho_n\}_{n=0}^{\infty}$ satisfy the following asymptotic formulae:

(5.1) $\sqrt{\lambda_n} = n + \frac{a_0}{n} + O(\frac{1}{n^3})$, and $\rho_n = \frac{\pi}{2} + \frac{b_0}{n^2} + O(\frac{1}{n^3})$.

We are also given the spectral characteristics $\{\mu_n,\sigma_n\}_{n=0}^{\infty}$ of $E(p(x);h,H)$ with unknown $p(x)\in C^1[0,\pi]$, which satisfy

(5.2) $\sqrt{\mu_n} = n + \frac{a_0'}{n} + O(\frac{1}{n^3})$, and $\sigma_n = \frac{\pi}{2} + \frac{b_0'}{n^2} + O(\frac{1}{n^3})$.

1°) In the first place, we consider the simple case $a_0=a_0'$.
We put

(5.3) $\alpha = 8(b_0'-b_0)/\pi^2$, $r(x) = q(x)+\alpha\cdot(x-\frac{\pi}{2})$, and $e(x) = r(x)-q(x)$.

We denote the spectral characteristics of $E(r(x);h,H)$ by $\{\nu_n,\tau_n\}_{n=0}^{\infty}$.
Then we obtain from Lemma 2.3 that

(5.4) $\sqrt{\nu_n} = n + \frac{a_0}{n} + O(\frac{1}{n^3})$ and $\tau_n = \frac{\pi}{2} + \frac{b_0'}{n^2} + O(\frac{1}{n^3})$,

i.e., $\sqrt{\nu_n} - \sqrt{\mu_n} = O(\frac{1}{n^3})$ and $\tau_n - \sigma_n = O(\frac{1}{n^3})$ hold.

Lemma 5.1.
If $|\alpha|$ is sufficiently small, then we have

(5.5) $\sum_{n=1}^{\infty}\{\sqrt{\nu_n}|\tau_n-\rho_n-\frac{b_0'-b_0}{n^2}|+|\nu_n-\lambda_n|\} \le C|\alpha|$,

where $C>0$ is a constant depending only on $q(x)$, h and H.

Proof.
We first show that

(5.6) $\sum_{n=0}^{\infty}|\nu_n-\lambda_n| \le C|\alpha|$.

To do this, it is sufficient to prove that

(5.7) $|\lambda_n-\nu_n| \le C\frac{|\alpha|}{n^2}$, for large n.

The eigenvalues $\{\lambda_n\}$ and $\{\nu_n\}$ are the roots of the equations

(5.8) $\phi'(\pi,\lambda) + H\phi(\pi,\lambda) = 0$,

and

(5.9) $\omega'(\pi,\lambda) + H\omega(\pi,\lambda) = 0,$

respectively, where $\omega(x,\lambda)$ is the solution of a differential equation

(5.10) $y'' + (\lambda - r(x))y = 0,$

with initial conditions

$$y(0) = 1, \quad y'(0) = h.$$

Moreover we obtain from Lemma 2.4 that

(5.11) $\omega(x,\lambda) = \phi(x,\lambda) + \int_0^x K_1(x,t)\phi(t,\lambda)dt,$

where $K_1(x,t)$ is the solution of a differential equation

$$K_{xx} - r(x)K = K_{ss} - q(s)K, \quad 0<s<x<\pi,$$

with boundary conditions

$$K(x,x) = \frac{1}{2}\int_0^x (r(t)-q(t))dt, \quad 0\le x\le\pi,$$

$$K_s(x,0) - hK(x,0) = 0, \quad 0<x<\pi.$$

By virtue of the asymptotic formulae of the eigenvalues (Lemma 2.3), the identity (2.4), (5.8), (5.9) and (5.11), the estimate (5.7) is shown to hold.
Let us now prove that

(5.12) $\sum_{n=1}^{\infty}\sqrt{\nu_n}\,|\tau_n - \rho_n - \frac{b_0'-b_0}{n^2}| \le C|\alpha|.$

We calculate

$$\tau_n - \rho_n = \int_0^\pi \{\omega(x,\nu_n)^2 - \phi(x,\lambda_n)^2\}dx$$

$$= \int_0^\pi \{\omega(x,\nu_n)^2 - \phi(x,\nu_n)^2\}dx + \int_0^\pi\{\phi(x,\nu_n)^2 - \phi(x,\lambda_n)^2\}dx \equiv I_3 + I_4.$$

It follows from (5.7) and Lemma 2.1 that

$$I_4 = O(\frac{|\alpha|}{n^3}), \quad \text{for large } n.$$

On the other hand, we obtain from Lemma 2.2 that

$$I_3 = \int_0^\pi(\omega(x,\nu_n)+\phi(x,\nu_n))(\omega(x,\nu_n)-\phi(x,\nu_n))dx$$

$$= \int_0^\pi\{2\cos\eta x + \frac{\sin\eta x}{\eta}(Q(x)+2h+\frac{1}{2}E(x)) + O(\frac{1}{\eta^2})\} \times$$

$$\times\, [\frac{\sin\eta x}{2\eta}E(x) + \frac{\cos\eta x}{4\eta^2}\{\alpha x - E(x)\cdot(2h+Q(x)+\frac{1}{2}E(x))\} + O(\frac{|\alpha|}{\eta^3})]dx$$

$$= \frac{\pi^2}{8\eta^2}\cdot\alpha + O(\frac{|\alpha|}{\eta^3}) = (b_0'-b_0)/n^2 + O(\frac{|\alpha|}{n^3}),$$

where $\eta = \sqrt{\nu_n}$, $Q(x) = \int_0^x q(t)dt$ and $E(x) = \int_0^x e(t)dt$.

So we have

$$|\tau_n - \rho_n - \frac{b_0'-b_0}{n^2}| \le C\cdot\frac{|\alpha|}{n^3}, \quad \text{for large } n,$$

while for small $n\ge 1$, we easily obtain that

$$|\tau_n-\rho_n-\frac{b_0'-b_0}{n^2}| \le C|\alpha|.$$

Therefore Lemma 5.1 is obtained.

2°) We may apply Theorem 1 to estimate the difference $|p(x)-r(x)|$.

If $\sum_{n=0}^{\infty}[\sqrt{\nu_n}|\sigma_n-\tau_n|+|\mu_n-\nu_n|]$ is sufficiently small, then we have

$$|p(x)-r(x)| \le \tilde{C}_1\cdot\sum_{n=0}^{\infty}[\sqrt{\nu_n}|\sigma_n-\tau_n|+|\mu_n-\nu_n|],$$

where $\tilde{C}_1$ is a constant depending only on $q(x)$, h and H.

3°) Finally, we consider the general case $a_0\neq a_0'$.

We put $q_1(x) = q(x)+2(a_0'-a_0)$. Clearly the spectral characteristics $\{\lambda_n^1,\rho_n^1\}_{n=0}^{\infty}$ of $E(q_1(x);h,H)$ satisfy

$$\sqrt{\lambda_n^1} = n + \frac{a_0'}{n} + O(\frac{1}{n^3}) \quad \text{and} \quad \rho_n^1 = \frac{\pi}{2} + \frac{b_0}{n^2} + O(\frac{1}{n^3}).$$

We obtain from 1°),2°) and 3°) that

$$|p(x)-q(x)| \le |p(x)-r(x)| + |r(x)-q(x)|$$

$$\le C\cdot\sum_{n=0}^{\infty}[\sqrt{\nu_n}|\sigma_n-\tau_n|+|\mu_n-\nu_n|]+|2(a_0'-a_0)+\frac{8}{\pi^2}(b_0'-b_0)(x-\frac{\pi}{2})|$$

$$\le C\{\sum_{n=1}^{\infty}[\sqrt{\lambda_n}|\sigma_n-\rho_n-\frac{b_0'-b_0}{n^2}|+|\mu_n-\lambda_n-2(a_0'-a_0)|]+$$

$$+|a_0'-a_0|+|b_0'-b_0|+|\sigma_0-\rho_0|+|\mu_0-\lambda_0|\},$$

if the right side is sufficiently small. Thus Theorem 2 is obtained.

§Appendix. Algorithm to solve the inverse Sturm-Liouville problem

To begin with, we state the well-known Gel'fand-Levitan's theorem.

Theorem[2].

Suppose that we are given two sequences of positive numbers $\{\lambda_n,\rho_n\}_{n=0}^{\infty}$ which satisfy the following asymptotic formulae:

(A.1) $$\sqrt{\lambda_n} = n + \frac{a_0}{n} + \frac{a_1}{n^3} + O(\frac{1}{n^4}), \quad \lambda_i<\lambda_j \text{ for } i\neq j,$$

(A.2) $$\rho_n = \frac{\pi}{2} + \frac{b_0}{n^2} + O(\frac{1}{n^4}),$$

where a_0, b_0 and a_1 are real constants.

Then there exist a function $q(x)\in C^1[0,\pi]$ and real numbers h and H such that the given data $\{\lambda_n,\rho_n\}_{n=0}^{\infty}$ are the spectral characteristics of the eigenvalue problem $E(q(x);h,H)$.

We introduce a new algorithm to prove the above Theorem.

The algorithm is as follows:

(A-1) There exist real numbers α, β and J such that the spectral characteristics $\{\nu_n,\tau_n\}_{n=0}^{\infty}$ of the eigenvalue problem $E(\alpha+\beta\cos 2x;0,J)$ satisfy the same asymptotic formulae as the given data$\{\lambda_n,\rho_n\}_{n=0}^{\infty}$, i.e.,

(A.3) $\sqrt{\nu_n} = n + \frac{a_0}{n} + \frac{a_1}{n^3} + O(\frac{1}{n^4})$, (A.4) $\tau_n = \frac{\pi}{2} + \frac{b_0}{n^2} + O(\frac{1}{n^4})$.

We put $r(x) = \alpha+\beta\cos 2x$.

(A-2) We denote by $\omega(x,\lambda)$ the solution of a differential equation

(A.5) $y'' + (\lambda-r(x))y = 0$,

which satisfies the initial conditions

(A.6) $y(0) = 1, \quad y'(0) = 0$.

and we put

(A.7) $$F(x,s) = \sum_{n=0}^{\infty}\Big[\frac{\omega(x,\lambda_n)\omega(s,\lambda_n)}{\rho_n} - \frac{\omega(x,\nu_n)\omega(s,\nu_n)}{\tau_n}\Big].$$

Then the function $F(x,s)$ is twice continuously differentiable in $0\le x,s\le\pi$.

(A-3) The integral equation

(A.8) $$K(x,s) + \int_0^x K(x,t)F(t,s)dt + F(x,s) = 0, \quad 0<s<x<\pi,$$

has a unique solution $K(x,s)$ for every fixed x, and the solution $K(x,s)$ is twice continuously differentiable in $0\le s\le x\le\pi$.

(A-4) We put

(A.9) $$\phi(x,\lambda) = \omega(x,\lambda) + \int_0^x K(x,s)\omega(s,\lambda)ds,$$

(A.10) $q(x) = 2\frac{d}{dx}K(x,x) + r(x)$, and (A.11) $h = K(0,0)$.

Then $\phi(x,\lambda)$ satisfies a differential equation

(A.12) $\phi''(x,\lambda) + (\lambda-q(x))\phi(x,\lambda) = 0$, for $\lambda>0$ and $0<x<\pi$,

and the initial conditions

(A.13) $\phi(0,\lambda) = 1, \quad \phi'(0,\lambda) = h$.

Furthermore it follows that

(A.14) $$\rho_n = \int_0^{\pi}\phi(x,\lambda_n)^2dx,$$

and that the ratio $\phi'(\pi,\lambda_n)/\phi(\pi,\lambda_n)$ is a constant independent of n.

Thus if we put $H=-\phi'(\pi,\lambda_n)/\phi(\pi,\lambda_n)$, then we find that the given numbers $\{\lambda_n,\rho_n\}_{n=0}^{\infty}$ are the spectral characteristics of the found eigenvalue problem $E(q(x);h,H)$.

Gel'fand-Levitan considered the fixed eigenvalue problem $E(0;0,0)$ as a starting approximation. In this case we have $\omega(x,\lambda)=\cos\sqrt{\lambda}x$.
According to the algorithm based on our method, it is much easier to verify the regularity of the function $q(x)$.
For the proof of (A-1)~(A-4), see Mizutani[17].

<u>References:</u>

[1] G.Borg, Eine Umkehrung der Stum-Liouvillischen Eigenwertaufgabe, Acta Math., 78(1946) 1-96.
[2] I.M.Gel'fand and B.M.Levitan, On the determination of a differential equation by its spectral function, Izv.Akad.Nauk SSSR, Ser.Mat.15(1951), 309-360. (English

Translation: Amer.Math.Soc.Translations,Ser.2,1(1955), 253-304.)
[3] H.Hochstadt, On the determination of a Hill's equation from its spectrum, Arch Rat.Mech.Anal.,19(1965) 353-362.
[4] H.Hochstadt, On the wellposedness of the inverse Sturm-Liouville problems, J. Diff.Eq.23(1977), 402-413.
[5] N.Levinson, Gap and Density Theorems, Colloq.Publ. vol.26. Amer.Math.Soc., New York(1940).
[6] N.Levinson, The inverse Sturm-Liouville problem, Mat.Tidsskr.B,(1940),25-30.
[7] B.M.Levitan and M.G.Gasymov, Determination of a differential equations by two of its spectra, Russian Math.Survey, 19-2(1964),1-63.
[8] B.M.Levitan and I.S.Sargsjan, Introduction to Spectral Theory: Selfadjoint Ordinary Differential Operators, Amer.Math.Soc.Transl.Math.Monographs(1975).
[9] V.A.Marchenko, Some problems in the theory of second-order differential operators, Dokl.Akad.Nauk.SSSR 72(1950), 457-560.
[10] R.Murayama, The Gel'fand-Levitan theory and certain inverse problems for the parabolic equation, J.Fac.Sci.Univ.Tokyo, 28(1981), 317-330.
[11] T.Suzuki, Uniqueness and nonuniqueness in an inverse problem for the parabolic equation, J.Diff.Eq.47(1983),296-316.
[12] T.Suzuki, Inverse problems for heat equations on compact intervals and on circles,I.
[13] T.Suzuki, On the inverse Sturm-Liouville problem for spatially symmetric operators.
[14] T.Suzuki, Inverse problems for the heat equation,(Japanese),Sūgaku,34(1982), 55-64.
[15] K.Yosida, Theory of Integral Equations, 2nd ed.(Japanese),Iwanami(1978).
[16] M.A.Naimark, Linear Differential Operators,I and II, Frederic Ungsr Publ.Co., New York(1968).
[17] A.Mizutani, On the inverse Sturm-Liouville problem, in preparation.

Computing Methods in Applied Sciences and Engineering, VI
R. Glowinski and J.-L. Lions (Editors)
Elsevier Science Publishers B.V. (North-Holland)

MODULEF AND COMPOSITE MATERIALS

D. BEGIS, S. DINARI, G. DUVAUT, A. HASSIM, F. PISTRE

INRIA, Rocquencourt, Le Chesnay.

This paper is concerned with the numerical treatment of composite materials. The proposed treatment is based on the homogeneization method [3, 4]. This method is original because of its possibility to give a macroscopic and microscopic field of stresses in the composite using asymptotic expansions. Numerical examples are finally presented. They have been carried out by using finite element code Modulef [2].

INTRODUCTION

Composite materials consisting of high tensile resin-impregnated fibers are being more and more frequently used in structures capable of high mechanical performance. Direct calculation of deformation of these structures using the finite elements method raises major difficulties due mainly to the very high number of heterogeneities in the material. Computation methods are, therefore, based on investigation of equivalent homogeneous materials, i.e. effective behavior moduli (Willis, Hashin ...).

In this paper we use the homogenization method. This method applies when the material being investigated has a periodic structure. It can then be shown that when the dimensions of the period tend homothetically to zero the fields of deformation and stresses tend to those corresponding to a homogeneous structure whose elastic properties can be computed precisely when a single period of the composite medium to be investigated is known. This boundary value structure is the homogenized structure and its behavior coefficients are the homogenized coefficients. This is the macroscopic equivalent structure. Furthermore by a localization procedure the method allows an easy computation of the microscopic field of stresses and, in particular, of stress forces at the boundaries between fibers and matrix. These stress-forces are particularly important because they can initiate cracks and delaminations. The overstresses at the microscopic level may produce fiber ruptures.

DESCRIPTION OF THE HOMOGENIZATION METHOD

FORMULATION OF THE PROBLEM

Let us consider an elastic body which occupies a region Ω related to a system of orthogonal axes $0x_1\ x_2\ x_3$. This body is subjected to a system of voluminal forces $\{f_i\}$ and surface forces $\{F_i\}$ on a portion Γ_F of boundary $\partial\Omega$. The other portion of the boundary is Γ_0, to which a zero movement condition is imposed.

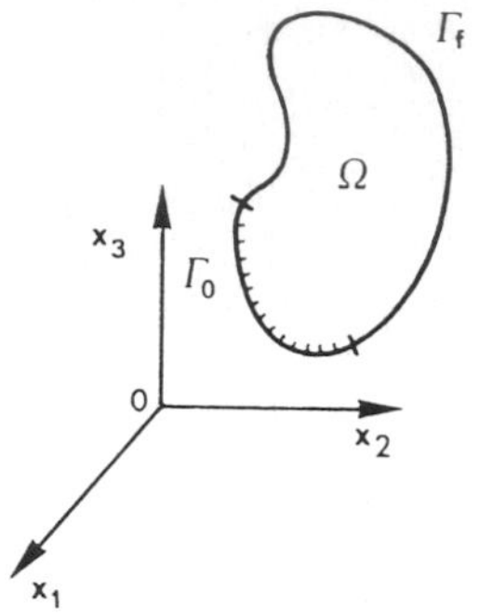

The field of stresses at equilibrium satisfies the equilibrium equations

1) $\frac{\delta \sigma_{ij}}{\delta x_j} + f_i = 0$ in Ω

2) $\sigma_{ij}\, n_j = F_i$ on Γ_F

Furthermore, the material is elastic with fine periodic structure, i.e. Ω is covered by a set of identical periods.

All the period forms must be such that opposing faces which correspond in a translation can be defined two by two.

In all cases we shall designate as Y a period characteristic of the material which has been enlarged by homothetics and fixed once and for all. ε then designates the homothetic ratio which is small and which takes us from Y to a period in the elastic material. The elastic structure of the material is then fully known if it is given over a single period, e.g. the enlarged period Y related to the orthonormal axis system $Oy_1\, y_2\, y_3$. Then let $a_{ijkh}(y)$ be the coefficients of elasticity on Y, which generally alter very quickly with respect to y, but satisfy in all respects the symmetry relation

$$a_{ijkh}(y) = a_{jikh}(y) = a_{khij}(y)$$

and positivity relation

$$\exists \alpha_0 > 0, a_{ijkh}(y)\, \tau_{ij}\, \tau_{kh} \geq \alpha_0\, \tau_{ij}\, \tau_{ij}, \forall \tau_{ij} = \tau_{ji}$$

The functions $y \to a_{ijkh}(y)$ defined on Y are extended by Y-periodicity to the entire space $Oy_1\, y_2\, y_3$ assumed to be covered by contiguous periods identical to Y.

The coefficients of elasticity in the material Ω are then $a^{\varepsilon}_{ijkh}(x)$ defined by :

$$a^{\varepsilon}_{ijkh}(x) = a_{ijkh}(y), \quad y = \frac{x}{\varepsilon}$$

For greater simplification in the text we shall write :

$$a(y) = \{a_{ijkh}(y)\}, \quad a^{\varepsilon}(x) = a(\tfrac{x}{\varepsilon}), \quad \sigma = \{\sigma_{ij}\}$$

and shall consider $a(y)$ or $a^{\varepsilon}(x)$ as known matrix 6x6 indexed by the symmetrical pairs (i,j).

The law of elasticity

3) $\sigma_{ij} = a^{\varepsilon}_{ijkh}(x) \quad e_{kh}(u)$

is written

$$\sigma = a^{\varepsilon}(x)\, e(u),$$

where

$$e(u) = \{e_{ij}(u)\}, \; e_{ij}(u) = \frac{1}{2}\left(\frac{\delta u_i}{\delta x_j} + \frac{\delta u_j}{\delta x_i}\right)$$

When an ambiguity is possible, either $e_x(u)$ or $e_y(u)$ will be specified depending on wheter the drift occurs with respect to x or y. The boundary conditions are finalized by :

4) $u = 0$ on Γ_0

The problem posed by (1) (2) (3) (4) has a unique solution which depends on ε and which we shall designate u^{ε} ; to this corresponds a field of stresses σ^{ε} given by :

5) $\sigma^{\varepsilon} = a^{\varepsilon}(x)\, e(u^{\varepsilon})$

Numerically it is very difficult when ε is small to calculate u^{ε} since there are a large number of heterogeneities in the elastic medium. We therefore try to obtain a limited expansion of the solution u^{ε}, σ^{ε}.

ASYMPTOTIC EXPANSIONS

The solution is affected by two factors :

i) The first is the scale of Ω and arises from the forces applied and the conditions at the boundaries.

ii) The second is due to the periodic structure ; it is on the same scale as the period and is repeated periodically.

This justifies looking for an asymptotic expansion of the form :

6) $u^{\varepsilon} = u^0(x,y) + \varepsilon u^1(x,y) + \varepsilon^2 u^2(x,y) + \ldots$

where the $u^{\alpha}=(x,y)$ are, for each $x \in \Omega$, Y-periodic functions with respect to the variable $y \in Y$. Then $y = \frac{x}{\varepsilon}$ is applied to (6). Associated with the expansion (6) in an expansion of the field of deformation $e(u^{\varepsilon})$.

7) $e(u^{\varepsilon}) = \frac{1}{\varepsilon}\, e_y(u^0) + e_x(u^0) + e_y(u^1) + \varepsilon[e_x(u^1) + e_y(u^2)] + \ldots$

and of the field of stresses σ^{ε}

8) $\sigma^{\varepsilon} = \frac{1}{\varepsilon}\, \sigma^0(x,y) + \sigma^1(x,y) + \varepsilon\sigma^2(x,y) + \ldots$

with

$$\sigma^0(x,y) = a(y)\, e_y(u^0)$$

$$\sigma^1(x,y) = a(y)\, [e_y(u^1) + e_x(u^0)]$$

$$\sigma^2(x,y) = a(y)\, [e_y(u^2) + e_x(u^1)]$$

The equilibrium equations (1) applied to σ^{ε} give

$$\frac{\delta}{\delta x_j} \sigma^{\varepsilon}_{ij} + f_i = 0$$

or in a more condensed form

9) $\quad \operatorname{div} \sigma^{\varepsilon} + f = 0$

Given the expansion (8) of σ^{ε} we have

10) $\frac{1}{\varepsilon^2} \operatorname{div}_y \sigma^0 + \frac{1}{\varepsilon}(\operatorname{div}_y \sigma^1 + \operatorname{div}_x \sigma^0) + \operatorname{div}_y \sigma^2 + \operatorname{div}_x \sigma^1 + f + \ldots = 0$

$x \in \Omega,\ y \in Y.$

where

$$\operatorname{div}_y \sigma^{(\alpha)} = \{\frac{\delta\sigma^{\alpha}_{ij}}{\delta y_j}\} \ ; \ \operatorname{div}_x \sigma^{(\alpha)} = \{\frac{\delta\sigma^{\alpha}_{ij}}{\delta x_j}\} \ .$$

The boundary conditions (2) are treated in the same way :

11) $\frac{1}{\varepsilon} \sigma^0.n + \sigma^1.n - F + \varepsilon \sigma^2.n + \ldots = 0 \qquad$ for $x \in \Gamma_F\ y \in Y.$

Finally the conditions (4) mean that

12) $u^0 + \varepsilon u^1 + \varepsilon^2 u^2 + \ldots = 0 \qquad$ for $x \in \Gamma_0,\ y \in Y.$

By making the various powers of ε zero we obtain :

13) $\begin{cases} \operatorname{div}_y \sigma^0 = 0 \\ \sigma^0 = a(y)\ e_y(u^0) \end{cases}$

14) $\begin{cases} \operatorname{div}_y \sigma^1 + \operatorname{div}_x \sigma^0 = 0 \\ \sigma^1 = a(y)\ [e_y(u^1) + e_x(u^0)] \end{cases}$

15) $\begin{cases} \operatorname{div}_y \sigma^2 + \operatorname{div}_x \sigma^1 + f = 0 \\ \sigma^2 = a(y)\ [e_y(u^2) + e_x(u^1)] \end{cases}$

The equations (11) and (12) will be used later.

RESOLUTIONS

The systems (13) (14) (15) contain differential operators in y. They therefore constitute equations with partial derivatives on the period of base Y, the unknown factors being the Y-periodic functions.

System (13) : This leads immediately to :

16) $\quad \sigma^0 = 0,\ u^0 = u^0(x)$

System (14) : In view of (16) it is reduced to :

17) $\quad \operatorname{div}_y \sigma^1 = 0,\ \sigma^1 = a(y)\ [e_y(u^1) + e_x(u^0)]$

The deformation $e_x(u^0)$ is a function only of x ; it therefore plays the role of a parameter with respect to the differential system in y. Due to the linearity, σ^1, u^1 may therefore be written in the form :

18) $$\begin{cases} \sigma^1 = s^{kh}(y)\, e_{kh}(u^0) \\ u^1 = \chi^{kh}(y)\, e_{kh}(u^0) \end{cases}$$

where

$$e_{kh}(u^0) = \frac{1}{2}\left(\frac{\delta u_k^0}{\delta x_h} + \frac{\delta u_h^0}{\delta x_k}\right)$$

19) $$\begin{cases} \mathrm{div}_y\, s^{kh} = 0 \\ s^{kh} = a(y)[\tau^{kh} + e_y(\chi^{kh})] \\ \chi^{kh} \text{ is } Y\text{-periodic} \end{cases}$$

The tensor τ^{kh} has components given by

$$\tau_{ij}^{kh} = \frac{1}{2}(\delta_{ik}\,\delta_{jh} + \delta_{ih}\,\delta_{jk})$$

It can be proved that the system (19) determines the vector $\chi^{kh}(y)$ to within an additive constant.

For any function $\phi = \phi(x,y)$, we define

$$<\phi> = \frac{1}{\mathrm{mes}\ Y} \int_Y \phi(x,y)\, dy$$

The solution σ^1 of (14) is given by,

20) $$\sigma^1(x,y) = a(y)\ [\tau^{kh} - e_y(\chi^{kh})]\ e_{kh}(u^0),$$

and taking the mean value, we obtain,

21) $$<\sigma_{ij}^1> = q_{ij}^{kh}\ e_{kh}(u^0)$$

where

22) $$q_{ij}^{kh} = <a_{ijkh}(y)> - <a_{ijpq}(y)\ e_{pq}(\chi^{kh}(y)>$$

System (15) : It suffices to take the mean on Y in the first equation to obtain :

23) $$\mathrm{div}_x <\sigma^1> + f = 0 \qquad \text{in } \Omega$$

If we introduce $\Sigma = <\sigma_1>$, we have

24) $$\begin{cases} \mathrm{div}_x\, \Sigma + f = 0 & \text{in } \Omega \\ \Sigma_{ij} = q_{ij}^{kh}\ e_{kh}(u^0) \end{cases}$$

Using equation (12) and taking the mean on Y in (11), we obtain :

$$(25)\quad \begin{cases} u^0 = 0 & \text{on } \Gamma_0 \\ \Sigma.n = F & \text{on } \Gamma_F. \end{cases}$$

The system (24) with boundary conditions (25) is a well posed elasticity problem ; the equilibrium equations are unchanged, as well as the boundary conditions. The elastic constitutive relation is

$$\Sigma_{ij} = q_{ij}^{kh}.e_{kh}(u^0)$$

It is homogeneous since the coefficients q_{ij}^{kh} given by (22) are independent of $x \in \Omega$. These coefficients define the equivalent homogeneous material. They are called homogenized coefficients. The stress field $\Sigma = (\Sigma_{ij})$ is called the macroscopic stress field and is defined by

$$\Sigma = \langle \sigma^1 \rangle$$

The strain field $E = e_x(u^0)$ is called the macroscopic strain field and satisfies

$$E = \langle e_x(u^0) + e_y(u^1) \rangle$$

It can be proved that the homogenized coefficients q_{ij}^{kh} satisfy

$$\begin{cases} q_{ij}^{kh} = q_{kh}^{ij} \ (= q_{ijkh}) \\ \exists \alpha_1 > 0,\ q_{ij}^{kh}\ s_{ij}\ s_{kh} \geq \alpha_1\ s_{ij}\ s_{ij},\ \forall s_{ij} = s_{ji} \end{cases}$$

This shows that (q_{ij}^{kh}) are reasonable elastic coefficients and that the macroscopic scale problem (24) (25) has a unique solution.

MICROSCOPIC FIELDS. LOCALIZATION

The stress field $\sigma^1(x,y)$ is the first term of the asymptotic expansion (8) of the stress field $\sigma^\varepsilon(x)$ solution of the initial exact problem. The field $\sigma^1(x,y)$ is called the microscopic stress field. If we imagine that at each point $x \in \Omega$, there is a small εY period with its composite structure, then $\sigma^1(x,y)$ gives, for x kept fixed in Ω, a stress field in this period.

It can be shown that $\sigma^\varepsilon(x) - \sigma^1(x,\frac{x}{\varepsilon})$ tends to zero in the $L^1(\Omega)$ norm when ε tends to zero. This proves that $\sigma^1(x,\frac{x}{\varepsilon})$ is a good approximation of $\sigma^\varepsilon(x)$ when ε is small. The microscopic stress field $\sigma^1(x,y)$, $y = \frac{x}{\varepsilon}$ can be calculated as follows :

i) First we obtain the six $\chi^{kh}(y)$ vector fields on Y, each one been associated with tensor $\tau^{kh} = \tau^{hk}$. These six vector-fields are solution of problem (19) which is an elastic type problem on the inhomogeneous period Y.

ii) From the vector fields $\chi^{kh}(y)$ we get the homogenized coefficients q_{ij}^{kh} by formula (22).

iii) We solve the macroscopic scale, homogenized elastic problem (24) (25) on Ω. It gives the macroscopic stress field $\Sigma(x)$ and the macroscopic strain field $e_x(u^0) = E(x)$, for $x \in \Omega$.

iv) Localization procedure : using formula (20) we can calculate $\sigma^1(x,y)$. For x fixed in Ω, this stress field on Y shows how the macroscopic stress $\Sigma(x) = <\sigma^1(x,y)>$ is localized in an εY period at $x \in \Omega$.

It can be proved that when ε tends to zero, the stress field $\sigma^\varepsilon(x)$ tends to $\Sigma(x)$ in the weak $L^2(\Omega)$ topology. Nevertheless $\sigma^1(x,\frac{x}{\varepsilon})$ is a better approximation of $\sigma^\varepsilon(x)$ than $\Sigma(x)$: the norm $L^1(\Omega)$ convergence implies that

$$\sigma^\varepsilon(x) - \sigma^1(x,\frac{x}{\varepsilon})$$

tends to zero for almost every point in Ω, while the weak $L^2(\Omega)$ convergence does not. The macroscopic stress field $\Sigma(x)$ is just a mean value while $\sigma^1(x,\frac{x}{\varepsilon})$ takes into account the fine periodic structure of the composite material.

APPLICATION TO A CROSS PLY LAMINATE

We consider a cross ply lamination consisting of n orthotropic layers staked with an alternating orientation of 90° between the layers. We assume that the x and the y axes are parellel to the fiber reinforcement in the odd and the even numbered layers respectively. All the odd layers are identique. The even layers also. Three examples of layers are considered :

Example 1 : each layer is reinforced by a single fiber row (Figure 1, 6)

Example 2 : each layer is reinforced by a double fiber rows (Figure 2,7)

Example 3 : each layer is reinforced by multiple fiber rows (Figure 3)

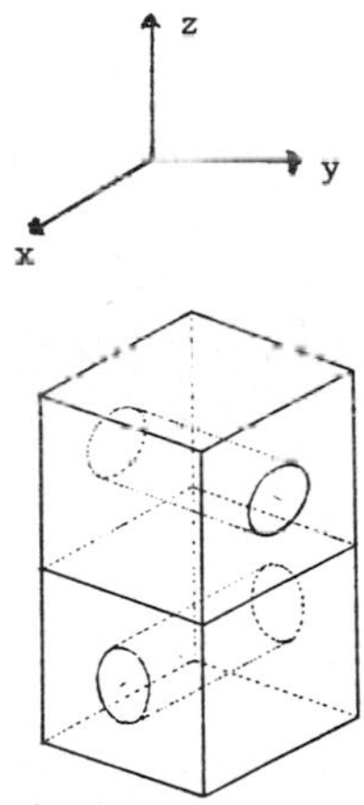

Figure 1 : Base period
Each layer is reinforced by a single row.

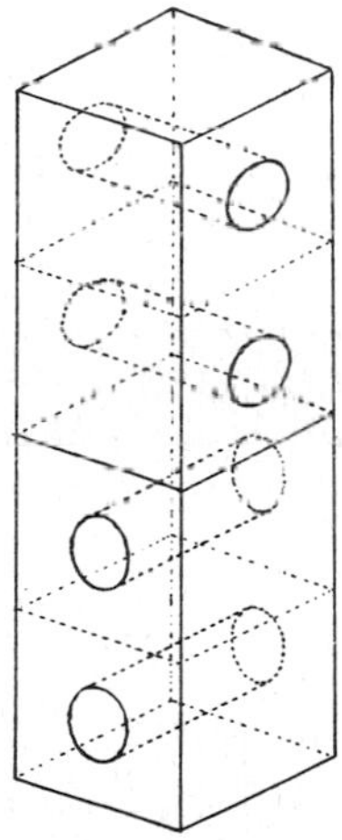

Figure 2 : Base period
Each layer is reinforced by a double fiber rows

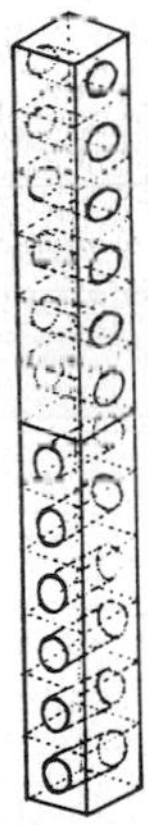

Figure 3 : Base period
Each layer is reinforced by a multiple fiber rows.

The two first examples are solved by using the three-dimensional homogenization techniques.

The last example (Figure 3) is solved in two steps by using a two-dimensional analysis :

i) homogenization of each layer reinforced by fibers running in the same direction. The problem (19) is bidimensional.

ii) homogenization of the cross ply lamination using the computed results for each layer given by the first step.

The numerical results are obtained by using the finite element code MODULEF.

THREE DIMENSIONAL HOMOGENIZATION OF THE CROSS PLY LAMINATE.

To obtain the homogenized moduli q_{ijkh}(22) it is necessary, first, to compute the functions $\vec{\chi}^{kh}$ solutions of elliptic boundary value problems on the basic period (19). As our purpose is to compute the functions $\vec{\chi}^{kh}$ by the finite element code Modulef, the variational formulation associated to (19) is :

$$\text{26)} \quad \begin{cases} \vec{\chi}^{kh} \in V = \{\vec{v} = (v_1, v_2, v_3),\ \vec{v} \in [H^1(Y)]^3, \vec{v}\ \text{Y-Periodic}\} \\ a_y(\vec{\chi}^{kh}, \vec{v}) = \int_Y a_{ijkh}(y)\ e_{ij}(v)\ dy, \quad \forall v \in V \end{cases}$$

where

$$a_y(\vec{u},\vec{v}) = \int_Y a_{ijkh}(y)\ e_{kh}(\vec{u})\ e_{ij}(\vec{v})\ dy.$$

The last integral of (26) can be written in term of surface load :

$$\int_Y a_{ijkh}(y)\ e_{ij}(\vec{v})\ dy = \int_\Gamma \vec{F}^{kh}.\vec{v}\ d\gamma$$

with

$$F_1^{11} = [a_{11\ell 1}]\ n_\ell \qquad F_2^{11} = [a_{11\ell 2}]\ n_\ell \qquad F_3^{11} = [a_{11\ell 3}]\ n_\ell$$

$$F_1^{22} = [a_{22\ell 1}]\ n_\ell \qquad F_2^{22} = [a_{22\ell 2}]\ n_\ell \qquad F_3^{22} = [a_{22\ell 3}]\ n_\ell$$

$$F_1^{33} = [a_{33\ell 1}]\ n_\ell \qquad F_2^{33} = [a_{33\ell 2}]\ n_\ell \qquad F_3^{33} = [a_{33\ell 3}]\ n_\ell$$

$$F_1^{23} = [a_{23\ell 1}]\ n_\ell \qquad F_2^{23} = [a_{23\ell 2}]\ n_\ell \qquad F_3^{23} = [a_{23\ell 3}]\ n_\ell$$

$$F_1^{13} = [a_{13\ell 1}]\ n_\ell \qquad F_2^{13} = [a_{13\ell 2}]\ n_\ell \qquad F_3^{13} = [a_{13\ell 3}]\ n_\ell$$

$$F_1^{12} = [a_{12\ell 1}]\ n_\ell \qquad F_2^{12} = [a_{12\ell 2}]\ n_\ell \qquad F_3^{12} = [a_{12\ell 3}]\ n_\ell$$

where $\vec{n} = (n_1, n_2, n_3)$ is the outward unit normal to the interface of the components (fiber-resin) and where the bracket [.] denotes the jump of a function through the interface.

Provided the numerical integration of the functions $\vec{\chi}^{kh}$, the homogenized coefficients Q_{ijkh} of the composite are the mean value, on Y, of the corresponding a_{ijkh} altered by a corrective term depending on the $\vec{\chi}^{kh}$ (22) :

$$\begin{cases} q_{ijkh} = \langle a_{ijkh}(y)\rangle - q^*_{ijkh} \\ \text{with} \\ q^*_{ijkh} = \frac{1}{|Y|} \int_{Y^*} a_{ijpq} \frac{\partial \chi^{kh}{}_p}{\partial y_q} \, dy. \end{cases}$$

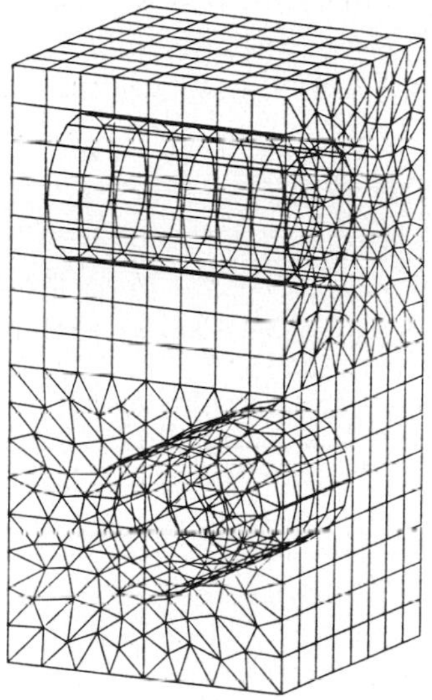

Figure 6 : Period Y (single fiber row)
Finite element mesh

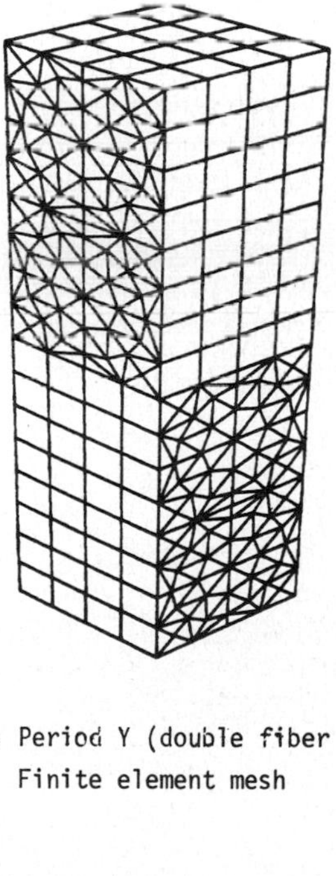

Figure 7 : Period Y (double fiber rows)
Finite element mesh

TWO-DIMENSIONAL HOMOGENIZATION OF THE CROSS PLY LAMINATE

The example 3 is solved in two steps :

Step 1 : Homogenization of a composite reinforced by fibers running in the same direction.

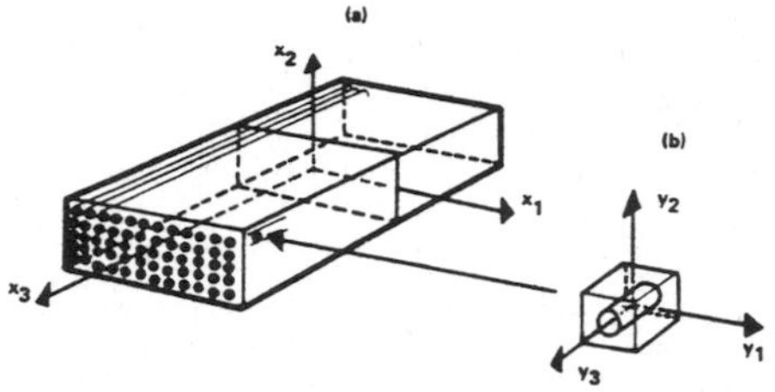

a) Structuration of fibers
b) Base period.

Calculation of the homogenized coefficients q_{ijkh} calls for the resolution of (19). In the present case the coefficients $a_{ijkh}(y)$ are independent of y_3 ; the result is that the fields $\vec{X}^{ij}(y)$ are also independent of y_3 ; in (19, 26) the various indices give a zero contribution when they refer to $\frac{\delta}{\delta y_3}$ making computation of $\vec{X}^{ij}(y)$ a bidimensional problem.

Morever, we have :

$$\vec{X}^{kh}(y) = \vec{X}^{kh}(y_1,y_2) = [X_1^{kh}(y_1,y_2),\ X_2^{kh}(y_1,y_2),\ 0]$$

$$\text{for } (k,h) = [(1,1),\ (2,2),\ (3,3),\ (1,2)]$$

These functions are therefore solution of plane strain elasticity problem.
And,

$$\vec{X}^{kh}(y) = \vec{X}^{kh}(y_1,y_2) = [0,0,\ X_3^{kh}(y_1,y_2)]$$

$$\text{for } (k,h) = [(1,3),\ (2,3)].$$

These two functions are solutions of a scalar problem in R^2.

For the details, refer to D. Begis, G. Duvaut, A. Hassim [1] and to the references in this publication.

Step 2 : Homogenization of the cross ply lamination.

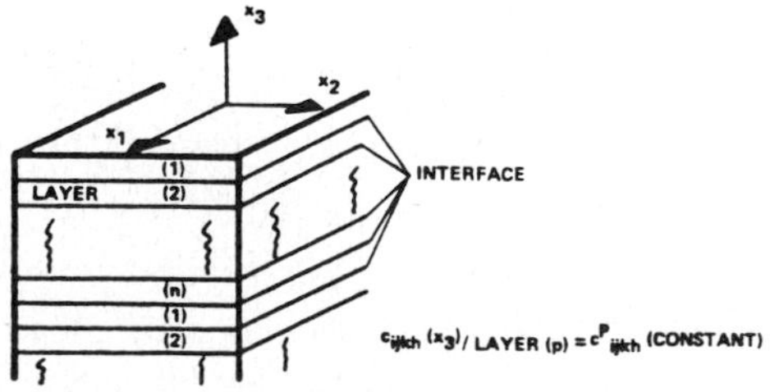

Multiple layers. Each layer possesses a plane of elastic symmetry normal to the x_3 axis (monoclinic symmetry).

Given the homogenized moduli of each layer (steep one), the homogenization formulae are considerably simplified. The problem (19) is reduced to a system of differential equations which may be solved explicitely. For the details see [7].

NUMERICAL RESULTS

The finite element code results obtained for the two first examples are compared with those obtained for the last example 3.

Each component is assumed elastic, homogeneous and isotrop, with :

E_f = 84 000 MPa ; ν_f = .22 for the fiber

E_r = 4 000 MPa ; ν_r = .34 for the resin.

Tables presented hereafter for 3 impregnations of resin (36%, 50%, 65%) show the "good" agreement of the bidimensional homogenization method.

Example	E1 (MPa)	E2 (MPa)	E3 (MPa)	G23 (MPa)	G13 (MPa)	G12 (MPa)	ν_{23}	ν_{13}	ν_{12}
1	37 700	37 700	18 900	5 600	5 600	6 775	.27	.27	.135
2	37 780	37 780	19 850	5 570	5 570	6 670	.259	.259	.137
3	37 750	37 750	20 614	5 298	5 298	6 304	.24	.24	.138

Table 1 : Comparative table for 36% of resin in volume.

Example	E1 (MPa)	E2 (MPa)	E3 (MPa)	G23 (MPa)	G13 (MPa)	G12 (MPa)	ν_{23}	ν_{13}	ν_{12}
1	28 340	28 340	12 996	3 800	3 800	4 300	.32	.32	.122
2	28 400	28 400	13 280	3 780	3 780	4 276	.308	.308	.124
3	28 801	28 801	13 563	3 598	3 598	4 168	.297	.297	.124

Table 2 : Comparative table for 50% of resin in volume.

Example	E1 (MPa)	E2 (MPa)	E3 (MPa)	G23 (MPa)	G13 (MPa)	G12 (MPa)	ν_{23}	ν_{13}	ν_{12}
1	20 500	20 500	9 471	2 880	2 880	3 076	.35	.35	.126
2	20 560	20 560	9 565	2 850	2 850	3 057	.35	.35	.128
3	20 237	20 237	9 243	2 657	2 657	2 910	.35	.35	.126

Table 3 : Comparative table for 65% of resin in volume.

NUMERICAL RESULTS OF MICROSCOPIC STRESS-FIELD

In the previous paragraph, we have shown that we get the homogenized moduli q_{ijkh} from the six vector fields $\vec{X}^{kh}(y)$. We can then solve the homogenized elastic problem (24), (25), on Ω, which gives the macroscopic stress field $\Sigma(x)$, for $x \in \Omega$. The computations of the microscopic stress-field and stress forces at the interface between fiber and resin, are particulary important because they can initiate cracks and delaminations. The localization procedure allows an easy computations of these microscopic stress-field and stress forces.

Some numerical results for an unidirectionally fiber-reinforced composite subjected to :
i) a uniform load parallel to the fibers
ii) a uniform load perpendicular to the fibers
iii) a shearing stress-field normal to the direction of the fibers
are shown in the following figures. For each applied load, the stress forces $\sigma.\vec{n}$ at the interface and the components of microscopic stress field are plotted. The density of lines plotted indicate the level of microscopic stress concentration.

The components elastic moduli are taken as

Fiber : E_f = 84 000 MPa ν_f = 0.22

Resin : E_r = 4 000 MPa ν_r = 0.34

The corresponding homogenized moduli for 60% of resin's impregnations are

E1 = 10141 MPa	ν_{32} = 0.287	G_{32} = 3106
E2 = 9685 MPa	ν_{31} = 0.281	G_{31} = 3386
E3 = 35655 MPa	ν_{12} = 0.353	G_{12} = 2606

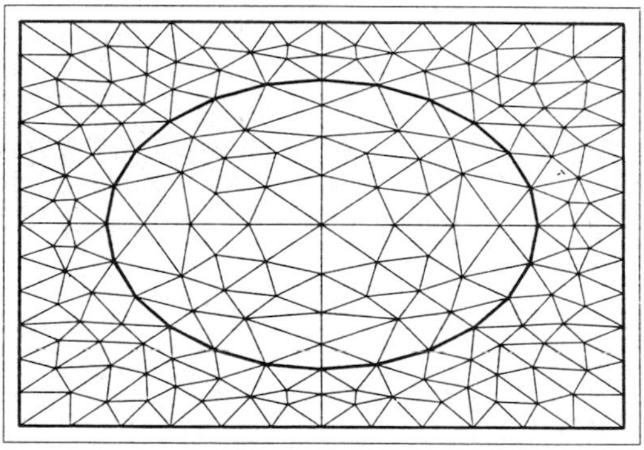

Period Y
Triangular finite element mesh of the cross section

UNIFORM LOAD PARALLEL TO THE FIBERS (FIBERS // OX_3)

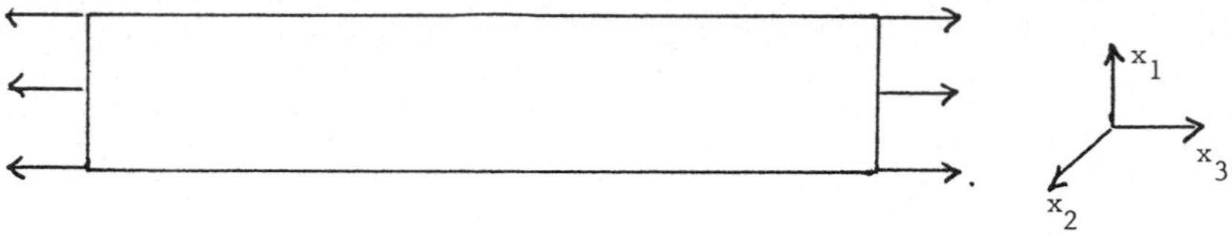

Macroscopic stress-field

$$\sigma^0 = \begin{pmatrix} 0 & 0 & 0 \\ 0 & 0 & 0 \\ 0 & 0 & \sigma^0_{33} \end{pmatrix}$$

$\sigma^0_{33} = 100$

Microscopic stress-field

$$\sigma(x,y) = \begin{pmatrix} \sigma_{11} & \sigma_{12} & 0 \\ \sigma_{12} & \sigma_{22} & 0 \\ 0 & 0 & \sigma_{33} \end{pmatrix}$$

σ_{11} max = 1,4 σ_{33} max = 235

σ_{22} max = 1,3 $|\vec{F}|$ max = 0,46

σ_{12} max = 0,72 F_3 = 0.

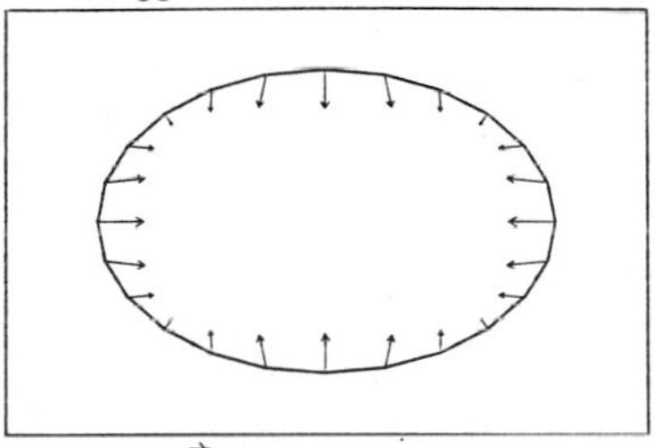

$|\vec{F}|$ max = 0,46
F_3 = 0

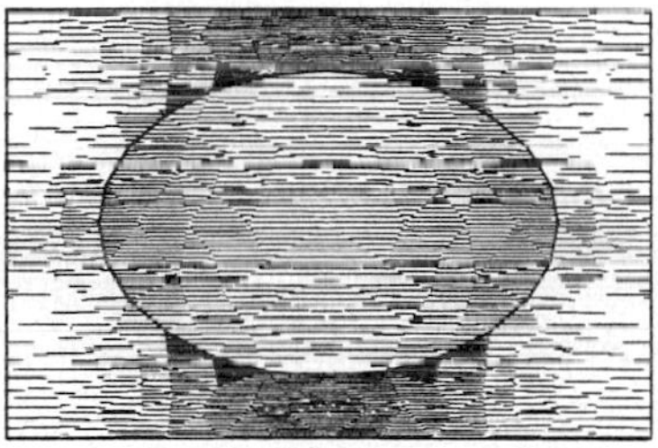

σ_{11} max = 1.4

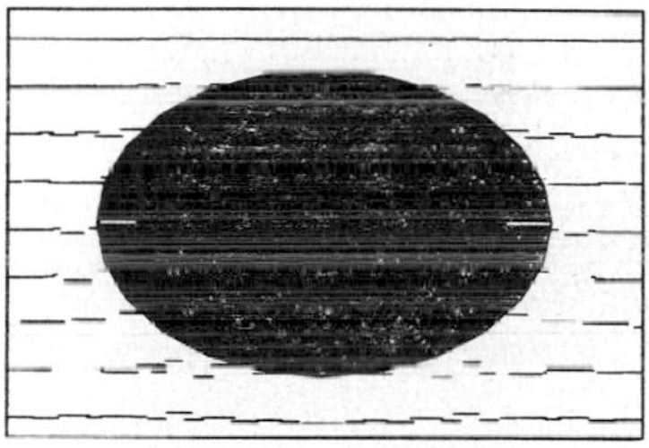

σ_{33} max = 235

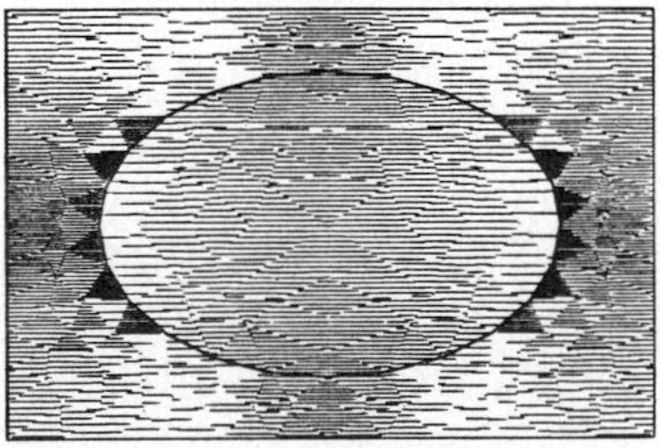

σ_{22} max = 1.3

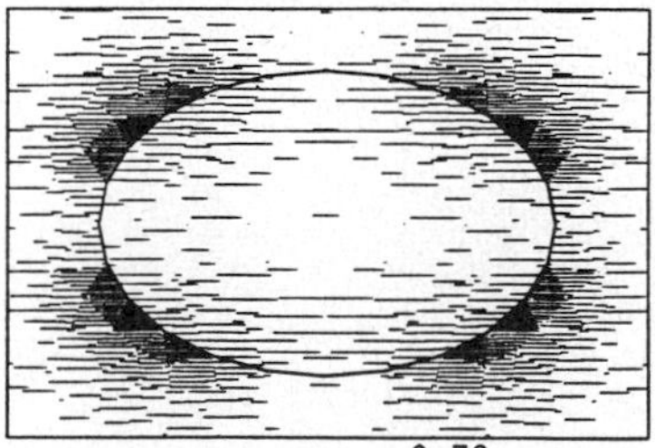

σ_{12} max = 0.72

UNIFORM LOAD PERPENDICULAR TO THE FIBERS (FIBERS // OX_3)

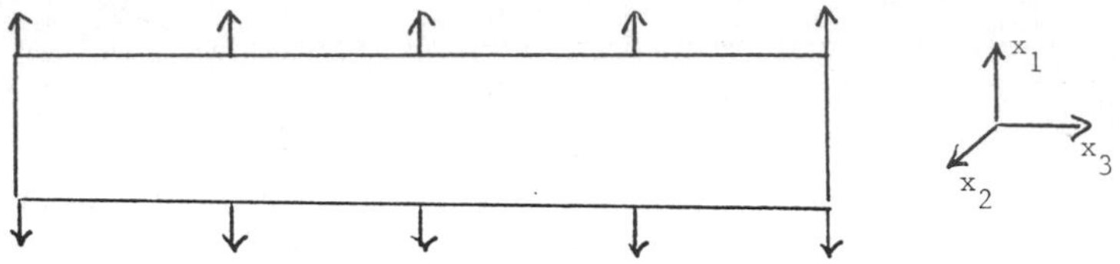

Macroscopic stress-field

$$\sigma^0 = \begin{pmatrix} \sigma^0_{11} & 0 & 0 \\ 0 & 0 & 0 \\ 0 & 0 & 0 \end{pmatrix}$$

$$\sigma^0_{11} = 100.$$

Microscopic stress-field

$$\sigma(x,y) = \begin{pmatrix} \sigma_{11} & \sigma_{12} & 0 \\ \sigma_{12} & \sigma_{22} & 0 \\ 0 & 0 & \sigma_{33} \end{pmatrix}$$

σ_{11} max = 157 σ_{33} max = 77

σ_{22} max = 73 $|\vec{F}|$ max = 74

σ_{12} max = 27 F_3 = 0.

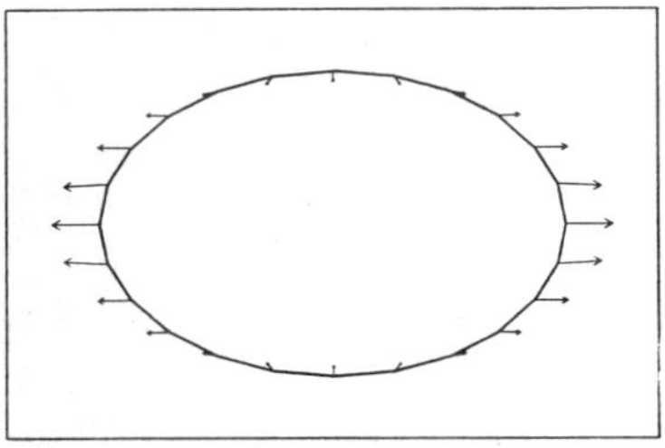

$|\vec{F}|$ max = 74
F_3 = 0

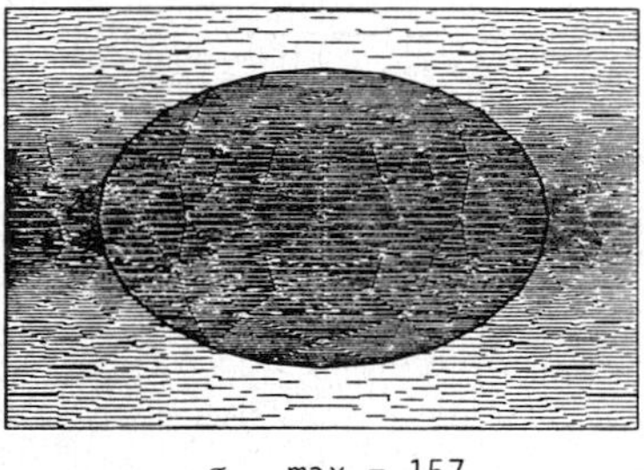

σ_{11} max = 157

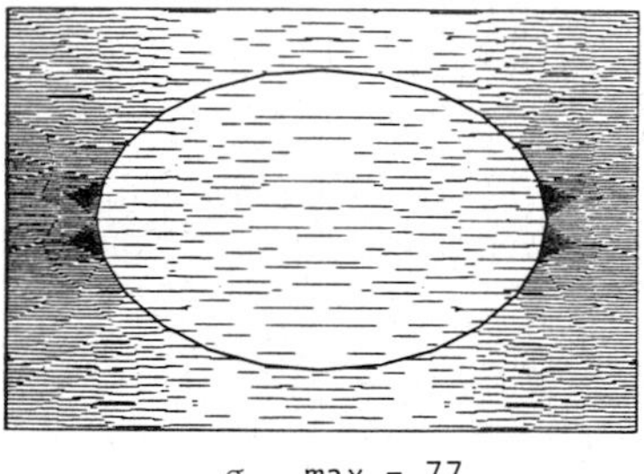

σ_{33} max = 77

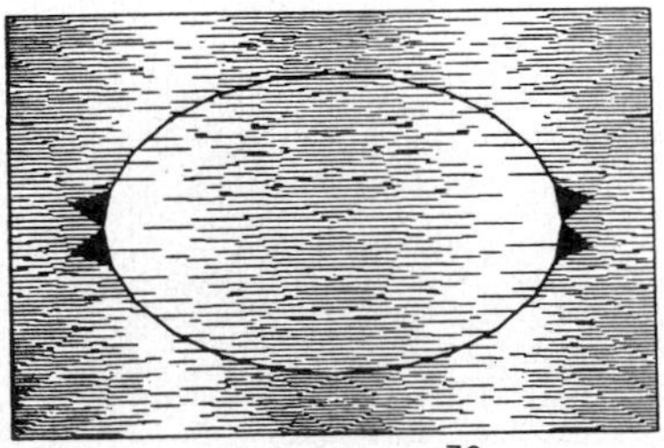

σ_{22} max = 73

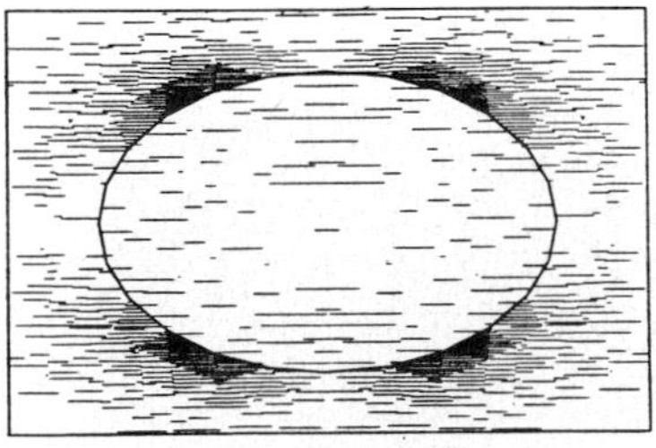

σ_{12} max = 27

SHEARING STRESS FIELD NORMAL TO THE DIRECTION OF THE FIBERS

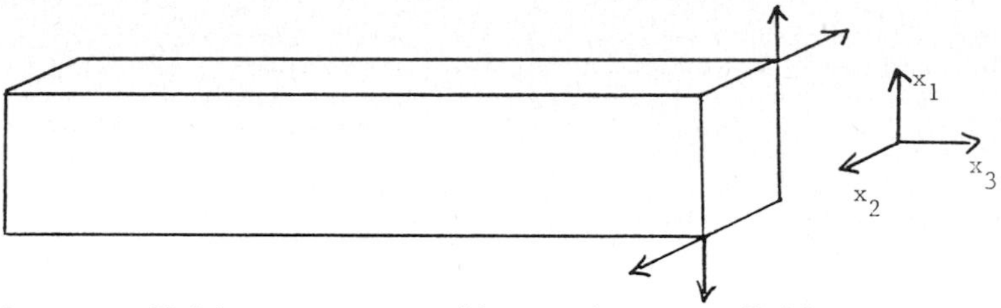

Macroscopic stress-field

$$\sigma^0 = \begin{pmatrix} 0 & \sigma^0_{12} & 0 \\ \sigma^0_{12} & 0 & 0 \\ 0 & 0 & 0 \end{pmatrix}$$

$\sigma^0_{12} = 100.$

Microscopic stress-field

$$\sigma(x,y) = \begin{pmatrix} \sigma_{11} & \sigma_{12} & 0 \\ \sigma_{12} & \sigma_{22} & 0 \\ 0 & 0 & \sigma_{33} \end{pmatrix}$$

σ_{11} max = 47.8 $\quad \sigma_{33}$ max = 35.3

σ_{22} max = 60 $\quad |\vec{F}|$ max = 35

σ_{12} max = 69.1 $\quad F_3$ = 0.

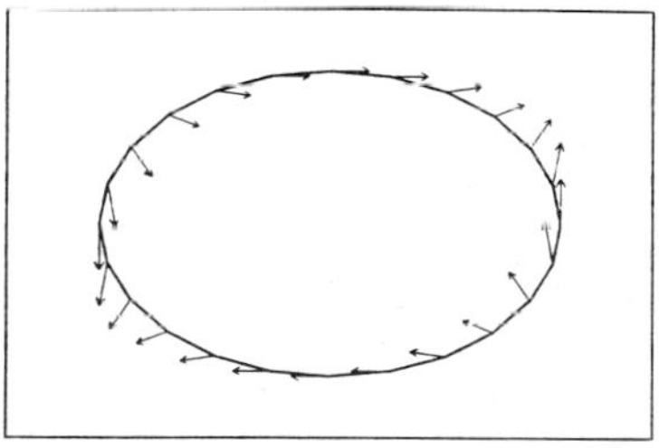

$|\vec{F}|$ max = 35
$F_3 = 0$

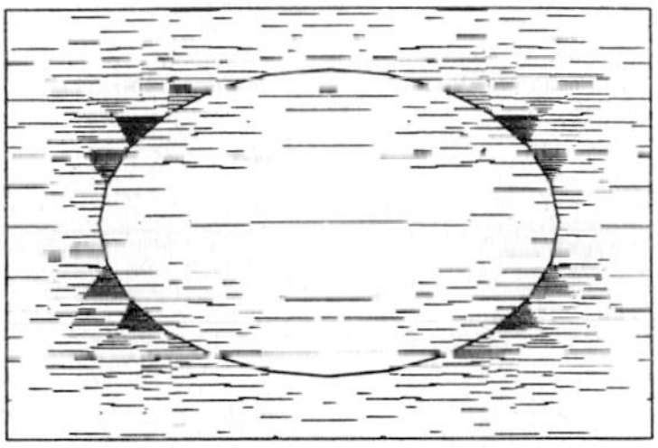

σ_{11} max = 47.8

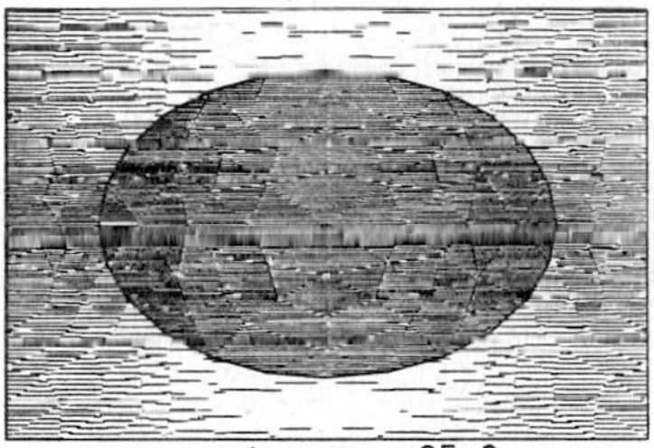

σ_{33} max = 35.3

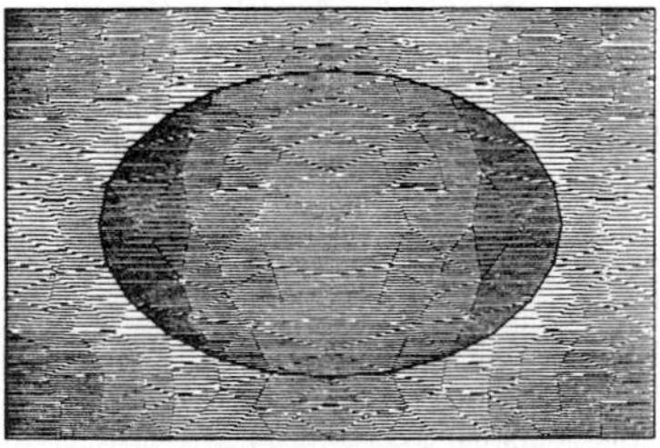

σ_{22} max = 60

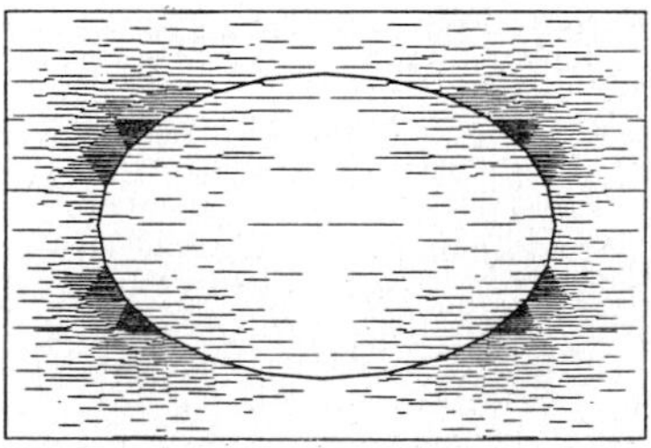

σ_{12} max = 69.1

CONCLUSION

The homogenization theory seems efficient to compute the mechanical characteristics of composite materials. Validity of the results is evidently subjected to the assumptions made on shapes and lay out of fibers (periodicity). Of course, a configuration taking into consideration random direction will probably be nearer to the truth. However, the undeniable advantage of this method aims at supplying complete and consistent sets of values, mutually coherent. The comparison with experimental characterization made by the Aerospace Industrie SNIAS Marignane shows that the estimates based on the homogenization theory seems nearer to those based on other methods.

BIBLIOGRAPHY

[1] D. BEGIS, G. DUVAUT, A. HASSIM, Homogénéisation par éléments finis des modules de comportements élastiques de matériaux composites. Rapport de Recherche n°101, INRIA.

[2] THE CLUB MODULEF : A library of computer procedures for finite element analysis. Publication Modulef n°73, June 1983, INRIA.

[3] A. BENSOUSSAN, J.L. LIONS, G. PAPANICOLAOU, Asymptotic analysis for periodic structures. North-Holland, Publishing Company, 1978.

[4] G. DUVAUT, Matériaux élastiques composites à structure périodique, Homogeneization. Theoretical and applied mechanics. W.T. Koiter, cd North-Holland publishing company, 1976.

[5] G. DUVAUT, Effective and homogenized coefficients Symposium on functional analysis and differential equations. Lisboa, Portugal, March 29-April 2, 1982.

[6] S. FLUGGE, Encyclopedia of Physics, Mechanics of solids II, Springer Verlag, volume 2.

[7] A. HASSIM, Homogénéisation par éléments finis d'un matériau élastique renforcé par des fibres. Thèse de 3ème cycle (Toulouse 1980).

[8] F. LENE, G. DUVAUT, Résultats d'isotropie pour des milieux homogénéisés. CRAS Paris, Tome 293, pp. 477-480 série 2 (Octobre 1981).

[9] F. LENE, D. LEGUILLON, Etude de l'influence d'un glissement entre les constituants d'un matériau composite sur ses coefficients de comportement effectifs Journal de Mécanique, vol. 20, n°2, 1981.

[10] F. PISTRE, Calcul des contraintes dans les matériaux composites. Rapport MRT action concertée - mécanique, 1981.

[11] A. PUCK, Grundlagen der Spannungs und verformungs Analyse. Dipl. Ing. Kunststoffe, Bd 57, Heft 4 (1967).

[12] E. SANCHEZ-PALENCIA, Non-homogeneous media and vibration theory. Lectures Notes in Physics 127, 1980.

[13] P. SUQUET, Une méthode duale en homogénéisation. CRAS Paris, Tome 291 (A), pp. 181-184, 1980.

[14] S.W. TSAI, J.C. HALPIN, N.J. PAGANO, Composite materials Workshop. Technomic Publishing Co. Inc. Connecticut (USA).

NONLINEAR ANALYSIS

ANALYSE NON LINEAIRE

Computing Methods in Applied Sciences and Engineering, VI
R. Glowinski and J.-L. Lions (Editors)
Elsevier Science Publishers B.V. (North-Holland)

NUMERICAL METHODS FOR HOPF BIFURCATION AND CONTINUATION OF PERIODIC SOLUTION PATHS

E.J. Doedel[1], A.D. Jepson[2], H.B. Keller[3]

Path following techniques for steady state problems are readily extended to apply to periodic solutions of autonomous systems. Fold points and bifurcations are easily determined. In addition paths of singular points - folds, simple bifurcations, Hopf bifurcatons - that occur in multiparameter problems are computed directly. This is basically a survey of known methods devised and applied by the authors.

Introduction; Steady State Problems.

Numerical methods for approximating paths and bifurcations of steady state problems

(1) $$F(x,\lambda) = 0, \qquad F: \mathcal{B}_1 \times \mathbb{R} \to \mathcal{B}_2$$

are well developed [7,13]. Indeed if the path or arc of solutions is parametrized by λ as

(2) $$\Gamma:\{x(\lambda),\lambda\}, \quad a \le \lambda \le b,$$

or in terms of an intrinsic parameter, s, as

(3) $$\Gamma:\{x(s),\lambda(s)\}, \quad s_a \le \lambda \le s_b$$

then predictor-solver continuation methods are in abundance; they can be quite efficient and can even be proven to yield accurate results if the solution path is regular (i.e. $F_x(x(\lambda),\lambda)$ is nonsingular for $a\leq\lambda\leq b$). At distinct points on the path where $F_x(x(\lambda_0),\lambda_0)$ may be singular, methods have been devised to circumvent the singularity by a) using a new parametrization as in (3), or b) jumping over the singular point. These ideas are discussed in [7,13] and elsewhere. Our goal here is to show how precisely these techniques can be used to study similar phenomena concerned with periodic solutions of autonomous differential systems. In addition we consider problems depending upon two (or more) real parameters and show how paths or families of singular solutions can be computed by the same techniques. In closing these introductory remarks we stress that for regular solutions it is easy to be assured that consistent, stable numerical methods yield good approximations [8]. Thus one basic technique is to reformulate singular cases so that they are regular - and we do this many times in this work. Then we do not have to spell out details of the numerical methods since any experienced numerical analyst can devise stable, consistent schemes.

[1]Computer Science Department, Concordia University, Montreal, Canada. Supported in part by NSERC (Canada) #A4274 and FCAC Quebéc (#EQ1438).

[2]Department of Computer Science, University of Toronto, Toronto, Ontario, Canada. Supported in part by USARO and NSERC Canada.

[3]Applied Mathematics, California Institute of Technology 217-50, Pasadena, California 91125. Supported in part under USARO contract No. DAAG29-81-K-0107 and DOE contract DE-AS03-76SF-00767, Project Agreement No. DE-AT03-7ER 72012.

Autonomous Time Dependent Problems.

Solutions of the autonomous system

(4) a) $\frac{du}{d\tau} = f(u,\lambda)$; $f: \mathcal{B}_3 \times \mathbb{R} \to \mathcal{B}_4$

are periodic of period T if

(4) b) $u(0) = u(T)$.

To cast such problems into the form (1) we rescale the time as: $\tau = Tt$, set $y(t) \equiv u(\tau)$ and get:

(5)

a) $\frac{dy}{dt} = Tf(y,\lambda)$, $0 \leq t \leq 1$;

b) $y(0) - y(1) = 0$.

Since the solution is translation invariant, i.e. $[y(t+\theta),T,\lambda]$ satisfies (5) for any θ if $[y(t),T,\lambda]$ is a solution, we can adjoin a scalar constraint to fix the phase; say

(5) c) $p(y,T,\lambda) = 0$, $p: \mathcal{B}_3 \times \mathbb{R}^2 \to \mathbb{R}$.

Some choices for $p(\cdot,\cdot,\cdot)$ are:

(6)

a) $p_I(y,T,\lambda) \equiv \zeta^* \cdot [y(0) - y_0(0)]$;

b) $p_{II}(y,T,\lambda) \equiv \int_0^1 [y(t) - y_0(t)]^* \cdot \frac{dy(t)}{dt}\, dt$;

c) $p_{III}(y,T,\lambda) \equiv \int_0^1 [y(t) - y_0(t)]^* \cdot \frac{dy_0(t)}{dt}\, dt$.

Here $y_0(t)$ is a (periodic) solution of (5) for $T=T_0$, $\lambda=\lambda_0$ and $\zeta^* \cdot dy_0(0)/dt \neq 0$. Condition (6a) is that due to Poincaré and (6b) is due to Doedel.

Now we can write (5a,b,c) as:

(7) $$F_B(x,\lambda) \equiv \begin{pmatrix} \frac{dy}{dt} - Tf(y,\lambda) \\ y(0) - y(1) \\ p(y,T,\lambda) \end{pmatrix} = 0, \quad x \equiv (y(t),T)$$

and the analogy with (1) is clear.

Another approach is to solve the initial value problem:

(8) $\frac{dz}{dt} = Tf(z,\lambda)$; $z(0) = \zeta$.

We denote the solution as $z \equiv z(t;\zeta,T,\lambda)$ and (5b,c) become

(9) $$F_S(x,\lambda) \equiv \begin{pmatrix} \zeta - z(1;\zeta,T,\lambda) \\ p(z,T,\lambda) \end{pmatrix} = 0, \quad x \equiv (\zeta,T).$$

This is also analogous to (1) and it is known as the "shooting" formulation. Whereas (7) and (9) are theoretically equivalent there are great differences in their direct numerical approximation. The latter is easier to formulate and implement while the former is more stable (when done correctly). Indeed the shooting formulation cannot yield unstable periodic solutions while the "boundary value" formulation (7) easily does so.

Poincaré Continuation.

When the Frechét derivative $F_x(x(\lambda),\lambda)$ is nonsingular at a solution of (1) the implicit function theorem yields a constructive proof of the continuation, in λ, of the solution. When this result is applied to the periodic case it is known as Poincaré continuation. Using either formulation (7) or (9) must be equivalent and indeed we have the basic:

THEOREM 10. Let $[y_0(t),T_0,\lambda_0]$ be a nontrivial solution of (5a,b,c). Let A^0 be the linear two point boundary operator:

(11) $$A^0\phi \equiv \begin{pmatrix} \frac{d}{dt}\phi - T_0 f_y(y_0(t),\lambda_0)\phi \\ \phi(0) - \phi(1) \end{pmatrix}.$$

Then:

A] $\dfrac{\partial F_B^0}{\partial(y,T)}$ is nonsingular iff:

i) $\text{Null}(A^0) = \text{span}\left\{\dfrac{dy_0(t)}{dt}\right\}$;

ii) $\begin{pmatrix} f(y_0(t),\lambda_0) \\ 0 \end{pmatrix} \notin \text{Range}(A^0)$;

iii) $p_y(y_0,T_0,\lambda_0)\dfrac{dy_0(t)}{dt} \neq 0$.

Let $V^0(t,\tau)$ be the fundamental solution satisfying:

(12) $$\frac{d}{dt}V^0(t,\tau) = T_0 f_y(y_0(t),\lambda_0)V^0(t,\tau), \quad V(\tau,\tau) = I.$$

Then:

B] $\dfrac{\partial F_S^0}{\partial(\zeta,T)}$ is nonsingular iff:

i) $\kappa = 1$ is a simple eigenvalue of $V^0(1,0)$;

The standard way to locate such Hopf bifurcations is to continue along the path Γ and to compute (approximate) the "entire" spectrum of $f_y(\lambda)$ along the path. Then a sign change in the real part of any eigenvalue usually signals Hopf bifurcation. This is an extremely complicated and inefficient procedure. Our technique is to locate Hopf biurcations directly, without generating entire paths, by solving the extended system:

$$(31) \qquad F(z) \equiv \begin{pmatrix} f(y,\lambda) \\ f_y(y,\lambda)a + \omega b \\ f_y(y,\lambda)b - \omega a \\ a^*a + b^*b - 1 \\ \xi^*b \end{pmatrix} = 0, \quad z \equiv \begin{pmatrix} y \\ a \\ b \\ \lambda \\ \omega \end{pmatrix} \varepsilon\ \mathscr{B}_3^3 \times \mathbb{R}^2$$

The justification for this is simply stated as:

THEOREM 32. At a Hopf bifurcation point z_0, $F(z_0)=0$ and

$$\frac{\partial F}{\partial z}(z_0) \text{ is nonsingular}$$

provided that

$$\xi^* \not\perp \text{span}\{a_0, b_0\}.$$

Proof. See [5].

Since Hopf bifurcations are isolated or regular solutions of (31), Newton's method and related techniques yield effective methods for approximating them. These ideas have been used in many such calculations. However the technique can even be extended for multiparameter problems to generate paths of Hopf bifurcations. Thus we consider generalizations of (5) in the form:

$$(33) \qquad \begin{aligned} &a) \quad \frac{dy}{dt} = Tf(y,\lambda,\tau), \quad f: \mathscr{B}_3 \times \mathbb{R}^2 \to \mathscr{B}_4 . \\ &b) \quad y(0) - y(1) = 0 \end{aligned}$$

For any fixed τ, a Hopf bifurcation for (33) is now directly determined by solving:

$$(34) \qquad F(z,\tau) \equiv \begin{pmatrix} f(y,\lambda,\tau) \\ f_y(y,\lambda,\tau)a + \omega b \\ f_y(y,\lambda,\tau)b - \omega a \\ a^*a + b^*b - 1 \\ p(a,b) \end{pmatrix} = 0; \quad z \equiv \begin{pmatrix} y \\ a \\ b \\ \lambda \\ \omega \end{pmatrix}.$$

Obviously our path following methods can be applied here to generate paths,

$\{z(\tau),\tau\}$, of Hopf bifurcations. The roles of λ and τ can be interchanged or we can use arclength-like parametrizations, $\{z(s),\tau(s)\}$, to generate these paths. As expected singularities such as folds or bifurcations develop along these paths and our previous techniques can be used to circumvent or to locate the singular points. We illustrate with an example.

PREDATOR-PREY MODEL.

This example is from a paper of E. Doedel [2] and uses his code AUTO [1] which incorporates most of the ideas discussed in this paper. The ODE systems involved are discretized by a high order accurate superconverging collocation procedure.

A dynamical system modeling plankton (u), fish (v) and sharks (w) is given by:

$$\text{a)}\quad \frac{du}{dt} = u[1-u] - c_{uv}uv \,,$$

$$(35)\qquad \text{b)}\quad \frac{dv}{dt} = -c_v v + c_{uv}uv - c_{vw}vw - \lambda\left(1 - e^{-c_e v}\right),$$

$$\text{c)}\quad \frac{dw}{dt} = -c_w w + c_{vw}vw \,.$$

The plankton grow under a logistics law but are eaten by the fish. The fish decay at a characteristic rate, they increase due to plankton, decrease due to sharks and (the last term in (35b)) they are harvested by fishermen. The sharks decay at their rate but prey upon the fish. In this simple model we set:

$$c_v = 0.25, \quad c_w = 0.5, \quad c_e = 5.0, \quad c_{vw} = 3.0$$

and consider the two free parameters λ, c_{uv} in:

$$0 \le \lambda \le 0.6, \qquad 2.5 \le c_{uv} \le 6.$$

The results of some calculations are shown in Figure 2. Steady state and periodic solution paths are shown in Figure 2a for fixed $c_{uv} = 3.0$. Two Hopf bifurcations and two simple bifurcations in steady states are shown. Stability has been determined along these paths by computing the Floquet multipliers of the linearization of the Poincaré return map. The elimination procedure employed in AUTO is designed to make these calculations routine.

In Figure 2b paths of Hopf bifurcations are obtained by allowing c_{uv} to vary. We observe folds or limit points as well as bifurcations on these branches of Hopf points. More detailed discussions are contained in [2].

References

[1] E.J. Doedel, AUTO: A program for the automatic bifurcation analysis of autonomous systems. Cung. NUM 30 (265-284) [Proc. 10th Manitoba Conf. on Num. Math & Computing, Winnipeg, Oct. 1980].

[2] E.J. Doedel, The computer aided bifurcation analysis of predator-prey models, submitted to J. Math Bio. (1983).

[3] J. Fier, Thesis in Applied Math., California Institute of Technology, Pasadena, CA 1984.

[4] J. Fier and H.B. Keller, Follow the folds, in preparation.

[5] A.D. Jepson, Numerical Hopf Bifurcation, Part II, Thesis in Applied Math., California Institute of Technology, Pasadena, CA 1981.

[6] A.D. Jepson and A. Spence, Folds in solutions of two parameter systems: Part I; Tech. Rept. NA-92-02, Comp. Sci. Dept., Stanford U., Stanford, CA 1982.

[7] H.B. Keller, Numerical solution of bifurcation and nonlinear eigenvalue problems; in: Applications of Bifurcation Theory (ed. P.H. Rabinowitz) Academic Press, New York, (1977) 359-384.

[8] H.B. Keller, Approximation methods for nonlinear problems with application to two-point boundary value problems, Math. of Comp. 29 (1975) 464-474.

[9] H.B. Keller and R.-K. Szeto, Calculation of flows between rotating disks, in Computing Methods in Applied Sciences and Engineering (eds. R. Glowinski, J.L. Lions) North-Holland (1980) 51-61.

[10] J.P. Keener and H.B. Keller, Perturbed bifurcation theory, Arch. Rat. Mech. Anal. 50 (1973) 159-175.

[11] G. Moore and A. Spence, The calculation of turning points of nonlinear equations, SIAM J. Num. Anal. 17 (1980) 567-576.

[12] W.C. Rheinboldt, Computation of critical boundaries on equilibrium manifolds, SIAM J. Num. Anal. 19, (1982) 653-669.

[13] W.C. Rheinboldt, Solution fields of nonlinear equations and continuation methods, SIAM J. Num. Anal. 17 (1980) 221-237.

[14] R.-K. Szeto, The flow between rotating coaxial disks, Thesis in Applied Math., California Institute of Technology, Pasadena, CA 1978.

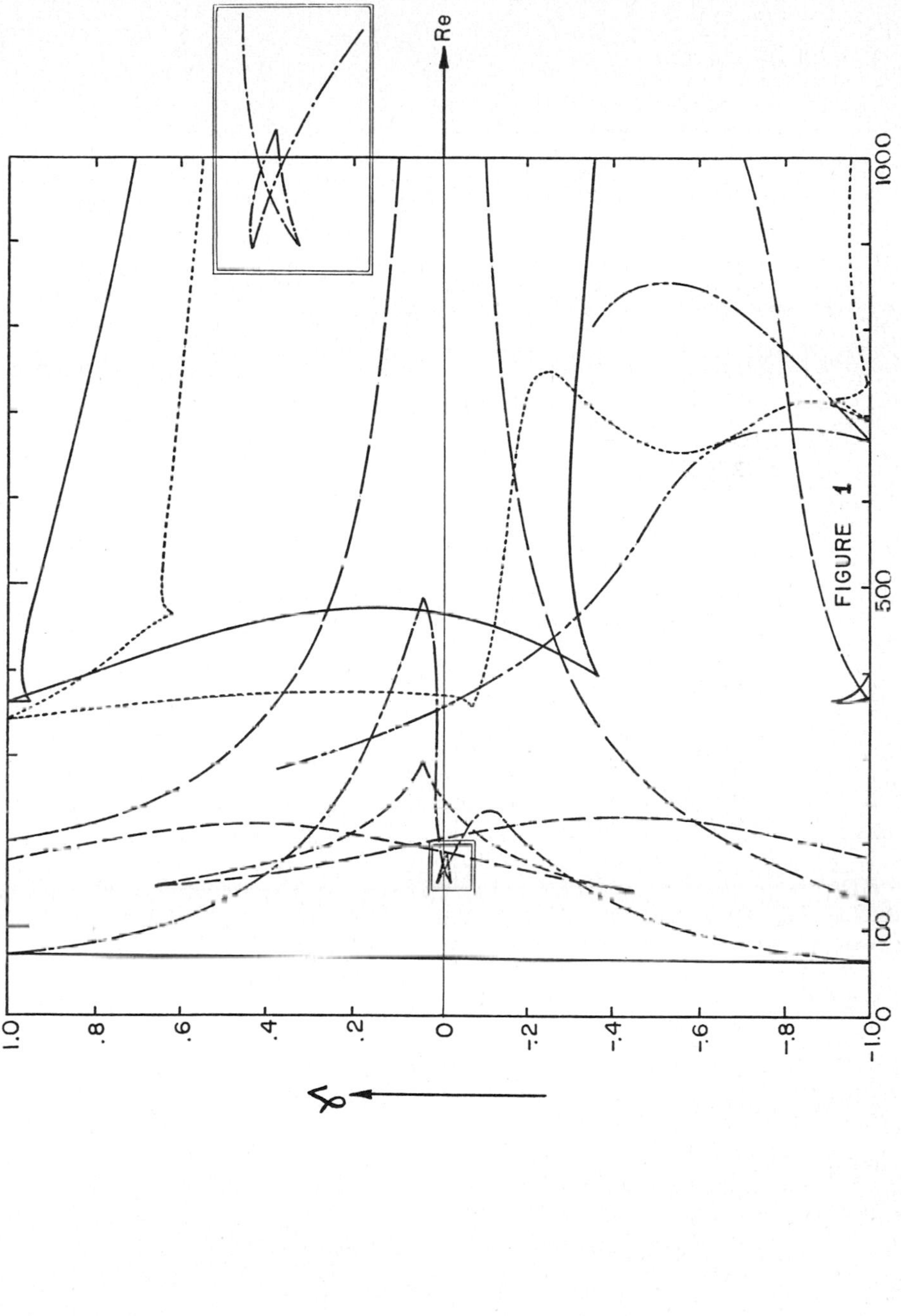

FIGURE 1

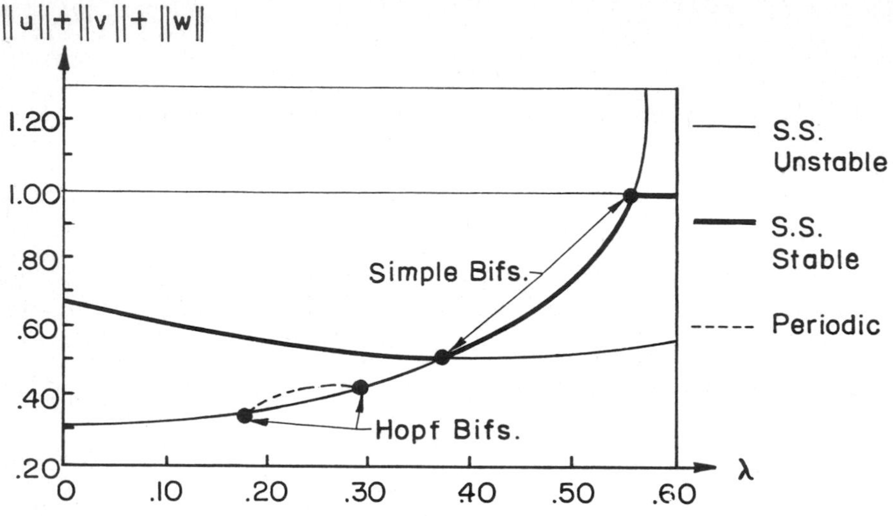

(A.) Bifurcation Diagram For C_{uv} = 3.00

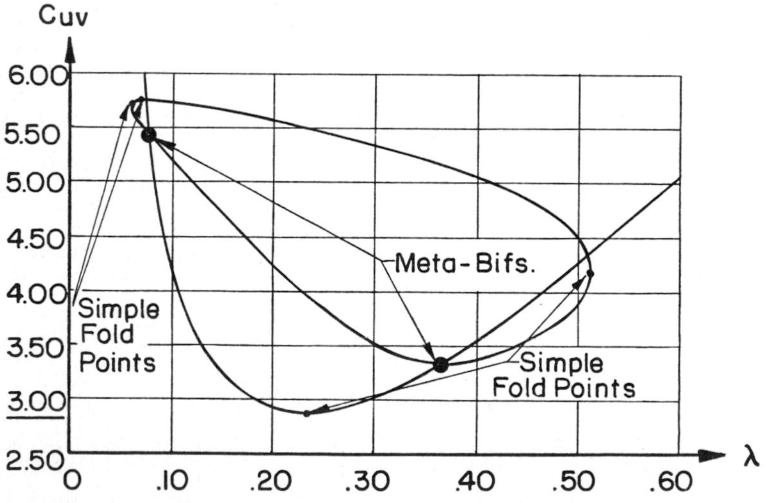

(B.) Curves Of Hopf Bifurcation Points

FIGURE 2

PREDATOR-PREY MODEL

Computing Methods in Applied Sciences and Engineering, VI
R. Glowinski and J.-L. Lions (Editors)
Elsevier Science Publishers B.V. (North-Holland)

On some multigrid finite difference schemes which describe everywhere non differentiable functions

Masaya Yamaguti, Masayoshi Hata

Department of Mathematics, Faculty of Sciences
Kyoto University
Kyoto 606
Japan

We propose a series of multigrid finite difference schemes which can describe everywhere non differentiable functions like Takagi's function and Lebesque's singular function. Physical meanings of these functions are explained. This study is related to the numerical solution of some singular perturbation problem.

1. Introduction

Recently we observed that the Weierstrass function which is continuous but everywhere non differentiable can be obtained as a solution of very simple functional equation [2]. This functional equation contains a one-dimensional dynamical system and an initially given function g. And then, by changing the dynamical system and the function g, we can get many such families of irregular continuous functions that include the Takagi and Van der Waerden function which was found by T. Takagi in 1903[1]. On the other hand, we noticed that C. De Rham had found some very simple functional equation which is satisfied by Lebesque's singular function. This functional equation is very much related to our functional equation. We clarified that these functional equations can be converted to some boundary value problems for multigrid finite difference schemes which are an analogue of singular perturbation. Using these result, we succeeded to get a very simple relation between Takagi's function and Lebesque's singular function. And by product, using de Rham's functional equation, we could compute the Fourier-Stieltjes coefficient of Lebesque's singular function and prove that it does not satisfy Riemann-Lebesque theorem.

In the last section, we will show the physical meanings of these functions, explained by H. Takayasu who is a physicist in Nagoya.

2. Functional equation which describe everywhere non differentiable continuous functions

We begin with some trivial remarks. The first example is Weierstrass's function:

$$W_{a,b}(x) = \sum_{n=0}^{\infty} a^n \cos(\pi b^n x) \qquad (0 \le x \le 1),$$

a, b real positive, $0 < a < 1$.

This can be represented, when $b = 2$,

(1) $$W_{a,2}(x) = \sum_{n=0}^{\infty} a^n \cos(\pi\phi^n(x))$$

where $\phi(x) = 2x$ $(0 \le x \le 1/2)$, $\phi(x) = 2(1-x)$ $(1/2 \le x \le 1)$ and $\phi^n(x)$ means n-th iterate of $\phi(x)$. (Specially, $\phi^o(x) \equiv x$.)

Next example is Takagi's function[1]

(2) $$T(x) = \sum_{n=1}^{\infty} (\tfrac{1}{2})^n \phi^n(x).$$

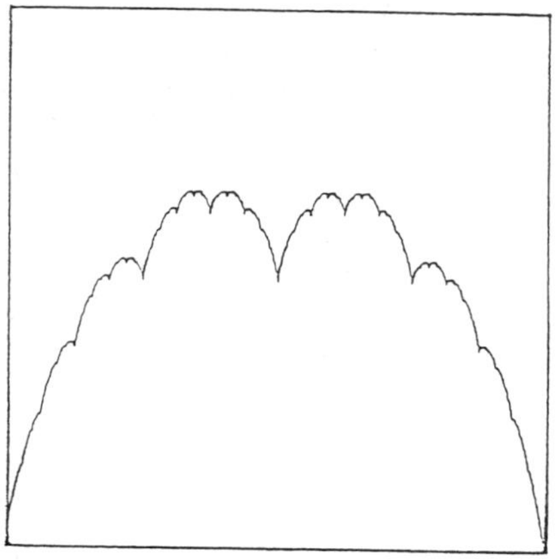

Figure 1. Graph of Takagi's function.

Remark. The above is not the original form of Takagi's function but we interprete the original definition using $\phi(x)$.

Both functions (1) and (2) satisfy the following functional equation:

(3) $$F(t,x) = tF(t,\psi(x)) + g(x), \quad (0 \le t < 1)$$

where $\psi(x)$ is a given mapping from [0, 1] to [0, 1], and $g(x)$ is a given bounded function.

It is easy to see that $F(a,x) = W_{a,2}(x)$ for $g(x) = \cos\pi x$ and $\psi(x) = \phi(x)$, and that $F(1/2,x) = T(x)$ for $g(x) = \phi(x)/2$ and $\psi(x) = \phi(x)$.

One can put the equation (3) an initial value problem:

(4) $$\begin{cases} \frac{\partial F}{\partial t}(t,x) = \frac{\partial}{\partial t}\{tF(t, \psi(x))\} & (0 < t \le 1), \\ F(0,x) = g(x). \end{cases}$$

Then we get the following theorem:

Theorem 1. Suppose that $g:[0,1] \longrightarrow \mathbb{R}$ is a bounded function and that $\psi:[0,1] \longrightarrow [0,1]$ is a dynamical system. Then $F(t,x)$, which satisfies (3) and is bounded with respect to x for each t, is uniquely determined and expressed by the following

(5) $$F(t,x) = \sum_{n=0}^{\infty} t^n g(\psi^n(x)).$$

We omit the proof because it is so easy.

Remark. This theorem is very general. For example, given functions $f(x)$ and $\psi(x)$ and a real value s ($0 < s < 1$), we can construct $g_s(x)$ such that the solution $F(t,x)$ of the initial value problem (4) with initial data $g_s(x)$ satisfies

(6) $$F(s,x) = f(x).$$

Thus, we can obtain an expansion of usual Cantor function:

(7) $$\sum_{n=0}^{\infty} \frac{1}{2^{n+1}} \chi_{[1/3,1]}(\phi^n(x))$$

where $\chi_{[1/3,1]}(x)$ is the characteristic function on the interval $[1/3,1]$.

Now, let us recall de Rham's functional equation. His original work[4] was more general but we mention here a special case which relates to our equation. $M(x)$ is unknown function. His equation is as follows:

(8) $$\begin{cases} M(x) = \alpha M(2x) & (0 \le x \le \frac{1}{2}) \\ M(x) = (1-\alpha)M(2x-1) + \alpha & (\frac{1}{2} \le x \le 1) \end{cases}$$

where α is a real number such that $0 < \alpha < 1$.

For comparison, we examine a special case of our equation for Takagi's function:

$$T(x) = \frac{1}{2}T(\phi(x)) + \frac{\psi(x)}{2}$$

which is rewritten in detail as below,

(9) $$\begin{cases} T(x) = \frac{1}{2}T(2x) + x & (0 \le x \le \frac{1}{2}) \\ T(x) = \frac{1}{2}T(2(1-x)) + 1 - x & (\frac{1}{2} \le x \le 1), \end{cases}$$

then (9) is very similar to (8).

The solution of (8) is Lebesque's singular function, which is strictly increasing continuous and has zero-derivatives almost everywhere for $\alpha \neq 1/2$. We denote this function $M_\alpha(x)$. Later on, we will see that there is a neat relation between $T(x)$ and $M_\alpha(x)$.

3. Schauder expansion

As an analogy of Fourier expansion, we have Schauder expansion of all continuous function on the closed interval [0, 1]. The basis function $F_{i/2^k,(i+1)/2^k}(x)$ is obtained from the function $F_{\alpha,\beta}(x)$ which is defined as follows:

(10) $$F_{\alpha,\beta} = \frac{1}{\beta - \alpha}\{|x - \alpha| + |x - \beta| - |2x - \alpha - \beta|\}.$$

Our bases are the following sequence of functions:

(11) $$1,\ x,\ F_{0,1},\ F_{0,1/2},\ F_{1/2,1},\ \cdots,\ F_{i/2^k,(i+1)/2^n},\ \cdots.$$

Theorem 2. Any continuous function $f(x)$ on $[0, 1]$ can be expanded uniquely as follows:

(12) $$f(x) = f(0) + [f(1) - f(0)]x + \sum_{k=0}^{\infty} \sum_{i=1}^{2^k-1} a_{i,k}(f) F_{i/2^k,(i+1)/2^n}(x)$$

where the coefficients $a_{i,k}$ are

(13) $$a_{i,k}(f) \equiv f\left(\frac{2i+1}{2^{k+1}}\right) - \frac{1}{2}\left\{f\left(\frac{i}{2^n}\right) + f\left(\frac{i+1}{2^k}\right)\right\}.$$

The proof is very elementary.

Now we can observe that

$$F_{i/2^k,(i+1)/2^k} = \chi_{[i/2^k,(i+1)/2^k]}(x)\phi^{k+1}(x).$$

We can replace (12) by

(14) $$f(x) = f(0) + [f(1) - f(0)]x + \sum_{k=0}^{\infty} b_k(x)\phi^{k+1}(x)$$

where $b_k(x) = \sum_{i=0}^{2^k-1} a_{i,k}(f)\chi_{[i/2^k,(i+1)/2^k]}(x)$.

Therefore we can say that Takagi's function $T(x)$ has a special expansion (14) because that $f(0) = f(1) = 0$ for $T(x)$, and that

(15) $$\forall k,\quad b_k(x) = \frac{1}{2^{k+1}} \qquad (0 \le x \le 1),$$

which means that $T(x)$ satisfies the following boundary value problem for a multigrid finite difference scheme because of (13).

(16) $$\begin{cases} T\left(\frac{2i+1}{2^{k+1}}\right) - \frac{1}{2}\left\{T\left(\frac{i}{2^k}\right) + T\left(\frac{i+1}{2^k}\right)\right\} = \frac{1}{2^{k+1}}, & (0 \le i \le 2^k - 1,\ \forall k) \\ T(0) = T(1) = 0. \end{cases}$$

Remark. If we replace $1/_2k+1$ at the right hand side in (16) by $(1/_{2^{k+1}})^2$, then we get usual smooth solution $x(1-x)$ of a Poisson equation $\partial^2 u/_{\partial x}2 = -2$ with boundary condition $u(o) = u(1) = 0$.

The following theorem suggests that Takagi's function can be generalized to some nice class.

Theorem 3. A function $f(x)$ is continuous on $[0, 1]$ and $f(0) = f(1) = 0$ if and only if it has expansion $\sum_{n=1}^{\infty} c_n \phi^n(x)$ whose coefficient c_n satisfies

$$\sum_{n=1}^{\infty} |c_n| < +\infty . \tag{17}$$

The proof of sufficiency is easy. The necessity is a little hard to prove. We only point out that the property of some orbit of the dynamical system $x_{n+1} = \phi(x_n)$ which pass near the mid point 1/2 plays an important role (See [1]). We think this theorem correspond to Sidon's theorem for lacunary Fourier series. Of course, the regularity depends on $\{c_n\}$. If $\{2^n c_n\} \in l_1$, then $f(x)$ is bounded variation. If $\overline{\lim_{n\to\infty}} |2^n c_n| > 0$, then $f(x)$ has no derivative everywhere. We call $f(x)$ the generalized Takagi function in this last case.

4. Multigrid finite difference schemes

As we have seen from (16) that $T(x)$ satisfies some multigrid finite scheme, the generalized Takagi function $T_G(x)$ satisfies the following boundary value problem:

$$\text{(18)} \quad \begin{cases} T_G(\frac{2i+1}{2^{k+1}}) = \frac{1}{2}\{T_G(\frac{i}{2^k}) + T_G(\frac{i+1}{2^k})\} + c_k \\ T_G(0) = T_G(1) = 0, \quad 0 \le i \le 2^k - 1, \quad \forall k \end{cases}$$

where $\{c_n\} \in l_1$, and $\overline{\lim_{n\to\infty}} |2^n c_n| > 0$.

Now we are going to look for some multigrid boundary value problem for the function $M_\alpha(x)$.

We proved that the following boundary problem is the right one:

$$\text{(19)} \quad \begin{cases} M_\alpha(\frac{2i+1}{2^{k+1}}) = (1-\alpha)M_\alpha(\frac{i}{2^k}) + \alpha M_\alpha(\frac{i+1}{2^k}) \\ M_\alpha(0) = 0, \quad M_\alpha(1) = 1, \quad 0 \le i \le 2^k - 1, \quad \forall k, \end{cases}$$

the proof is easy using the equation (8).

This boundary value problem is closely related to some singular perturbation problem:

$$(*) \quad \begin{cases} \varepsilon \dfrac{d^2u}{dx^2} - \dfrac{du}{dx} = 0 \qquad \varepsilon \text{ small} > 0 \\ u(0)=0, u(1) = 1, \end{cases}$$

because our problem (18) can be rewritten as follows:

$$(**) \quad \begin{cases} \dfrac{1}{4\cdot 2^{K+1}} \dfrac{M_\alpha((i+1)/2^k) - 2M_\alpha((2i+1)/2^{k+1}) + M_\alpha(i/2^k)}{(1/2^{k+1})^2} \\ + (\alpha - \frac{1}{2}) \dfrac{M_\alpha((i+1)/2^k) - M_\alpha(i/2^k)}{\frac{1}{2^k}} = 0, \\ M_\alpha(0) = 0, \quad M_\alpha(1) = 1, \quad 0 \le i \le 2^k - 1, \quad \forall k \text{ and } 0 < \alpha < \frac{1}{2}. \end{cases}$$

That is, (**) is some kind of discretization of (*) but the mesh size depends on ε. Thus we can regard $M_\alpha(x)$ as a kind of boundary layer solution.

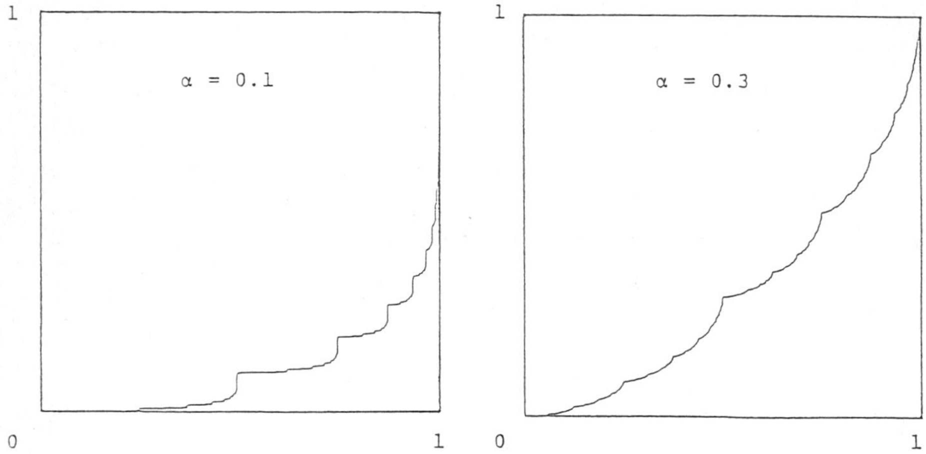

Figure 2. Graphs of Lebesque's singular function M_α.

5. A relation between $T(x)$ and $M_\alpha(x)$

Using (13) and (19), we get the coefficient $a_{i,k}(M_\alpha)$ in Schauder expansion of M_α:

$$a_{i,k}(M_\alpha) = (\alpha - \tfrac{1}{2})[M_\alpha(\tfrac{i+1}{2^k}) - M_\alpha(\tfrac{i}{2^k})]. \tag{20}$$

With this equality, we can obtain the relation

$$\begin{cases} a_{2m,k} = \alpha a_{m,k-1} \\ a_{2m+1,k} = (1 - \alpha) a_{m,k-1}. \end{cases} \tag{21}$$

Theorem 4.

$$a_{i,k}(M_\alpha) = (\alpha - \tfrac{1}{2})\alpha^p (1 - \alpha)^q$$

where $p + q = k$, p is the number of 0's in the binary expansion of

$$i = \sum_{j=1}^{k} w_j 2^{j-1} \qquad (0 \le i \le 2^k - 1),$$

q is the number of 1's.

Because of this theorem, the series of Schauder expansion for M_α is holomorphic function of α in neighbourhood of 1/2. We can differentiate (19) with respect to α in some neighbourhood of 1/2, and if we put $\alpha = 1/2$, then we get

$$2T(X) = \frac{\partial M_\alpha(x)}{\partial \alpha}\Big|_{\alpha=\frac{1}{2}} .$$

<u>Theorem 5.</u>

$$\frac{\partial M_\alpha(x)}{\partial \alpha}\Big|_{\alpha=\frac{1}{2}} = 2T(x).$$

6. Physical meaning of $M_\alpha(x)$ and $T(x)$

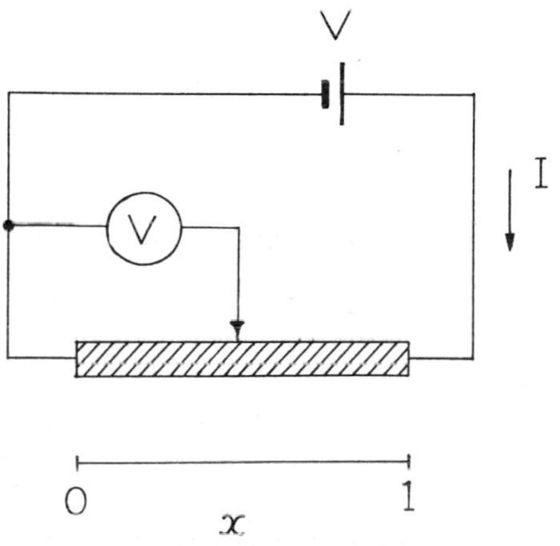

Figure 3.

We are thinking of an electric circuit in which constant voltage V is applied to a 1-dimensional resistance of length unity (Fig. 3). The Ohm's law is the following.

$$(22) \qquad E(x) = R(x)I$$

where I is the electric current, $E(x)$ and $R(x)$ are the electric field and the resistivity respectively.

If we assume that the resistivity is proportional to the density of impurity $\rho(x)$, namely

$$R(x) = \kappa\rho(x) \qquad \kappa : \text{constant},$$

then $V(x)$ voltage at point x, becomes

$$(23) \qquad V(x) = \int_0^x E(x')dx' = V\kappa\int_0^x \rho(x')dx'$$

where ρ is normalized as

$$\int_0^1 \rho(x')dx' = 1.$$

Now, we consider the case where $\rho(x)$ is de Wij's fractal [6]. De Wij's fractal $\rho_\alpha(x)$ is a self-similar function specified by only one real parameter α $(0 \le \alpha \le 1)$. It is defined by a limit of cascade of the coarse grained distribution $\rho_\alpha^{(k)}(x)$ which are defined by

$$(24) \qquad \begin{cases} \rho_\alpha^{(k+1)}\left(\frac{2i}{2^{k=1}}\right) = \alpha\rho_\alpha^{(k)}\left(\frac{i}{2^k}\right) \\ \rho_\alpha^{(k+1)}\left(\frac{2i+1}{2^{k+1}}\right) = (1-\alpha)\rho_\alpha^{(k)}\left(\frac{i}{2^k}\right) & (0 \le i \le 2^k - 1) \\ \rho_\alpha^{(0)}(0) = 1. \end{cases}$$

Fractal dimension D of ρ_α is known to be

$$D = -\{\alpha\log_2\alpha + (1-\alpha)\log_2(1-\alpha)\}.$$

Now the voltage $V(x)$ is obtained from (23),

$$(25) \qquad V(x) = V\kappa\int_0^x \rho_\alpha(x')dx' = V\, M_\alpha(x)$$

where $M_\alpha(x)$ is Lebesque's singular function appeared in preceding sections.

Next we treat the case where the density $\rho_\alpha(x)$ of the impurity changes with time.

From the conservation of impurity, the flux of the impurity $j(t,x)$ is determined as

$$j(t,x) = -\int_0^x \frac{\partial}{\partial t}\rho(x',t)dx'.$$

If we assume that the density is uniform at $t = 0$, namely $\rho(x,0) = \rho_{1/2}(x) = 1$, and that it becomes De Wij's fractal after a short time Δt, namely $\rho(x,\Delta t) = \rho_{1/2+\alpha'\Delta t}(x)$,

then the flux at $t = 0$ can be computed as

$$\begin{aligned} j(x,0) &= -\lim_{\Delta t\to 0}\int_0^x \frac{\rho_{1/2+\alpha'\Delta t}(x') - \rho_{1/2}(x')}{\Delta t}dx' \\ &= -\alpha'\frac{\partial}{\partial\alpha}M_\alpha(x)\Big|_{\alpha=\frac{1}{2}} \\ &= -2\alpha' T(x) \qquad\qquad \text{(Theorem 5)} \end{aligned}$$

where $T(x)$ is Takagi function.

The above discussion is also valid to other cases, for example, laminar shear flow ρ, V and j represent density of vorticity, velocity of the fluid and flux of vorticity respectively. The third case is just concerning about density of change, electric field and electric current respectively.

<u>Appendix.</u> The Fourier Stieltzes coefficient of $M_\alpha(x)$ can be computed using (7). Let

$$I(t) = \int_0^1 e^{itx} dM_\alpha(x)$$

be this coefficient, then we get

$$I(t) = \prod_{n=1}^{\infty} \{ \alpha + (1 - \alpha) e^{\frac{it}{2^n}} \}$$

for all integer $p \geq 2$,

$$I(2^p \pi) = (2\alpha - 1) I(\pi).$$

If $\alpha \neq 1/2$, $I(t)$ never vanish as t tends to $+\infty$. The other expression of $I(t)$ is

$$I(t) = e^{it} \prod_{n=1}^{\infty} \{\cos\frac{t}{2^{n+1}} + (1 - 2\alpha) i \sin\frac{t}{2^{n+1}}\}.$$

Refferences

[1] Takagi, T., A simple example of the continuous function without derivative, Proc. Phys.-Math. Japan 1 (1903) 176-177.

[2] Yamaguti, M. and Hata, M., Weierstrass's function and chaos, Hokkaido Math. Jrnl. (1983) Vol. XII. No. 3

[3] Hata, M. and Yamaguti, M., On Takagi's function , Japan Jrnl. of Applied Math. (New Jrnl.) to appear.

[4] De Rham, C., Sur quelques courbes définies par des équations functionnelles, Rend. Semi. Math. Torino 16 (1957) 101-113.

[5] Faber, G., Über stetige Funktionen, Math. Ann. 69 (1910) 372-443.

[6] Mandelbrot, B., The fractal geometry of nature, Freeman (1982) 376.

Computing Methods in Applied Sciences and Engineering, VI
R. Glowinski and J.-L. Lions (Editors)
Elsevier Science Publishers B.V. (North-Holland)

Partial Differential Equations with Hysteresis Functionals

Augusto Visintin

Istituto di Analisi Numerica del C.N.R. - C.so C. Alberto, 5
27100 Pavia (Italy)

The concept of hysteresis functional is introduced and examples are shown; a weak differential formulation is given for discontinuous functionals. Relations involving such functionals are coupled with partial differential equations; results of existence and uniqueness of weak solutions are given. A model of the evolution of ferromagnetic materials is introduced.

INTRODUCTION

Let us consider a constitutive relation establishing a dependence between two real variables: $u \mapsto w$. Hysteresis appears when the "output" $w(t)$ is not uniquely determined by the "input" $u(t)$ at the same instant $t \in [0,T]$, but instead $w(t)$ depends on the evolution of u in $[0,t]$ and on the initial value w^0 :

$$w(t) = [\mathbf{F}(u(.), w^0)](t) ,$$

where $u(.)$ denotes the function $u:[0,T] \to \mathbf{R}$ and $\mathbf{F}$ is a Volterra (i.e. causal) functional.

In order to exclude other memory effects, as viscosity e.g., we require that this dependence be "rate-independent", i.e. that w depends just on the range of u in $[0,t]$ and on the order in which these values are taken, not on its velocity (see (1.6)).

Ferromagnetism (see fig. 1), plasticity, supercooling and superheating effects are classical examples of hysteresis phenomena; hysteresis appears also in filtration through porous media, in biology, in chemistry and in many other fields. Though these phenomena have been intensively studied from the applicative viewpoint, the mathematical literature on hysteresis seems little (with the exception of works on plasticity); as far as this author knows, the only systematic mathematical research is due to Krasnosel'skiǐ, Pokrovskiǐ and co-workers in the 1970's (see [8] and its references).

The present author has recently started to study hysteresis especially in connection with partial differential equations, with the aim of investigating some physical problems, ferromagnetism in particular, and other applicative situations. Here some of these results are reviewed.

Fig. 1. For $-u_1 < u < u_1$, w is not uniquely determined. If $t \mapsto u(t)$ is as in (a), then $t \mapsto (u(t), w(t))$ follows the path drawn in (b). Remark that (u,w) moves along monotone curves ("piecewise monotonicity"). Picture (b) corresponds to ferromagnetism if u is interpreted as the magnetic field H and w as the magnetization M.

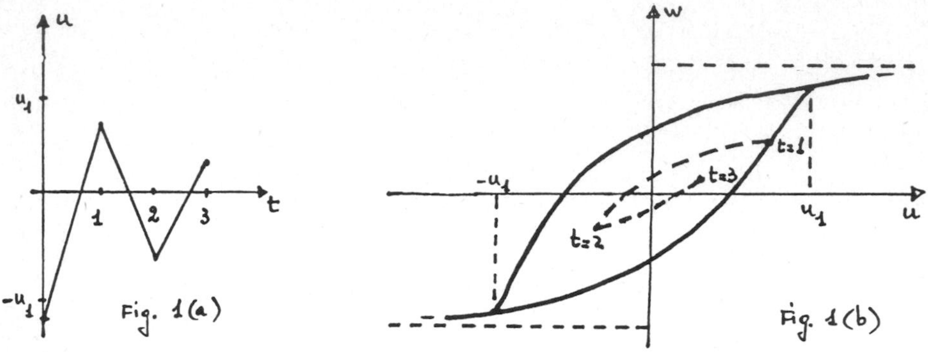

Fig. 1(a) Fig. 1(b)

1. HYSTERESIS FUNCTIONALS

1.1 Definitions. (Z,L,F) (or more shortly **F**) is named a *memory functional* iff

$$\mathbf{Z} \textit{ is a real Banach space}; \ \mathbf{L} \subset \mathbf{Z} \times \mathbf{Z}' \ (\mathbf{Z}' := \textit{dual of } \mathbf{Z}) \tag{1.1}$$

$$\begin{cases} Dom\,(\mathbf{F}) = \{(v,\xi) \in C^\circ([0,T]); Dom\,(\mathbf{L})) \times \mathbf{Z}' \mid (v(0),\xi) \in \mathbf{L}\} \\ (Dom\,(\mathbf{L}) := \{v \in \mathbf{Z} \mid \exists w \in \mathbf{Z}' : (v,w) \in \mathbf{L}\}) \end{cases} \tag{1.2}$$

$$\begin{cases} \forall (v,\xi) \in Dom\,(\mathbf{F}),\ \mathbf{F}(v,\xi) \in C^\circ([0,T]; \mathbf{Z}'),\ [\mathbf{F}(v,\xi)](0) = \xi \\ \text{and } \forall t \in [0,T],\ (v(t),\ [\mathbf{F}(v,\xi)](t)) \in \mathbf{L} \end{cases} \tag{1.3}$$

$$\begin{cases} \forall (v_1,\xi),(v_2,\xi) \in Dom\,(\mathbf{F}),\ \forall \bar{t} \in [0,T], \\ \text{if } v_1 = v_2 \textit{ in } [0,\bar{t}] \textit{ then } [\mathbf{F}(v_1,\xi)](\bar{t}) = [\mathbf{F}(v_2,\xi)](\bar{t}) \quad (\textit{Causality}). \end{cases} \tag{1.4}$$

(Z,L,F) is named a *hysteresis functional* iff (1.1),...,(1,4) hold and

$$\begin{cases} \forall (v,\xi) \in Dom\,(\mathbf{F}),\ \forall \tau \in [0,T],\ \textit{setting } \xi_\tau := [\mathbf{F}(v,\xi)](\tau) \text{ and} \\ v_\tau(.) := v(.+\tau), \forall t \in [\tau,T].\ [\mathbf{F}(v_\tau,\xi_\tau)](t-\tau) = [\mathbf{F}(v,\xi)](t) \\ (\textit{Semigroup property}) \end{cases} \tag{1.5}$$

$$\begin{cases} \forall (v,\xi) \in Dom\,(\mathbf{F}),\ \forall s : [0,T] \to [0,T] \ \textit{increasing homeomorphism}, \\ \forall t \in [0,T],\ [\mathbf{F}(v \circ s,\xi)](t) = [\mathbf{F}(v,\xi)](s(t)) \quad (\textit{Rate independence}). \end{cases} \tag{1.6}$$

This last appears as the distinguishing property of hysteresis; for instance it is not fulfilled by time convolution functionals representing viscosity effects.

We introduce a further property, which appears as a natural generalization of monotonicity to hysteresis functionals:

$$\left.\begin{array}{l} \forall t',t'' \in [0,T]\,(t' < t''),\ \forall v \in C^\circ([0,t'];\mathbf{Z}),\ \forall \xi \in \mathbf{Z}' \textit{ such that } (v(0),\xi) \in \mathbf{L},\ \forall z \in \mathbf{Z}, \textit{set} \\ v_z := \begin{cases} v & \textit{in } [0,t'] \\ v(t') + \dfrac{t-t'}{t''-t'}[z - v(t')] & \textit{in } [t',t''] \\ z & \textit{in } [t'',T] \end{cases} ;\ \Phi_v(z) := [\mathbf{F}(v_z,\xi)](t''). \\ \textit{It is required that } \Phi_v : \mathbf{Z} \to \mathbf{Z}' \textit{ be cyclically maximal monotone} \end{array}\right\} \tag{1.7}$$

(*Piecewise monotonicity*)

i.e., roughly speaking, when v increases (decreases), then $F(v,\xi)$ increases (decreases , respect.).

These definitions and properties can be generalized allowing an explicit dependence of **F** on t and accordingly requiring (1.5),....,(1.7) for any fixed t .

1.2 - An example. Let $\mathbf{Z} = \mathbf{R}$, $\mathbf{L} \subset \mathbf{R}^2$, $g_1, g_2 \in C^0(\bar{\mathbf{L}})$ and $(u^0, w^0) \in \mathbf{L}$; the functional $\Phi_{(u^0,w^0)}: \{u \in W^{1,1}(0,T) \mid u(0) = u^0\} \mapsto W^{1,1}(0,T) : u \to w$ is defined by means of the following Cauchy problem:

$$\begin{cases} \dfrac{dw}{dt} = g_1(u,w).(\dfrac{du}{dt})^+ - g_2(u,w).(\dfrac{du}{dt})^- \quad a.e. \ in \ ]0,T[\\ w(0) = w^0; \end{cases} \tag{1.8}$$

in particular g_1 and g_2 have to be such that (u,w) does not leave **L**. Using Φ it is easy to construct a functional **F** such that (**R** , **L** , **F**) fulfills (1.1),...,(1.6) (and also (1.7) if $g_1, g_2 \geq 0$) but with $C^0([0,T])$ replaced by $W^{1,1}(0,T)$. The continuity properties of **F** are studied in [19]; in general **F** cannot be extended by continuity to continuous input functions; the cases in which this is possible and **F** is continuous w.r.t. the uniform topology have been characterized by Krasnosel'skiĭ (see [8]).

1.3 - Relays (see [13,15,16] - Let $\rho := (\rho_1, \rho_2) \in \mathbf{R}^2$, $\rho_1 < \rho_2$. $\forall u \in C^0([0,T])$, $\forall w^0 \in \{-1,1\}$ such that

$$w^0 = -1 \text{ if } u(0) \leq \rho_1 \ , \quad w^0 = 1 \text{ if } u(0) \geq \rho_2 \ , \tag{1.9}$$

we define $w := f_\rho(u,w^0) : [0,T] \to \{-1,1\}$ as follows: $w(0) = w^0$ and

$$\begin{cases} \forall t \in]0,T] \ , \text{ if } A_t := \{\tau \in]0,t] \mid u(\tau) = \rho_1 \ or \ \rho_2\} = \phi \ , then \\ w(t) = w^0; \text{ if } A_t \neq \phi \text{ and } u(\max A_t) = \rho_1 \ (= \rho_2, respect.) \ , \\ then \ w(t) = -1 \ (= 1, respect.). \quad (see fig. 2). \end{cases} \tag{1.10}$$

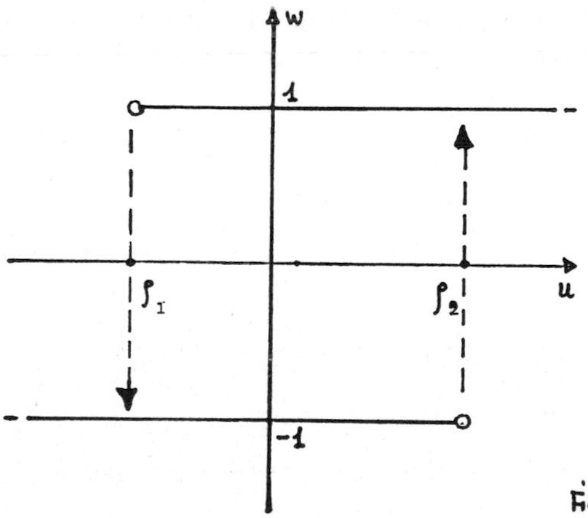

Fig. 2

These conditions define a unique measurable function w. Moreover w has bounded total variation; indeed w has a variation equal to 2 each time that u reaches one of the critical values ρ_1,ρ_2 and the uniformly continuous function u can have just a finite number of oscillations between ρ_1 and ρ_2. Therefore, if compared with the jump function, the functional f_ρ has a regularizing effect. The functional $\{u \in C^\circ([0,T]) |\ (1.9)\ \textit{holds}\} \to L^\infty(0,T)$: $u \mapsto f_\rho(u,w^0)$ is not closed w.r.t. the strong topology of $C^\circ([0,T])$ and the weak star topology of $L^\infty(0,T)$. We shall denote its closure by $\mathcal{F}_\rho$. We introduce some auxiliary functions:

$$\begin{cases} \alpha_\rho(\xi) := (\xi-\rho_2)^+ - (\xi-\rho_1)^- , \quad \beta_\rho(\xi) := \xi - \alpha(\xi) \quad \forall \xi \in \mathbf{R} \\ S(\xi) := \begin{cases} \{-1\} & \text{if } \xi < 0 \\ [-1,1] & \text{if } \xi = 0 \\ \{1\} & \text{if } \xi > 0 \end{cases} , \quad R_\rho(\xi) := \begin{cases}]-\infty,0] & \text{if } \xi = \rho_1 \\ \{0\} & \text{if } \rho_1 < \xi < \rho_2 \\ [0,+\infty[& \text{if } \xi = \rho_2 \end{cases} \end{cases} \tag{1.11}$$

$\forall u \in C^\circ([0,T])$, $\forall w^0$ such that

$$-1 \le w^0 \le 1 \text{ if } \rho_1 \le u(0) \le \rho_2 , \quad w^0 = -1 \text{ if } u(0) < \rho_1 , \quad w^0 = 1 \text{ if } u(0) > \rho_2 \tag{1.12}$$

$w \in \mathcal{F}_\rho(u,w^0)$ *iff* $w \in BV(0,T)$ (Banach space of functions $[0,T] \to \mathbf{R}$ with bounded total variation) and

$$w(0) = w^0 \tag{1.13}$$

$$w \in S(\alpha_\rho(u)) \quad \text{a.e. in }]0,T[\tag{1.14}$$

$$\frac{\partial w}{\partial t} \in R_\rho(\beta_\rho(u)) \text{ in } C^\circ([0,T])' ; \tag{1.15}$$

the latter can be written in the form

$$_{C^\circ([0,T])'}\!\left< \frac{\partial w}{\partial t}, \beta_\rho(u) - v \right>_{C^\circ([0,T])} \ge 0 \quad \forall v \in C^\circ([0,T]) : \rho_1 \le v \le \rho_2 . \tag{1.16}$$

$\mathcal{F}_\rho$ is sketched in fig. 3. A natural approximation of the (discontinuous) relay functional $\mathcal{F}_\rho$ is given by a sequence of (continuous) hysteresis functionals $\{\mathrm{F}_\rho^n\}_{n\in\mathbf{N}}$ as sketched in fig. 4.

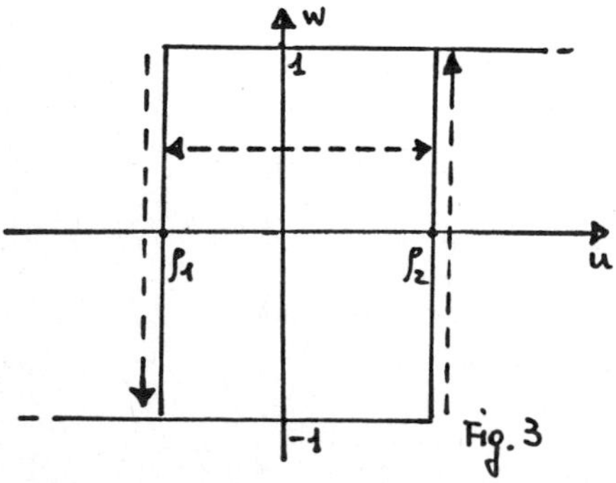

Fig. 3

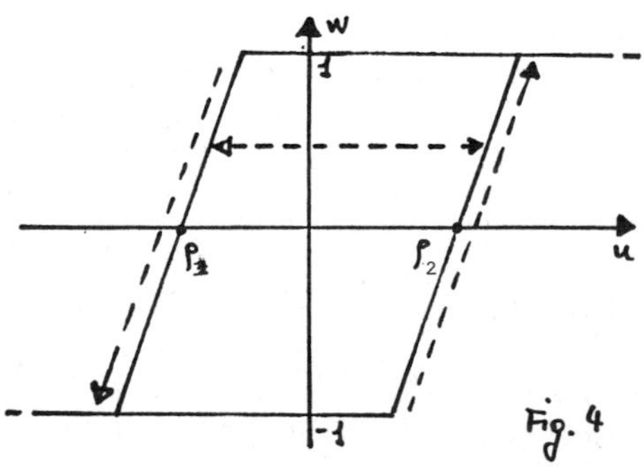

Fig. 4

1.4 Preisach Model (see [8,14]) - Let μ be a density over the so-called Preisach plane $P: = \{\rho = (\rho_1,\rho_2) \in \mathbf{R}^2 \mid \rho_1 < \rho_2\}$; let $u \in C^\circ([0,T])$ and $\{w_\rho^0\}_{\rho \in P}$ denote an application $P \to \{-1,1\}: \rho \mapsto w_\rho^0$ such that (1.9) holds μ -a.e. in P . We set

$$\mathbf{F}_\mu(u,\{w_\rho^0\}_{\rho \in P}): = \int_P f_\rho(u,w_\rho^0)d\mu_\rho \quad in\,[0,T]\,. \tag{1.17}$$

In general this function is not continuous in [0,T]; however if μ has no masses concentrated either on segments parallel to the axes or in points, then $\mathbf{F}_\mu(u,\{w_\rho^0\}_{\rho \in P}) \in C^\circ([0,T])$ and $\mathbf{F}_\mu$ is continuous w.r.t. u from $C^\circ([0,T])$ endowed with the strong topology into itself. If $\mu \geq 0$ then $\mathbf{F}_\mu$ is piecewise monotone; other properties of $\mathbf{F}_\mu$ can be easily deduced using the representation of the Preisach plane (see [14]). If μ does not fulfill the previous non-singularity property, then $\mathbf{F}_\mu$ is not continuous w.r.t. u and its closure $\tilde{\mathbf{F}}_\mu$ corresponds to replacing f_ρ by $\tilde{f}_\rho$ in (1.17):

$$\begin{cases} w \in \tilde{\mathbf{F}}_\mu(u,\{w_\rho^0\}_{\rho \in P}) \quad in\,[0,T]\; iff\;\; w(t) = \int_P w_\rho(t)d\mu_\rho \; in\; [0,T], \\ where\; w_\rho\; fulfills\; (1.13),...,(1.15)\; \mu_\rho - a.e.\; in\; P. \end{cases} \tag{1.18}$$

Vectorial generalizations of the Preisach model have been proposed in [2].

An important question is the identification of hysteresis functionals, see [8].

1.5 Rheological models (see [20]) - In continuum mechanics it is usual to represent the properties of materials by means of so-called rheological models (see [10], e.g.). We shall denote the stress and strain tensors by σ and ε , respect; here we confine ourselves to the univariate case, in order to avoid the distinction between the spheric and isotropic components of σ and ε .

Non-linear viscosity and elasticity correspond to constitutive relations of the form

$$\dot{\varepsilon} \in \alpha(\sigma) \quad (\textit{the dot denotes the time derivative}) \text{ and} \tag{1.19}$$

$$\varepsilon \in \beta(\sigma) \tag{1.20}$$

respectively; α and β are maximal monotone graphs $\mathbf{R} \to \mathbf{R}$. As a particular case of (1.19) we have the law of a rigid perfectly plastic material

$$\dot{\varepsilon} \in \partial I_K(\sigma), \tag{1.21}$$

where $K = [a,b]$ with $a < 0 < b$ corresponds to the yield criterium and I_K denotes the indicator function of K (i.e. $I_K(\xi) = 0$ if $\xi \in K$, $I_K(\xi) = +\infty$ if $\xi \notin K$). (1.21) fulfills the property of rate-independence, hence we can regard plasticity as a hysteresis effect.

Large classes of rheological models can be obtained by combining the basic viscous and elastic elements in series and in parallel; arrangements of elastic and plastic elements fulfill the property of rate-independence. For instance joining a linear elastic element in parallel with a plastic one we get a law of the form

$$\sigma \in \lambda\varepsilon + (\partial I_K)^{-1}(\dot{\varepsilon}) \quad (\lambda \in \mathbf{R}^+)\,, \tag{1.22}$$

whereas the combination in series of the same elements yields

$$\dot{\varepsilon} \in \lambda^{-1}\dot{\sigma} + \partial I_K(\sigma)\,; \tag{1.23}$$

as it is easy to check (see [20]), (1.22) and (1.23) correspond to

$$\varepsilon = \mathbf{F}(\sigma(.), \varepsilon^0) \text{ and } \sigma = \mathbf{G}(\varepsilon(.), \sigma^0) \quad (\textit{respect.}), \tag{1.24}$$

F and **G** being hysteresis functionals.

The multivalued relay functional $\tilde{f}_\rho$ introduced above corresponds to the arrangement in parallel of a plastic element with yield criterium $K = [\rho_1, \rho_2]$ and a non-linear elastic one, characterized by $\beta = (\partial I_{[-1,1]})^{-1}$: $\varepsilon \in \tilde{f}_\rho(\sigma)$ iff

$$\sigma \in \partial I_{[-1,1]}(\varepsilon) + (\partial I_{[\rho_1,\rho_2]})^{-1}(\dot{\varepsilon}) \ . \tag{1.25}$$

Moreover combining in series several (possibly infinite) models of this form corresponding to different $\tilde{K}$ and K, rheological representation of the Preisach model is obtained (see [20]).

2. Partial differential equations with hysteresis functionals

2.1 A model problem (see [12.14]). Let Ω be a bounded domain of $\mathbf{R}^N$ $(N \geq 1)$ and $T > 0$; set $Q := \Omega \times]0, T[$. Let V be a dense subspace of $L^2(\Omega)^M$ $(M \geq 1)$; identifying the latter with its dual we get $V \subset L^2(\Omega)^M = (L^2(\Omega)^M)' \subset V'$, with dense inclusions. Let $(\mathbf{R}^M, \mathrm{R}, \mathbf{F})$ be a memory functional, in the sense of (1.1),...(1.4), $A \in \mathbf{L}(V, V')$, $f \in L^1(0,T;V')$ and

$$u^0, w^0 : \Omega \to \mathbf{R}^M \ \textit{measurable} \ ; \ (u^0(x), w^0(x)) \in R \quad \textit{a.e. in } \Omega. \tag{2.1}$$

We introduce a weak problem

(P1) Find $u \in L^1(0,T;V)$ such that

$$u(x,.) \in C^\circ([0,T])^M \ , \quad u(x,0) = u^0(x) \quad \textit{a.e. in } \Omega \tag{2.2}$$

and such that, setting

$$w(x,t) := [\mathbf{F}(u(x,.), w^0(x))](t) \quad \forall t \in [0,T], \ \textit{a.e. in } \Omega, \tag{2.3}$$

then $w(.,t)$ is measurable in Ω $\forall t$, $u + w \in W^{1,1}(0,T;V')$ and

$$\frac{\partial}{\partial t}(u + w) + Au = f \quad \textit{in } V', \ \textit{a.e. in }]0,T[\tag{2.4}$$

$$(u + w)|_{t=0} = u^0 + w^0 \quad \textit{in } V'. \tag{2.5}$$

Theorem 1 - Assume that $(\mathbf{R}^M, R, \mathbf{F})$ fulfills (1.1),...,(1.4),(1.7), that (2.1) holds and that

$$R \subset \{ (u,w) \in (\mathbf{R}^M)^2 \mid |w| \leq c_1 |u| + c_2 \} \quad (c_1, c_2 : \textit{positive constants}) \tag{2.6}$$

$$\left\{ \begin{array}{l} \forall \{(v_n, \xi) \in \mathit{Dom}(\mathbf{F})\}_{n \in \mathbf{N}}, \text{ if } v_n \to v \ \textit{strongly in } C^\circ([0,T])^M , \\ \textit{then } (v, \xi) \in \mathit{Dom}(\mathbf{F}) \text{ and } \mathbf{F}(v_n, \xi) \to \mathbf{F}(v, \xi) \ \textit{strongly in } C^\circ([0,T])^M \end{array} \right. \tag{2.7}$$

$$\textit{The injection } V \to L^2(\Omega)^N \textit{ is compact} \tag{2.8}$$

$$\left\{ \begin{array}{l} A : V \to V' \ \textit{is linear, continuous, symmetric} \text{ and } \textit{coercive}, \textit{i.e.} \\ \forall v \in V, \ {}_{V'}\langle Av, v \rangle_V + \mu_1 \|v\|^2_{L^2(\Omega)^M} \geq \mu_2 \|v\|^2_V \ (\mu_1, \mu_2 : \textit{constants} > 0) \end{array} \right. \tag{2.9}$$

$$u^0 \in V, \ f \in L^2(Q)^M + W^{1,1}(0,T;V') . \tag{2.10}$$

Then problem **(P1)** has at least one solution such that moreover

$$u \in H^1(0,T;L^2(\Omega)^M) \cap L^\infty(0,T;V) . \tag{2.11}$$

Remark - The piecewise monotonicity property (1.7) gives a parabolic character to **(P1)**.

Sketch of the proof - (see §2 of [12], §2 of [14]). Let $m \in \mathbf{N}$, $k = \dfrac{T}{m}$.

$(\text{P1})_m$ - Find $u_m^n \in V$ for n=1,...,m, such that, setting

$$\begin{cases} u_m(x,.) := \textit{linear interpolate in } [0,T] \textit{ of } u_m(x,nk) := u_m^n(x) \\ (u_m^0 := u^0), \quad a.e. \ in \ \Omega \end{cases} \tag{2.11}$$

$$w_m^n(x) := [\mathrm{F}(u_m(x,.), w^0(x))](nk), \ a.e. \ in \ \Omega \ (n = 0,...,m), \tag{2.12}$$

then w_m^n is measurable in Ω, $w_m^n \in V'$ and

$$\frac{u_m^n - u_m^{n-1}}{k} + \frac{w_m^n - w_m^{n-1}}{k} + Au_m^n = f_m^n \quad in \ V', for \ n = 1,...,m \tag{2.13}$$

(with obvious definition of f_m^n). This problem can be solved step by step. For any n, if $u_m^0, \ldots, u_m^{n-1}$ are known, then by (2.12) w_m^n is a monotone function of u_m^n; therefore (3.13) is equivalent to a standard minimization problem and has a unique solution, which can also be easily numerically approximated. Multiplying (2.13) against $u_m^n - u_m^{n-1}$ and summing in n, using (1.7) and (2.9), we get

$$\|u_m\|_{H^1(0,T;L^2(\Omega)^M) \cap L^\infty(0,T;V)} \le Constant \tag{2.14}$$

and then by (2.6), denoting the time-interpolate of the w_m^n's by w_m,

$$\|w_m\|_{L^\infty(0,T;L^2(\Omega)^M)} \le Constant . \tag{2.15}$$

Therefore there exists u such that, possibly taking a subsequence,

$$u_m \to u \quad weakly \ star \ in \ H^1(0,T;L^2(\Omega)^M) \cap L^\infty(0,T;V) ; \tag{2.16}$$

this space is included with compactness into $L^2(\Omega; C^\circ([0,T]^M)$; hence, possibly extracting a further subsequence,

$$u_m(x,.) \to u(x,.) \quad strongly \ in \ C^\circ([0,T])^M, \ a.e. \ in \ \Omega \tag{2.17}$$

and then by (2.7)

$$\mathrm{F}(u_m(x,.), w^0(x)) \to \mathrm{F}(u(x,.), w^0(x)) \quad strongly \ in \ C^\circ([0,T])^M \ a.e. \ in \ \Omega, \tag{2.18}$$

whence (2.3). ■

The uniqueness of the solution of **(P1)** is an open question.

We introduce another variational problem:

(P2) As problem **(P1)** with (2.4) and (2.5) replaced by

$$\frac{\partial u}{\partial t} + Au + w = f \quad in \ V', \ a.e. in \ ]0,T[. \tag{2.19}$$

THEOREM 2 - Assume that the hypotheses of theorem 1 are fulfilled, with the possible exception of (1.7) (piecewise monotonicity). Then problem **(P2)** has at least one solution such that moreover

$$u \in H^1(0,T;L^2(\Omega)^M) \cap L^\infty(0,T;V) \,. \tag{2.20}$$

Proof - (see [15]) - Quite similar to that of theorem 1. ∎

THEOREM 3 - Assume that $(\mathbf{R}^M, R, \mathbf{F})$ fulfills (1.1),...,(1.4), that (2.1) and (2.9) hold as well as the following property of Lipschitz-continuity

$$\begin{cases} \exists \sigma \in C^\circ(\mathbf{R}^+) \text{ such that } \sigma(0) = 0 \text{ and} \\ \forall t', t'' \in [0,T]\ (t' < t'')\ , \forall (v_1,\xi),(v_2,\xi) \in Dom(\mathbf{F}), \\ \text{if } v_1 = v_2 \text{ in } [0,t'], \\ \text{then } \|\mathbf{F}(v_1,\xi) - \mathbf{F}(v_2,\xi)\|_{L^2(t',t'')} \le \|v_1 - v_2\|_{H^1(t',t'')}\ \sigma(t''-t')\,. \end{cases} \tag{2.21}$$

Then problem **(P2)** has at most one solution fulfilling the further regularity (2.20).

Proof - Let $u_1, u_2 \in H^1(0,T;L^2(\Omega)^M) \cap L^\infty(0,T;V)$ be two solutions of **(P2)**. We take the difference between the two corresponding equations (2.19) and multiply against the incremental ratio w.r.t. time of $u_1 - u_2$; thus we get

$$\int_0^t \|\frac{\partial}{\partial t}(u_1-u_2)\|^2_{L^2(\Omega)^M} d\tau + \frac{\mu_2}{2}\|(u_1-u_2)(t)\|^2_V - \frac{\mu_1}{2}\|(u_1-u_2)(t)\|^2_{L^2(\Omega)^M} \le$$

$$\le \sigma(t)\Big(\int_0^t \|u_1-u_2\|^2_{L^2(\Omega)^M} d\tau\Big)^{\frac{1}{2}} \cdot \Big(\int_0^t \|\frac{\partial}{\partial t}(u_1-u_2)\|^2_{L^2(\Omega)^M} d\tau\Big)^{\frac{1}{2}}.$$

If t is small enough, then $u_1 = u_2$ in $\Omega \times]0,t[$. Repeating this argument step by step in time we get the thesis. ∎

2.2 Ferromagnetism (see [9,17,18]) - For slowly varying fields, the displacement current term can be neglected; accordingly the system of Maxwell's equations is equivalent to the single equation

$$\frac{\partial B}{\partial t} + \nabla \times \nabla \times H = f \tag{2.22}$$

(with normalized constants and f datum); for a ferromagnetic medium the constitutive law can be assumed of the form

$$B = \mathbf{F}(H, B^0) \tag{2.23}$$

with $\mathbf{F}$ hysteresis functional. These equations can be set in the form of **(P1)**, with N=M=1 and $V = \{v \in L^2(\Omega)^3 \mid \nabla \times v \in L^2(\Omega)^3\}$; but theorem 1 cannot be applied, as (2.8) does not hold. Only in the univariate case V reduces to $H^1(\Omega)$ and (2.8) is fulfilled. As for the assumptions on $\mathbf{F}$, (2.6) is satisfied as $|M| \le M^0$: *constant* ; (1.7) has thermodynamic justifications;(2.7) is a strong limitation on the choice of the hysteresis functional: for instance $\mathbf{F}$ cannot be taken as in section 1.2 (this should be equivalent to replace (2.3) by (1.8) a.e. in Ω), thus in this last case even existence of a solution is an open question.

In order to deal with the multivariate case, we consider a classical physical model. According to the classical theory of Weiss, below a critical temperature on a microscopic scale a ferromagnetic body is magnetically saturated, that is denoting the microscopic magnetic field by m

$$|m(x,t)| = \mathcal{M}\ (positive\ constant)\quad in\ \Omega \tag{2.24}$$

The evolution of m is governed by Landau-Lifshitz' equations

$$\frac{\partial m}{\partial t} = \lambda_1 m \times h^e - \lambda_2 m \times (m \times h^e) \quad in\ \Omega \tag{2.25}$$

$$h^e = h + \nabla.(F.\nabla m) - G.m \quad in\ \Omega, \tag{2.26}$$

where λ_1 and λ_2 are constants, $\lambda_2 > 0$; h^e is the "effective magnetic field" acting on the elementary magnets, h is the magnetic field (the field appearing in Maxwell's equations), F and G are positive definite 3^2 - tensors (see [9]). (2.25) and (2.26) are to be coupled with the complete set of Maxwell's equations written for the microscopic fields m, h, etc. The mathematical aspects of this problem have been studied in [17], where in particular the existence of a weak solution has been proved.

In order to get a model for the macroscopic situation, following a standard procedure we introduce the space average operator $< >$ and we set $M := <m>$, $H := <h>$, $H^e := <h^e>$, etc. Thus (2.24) and (2.26) yield

$$|M(x,t)| \le \mathcal{M} \quad in\ \Omega \tag{2.27}$$

$$H^e := H + \nabla\cdot(F.\nabla M) - G.M \quad in\ \Omega. \tag{2.28}$$

Difficulties arise averaging (2.25); then we rather use an experimental relationship between H^e and M, which for the quasi-stationary evolution can be assumed of the form

$$M = \mathfrak{F}(H^e, M^0) \quad in\ \Omega \tag{2.29}$$

with $\mathfrak{F}$ hysteresis functional; the presence of hysteresis is strictly related to the multiplicity of solutions for the stationary microscopic problem (see §2 of [17]). Notice that for a non-distributed system $H^e = H$ and $\mathfrak{F}$ can be experimentally evaluated (at least in principle). Assuming that $\mathfrak{F}$ can be inverted (a concept to be precised!), (2.29) can be rewritten in the form

$$H^e = \mathfrak{G}(M, H^{e0}) \quad in\ \Omega, i.e. \tag{2.30}$$

$$H = \mathfrak{G}(M, H^{e0}) - \nabla\cdot(F.\nabla M) + G.M := \mathbf{G}(M) \tag{2.31}$$

and if also $\mathbf{G}$ can be inverted, then we get

$$M = \mathbf{F}(H, M^0) \quad in\ \Omega; \tag{2.32}$$

$\mathfrak{F}, \mathfrak{G}, \mathbf{F}, \mathbf{G}$ are all ysteresis functional; $\mathfrak{F}$ and $\mathfrak{G}$ are set pointwise in space whereas $\mathbf{F}$ and $\mathbf{G}$ have a global character. Each of the equivalent constitutive relations (2.29),...,(2.32) can be coupled with (2.1).

In the case of fast evolution, H^e must be replaced by

$$\tilde{H}^e := H^e - \gamma \frac{\partial M}{\partial t} \quad in\ \Omega, \tag{2.33}$$

with γ viscosity coefficient. Thus instead of (2.30) we get

$$H = \mathbf{G}(M, H^0) + \gamma \frac{\partial M}{\partial t} \quad in\ \Omega \tag{2.34}$$

Existence results for weak formulations in Sobolev spaces hold for the following settings (see [18]): (2.31) or (2.34) coupled with Maxwell's equations with no displacement current, (2.34) coupled with Maxwell's equations with displacement current, (2.31) or (2.34) coupled with the equations of magnetostatics: $\nabla \times H = 0$ and $\nabla . B = 0$ in Ω . The basic procedure is as for theorem 1: approximation by time discretization, use of piecewise monotonicity for deducing a priori estimates and limit in the hysteresis term by compactness. In all of these cases the question of uniqueness is open. The excluded case of (2.31) coupled with Maxwell's equations with displacement current is not physical, since the viscosity term $\frac{\partial M}{\partial t}$ in (2.32) is more important than the displacement current term.

Another question concerns periodicity: in the previous problems, if the boundary data and the source term are periodic in time, do periodic solutions exist?

The previous approach rises some further questions. Let us assume that the measure μ has the non-singularity properties indicated in section 1.4, so that the corresponding functional $\mathbf{F}_\mu$ defined in (1.17) is continuous from $C^\circ([0,T])$ into $C^\circ([0,T])$ w.r.t. the uniform topology; moreover assume that $\mu > 0$ (i.e. $\mu(A) > 0$ for any $A \subset P$ with non-vanishing Lebesgue measure). Can $\mathbf{F}_\mu$ be inverted? if so, what are the properties of its inverse?

2.3 Equations with discontinuous hysteresis functionals (see [13,14,15,16]) - As a model situation we consider a weak formulation of equation (2.4) coupled with the constitutive relation (defined in section 1.3)

$$w \in \tilde{\mathcal{F}}_\rho(u,w^0) \quad in\ [0,T] \tag{2.35}$$

We shall use the same notations as in section 1.2.We assume that

$$u^0 \in L^2(\Omega)\ ,\ w^0 \in L^\infty(\Omega)\ ,\ u^0(x) \text{ and } w^0(x)\ fulfill\ (1.12)\ \ a.e.\ in\ \Omega\ . \tag{2.36}$$

(P3) - Find $u \in L^1(0,T;V)$, $w \in L^\infty(Q)$ such that $u + w \in W^{1,1}(0,T;V')$, $\beta_\rho(u(x,.)) \in C^\circ([0,T])$, $w(x,.) \in BV(0,T)$ a.e. in Ω and

$$w \in S(\alpha_\rho(u)) \quad a.e.\ in\ Q \tag{2.37}$$

$$\begin{cases} {}_{C^\circ([0,T])'}< \frac{\partial w}{\partial t}(x,.), \beta_\rho(u(x,.)) - v >_{C^\circ([0,T])} \geq 0 \\ \forall v \in C^\circ([0,T]): \rho_1 \leq v \leq \rho_2\ ,\ a.e.\ in\ \Omega \end{cases} \tag{2.38}$$

$$w(x,0) = w^0(x) \quad a.e.\ in\ \Omega \tag{2.39}$$

$$\frac{\partial}{\partial t}(u+w) + Au = f \quad in\ \ V'\ ,\ a.e.\ in\ ]0,T[\tag{2.40}$$

$$(u+w)\,|_{t=0} = u^0 + w^0 \ \ in\ \ V'\ . \tag{2.41}$$

THEOREM 4 - Assume that

$$u^0 \in V\ ,\ \ f \in L^2(Q) + W^{1,1}(0,T;V') \tag{2.42}$$

Then problem **(P3)** has at least one solution such that moreover

$$u \in H^1(0,T;L^2(\Omega)) \cap L^\infty(0,T;V)\ ,\ w \in L^2(\Omega;BV(0,T))\ . \tag{2.43}$$

Sketch of the proof (see §3 of [14]) - As for **(P1)**, an implicit time discretization is used. At every time step the hysteresis relation can be written in the form

$$w_m^n \in \Phi^\rho_{w_m^{n-1}}(u_m^n) \quad in\ \Omega\ , \tag{2.44}$$

$\Phi^{\rho}_{w_m^{n-1}}$ being a maximal monotone graph, as for $(\mathbf{P1})_m$. The a priori estimates (2.14) can be obtained, hence by the properties of the hysteresis relation (cf. section 1.3) we have

$$\|w_m\|_{L^2(\Omega;W^{1,1}(0,T))} \le \frac{2}{\rho_2-\rho_1}\|u_m\|_{L^2(\Omega;W^{1,1}(0,T))} + Const. \le Const. \tag{2.45}$$

Then besides (2.16) we get (possibly extracting a subsequence)

$$w_m \to w \text{ weakly star in } L^\infty(Q)\cap L^2(\Omega;BV(0,T))\,. \tag{2.46}$$

Notice that the graphs S and R_ρ (defined in (1.11)) are maximal monotone and that by (2.16)

$$\alpha_\rho(u_m)\to\alpha_\rho(u),\ \beta_\rho(u_m)\to\beta_\rho(u) \text{ strongly in } L^2(Q); \tag{2.47}$$

then by (2.44) we get (2.37) and (2.38).

Another obvious approximation procedure consists in approximating the discontinuous and multivalued functional $\mathcal{F}_\rho$ by means of a sequence $\{\mathrm{F}^n_\rho\}_{n\in\mathbf{N}}$ of continuous hysteresis functionals, as indicated in section 1.3 (see fig. 4). By theorem 1 the corresponding approximated problems have at least one solution. The rest of the proof follows as before. This existence result can be easily extended to the case of the Preisach model, requiring $\mu>0$ in order to have the piecewise monotonicity property (see [14]), and also to the case that $\rho = (\rho_1,\rho_2)$ depends on (x,t) (see [16], where weaker a priori estimates are used).

We introduce a last variational problem:

(P4) - As problem **(P3)** with (2.40) and (2.41) replaced by (2.19) and an initial condition for u .

Also for this last problem an existence result holds. For **(P3)** and **(P4)** uniqueness of the solution is an open question.

If u and w are interpreted as temperature and enthalpy density (respect.), then **(P3)** can be regarded as a generalization of the classical Stefan problem for phase transition including supercooling and superheating effects (see [13]). Its extension to the Preisach model corresponds for instance to the quasi-stationary evolution of a univariate ferromagnetic body.

A parabolic system with a hysteresis term in the source has been proposed by Hoppensteadt and Jäger as a model for certain biological and chemical phenomena exhibiting pattern formation (see [7]). If the hysteresis term is represented by means of the functional $\mathcal{F}_\rho$ as above, then existence of a weak solution can be proved (see 6 of [15]); also here uniqueness is an open question.

2.4 Numerical approximation - The proofs of the previous existence theorems are based on (implicit) time discretization; it is clear that the discretization in space (by finite elements or by finite differences) can be introduced as well; then the same a priori estimates hold and the limit procedure can be carried out in the same way as above.

Some numerical tests have been performed by the author with the collaboration of Dr. Claudio Verdi, who has elaborated the programs and has taken care of the implementation on the computer. An especially simple situation has been considered, with Ω reduced to a bounded interval; problems **(P1)** and **(P2)** have been taken into account with **F** as sketched in fig. 4 and problems **(P3)** and **(P4)** with the hysteresis graph $\mathcal{F}_\rho$ of fig. 3. In these cases the quantity $\rho_2 - \rho_1$ can be regarded as a measure of the "amount of hysteresis"; another important parameter is the Lipschitz-constant of **F**, corresponding to the slope of the oblique part of the graph of fig. 4.

The numerical tests lead to the following conclusions (see [21]). As the discretization steps tend to zero, the whole sequence of approximate solutions converges; this heuristically supports the guess of uniqueness of a stable solution. For (P1) and (P3) the processing time (i.e. the cost) increases with the "amount of hysteresis" $\rho_2-\rho_1$, for a fixed discretization and a prescibed error test; this effect does not appear for (P2) and (P4).

Finally, both costs and errors do not seem to depend on the Lipschitz-constant of **F**, hence they are approximately the same for **(P1)** and **(P3)**, for **(P2)** and **(P4)**; the same happens in the case without hysteresis.

Other numerical investigations on hysteresis problems can be found in [3], e.g. .

REFERENCES

[1] R. BOUC - Modèle mathématique d'hystérésis - Acustica, 24(1971), 16-25.

[2] A. DAMLAMIAN, A. VISINTIN - Une généralisation vectorielle du modèle de Preisach pour l'hystérésis (to appear on C.R. Acad. Sci. Paris).

[3] R.M. DEL VECCHIO - The inclusion of hysteresis in a special class of electromagnetic finite element calculations - I.E.E.E. Trans. on Magn., vol. Mag-18, No.1(1982), 275-284.

[4] M. FABRIZIO, C. GIORGI - L'isteresi per un sistema elettromagnetico - Fisica Matematica, Suppl. B.U.M.I., vol. 1(1981), 33-45.

[5] N. GERMAY, S. MASTERO - Etude des cycles d'hystérésis magnétique et application à la ferrorésonance - Report of LABORELEC(1973).

[6] U. HORNUNG - Mathematical models for hysteresis effects - Zeitschr. Angew. Math. Mech. 63(1983), 324-325.

[7] W. JÄGER - A diffusion reaction system modelling spatial patterns - Equadiff, Bratislava 1981, Teubner Texte zur Math., 47(1982).

[8] M.A. KRASNOSEL'SKIǏ - Equations with nonlinearities of hysteresis type - (Russian), VII Int. Konf. Nichtlin. Schwing, Berlin 1975; Abh. Akad. Wiss. DDR, Jahrg. 1977, pp. 437-458 (English abstract Zbt. fur Math. 406-93032).

[9] L. LANDAU, E. LIFSHITZ - Electrodynamique des milieux continues. Cours de physique theorique, tome VIII (MIR, Moscou 1969).

[10] M. REINER - Rheology, in: H.P.J. Wijn (ed.), Encyclopedia of Physics, vol XVIII/2: Ferromagnetism (Springer Verlag, Berlin, 1966).

[11] A. VISINTIN - Hystérésis dans les systèmes distribués - C.R. Acad. Sci. Paris, 293(14 Déc. 1981), 625-628.

[12] A. VISINTIN - A model for hysteresis of distributed systems - Ann. Mat. Pura Appl., 131(1982), 203-231.

[13] A. VISINTIN - A phase transition problem with delay - Control and Cybernetics, 1-2(1982), 5-18.

[14] A. VISINTIN - On the Preisach model for hysteresis - Preprint No. 189 of SFB 123, Heidelberg (1982); to appear on Non linear Analysis T.M.A.(1983).

[15] A. VISINTIN - Evolution problems with hysteresis in the source term - Preprint No. 173 of SFB 123, Heidelberg (1982).

[16] A. VISINTIN - On variable hysteresis operators - Preprint of SFB123, Heidelberg (1983). (To appear on Bull. Un. Mat. Ital.).

[17] A. VISINTIN - On Landau-Lifshitz equations for ferromagnetism - Preprint of SFB123, Heidelberg (1983) .

[18] A. VISINTIN - On the evolution of ferromagnetic media - Preprint of I.A.N. of C.N.R., Pavia (1983) . (To appear on Math. Modelling).

[19] A. VISINTIN - Continuity properties of a class of hysteresis functionals - Preprint of I.A.N. of C.N.R., Pavia(1983).

[20] A. VISINTIN - Rheological models and hysteresis effects - (in preparation).

[21] C. VERDI, A. VISINTIN - Numerical approximation of hysteresis problems - Preprint of I.A.N. of C.N.R., Pavia (1983).

Computing Methods in Applied Sciences and Engineering, VI
R. Glowinski and J.-L. Lions (Editors)
Elsevier Science Publishers B.V. (North-Holland)

NORMAL FORMS for SINGULARITIES of NONLINEAR ORDINARY DIFFERENTIAL EQUATIONS

Shigehiro USHIKI

Institute of Mathematics, Yoshida College, Kyoto University
Kyoto 606, Japan

§0. Abstract

A mathematical method to compute the normal forms for systems of ordinary differential equations is described. The normal forms obtained here are simpler than those given by Poincaré, Arnold and Takens. Some implication of this theory to the numerical and analytical study of bifurcation problems of ODEs and PDEs is discussed.

§1. Introduction

Recently, much effort has been focused upon the theoretical and numerical study of bifurcation problems. When we study the bifurcation of ODEs or of PDEs, we first try to reduce the problem into a simpler form. As an example, let us consider the case of the reaction-diffusion equation. We formulate the system so that it defines an evolution equation on an appropriate function space. This function space is, in general, of infinite dimension. Suppose that there exists an equilibrium solution, say u_*. If the spectrum of the linearized operator of the evolution system has only finite number of points with zero real part and all other points in the spectrum has negative real part, then, by virtue of the center manifold theorem, the essential part of the dynamics of the evolution equation can be described by a system of ordinary differntial equation on a finite dimensional manifold.

The Lyapounov-Schmidt method is usually employed to study the bifurcation of the equilibrium states. In some cases we are to study non-equilibrium solutions, for example, periodic solutions or non-periodic but steady solutions, i.e., a solution of the evolution equation which does not converge to an equilibrium state nor diverges to the infinity nor breakes down to non-differentiable one, but oscillates aperiodically forever. In this case, the usual Lyapounov-Schmidt method and the singularity theory of mappings is not appropriate, since these theories detects only stationary solutions.

We analyse the bifurcation of the steady solution u_* by computing the behavior of the solutions for the evolution equation which bifurcates from the steady solution. In general, we look for a coordinate transformation near the bifurcation point to transform the equation to a simpler one. For this purpose, the theory developed by POINCARE[16][17][18], BIRKHOFF[6], ARNOLD[2][3][4][5], TAKENS[21] and BROER[8] is employed. See ARNOLD[4][5], BOGDANOV[7], LANGFORD[13], LANGFORD-IOOSS[14], GUCKENHEIMER[9], HOLMES[10][11], and ARNEODO-COULLET-SPIEGEL-TRESSER[1] for some applications of the normal forms theory to bifurcation problems. Normal forms thory mentioned above is, however, insufficient for the study of bifurcation of degenerate dynamics around the singular point of the system by the following reasons. First, it contains too many parameters even for truncated family of normal forms. Second, the normal forms has superfluous terms and the corresponding normal form for a given system of ODEs is not determined uniquely. We employ the normal form and versal family theory developed in USHIKI[22], which is an improved version of classical normal form theory for singularities of vector fields. The classical theory for normal forms are known and has been employed in many authors to study the bifurcation of vector fields.

The normal forms theory, started by H.Poincaré, aims to construct a "universal" family of vector fields. Suppose the following situation: we are given a system of ODEs on the euclidean space $\mathbb{R}^n$; we are to find the simplest form of the given system which can be obtained by a change of coordinates; if we had a list of normal forms of vector fields and we could tell to which simplest form a system can be transformed, then the above problem is easy to solve. The thory of Jordan normal forms for matrices can be regarded as the first order part of such normal forms theory of singularities of vector fields.

The author[22] improved the normal form theory by computing the coordinate transformations considered as a Lie group action acting over the Lie algebra of jets in an exaustive manner, so that the obtained normal forms has the uniqueness, i.e., there corresponds only one normal form for any given system in the list of normal forms of vector fields up to certain higher order terms. The list also provides the versal deformation for these singularities.

As an example of the application of our normal form theory to bifurcation problem, we apply it to the Lorenz family of ordinary differential equations [15] and the Rössler's family[19]. These families are embedded in the family of normal forms by a suitable smooth changes of coordinates. By rescaling the time and by renormalizing the family to the normal form, we shall obtain a one parameter family of families of systems, which connects the original family having quite complicated dynamics studied by many people, to a very simple, but degenerate, "integrable" system. Here, "integrable" means that it has explicit formula for the solutions. The result can be interpreted as follows. There exists "Lorenz attractor" in the neighborhood of the origin in some perturbation of the system $\ddot{x} = -x^3$, $\dot{z} = x^2$. Similar result is obtained for the Rössler's family of ODEs. See USHIKI-OKA-KOKUBU[23] for the detail. In the case of Rössler's attractor, the rescaled attractor does not shrink in all directions. It becomes thin and finally almost one-dimensional in the limit.

Another application of the normal forms theory is obtained by H.KOKUBU(in preparation). He detected the appearance of periodic solution via a secondary Hopf bifurcation for some reaction-diffusion equation.

§2. Jordan normal form and Poincaré's normal form

Consider a system of ordinary differential equations :

$$\dot{x}_i = f_i(x_1,\ldots,x_n) \qquad i=1,\ldots,n \tag{2.1}$$

and suppose that the origin is singular, i.e., $f_i(0) = 0$, $i=1,\ldots,n$. We write (2.1) in the vector form :

$$\dot{x} = F(x) = F_1(x) + F_2(x) + \ldots, \tag{2.2}$$

where $F(x)$ denotes the right hand side of (2.1) and $F_k(x)$ denotes the k-th order homogeneous part of $F(x)$. Let A be the n x n matrix corresponding to the linear part $F_1(x)$, i.e.:

$$Ax = F_1(x). \tag{2.3}$$

Let us see how a linear change of coordinates transforms (2.1). Let P be a non-singular n x n matrix. Define a linear change of coordinates by $y = Px$. Then in the y coordinates, (2.2) can be written in the form :

$$\dot{y} = P\dot{x} = PF(P^{-1}y) = PF_1(P^{-1}y) + PF_2(P^{-1}y) + \ldots \tag{2.4}$$

Linear transformation of coordinates P transforms the matrix A into PAP^{-1}. Hence (2.1) can always be transformed into the form whose linear part is of Jordan normal form. Any system (2.1), which is singular at the origin can be transformed into the Jordan normal form by a coordinate change up to terms of order two.

Now let us recall the Poincaré's normal forms theory. Suppose F(z) is a complex vector of power series in complex vector $z = (z_1,\ldots,z_n) \in \mathbb{C}^n$, which vanishes at the origin. Let w be the vector of eigenvalues of the linear part of F. Let $m = (m_1,\ldots,m_n)$ denote an n-tuple of non-negative integers. We call m a multi-index. We set

$$z^m = z_1^{m_1} \cdot z_2^{m_2} \cdot \ldots \cdot z_n^{m_n}, \quad |m| = m_1 + \ldots + m_n, \tag{2.5}$$

and $(m,w) = m_1w_1 + m_2w_2 + \ldots + m_nw_n$. A monomial vector field

$$\dot{x}_i = z^m, \ \dot{x}_j = 0 \quad (j=1,\ldots,n, \ j\neq i)$$

is said to be resonant if $w_i = (m,w)$ holds and the length $|m| \geq 2$. H. Poincaré proved the following theorem.

<u>Theorem 2.1</u> (Poincaré) System of ordinary differential equations $\dot{z} = F(z)$ defined by a vector of formal power series can be transformed by a formal transformation $z = y + \ldots$ into the form

$$\dot{y} = F_1 + W(y), \tag{2.6}$$

where all monomial terms in the power series W(y) are resonant.

We call (2.6) the Poincaré's normal form. TAKENS[21] improved the Poincaré's normal forms into the following form.

<u>Theorem 2.2</u> (Takens) Let H_k denote the linear space of homogeneous polynomial vector fields of degree k on $\mathbb{R}^n$. Let X be a vector field on $\mathbb{R}^n$ which vanishes at the origin (X(O) = 0). Let X_k denote the k-jet at the origin of X, i.e., the Taylor expansion of X at the origin truncated at order k. Let Y be a vector field belonging to H_k. The linear part X_1 of X defines a linear map L_k of H_k into itself by $L_k(Y) = [X_1,Y]$. Denote the range of L_k as B_k. Then, if $V_k - X_k$ belongs to R_k, there exists a local change of coordinates around the origin such that the coordinate change transforms the vector field X into a vector field which has the same k-jet as V_k.

He and ARNOLD[5] computed the explicit forms of normal forms and they studied in detail the global phase portraits of these singularities. The normal forms computed by Arnold and Takens are, however, unsatisfactory by the following reason. The family of normal forms can be considered as the set of representatives for each equivalence class of vector fields under the coordinate transformations. If a certain family is a family of normal forms, the family derived from this family by applying a coordinate transformation simultaneously to all systems of the family is also a family of normal forms. We cannot claim the family to be unique. We can expect, however, that the normal form (2.6) should be determined uniquely from the original system $\dot{z} = F(z)$. Unfortunately, Poincaré-Arnold-Takens theory does not give the uniqueness of the normal form in the above sense.

§3. Versal family of vector fields

Consider a family of systems :

$$\dot{x} = F_r(x),\ x \in \mathbb{R}^n,\ r \in \mathbb{R}^p \tag{3.1}$$

which has a singular point at the origin for all values of parameter r. How can we transform the family into a normal form by a family of coordinate transformations smoothly depending upon the parameter r ? A family of systems $\dot{y} = G_s(y)$ on $\mathbb{R}^n$ with parameter $s \in \mathbb{R}^q$ is said to be k-versal if for any family (3.1) such that $F_0(x) = G_0(x)$, there exists a family of coordinate transformations $y = T_r^0(x)$, smoothly depending on the parameter r and a smooth mapping of parameters $s = t(r)$, such that the system $\dot{x} = F_r(x)$ transformed by $y = T_r(x)$ and the system $\dot{y} = G_{t(r)}(y)$ coinsides up to terms of order higher than k. ARNOLD[3] gave the solution for this problem in the case of matrices. He gave the versal family of matrices, which can be considered as the normal forms for families of matrices. The versal family is also the vesal family of first order for ordinary differential equations with a singular point at the origin.

It is known that most of degenerate singularities must have an infinite number of unfolding parameters for the unfolding family to be versal, if we don't neglect any higher order terms (see ICHIKAWA [12]).

§4. Normal forms for some degenerate singularities

In this section, we reproduce the normal forms obtained in [22] for the systems of ODEs on $\mathbb{R}^2$ and $\mathbb{R}^3$, whose linear part include zero or pure imaginary eigenvalues. These results for truncated normal forms are obtaind by computing, up to higher order terms, the effect of all the coordinate transformations exaustively. See [22] for the detail. The coordinates are represented by (x,y) for $\mathbb{R}^2$ and (x,y,z) for $\mathbb{R}^3$. Polar coordinate (r,θ,z) is also used for $\mathbb{R}^3$. The folowing is a list of simplest normal forms with non-hyperbolic linear parts in $\mathbb{R}^3$ (or $\mathbb{C}^3$). Non-generic cases are omitted. All parameters are uniquely determined from the original system.

Case (A). Let X be a smooth vector field on $\mathbb{R}^2$ (or $\mathbb{C}^2$). Assume that X vanishes at the origin and that the linear part of X is $\dot{x} = y$, $\dot{y} = 0$. The matrix corresponding to this linear part is of the form

$$\begin{bmatrix} 0 & 1 \\ 0 & 0 \end{bmatrix}$$

Then X can be transformed by a transformation of coordinates into one of the following forms up to terms of order 5.

(a) $\dot{x} = y,\quad \dot{y} = \pm x^2 + uxy + wx^3 + qx^3,$

(b) $\dot{x} = y, \quad \dot{y} = \pm xy + v_1 x^3 + v_2 x^2 y + qx^3 + sx^4,$

or (c) $\dot{x} = y, \quad \dot{y} = w_1 x^3 + w_2 x^2 y + qx^3 + sx^4$, with $w_1^2 + w_2^2 = 1$.

Parameters u, v_1, v_2, w, q, s, w_1, w_2 are uniquely determined from the 4-th order Taylor expansion of X at the origin.

Versal families for these singularities are obtained by adding the supplementary parameters. For instance, a versal family of fourth order for a system of type (a) above, say X_* for $u = u_0$, $w = w_0$ and $q = q_0$, is given by adding the Arnold's versal family for linear part to the family of normal forms :

$$\text{(a')} \quad X_* + \begin{pmatrix} 0 \\ p_1 x + p_2 y + uxy + wx^3 + qx^3 \end{pmatrix}.$$

Case (B) Linear part is of the form:

$$\dot{\theta} = 1, \quad \dot{r} = 0, \quad \dot{z} = 0.$$

The matrix for this linear part is

$$\begin{bmatrix} 0 & -1 & 0 \\ 1 & 0 & 0 \\ 0 & 0 & 0 \end{bmatrix}$$

in (x,y,z) coordinates. Fifth order normal form is

$$\begin{aligned} \dot{\theta} &= 1 + bz + ez^2 + gr^4 \\ \dot{r} &= azr + dz^2 r + fr^5 \\ \dot{z} &= \pm r^2 \pm z^2 + cz^3. \end{aligned} \tag{4.1}$$

Case (C) Linear part is of the form :

$$\dot{x} = y, \quad \dot{y} = 0, \quad \dot{z} = 0.$$

The matrix for this linear part is

$$\begin{bmatrix} 0 & 1 & 0 \\ 0 & 0 & 0 \\ 0 & 0 & 0 \end{bmatrix}$$

in (x,y,z) coordinates. Third order normal form is

$$\begin{aligned} \dot{x} &= y \\ \dot{y} &= \pm x^2 + axy \pm z^2 + cyx + ex^3 + hxyz + iyz^2 \\ \dot{z} &= bxz + dz^2 + fx^3 + gz^3. \end{aligned} \tag{4.2}$$

Case (D) Linear part is of the form :

$$\dot{x} = y, \quad \dot{y} = z, \quad \dot{z} = 0.$$

The matrix for this linear part is

$$\begin{bmatrix} 0 & 1 & 0 \\ 0 & 0 & 1 \\ 0 & 0 & 0 \end{bmatrix}$$

in (x,y,z) coordinates. Third order normal form is

$$\dot{x} = y$$

$$\dot{y} = z$$
$$\dot{z} = \pm x^2 + axy + bxz + cx^2y + dxz^2 + ex^3. \quad (4.3)$$

§5. Comparison of normal forms with those of Poincaré and Arnold-Takens

Let us compare, in this section, the number of parameters included in the family of normal forms by Poincaré-Dulac, by Arnold-Takens and ours. For each typical type of Jordan normal form of matrices, we computed the families of normal forms of vector fields whose linear part is of the Jordan normal form. We compute the number of resonant monomials for the Poincaré's family of normal forms. For Arnold and Takens normal forms, we also computed the number of parameters included in the family. For our normal forms, we computed the number of parameters assuming the system to be generic, i.e., assuming one of the nonlinear terms to be non-zero.

In these tables, numbers in brackets as [1] represent the number of terms whose coefficients can be normalized to 0 or ± 1. The number of parameters listed in the following tables are those vector fields which have non-vanishing non-linear term indicated above and hence normalized to ± 1. Numbers in round brackets as (1) represent the number of terms normalized to zero (degenerate case). Columns marked as * are not computed yet.

Table 5.1. Number of parameters in the family of normal forms for the Hopf singularity on the plane

Linear part of the system $\begin{bmatrix} 0 & -1 \\ 1 & 0 \end{bmatrix}$

Number of parameters in the linear versal family of Arnold : 2

Degree	2nd	3rd	4th	5th	6th	7th	8th	9th
Poincaré's normal form	0	1	0	1	0	1	0	1
Arnold-Takens	0	[1]	0	1	0	1	0	1
Ours	0	[1]	0	1	0	0	0	0

Table 5.2. Number of parameters in the family of normal forms for the Bogdanov singularity on the plane --- Case (A)

Linear part of the system $\begin{bmatrix} 0 & 1 \\ 0 & 0 \end{bmatrix}$

Number of parameters in the linear versal family of Arnold : 2

Degree	2nd	3rd	4th
Poincaré's normal form	6	8	10
Arnold-Takens	2	2	2
Ours	1+[1]	1	1
	(1)+[1]	2	2
	(2)	1	2

Table 5.3. Number of parameters in the family of normal forms for the Hopf-saddle-node singularity in $\mathbb{R}^3$ --- Case (B)

Linear part of the system $\begin{bmatrix} 0 & -1 & 0 \\ 1 & 0 & 0 \\ 0 & 0 & 0 \end{bmatrix}$

Number of parameters in the linear versal family of Arnold : 3

Degree	2nd	3rd	4th	5th
Poincaré's normal form	4	6	7	9
Arnold-Takens	4	6	7	9
Ours	2+[2]	3	0	2

Table 5.4. Number of parameters in the family of normal forms for the Bogdanof-saddle node singularity in $\mathbb{R}^3$ --- Case (C)

Linear part of the system $\begin{bmatrix} 0 & 1 & 0 \\ 0 & 0 & 0 \\ 0 & 0 & 0 \end{bmatrix}$

Number of parameters in the linear versal family of Arnold : 5

Degree	2nd	3rd	4th	5th
Poincaré's normal form	9	12	15	18
Arnold-Takens	8	11	14	17
Ours	4+[2]	5	*	*

Table 5.5. Number of parameters in the family of normal forms for the triply degenerate singularity in $\mathbb{R}^3$ --- Case (D)

Linear part of the system $\begin{bmatrix} 0 & 1 & 0 \\ 0 & 0 & 1 \\ 0 & 0 & 0 \end{bmatrix}$

Number of parameters in the linear versal family of Arnold : 3

Degree	2nd	3rd
Poincaré's normal form	9	12
Arnold-Takens	4	6
Ours	2+[1]	3

§6. Normalization and rescaling of Lorenz system

In this section we construct a one-parameter family of systems of ODEs, which connects the Lorenz system to an "integrable" system. The Lorenz system is :

$$\dot{x} = sy - sx, \ \dot{y} = -xz + rx - y, \quad \dot{z} = xy - bz. \tag{6.1}$$

This system has a symmetry: the system takes exactly the same form when it is transformed by coordinate transformation

$$(x,y,z) \to (-x,-y,z).$$

The normal form theory works also for such equivariant systems taking only equivariant systems and equivariant coordinate transformations. The third order versal family of vector field $\dot{x} = y$, $\dot{y} = 0$ under the symmetry above is given by

$$\dot{x} = y,\ \dot{y} = Ax+By+Eyz+Fxz+Hx^2y+Iyz^2+Jx^3,\ \dot{z} = Cz+x^2+Gz^2+Kz^3,$$

where F is a non-zero constant and A, B, C, E, G, H, I, J, K are unfolding parameters. For a degenerate case

$$\dot{x} = y,\ \dot{y} = -x^3,\ \dot{z} = x^2, \tag{6.2}$$

versal family for this system is computed as :

$$\dot{x} = y,\ \dot{y} = Ax+By+Eyz+Fxz+Hx^2y+Iyz^2+Jx^3+Lxz^2,$$

$$\dot{z} = Cz+x^2+Gz^2+Kz^3, \tag{6.3}$$

here all parameters are unfolding parameters.

The linear part of (6.1) at the origin has triple zero eigenvalues when $(s,r,b) = (-1,1,0)$. Assume $s \neq 0$ and $b \neq 2s$ and execute the change of variables for (6.1): $X = \sqrt{2}x$, $Y = \sqrt{2}\, s(y-x)$, $Z = Q(2sz-x^2)$, where $Q = 1 - b/2s$. Put $A = s(r-1)$, $B = -(1+s)$, $C = -b$, and $S = 2Qs$. We obtain :

$$\dot{x} = y,\ \dot{y} = Ax+By-Sxz-x^3,\ \dot{z} = Cx+x^2, \tag{6.4}$$

which is a subfamily of (6.3). Observe that (6.4) has four parameters whereas (6.1) contains only three parameters.

Now change the time scale from t to $t' = t/p$ and renormalize the obtained system by coordinate change $X = px$, $Y = p^2y$, $Z = pz$. The renormalized system is of the form

$$\dot{x} = y,\ \dot{y} = p^2Ax+pBy-pSxz-x^3,\ \dot{z} = pCx+x^2, \tag{6.5}$$

which shows that there exists a "miniature" of Lorenz attractor in the neighborhood of the origin for small parameter p, since for any positive value p, the system is analytically conjugate to the Lorenz system.

REFERENCES

[1] Arneodo,A., Coullet,P.H., Spiegel,E.A. and Tresser,C.: Asymptotic chaos, to appear in PHYSICA D.
[2] Arnold,V.I.: On matrices depending on a parameter, Russ. Math. Sueveys 26 (1971) 29-43.
[3] Arnold,V.I.: Lectures on Bifurcation in versal families, Russ. Math. Surveys 27 (1972) 54-123.
[4] Arnold,V.I.: Loss of stability of self oscillations close to resonance and versal deformations of equivariant vector fields, Funct. Anal. Appl., 11 (1977) 85-92.
[5] Arnold,V.I.: Chapitres supplémentaires de la théorie des équations différentielles ordinaires, Editions Mir, Moscow, 1980.
[6] Birkhoff,G.D.: Dynamical systems, Amer. Math. Soc. Colloquium Publications, New York (1927).
[7] Bogdanov,R.I.: Versal deformation of a singular point of a vector field on a plane in the case of zero eigenvalues, Proceedings

of I.G.Petrovskii Seminar, 2 (1976) 37-65.
[8] Broer,H.: Formal normal form theorems for vector fields and some consequences for bifurcations in the volume preserving case, Lecture Notes in Math. 898, Dynamical Systems and Turbulence, Warwick 1980, Springer, (1981) 54-74.
[9] Guckenheimer,J.: On a codimension two bifurcation, Lecture Notes in Math. 898, Dynamical Systems and Turbulence, Warwick 1980, Springer, (1981) 99-142.
[10] Holmes,P.J.: Center manifolds, normal forms and bifurcation of vector fields, Physica 2D (1981) 449-481.
[11] Holmes,P.J.: A strange family of three-dimensional vector fields near a degenerate singularity, J. Diff. Eq. 37 (1980) 382-403.
[12] Ichikawa,F.: Finitely determined singularities of formal vector fields, Invent. Math. 66 (1982) 199-214.
[13] Langford,W.: Peridic and steady-state mode interactions lead to tori, SIAM J. Appl. Math., 37 (1979) 22-48.
[14] Langford,W., Iooss,G.: Interactions of Hopf and pitchfork bifurcations, Workshop on Bifurcation Problems, Birkhäuser Lecture Notes, (1980).
[15] Lorenz,E.N.: Deterministic nonperiodic flows, J. Atmospheric Sci., 20, (1963), 130-141.
[16] Poincaré,H.: Thèse (1879), Oeuvre I, Gauthier-Villars (1928) 69-129.
[17] Poincaré,H.: Mémoire sur les courbes définies par une équation différentielle, I, II, III, and IV, J.Math. Pures Appl. (3) 7 (1881) 375-422, (3) 8 (1882) 251-296, (4) 1 (1885) 167-244, (4) 2 (1886) 151-217.
[18] Poincaré,H.: Les méthodes nouvelles de la mécanique céleste, I (1892), III (1899).
[19] Rössler,O.E.: Continuous chaos - four prototype equations, Ann. New York Acad. Sci., 316, (1979), 376-392.
[20] Rössler,O.E.: Different types of chaos in two simple differential equations, Z. Naturforsch, 31a, (1976), 1664-1670.
[21] Takens,F.: Singularities of vector fields, Publ. Math. IHES, 43 (1973) 47-100.
[22] Ushiki,S.: Normal forms for singularities of vector fields, preprint.
[23] Ushiki,S., Oka,H., and Kokubu,H.: Attracteurs étranges engendré par une singularité des systèmes intégrables, preprint.

Computing Methods in Applied Sciences and Engineering, VI
R. Glowinski and J.-L. Lions (Editors)
Elsevier Science Publishers B.V. (North-Holland)

DISCRETE ASYMPTOTIC BEHAVIOR FOR A NONLINEAR DEGENERATE DIFFUSION EQUATION

Tsutomu Ikeda

Department of Mathematics
Faculty of Science
Kyoto University
Kyoto, Japan

We are concerned with a finite difference approximation for a spatially aggregating population model which includes a nonlinear diffusion process and a nonlocal interaction. We propose a finite difference scheme involving a nonlinear artificial viscosity term. After studying how the nonlinear artificial viscosity works, we show that the scheme preserves such properties of the model as finite speed of propagation and formation of solitary wave patter for large time.

1. Introduction

In this paper, we study the asymptotic behavior of a solution of a finite difference approximation for a nonlinear degenerate diffusion equation. From ecological points of view, there have been proposed a number of spatially spreading population models which yield several traveling wave patterns or several stable stationary patterns for large time. We deal with one of such models that has infinitely many stationary solutions. (It is relatively easy to show the asymptotic behavior of a finite difference solution for linear problems and nonlinear problems having a unique stationary solution.)

We consider the following nonlinear degenerate diffusion equation

$$(1.1)\qquad \begin{cases} U_t = (U^2)_{xx} + [U(\int_{-\infty}^{x} U dy - \int_{x}^{\infty} U dy)]_x & \text{in } \mathbb{R} \times (0, \infty), \\ U(x,0) = U^0(x) \geq 0 & \text{for } x \in \mathbb{R}, \end{cases}$$

where $m > 1$ and $U(x,t) \geq 0$ denotes the population density at position $x \in \mathbb{R}$ and time $t > 0$. Equations of this type, proposed by Nagai and Mimura[8], represent a spatially spreading population model for a class of aggregating phenomena of individuals. The first term on the right-hand side of (1.1) ecologically corresponds to the transport of population through a nonlinear diffusion process called density-dependent diffusion ([4] and [5]). The second term provides the aggregative mechanism that moves individuals to the right (resp. left) direction when

$$\int_{x}^{\infty} U(y,t)dy > \int_{-\infty}^{x} U(y,t)dy \qquad \text{(resp. <)}.$$

The solution of (1.1) has the following interesting properties:

[A] (finite propagation) the speed of propagation of $U(x,t)$ is finite when U^0 has a compact support,

[B] (asymptotic behavior) a solution $U(x,t)$ forms a stationary solitary wave pattern for large time, which is caused by a suitable balance between the nonlinear diffusion process and the aggregative effect.

From [A], one can define "interface curves" by

$$\partial\{(x,t) \in \mathbb{R}\times(0,\infty);\ U(x,t) > 0\}\backslash(\text{support of } U^0)$$

([1], [2] and [7] for the porous medium equation).

The total distribution $\int_{-\infty}^{\infty} U(y,t)dy$ is kept constant when $U(x,t)$ is a solution of (1.1). Hence the problem (1.1) is transformed into the following Cauchy problem, including no nonlocal interaction,

$$\text{(1.2)}\quad \begin{cases} u_t = [c^{m-1}u_x^m + cu^2]_x & \text{in } \mathbb{R}\times(0,\infty), \\ u_x \geq 0 & \text{in } \mathbb{R}\times(0,\infty), \\ u(-\infty,t) = -1 \quad \text{and} \quad u(\infty,t) = 1 & \text{for } t > 0 \end{cases}$$

subject to a non-decreasing initial condition

$$\text{(1.3)}\quad u(x,0) = u^0(x) = \frac{1}{c}\int_{-\infty}^{x} U^0(y)dy - 1 \qquad \text{for } x \in \mathbb{R}$$

through

$$u(x,t) = \frac{1}{c}\int_{-\infty}^{x} U(y,t)dy - 1\ , \qquad c = \frac{1}{2}\int_{-\infty}^{\infty} U^0(y)dy.$$

The one-to-one correspondence between two solutions of (1.1) and (1.2) subject to (1.3) has been shown by Nagai and Mimura[8]. We introduce a set V of fuctions given by

$$V = \{v \in C(\mathbb{R});\ \text{there exist constants a and b such that } v(x) = -1 \text{ for } x \leq a \text{ and } v(x) = 1 \text{ for } x \geq b\}.$$

The properties [A] and [B] of the solution of (1.1) are described as follows.

Theorem (Nagai-Mimura[9]).

(1) Let $u^0 \in V$ be a non-decreasing function. Then there exists a unique stationary solutions $w(x)$ of (1.2) that satisfies

$$\text{(1.4)}\qquad \int_{-\infty}^{\infty}(w(x)-u^0(x))dx = 0\ .$$

(2) Let v^0 and w^0 be non-decreasing fuctions belonging to V, and let v and w be solutions of (1.2) subject to initial conditions $v = v^0$ and $w = w^0$, respectively. If $v^0(x) \geq w^0(x)$ for $x \in \mathbb{R}$, then $v(x,t) \geq w(x,t)$ for $x \in \mathbb{R}$ and $t > 0$.

(3) Let $u^0 \in V$ be a non-decreasing function. A solution $u(x,t)$ of (1.2) subject to the initial condition $u = u^0$ tends to the stationary solution $w(x)$ uniquely determined by (1.4) as t tends to infinity.

We now proceed to finite difference schemes for (1.2). An approximate solution for the original problem(1.1) is obtained by differencing that for (1.2). It is desirable for a finite difference scheme for (1.2)

(i) to be stable,
(ii) to preserve the monotone non-decreasing property of $u(x,t)$,
(iii) to have "sufficiently many" stationary solutions,
(iv) to satisfy a kind of comparison theorem corresponding to (2) in the above theorem,
(v) to give a solution tending to a stationary solution as time t

tends to infinity for any spatial mesh size h,

(vi) to approximate interface curves as well as the value of exact solution $u(x,t)$ ([3], [6] and [11] for the porous medium equation).

In the above, "sufficiently many" means that for any non-decreasing function w belonging to V, there exists a stationary solution w_h of a finite difference scheme that satisfies

$$\int_{-\infty}^{\infty}(w_h(x)-\bar{w}(x))dx = 0 ,$$

where $\bar{w}$ is a projection of w on the set of functions admissible to the scheme. Hence the existence of countable set of stationary solutions does not always fulfil the requirement (iii). For instance, assume that there exists a finite difference scheme in conservation form having a stationary solution $w_h(x)$. Then, for $i = \pm 1, \pm 2, \ldots$, the translation $w_{h,i}$ of w_h by ih is a stationary solution. None of $w_{h,i}$'s, however, satisfy the equality

$$\int_{-\infty}^{\infty}(w_{h,i}-v_h)dx = 0 \quad \text{for } v_h \neq w_{h,0},\ v_h \neq w_{h,1},\ w_{h,1} \geq v_h \geq w_{h,0} .$$

(See Figure 1.1.)

We propose in the next section a finite difference scheme for (1.2) with $m = 2$ that fulfils the requirements (ii), (iii), (iv) and (v). Our scheme involves a nonlinear degenerate artificial viscosity term which permits the existence of "sufficiently many" stationary solutions.

In Section 2, we state that the scheme preserves the monotone non-decreasing property and satisfies a comparison theorem under suitable conditions on the time increment. In section 3, we give a procedure to obtain all stationary solutions of the scheme, and show that the scheme has sufficiently many stationary solutions. Section 4 deals with the asymptotic behavior of a solution obtained by the scheme.

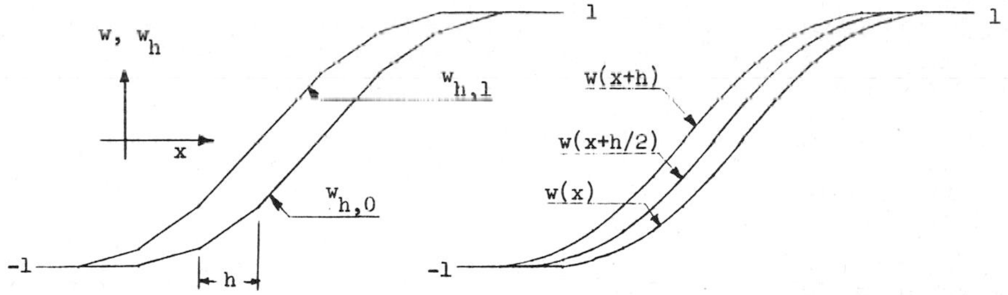

Figure 1.1. $w_{h,0}$ and $w_{h,1}$ correspond to $w(x)$ and $w(x+h)$, respectively. What corresponds to $w(x+h/2)$?

2. A finite difference scheme

In this section, we propose a finite difference scheme for (1.2) with $m = 2$:

$$u_t = [u_x^2 + u^2]_x \qquad \text{in } \mathbb{R} \times (0, \infty) \tag{2.1}$$

(one may put $c = 1$ without loss of generality), and study the behavior of a nonlinear artificial viscosity term involved in the scheme. We then state that the scheme obeys a conservation law, and that it preserves the monotone non-decreasing property of a solution of (1.2) and satisfies a comparison theorem under suitable conditions on the time increment.

With $0 < h \le \sqrt{2}$, we associate $V_h = \{u_h \in C(\mathbb{R});\ u_h$ is linear on each interval $(ih, (i+1)h)$, $i = 0, \pm 1, \ldots$, and there exist integers M and N such that $u_h(x) = -1$ for $x \le Mh$ and $u_h(x) = 1$ for $x \ge Nh\}$, and introduce a finite difference operator $L_h : V_h \to V_h$ defined by

$$\begin{aligned}(L_h u_h)(ih) = {} & \frac{1}{h}[\{\frac{1}{h}(u_{i+1}-u_i)\}^2 - \{\frac{1}{h}(u_i - u_{i-1})\}^2] + \frac{1}{2h}[u_{i+1}^2 - u_{i-1}^2] \\ & + [a(u_i, u_{i+1})(u_{i+1}-u_i) - a(u_{i-1}, u_i)(u_i - u_{i-1})]/h^2\end{aligned} \tag{2.2}$$

for $i = 0, \pm 1, \ldots$, where u_i denotes the value of u_h at position $x = ih$ and $a(v,w)$ is a non-negative continuous function defined on $\mathbb{R}^2$ by

$$a(v,w) = \max\{\frac{h}{2}|v+w| - \frac{w-v}{h},\ 0\} \ . \tag{2.3}$$

Using L_h, our finite difference scheme for (2.1) is defined by

$$\begin{cases} \text{Find } \{u_h^n\}_{n=0}^{\infty} \subset V_h \text{ such that} \\ u_h^{n+1} = u_h^n + \tau^n L_h u_h^n & \text{for } n = 0, 1, \ldots, \\ u_i^0 = u^0(ih) & \text{for } i = 0, \pm 1, \ldots, \end{cases} \tag{2.4}$$

where $\tau^n > 0$ is the time increment. The first and second terms on the right-hand side of (2.2) are finite difference approximations for $(u_x^2)_x$ and $(u^2)_x$, respectively; the third is the nonlinear artificial viscosity term that permits the existence of sufficiently many stationary solutions of the scheme (2.4).

We use the following two different expressions of L_h:

$$(L_h u_h)(ih) = \frac{1}{h}[P(u_i, u_{i+1}) - P(u_{i-1}, u_i)] \ , \tag{2.5}$$

$$(L_h u_h)(ih) = [Q_+(u_i, u_{i+1})(u_{i+1}-u_i) - Q_-(u_{i-1}, u_i)(u_i - u_{i-1})]/h^2 \tag{2.6}$$

where $P(v,w)$, $Q_+(v,w)$ and $Q_-(v,w)$ are given by

$$\begin{cases} P(v,w) = (\frac{w-v}{h})^2 + \frac{1}{2}(v^2+w^2) + \frac{1}{h}\, a(v,w)(w-v), \\ Q_+(v,w) = \frac{w-v}{h} + \frac{h}{2}(v+w) + a(v,w), \\ Q_-(v,w) = \frac{w-v}{h} - \frac{h}{2}(v+w) + a(v,w). \end{cases} \tag{2.7}$$

We here demonstrate the behavior of the functions $a(v,w)$, $P(v,w)$, $Q_+(v,w)$ and $Q_-(v,w)$ on the half plane $D = \{(v,w) \in \mathbb{R}\times\mathbb{R};\ v < w\}$.
Let D_+ and D_- be subdomains of D defined by

$$\begin{cases} D_+ = \{(v,w) \in D;\ v < w < v(2-h^2)/(2+h^2) < 0\}, \\ D_- = \{(v,w) \in D;\ 0 < v < w < v(2+h^2)/(2-h^2)\}, \end{cases}$$

respectively. See Figure 2.1, in which the values of the functions a, Q_+ and Q_- are shown. We note that

(2.8) $a(v,w)$, $Q_+(v,w)$ and $Q_-(v,w)$ are non-negative on D,

(2.9) $a(v,w)$ vanishes on $D\setminus(D_+\cup D_-)$ while Q_+ and Q_- vanish on D_+ and D_-, respectively,

(2.10) $a(v,w)$ is non-decreasing with v and non-increasing with w,

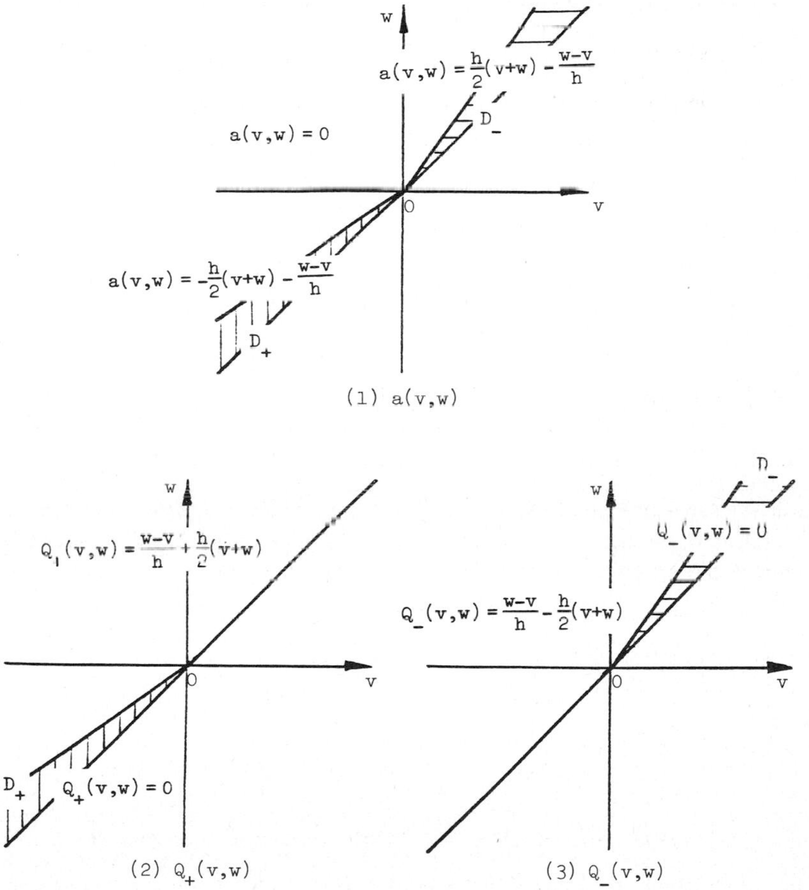

Figure 2.1. Behavior of $a(v,w)$, $Q_+(v,w)$ and $Q_-(v,w)$ on D.

and the following estimates holds:

$$\begin{cases} 0 \le a(v,w) - a(u,w) \le (\frac{1}{h} + \frac{h}{2})(v-u) , \\ 0 \le a(u,v) - a(u,w) \le (\frac{1}{h} + \frac{h}{2})(w-v) \quad \text{for } u \le v \le w , \end{cases}$$

(2.11) $P(-1,v) = 1$ for $(-1,v) \in D_+$ and $P(v,1) = 1$ for $(v,1) \in D_-$.

The expression (2.5) of L_h implies

Proposition 2.1 (Conservation law). It holds that

$$\int_{-\infty}^{\infty} [(u_h + \tau L_h u_h)(x) - u_h(x)]dx = 0 \qquad \text{for all } u_h \in V_h \text{ and } \tau > 0.$$

We have obtained the following two propositions by observing the behavior of the nonlinear artificial viscosity and by using (2.6). (The scheme (2.4) appears to be stable under the condition (2.13) below. A complete proof of stability, however, has not obtained.)

Proposition 2.2 (Preservation of non-decreasing property). Let $u_h \in V_h$ be a non-decreasing function. Then, $v_h = u_h + \tau L_h u_h$ becomes a non-decreasing function under the condition

$$(2.12) \qquad 2\tau[\frac{1}{h}(u_{i+1} - u_i) + a(u_i, u_{i+1})] \le h^2 \quad \text{for } i = 0, \pm 1, \dots .$$

Proposition 2.3 (Comparison theorem). Let u_h and v_h be non-decreasing functions belonging to V_h, and put

$$u_h' = u_h + \tau L_h u_h \quad \text{and} \quad v_h' = v_h + \tau L_h v_h .$$

If $u_h \ge v_h$, then $u_h' \ge v_h'$ under the condition

$$(2.13) \qquad \begin{cases} 2\tau[\frac{2}{h}(u_{i+1} - u_i) + a(u_i, u_{i+1})] \le h^2 , \\ 2\tau[\frac{2}{h}(v_{i+1} - v_i) + a(v_i, v_{i+1})] \le h^2 \quad \text{for } i = 0, \pm 1, \dots . \end{cases}$$

3. Stationary solutions of finite difference scheme

In this section, we give a procedure to obtain all stationary solutions of the scheme (2.4), and then show that (2.4) fulfils the requirement (iii) described in Section 1.

Definition. A stationary solution of (2.4) is defined to be a monotone non-decreasing function w_h belonging to V_h that satisfies $L_h w_h = 0$, or in other words,

(3.1) $P(w_i, w_{i+1}) = 1 \qquad$ for $i = 0, \pm 1, \dots$,

(3.2) $Q_+(w_i, w_{i+1})(w_{i+1} - w_i) = Q_-(w_{i-1}, w_i)(w_i - w_{i-1}) \quad$ for $i = 0, \pm 1, \dots$.

We first note that if (2.3) is replaced by others, then the scheme may have no stationary solution. For instance, let us replace (2.3) by $\max\{s[h|v+w|/2 - (w-v)/h], 0\}$ $(s > 0)$. If $s > 1$, then the scheme has no stationary solution in the sense of the above definition. On the other hand, if $s < 1$, then the scheme violates the monotone non-decreasing property and gives a solution that is meaningless from both mathematical and ecological points of view. When we replace $a(v,w)$ by a linear artificial viscosity, the scheme has no stationary solution while the stability of the scheme is shown under a condition similar to (2.13).

We now return to our scheme (2.4) with (2.2) and (2.3), and begin with necessary conditions for a stationary solution.

Proposition 3.1. If w_h is a stationay solution of (2.4) such that $\ldots = w_0 = -1 < w_1 \le w_2 \le \ldots \le w_{n-1} < 1 = w_n = \ldots$, then it holds that

$$(3.3)\quad \begin{cases} (w_0,w_1) \in D_+ , \quad (w_{n-1},w_n) \in D_- \\ (w_i,w_{i+1}) \in D\setminus(D_+ \cup D_-) \quad \text{for } i = 1,2,\ldots,n-2 , \end{cases}$$

or in other words, w_h satisfies

$$(3.4)\quad \begin{cases} -1 < w_1 \le -(2-h^2)/(2+h^2), \quad (2+h^2)/(2-h^2) \le w_{n-1} < 1, \\ w_i < w_{i+1} \text{ and } a(w_i,w_{i+1}) = 0 \quad \text{for } i = 1,\ldots,n-2 . \end{cases}$$

Lemma 3.2. Let $-1 \le u \le (2-h^2)/(2+h^2) = d(h)$, then there exists a unique solution $v(u)$ of $P(u,v) = 1$ subject to $a(u,v) = 0$ and $u < v < 1$. Moreover, it holds that
(1) v increases as u in the interval $[-1,d(h)]$,
(2) $v(d(h)) = 1$ and $v(-1) = -d(h)$.

Proposition 3.1 together with (2.11) and Lemma 3.2 gives the following procedure to obtain all the stationary solutions of (2.4):

(3.5)
```
fix an integer I and a positive constant e ≤ 1 - d(h),
set w_i = -1 for all i ≤ I,
set i = I+1,
set w_i = -1+e,
do while w_i < 1
 if w_i ≥ d(h) then set w_j = 1 for all j ≥ i+1,
 else set w_{i+1} = [a unique solution of P(w_i,w_{i+1}) subject to
 a(w_i,w_{i+1}) = 0 and w_i < w_{i+1} < 1],
 set i = i+1,
end.
```

With $r \in \mathbb{R}$, we associate a stationay solution $w_h(\cdot;r)$ of (2.4) that satisfies the conditions

$$\begin{cases} w_h(ih;r) = -1 \quad \text{for all integers } i \le I , \\ w_h((I+1)h;r) = -1 + [1-d(h)][(I+1)h-r]/h > -1 , \end{cases}$$

where $I = [r/h]$ ([ ] indicates the Gauss symbol). This gives a one-to-one correspondence between $\mathbb{R}$ and the set of all stationary solutions of (2.4). Figures 3.1 and 3.2 represent stationary solutions of (2.4) for various values of spatial mesh size h. It follows from (1) and (2) of Lemma 3.2 that

$$\begin{cases} w_h(\cdot;s) \ge w_h(\cdot;r) \quad \text{for } s \le r , \\ \int_{-\infty}^{\infty} [w_h(x;s)-w_h(x;r)]dx \quad \text{tends to zero as } s \to r . \end{cases}$$

Thus, we obtain:

Theorem 3.3 (Existence of stationary solution). For any non-decreasing function v belonging to V, there exists a unique stationary solution $w_h(.;r)$ of (2.4) that satisfies the equality

$$\int_{-\infty}^{\infty}[w_h(x;r)-v(x)]dx = 0 . \tag{3.6}$$

Moreover, when h tends to zero, $w_h(.;r)$ tends to the stationary solution w of (1.2) satisfying

$$w(x) = -1 \quad \text{for} \quad x \leq r \quad \text{and} \quad w(x) > -1 \quad \text{for} \quad x > r .$$

(The length of the support of $\frac{1}{h}(w_h(x+h;r)-w_h(x;r))$ is bounded from the below and above uniformly in h and r, which is shown by (3.1) and a careful observation of the procedure (3.5).)

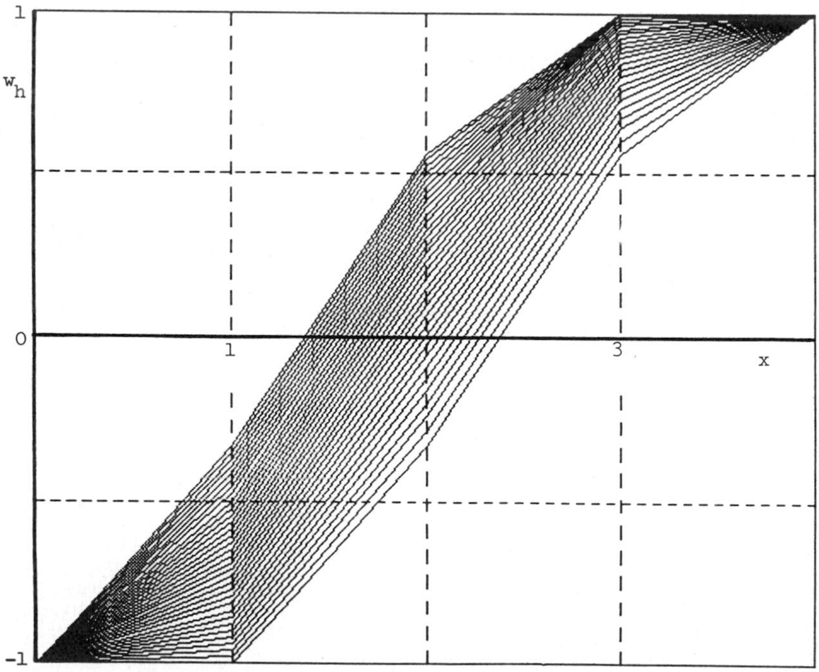

Figure 3.1. Stationary solutions of (2.4) for h = 1. Stationary solutions $w_h(x;i/32)$ i = 0,1,...,32, are shown.

4. Asymptotic behavior of finite difference solution

Figure 4.1 represents an approximate solution for (1.1) with m = 2 obtained by differencing a solution of the scheme (2.4).

Theorem 4.1 (Asymptotic behavior). Let $w_h = w_h(\cdot;r)$ be the stationary solution of (2.4) that satisfies (3.6) with $v = u_h^0$. Then, the solution u_h^n of (2.4) tends to w_h as $n \to \infty$ under the condition

$$(4.1) \qquad \begin{cases} 2\tau^n[\frac{2}{h}(u_{i+1}^n - u_i^n) + a(u_i^n, u_{i+1}^n)] \le h^2, \\ 2\tau^n[\frac{2}{h}(w_{i+1} - w_i) + a(w_i, w_{i+1})] \le h^2 \end{cases}$$

for $i = 0, \pm 1, \ldots$ and $n = 0, 1, \ldots$. Moreover, it holds that

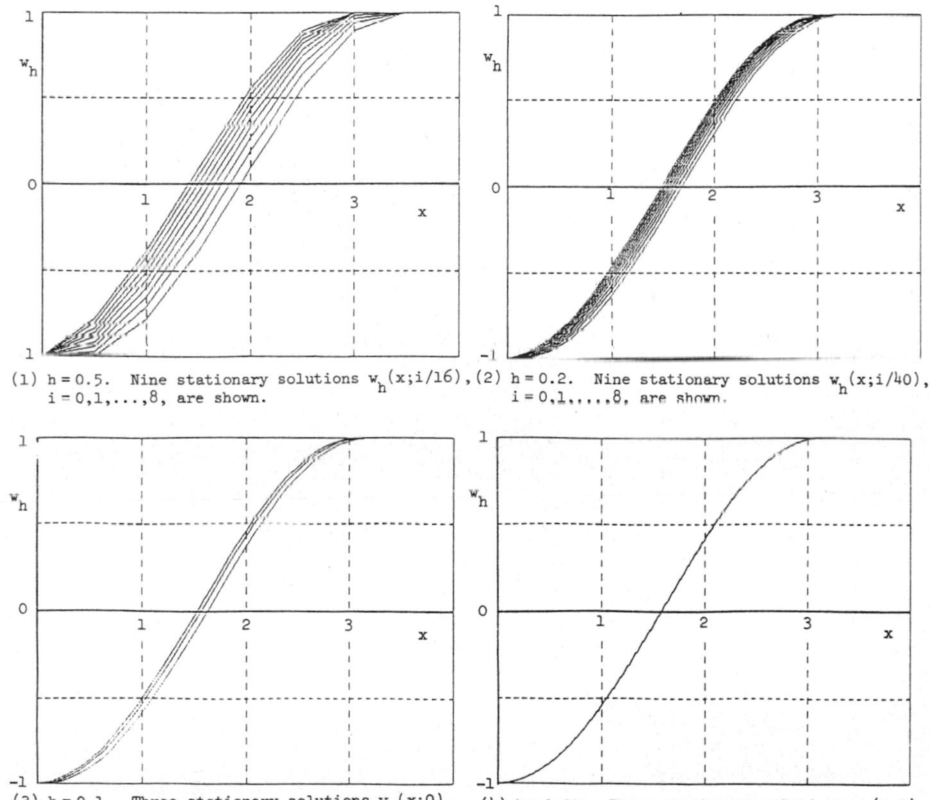

(1) h = 0.5. Nine stationary solutions $w_h(x;i/16)$, i = 0,1,...,8, are shown.
(2) h = 0.2. Nine stationary solutions $w_h(x;i/40)$, i = 0,1,...,8, are shown.
(3) h = 0.1. Three stationary solutions $w_h(x;0)$, $w_h(x;0.05)$ and $w_h(x;0.1)$ are shown.
(4) h = 0.01. Three stationary solutions $w_h(x;0)$, $w_h(x;0.005)$ and $w_h(x;0.01)$ are shown.

Figure 3.2. Stationary solutions of (2.4).

(4.2) $$\sup_{x\in\mathbb{R}}|u_h^n(x)-w_h(x)| \le C\exp(-Kt_n)$$

where C and K are positive constants and $t_n = \tau^0+\ldots+\tau^{n-1}$.

Sketch of the proof. Let I and J be integers such that

$$w_h(x+Ih) \le u_h^0(x) \le w_h(x+Jh) \qquad \text{for all } x \in \mathbb{R},$$

then, under the condition (4.1), Proposition 2.3 yields

$$w_h(x+Ih) \le u_h^n(x) \le w_h(x+Jh) \qquad \text{for all } x \in \mathbb{R} \text{ and } n = 1,2,\ldots .$$

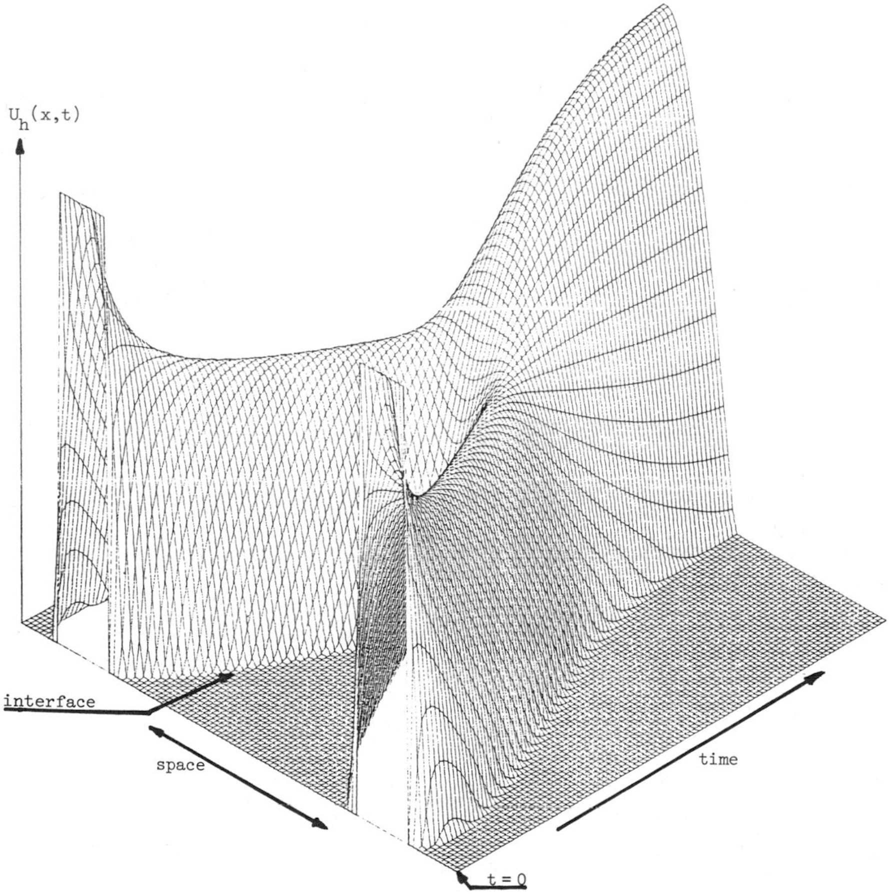

Figure 4.1. An approximate solution for (1.1) with m = 2 obtained by differencing a solution of (2.4).

In order to show the exponential decay of $|u_h^n - w_h|$, we deduce a finite difference equation for

$$v_h^n(x) = \int_{-\infty}^{x} (u_h^n(y) - w_h(y))dy$$

from (2.4) and construct a sequence $\{z_h^n\}$ of piecewise linear functions on $\mathbb{R}$ that satisfies

$$z_h^n(x) \geq |v_h^n(x)| \quad \text{and} \quad 0 \leq z_h^n(x) \leq C\exp(-Kt_n)$$

for all $x \in \mathbb{R}$ and $n = 0,1,\ldots$. This idea of the proof is similar to that for the continuous problem (Nagai and Mimura[9]). For the construction of $\{z_h^n\}$, however, we need an elaborate technique which is different from that for the continuous problem.

References

[1] D. G. Aronson, Regularity property of flows through porous media: The interface, Arch. Rational Mech. Anal. v.37(1970), pp.1-10.

[2] D. G. Aronson, L. A. Caffarelli & S. Kamin, How an initially stationary interface begins to move in porous medium flow, Univ. of Minnesota Mathematics Report(1981), pp.81-113.

[3] E. DiBenedetto & D. Hoff, An interface tracking algorithm for the porous medium equation, to appear in Trans. A.M.S.

[4] W. S. C. Gurney & R. M. Nisbet, The regulation of inhomogeneous populations, J. Theoret. Biol., v.52(1975), pp.441-457.

[5] M. E. Gurtin & R. C. MacCamy, On the diffusion of biological populations, Math. Biosci., v.33(1979), pp.35-49.

[6] D. Hoff, A linearly implicit finite difference scheme for the one-dimensional porous medium equation, (preprint).

[7] B. F. Knerr, The porous medium equation in one dimension, Trans. Amer. Math. Soc., v.234(1977), pp.381-415.

[8] T. Nagai & M. Mimura, Some nonlinear degenerate diffusion equations related to population dynamics, J. Math. Soc. Japan, v.35 (1983), pp.539-562.

[9] T. Nagai & M. Mimura, Asymptotic behavior for a nonlinear degenerate diffusion equation in population dynamics, SIAM J. Appl. Math., v.43(1983), pp.449- 464.

[10] T. Nagai, Some nonlinear degenerate diffusion equations with a nonlocally convective term in ecology, Hiroshima Math. J., v.13 (1983), pp.165-202.

[11] K. Tomoeda & M. Mimura, Numerical approximations to interface curves for a porous media equation, Hiroshima Math. J., v.13 (1983), pp.273-294.

MULTIGRID METHODS

METHODES MULTIGRILLES

Computing Methods in Applied Sciences and Engineering, VI
R. Glowinski and J.-L. Lions (Editors)
Elsevier Science Publishers B.V. (North-Holland)

SOFTWARE DEVELOPMENT BASED ON MULTIGRID TECHNIQUES

Klaus Stüben, Ulrich Trottenberg, Kristian Witsch

Gesellschaft für Mathematik und Datenverarbeitung
Postfach 1240, 5205 St.Augustin 1, West-Germany

Summary *)

One of the most important obstacles to the spreading of multigrid ideas and techniques among PDE users certainly lies in the fact that multigrid components usually can hardly be incorporated into existing non-multigrid PDE software. In many cases this may have "only" technical reasons (multigrid-incompatible data structures etc.). But even if such technical difficulties are not present - several multigrid experts claim that this way of developing multigrid software will not at all lead to the optimal, not even to really efficient multigrid codes. One argument that is emphazised in this context is the following: Usual PDE packages are modularly structured in particular with respect to the discretization and the solution procedures, and this modular structure requires a clear distinction between both processes. Exactly this sharp distinction, however, is in contradiction to a central element of the multigrid philosophy. According to Achi Brandt, efficient realization of multigrid techniques requires, on the contrary, an as strong as possible connection of both processes.

In this paper we want to contribute to the discussion of this question, namely whether there are natural (technical and/or more fundamental) requirements for the design of multigrid software.

One result of our experiences is that the answer on this question strongly depends on the objectives and on the purpose the concrete software product is designed for. Indeed, these objectives which are different for the different codes seem to have more influence on the design question than the common "multigrid requirements". In particular, we want to report on three activities in the area of multigrid software development which have been persued by the GMD multigrid group:

- (1) developing highly efficient codes for special scalar second order elliptic equations (MG00, MG01...)

- (2) writing a black-box (test-)program for the solution of a large class of sparse linear systems, based on the algebraic multigrid approach (AMG01)

- (3) participating in the further development of the GRIDPACK-system: providing tools for the convenient definition and handling of grid-structures and grid-routines (such as interpolation, restriction etc.).

The emphasis of our presentation lies on the MG00 package and the AMG01 code. For both products, we describe in detail its purpose and user interfaces, the underlying data structure and memory management, the modularity of subroutines for subtasks etc. We will recognize that the very different objectives of MG00 and AMG01 lead to very different realization of software design principles.

---

*) The full paper is contained in: Proceedings of the IFIP Working Conference "PDE Software: Modules, Interfaces and Systems". Söderköping, 22.-26. August 83

Computing Methods in Applied Sciences and Engineering, VI
R. Glowinski and J.-L. Lions (Editors)
Elsevier Science Publishers B.V. (North-Holland)

PARABOLIC MULTI-GRID METHODS

Wolfgang Hackbusch

Institut für Informatik und Praktische Mathematik
Christian-Albrechts-Universität
Olshausenstr 4o
D-23oo Kiel 1, RFA

A multi-grid iteration for solving parabolic partial differential equations is presented. It is characterized by the simultaneous computation of several time steps in one step to the computational process.

1. INTRODUCTION

1.1 MULTI-GRID METHODS FOR STATIONARY PROBLEMS

Multi-grid methods are very efficient solvers for elliptic equations. The iteration consists of a 'smoothing step' and a 'coarse-grid correction' involving a sequence of coarser grids (cf [1],[4],[5]). Let $\{h_l\}_{l\geq o}$ be a sequence of grid sizes (eg, $h_{l-1}=2h_l$) and denote the linear discrete problem by

$$L_l\, u_l = f_l\,. \tag{1.1}$$

The multi-grid iteration al level l for solving Eq (1.1) is defined by the following recursive procedure performing one iteration step:

```
procedure MGM(l,u,f); integer l; array u,f;
if  l=o  then  u:=L_o^-1·f  else
begin integer j; array v,d;
      for j:=1 step 1 until ν do u:=S_l(u,f);
      d:=r·(L_l·u-f);                                   (1.2)
      v:=o;
      for j:=1 step 1 until γ do MGM(l-1,v,d);
      u:=u-p·v
end;
```

S_l denotes a 'smoothing iteration' (eg, the Gauß-Seidel iteration). The number ν of smoothing iterations is independent of l. p is a prolongation (interpolation) from the coarse grid of level $l-1$ into the fine one of level l. For elliptic problems of second order the usual choice is a piecewise linear interpolation. The restriction r from the fine into the coarse grid may be chosen as adjoint of p: $r=p^*$ (cf [4]).

The typical multi-grid convergence speed is uniform with respect to the discretization parameter (cf [4]).

Usually, the coarser grids are also used to provide good starting values. The resulting 'nested iteration' reads as follows:

```
u_o := L_o^{-1} f_o ;
for k:=1 step 1 until l do
begin u_k:=p̃·u_{k-1};                                        (1.3)
      for j:=1 step 1 until i do MGM(k,u_k,f_k)
end;
```

Here, the number i of iterations is often very small (eg, $i = 1$).

1.2 PARABOLIC EQUATIONS

A linear parabolic problem is given by

$$u_t + Lu = f \tag{1.4}$$

(L: elliptic operator) together with initial and boundary conditions. Assume that again different levels of discretizations are given characterized by a spatial grid size Δx_l. In addition there are time steps Δt_l. A simple discretization is the implicit Euler formula

$$P_l(t,u_l(t),u_l(t-\Delta t_l)) = f_l(t), \tag{1.5a}$$

where the discrete parabolic operator is defined by

$$P_l(u_l,u_l',t) := (u_l-u_l')/\Delta t_l + L_l(t)\, u_l \,. \tag{1.5b}$$

The advantage of the implicit scheme (1.5b) is the stability for all ratios $\Delta t_l/\Delta x_l^2$. One might prefer other implicit schemes of higher order of consistency (eg, the Crank-Nicholson scheme), but for the sake of simplicity we shall always consider (1.5a,b) in the sequel.

Also for simplicity we shall use fixed time steps Δt_l. Of course, a good parabolic solver has to use variable time steps, but we do not want to mix the discussion of time stepping strategies with the explanation of the multi-grid procedure for solving the implicit equations (1.5a,b).

1.3 EXISTING MULTI-GRID APPROACHES

Eq (1.5a,b) is a discrete elliptic problem for the unknown grid function $u_l(t)$. Hence, it can be solved by one or more iterations of the usual multi-grid methods for solving stationary problems (cf [3]).

A second approach is the 'frozen τ-technique' of Brandt [2]. It is restricted to the case, where $u(t)$ approaches the stationary limit $u_\infty = \lim\{u(t): t \to \infty\}$ and exploits the smoothness of the remainder $u(t)-u_\infty$. Numerical results are reported by Kroll [6].

The approach explained below will be able to converge to the discrete solution of (1.5) regardless of the smoothness or non-smoothness of the solution. But if $u_l(t)$ changes smoothly in time, it will be possible to save a part of the computational work.

2. PARABOLIC MULTI-GRID ITERATION

The conventional approach is to solve Eq (1.5) time step by time step; $u_l(t)$ is computed from $u_l(t-\Delta t_l)$, then $u_l(t+\Delta t_l)$ from $u_l(t)$ etc. The following process will be different. Assume that $u_l(t)$ is already computed or given as initial state. Simultaneously, we shall solve for $u_l(t+\Delta t_l)$, $u_l(t+2\Delta t_l)$, ..., $u_l(t+k\Delta t_l)$ in one step of the algorithm. The following step of the algorithm will yield $u_l(t+(k+1)\Delta t_l)$, ..., $u_l(t+2k\Delta t_l)$ etc. The number k may depend on l if Δt_l does so:

$$k = k_l = \delta t \,/\, \Delta t_l, \tag{2.1}$$

where δt is the time length by which the computation proceeds per step. In §2.1 we shall consider the case of $\Delta t_l = \Delta t_{l-1} = \ldots = \Delta t_o$, while the coarsening in time ($\Delta t_l < \Delta t_{l-1}$) is discussed in §2.2.

2.1 CASE OF EQUAL TIME STEPS

For simplicity consider only the two levels l and l-1 with time steps $\Delta t_l = \Delta t_{l-1}$, whereas the spatial grid size is coarsened, eg by $\Delta x_{l-1} = 2\Delta x_l$. Fig 2.1a,b shows the situation for k=4 in the 1-D case.

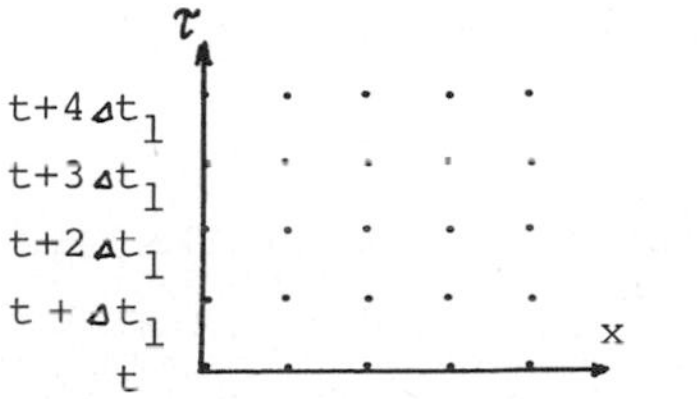

Fig 2.1a: Grid at level l

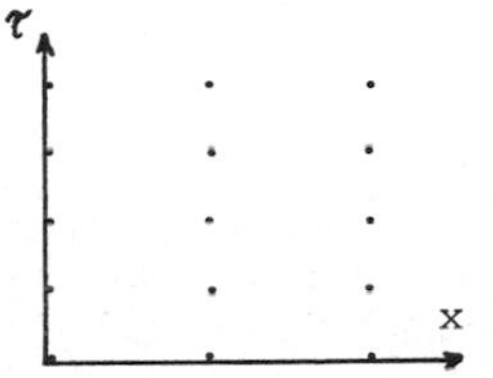

Fig 2.1b: Grid at level l-1

Let $u_l^{(\nu)} = S_{l,stat}(t, u_l^{(\nu-1)}, f_l)$ be a stationary smoothing iteration as used in (1.2), where L_l is replaced by $L_l(t) + I/\Delta t_l$. Eg, $S_{l,stat}$ may be the Gauß-Seidel iteration. The corresponding iteration applied to Eq (1.5) is $S_l = S_l(t, u_l(t), u_l(t-\Delta t_l), f_l(t))$ defined by

$$S_l(t, u_l, u_l', f_l) = S_{l,stat}(t, u_l, f_l + u_l'/\Delta t_l). \tag{2.2}$$

$u_l^{(\nu)} = S_l(t, u_l^{(\nu-1)}, u_l', f_l)$ is an iteration for $u_l = u_l(t)$, whereas $u_l' = u_l(t-\Delta t_l)$ is fixed.

We recall that the multi-grid algorithm is defined by the smoothing step and the coarse-grid correction. In the parabolic case the smoothing step consists of ν applications of S_l to $u_l(t+\Delta t_l), \ldots, u_l(t+k\Delta t_l)$; eg by

$$\begin{aligned}
&\text{for } \tau := t+\Delta t_l \text{ step } \Delta t_l \text{ until } t+k\Delta t_l \text{ do}\\
&\text{for } j := 1 \text{ step } 1 \text{ until } \nu \text{ do}\\
&u_l(\tau) := S_l(\tau, u_l(\tau), u_l(\tau-\Delta t_l), f_l(\tau))
\end{aligned} \tag{2.3}$$

The coarse-grid correction of the parabolic multi-grid iteration reads as follows:

1) compute the defects

$$d_l(\tau) := P_l(\tau, u_l(\tau), u_l(\tau-\Delta t_l)) - f_l(\tau) \tag{2.4a}$$

for $\tau = t+j\Delta t_l$, $1 \le j \le k$ (d_l is defined on the grid of Fig 2.1a);

2) restrict the defects to grid functions of level l-1:

$$d_{l-1}(\tau) := r d_l(\tau), \quad \tau = t+j\Delta t_l, \quad 1 \le j \le k, \tag{2.4b}$$

where r is chosen as mentioned in §1.1 (d_{l-1} is defined on the grid of Fig 2.1b);

3) solve the following coarse-grid equation (parabolic initial-value problem):

$$\begin{aligned}
&v_{l-1}(t) = o,\\
&P_{l-1}(\tau, v_{l-1}(\tau), v_{l-1}(\tau-\Delta t_l)) = d_{l-1}(\tau), \quad \tau = t+j\Delta t_l, \quad 1 \le j \le k;
\end{aligned} \tag{2.4c}$$

4) prolongate v_{l-1} by

$$v_l(\tau) := p\cdot v_{l-1}(\tau), \quad \tau=t+j\Delta t_l, \quad 1\le j\le k \tag{2.4d}$$

with p from §1.1;

5) correct $u_l(\tau)$ by

$$u_l(\tau) := u_l(\tau) - v_l(\tau), \quad \tau=t+j\Delta t_l, \quad 1\le j\le k. \tag{2.4e}$$

The different time levels are coupled by (2.4c). The following note applies to the other parts of the coarse-grid correction:

NOTE 2.1 The steps (2.4a,b,d,e) of the coarse-grid correction can be performed by k parallel processors.

In the two-grid case the coarse-grid equation (2.4c) is solved exactly. If we replace the exact solution of (2.4c) by γ (cf (1.2)) iterations of the multi-grid process for the parabolic problem at level $l-1$ we obtain the multi-grid iteration for solving (1.5) at $\tau=t+j\Delta t_l$, $1\le j\le k$.

In the process (2.5a-e) we did not mention the boundary conditions explicitly. If $u_l(\tau)$ satisfies some Dirichlet conditions, $v_{l-1}(\tau)$ from (2.4c) has to fulfil homogeneous Dirichlet values. Then, the correction (2.4e) does not destroy the boundary conditions.

REMARK 2.2 It can be proved that the two-grid iteration (2.3), (2.4a-e) as well as the corresponding multi-grid iteration converge with the typical speed of multi-grid iterations. The contraction number is bounded by $\rho \ll 1$ independently of the step sizes Δt_l and Δx_l. This holds in particular for large ratios $\Delta t_l/\Delta x_l^2$, where eg the Gauß-Seidel iteration is very slow.

2.2 CASE OF COARSER TIME STEPS

Consider the case of Fig 2.2, where $\Delta t_{l-1}=4\Delta t_l$. A coarsening with

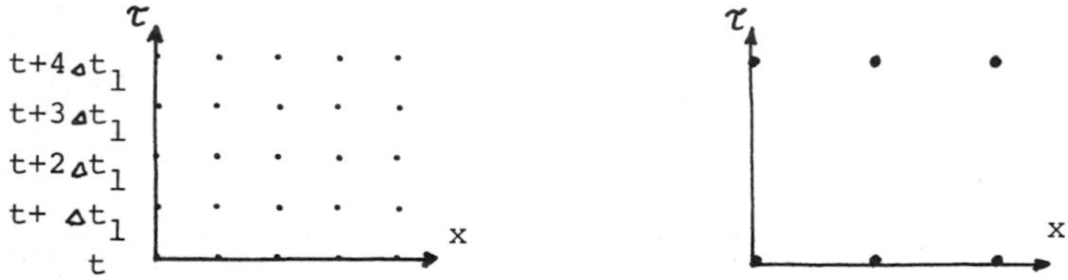

Fig 2.2a: Grid at level l   Fig 2.2b: Grid at level l-1

respect to all (space and time) variables seemes to be natural for a multi-grid algorithm. The smoothing step is again (2.3). The coarse-grid correction (2.4) has rather to be changed. (2.4b) becomes

$$d_{l-1}(\tau) := rd_l(\tau), \quad \tau=t+j\Delta t_{l-1}, \quad 1\le j\le k_{l-1}, \tag{2.4b'}$$

where r is a restriction also in time. k_l and k_{l-1} are connected by $k_l\Delta t_l=k_{l-1}\Delta t_{l-1}=\delta t$ (cf (2.1)). In case of Fig 2.2 we have $k_l=4$ and $k_{l-1}=1$. Analogously, the coarse-grid equation has to be solved in steps of width Δt_{l-1}:

$$v_{l-1}(t) = o, \tag{2.4c'}$$
$$P_{l-1}(\tau, v_{l-1}(\tau), v_{l-1}(\tau-\Delta t_{l-1})) = d_{l-1}(\tau), \quad \tau=t+j\Delta t_{l-1}, \quad 1\le j\le k_{l-1}.$$

Step (2.4d) is formally the same as before, but p is a prolongation also in time direction. (2.4e) remains unchanged.

If $k_{l-1}=1$, a very simple restriction is given by

$$d_{l-1}(t+\delta t) := r_{spatial}\cdot d_l(t+\delta t),\quad \delta t=\Delta t_{l-1}=k_l\Delta t_l, \tag{2.4b''}$$

with no weighting in time direction. In that case the defect (2.4a) has to be determined in $t+\delta t$, only.

REMARK 2.3 The two-grid iteration (2.3), (2.4a,b',c',d,e) does not converge with the usual multi-grid speed.

The reason of slow convergence (or divergence) are errors which are smooth in the spatial directions but nonsmooth in time. Assume, eg, in case of Fig 2.2 that the errors at level l are $e_l(t+j\Delta t_l)=o$ for $j=o$ and $j=2,3,4$, but $e_l(t+\Delta t_l)=v_l$ ($o\neq v_l$ smooth). Since the defects are smooth in space, the smoothing step does not improve the solution, significantly. On the other hand the coarse-grid correction does not provide a good correction because of $\Delta t_{l-1} > \Delta t_l$.

However, if the defects $d_l(\tau)$ are smooth in time, the coarse-grid correction (with or without preceding smoothing step) yields an efficient improvement.

REMARK 2.4 Assume that the exact discrete solution $u_l(\tau)$, $\tau=t+j\Delta t_l$, $o\le j\le k_l$, as well as the starting guess $u_l^o(\tau)$ is smooth in time. Then one step of the two-grid iteration (2.3), (2.4a,b',c',d,e) (with coarser time step $\Delta t_{l-1}>\Delta t_l$) yields a much better next iterate $u_l^{(1)}(\tau)$.

Proof. By assumption the error $u_l(\tau)-u_l^{(o)}(\tau)$ and thereby the defect is smooth.

After one iteration nonsmooth error components may dominate so that further iterations does not give the same success as the first one (cf Remark 2.3).

So far we only mentioned the two-grid iterations. There are different possibilities for the multi-grid iteration depending on the choice of $\Delta t_{l-2},\ldots,\Delta t_o$. One may proceed with $\Delta t_o=\ldots=\Delta t_{l-2}=\Delta t_{l-1}$ (this is the only choice if $\Delta t_{l-1}=\delta t$, ie $k_{l-1}=1$). Then, the iterations at lower levels are multi-grid iterations as explained in §2.1. The choice $\Delta t_{l-2}>\Delta t_{l-1}$ might give rise to slow convergence because of Remark 2.3 unless the defects $d_{l-1}(\tau)$ are smooth in time.

2.3 STARTING VALUES

The multi-grid iteration (either from §2.1 or 2.2) requires starting iterates $u_l^o(\tau)$, $\tau=t+j\Delta t_l$, $o\le j\le k_l$. A possible simple choice is

$$u_l^{(o)}(t+j\Delta t_l) = u_l(t),\quad o\le j\le k_l, \tag{2.5}$$

where $u_l(t)$ is the initial value given either by the previous computation or by the initial value of the given parabolic problem. The choice (2.5) requires that the boundary values are constant in time. Otherwise, the boundary values have to be corrected in (2.5).

Another cheap approach corresponds to the technique of the nested iteration (1.3). By the simple injection to the coarser grid we can define initial values

$$v_{l-1}(t) := r_{injection}\cdot u_l(t) \tag{2.6a}$$

at time t. Solve the discrete parabolic problem at level l-1 by the multi-grid process:

Let $v_{l-1}(\tau)$, $\tau=t+j\Delta t_{l-1}$, $1\le j\le k_{l-1}$, be an approximate solution of $P_{l-1}(\tau,v_{l-1}(\tau),v_{l-1}(\tau-\Delta t_{l-1}))=f_{l-1}(\tau)$, (2.6b)

where possibly $\Delta t_{l-1}>\Delta t_l$ (cf Fig 2.1 or Fig 2.2, resp). Calculate

$$\left(v_l(t+\Delta t_l),\dots,v_l(t+\delta t)\right) := \tilde{p}\left(v_{l-1}(t+\Delta t_{l-1},\dots,v_{l-1}(t+\delta t)\right), \quad (2.6c)$$

where $\tilde{p}$ describes a suitable interpolation of the coarse-grid function v_{l-1} in the fine grid. If $\Delta t_l=\Delta t_{l-1}$, $\tilde{p}$ involves only a spatial interpolation. v_l is almost the starting guess at level l. However, $v_l(t)$ is not equal to $u_l(t)$ in general. Therefore, the final definition reads as

$$u_l^{(o)}(\tau) := v_l(\tau) + u_l(t) - v_l(t), \quad \tau=t+j\Delta t_l, \quad o\le j\le k_l. \quad (2.6d)$$

If v_l satisfies the Dirichlet boundary conditions, then $u_l^{(o)}$ does so, too. In more complicated cases one has to correct the boundary conditions suitably.

2.4 COMPLETE ALGORITHM

Combining the various parts explained above we obtain the following algorithms for the computation of the solution in $[t,t+\delta t]$.

first variant: (2.7)
a) compute the starting iterate $u_l^{(o)}$ by (2.6a-d),
b) apply i multi-grid iterations with $\Delta t_l=\Delta t_{l-1}$ (cf §2.1).

Due to Remark 2.2 the iteration converges fast. The computational work can be reduced by choosing $\Delta t_{l-1}>\Delta t_l$ in the first step of the iteration. According to Remark 2.4, the error of $u_l^{(o)}$ can greatly be reduced by one step with coarser time width Δt_{l-1}, provided that $u_l^{(o)}$ and u_l are smooth in time. The resulting algorithm is the

second variant: (2.8)
a) compute the starting iterate $u_l^{(o)}$ by (2.6a-d),
b) apply one multi-grid iteration with $\Delta t_{l-1}>\Delta t_l$ (cf §2.2),
c) apply i multi-grid iterations with $\Delta t_{l-1}=\Delta t_l$ (cf §2.1).

The number i of iterations should be as small as possible. The smallest choice i=o yields the

third variant: (2.9)
a) compute the starting iterate $u_l^{(o)}$ by (2.6a-d),
b) apply one multi-grid iteration with $\Delta t_{l-1}>\Delta t_l$ (cf §2.2).

In contrast to algorithms (2.6) and (2.7), the latter version is no iteration. It yields a result $u_l^{(1)}$, which is different from the exact solution of the discrete problem (1.5). Algorithm (2.9) defines a new discretization, which has almost the accuracy of the discretization (1.5) and is much more accurate than the algorithm (2.7) with i=o. Note that algorithm (2.7) with i=o yields the solution corresponding to the time step size Δt_{l-1} instead of Δt_l.

2.5 MODIFICATIONS

REMARK 2.5 The algorithms described above apply also to other discretizations, eg to the Crank-Nicholson scheme.

After some obvious modifications the algorithm can also be applied to multi-step discretizations.

In Note 2.1 we mentioned that the most parts of the coarse-grid correction can be performed in parallel. This does not hold for the smoothing iteration defined by (2.3). However, if the smoothing is performed by

$$\begin{aligned}&\text{for } \tau:=t+\delta t \text{ step } -\Delta t_l \text{ until } t+\Delta t_l \text{ do}\\ &\text{for } j:=1 \text{ step } 1 \text{ until } \nu \text{ do}\\ &u_l(\tau):=S_l(\tau,u_l(\tau),u_l(\tau-\Delta t_l),f_l(\tau))\end{aligned} \qquad (2.1o)$$

instead of (2.3), the computations at all k different time steps $t+j\Delta t_l$ are independent:

REMARK 2.6 In case of (2.1o) the smoothing step can be performed by k parallel processors.

Does the multi-grid iteration converge if (2.3) is replaced with (2.1o)? Assume that the errors are $e_l(t+j\Delta t_l)=\delta_{jm}v_l$ (δ_{jm}: Kronecker's symbol; $1\leq m\leq k_l$). Let v_l be an eigenfunction of $L_l(t)$. If v_l is of low frequency (ie, if the corresponding eigenvalue is small), this error will be reduced by the subsequent coarse-grid correction. Therefore, assume that v_l is of high frequency and that the sequential smoothing iteration (2.3) at time $t+m\Delta t_l$ reduces the error v_l to ρv_l with $\rho \ll 1$. The errors at $t+(m+1)\Delta t_l,\ldots,t+\delta t$ generated by the further steps of (2.3) are even smaller. The result of the parallel smoothing (2.1o) is different. Again, the new error at time $t+m\Delta t_l$ is ρv_l. The starting defect at time $t+(m+1)\Delta t_l$ is $-v_l/\Delta t_l$, which is also reduced by the factor ρ. Hence, the new error at time $t+(m+1)\Delta t_l$ is αv_l with $\alpha=(1-\rho)/(1+\Delta t_l\lambda)$, where $\lambda v_l=L_lv_l$. Since ρ and λ are related by $1-\rho\approx \text{const}\cdot\lambda\cdot\Delta x_l^2$, one concludes that $\alpha = O(\Delta x_l^2/\Delta t_l)$.

REMARK 2.7 If $\Delta t_l/\Delta x_l^2 \gg 1$ the parallel smoothing (2.1o) is successful. Note that $\Delta t_l/\Delta x_l^2 \gg 1$ is just the case, where implicit schemes are interesting because of their stability.

Although we want to avoid the discussion of possible time stepping strategies, we state that the present multi-grid algorithms yield the necessary dates.

REMARK 2.8 The starting step a) of (2.7), (2.8), or (2.9) yields $u(x,\tau;\Delta x_{l-1},\Delta t_{l-1})$. Eg, we can first compute $u(x,\tau;\Delta x_{l-1},\delta t)$ for $\Delta t_{l-1}=\delta t$, which serves as starting guess for the computation of $u(x,\tau;\Delta x_{l-1},\Delta t_l)$ with $\Delta t_l<\delta t$. Then, the remaining part of the algorithm yields $u(x,\tau;\Delta x_l,\Delta t_l)$. By these three values one can determine how the error depends on Δx and Δt. Hence, new adapted grid sizes can be guessed.

Stationary multi-grid methods can be applied not only to the linear problem (1.1) but also to nonlinear equations $L_l(u_l)=f_l$. The nonlinear multi-grid iterations is described, eg, in [2] and [4].

REMARK 2.9 Replacing the linear multi-grid scheme (1.2) by the nonlinear iteration, one can formulate nonlinear versions of the algorithms (2.7), (2.8), and (2.9).

3. NUMERICAL EXAMPLE: BUOYANCY-DRIVEN FLOW

We consider the time-dependent natural convection in a square cavity, which can be described by

$$-\Delta\psi_t + \Pr\Delta^2\psi + \psi_y\Delta\psi_x - \psi_x\Delta\psi_y - \text{Ra}\cdot\Pr\cdot T_x = 0, \qquad (3.1a)$$

$$T_t - \Delta T + \psi_x T_y - \psi_y T_x = 0 \qquad (3.1b)$$

in $\Omega := \{(x,y): 0<x<1,\ 0<y<1\}$ and $0\le t\le 1$, where ψ: stream function, T: temperature, Ra=1000: Rayleigh number, Pr=0.71: Prandtl number. The boundary conditions are

$$\psi = \partial\psi/\partial n = 0 \quad \text{on } \Gamma = \text{boundary of } \Omega,\ 0\le t\le 1, \tag{3.2}$$

$$T=0 \text{ at } x=0, \quad \partial T/\partial n = 0 \text{ at } y=0 \text{ and } y=1. \tag{3.3a}$$

The right wall is heated according to

$$T(1,y,t) = t(2-t) \qquad \text{for } 0\le y\le 1,\ t\ge 0. \tag{3.3b}$$

The initial value is

$$\psi(x,y,0) = T(x,y,0) = 0 \quad \text{for } (x,y)\in\Omega. \tag{3.4}$$

Although Eq (3.1) is not of the form of Eq (1.4), the equations (3.1 - 4) describe a well-behaving parabolic initial-boundary value problem.

The equations (3.1a,b) are discretized by central differences, Δ^2 is replaced by the square of the five-point scheme. The details of the multi-grid algorithm are as follows:

- smoothing iteration S_1: Gauß-Seidel iteration with chequer-board ordering of the grid points;
- p: piecewise linear interpolation for T (socalled nine-point prologation, cf [4]), piecewise cubic interpolation for ψ;
- r: for both T and ψ the restriction is chosen as nine-point restriction (adjoint of nine-point prolongation; cf [4]);
- $\nu=2$: number of smoothing iterations;
- $\gamma=1$: socalled V-cycle;
- $\Delta x_0=1/2$, $\Delta x_1=1/4$, $\Delta x_2=1/8$, $\Delta x_3=1/16$; $\Delta t_0=\ldots=\Delta t_3=1/4$; k=4;
- the nonlinearity is treated according to Remark 2.9.

The following table contains the values of ψ, T, and of the vorticity $\zeta=-\Delta\psi$ at the midpoint (1/2,1/2). The last column indicates the computational work measured by the unit 1W, which is one Gauß-Seidel iteration or one evaluation of the defects at the finest grid with $\Delta x=1/16$. Rows 1 to 3 show the exact values for different grid sizes. One concludes that the discretization error with respect to Δt is smaller than the error arising from the spatial grid width. The further rows correspond to the finest grid parameters $\Delta t=1/4$, $\Delta x=1/16$ and different numbers of iterations:

- run A: multi-grid version of (2.7) with i=2;
- run B: same with i=3;
- run C: multi-grid version of (2.8) with two iterations in b) and two iterations in c);
- run D: same as C with three iterations in b) and c);
- run E: same as D, but the sequential smoothing (2.3) is replaced by the parallel smoothing (2.10).

One observes that the parallel smoothing (run E) yields almost the same results as the sequential smoothing (2.3). Even with a smaller number of iterations one obtaines good approximations to T.

parameters	$\psi(1/2,1/2)$	T(1/2,1/2)	$\zeta(1/2.1/2)$	W
1 $\Delta t=1/4$, $\Delta x=1/16$; exact	1.203050	0.473829	31.86	
2 $\Delta t=1/8$, $\Delta x=1/16$; exact	1.208916	0.480620	32.04	
3 $\Delta t=1/4$, $\Delta x=1/8$; exact	1.317663	0.473752	32.55	
4 $\Delta t=1/4$, $\Delta x=1/16$; run A	1.3398	0.4761	-	39
5 $\Delta t=1/4$, $\Delta x=1/16$; run B	1.2226	0.4749	27.7	56
6 $\Delta t=1/4$, $\Delta x=1/16$; run C	1.1260	0.4739	22.1	44
7 $\Delta t=1/4$, $\Delta x=1/16$; run D	1.2090	0.4739	32.2	66
8 $\Delta t=1/4$, $\Delta x=1/16$; run E	1.2121	0.4738	32.2	66

REFERENCES

[1] Brandt, A., Multi-level adaptive solutions to boundary-value problems, Math. Comp. 31 (1977) 333-390.

[2] Brandt, A., Multi-level adaptive finite-element methods: variational problems, in: Frehse, J., Pallaschke, D., and Trottenberg, U. (eds.), Special topics of applied mathematics (North-Holland, Amsterdam, 1980).

[3] Hackbusch, W., A fast numerical method for elliptic boundary value problems with variable coefficients, in: Hirschel, E. H. and Geller, W. (eds.), Second GAMM conference on numerical methods in fluid mechanics (DFVLR, Köln-Porz, 1977).

[4] Hackbusch, W., Introduction to multi-grid methods for the numerical solution of boundary value problems, in: Essers, J.A. (ed.), Computational methods for turbulent, transonic, and viscous flows (Hemisphere, Washington, 1983).

[5] Hackbusch, W., and Trottenberg, U. (eds.), Multigrid methods (Springer, Berlin, 1982).

[6] Kroll, N., Direkte Anwendung von Mehrgittertechniken auf parabolische Anfangsrandwertaufgaben, Diplomarbeit, Bonn (1981).

[7] van der Houwen, P.J., and de Vries, H.B., Preconditioning and coarse grid corrections in the solution of the initial value problem for nonlinear partial differential equations, SIAM J. Sci. Stat. Comput. 3 (1982) 473-485.

Computing Methods in Applied Sciences and Engineering, VI
R. Glowinski and J.-L. Lions (Editors)
Elsevier Science Publishers B.V. (North-Holland)

MULTIGRID ACCELERATION OF AN ITERATIVE METHOD WITH APPLICATION TO TRANSONIC POTENTIAL FLOW

Z. Nowak
Politechn. Warsz. ITLiMS
00-665 Warsaw, Nowowiejska 24, Poland
and
P. Wesseling
Dept. of Math. and Inf., Delft Un. of Technology
P.O. Box 356, 2600 AJ Delft, The Netherlands

The full transonic potential equation is solved numerically with the aid of a multigrid method. The multigrid part of the method is completely divorced from the handling of the nonlinearity, which makes the method relatively simple. A brief outline is given of algebraic aspects of multigrid acceleration of iterative methods for linear problems. The potential equation is discretized with the finite volume technique, using the retarded density concept. The physical domain is mapped numerically onto a rectangle. Results are presented for flows around the NACA 0012 airfoil.

INTRODUCTION

The purpose of this paper is to describe an application of a multigrid method to the full transonic potential equation. The multigrid method used is based essentially on the MGD1 method described in [15, 16] and further developed and tested in [5, 6, 9]. This is a method to accelerate the convergence of a stationary iterative method for a linear algebraic system arising from the discretization of a second order partial differential equation. It is easy to code such a multigrid method such that the user needs to provide only the matrix and the right-hand side, so that the code is perceived by the user just like any other standard subroutine for solving linear systems. The user remains unaware of the underlying multigrid algorithm.

One may well ask whether such a multigrid code would be versatile enough to deal effectively with difficult nonlinear problems, such as for example the full transonic potential equation, because the user cannot adapt the multigrid part of his code to and tune it for the problem at hand. This is what has (sometimes painstakingly) been done in existing successfull multigrid methods for transonic potential flow, e.g. [1, 8]. The present paper aims to investigate this question.

LINEAR MULTIGRID METHODS

By linear multigrid methods we mean multigrid methods for solving linear algebraic systems, denoted by

$$A\phi = b. \tag{1}$$

We will give a brief outline of a theoretical framework for linear multigrid methods.

The work of the first author was sponsered by the Netherlands Organization for the Advancement of Pure Research (Z.W.O.).

Equation (1) is assumed to represent a discretization of a partial differential equation on a computational grid G. The set of grid-functions $G \to \mathbb{R}$ is denoted by Φ. We start by discussing two-grid methods. There is also given a coarser grid $\bar{G}$; the corresponding set of grid-functions is denoted by $\bar{\Phi}$. The coarse grid approximation of (1) is

$$\bar{A}\bar{\phi} = \bar{b}. \tag{2}$$

Furthermore, let there be given a prolongation operator P and a restriction operator R:

$$P : \bar{\Phi} \to \Phi \; ; \quad R : \Phi \to \bar{\Phi}. \tag{3}$$

A linear two-grid method can then be described as follows. Let u^j be the current iterand, and $u^{j+\frac{1}{2}}$ the result of applying a coarse grid correction to u^j, as follows:

$$\phi^{j+\frac{1}{2}} = \phi^j - P\,\bar{A}^{-1} R\,(A\phi^j - f)\;, \tag{4}$$

where we assume that the coarse grid problem is solved exactly. For the residue $r^j = A\phi^j - b$ we find:

$$r^{j+\frac{1}{2}} = (I - AP\bar{A}^{-1}R)\; r^j \;. \tag{5}$$

We now make the following choice for $\bar{A}$ (Galerkin approximation):

$$\bar{A} = RAP. \tag{6}$$

Then it follows from (5) that

$$r^{j+\frac{1}{2}} \in \mathrm{Ker}(R) \tag{7}$$

as noted in [12]. In other words, $r^{j+\frac{1}{2}} \perp \mathrm{Ker}^{\perp}(R)$, which justifies the appellation "Galerkin approximation" for (6). We now define:

<u>Definition 1</u>. The set of R-smooth grid-functions is $\mathrm{Ker}^{\perp}(R)$.

Whether the grid-functions just defined are also what one would call physically smooth, or, a little more precise, smooth in the sense of the Fourier analysis proposed in [2], will depend on the choice made for R.

It remains to annihilate the non-smooth part of r, and this is done in the second part of the two-grid iteration, called smoothing. The smoothing process usually is some classical stationary iterative method, which can generally be denoted as

$$\phi^{j+1} = (I - BA)\phi^{j+\frac{1}{2}} + Bb, \tag{8}$$

where B depends on the choice one makes for the smoothing process. For the residue we find

$$r^{j+1} = (I - AB)\; r^{j+\frac{1}{2}}. \tag{9}$$

The projection operator on Ker(R) is given by $I - R^T(RR^T)^{-1}R$, and we may conclude that

$$||r^{j+1}|| \leq ||(I-AB)(I-R^T(RR^T)^{-1}R)||\; ||r^{j+\frac{1}{2}}|| \;. \tag{10}$$

This leads us to the following definition:

Definition 2. The R-smoothing factor of the smoothing process (8) is

$$\mu_R := ||(I-AB)(I-R^T(RR^T)^{-1}R||\ .$$

Whether this smoothing factor will be approximately equal to the smoothing factor defined in [2] depends on whether Ker(R) corresponds to non-smooth functions in Fourier-space, and on the applicability of the Fourier-analysis on which Brandt's smoothing analysis is based. This Fourier analysis presupposes constant or at least locally smooth coefficients in the given differential problem, and, furthermore, that the influence of the boundary conditions is not felt in the interior of the domain. Therefore the smoothing factor of definition 2 is more accurate, but it is also more difficult to compute.

We now take the dual viewpoint of considering the error instead of the residue. We define

Definition 3. The set of P-smooth grid-functions is Range (P).

Let the error e^j be defined by $e^j = \phi^j - \phi$. Then it follows from (4) that

$$e^{j+\frac{1}{2}} = (I-P\bar{A}^{-1}RA)\ e^j\ . \tag{11}$$

A streamlined line of reasoning is obtained if we now assume that smoothing precedes coarse grid correction, so that (8) is replaced by

$$\phi^j = (I-BA)\phi^{j-\frac{1}{2}} + Bb\ . \tag{12}$$

Let $e^j = e_1^j + e_2^j$ with $e_1^j \subset$ Range(P), $e_2^j \subset \text{Range}^{\perp}(P)$. Again choosing $\bar{A}_j$ according to (6) we see that $(I-P\bar{A}^{-1}RA)e_1^j = 0$ (write $e_1^j = P\hat{e}$ for some $\hat{e}$), so that

$$e^{j+\frac{1}{2}} = (I-P\bar{A}^{-1}RA)e_2^j \tag{13}$$

This will in general be small only if e_2^j is small, which motivates the following

Definition 4. The P-smoothing factor of the smoothing process (12) is

$$\mu_p = ||(I-P)P^TP)^{-1}P^T)(I-BA)||\ .$$

(Note that the projection operator on $\text{Range}^{\perp}(P)$ is given by $I-P(P^TP)^{-1}P^T$).

In the special case that $R = P^T$ we have that the sets of R-smooth and of R-smooth grid functions are identical, since $\text{Ker}^{\perp}(R) = \text{Range}(R^T)$, but μ_R and μ_P will in general not be identical. The quantity μ_P is defined and studied in [12].

The two-grid method defined by (4), (8) or by (4), (12) can be regarded as an acceleration method for the iterative process (8). The striking efficiency of multigrid methods is due to the fact that there exist simple iterative methods of type (8) for which μ_R and μ_P are well below 1 for large classes of problems.

DETAILS OF THE MULTIGRID METHOD USED

The multigrid method to be used here is a modification of the method MGD1, called MGD4. For details on MGD1 we refer to [5, 6, 15, 16]. MGD4 assumes a 9-point finite difference molecule instead of a 7-point molecule as in MGD1, for reasons of symmetry: with a 7-point molecule we obtained lift differing slightly from zero for symmetric flows.

MGD4 is based on (4), (8), with coarse grid approximation according to (6). The coarse grid equation is not solved exactly of course, but approximately, with one two-grid iteration employing an additional coarser grid, and so on recursively till the coarsest grid is reached, where we solve approximately with three iterations (8). The resulting multigrid method is said to be of sawtooth type, because its schedule (the switching strategy between the grids) is represented in a natural way by the diagram given in fig. 1, exhibiting a sawtooth curve.

The computational grid is assumed to be the rectangle $[0,a] \times [0,b]$ in the ξ-η plane. On $\eta = 0,b$ we have Dirichlet or Neumann conditions. On $\xi = 0,a$ we have periodicity: either

$$\phi(\xi+a,\eta) = \phi(\xi,\eta) \tag{14}$$

or

$$\phi(\xi+a,\eta) = \phi(\xi,\eta) + c \tag{15}$$

(periodicity modulo a constant increment). Coarser grids are obtained by removing every other horizontal and vertical grid line. The lines $\eta = 0,b$ and $\xi = 0,a$ are common to all grids; the unknowns at $\xi = a$ are eliminated with (14) or (15). P and R are bilinear interpolation in grid cells and its adjoint (full weighting) respectively; for exact details see for example [15] (9-point prolongation and restriction). As a result the coarse grid operators generated by (6) have the same 9-point structure as the finest grid operator, which will be discussed in more detail in the sequel. Where necessary all operators are periodically extended near $\xi = 0,a$. As in MGD1, smoothing is done by means of an incomplete LU-decomposition, modified to take periodicity into account. Details are given in the following section.

In order to get some superficial insight in the computational cost of one MGD4 iteration we counted arithmetic operations. Application of P, R, C or $(LU)^{-1}$ to a grid-function on a given grid takes 8, 11, 11, 23 operations per grid-point; we chose not to store C. For the total cost of one MGD4 iteration an operation count of 66 per finest grid-point was obtained, assuming a large number of grids. One execution of (6) takes 339 operations per point of $\bar{G}$; hence the total cost of generating all coarse grid operators is 339 (1/4+1/16+...) = 113 operations per point of the finest grid. The generation of all ILU decompositions in the way described in the next section takes 37 operations per finest grid-point.

PERIODIC INCOMPLETE LU DECOMPOSITIONS

Incomplete LU (ILU) decomposition will be used in the smoothing process (8). Lower and upper triangular matrices L and U are constructed such that

$$LU = A + C \tag{16}$$

with C the so-called error matrix. The smoothing process (8) is chosen as follows:

$$u^{j+1} = (I-(LU)^{-1}A)u^{j+\frac{1}{2}} + (LU)^{-1}f ,$$

or

$$LUu^{j+1} = Cu^{j+\frac{1}{2}} + f . \tag{17}$$

For the construction of ILU decompositions and their use in multigrid methods, see for example [5, 6, 9, 13, 15, 16]. In the application that we wish to consider periodic boundary conditions occur. Therefore we present here a brief discription of periodic ILU decompositions. For a sparsity pattern corresponding to the standard 5-point molecule, [13] gives detailed formulae for the construction of L and U in the periodic case. We will briefly describe a periodic ILU decomposition for a 9-point case.

Let the computational grid be the unit square in the ξ-η plane, with uniform mesh sizes $\Delta\xi$ and $\Delta\eta$. In order to obtain an elegant presentation, the grid is thought to be extended in the ξ-directions, so that we obtain the infinite strip $(-\infty,\infty)\times[0,1]$. The periodicity condition is given by (14) or (15).

Let $S_\xi^{\pm}$, $S_\eta^{\pm}$ be the shift operators defined by

$$S_\xi^{\pm}\phi(\xi,\eta) = \phi(\xi\pm\Delta\xi,\eta),$$
$$S_\eta^{\pm}\phi(\xi,\eta) = \phi(\xi,\eta\pm\Delta\eta). \tag{18}$$

Let the equation to be solved $A\phi = b$ arise from a 9-point discretization. Instead of the usual matrix notation the operator A can with the aid of the shift operators just defined also be represented as

$$A = a_1 S_\xi^- S_\eta^- + a_2 S_\eta^- + a_3 S_\xi^+ S_\eta^- + a_4 S_\xi^- + a_5 I$$
$$+ a_6 S_\xi^+ + a_7 S_\xi^- S_\eta^+ + a_8 S_\eta^+ + a_9 S_\xi^+ S_\eta^+ , \tag{19}$$

with I the identity operator, and $a_1, a_2, \ldots, a_9$ grid-functions. The domain of A is the set of grid-functions defined on the infinite strip introduced before. The L and U factors are chosen as follows:

$$L = \ell_1 S_\xi^- S_\eta^- + \ell_2 S_\eta^- + \ell_3 S_\xi^+ S_\eta^- + \ell_4 S_\xi^- + I , \tag{20}$$

$$U = u_1 I + u_2 S_\xi^+ + u_3 S_\xi^- S_\eta^+ + u_4 S_\eta^+ + u_5 S_\xi^+ S_\eta^+ . \tag{21}$$

We now require

$$LU = A + C , \tag{22}$$

with C an operator in which the coefficients of those (combinations of) shift operators that occur in (19) are zero. In matrix terminology, C has zero elements in the sparsity pattern of A. This leads easily to the following relations:

$$\ell_1 = a_1 / S_\xi^- S_\eta^- u_1 ,$$

$$\ell_2 = (a_2 - \ell_1 S_\xi^- S_\eta^- u_2) / S_\eta^- u_1 ,$$

$$\ell_3 = (a_3 - \ell_2 S_\eta^- u_2) / S_\xi^+ S_\eta^- u_1 ,$$

$$\ell_4 = (a_4 - \ell_1 S_\xi^- S_\eta^- u_4 - \ell_2 S_\eta^- u_3) / S_\xi^- u_1 ,$$

$$u_1 = a_5 - \ell_1 S_\xi^- S_\eta^- u_5 - \ell_2 S_\eta^- u_4 - \ell_3 S_\xi^+ S_\eta^- u_3 - \ell_4 S_\xi^- u_2 \;, \tag{23}$$

$$u_2 = a_6 - \ell_2 S_\eta^- u_5 - \ell_3 S_\xi^+ S_\eta^- u_4 \;,$$

$$u_3 = a_7 - \ell_4 S_\xi^- u_4 \;,$$

$$u_4 = a_8 - \ell_4 S_\xi^- u_5 \;,$$

$$u_5 = a_9 \;.$$

For the operator C we find

$$C = -(\ell_1 S_\xi^- S_\eta^- u_3) S_\xi^- S_\xi^- - (\ell_3 S_\xi^+ S_\eta^- u_2) S_\xi^+ S_\xi^+ S_\eta^-$$
$$-(\ell_3 S_\xi^+ S_\eta^- u_5) S_\xi^+ S_\xi^+ - (\ell_4 S_\xi^- u_3) S_\xi^- S_\xi^- S_\xi^+ \;. \tag{24}$$

The grid-functions $\ell_1, \ell_2, \ldots, u_5$ can be calculated recursively in the unit square in the ξ-η plane by sweeping along η=constant lines in increasing ξ-direction, and taking the lines in order of increasing η. The recursion is started as follows. At $\eta = 0$ we have $a_1 = a_2 = a_3 = 0$, hence $\ell_1 = \ell_2 = \ell_3 = 0$. The boundaries $\xi = 0$ and $\xi = 1$ are handled by requiring that $\ell_1, \ell_2, \ldots, u_5$ are periodic, i.e. satisfy (14); $a_1, a_2, \ldots, a_9$ are also periodically extended according to (14). Along the grid-line $\eta = 0$ we can write

$$\ell_4 = a_4 / S_\xi^- u_1 \;, \tag{25}$$

$$u_1 = a_5 - \ell_4 S_\xi^- a_6 \;. \tag{26}$$

Eliminating ℓ_4 we obtain

$$u_1 = a_5 - a_4 (S_\xi^- a_6) / S_\xi^- u_1 \;, \tag{27}$$

or more precisely, using periodicity

$$u_1(0,0) = a_5(0,0) - a_4(0,0)\; a_6(1-\Delta\xi,0)/u_1(1-\Delta\xi,0),$$
$$u_1(k\Delta\xi,0) = a_5(k\Delta\xi,0) - a_4(k\Delta\xi,0)\; a_6((k-1)\Delta\xi,0)/u_1((k-1)\Delta\xi,0) \;. \tag{28}$$

In [13] is explained how $u_1(0,0)$ can be determined such that the periodicity condition

$$u_1(0,0) = u_1(1,0) \tag{29}$$

is satisfied. We choose a cheaper, approximate method not leading to exactly periodic L and U operators by taking at $(0,\eta)$:

$$\ell_4(0,\eta) = \left\{(a_4 - \ell_1 S_\xi^- S_\eta^- u_4 - \ell_2 S_\eta^- u_3)/S_\xi^- a_5)\right\}\Big|_{(0,\eta)} \tag{30}$$

At other grid-points (23) is applied. As a consequence, the error operator C is at grid-points on $\xi = 0$ enlarged with a term

$$(a_4 - \ell_1 S_\xi^- S_\eta^- u_4 - \ell_2 S_\eta^- u_3 - \ell_4 S_\xi^- u_1) S_\xi^- . \tag{31}$$

It was found that this deviation from exact periodicity in L and U at the cost of an extra term in C did not affect the rate of convergence adversely. Smoothing takes place by solving (17) with simple back-substitution, using at $\xi = 0$ eq. (14) or (15), as the case may be. Along each horizontal line η=constant an auxiliary forward substitution is executed, solving bi-diagonal equations with the periodicity conditions (14) or (15).

THE TRANSONIC FULL POTENTIAL EQUATION

We study two-dimensional transonic potential flow. Let ϕ and ρ be the velocity potential and the density. Let units of velocity and density be the critical sound speed and the stagnation density. Then the mass conservation principle results in the following potential equation in the physical (x,y) plane:

$$(\rho\phi_x)_x + (\rho\phi_y)_y = 0 , \tag{32}$$

with

$$\rho = \{1 - \frac{\gamma-1}{\gamma+1} (\phi_x^2+\phi_y^2)\}^{1/(\gamma-1)} , \tag{33}$$

where γ is the ratio of specific heats, and a subscript denotes differentiation. Let us consider unbounded flow around an airfoil. On the airfoil we have the following Neumann boundary condition:

$$\phi_n = 0 . \tag{34}$$

Furthermore, the velocity magnitude must tend to equal limits as the trailing edge is approached along the upper and the lower side of the airfoil (Kutta condition). At infinity ϕ must tend asymptotically to uniform subsonic stream conditions with (nondimensional) velocity λ_∞ and Mach number M_∞:

$$\phi \cong \lambda_\infty \bar{x} + \frac{\Gamma}{2\pi} \arctan (\sqrt{1 - M_\infty^2}\, \bar{x}/\bar{y}) \tag{35}$$

Here $(\bar{x},\bar{y})$ are Cartesian coordinates with the $\bar{x}$-axis pointing in the direction of the free stream velocity vector, and Γ is the circulation.

TRANSFORMATION TO COMPUTATIONAL RECTANGLE

In order to facilitate accurate implementation of boundary conditions, boundary fitted coordinates are used. This also makes application of the multigrid subroutine MGD4 possible, since this assumes a rectangular computational region with a uniform mesh.

Boundary condition (35) is applied not at infinity but at a circle with radius 6 chord lengths and origin inside the airfoil. The doubly connected physical domain in the (x,y) plane is mapped onto the rectangle $[0,a] \times [0,b]$ in the (ξ,η) plane. The airfoil surface is mapped on $\{[0,a],0\}$, whereas the outer circle is mapped on $\{[0,a],b\}$. The situation is depicted in fig. 2.

The mapping $\xi(x,y)$, $\eta(x,y)$ is constructed with the method described in [14], namely as the solution of the following system of quasilinear elliptic equations:

$$\alpha x_{\xi\xi} - 2\beta x_{\xi\eta} + \gamma x_{\eta\eta} = 0 , \tag{36}$$

$$\alpha y_{\xi\xi} - 2\beta y_{\xi\eta} + \gamma y_{\eta\eta} = 0 \,, \tag{37}$$

where α, β, γ are the following geometrical quantities:

$$\alpha = x_\eta^2 + y_\eta^2 \,, \quad \beta = x_\xi x_\eta + y_\xi y_\eta \,, \quad \gamma = x_\xi^2 + y_\xi^2 \,. \tag{38}$$

At $\eta = 0,b$ Dirichlet boundary conditions are applied, at $\xi = 0,a$ periodicity is required.

Eqs. (36), (37) are discretized with a 7-point finite difference molecule as described for example in [15]. As initial guess we take $\alpha=\gamma=1$, $\beta=0$. The resulting linear system is solved, α, β, γ are updated and so on. The number of iterations needed may be quite large. For the generation of the 128×41 grid shown in fig. 3, 30 iterations were used, at a cost of 2 min. 43 sec. CP time on an Amdahl 470. Convergence was slow near the trailing edge. Probably it would be better to let the numerical mapping be preceded by a simple analytic mapping rectifying the trailing edge. The linear systems were approximately solved with one iteration with the multigrid code MGD3, which is an adaptation of the code MGD1 for periodic boundary conditions.

In the (ξ,η) plane the problem (32), (33) takes the following form (cf.[7]):

$$(\rho U/J)_\xi + (\rho V/J)_\eta = 0 \,, \tag{39}$$

$$\rho = \{1 - \frac{\gamma-1}{\gamma+1} \lambda^2\}^{1/(\gamma-1)} \,, \tag{40}$$

where $U = A_1\phi_\xi + A_2\phi_\eta$, $V = A_2\phi_\xi + A_3\phi_\eta$, $\lambda = |U\phi_\xi + V\phi_\eta|^{\frac{1}{2}}$, $J = 1/(x_\xi y_\eta - y_\xi x_\eta)$,

$A_1 = J^2(x_\eta^2+y_\eta^2)$, $A_2 = -J^2(x_\xi x_\eta + y_\xi y_\eta)$, $A_3 = J^2(x_\xi^2+y_\xi^2)$. U and V are the contravariant components of the velocity, λ is the magnitude of the velocity and J is the Jacobian of the transformation.

The Dirichlet condition (35) is imposed on $\eta = b$. The Neumann condition (2), imposed on $\eta = 0$, takes the following form:

$$A_2\phi_\xi + A_3\phi_\eta = 0 \,. \tag{41}$$

Along $\xi = 0,a$ the following periodicity condition (modulo a constant) is imposed:

$$\phi(a,\eta) - \phi(0,\eta) = \Gamma \,. \tag{42}$$

DISCRETIZATION.

Discretization is done with the familiar control volume method. We give a brief description.

The computational rectangle is covered by a uniform mesh, part of which is shown in fig. 4. Each grid point is the center of a rectangular control volume with sides $\Delta\xi$ and $\Delta\eta$, cf. fig. 4. The first derivatives of ϕ, x, y occurring in (39) are evaluated in the centers of the sides of the control volume (points A, B, C, D in fig. 4), using second order accurate finite difference formulae, containing values of ϕ, x, y at six surrounding grid-point, for example grid-points 2, 3, 5, 6, 8, 9 for A. Hence, at the centers of control volume sides we also have available ρ, λ and the Mach-number, given by

$$M^2 = 2\lambda^2/\{\gamma+1-\lambda^2(\gamma-1)\} \ . \tag{43}$$

In order to satisfy the entropy condition and capture shock waves accurately we use the concept of retarded density [3, 7]. It turns out that the (ξ,η) coordinates are shaped such that the flow is roughly aligned with the ξ-direction, so that we will use density retardation only in the ξ-direction. The density ρ is replace by the retarded density defined by

$$\tilde{\rho} = \rho - \nu(M)\ \Delta\xi\rho_{\overleftarrow{\xi}} \ , \tag{44}$$

with $\rho_{\overleftarrow{\xi}}$ the upwind difference of ρ, and ν the following smooth switching function:

$$\begin{aligned} \nu &= 0 \ , \quad M/M_c < (1+\varepsilon)^{-\frac{1}{2}} \ ; \\ \nu &= ((M_c/M)^2-1-\varepsilon)^2/4\varepsilon \ , \quad (1+\varepsilon)^{-\frac{1}{2}} \leqslant M/M_c < (1-\varepsilon)^{-\frac{1}{2}} \ ; \\ \nu &= 1 - M_c^2/M^2 \ , \quad M/M_c \geqslant (1-\varepsilon)^{-\frac{1}{2}} \ , \end{aligned} \tag{45}$$

where M_c and ε are to be chosen: M_c slightly below unity, ε a small positive constant. We found that better shock resulution is obtained with ν evaluated at grid-points instead of at the centers of control volume sides. For this purpose the Mach number at grid-points is computed according to

$$M_5^2 = (M_A^2 + M_B^2)/2 \ , \tag{46}$$

cf. fig. 4. Eq. (39) is replaced by

$$(\tilde{\rho}U/J)_\xi + (\rho V/J)_\eta = 0 \ . \tag{47}$$

Using the control volume method we obtain the following discretization:

$$(\tilde{\rho}U/J)_A - (\tilde{\rho}U/J)_C + (\rho V/J)_B - (\rho V/J)_D = 0 \ , \tag{48}$$

cf. fig. 4. When $U > 0$, $\tilde{\rho}_A$ and $\tilde{\rho}_C$ are computed using ν_5 and ν_4, respectively; when $U < 0$, ν_6 and ν_5 are used.

When the central point 5 of the control volume under consideration belongs to the airfoil surface $\eta = 0$, the term $(\rho V/J)_D$ is deleted from (48), and at the surface points A and C ρ is calculated from (40), where we put

$$\lambda^2 = \phi_\xi^2/(x_\xi^2+y_\xi^2) \ . \tag{49}$$

The Kutta condition is handled as follows. Let the central grid-point 5 of the control volume under consideration lie at the trailing edge (cf. fig. 5). The Kutta condition is $\lambda_A=\lambda_C$, or

$$|\phi_6-\phi_5|(x_\xi^2+y_\xi^2)_A^{-\frac{1}{2}} = |\phi_5-\phi_4|(x_\xi^2+y_\xi^2)_C^{-\frac{1}{2}} \ . \tag{50}$$

ITERATIVE SOLUTION METHOD

In general, the left-hand side of (48) is a nonlinear function $f(\phi_1,\phi_2,\cdots,\phi_{12})$, where the subsripts correspond to the grid-points indicated in fig. 4. At subsonic grid-points f does not depend on ϕ_{10}, ϕ_{11}, ϕ_{12}. Fig. 4 represents the case U > 0. When U < 0, the grid-points 10, 11, 12 must be chosen to the right of the control volume. Eq. (48) is Newton-linearized. The equation for the n-th Newton correction $\delta\phi^{(n)}$ is given by

$$\sum_{i=1}^{12} a_i^{(n-1)} \delta\phi_i^{(n)} = - f^{(n-1)} , \tag{51}$$

where $a_i^{(n-1)} := \partial f/\partial\phi_i$ and $f^{(n-1)}$ are to be evaluated at the previous iterand $\phi^{(n-1)} = \phi^{(n-2)}+\delta\phi^{(n-1)}$. Each control volume gives rise to an equation of type (51). The system of equations is represented symbolically as

$$A^{(n-1)} \delta\phi^{(n)} = - f^{(n-1)} . \tag{52}$$

As initial guess we have always taken uniform flow:

$$\phi^{(0)} = \lambda_\infty \bar{x} . \tag{53}$$

The boundary conditions for $\delta\phi^{(n)}$ are (41), and

$$\delta\phi^{(n)}(\xi,b) = \frac{\delta\Gamma^{(n)}}{2\pi} \arctan (\sqrt{1 - M_\infty^2}\ \bar{x}/\bar{y}) , \tag{54}$$

and

$$\delta\phi^{(n)}(a,\eta) - \delta\phi^{(n)}(0,\eta) = \delta\Gamma^{(n)}. \tag{55}$$

The circulation correction $\delta\Gamma^{(n)}$ is to be chosen such that the updated potential $\phi^{(n)} = \phi^{(n-1)}+\delta\phi^{(n)}$ satisfies the Kutta condition (50). This is achieved as follows. We seek $\delta\phi^{(n)}$ in the following form

$$\delta\phi^{(n)} = \delta\Gamma^{(n)}\delta\hat{\phi}^{(n)} + \delta\tilde{\phi}^{(n)} , \tag{56}$$

with $\delta\hat{\phi}^{(n)}$, $\delta\tilde{\phi}^{(n)}$ satisfying (51) and the boundary conditions, but $\delta\hat{\phi}^{(n)}$ with $\delta\Gamma^{(n)} = 1$ and zero right hand side, and $\delta\tilde{\phi}^{(n)}$ with $\delta\Gamma^{(n)} = 0$. The correct value for $\delta\Gamma^{(n)}$ follows by substitution of (56) in the Kutta condition.

For subcritical flow, with the density bias switched off, the operator $A^{(n-1)}$ in (52) corresponds to a symmetric 9-point finite difference molecule of the type assumed by the multigrid code MGD4. It was found that MGD4 solves (52) efficiently in the subcritical case, with a residue reduction factor around 0.1 per iteration. The Newton process converged rapidly and quadratically.

For supercritical flows (52) could be solved by a combination of MGD4 and Picard iteration, although with a slower rate of convergence. A major problem encountered in the supercritical case was erratic convergence of the Newton process. Especially for fine meshes the Newton process did not converge for certain cases. The Newton process was therefore modified as follows. The operator $A^{(n)}$ is split into two parts:

$$A^{(n)} = B^{(n)} + C^{(n)} , \tag{57}$$

with

$$\begin{aligned} B^{(n)} = {} & (a_1^{(n)}+a_{10}^{(n)})S_\xi^- S_\eta^- + a_2^{(n)} S_\eta^- + a_3^{(n)} S_\xi^+ S_\eta^- + (a_4^{(n)}+(1-\sigma)a_{11}^{(n)})S_\xi^- + \\ & (a_5^{(n)}+\sigma a_{11}^{(n)})I + a_6^{(n)} S_\xi^+ + (a_7^{(n)}+a_{12}^{(n)})S_\xi^- S_\eta^+ + a_8^{(n)} S_\eta^+ + a_9^{(n)} S_\xi^+ S_\eta^+ , \end{aligned} \tag{58}$$

$$C^{(n)} = \{a_{10}^{(n)} S_\xi^- S_\eta^- + a_{11}^{(n)} (S_\xi^- + \sigma I) + a_{12}^{(n)} S_\xi^- S_\eta^+\}(S_\xi^- - I) , \tag{59}$$

where σ is a parameter to be chosen. The iterative method that we will use is given by

$$B^{(n-1)} \delta\hat{\phi}^{(n)} = - C^{(n-1)} \delta\hat{\phi}^{(n-1)} , \tag{60}$$

with $\delta\hat{\phi}^{(n)}$ satisfying the boundary conditions with $\delta\Gamma^{(n)} = 1$;

$$B^{(n-1)} \delta\tilde{\phi}^{(n)} = r^{(n-1)} , \tag{61}$$

with $\delta\tilde{\phi}^{(n)}$ satisfying the boundary conditions with $\delta\Gamma^{(n)} = 0$. The iteration is completed by forming

$$\phi^{(n)} = \phi^{(n-1)} + \delta\Gamma^{(n)} \delta\hat{\phi}^{(n)} + \delta\tilde{\phi}^{(n)} , \tag{62}$$

and choosing $\delta\Gamma^{(n)}$ such that the Kutta condition is satisfied.

Because the matrix associated with the operator $B^{(n)}$ has the sparsity pattern assumed by the multigrid code MGD4, we can apply MGD4 to the solution of (60) and (61). The main diagonal of the matrix associated with $B^{(n)}$ can be increased by increasing σ, because $A_5^{(n)}$ and $A_{11}^{(n)}$ have the same sign. This enhances the rate of convergence of MGD4. On the other hand, $C^{(n)}$ gets bigger. For the calculations to be reported we have always chosen $\sigma = 1$. In practice only one MGD4 iteration was found to give a sufficiently accurate solution of (60) and (61).

Obvious changes must be made in the definition of $B^{(n)}$ and $C^{(n)}$ when the grid-points 10, 11, 12 lie at the other side of the control volume under consideration ($U < 0$).

In subcritical flow $C^{(n)} \equiv 0$, and the iteration method described is in fact the Newton method.

In (61) the operator $C^{(n-1)}$ has been neglected, unlike (60), in order to damp oscillations in the supersonic zone. If we consider symmetric flow we have to deal only with (61). Interpreting the iteration method as a time-like process, we find that (61) is consistent with the following continuous interpretation:

$$(A-C)\partial\phi/\partial t = - f , \tag{63}$$

where $C\partial\phi/\partial t$ contains a leading term proportional to $\partial^2\phi/\partial\xi\partial t$, which is known to have a damping effect on oscillations in the supersonic region, cf. [7, 8]. We found that for nonsymmetric flow it is better to retain the operator $C^{(n)}$ in (60).

RESULTS

Figure 6 gives the convergence history for the C_p distribution for the flow around the NACA 0012 airfoil at angle of attack $\alpha=2^0$ for $M_\infty = 0.63$. This flow is subcritical. The convergence is quadratic. The final solution does not differ perceptibly from the solution found in [7], except for the slight oscillations near the peak value. These oscillations are present only when the velocity is calculated by means of a finite difference approximation to ϕ_ξ in (49) in the middle between two grid-points, as we have done. The lift coefficient C_L=0.33096 obtained by numerical integration of the pressure distribution, and the value C_{Lt}=0.33117 found from $C_{Lt}=-2\Gamma/\lambda_\infty$, agree well with the values C_L=0.334 and C_L=0.335 obtained in [7] and [10], respectively. Figure 7 gives the evolution of the maximum norm of the residue f for the nonlinear system and the correction to C_p, as well as the ℓ_2-norms of the residues of the linear systems (60) and (61) before and after one MGD4 iteration. We observe quadratic convergence for f and C_p. MGD4 used four grids. The finest grid contained 128 points in the ξ-direction and 41 nodes in the η-direction. The overall calculation required 37 CP seconds on an Amdahl 470.

Figure 8 gives results for the flow around NACA 0012 at $\alpha=2^0$ for M_∞=0.75. This flow is supercritical. The C_p-distribution after 20, 40, 60 and 72 iterations (60), (61) is shown. The final values C_L=0.55958 (C_{L_t}=0.56106) and C_D=0.01625 are slightly less than the values C_L=0.571, C_D=0.0175 reported in [7] and C_L=0.5878, C_D=0.0182 reported in [8]. This could be ascribed to differences in region and grid geometry; we remind the reader that in our calculations infinity is at 6 chord lengths distance. The maximum Mach number is 1.374. Figure 10 gives details of the convergence history. Initial rapid convergence is followed by a more erratic and slower convergence behaviour, especially around the 35-th iteration, when the shock arrives at its final position. Convergence is delayed because of a weak dependence of the flow in the supersonic region on the flow in the subsonic region downstream of the shock wave; it is as if the flow is partially choked. Also, any changes in shock strength and position strongly affect the circulation, which in turn influences the shock, a "noisy" event producing near the shock local jumps of the flow properties and the residue. The rate of convergence speeds up again when the circulation stabilizes around its final value. The maximum norm of the residue is reduced by a factor 10^{-3} in 65 iterations, taking 8 min. 23 sec. CP time on a Amdahl 470. The rapid convergence of MGD4 is illustrated in fig. 9 during the first 20 iterations.

DISCUSSION

A multigrid method for transonic flow computations has been presented in which an autonomous (black box) multigrid subroutine is used as a linear equation solver. Hence, the multigrid part of the code is completely divorced from the handling of the nonlinearity. This makes the method simpler than other multigrid methods for transonic flow that have been proposed.

We do not know how the efficiency of the present method compares with other multigrid methods for transonic potential flow. Obviously, there is room for improvement. The efficiency can perhaps be improved by intertwining the multigrid part of the handling of the nonlinearity, as in [8]. Of course, this leads to a more complicated method. Because of its rapid initial convergence (cf. fig. 9), (60), (61) together with ILU decomposition would probably be a very efficient smoothing process in a full approximation storage multigrid algorithm of the type proposed in [2]. Another, simpler way to gain efficiency would probably be to use a converged solution on a coarse grid, interpolated to a finer grid, as initial iterand on the finer grid, while preserving the present method, and so on, until the finest grid is reached.

REFERENCES

[1] Boerstoel, J.W., Kassies, A., Integrating multigrid relaxation into a robust fast solver for transonic potential flows around lifting airfoils, NLR Report MP 83021 U, Amsterdam, 1983; AIAA 6th Computational Fluid Dynamics Conference, Danvers, Mass., 1983.

[2] Brandt, A., Multi-level adaptive solutions to boundary-value problems. Math. Comp. 31 (1977) 333-390.

[3] Hafez, M., South, J., Murman, E., Artificial compressibility methods for numerical solution of transonic full potential equation, AIAA Journal 17 (1979) 838-844.

[4] Hackbusch, W. and Trottenberg, U. (eds.), Multigrid Methods, Proceedings, Köln-Porz, 1981. Lecture Notes in Mathematics 960 (Springer-Verlag, Berlin etc., 1982).

[5] Hemker, P.W., Kettler, R., Wesseling, P., and de Zeeuw, P.M., Multigrid methods: development of fast solvers. Report 83-05, Dept. of Mathematics and Informatics, Delft University of Technology, 1983. To appear in the Proceedings of the International Multigrid Conference, Copper Mountain, U.S.A., April 1983.

[6] Hemker, P.W, Wesseling, P. and de Zeeuw, P.M., A portable vector-code for autonomous multigrid modules. Report NW 154/83, Mathematical Center, Amsterdam 1983.

[7] Holst, T.L., An implicit algorithm for the conservative, transonic full potential equation using an arbitrary mesh, AIAA Paper 78-1113, 1978.

[8] Jameson, A., Acceleration of transonic potential flow calculations on arbitrary meshes by the multiple grid method, AIAA Paper 79-1458 CP, 1979.

[9] Kettler, R., Analysis and Comparison of Relaxation Schemes in Robust Multigrid and Preconditioned Conjugate Gradient Methods, in [4], pp. 502-534.

[10] Lock, R.C., Test cases for numerical methods in two-dimensional transonic flows, AGARD Report No. 575, 1970.

[11] McCarthy, C.J., Investigation into the Multigrid Code-MGD1. Report R10889, AERE Harwell, Oxfordshire, April 1983.

[12] McCormick, S.F., An algebraic interpretation of multigrid methods, SIAM J. Numer. Anal. 19 (1982) 548-560.

[13] Meijerink, J.A. and van der Vorst, H.A., Guidelines for the usage of incomplete decompositions in solving sets of linear equations as they occur in practical problems, J. Comp. Phys. 44 (1981) 134-155.

[14] Thompson, J.F., Thames, F.C. and Mastin, C.M., TOMCAT - a code for numerical generation of boundary-fitted curvilinear coordinate systems on fields containing any number of arbitrary two-dimensional bodies, J. Comp. Phys. 24 (1977) 274-302.

[15] Wesseling, P., Theoretical and practical aspects of a multigrid method, SIAM J. Sci. Stat. Comput. 3 (1982) 387-407.

[16] Wesseling, P., A robust and efficient multigrid method, in [4], pp. 614-630.

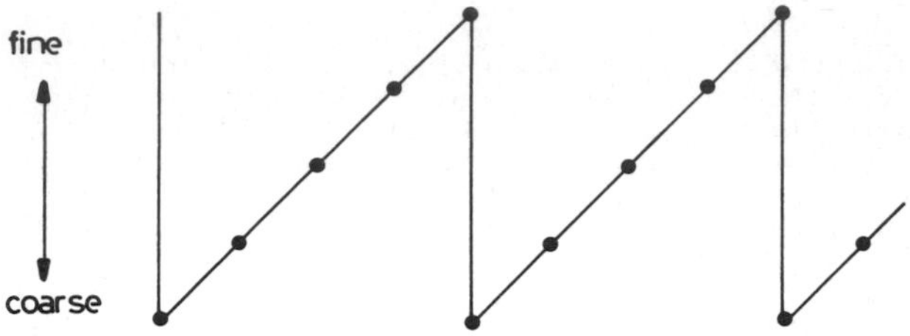

Figure 1. Multigrid sawtooth schedule. A dot represents application of the smoothing process (8).

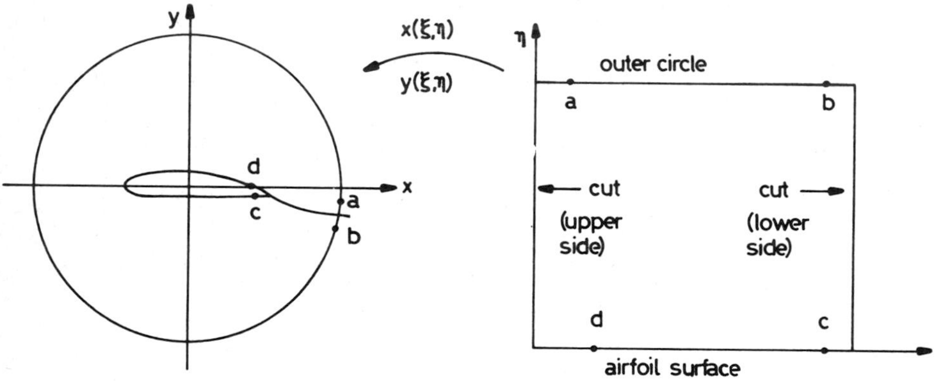

Figure 2. Mapping between physical and computational plane.

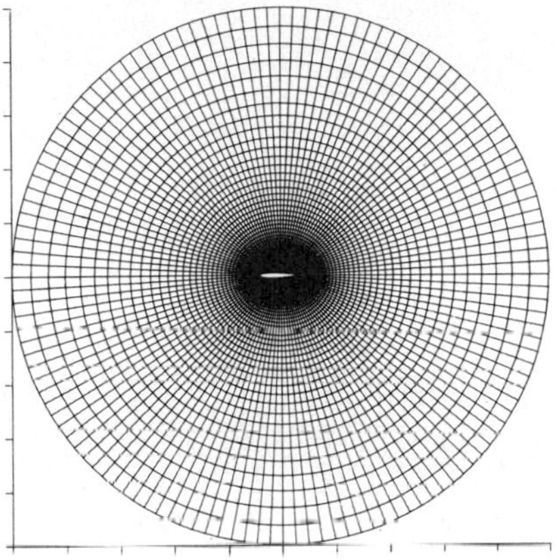

Figure 3. Example of image of computational grid in physical plane.

Figure 4. Portion of computational grid, with control volume (dotted line).

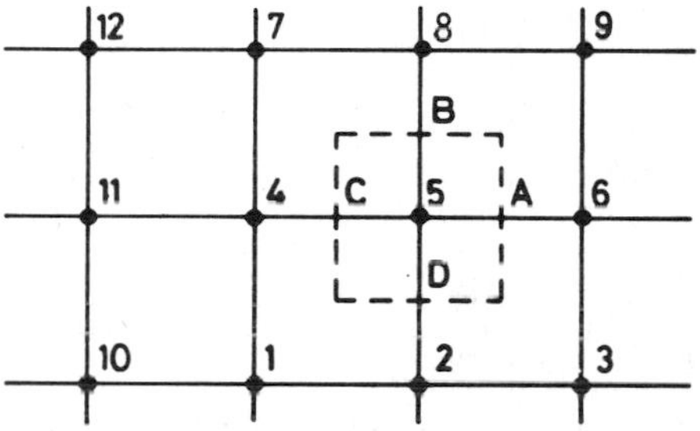

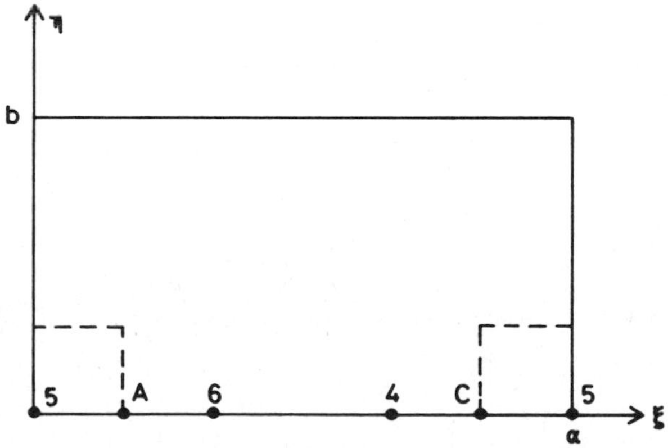

Figure 5. Control volume at trailing edge.

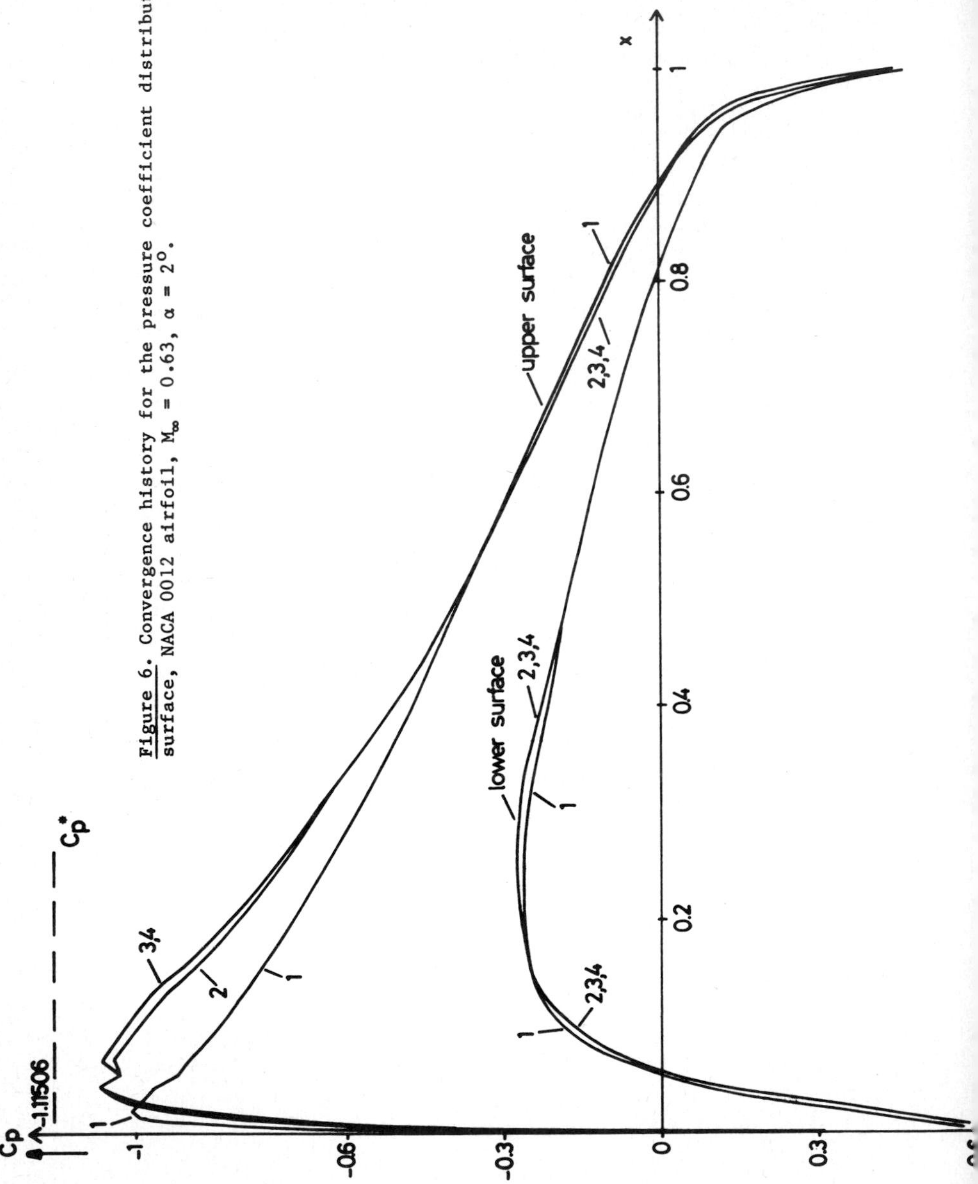

Figure 6. Convergence history for the pressure coefficient distribution on the surface, NACA 0012 airfoil, $M_\infty = 0.63$, $\alpha = 2^o$.

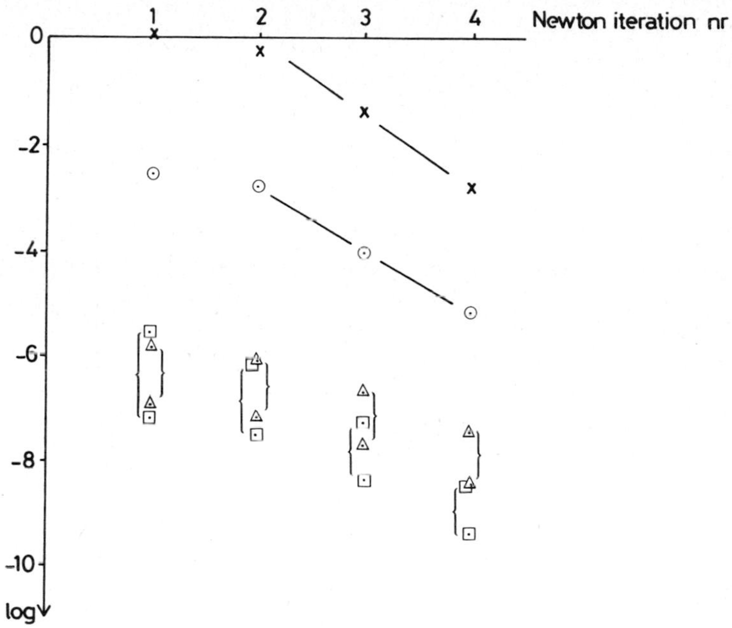

Figure 7. Convergence history, same flow as fig. 6.
⊙ : maximum norm of the residue for the nonlinear system.
x : maximum norm of C_p correction.
△,⊡ : ℓ_2-norm of residue before and after one MGD4-iteration on (60), (61) respectively.

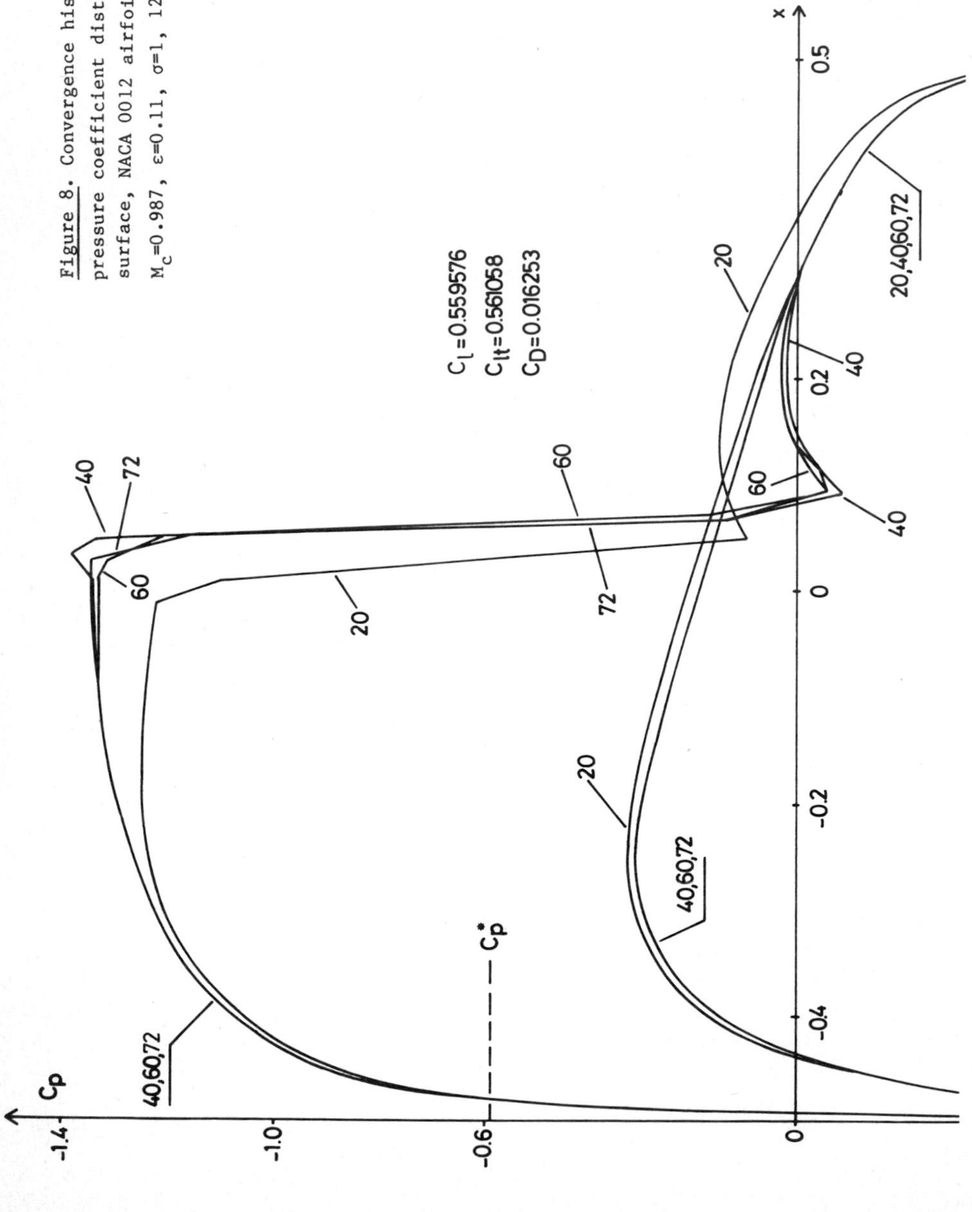

Figure 8. Convergence history for the pressure coefficient distribution on the surface, NACA 0012 airfoil, M_∞=0.75, α=2°, M_c=0.987, ε=0.11, σ=1, 128×41 grid.

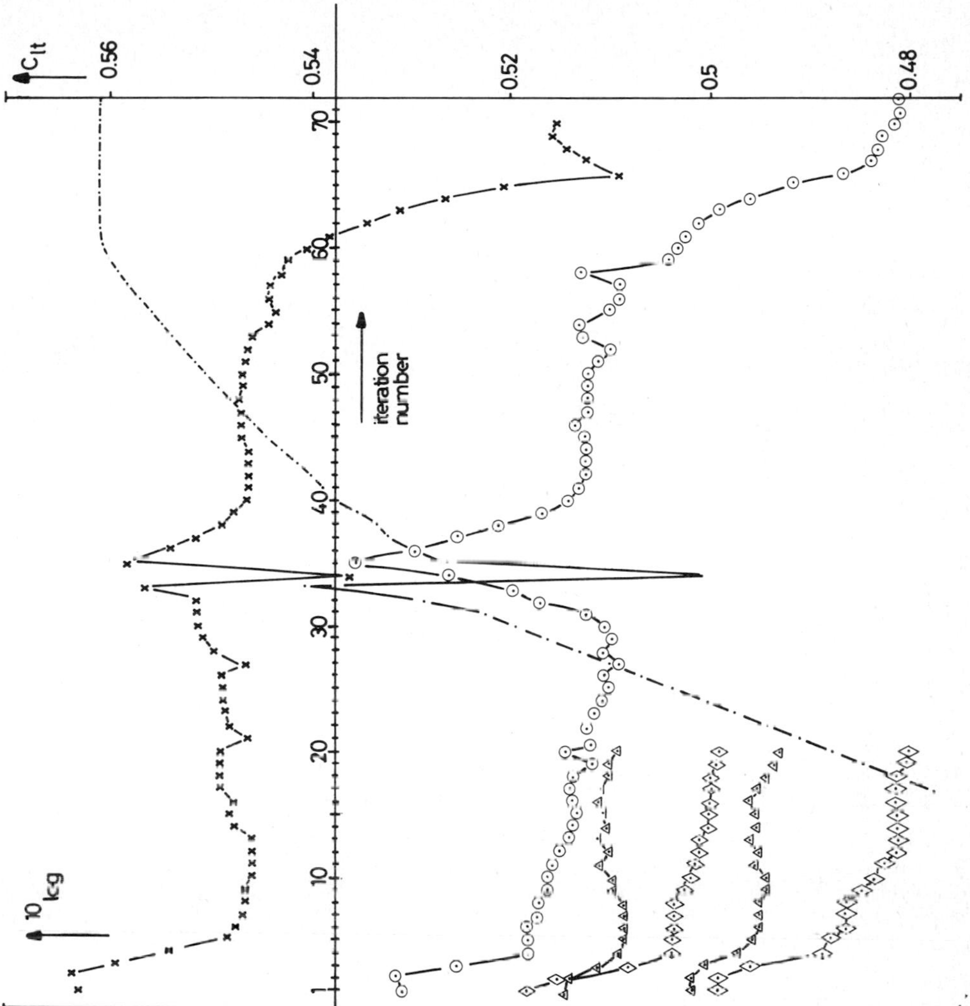

Figure 9. Convergence history, same flow as fig. 8.
⊙ : maximum norm of residue for nonlinear system.
× : maximum norm of C_p correction.
-.-: value of C_{Lt} : = $-2\Gamma/\lambda_\infty$.
Δ,◊ : 100 × ℓ_2-norm of residue before and after one MGD4-iteration on (60), (61) respectively.

PARALLEL COMPUTING

CALCUL PARALLELE

Computing Methods in Applied Sciences and Engineering, VI
R. Glowinski and J.-L. Lions (Editors)
Elsevier Science Publishers B.V. (North-Holland)

THE FIFTH GENERATION COMPUTER SYSTEMS AND ITS COMPUTER ARCHITECTURE

Hidehiko Tanaka

Department of Electrical Engineering
The University of Tokyo
Hongo, Bunkyo-ku, Tokyo 113
JAPAN

A project called Fifth Generation Computer Systems (FGCSs) is currently underway in Japan. The objective of this project is to develop the basic technology that is required to realize the FGCSs which will be used in the 1990s. These systems are oriented towards knowledge information processing system, the basic elements of which are the problem-solving and inference systems and the knowledge base system. A 10-year research period has been set up, and this commenced on June 1, 1982. This paper presents the general view of the FGCS, the architectural consideration, and the description of a few architectural projects.

1. INTRODUCTION

The progress of computer technology is striking nowadays. For example, we now have single-chip computers whose processing power far exceeds that of the First Generation computers. We have also various kinds of high-level languages, operating systems, and database systems. As a result, we can make programs for almost any kind of applications, provided that their algorithms can be described explicitly. This means that computers can take the place of people in many areas because of their high-speed processing ability and their large memory capability. However, there are still some application fields in which computers have not been able to demonstrate their power satisfactorily; these are pattern recognition, natural language translation, summarizing capability of sentences, learning capability of past experiences, global decision making through various kinds of information, and so on.

On the other hand, we are also facing a serious problem, namely the 'software crisis'. This arises because of the gap between the logic level required by applications and the one offered by high-level languages. As the application field expands more, the gap becomes broader and broader. This means that computers should support a higher logic level which is at the same time friendly and familiar for humans. We call such computer systems Fifth Generation Computer Systems (FGCSs) that are expected to solve these problems. In other words, the FGCSs are Knowledge Information Processing Systems (KIPSs); they have knowledge bases and are able to infer from the knowledge and solve problems as humans do. Only when computers gain such capability, they can be truly useful and tide us over the software crisis. To realize such high-level computers, various kinds of technologies are required such as knowledge representation, inference operation, knowledge acquisition, very efficient parallel execution mechanism, VLSI technology and so on. These technologies are not completed yet, but the time is now ripe to start a project to develop Fifth Generation computers which are based on artificial intelligence.

In Japan, a few study groups were formed in 1979 made up of university professors, and researchers of national laboratories and industries. In the course of three years they fixed the image of Fifth Generation computers[1], and drew up a 10-year research plan divided into three term of three, four and three years. A laboratory to lead this research and development was founded in June 1982 in Tokyo and funded by the Ministry of International Trade and Industry.

Section 2 of this paper summarizes the overall view of Fifth Generation Computer Systems. Section 3 analyzes the processing model of logic programming and shows the models of logic programming oriented computer architecture. Section 4 reports the present status of computer architecture research projects which are underway at and around the Institute of New Generation Computer Technology (ICOT). Section 5 concludes this paper and shows some of the future works.

2. IMAGES OF THE FIFTH GENERATION COMPUTER SYSTEMS [2]

The FGCSs are oriented to knowledge information processing that is based on the innovative theories and technologies which offer the intelligent dialogue functions and the inference mechanism for knowledge bases that will be required in the 1990s. The functions of FGCSs can be divided into three kinds as follows:

1. Problem-solving and inference function.
2. Knowledge base management function.
3. Intelligent interface function.

The problem-solving and inference function corresponds to the basic four arithmetic operations (plus, minus, multiply, divide) and the control functions needed to perform arithmetic calculations. The system which realizes this function will be organized by the hardware inference mechanism, the control mechanism of parallel processing, the logic programs made up on these mechanisms, and the software system for higher-order predicate calculus. A language which includes the language PROLOG will be used as the interface between software and hardware. The knowledge-base management function is an innovative database management function which is enhanced by the processing function of semantic information. The relational data model is selected as the basic data model, which matches to predicate calculus very well. The software of knowledge-base management will be implemented over the hardware function which processes the relational algebra with high speed.

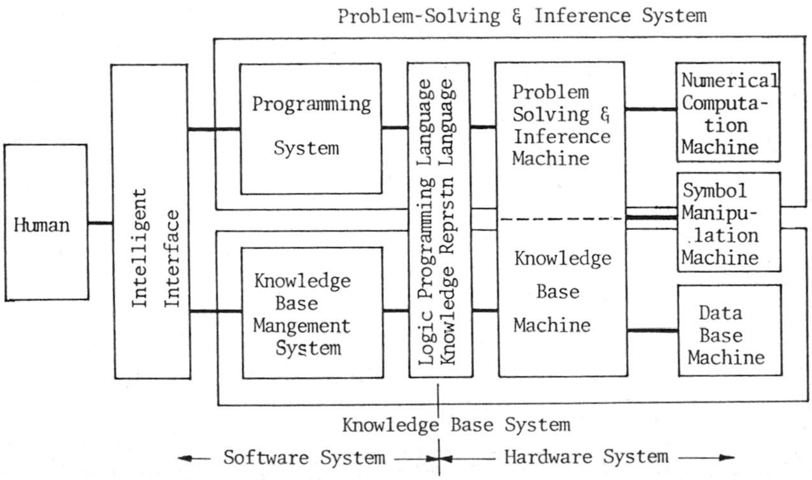

Fig.1 Image of the Fifth Generation Computer System

Intelligent interface function, which corresponds to the input/output function of conventional computers at present, will consist of a collection of systems which process, recognize or synthesize many kinds of input/output media such as characters, voice, figures, and images. It is very important to offer some friendly interfaces for humans to exchange information with FGCSs. These functions are realized through corresponding software and hardware systems. The conceptual image of FGCSs is shown in figure 1. The right upper half of the system corresponds to the problem-solving and inference function, and the right lower half to the knowledge-base function. The left area corresponds to the intelligent interface function, which depends heavily on the previous two functions as shown in the figure. This figure shows that computers will closely approach the human system through a remarkable improvement of the logical level of the hardware system and by placing the modeling system between the hardware system and the human system.

In the knowledge-base system, three kinds of knowledge are stored: knowledge about problem domain; knowledge regarding languages, images etc. which are used by the intelligent interface function; and knowledge about the machine system and knowledge representation.

When some problem is presented by the application system through some end-user language that uses voice, figures and images, this problem is analyzed and recognized by using knowledge about the language and the images, and is transformed into intermediate specifications which are given to the programming system. The problem is understood here using the knowledge about the problem domain, and the processing specification is made up as the result. This specification is transformed into a program and optimized through referencing the knowledge about the machine system and knowledge representation. The program, written in a logic programming language, is processed by the problem-solving and inference machine and the knowledge-base machine. We think that such hardware as the symbol manipulation machine, scientific calculation machine and database machine can be regarded as machines which undertake to subcontract from the problem-solving and inference machine and the knowledge-base machine. As all of the computer hardware up to the Fourth Generation play the role of the subcontractor, we can say that the functions of hardware and software of FGCSs make the semantic gap between human and machine very small.

The inference machine which processes logic programming languages such as PROLOG is equipped with a high-power inference mechanism of hardware and a data-flow oriented control mechanism for parallel processing. The innovative von Neumann mechanism will be provided to accept the conventional programs effectively. The abstract data type support mechanism is required from the viewpoint of module programming and data security as well.

The knowledge-base machine will be realized by the combination of the relational database mechanism and the parallel processing machine of relational algebra operations, and be integrated with the inference machine in future.

These machines, of which machine language is a common kernel language for FGCSs, can be connected to each other to organize a distributed processing system. This kernel language interface the software system with the hardware system. All software systems are represented by this kernel language which the hardware system executes directly. For the implementation of this hardware, VLSI technology is indispensable as the number of gates required will be very large.

3. FIFTH GENERATION COMPUTER ARCHITECTURE

3.1. GENERAL VIEW

The major functions supported by the fifth generation computer hardware are the problem solving and inference function and knowledge base function. These functions are implemented by the inference machine and knowledge base machine, and provided through a language interface which will be developed on the basis of logic programming and knowledge representation. Though these two machines are expected to be fused into a single machine in future, we treat these machines individually at present so that the research of architecture goes at ease.

The roles of architecture support for the fifth generation computer systems can be summarized as follows.

(1) To shorten the semantic gap through supporting directly the high level operations.
(2) To get high speed processing capability, as the processing complexity will expand more and more, and the needs of high speed processing will be developed more as the application field emerges.
(3) To make it feasible to handle enormous amount of and growing knowledge data.

3.2. INFERENCE MACHINE ARCHITECTURE

A. INFERENCE PROCESS

Assume a prolog program as follows.

```
sister-of(X,Y) :- female(X),parent(X,M,F),parent(Y,M,F).    (1)
parent(alice,albert,victoria).                                (2)
parent(edward,albert,victoria).                               (3)
female(alice).                                                (4)
female(victoria).                                             (5)
```

When a question "?- sister-of(X,edward). (Who is the sister of edward?)" arises, the process to get the answer is as follows.

(a) (1) matches to the question, then,
sister-of(X,edward) :- female(X),parent(X,M,F),parent(edward,M,F).
(b) A new goal is generated.
?- female(X),parent(X,M,F),parent(edward,M,F).
(c) A literal "female(X)" is tested in the program.
(d) It is found that there are 2 candidates for X from (4) and (5).
X = alice, X = victoria.
(e) X = alice is applied to the goal.
?- parent(alice,M,F),parent(edward,M,F).
(f) The first literal "parent(alice,M,F)" is tested, and found to match when variable M and F is set to be albert and victoria, respectively.

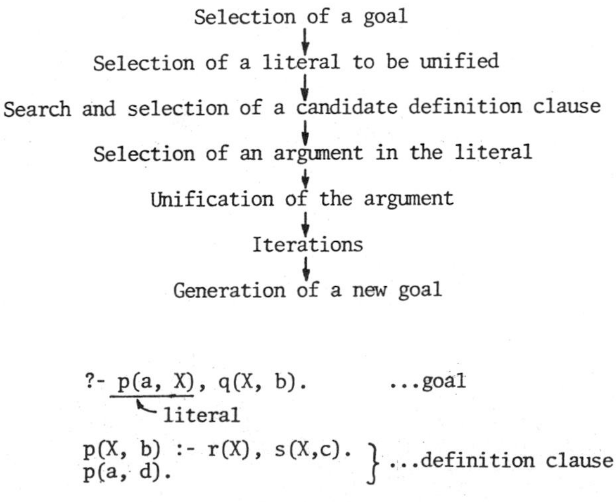

Fig.2 The procedure of inference process

Generally speaking, these are 2 kinds of inference. One is deductive inference and another is inductive inference. The process above is an example of deductive inference. As the deductive inference matches well to the programming, it can be set as the basic operation of inference machine. The inductive inference is expected to play an important role when knowledge acquisition is considered. Accordingly, it will be taken into account for the architecture design as the research of knowledge base progresses. So far as the deductive inference of first order predicate logic is concerned, the details of inference process can be written as a procedure of fig. 2. The procedure of fig. 2 corresponds to the machine language interpretation of conventional computers by microprogram. When we see the process from the viewpoint of logic, it can be regarded as a search process on a AND-OR tree like fig. 3. That is, a goal (?- p,q.) is expressed as an AND condition (p AND q). To satisfy a part (for example, p) of the AND condition, a few OR conditions can be found (p at node 3, or p at node 4). To satisfy each of the OR conditions, an AND condition (for example, r AND s) is required by a definition clause (for example, p :- r,s.), and so on.

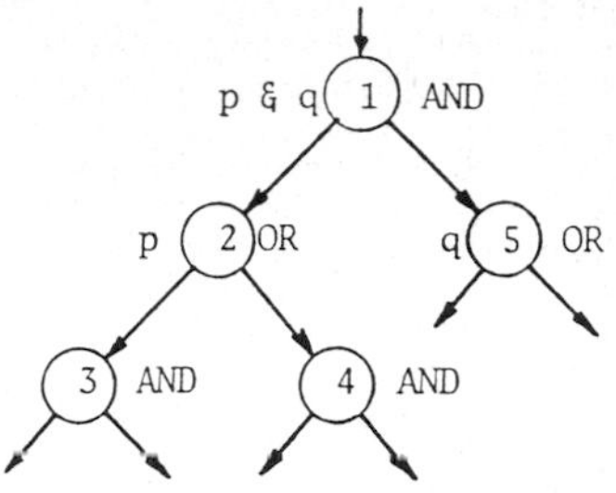

Fig.3 AND-OR tree

B. GOAL SEARCH ALGORITHMS

In logic programming based on the predicate logic, unification is the key procedure for inference. The unification is composed of a selection of a definition clause which matche to the goal and the correspondence among arguments. When a goal succeeds, the set of correspondence among arguments defines values of variables, which can be used as answer to the question. In order to make the inference process efficient, many kinds of schemes can be proposed. They can be grouped into four kinds as follows.

(1) Parallel search.
(2) Optimization of search sequence.
(3) Usage of designation by programmer.
(4) Elimination of recomputation.

PARALLEL SEARCH

On conventional computers, the inference process i.e. tree search is executed sequentially. The tree is traversed node by node. Accordingly, when a goal is failed at a node, the downward search operation is stopped there, and returned to a upper node to test the other alternatives. This return operation is called backtrack. Parallel search is a straightforward approach to speed up the search procedure. Examining the inference process of fig.2, we can find tree kinds of parallelism. First is the OR parallelism, which comes from the fact that there are several definition clauses of which head literal matches to the goal. Second is the AND parallelism among literals of a goal. Third is the intra-literal parallelism which

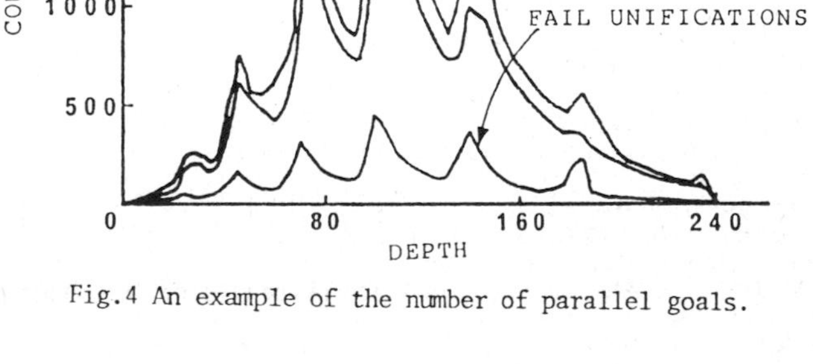

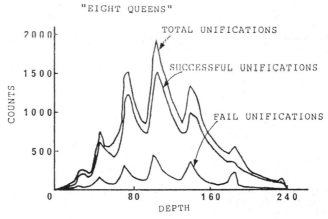

Fig.4 An example of the number of parallel goals.

performs the unification of argument in parallel with every argument. Fig. 4 is a measurement of OR parallelism. The sample program is eight queens. The horizontal axis denotes the depth of the tree. The vertical axis is the number of OR nodes at the depth. This shows that we can have a lot of parallelism easily and that the number of parallel nodes grows exponentially. Accordingly, we can expect enough parallelism through the OR parallelism. Regarding AND parallelism, the sharing of variables among literals makes the problem complicated, though AND parallelism without shared variables can be handled easily. The intra-literal parallelism can be used effectively for the design of unification executer (i.e. unifier). We can gain the speed up factor of the number of arguments through placing many unifiers of single argument in parallel.

OPTIMIZATION OF SEARCH SEQUENCES

When we want to get an answer as fast as possible, we can make the search process efficient by searching the most possible candidate first. Most popular algorithms are the depth-first one and the breadth-first one. The depth- first search selects the child node first and the breadth-first search the brother node first. Besides these, we can optimize the search sequence in terms of clauses, literals and arguments. Table 1 is an example of time measurements by changing the literal selection algorithm.

Table 1. Total number of unifications of 2 sample programs

	program JE	program EJ
scheme1	145	27
scheme2	36	27
scheme3	192	34

,where program JE(EJ) is a small program of Japanese to English (vice versa) translation.

USAGE OF DESIGNATION BY PROGRAMMER

As a programmer knows best about the behavior of his program, we can expect the efficient execution through making programmer designate the control of his program execution by introducing a few syntax into the programming languages. Examples are as follows.

a. Execution order : sequential OR, parallel OR, sequential AND, parallel AND

b. search region : To restrict the search region by removing unnecessary subtrees when some conditions met. This examples are guard[3], remote-cut[4] and success-throw[4].

c. Input/Output direction of variables. When a goal has shared variables, we face to the ambiguity which appearance of the variable defines the value. This ambiguity makes the AND parallelism difficult. Accordingly, if the ambiguity is removed with the help of programmer, it makes the AND parallel operation very easy. An example is the read-only-annotation of concurrent Prolog[3].

d. Lazy evaluation : By restricting the number of repetitive usages of some clauses, we can protect against the occurrence of unnecessary tasks.

ELIMINATION OF RECOMPUTATION

There arises a lot of unifications and interim goals during the search process. However, there may be many duplicated goals and same kind of unifications among these. We can expect to make the search efficient by eliminating these unnecessary duplicated processing. Some memory mechanism is needed to implement this operation. Regarding the duplicated goals, memorizing some middle goals as if they are predefined theorems will improve the inference speed much. As this memorizing process changes the original program (add new knowledge), it is closely related to the learning mechanism, and further research is required. One example of this scheme is the proposal of applicative cache [7] by Keller.

There may happen the duplicate processing in a goal. For example, assume a definition of Fibonacci number as follows.

```
f(N,V):-f(N-1,V1),f(N-2,V2),add(V1,V2,V).
f(Ø,Ø).
f(1,1).
```

When a goal ?-f(6,V). is given, the first unification gives a next goal.

```
?-f(5,V1),f(4,V2),add(V1,V2,V).
```

Then, the second unification gives,

```
?-f(4,V1'),f(3,V2'),add(V1',V2',V1),f(4,V2),add(V1,V2,V).
```

In this goal, there arises literals f(4,V1') and f(4,V2). If we know that the function f is a unique function, we can remove one of them. Therefore, the goal can be changed as follows.

```
?-f(4,V2),f(3,V2'),add(V2,V2',V1),add(V1,V2,V).
```

While the original Fibonacci number calculation requires exponential order of processing, the above calculation needs linear order of processing. These elimination scheme may give substantial improvement for the inference processing.

C. PROCESSING MODELS

There are several processing models for the implementation of parallel inference machine as follows.

(1) Multi-process model.
(2) Data-flow model.
(3) Functional model.
(4) Logic model.

The models from 1 to 3 are such models that logic programs are translated into some intermediate structures which are different from model by model. On the other hand, the model 4 is a direct execution model of logic programs.

MULTI PROCESS MODEL[8]

This is a model which uses the conventional multi-processes structure for the implementation of inference processing. Processes which communicate each other are created and destroyed dynamically. Usually, each node of AND-OR tree is implemented by a process. This model is a very natural application of multi-processes structure, and can be implemented easily on conventional multi-processor computers.

DATA FLOW MODEL

Logic programs are translated into data flow graphs which are executed directory by a data flow machine. Data driven structure, which controls the start of each operation through the availability of required data, can be thought to be the basic principle of parallel processing. There are 2 research themes. One is to show the feasibility of data flow architecture of high degrees of parallelism. The other is to find out the algorithm which maps the logic programs to the data flow graphs efficiently. This algorithm is the key point of this model, which determines the parallelism of logic program execution. To implement OR parallelism on data flow model, all OR node processes are driven in parallel. The only results of processes whose unification succeed are merged into a stream and returned to the upper AND process. The sequence of this stream data is non deterministic. Regarding AND parallelism, the consistency check among the solutions is required generally. However, this may incur a lot of overhead especially in the case that the size of solution space is large. Accordingly, the pipeline schemes which executes the processing of goal literals with pipeline fashion are suggested for the AND parallelism. This is shown to be feasible through using the stream concept stated above by the PIM-D machine of ICOT.[13]

FUNCTIONAL MODEL

In this model, logic programs are translated into an intermediate form which is suitable for the execution of some functional programming languages. Examples are ALICE [9] of Imperial College of London and M3L [10] of Toulouse Univ. This model focalizes on the input/output relation of each operation. The output of function is determined by the input only. This means that the function has no side effect. This characteristics is fit for preserving the locality of each process, so that the parallel execution becomes easy. The parallelism of this model comes from the parallel evaluation of an expression. A program is represented by a graph whose each node expresses a function. When some function is reducible using rewrite rules, the rewrite operation is performed independently of others. To implement logic programs, some new rewrite rules should be added considering the variable instantiation of unifications.

LOGIC MODEL

This model simulates the behavior of logic program execution directory by hardware. Therefore, this one corresponds to the high level language machine of predicate calculus. The abstract image of this model is shown in fig. 5. In the goal pool, a lot of alternative goals are stored. Each goal data is transferred to unify-processor and performed unification. The result generats new goals and is returned to the goal pool. In this case, there are several alternatives for the implementation. For example, we can select either all data of a goal or a part data of a goal as the transfer unit between unify processors and goal pool. Though all goals are logically independent, they can be implemented by data-sharing for reasons of efficiency. OR parallelism generates several goals from a goal. Intra literal parallelism can be used to speed up the unification in the unify processor. Examples of this model is PIE [11],[12] of university of Tokyo and PIM-R [6] of ICOT. Fig. 4 is a simulation data of PIE, which shows a lot of parallelism. The machines of this model try to realize this kind of parallelism directory. As the behavior of this machine directly reflects the behavior of logic program, it is very easy to understand the machine behavior.

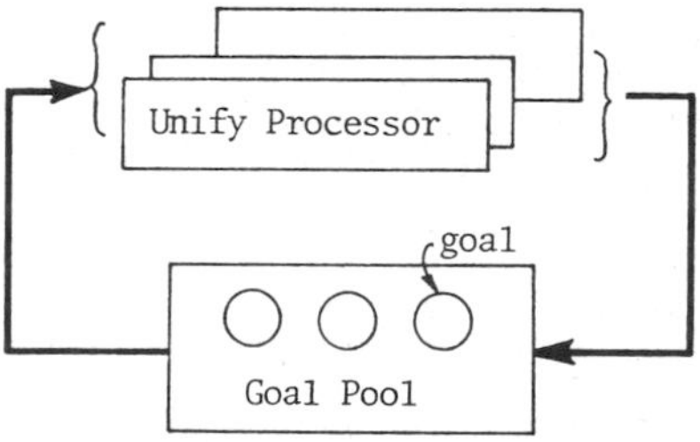

Fig.5 The abstract image of logic model

D. CONSIDERATION OF PARALLEL MACHINE IMPLEMENTATION

Followings are the consideration of parallel machine implementation by taking the logic model machine of fig. 5 as an example. The same kind of consideration can be applied to the other models as well. Assume that all of the unify processors are equipped with definition memories which store all the definition clauses. Accordingly, access to the definition memory doesn't make any influence to the behavior of major processing loop, that is the loop between unify processors and goal pool. For the detail design of this machine, 6 design points can be distinguished as follows.

(1) Transfer data unit between unify processor and goal pool.
(2) Definition memory...Shared or dedicated.
(3) Goal data...Shared or copied.
(4) Structure memory...Separated or integrated.
(5) Cache.
(6) Pipeline control.

TRANSFER DATA UNIT

Alternatives are all data of a goal and a part data of a goal. When all data of a goal are used as the unit, the behavior of machine becomes very simple. Each unify processor fetches a goal, preforms one unification and returns the new generated goals into the goal pool. All unify processor can operate independently.

On the other hand, the volume of the unit may be very large. This may reduce the speed of one round of unification. By an initial measurement result, the volume of the goal is about a few hundred bytes. When a part data of a goal is used as the unit, the transfer data between unify processor and goal pool can be reduced to only the one required. So the data traffic may be reduced. However, the access to the goal pool becomes random rather than burst, and the control can be complicated more.

DEFINITION MEMORY

The definition clauses should be accessible from all unify processors. There are 2 alternatives for the definition memory implementation. One is to use a shared memory among unify processors. The other is to equip all unify processors with their own local definition memory. To evaluate these schemes, think the model of fig. 6. Each processor has a local memory and an access path to the shared memory. Assume that one cycle of processing (unification) needs m memory accesses, within which $m\rho$ accesses are the shared memory accesses. When the total throughput of this system is H cycles/sec, the throughput restraints of memories can be expressed as

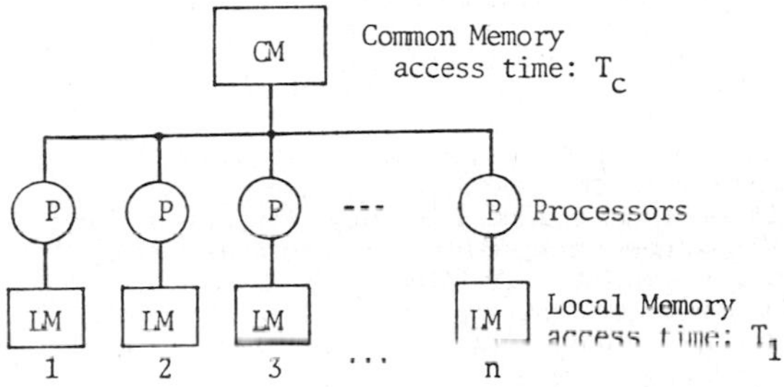

Fig.6 Shared memory model

$$\frac{H}{n} m\rho \cdot n \cdot T_c < 1 \text{ , and} \tag{6}$$

$$\frac{H}{n} m(1-\rho) \cdot T_1 < 1 \text{ ,} \tag{7}$$

where, n is the number of processors, and Tc and Tp are the access time of shared (common) memory and of local memory, respectively. From these constraints, the maximum throughput Hmax is, by changing the ρ as the parameter,

$$H_{max} = \frac{1}{m \cdot \rho_0 \cdot T_c} \text{ ,} \tag{8}$$

$$\text{where } \rho_0 = \frac{T_1}{nT_c + T_1} \tag{9}$$

From the equation (9), the effective degree of parallelism ne is given as follows.

$$n_e = \frac{1-\rho}{\rho} \cdot \frac{T_1}{T_c} \tag{10}$$

This means that it is useless to have more processors than ne when the shared memory access probability is given. If we require more than 100 of parallelisms, the following expression should be satisfied.

$$100 \ll \frac{1-\rho}{\rho} \frac{T_1}{T_c} \tag{11}$$

that is,

$$\rho \ll \frac{1}{100 \; (T_c / T_1) + 1} \tag{12}$$

When $T_c/T_p = 2$, ρ must be less than 0.005. Therefore, shared access should be less than 0.1%, if we require the parallelism of more than a few hundreds. This means that fairly large amount of local memory should be provided as the definition memory of each processor.

GOAL DATA STORAGE
While several new goals are generated from a goal by OR parallelism, the difference among data of these new goals comes from the difference among definition clauses which are the candidates of unification between the head literal of the old goal and definition clauses. The rest data are the same. Usually, a goal data includes many environment data such as variable instantiation. Accordingly, changed data by a unification is rather less than the unchanged one. This means that shared data scheme among goals can be very efficient. However, sharing may introduces dependency among goals, which is harmful for the highly parallel operation and makes the operation complicated. Some compromised schemes are needed, which preserve logical independence while some data are shared physically.

SEPARATION OF STRUCTURE DATA
Logic programs need to process large structure data. At that time, if all data of the structure should be always transferred from goal memory to unify processor, the overhead of transmission cannot be ignored. Instead, the operation code could be transferred to the structure, where the operation is executed. This means that it can improve the throughput of the system to provide a structure memory which is optimized for the structure manipulation and structure storage.

USAGE OF CACHE
Since cache system (buffer memory) is very effective to reduce the memory access time, it is very natural to use cache for the implementation of inference machine which is rather memory speed necked system. The memories which can be equipped with cache are the definition memory and goal memory. The definition memory can be regarded as a kind of cache for the knowledge base. Before starting some job, all necessary definition clauses are assumed to be loaded into all definition memories, in the inference machine discussed previously. The interface between definition memory and knowledge base is one of the knowledge base machine interface. This will be clarified as the research of knowledge base progresses. One more role of the definition memory is to store the intermediate results as a kind of theorems. A simple way to do this is to store all intermediate results into the capacity bounded cache. By applying "least recently used (LRU) algorithm", results of high usage-probability would be maintained in the cache. Regarding goal pool, the goal buffer memory in the unify processor acts like cache when all data of a goal is used as the transfer unit between unify processors and goal pool.

PIPELINE CONTROL
In one cycle of unification, a goal is fetched by a unify processor, the unification is executed, new goals are generated and they are returned into the goal memory. This sequence could be controlled with pipeline fashion as fig.7.

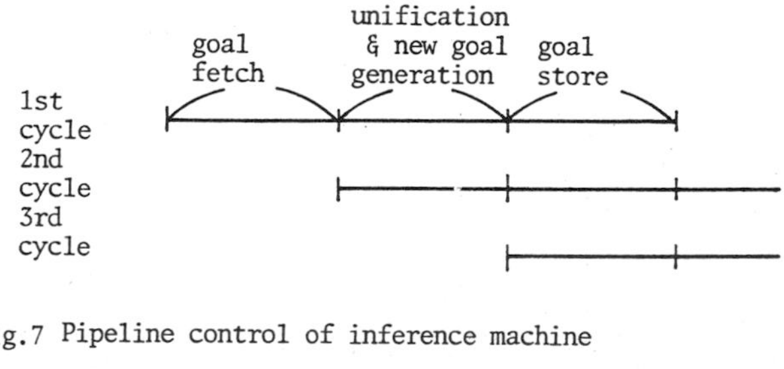

Fig.7 Pipeline control of inference machine

3.3. KNOWLEDGE BASE MACHINE ARCHITECTURE
A. THE CONCEPT OF KNOWLEDGE BASE
Knowledge information processing systems handle all information as knowledge. This includes simple data such as country population, rules such as mathematical theorems, programs, and common knowledge. From the operation point of view, KIPS can be seen as inference systems. On the other hand, KIPS can be seen as

knowledge base systems from the data point of view. As stated previously, the future FGCS will be a fused version of inference machine and knowledge base machines. However, at present, they should be investigated separately as their characteristics differ much from each other. Knowledge has many sided views. One aspect is rules and facts. The other aspect is permanent or temporary. Though mathematical theory can be regarded as permanent, knowledges expressed by some application programs are rather temporal (i.e. only within the scope of the program).

B. THE ROLES OF KNOWLEDGE BASE MACHINE
The machine which supports knowledge base is the knowledge base machine. The roles of knowledge base machine are,

(1) knowledge manipulation,
(2) knowledge storage, and
(3) knowledge acquisition.

KNOWLEDGE MANIPULATION
This includes the search operation for some stored facts or rules, and the search operation for such knowledge that are gotten as the results of the combination of stored facts and rules. The latter one is to get the results from inference operations, and corresponds to inference itself. Though any inference operation can be executed on the inference machine stated previously, there are such knowledge that are fit better for set operation like relational algebra. An example is relational data. Though relational data can be expressed by predicate calculus languages such as PROLOG, its manipulation could be better performed on some relational data base machine. This means that some part of inference should be done on knowledge base machine.

KNOWLEDGE STORAGE
Knowledge base machine should store a large amount of knowledge which will grow time by time. The same kind of functions as data base are required to maintain knowledge. Furthermore, some innovative background operation will be needed to detect inconsistency in the knowledge and to improve the efficiency of storage and of manipulation.

KNOWLEDGE ACQUISITION
Knowledge acquisition is to make up new knowledge, to test inconsistency of the knowledge world after adding the new one, and to remove unnecessary redundant knowledge. Up to this time, knowledge acquisition is done in such systems as expert systems with the help of human. Though it will be very hard to do this with full automatic, this function will be most important in future. To make up new knowledge, induction inference will be needed. Deductive inference function will be used for the inconsistency check.

C. MACHINE ARCHITECTURE
The basic architecture for the knowledge base machine is the relational data base machine. The relational data base machine will be added by knowledge operation mechanism and evolve toward the knowledge base machine. At the beginning, only facts data will be stored in the relational data base machine as they matches well to the relational data, and rules will be transferred to inference machine when they are used for execution. Typical interface to relational data base machine is relational algebra or relational calculus. As relational algebra is directly fit for the implementation of actions, it will be better to use relational algebra as the basic interface. As the relational calculus which can be mapped to relational algebra is a kind of predicate calculus, relational data base machines seem to match very well to predicate calculus languages such as PROLOG.

4. COMPUTER ARCHITECTURE RESEARCH PROJECTS

4.1. RESEARCH ORGANIZATIONS AND PROJECTS

Fifth generation computer systems project started in April 1982 with the foundation of ICOT. In June 1982, a research center inaugurated at ICOT as the kernel part for research and development of the Fifth Generation Computer Systems. In addition, the advisory groups : the Project Promotion Committee and 5 working groups, were formed to advice and encourage the activities of ICOT. These groups are organized by many researchers and experts of the organizations such as University of Tokyo, Electro Technical Laboratory of MITI, Electrical Communication Laboratory of NTT and so on.

Many research projects are going on in parallel. Among those, the computer architecture related projects are summarized as follows.

1. Sequential inference machine (SIM) project
2. Parallel inference machine (PIM) project
3. Knowledge base machine (KBM) project

The sequential Inference Machine (SIM) is a pilot model for the efficient development of software for the Fifth Generation Computer Systems. The architecture of SIM is basically the one of conventional computers and tuned for the logic programming. The object of Parallel Inference Machine project is to clarify the architecture which supports the parallel execution of inference operations. Knowledge Base Machine project is the research of such architecture that supports the execution of basic knowledge base operations. This project includes a few items. One is the research and development of relational data base machine. Second is the research of parallel knowledge operations which provide speedy knowledge accumulation, retrieval and updating, data conversion, etc. As very large scale integration technology is one of the key technology for architecture design, a few VLSI related projects are going on, such as VLSI-CAD development, and experimental fabrication of a few machine components by VLSI. The projects stated above are going at ICOT or a few organization closely related to ICOT. However, besides these, many projects which are related to the knowledge information processing systems are going on at other organizations than ICOT, as well. Followings are the introduction of computer architecture related projects which are being carried out mainly at ICOT.

4.2. SEQUENTIAL INFERENCE MACHINE PROJECTS [14]

SIM is a pilot model for software development. The kernel language version Ø, KLØ, plays the role of interface between SIM hardware and software. That is, the KLØ can be regarded as the machine language of SIM. KLØ is a predicate calculus language which includes and extends PROLOG. Each predicate in the kernel language is converted to the internal representation format of the SIM, and then directly interpreted and executed by microprogram.

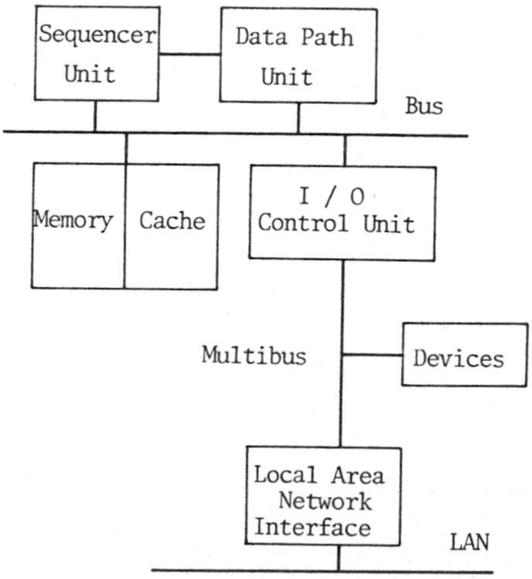

Fig.8 The configuration of the Sequential Inference Machine-PSI

The basic configuration of personal SIM (PSI) is shown in fig. 8. The data path units includes a few buses, some registers and an ALU. Some registers are specialized for resolution and unification. Resolution is a stack operation which accompanies the call/return of a clause. Unification performs the data comparison with the tag check. The memory module consists of a cache and address translation mechanism, and memory board. SIM doesn't support virtual memory. The memory board will accommodate 16MW. One word is 4Ø bits, of which format is shown in fig. 9. The tag part of 6 bits represents data types such as

symbolic atom, integer number, real number, vector, string, variable, undefined, reference, built in code, code, control mark, etc.

SIM provides a bit-map display and pointing device, a multibus (IEEE-796) for standard input/output, and a local area network interface. The instruction expression of SIM is shown in fig. 1Ø. This expression represents a clause in KLØ, and consists of a clause header part, an argument part of the head, and two or more body-goal parts. A body-goal may contain a user defined predicate (pointer and arguments), or a built-in predicate. KLØ has four stacks : local, global, control and fail stacks. KLØ runs using these stacks and a heap area where instruction codes are stored.

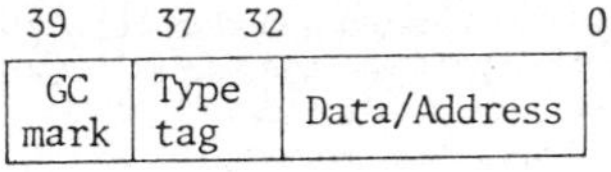

Fig.9 Word format

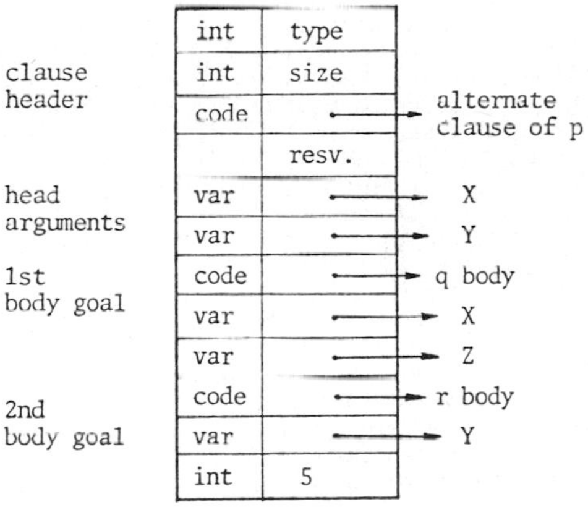

, where p(X,Y) :- q(X,Z), r(Y,5).

Fig 10 Clause representation of p.

<u>4.3. PARALLEL INFERENCE MACHINE PROJECT</u>

PIM project aims to clarify and to show the feasibility of the architecture which supports the parallel execution of inference operations. The research of a few processing models for parallel inference machine is being carried out in parallel. Followings are the brief description of 2 machines, data-flow model PIM and logic model PIM, among them.

A. DATA FLOW MODEL PIM

Among 3 types of parallelism (OR, AND, argument parallelism), OR parallelism and argument parallelism are realized on this machine. The argument parallelism is achieved by decomposing a unification process for a goal literal into several unification operations for each argument. The decomposition is represented by a data flow graph, through which efficient parallel processing is realized. Regarding AND parallelism, AND pipeline processing is realized instead of it as it may introduce great complexity.

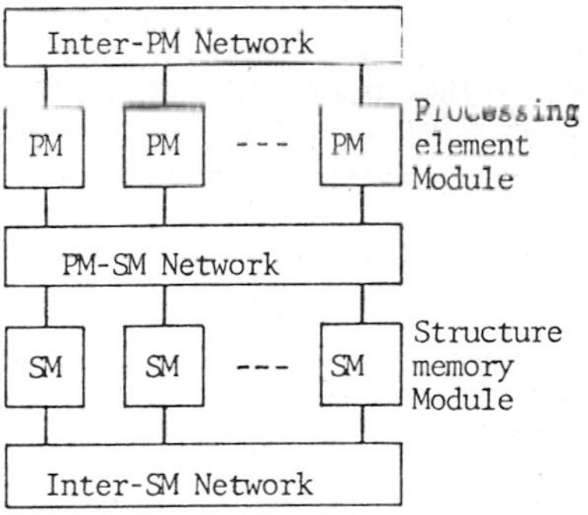

Fig.11 The configuration image of data-flow Parallel Inference Machine

The structure image of this machine is shown in fig. 11. [13] The machine is made of processing modules (PMs), structure memory modules (SMs), and three networks. PM which executes the unification processes, consists of an instruction control module and an execution module. These 2 modules form a circular pipeline structure. The instruction control module detects the firable operations of which operands are ready to be executed, and generates the instruction packets. The execution module receives the packets, interprets the instructions, and executes the corresponding functions.

The SM stores, manages and manipulates structured data such as list, vector and stream. To avoid the access conflicts on some specific SM, structures are distributed among the SMs. Each memory cell is provided with reference counter of 8 bits a read-only tag and a garbage tag of one bit so as to allow structure sharing and to control garbage collection. There are 3 networks, Inter-PM network, PM-SM network, and Inter-SM network. These networks support asynchronous packet communications. A compiler and a simulator are being developed regarding the software of this machine. The compiler compiles programs in KLØ to data flow graphs.

B. LOGIC MODEL PIM

The parallel inference machines of this model directly execute the inference operations. As shown in fig. 5, many goals are stored in a goal pool. Each unify processor fetches a goal at a time, performs unification with definition clauses and generates several new goals if the unifications succeed. Project PIE (parallel inference engine [11]) of this type is based on OR parallelism, and goals are independent each other. As each goal includes all the necessary information from the start of the inference to the processing step, goals may include a lot of same information each other. However, this doesn't always mean that each goal should use separate memory area. It is enough for each goal to be regarded as independent from logical point of view. On the other hand, you may fear at the first glance that the size of goal grows exponentially as the inference steps elapse. However, it is shown from fig. 12 that the size doesn't grow so much. The example program of fig. 12 is eight queens, which is the same program for fig. 4. (As this characteristics depend on the programs, we need more measurement results, of course.)

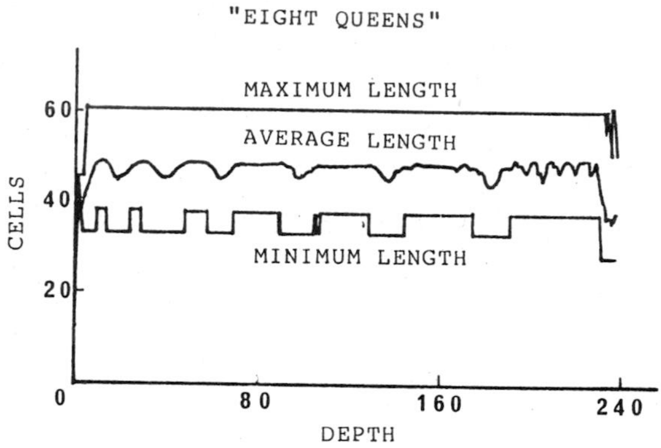

Fig.12 The change of average size of goals.

The configuration image of PIE is shown in fig. 13. Each unify processor is served by an individual definition memory, which stores the same copy of definition clauses needed for the processing. The loading of definition clauses is controlled by a system manager. Memory modules act as a goal pool on the whole. At the initial stage of research, this memory modules were assumed to memorize each goal physically independently due to its simplicity. However, data sharing schemes are being investigated now. By this scheme, each cell of memory modules is equipped with reference counter, and makes up the structure memory modules. Activity controllers control the search strategy of solutions. The examples are the "guard" operation to select only one success solution at an OR parallel execution, and "remote-cut" operation to discard some goals which turned out to be

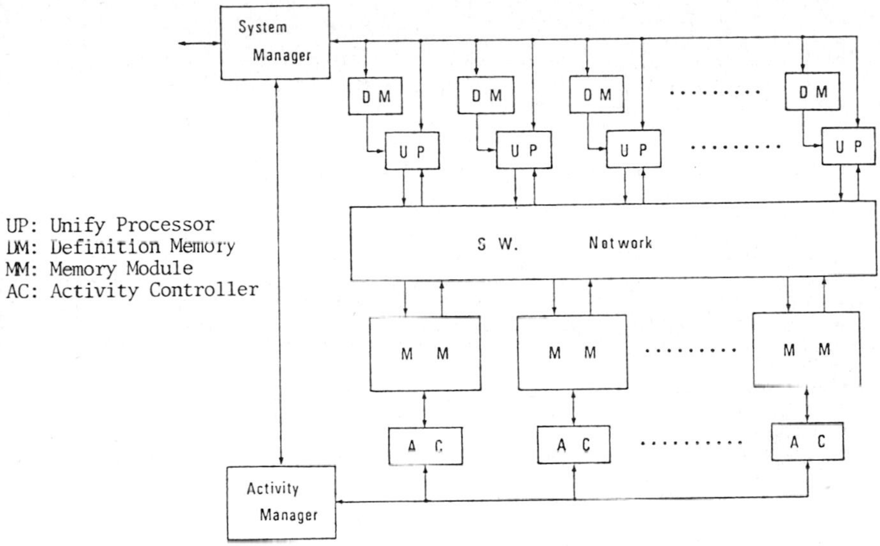

Fig.13 The configuration of Parallel Inference Engine-PIE

unneeded any more. For this purpose, each activity controller keeps the inference tree information in its memory. Following this tree, activity controllers send and receive commands each other, and control the activity of the system through cutting the tree connections or setting the status of goals "wait". The Activity Manager is the central controller of this system, while the activity controller is the local one.

For the investigation of PIE, we wrote a few software simulators such as,

1. Simulators for OR parallelism measurement,
2. Time simulator of PIE model, and
3. Structure memory simulator.

Besides these software simulator, we are building a hardware simulator and a unify processor board. The hardware simulator which will be made of a few work stations can simulate the parallel activity of PIE model. The unify processor board which will be implemented by TTL ICs is an experimental product for unification hardware.

4.4. RELATIONAL DATA BASE MACHINE PROJECT [5]

At the initial stage of FGCS project, the knowledge base system hardware is compose of a relational data base machine and a sequential inference machine. The software module for knowledge base management which runs on the sequential inference machine transforms the queries for external

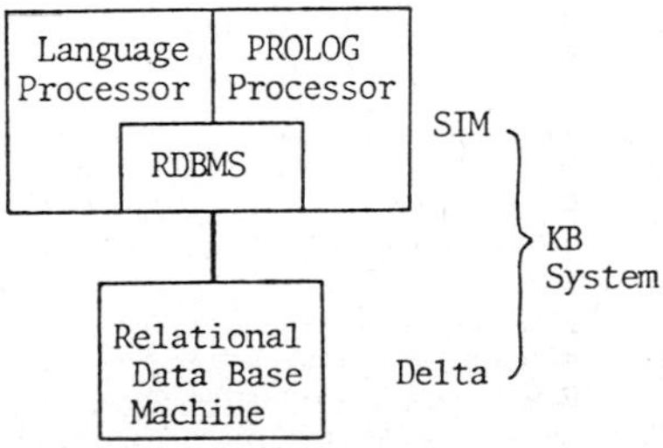

Fig.14 The experimental system of Knowledge Base System

data base into relational algebra commands, accesses to the relational data base machine, and returns the result information to the originator of queries. The architectural basic design data for the knowledge base subsystem will be gathered through these experiences. This configuration is shown in fig. 14. The relational data base machine for this experiment is called "Delta". Fig. 15 shows the internal organization of Delta. Control processor (CP) analyzes command trees given by SIM, breakdowns them into subcommands for hierarchical memory and relational data base engine, and manages the dictionary/directory. Relational data base engine (RDBE) is a hardware to execute retrieval subcommands in relational algebra. RDBE sorts the input attributes, and input the output into a relational operation unit. As Delta can be provided several RDBEs connected through a network, it can process the command tree in parallel.

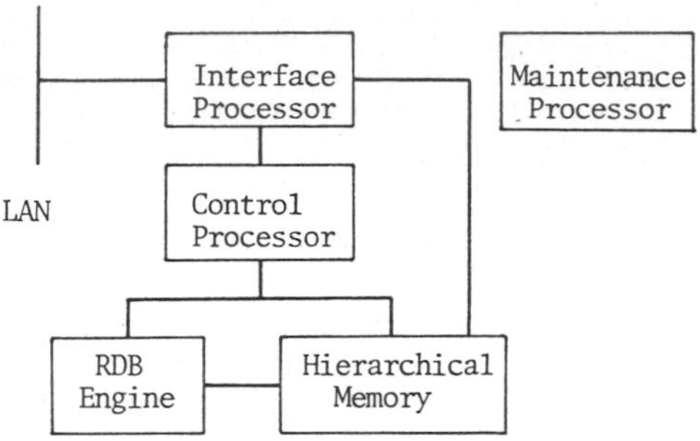

Fig.15 The internal organization of "Delta".

Hierarchical memory (HM) is a physical storage device of data. It retrieves data with high speed, and transfers the requested data to the RDBE, CP and IP. It also stores incoming data, insert, delete and modifies the stored data. The hardware components of (HM) are the working memory of high speed RAM, the silicon disk which is made of medium speed semiconductor memory to simulate a high speed disk memory, and the group of moving head disks. These 3 components form a memory hierarchy, and realize the high speed secondary storage of large volume. The working memory contains also the directory/dictionary information. The maintenance processor is a kind of supervisor processor. The roles are the bootstrapping of Delta, diagnosis of failures, recovery from the failure, loading control of existing data base, and data gathering for system evaluation.

Besides this hardware design and implementation, a software simulator is developed, whose results will be used to fix the detailed parameter and to find the system bottleneck.

5. CONCLUSION

In this paper, I summarized the fifth generation computer systems project which is now being carried out in Japan since 1982, gave consideration especially about the computer architecture for inference operations and knowledge operations, and described a few architectural projects which are going on at and around the institute of new generation computer technology.

At the end of 1984, several experimental systems are coming out such as personal sequential inference machines, parallel inference machines of data flow model and logic model, and relational data base machine. The middle term for FGCS project will begin at the fiscal year of 1985. Regarding the architectural research, the first year of the middle term will be devoted to the full scale evaluation of a few processing models. At the same time, the feasibility of parallel architecture will be evaluated in terms of the degree of parallelism. The major themes of the middle term are to built 2 subsystems, the inference machine subsystem and the knowledge base subsystem. These machines will be developed using the result of the initial term. The VLSI technology will play an important role for the implementation of hardware.

During the initial term, the kernel language Ø (KLØ) is used for the research. However this language is oriented to the conventional sequential machine. Though the second version of kernel language (KL1) is developed, which is oriented to the parallel processing and the object concept, the full scale usage of KL1 will begin at the middle term. The architectural research of the initial term is based on the subset of KLØ added by a few parallel processing feature.

[1] Proceedings of International Conference on Fifth Generation Computer Systems, Japan Information Processing Development Center, Oct.1981.

[2] Moto-oka,T. and Tanaka,H., The Fifth Generation -- Progress in Japan, Supercomputer Systems Technology, Series 1Ø, No.6, Pergamon Infotech Limited, 1982.

[3] Shapiro,E.Y., and Takeuchi, Object Oriented Programming in Concurrent Prolog, ICOT Technical Report TR-ØØ4, 1983.

[4] Chikayama,T., ESP as Preliminary Kernel Language of Fifth Generation Computers, ICOT Technical Report TR-ØØ5, 1983.

[5] Shibayama,S., Kakuta,T., Miyazaki,N., and Yokota,H., A Relational Database Machine "Delta", ICOT Technical Memorandum TM-ØØØ2, 1982.

[6] Research Report on Fifth Generation Computer Systems Project, Institute for New Generation Computer Technology, March 1983.

[7] Keller,R.M., and Sleep,M.R., Applicative Cashing: Programmer Control of Object Sharing and Lifetime in Distributed Implementations of Applicative Languages, Proc. 1981 Conference on Functional Programming Languages and Computer Architecture, 1981.

[8] Conery,J.S., and Kibler,D.F., Parallel Interpretation of Logic Programs, Proc. 1981 Conference on Functional Programming Languages and Computer Architecture, 1981.

[9] Darlington,J., and Reeve,M., Alice: A Multi-Processor Reduction Machine for the Parallel Evaluation of Applicative Languages, Proc. 1981 Conference on Functional Programming Languages and Computer Architecture, 1981.

[1Ø] Sansonnet,J.P., Castan,M., and Percebois,C., M2L: A List-Directed Architecture, International Symposium on Computer Architecture, La Baule, May 198Ø.

[11] Goto,A., Aida,H., Yamazaki,A., Maruyama,T. Yuhara,M., Tanaka,H., and Moto-oka,T., On the Efficient Parallel Processing of the Highly Parallel Inference Engine — PIE, Proc. Electronic Computer Society of IECE of Japan, EC83-9, 1983.

[12] Goto,A., Aida,H., Maruyama,T. Yuhara,M., Tanaka,H., and Moto-oka,T., Proc. of The Logic Programming Conference '83, Tokyo, March 1983 (in Japanese).

[13] Ito,T., Onai,R., Masuda,Y., and Shimizu,H., Prolog Machine based on the Data Flow Mechanism, Proc. of The Logic Programming Conference '83, Tokyo, March 1983 (in Japanese).

[14] Uchida,S., Yokota,M., Yamamoto,A., Taki,K., and Nishikawa,H., Outline of the Personal Sequential Inference Machine: PSI, New Generation Computing Vol.1, No.1, 1983.

Computing Methods in Applied Sciences and Engineering, VI
R. Glowinski and J.-L. Lions (Editors)
Elsevier Science Publishers B.V. (North-Holland)

Increasing the Performance of Mathematical Software Through High-Level Modularity

Jack J. Dongarra†

Mathematics and Computer Science Division
Argonne National Laboratory
Argonne, Illinois 60439
U.S.A.

Abstract — This paper describes the performance of some algorithms in linear algebra based on high-level modules. High-level modularity facilitates portability and aids in attaining performance efficiency on a wide variety of environments spanning scalar, vector, and certain parallel computers.

Motivation for the Study

This work is motivated by a number of different but related concerns. The first of these involves the performance of standard software for dealing with dense matrix problems. Although the software implementing the algorithms was almost fully vectorized, and the although vectorized part was performed by highly tuned assembly language routines, the performance on certain vector machines was a factor of three lower than possible.

The second concern centers on the desire to produce algorithms that perform well on scalar, vector, and certain parallel machines. The third concern involves the desire to develop such routines in a portable, Fortran implementation. These two points have obvious advantages when software is being moved from one computer to another and also when new computer architectures are implemented.

Timing Study with LINPACK

During the LINPACK [1] project, various timing studies were performed in an attempt to measure the speed of certain well-used routines in the package. In particular, the set of routines used to factor and solve a dense general matrix was run in a number of different environments. The software and test problem remained the same; only the machine, compiler, and operating system varied.

The LINPACK programs can be characterized as having a high percentage of floating point arithmetic operations. The routines involved in this timing study, SGEFA and SGESL, use column-oriented algorithms. By column orientation we mean that the programs usually reference array elements sequentially down a column, not across a row. Column orientation is important in increasing efficiency in a Fortran environment because of the way in which arrays are stored. Most of the floating point operations in

†Work supported in part by the Applied Mathematical Sciences Research Program (KC-04-02) of the Office of Energy Research of the U. S. Department of Energy under Contract W-31-109-Eng-38.

T. Jordan and K. Fong [3] have described an algorithm that accomplishes the same procedure. At the jth step multiples of parts of the first $j-1$ columns are added to a segment of the jth column.

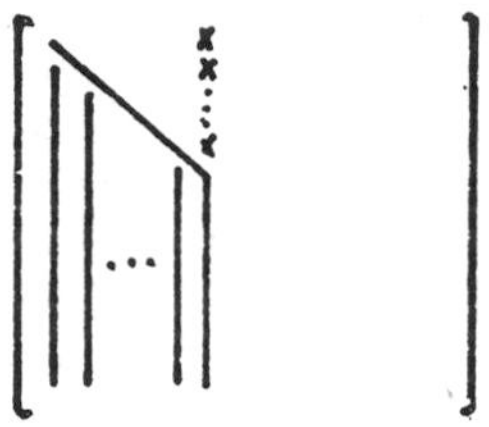

As before, the basic operation is $y \leftarrow \alpha x + y$, but the vector y can be retained in a register and is not stored during each operation. The vector x is loaded and used but not changed. Thus it is possible to achieve an execution rate of around 120 MFLOPS (super-vector speeds).

The problem with this approach is that it cannot achieve this rate in Fortran. Cray assembly language is necessary for two reasons: 1) irregular vector lengths are created during the main reduction loop, and 2) information just undated is needed for the next vector operations. The limitations of the Jordan-Fong approach led us to search for an alternative algorithm that used vectors of the same size throughout a given step and would be capable of producing high execution rates in Fortran.

Algorithm Design Based on Matrix-Vector Operations

The *LU* factorization based on the Crout method met our requirements. The algorithm is based on matrix-vector operations. Here at the jth step, a matrix formed from columns 1 through $j-1$ and rows j through n is multiplied by a vector constructed from the jth column, rows 1 through $j-1$, with the results added to the jth column, rows j through n. The second part of the jth step involves a vector-matrix product, where the vector is constructed from the jth row, columns 1 through $j-1$, and a matrix constructed from rows 1 through $j-1$ and columns $j+1$ through n, with the results added to the jth row, columns $j+1$ through n.

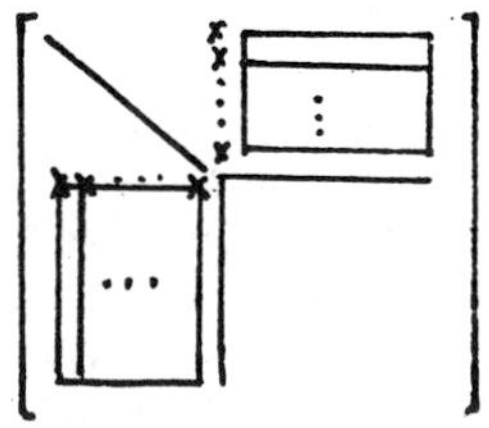

The vectors in this algorithm remain the same size throughout the step. The matrix-vector operation can be thought of as being performed in a subroutine, e.g.,

```
do j = 1,n
  do i = 1,m
    y(i) = y(i) + x(j)*M(i,j)
  continue
continue
```

The Cray CFT compiler does not detect the fact that the result can be accumulated in a register (and not stored between successive vector operations). Thus, the rate of execution is limited to vector speeds.

If, however, we unroll [4] the outer loop (in this case to a depth of four) and insert parentheses to force the arithmetic operations to be performed in the most efficient order, then the innermost loop becomes

```
do j = 1,n,4
  do i = 1,m
    y(i) = y(i) + x(j)*M(i,j)
                + x(j+1)*M(i,j+1)
                + x(j+2)*M(i,j+2)
                + x(j+3)*M(i,j+3)
  continue
continue
```

Now the code generated by CFT has *six* memory references for every *eight* floating-point operations. Thus the maximum rate of execution is ~100 MFLOPS (*super-vector performance from Fortran.*)

The same ideas for use of high-level modules can be applied to other algorithms, including matrix multiply, Cholesky decomposition, QR factorization, and Gram-Schmidt orthogonalization. These algorithms are based on standard procedures in linear algebra. They have been written to retain much of the original mathematical formulation and are based on matrix-vector operations that are isolated through calls to subroutines. Designing the algorithms in terms of such operations is really the difficult part of an implementation. When programming from an algorithmic description, one has a tendency to focus on small details during an implementation. To unveil the desired structure, one must look higher, avoiding the details, and must concentrate on operations that embody the computational components of an algorithm.

Just three modules are needed: matrix-vector multiplication ($y = y + Ax$), vector-matrix multiplication ($y^T = y^T + x^TA$), and a rank one update to a matrix ($A = A + xy^T$). These modules represent a high level of granularity in the algorithm in the sense that they are based on matrix-vector operations, $O(n^2)$ work, not just vector operations, $O(n)$ work. The granularity holds the key to production of efficient and portable routines. By isolating the computationally intense parts of an algorithm in high-level modules, one can treat the modules separately, perhaps retargeting them for quite different architectures, and making the overall algorithms efficient on the targeted architecture.

When the architecture changes, the basic algorithm is retained and just the matrix-vector modules are reprogrammed. One of our goals is to avoid locking the algorithms into one computer architecture, however fast that one may be. Another goal is to design the algorithms at a level that only the fundamental modules need be replaced to gain the desired performance.

Denelcor HEP

The HEP [5] computer is an MIMD computer consisting of one or more pipelined MIMD Process Execution Modules (PEMs) that share memory. Several separate pipelines interact on each PEM, but the major flavor can be given by considering only the main execution pipeline. Parallelism is simulated by having different pairs of operands occupying different stages of the pipeline simultaneously. On the HEP the execution pipeline that most instructions use is broken down into eight stages. Thus, independent instructions flow through the pipeline, with instructions finishing execution in eight steps. Independent processes are issuing instructions from independent instruction streams. Copies of the process state and program counters are kept for each of the processes. The diagram below shows the flow of information.

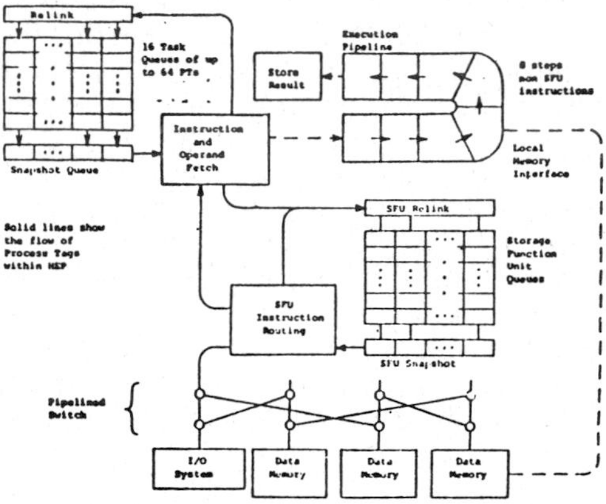

To produce on the HEP a parallel version of the algorithm described in the previous section, we replaced only the three basic matrix-vector modules. On the Cray the matrix-vector multiplication was performed by taking multiples of the columns and accumulating the results in a vector.

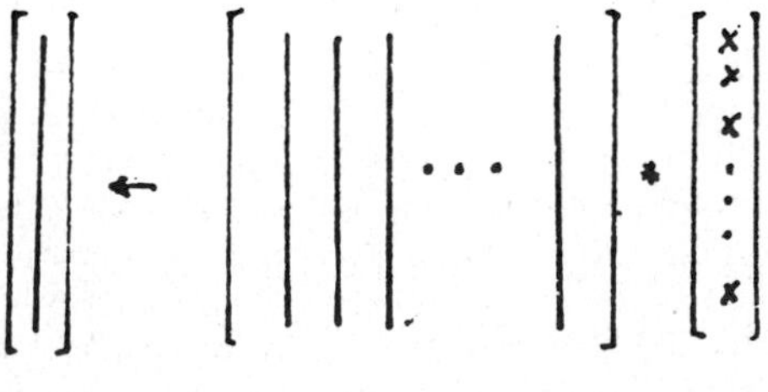

This approach allowed for vector operations in such a way that the result was maintained in a vector register throughout the calculation and super-vector rates were achieved. For the HEP, the parallelism in matrix-vector multiplication was obtained by performing m independent inner products with a matrix of size $m \times n$ and a vector of length n. For the vector-matrix multiplication, n independent inner products were performed with a vector of length m and a matrix of size $m \times n$:

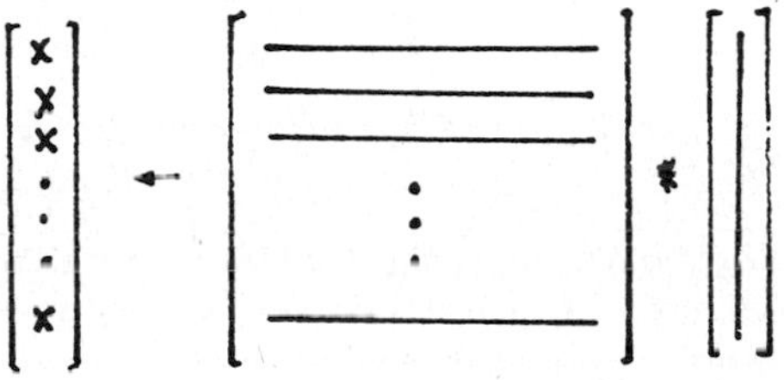

The parallelism in the rank one update was obtained by performing n of the following operations, a scalar times a vector added to a column of the matrix.

The graphs that follow compare the performance of the parallel algorithms and their sequential counterparts on the HEP. The routines were run first with straight sequential Fortran 77 versions of the modules, using no parallel constructions, and then with the sequential modules replaced by their parallel counterparts. The parallel algorithms, written in an extended version of Fortran 77, require the same number of floating point operations and have identical properties with respect to roundoff errors as their sequential counterparts. What is plotted is the ratio of execution time for the sequential program divided by the execution time for the parallel program as the matrix order is varied. These runs were made on a HEP that had one PEM. Since the instruction pipeline is segmented into eight pieces, one may expect the ratio of sequential to parallel times to reach a maximum at eight. In fact, however, the speedups can actually be slightly greater than eight as the result of memory references, which do not go through the instruction execution pipeline.

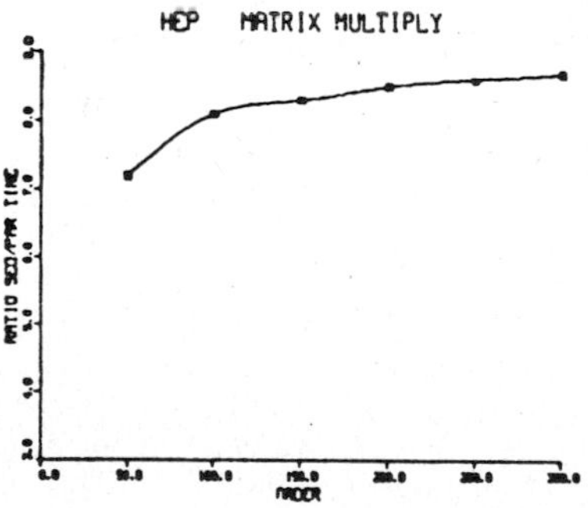

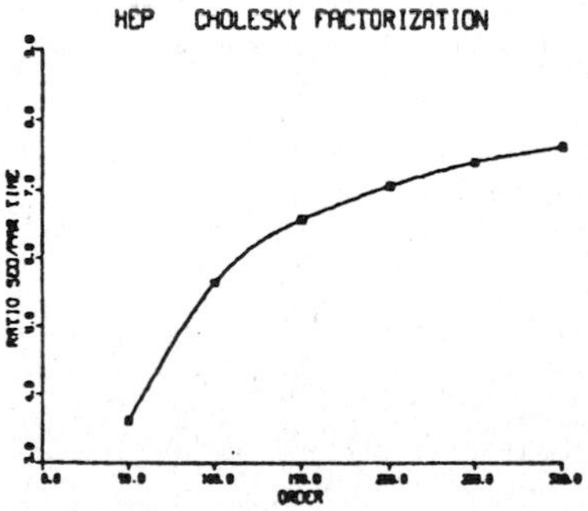

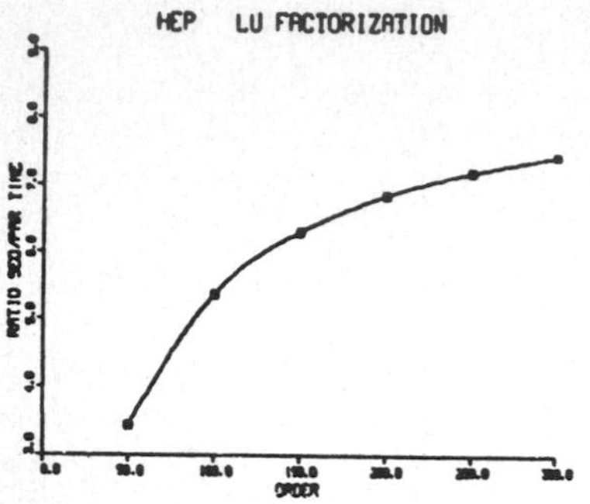

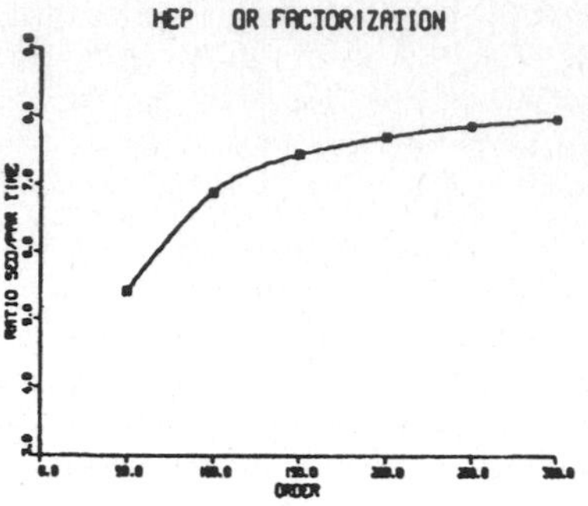

I include some recent results obtained on the CRAY X-MP using two processors. The algorithm involves *LU* decomposition with partial pivoting and is carried out as outlined above. The algorithm can be described as having a loop around three sections: ***perform a matrix-vector product***, ***perform a pivot search***, and ***perform a vector-matrix product***. The matrix-vector and vector-matrix products are independent in the sense that one does not require the results of the other. Thus, the algorithm may be able to take advantage of both processors on the CRAY X-MP. The picture below illustrates the operations:

TASK 1

$$y_1 \leftarrow y_1 + M_1 x_1$$

TASK 2

$$y_2^T \leftarrow y_2^T + x_2^T M_2$$

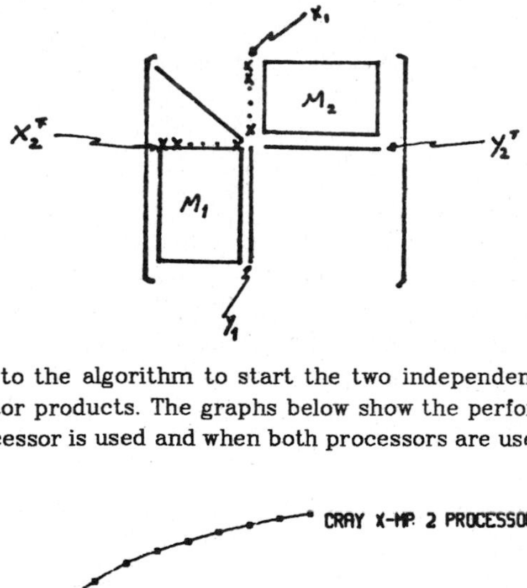

Small changes were made to the algorithm to start the two independent tasks associated with the matrix-vector products. The graphs below show the performance for the algorithm when one processor is used and when both processors are used.

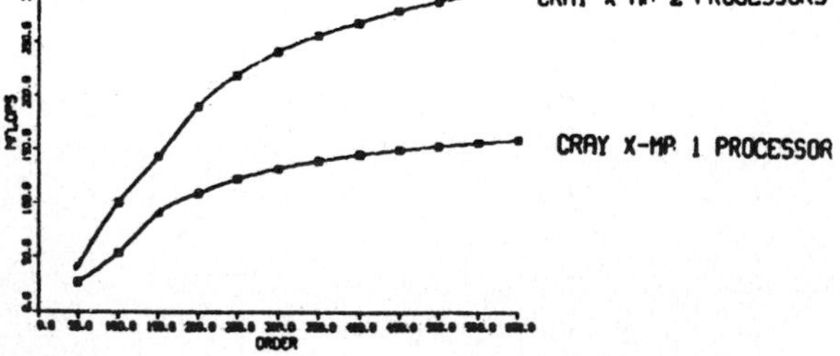

When both processors are used, the speedup is a factor 1.9 over the time for a single processor.

Conclusion

With such a wide variety of computer systems and architectures in use or proposed, there is a very real challenge for people designing algorithms, namely, how to write software that is both efficient and portable. The solution lies in the granularity of the task. Programs expressed in terms of modules with a high level of granularity reflect less of the detail and retain more of the basic mathematical formulation. This allows for a wider range of efficient implementations, since the computationally intense parts are isolated in high-level modules.

The module concept offers a further advantage. It allows one to divide a large problem into small, easily understood pieces that can be programmed separately and verified at each step of the development process. These pieces are then chosen, perhaps repeated, to solve various aspects of the larger problem. The success of this approach in efficiently solving problems across a wide spectrum of computers depends on how well the modules can be chosen. The modules must be at a high enough level to allow a significant number of arithmetic operations to be performed.

REFERENCES

[1] J.J. Dongarra, J.R. Bunch, C.B. Moler, and G.W. Stewart, *LINPACK Users' Guide*, SIAM Publications, Philadelphia, 1979.

[2] C. Lawson, R. Hanson, D. Kincaid, and F. Krogh, "Basic Linear Algebra Subprograms for Fortran Usage," *ACM Trans. Math. Software*, 5(3), 308-371, 1979.

[3] K. Fong and T. L. Jordan, *Some Linear Algebra Algorithms and Their Performance on CRAY-1*, Los Alamos Scientific Laboratory, UC-32, June 1977.

[4] J.J. Dongarra and S.C. Eisenstat, *Squeezing the Most out of an Algorithm in Cray Fortran*, Argonne National Laboratory, Argonne, Illinois, ANL Tech Memo ANL/MCS-TM-9, May 1983.

[5] *HEP Fortran 77 User's Guide*, Denelcor Inc., Aurora, Colorado, 1982.

Computing Methods in Applied Sciences and Engineering, VI
R. Glowinski and J.-L. Lions (Editors)
Elsevier Science Publishers B.V. (North-Holland)

QUELQUES PROGRES EN CALCUL PARALLELE ET VECTORIEL

Jocelyne ERHEL (*) ; William JALBY (*)
Alain LICHNEWSKY (**) ; François THOMASSET (*)

RESUME

Cet article se compose de deux parties, toutes deux liées à l'algorithmique sur calculateurs parallèles.

Dans une PREMIERE PARTIE, nous décrivons quelques adaptations pour des calculateurs parallèles ou vectoriels de l'algorithme du gradient conjugué préconditionné. Notre objectif était tout d'abord de tester différentes alternatives et de vérifier les bonnes propriétés de convergence de l'algorithme. Nous avons également pu mettre en évidence quelques heuristiques permettant d'accroitre le degré de parallélisme disponible et exploitable.

Nous objectif fut d'étudier plus en détail les variantes valables soit pour des multiprocesseurs (MIMD), soit pour des calculateurs vectoriels. Nous allons donner quelques résultats sur les possibilités d'application de nos techniques dans diverses situations. Comme nous envisageons le cas d'une matrice creuse résultant de la discrètisation par différences ou éléments finis d'un domaine 2D ou 2D, les modèles présentés montrent l'influence de la complexité du domaine sur l'accélération obtenue. Concernant l'approche MIMD nous montrons d'une part des résultats d'évaluations au moyen de modèles simples, d'autre part des résultats expérimentaux obtenus sur le CRAY-XMP.

(*) INRIA
(**) INRIA et Université Paris Sud

La majeure partie de l'algorithme du gradient conjugué consiste en multiplications matrice x vecteur, en combinaisons linéaires de vecteurs, et en produits scalaires, toutes opérations qui s'effectuent avec une bonne efficacité sur des calculateurs parallèles. Nous ne détaillerons donc que la résolution de systèmes linéaires qui est nécessitée par la présence de l'opérateur de préconditionnement.

Les techniques que nous présentons ici sont basées sur des renumérotations des inconnues permettant d'obtenir la structure désirée pour le graphe de connection de la matrice. Un algorithme de théorie des graphes permet d'exhiber une partition en sous-domaines utilisable dans une implémentation multiprocesseur. D'autre part des variantes des techniques de "coloriage" donnent un parallélisme de type vectoriel avec de longs vecteurs. Nous proposons des adaptations aux calculateurs multi-vectoriels (tels que le CRAY-XMP), combinant les deux heuristiques précédentes en une approche unifiée. Les nouvelles techniques de coloriage étudiées par [SWP 1983] sont un pas dans la même direction ; toutefois nous ne connaissons actuellement guère de résultat sur la numérotation "optimale".

L'efficacité des gradients conjugués préconditionnés et de leurs adaptations aux calculateurs vectoriels a été l'objet de nombreux travaux : on peut citer [MVV 1977, MVV 1981] ; [DGR 1979] ; [RoW 1982] ; [Meu 1982] ; [CGM 1983] ; [JMP 1982] ; [SMP 1983]; [Erh 1983a]. Notre approche est dérivée de [Lic 1981], [Lic 1982]. On présente d'abord le cas de maillages réguliers de type "différences finies" ; on étend ensuite la technique à des maillages d'éléments finis généraux.

La SECONDE PARTIE traite des problèmes d'accès aux données dans un calculateur SIMD. En effet l'une des difficultés majeures dans la parallèlisation d'algorithmes pour de telles machines est liée à la nécessité d'accéder en parallèle aux opérandes et de les amener aux unités de traitement à travers un réseau d'interconnexion.

Nous exposons des méthodes de résolution de ce problème dans le cas des méthodes multi-grilles ; nous évaluons les performances d'une machine SIMD munie d'un réseau d'interconnexion de type Ω sur de tels algorithmes [Jal 1983].

REFERENCES

O. AXELSSON [Axe 1976] : "Solution of linear systems of equations : iterative methods", in Sparse Matrix Techniques, V.A. Barker Ed., Springer.

P. CONCUS, G. GOLUB & G. MEURANT [CGM 1982] : "Block preconditionning for the conjugate method", Report LBL-14856, Stanford University, Stanford CA.

CRAY Research Inc. [CRI 1983] : "Proposed Multi-Tasking Facility for CFT".

P.F. DUBOIS, A. GREENBAUM & G. RODRIGUE [DGR 1979] : "Approximating the inverse of a matrix for use in iterative algorithms on vector processes", Computing, N°. 2, pp. 257-268.

J. ERHEL [Erh 1983a] : "Parallèlisation d'un algorithme de gradient conjugué préconditionné", Rapport de Recherche INRIA N° 189, Février 1983.

J. ERHEL [Erh 1983b] : "CREM : User's manuel", Rapport technique INRIA n° 25, Mai 1983.

J. ERHEL, A. LICHNEWSKY & F. THOMASSET [ELT 1982] : "Parallelism in finite element computation", Proceedings of IBM Symposium on parallel processing, Rome.

W.M. GENTLEMAN [Gen 1981] : "Design of Numerical Algorithms for Parallel Processing", CREST course, Bergamo, Italie, Juin 1981.

J.A. GEORGE [Geo 1977] : "Solution of linear systems of equations : direct methods for finite element problems", in Sparse Matrix Techniques, V.A. Barker Ed., Springer.

J.R. GILBERT [Gil 1980] : "Graph separator theorems and sparse gaussian elimination", Ph. D. Thesis, Stanford.

W. JALBY [Jal 1983] : These de 3ème cycle, Université Paris-Sud, Orsay, Juin 1983.

O.G. JOHNSON, C.A. MICHELLI & G. PAUL [JMP 1982] : "Polynomial preconditionning for conjugate gradient calculations", IBM Thomas J. Watson Research Center, Yorktown Heights, NY, 1982.

A. LICHNEWSKY [Lic 1981] : "Sur la résolution de systèmes linéaires issus de la méthode des éléments finis par une machine multi-processeurs", Rapport de Recherche INRIA, n° 119.

A. LICHNEWSKY [Lic 1982] : "Solving some linear systems arising in finite element methods on parallel processors" SIAM National Meeting, Stanford.

R.J. LIPTON & R.E. TARJAN [LiT 1979] : "A separation theorem for planar graphs", SIAM J. on Appl. Maths., 36 (1979), pp. 177-189.

T.A. MANTEUFFEL [Man 1978] : "The shifted incomplete Cholesky factori zation", SAND/78/8226, Sandia Lab., Albuquerque, NM.

J.A. MEIJERINK & H.A. VAN DER VORST [MVV 1977] : "An iterative solution method for linear systems of which the coefficient matrix is a symmetric M-matrix", Math. Comp., Vol. 31, pp. 148-162.

J.A. MEIJERINK & H.A. VAN DER VORST [MVV 1981] : "Guidelines for the usage of incomplete decomposition in solving sets of linear equations as they occur in practical problems", J. Comp. Ph., Vol. 44, pp. 134-155.

G. MEURANT [Meu 1983] : "Vector preconditionnings for the conjugate gradient method", to appear.

R. SCHREIBER & WEI-PEI-TANG [SWP 1983] : "Vectorizing the conjugate gradient method", to appear.

G. RODRIGUE & D. WOLITZER [RoW 1982] : "Preconditionning by incomplete block cyclic reduction", Research Report UCID-19502, LLNL, Livermore, CA.

H.A. VAN DER VORST [VDV 1983] : "On the vectorization of simple ICCG methods", First Int. Coll. on Vector & Parallel Computation, AFCET-GAMNI ISINA, Paris.

Computing Methods in Applied Sciences and Engineering, VI
R. Glowinski and J.-L. Lions (Editors)
Elsevier Science Publishers B.V. (North-Holland)

VECTOR PRECONDITIONING FOR THE CONJUGATE GRADIENT ON CRAY-I AND CDC CYBER 205

Gérard MEURANT

CEA Centre d'Etudes de Limeil

Conjugate gradient is a very attractive method to solve linear systems on vector computers like the Cray-I and the CDC Cyber 205.

Most of the algorithm is trivially vectorizable. However it is known that for the method to be efficient one needs a good preconditioner (Concus,Golub and O'Leary /2/). Unfortunately most of the well known good preconditioners are not fully vectorizable. So recently an active research has been conducted on finding good vector preconditioning methods for the conjugate gradient.

Dubois, Greenbaum and Rodrigue /3/ used a truncated Neumann series, Rodrigue and Wolitzer /II/ described an incomplete block cyclic reduction, Lichnewsky /8/ suggested to use nested dissection techniques, Johnson, Michelli and Paul /6/ show how to use efficient polynomial preconditioners, Jordan /7/ compared several of these approaches with Chebychev polynomials, Schreiber and Wei Pai Tang /I3/ introduced graph coloring techniques.

In this paper we show results of numerical experiments for three different preconditioners:

a) The classical incomplete Choleski decomposition IC(I,I) as vectorized by H.A. van der Vorst /I4/

b) A vector version of the block preconditioner introduced by Concus, Golub and Meurant /I/

c) A polynomial preconditioner used by Saad /I2/.

The numerical results are given for two vector computers Cray I-S and CDC Cyber 205.

In section 2 we recall for convenience some technical details about the computers used.

In section 3 we identify the basic operations needed for the conjugate gradient algorithm and we give figures showing the speed of computation we obtain for these operations.

Section 4 contains a brief description of the preconditioners Numerical results for a model problem are given in section 5.

As a conclusion we will see that the concept of "best preconditioner" is problem and machine dependant. Further results for the Cray X-MP will appear in a forthcoming paper /I0/.

2-Vector computers

We briefly recall here, for convenience, the main features of the Cray I-S and CDC Cyber 205. For more details we refer to the manufacturers manuals or to Hockney and Jesshope /5/.

2.I- The Cray I-S used is a 2300 model. It has a two millions words (64 bits) memory in eight banks. The clock period is I2.5 ns. As everyone knows the main feature of the Cray-I is eigth vector registers, each one being of 64 words.

The instruction set includes floating point vector operations. The data is read from memory, stored in the vector registers then flows to functional units, the result goes back to a register and then is written back to memory. However there is only one data path between memory and registers, so one cannot chain two loads from memory. This considerably slow down the potential computational speed. To code efficiently one has to avoid memory references. With Fortran it is not easy to have a computational speed greater than 50 Mflops (million of floating point operations per second).
The compiler used was CFT I.IO
2.2- The CDC Cyber 205 has a completely different architecture. The one we used was a 2 pipes machine with a I million word (64 bits) memory. It has a 20 ns clock period. The vectors have to be made of contiguous memory words (this is not a drawback for the conjugate gradient method).
The floating point operations are from memory to memory. For an operation the two vectors are read from memory, flow through a stream unit. The floating point operations are done into two identical "pipes", each of one has an additioner and a multiplier. Then data moves back to memory following an another path.
The maximum speed is 50 Mflops per pipe unless one has to perform a so called "linked triad" (a vector operation

involving two vectors and a scalar) in which case one can reach 100 Mflops per pipe that is to say 200 Mflops.
The start up time is usually long, hence the vector length has to be large enough to obtain the asymptotic speed.
Moreover the code must be written in CDC vector Fortran (no portability).

3-The conjugate gradient algorithm

We want to solve a linear system Ax=b whose matrix is a block tridiagonal symmetric Stieltjes matrix.

$$A = \begin{pmatrix} D_1 & A_2^T & & & \\ A_2 & D_2 & A_3^T & & 0 \\ & \ddots & \ddots & \ddots & \\ & 0 & A_{N-1} & D_{N-1} & A_N^T \\ & & & A_N & D_N \end{pmatrix}$$

N being the order of the blocks.
The model problem is coming from the standard 5 point discretization of the Laplacian operator with Dirichlet boundary conditions on the square $]0,1[\times]0,1[$.
The preconditioned conjugate gradient algorithm is the following (see /2/ or /4/)

$$x^0 \text{ random}, \quad r^0 = b - Ax^0, \quad p^{-1} \text{ arbitrary}$$

$$Mz^k = r^k$$

$$\beta_k = \frac{(Mz^k, z^k)}{(Mz^{k-1}, z^{k-1})}, \quad \beta_0 = 0$$

$$p^k = z^k + \beta_k p^{k-1}$$

$$\alpha_k = \frac{(Mz^k, z^k)}{(Ap^k, p^k)}$$

$$x^{k+1} = x^k + \alpha_k p^k$$

$$r^{k+1} = r^k - \alpha_k Ap^k$$

The choosen convergence criteria is $\frac{\|r^k\|_\infty}{\|r^0\|_\infty} \leq 10^{-6}$.

M is the preconditioning matrix. Apart from the solution of $Mz^k = r^k$, the whole algorithm is trivially vectorizable.

For each iteration we have to perform

a) 2 dot products

b) I sparse matrix $\times$ vector product

c) 3 vector+scalar $\times$ vector

What is the speed that we can reach for these operations ? Answers are given on figures I,2 and 3 where we show the speed as a function of vector length. We summarize them in Table I which gives for vector length 500 (upper number) and I0000 (lower number) the speed of computation for the three operations. BLAS is using the CAL coded routines SAXPY and SDOT on the Cray.

	Cray-I Fortran	Cray-I BLAS	CDC Cyber 205
a)	24 29	52 74	56 90
b)	54 60	-- --	70 90
c)	30 36	37 45	70 I65

Table I

Cyber 205 gives better results than the Cray-I for vector lengths greater than 300. Remark that c) is a linked triad.

4-Preconditioners

4-I The vectorized IC(I,I) (denoted hereafter by ICVDV) is described in a paper by H.A. van der Vorst /I4/. He used truncated Neumann series (with 4 terms) to solve the bidiagonal systems one gets during the forward and backward substitution. In this algorithm the vector length is N.

4-2 The block preconditioning INV(I) was introduced by Concus, Golub and Meurant /I/. M is choosen as

$$M = (\Delta + L)\Delta^{-1}(\Delta + L^T)$$

$$L = \begin{pmatrix} 0 & & & & \\ A_2 & 0 & & 0 & \\ & A_3 & 0 & & \\ & 0 & \ddots & \ddots & \\ & & & A_N & 0 \end{pmatrix} \qquad \Delta = \begin{pmatrix} \Delta_1 & & & \\ & \Delta_2 & & 0 \\ & & \ddots & \\ & 0 & & \Delta_N \end{pmatrix}$$

where the matrices Δ_i of order N are computed by

$$\Delta_1 = D_1$$

$$\Delta_i = D_i - A_i \Omega_{i-1}(3) A_i^T \quad , \quad 2 \leq i \leq N$$

$\Omega_{i-1}(3)$ is a tridiagonal matrix whose three main diagonals are the same than those of Δ_{i-1}^{-1}. This can be computed easily with the Choleski decomposition of Δ_{i-1}.

During the block forward and backward substitution one has to solve systems like

$$\Delta_i y_i = c_i$$

Δ_i is a tridiagonal matrix, this is not directly vectorizable. So we change the preconditioner by solving these systems setting

$$y_i = \Omega_i(j) c_i$$

where $\Omega_i(j)$ is a banded matrix with 2j+1 diagonals, whose

elements within the band are the same as those of Δ_i^{-1} .
More details are given in /9/ and /IO/. Numerical experiments show that we have to take j=3 to have the same rate of convergence as when we solve exactly $\Delta_i y_i = c_i$.
This preconditioner is called INVV3(I), the second V standing for "Vector". The vector length is N.

4-3 Polynomial preconditioners

Other polynomials than the truncated Neumann series were introduced by Johnson, Michelli and Paul /6/. Here we used a polynomial suggested by Saad /I2/.

$$M^{-1} = \alpha_0 + \alpha_1 A + \cdots + \alpha_k A^k = p_k(A)$$

The polynomial coefficients are computed to minimize

$$\| 1 - \lambda p_k(\lambda) \|_w^2 = \int_a^b (1 - \lambda p_k(\lambda))^2 w(\lambda) d\lambda$$

In the numerical experiments we choose (see /I2/)

$$a = 0 , \quad b = 8 \quad , \quad w(\lambda) = \lambda^{-\frac{1}{2}} (8 - \lambda)^{-\frac{1}{2}}$$

This method will be denoted by LSP(k) where k is the degree of the polynomial. The vector length is N^2.

5- Numerical results

We solve the model problem for various values of N, the mesh size being $h = \frac{1}{N+1}$.
We have to choose the degree of the polynomial. Figure 4 shows the number of iterations and the computing time as a function of the degree k. In view of the results we choose LSP(IO).

The number of iterations as a function of N is given on figure 5 for the three methods. LSP(IO) gives the smallest number of iterations. Figures 6, 7 and 8 give Mflops rates for the three preconditioners.

ICVDV reach 38.8 Mflops on the Cray-I, 74.5 for the Cyber 205.

INVV3(I) gives 49 Mflops on the Cray-I, 7I.5 on the 205.

With LSP(IO) (whose vector length is longer) we obtain 5I.8 Mflops on the Cray-I and I00 Mflops on the 205.

But, of course, what is of interest are the computing times. They are given on figures 9 and I0. Some of them are shown on Tables 2 and 3 below for N=50 and N=I50.

TABLE 2

	ICVDV	INVV3(I)	LSP(IO)
CRAY-I	0.098 sec 32.9 Mflops	0.046 sec 42 Mflops	0.083 sec 5I Mflops
CYBER 205	0.069 sec 45 Mflops	0.046 sec 42 Mflops	0.04I sec I02 Mflops

N=50 (order 2500)

TABLE 3

	ICVDV	INVV3(I)	LSP(IO)
CRAY-I	I.74I sec 37.7 Mflops	0.83I sec 47.7 Mflops	I.707 sec 5I.8 Mflops
CYBER 205	0.993 sec 70.4 M flops	0.634 sec 67.8 Mflops	0.955 sec 98.7 Mflops

N=I50 (order 22500)

It appears that the best results are obtained on Cyber 205, whatever the problem size is.

One can say that there is no "best method" since on Cyber 205 LSP(IO) is the best for $N \leqslant 60$ and then INVV3(I) becomes to be better. On the Cray-I INVV3(I) is always the best.

So the conclusion is that on this kind of problem with a very regular pattern of data Cyber 205 could give very good results. Moreover we have to be very careful when we compare algorithms for vectors computers. The results are not only problem dependant but also machine dependant. This is a parameter numerical analysts would have to include in their studies.

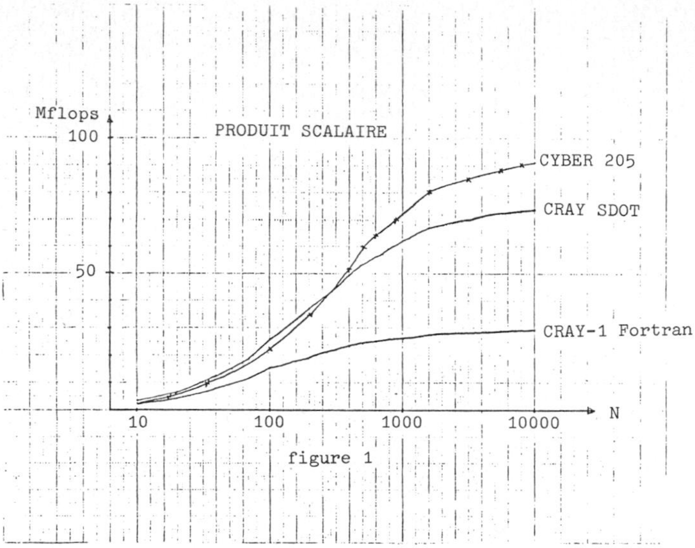

figure 1

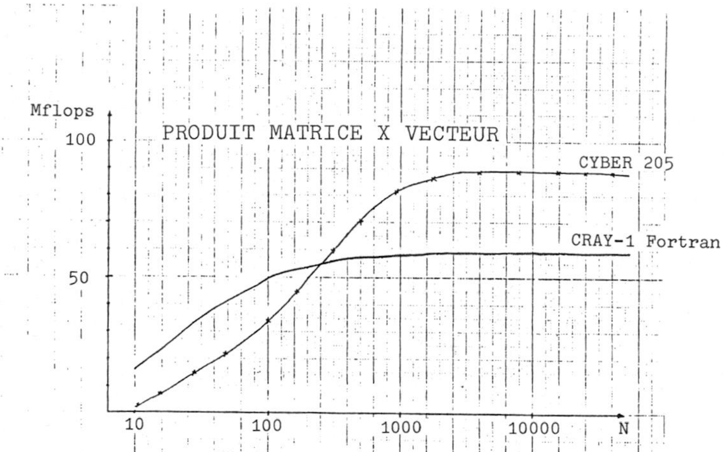

figure 2

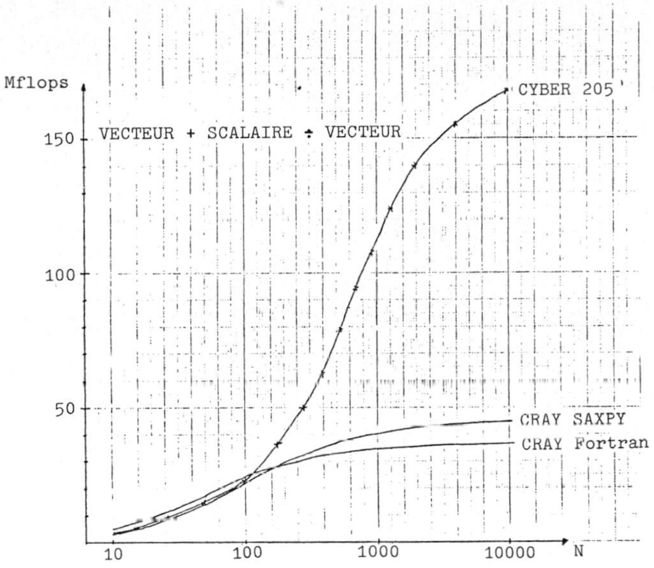

figure 3

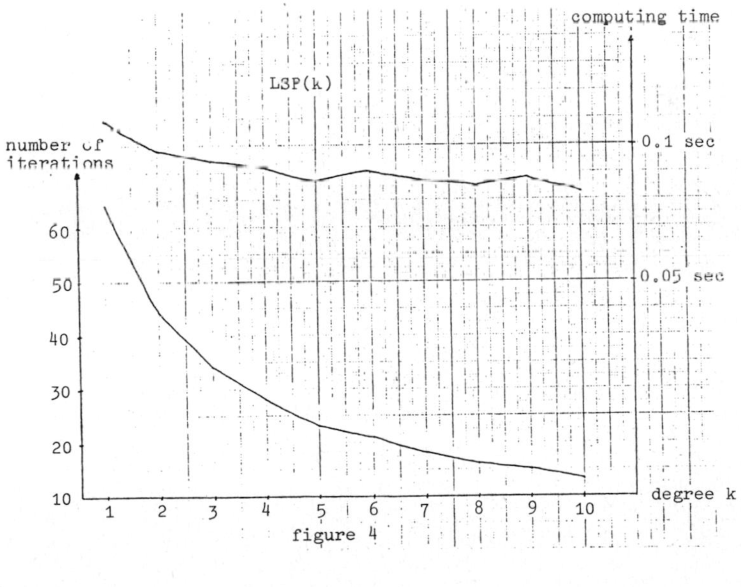

figure 4

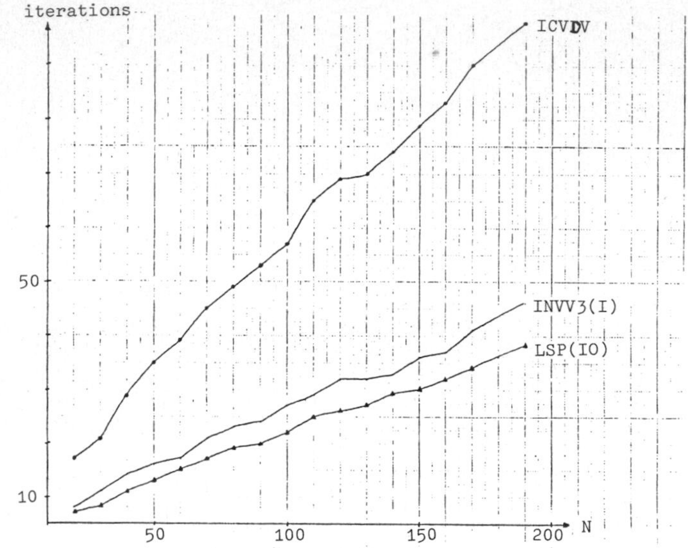

figure 5

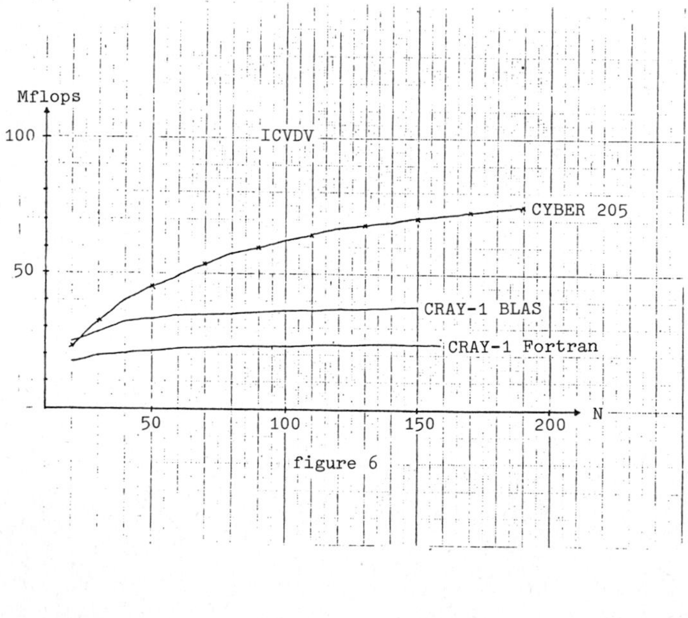

figure 6

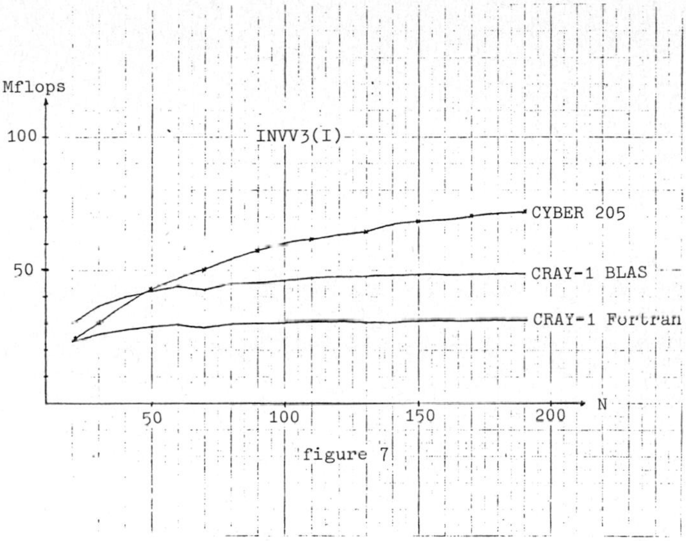

figure 7

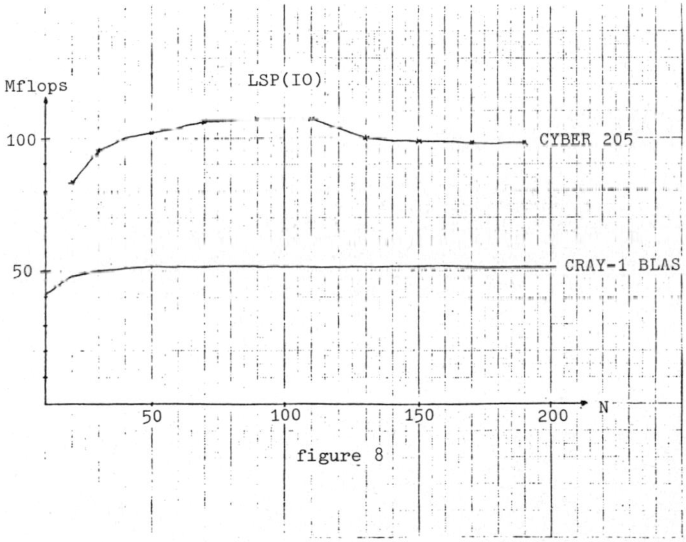

figure 8

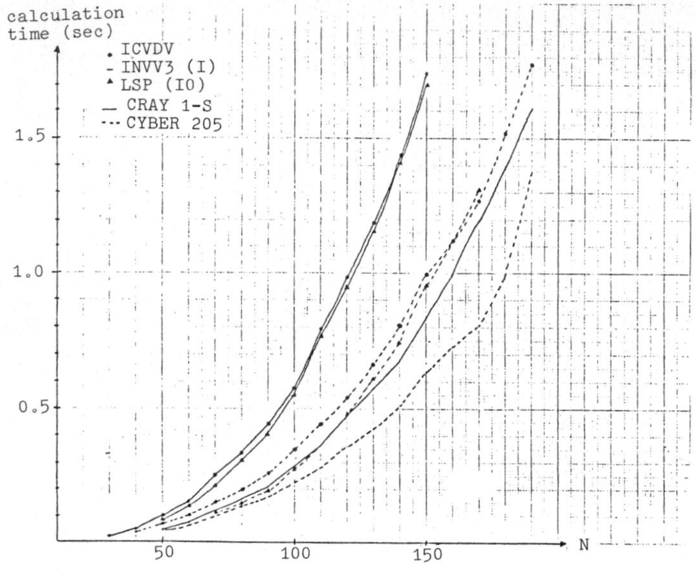

figure 9

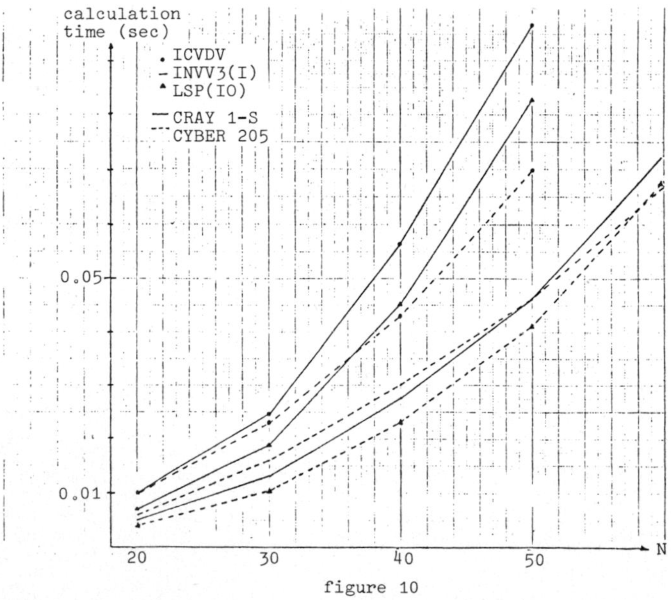

figure 10

REFERENCES

/1/ P.CONCUS, G.H. GOLUB and G. MEURANT
Block preconditioning for the conjugate gradient method
Lawrence Berkeley Laboratory LBL 14856 (1982)

/2/ P.CONCUS, G.H. GOLUB and D.P. O'LEARY
A generalized conjugate gradient method for the numerical solution of elliptic partial differential equations
in Sparse Matrix Computations J.R. Bunch and D.J. Rose eds
Academic Press, New York (1976) pp. 309-332

/3/ P.F.DUBOIS, A. GREENBAUM and D. RODRIGUE
Approximating the inverse of a matrix for use in iterative algorithms on vector computers
Computing v. 22 (1979) pp. 257-268

/4/ G.H.GOLUB and G. MEURANT
Résolution numérique des grands systèmes linéaires
vol. 49 Collection Etudes et Recherches EDF
Eyrolles, Paris (1983)

/5/ R.W. HOCKNEY and C.R. JESSHOPE
Parallel Computers
Adam Hilger, Bristol (1981)

/6/ O.G. JOHNSON, C.A. MICHELLI and G. PAUL
Polynomial preconditioning for conjugate gradient calculations
Siam J. on Numer. Anal. v. 20 (1983) pp. 362-376

/7/ T. JORDAN
Conjugate gradient preconditioners for vector and parallel processors
in Elliptic problems solvers, G. Birkhoff ed.
Academic Press, New York (I984)

/8/ A. LICHNEWSKY
Solving some linear systems arising in finite elements methods on parallel processors
Siam National Meeting, Stanford CA (I982)

/9/ G. MEURANT
Comparaisons numériques de différents préconditionnements pour la méthode du gradient conjugué sur Cray-I et Cyber 205
Note interne CEA (I983)

/I0/ G. MEURANT
The preconditioned conjugate gradient method on vector computers
to appear

/II/ G. RODRIGUE and D. WOLITZER
Preconditioning by incomplete block cyclic reduction
Lawrence Livermore Laboratory UCID I9502 (I982)

/I2/ Y. SAAD
Practical use of polynomial preconditioning for the conjugate gradient method
Research Report YALE U/DCS/RR-282 (I983)

/I3/ R.SCHREIBER and WEI PAI TANG
Vectorizing the conjugate gradient method
Report Dept. of Comp. Science Stanford Univ. (I983)

/I4/ H.A. van der VORST
A vectorizable version of some ICCG methods
Siam J. on Stat. and Sci. Comp. v.3 (I982) pp. 350-356

Computing Methods in Applied Sciences and Engineering, VI
R. Glowinski and J.-L. Lions (Editors)
Elsevier Science Publishers B.V. (North-Holland)

A GENERALIZATION OF THE NUMERICAL SCHWARZ ALGORITHM

Garry Rodrigue
and
Jeff Simon

Computing Research Group
Lawrence Livermore National Laboratory
University of California
Livermore, California
U.S.A.

The Schwarz alternating method of decomposing the domain has lately been found to be an effective means for solving elliptic partial differential equations on a multiprocessing computing system. In this paper, the method is recast into numerical linear algebra so that classical techniques of acceleration can be applied. Computational results using an incomplete-factorization as a pre-conditioner are given. A study of the effects of convergence by varying the subdomains is made.

1. The Fundamental Linear Systems

We consider the solution of the linear system

$$Tx = b \tag{1.1}$$

where T is a matrix of the form

$$T = \begin{bmatrix} F_1 & G_1 & & & \\ E_2 & F_2 & G_2 & & \\ & \ddots & \ddots & \ddots & G_{p-1} \\ & & & E_p & F_p \end{bmatrix} \tag{1.2}$$

$$F_1 = \begin{bmatrix} T_1 & R_1 \\ R_1^t & T_2 \end{bmatrix}, \qquad F_p = \begin{bmatrix} T_{2p-2} & R_{2p-2} \\ R_{2p-2}^t & T_{2p-1} \end{bmatrix}$$

$$(1.3) \qquad F_{i+1} = \begin{bmatrix} T_{2i} & R_{2i} & \\ R_{2i}^t & T_{2i+1} & R_{2i+1} \\ & R_{2i+1}^t & T_{2i+2} \end{bmatrix}, \quad 1 \le i \le p-2,$$

$$(1.4) \qquad G_i = \begin{bmatrix} 0 & 0 \\ 0 & R_{2i} \end{bmatrix}, \quad 2 \le i \le p$$

$$(1.5) \qquad E_i = \begin{bmatrix} R_{2i-1}^t & 0 \\ 0 & 0 \end{bmatrix}, \quad 2 \le i \le p$$

The source vector $b = [b_1, b_2, .., b_p]^t$ where

$$b_1 = \begin{bmatrix} \beta_1 \\ \beta_2 \end{bmatrix}, \qquad b_p = \begin{bmatrix} \beta_{2p-2} \\ \beta_{2p-1} \end{bmatrix},$$

$$(1.6) \qquad b = \begin{bmatrix} \beta_{2i} \\ \beta_{2i+1} \\ \beta_{2i+2} \end{bmatrix}, \qquad 1 \le i \le p-2.$$

The matrices defined in (1.3) - (1.5) may be of different order. We assume that the matrices $T_1, T_2, \ldots, T_{2p-1}$ are non-singular.

Associated with system (1.1) is

$$(1.7) \qquad T'x' = b'$$

where

$$(1.8) \qquad T' = \begin{bmatrix} T_1 & R_1 & & & \\ R_1^t & T_2 & R_2 & & \\ & \ddots & \ddots & \ddots & \\ & & & & R_{2p-2} \\ & & & R_{2p-2}^t & T_{2p-1} \end{bmatrix}$$

$$(1.9) \qquad b' = \begin{bmatrix} \beta_1 \\ \beta_2 \\ \cdot \\ \cdot \\ \cdot \\ \beta_{2p-1} \end{bmatrix}$$

The connection between the systems (1.1) and (1.7) is contained in the following theorem:

Theorm 1.1 Suppose from (1.1) we have Tx = b where $x = [x_1, x_2, \ldots, x_p]^t$ and

$$x_1 = \begin{bmatrix} X_1 \\ X_2 \end{bmatrix} \qquad x_p = \begin{bmatrix} X_{3p} \\ X_{3p+1} \end{bmatrix}$$

$$x_i = \begin{bmatrix} X_{3i} \\ X_{3i+1} \\ X_{3i+2} \end{bmatrix}, \quad 1 \le i \le p-1$$

Then the vector

$$x' = [X_1, X_3, X_4, X_6, \ldots, X_{3i}, X_{3i+1}, \ldots, X_{3p}, X_{3p+1}]^t$$

is the solution of the associated system (1.8).

Conversely, if the matrix T' in (1.7) is non-singular and the matrices T_i, $1 \le i \le 2_{p-1}$, are non-singular, then the matrix T in (1.1) is non-singular.

Proof:

Tx = b implies for $1 \le i \le p$

$$R^t_{2i-1} x_{3i-2} + T_{2i} x_{3i-1} + R_{2i} x_{3i+1} = \beta_{2i}$$

$$R^t_{2i-1} x_{3i-2} + T_{2i} x_{3i} + R_{2i} x_{3i+1} = \beta_{2i}$$

By the assumption of non-singularity of T_{2i}, we have $x_{3i} = x_{3i-1}$. The first result then follows.

Now, suppose $Tx = 0$ for some non-zero vector $x = [x_1, x_2, \ldots, x_{2p+1}]^t$. Using the same argument as above, we obtain $x_{3i-1} = x_{3i}$ for $1 < i < p$. It follows that the non-zero vector

$$x' = [x_1, x_3, x_4, x_6, \ldots, x_{3i}, x_{3i+1}, \ldots, x_{3p}, x_{3p+1}]^t$$

solves the equation $T'x' = 0$ which is a contradiction.

2. Relationship to the Schwarz Method

In this paper, we will be concerned with solving systems of the form (1.8) that arise from the numerical solution of elliptic and parabolic partial differential equations. To obtain the solution of (1.8), iterative methods on the associated system (1.1) will be studied. Note that if the inverse-positivity of T can be guaranteed then any regular splitting of T will yield a convergent iterative method, cf. Varga [4].

A variety of convergent splittings appropriate for parallel computing can be obtained by the domain decomposition method first put forth by H. Schwarz, [1]. In this method the numerical solution of the 2-dimensional elliptic problem

(2.1) $$Lu = \nabla \cdot K \nabla u = f,$$

$$(x,y) \in \Omega = [0,1] \times [0,1],$$

$$K(x,y) > 0.$$

f is a given function on Ω. u must satisfy the Dirichlet boundary condition $u(x,y) = g(x,y)$, $(x,y) \in \partial\Omega$, the boundary of Ω. Let n be a positive integer and define the set grid points $\{k\Delta x, \ell\Delta y\}$, $0 < k, \ell < n+1$, on Ω with $\Delta x = \Delta y = 1/n+1$. A discretization of the operator L yields a block tridiagonal system of equations $T'x' = b'$ where

$$(2.2) \qquad T' = \begin{bmatrix} A_1 & B_1 & & \\ B_1^t & A_2 & B_2 & \\ & \ddots & \ddots & B_{n-1} \\ & & B_{n-1}^t & A_n \end{bmatrix},$$

A_i, B_i = tridiagonal matrices, $1 \leq i \leq n$, cf. Rodrigue et al [6], Varga [4], Ciarlet [5]. The matrix T' is symmetric and positive definite.

Let

$$0 = k_1 < k_2 < .. < k_p < n$$

$$0 < m_1 < m_2 < ... < m_p = n$$

be two sequences of integers such that

$$k_1 < k_2 < m_1 < k_3 < m2 < ... < k_p < m_{p-1} < m_p.$$

Define the system (1.1) where for $0 \leq i \leq n-1$

$$(2.3) \qquad T_{21} = \begin{bmatrix} A_{k_i+1} & B_{k_i+1} & & \\ B_{k_i+1}^t & \ddots & \ddots & \\ & \ddots & \ddots & B_{k_{i+1}-2} \\ & & B_{k_{i+1}-2}^t & A_{k_{i+1}-1} \end{bmatrix}$$

$$(2.4) \qquad T_{2i+1} = \begin{bmatrix} A_{k_{i+1}} & B_{k_{i+1}} & & \\ B_{k_{i+1}}^t & \ddots & \ddots & \\ & \ddots & \ddots & B_{m_i-2} \\ & & B_{m_i-2}^t & A_{m_i-1} \end{bmatrix}$$

$$(2.5) \qquad T_{2i+1} = \begin{bmatrix} A_{m_i} & & B_{m_i} & & \\ & \ddots & & \ddots & \\ B^t_{m_i} & & \ddots & & B_{m_{i+1}-1} \\ & \ddots & & \ddots & \\ & & B^t_{m_{i+1}-1} & & A_{m_{i+1}} \end{bmatrix}$$

and

$$(2.6) \qquad R_k = \left[\begin{array}{c|c} & 0 \\ \hline B_\alpha & 0 \end{array}\right]$$

$$\alpha = \begin{cases} k_{i+1} - 1 , & k \text{ even} \\ m_i - 1 , & k \text{ odd} \end{cases} .$$

It follows from Theorem 1.1 that convergent iterative methods can be used on the system (2.4) - (2.6) to obtain solutions of (2.2). In fact, the block Jocobi splitting $T = M - N$ where

$$(2.7) \qquad M = \begin{bmatrix} F_1 & & \\ & \ddots & \\ & & F_p \end{bmatrix}$$

corresponds to solving a sequence of problems of the form (2.1). More specifically, for $1 \le i \le p$ define the sets

$$r_i = \{ (x,y) \in \Omega : y = m_i / n+1 \}$$

$$(2.8) \qquad \ell_i = \{ (x,y) \in \Omega : y = k_i / n+1 \}$$

$$\Omega_i = \{ (x,y) \in \Omega : k_i / n+1 \le y \le m_i / n+1 \}.$$

Then $\Omega = \bigcup_{i=1}^{p} \Omega_i$. Now consider the problems, $1 \le i \le p$,

$$(2.9) \qquad Lu_i = f_i,$$

$$(x,y) \in \Omega_i .$$

f_i is the restriction of f on Ω_i. The functions u_i are to satisfy the boundary conditions

$$(2.10) \qquad u_i\,(x,y) = g \text{ on } \gamma_i = \partial\Omega_i - r_i - \ell_i$$

$$(2.11) \qquad u_i\,(x,y) = u(x,y) \text{ on } \begin{cases} r_1 & , i = 1, \\ r_i \cup \ell_i, & 1 < i < p, \\ \ell_p & , i = p. \end{cases}$$

Since the boundary condition (2.11) is not known, the numerical solutions of (2.9) can be obtained via the block-Jacobi method (2.7). In fact, the block-Gauss-Siedel splitting of (2.4) - (2.6) is precisely the numerical Schwarz alternating procedure described in Miller [2], Kang [3] and Glowinski [9]. However, in both this and the block-Jacobi method an efficient means of inverting each of the F_i matrices must be done. An alternative would be to use an incomplete-factorization of the F_i matrices, cf. Rodrigue and Wolitzer [7] or Reiter and Rodrigue [8]. That is, if for $1 \le i \le p$

$$F_i = L_i L_i^t + Q_i$$

is an incomplete factorization of F^i, then a splitting $T = M - N$ given by

$$(2.11) \qquad M = \begin{bmatrix} L_1L_1^t & & & \\ E_2 & L_2L_2^t & & \\ & \ddots & \ddots & \\ & & E_p & L_pL_p^t \end{bmatrix}$$

would yield an easily invertible system. If the system T is an M-matrix, then the splitting (2.11) is regular and hence convergent.

3. Computational Results

We use the method (2.11) to numerically solve Laplace's equation

$$\frac{\partial^2 u}{\partial x^2} + \frac{\partial^2 u}{\partial y^2} = 0$$

on $R = [0,1] \times [0,1]$ with boundary conditions $u = 2$. A 5-point central difference approximation with $\Delta x = 1/31$, $\Delta y = 1/65$ being used to yield the standard Laplace's matrix of the form (2.2).

The first set of calculations were done to study the effect of increasing the distances between the sets r_i and ℓ_i of (2.8). That is, the effect of the overlap of the R_i regions on the overall convergence rate. For simplicity $p = 2$ is used. See Table 1.

The second set of calculations studies the effects of the number of regions (i.e. the magnitude of the integer p) on the convergence. Here, $m_i - k_{i+1} = 2$ so that all of the subregion overlaps have one grid line in common. See Table 2.

The third set of calculations is similar to the first set. Instead of passing only boundary information to the adjacent regions (the usual implementation of the Schwarz algorithm), all of the data in common to both regions is passed. See Table 3.

5. References

[1] Schwarz, H.A., "Gesammelte Mathematische Abhandlungen", Vol. 2, pp. 133-134, Berlin, Springer 1890.

[2] Miller, K., "Numerical analogs to the Schwarz alternating procedure", Numer. Math., 7 (1965), pp 91-103.

[3] Kang, L.S., "The Schwarz Algorithm", Wuhan University Journal, National Science Edition, Special Issue of Mathematics, China, 1981, pp 77-88.

[4] Varga, R.S., Matrix Iterative Analysis, Prentice Hall, New Jersey, 1962.

[5] Ciarlet, P.G., The Finite Element Method for Elliptic Problems, North-Holland, Amsterdam, 1978.

[6] Rodrigue, G., Hendrickson, C., Pratt, M., "An Implicit Solution of the Two-Dimensional Diffusion Equation and Vectorization Experiments", Parallel Computations, G. Rodrigue, ed., Academic Press, 1982.

[7] Rodrigue, G., Wolitzer, D., "Preconditioning by Incomplete Block Cyclic Reduction", to appear Mathematics of Computation.

[8] Rodrigue, G., Reiter, E., "An Incomplete Cholesky Factorization by a Matrix Partition Algorithm", to appear Elliptic Problem Solvers II, G. Birkhoff, ed., Academic Press.

[9] Glowinski, R., "Domain Decomposition Methods for Nonlinear Problems in Fluid Dynamics", INRIA Report, No. 147, July, 1982.

5. Acknowledgement

This work performed under the auspices of the U.S. Department of Energy by Lawrence Livermore National Laboratory under contract No. W-7405-Eng-48.

This work was performed under the auspices of the U.S. Department of Energy Office of Basic Energy Sciences, Applied Mathematics and Statistics Division.

Domain Definition (see (2.8))				No. of Iterations
m_1	k_1	m_2	k_2	
0	32	32	64	648
0	29	34	64	644
0	24	39	64	642
0	19	44	64	642
0	14	49	64	642
0	9	54	64	642

Table 1

No. of Subregions	Origin of Subregions	No. of Iterations
2	(0,0), (0,32)	648
3	(0,0), (0,21) (0,42)	657
4	(0,0), (0,15) (0,30), (0,46)	654
8	(0,0), (0,9) (0,20), (0,27) (0,35), (0,43) (0,50), (0,56)	666

Table 2

Domain Definition (see (2.8))				No. of Iterations
m_1	k_1	m_2	k_2	
0	32	32	64	638
0	29	34	64	595
0	24	39	64	523
0	19	44	64	456
0	14	49	64	392
0	9	54	64	343

Table 3

Computing Methods in Applied Sciences and Engineering, VI
R. Glowinski and J.-L. Lions (Editors)
Elsevier Science Publishers B.V. (North-Holland)

COMPUTING GENERALIZED INVERSES AND EIGENVALUES OF SYMMETRIC MATRICES USING SYSTOLIC ARRAYS

Robert Schreiber

Computer Science Department
Stanford University
Stanford, California
U. S. A.

New systolic array methods are presented for tridiagonalization of a dense, symmetric n x n matrix in $O(n^{3/2})$ time, tridiagonalization of a matrix of bandwidth b in O(bn) time, and finding the eigenvalues of a symmetric tridiagonal matrix in O(n) time. Systolic arrays for implementing an iterative method for the generalized inverse are given. They allow computation of the generalized inverse of an m x n matrix in O(m) time.

INTRODUCTION

This paper has two parts. First we describe a method for parallel computation of all the eigenvalues of a symmetric matrix that uses $n^{3/2} + O(n)$ processors and $O(n^{3/2})$ time. If the matrix has bandwidth $b \ll n$ then the method requires O(n) processors and O(bn) time. Second we examine the parallel computation of a method of Ben-Israel and Cohen [1] for computing the generalized inverse. We show that it requires O(m) time using $3n^2/2$ processors to compute the generalized inverse of an m x n matrix.

The parallel architectures we describe are planar arrays of simple, uniform computing cells that operate in unison on streams of data. These are called systolic arrays; they are of growing importance in numerical linear algebra [3,7] and its applications to signal processing [10].

COMPUTING THE EIGENVALUES IN $O(n^{3/2})$ TIME

Brent and Luk [2] describe a processor array implementation of a cyclic Jacobi method that requires O(n log n) time using $n^2/4$ processors. If the matrix is banded, their method takes no advantage of the fact. Schreiber [6] shows how to reduce a symmetric n x n matrix A to an orthogonally similar matrix A(b) of bandwidth b ($a_{ij} = 0$ if $|i - j| > b$) in time $O(n^2/b)$. A b x n trapezoidal systolic array is used: there are b(n - (b-1)/2) processors. He also shows how to then compute the QR iterates using a b x (2b+1) array. But, because of the need to compute good shifts, it appears that only O(1) QR iterations can usefully be performed in parallel. Finding all eigenvalues by the QR method takes, therefore, $O(n^2)$ time. Thus we seek other approaches.

It is possible to carry the reduction in bandwidth all the way to b=1 (tridiagonal) with a second array in an additional O(bn) time. The algorithm for bandwidth reduction is the usual one. A nonzero element is eliminated by symmetric plane rotations. This creates a new nonzero further toward the lower right corner of the matrix. The undesired nonzero is "chased" by plane rotations until it leaves the matrix. This process is illustrated in Figure 1. Note that O(n) operations are required to remove one nonzero. We suppose a processor is available that carries out this process in O(n) time; moreover, it takes in elements of the matrix from top left to bottom right, one codiagonal per clock, and sends out elements of the transformed matrix, in the same format, O(1) clocks later -- see Figure 2. Clearly a pipeline of n such processors can remove an entire diagonal in O(n) time. To reduce A(b) to a tridiagonal matrix A(1) one may use b-1 passes through the array, which takes O(bn) time. If we choose $b = n^{1/2}$, then the complete tridiagonalization process takes $O(n^{3/2})$ time and uses $n^{3/2} + O(n)$ processors.

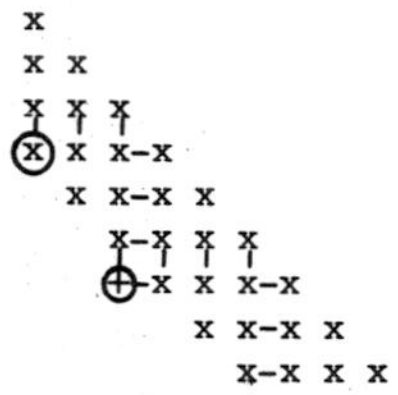

x-x a rotated pair

⊗ the zeroed element

⊕ nonzero created then removed

Figure 1
Chasing a nonzero

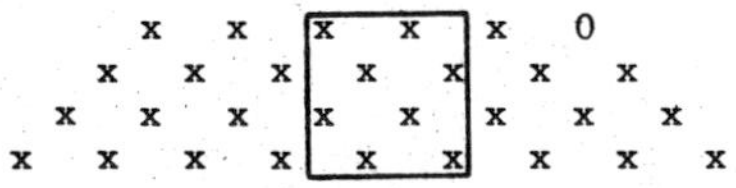

Figure 2
The bandwidth reduction processor

EIGENVALUES OF A SYMMETRIC TRIDIAGONAL MATRIX

We propose to find all n eigenvalues in O(n) time using an n-processor array. Either the bisection method, or Newton's method, or a combination of them can be used. Let the tridiagonal matrix

$$A(1) = \begin{bmatrix} a_1 & b_2 & & & \\ b_2 & \cdot & \cdot & & \\ & \cdot & \cdot & \cdot & \\ & & \cdot & \cdot & b_n \\ & & & b_n & a_n \end{bmatrix}$$

For $1 \leq j \leq n$, let $p_j(z)$ be the characteristic polynomial of the leading $j \times j$ submatrix. We seek the roots of $p_n(z)$. It is well known [5] that (with $p_0 \equiv 1$ and $p_{-1} \equiv 0$)

$$(1) \qquad p_j(z) = (a_j - z)\, p_{j-1}(z) - b_j^2\, p_{j-2}(z), \qquad j = 1, 2, \ldots, n$$

and that if $s_j(z)$ is the number of strong sign changes in the sequence $\{p_0(z), p_1(z), \ldots, p_j(z)\}$ then $s_n(z)$ is the number of eigenvalues less than z. From (1) we have that

$$p_j'(z) = (a_j - z)\, p_{j-1}'(z) - b_j^2\, p_{j-2}'(z) - p_{j-1}(z), \qquad j = 1, 2, \ldots, n$$

These recurrences can be used as the basis of a bisection or Newton's method for finding the eigenvalues.

We propose two n-processor pipelines, one shown in Figure 3 for the bisection method and the other, shown in Figure 4, for the Newton method. While O(n) clocks are needed for one iteration of either method for a single eigenvalue, it is possible to carry out an iteration on all n eigenvalues in the same amount of time. A controlling processor would have to update the iterates and test for convergence. Since O(1) iterations suffice for convergence, the total time is O(n). These procedures, moreover, can take advantage of good approximate eigenvalues, which are available, for example, in signal processing applications where A is an estimate for a covariance matrix that is frequently updated [9].

It is straightforward to devise similar arrays implementing inverse iteration for the eigenvectors of the tridiagonal matrix. Unfortunately, it would be excessively expensive to accumulate the transformations used to tridiagonalize A(b).

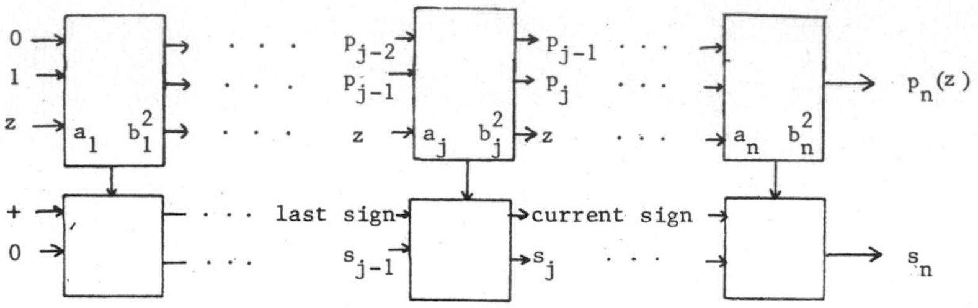

Figure 3
The bisection processor

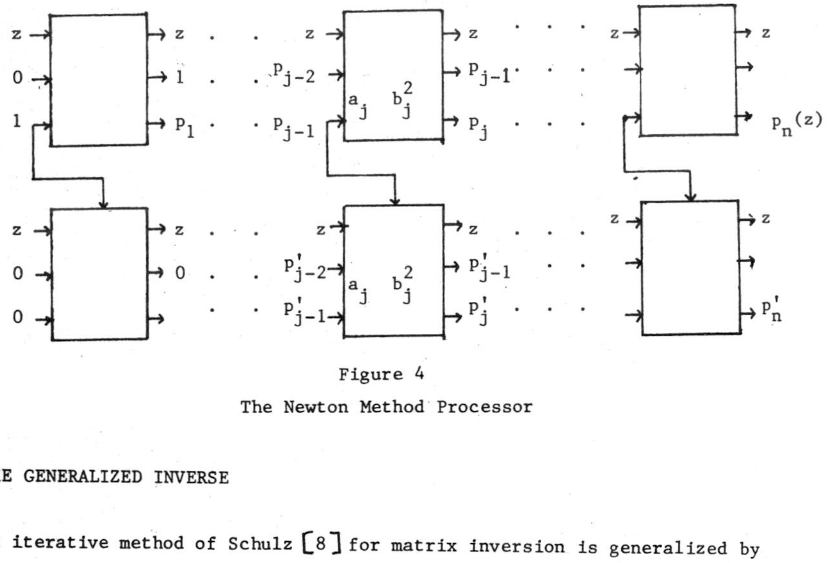

Figure 4
The Newton Method Processor

THE GENERALIZED INVERSE

An iterative method of Schulz [8] for matrix inversion is generalized by Ben-Israel and Cohen for genrealized inversion. They prove [1]

<u>Theorem</u>: Let A be an m x n matrix and let $s_1 \geq s_2 \geq \ldots \geq s_r$ be the nonzero singular values of A. If $0 < a_0 < 2/s_1^2$ then the sequence given by

(2) $$X_0 = a_0 A^T$$

(3) $$X_{k+1} = (2I - X_k A)X_k, \quad k = 0, 1, \ldots$$

converges to the generalized inverse A^+ of A. Convergence is ultimately quadratic.

Ben-Israel and Cohen determine the optimal a_0

(4) $$a_0 = \frac{2}{s_1^2 + s_r^2}.$$

This method is not usually taken seriously because the alternative of computing the singular value decomposition of A,

$$A = U \begin{bmatrix} S & 0 \\ 0 & 0 \end{bmatrix} V^T$$

where $S = \text{diag}(s_1, s_2, \ldots, s_r)$ and U and V are orthogonal, and then forming

$$A^{\dagger} = V \begin{bmatrix} S^{-1} & 0 \\ 0 & 0 \end{bmatrix} U^T$$

is more efficient. But because the iteration (2)-(3) is so conveniently implemented by a systolic array, we reconsider it here.

An analysis of the method is given by Söderström and Stewart [9]. Their general comclusion is that the method is neither efficient nor very robust. Some of their criticisms can be rebuffed, however, by improvements to the method. These are discussed in the following sections.

SYSTOLIC IMPLEMENTATION

Two systolic arrays can be used to efficiently implement the iteration (3). The overall system is shown in Figure 5. An n x n triangular array that accumulates the (symmetric) product $S_k = X_k A$ is shown in Figure 6. Each cell computes the sum of the products of the elements of X and A that flow through it. These elements flow through the array, one cell per clock.

When S_k is complete, it is unloaded from the triangular array and sent into a square n x n array that computes $X_{k+1} = 2X_k - S_k X_k$. The process of unloading S_k is shown in Figure 7. Elements of S_k begin to move out along the paths shown as soon as they are completed. Their motion can be triggered by a control signal that travels with the last column of X_k. Off-diagonal elements move both left and up while diagonal elements move left.

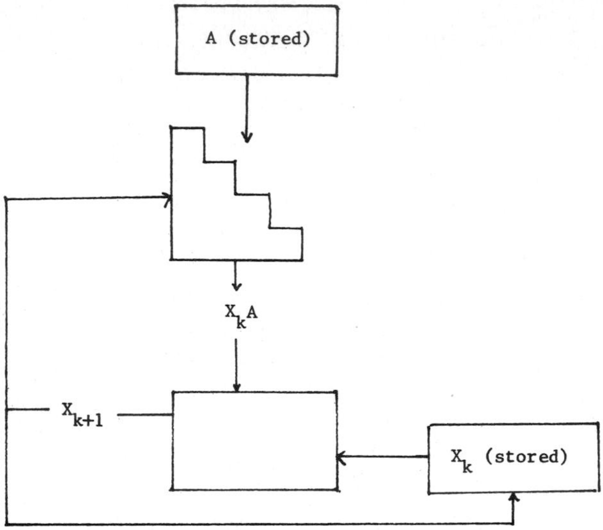

Figure 5
System Overview

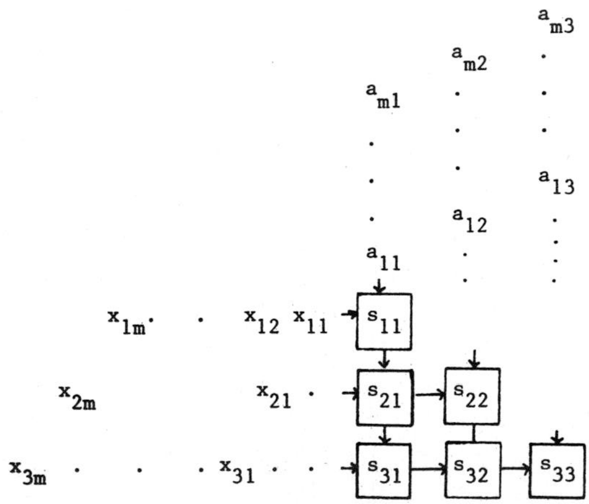

Figure 6
Computing $S_k = X_k A$

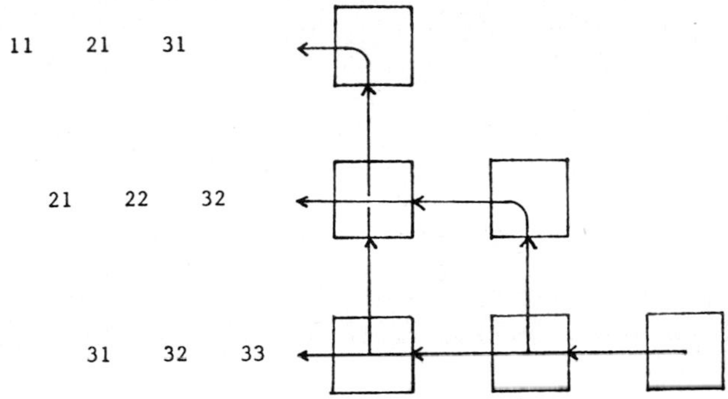

Figure 7
Unloading the Triangular Array

As they leave, elements of S_k can move directly into the second array, which is shown in Figure 8. The pattern of output from the triangular array matches the pattern of input to the square array.

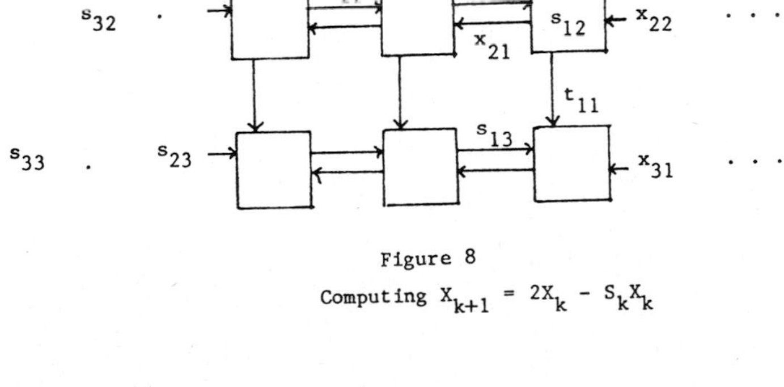

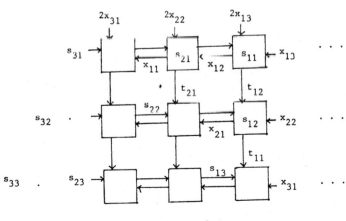

Figure 8
Computing $X_{k+1} = 2X_k - S_kX_k$

In the square array, which computes $T = X_k - S_kX_k$, elements of S_k move from left to right until stopped by a control signal. These signals travel with the first column of X_k. Elements of $2X_k$ enter at the top and move downward one cell per clock. The product $s_{ir}x_{rj}$ is subtracted from t_{ij} as it passes the rth cell in its traversal. The output, the new iterate X_{k+1}, can be sent directly into the triangular array; the output and input formats match. It must also be stored in memory for later use in the square array.

IMPROVEMENTS TO THE METHOD

We first describe an acceleration procedure that doubles the rate of convergence. Following Söderström and Stewart, we base the analysis on the eigendecomposition

$$X_kA = V\ T_k\ V^T,$$

$T_k = \text{diag}\ (t_1^{(k)}, t_2^{(k)}, \ldots, t_r^{(k)})$, of the symmetric matrix X_kA. The new iteration is

$$(5) \qquad X_{k+1} = a_{k+1}\ (2I - X_kA)X_k.$$

We show below how to choose the acceleration parameters a_k. From (2) and (5) it follows that

$$(6) \qquad t_j^{(0)} = 2s_j^2\ /\ (s_1^2 + s_r^2)$$

and that

$$(7) \qquad t_j^{(k+1)} = a_{k+1}\ (2 - t_j^{(k)})t_j^{(k)}.$$

Convergence occurs when all $t_j^{(k)}$ are sufficiently close to 1. Note that $g: x \mapsto (2 - x)x$ maps $(0,2)$ into $(0,1)$ and that $g(x) = g(2-x)$. Thus, with $a_k = 1$, all the eigenvalues $t_j^{(k+1)}$ lie in $(0,1)$. By choosing $1 < a_{k+1} < 2$, we move the leftmost eigenvalue, $t_r^{(k+1)}$ closer to 1. We must not, however, move any other eigenvalue further from 1 than is t_r. So, to choose a_{k+1}, we maintain estimates $\underline{t}$ of t_r and $\bar{t}$ of $\max(t_j : 1 \leq j \leq r)$. Let

$$\underline{t}^{(0)} = 2s_r^2\ /\ (s_1^2 + s_r^2),$$

and

$$\bar{t}^{(0)} = 2s_1^2 / (s_1^2 + s_r^2),$$

(of course, estimates of s_1 and s_r must be used) and, for $k = 0, 1, \ldots,$

$$\underline{t}^{(k+1)} = a_{k+1} (2 - \underline{t}^{(k)}) \underline{t}^{(k)}$$

$$\bar{t}^{(k+1)} = a_{k+1} .$$

Finally, choose a_{k+1} such that $\underline{t}^{(k+1)} = 2 \quad \bar{t}^{(k+1)}$, that is,

$$a_{k+1} = \frac{2}{1 + (2 - \underline{t}^{(k)})\underline{t}^{(k)}}$$

Without acceleration, convergence requires approximately $2 \log_2 (\kappa(A))$ iterations, where $\kappa(A) \equiv s_1/s_r$ [9]. The accelerated procedure takes roughly half this many.

As Söderström and Stewart point out, it may often be better to compute the generalized inverse of the matrix A(e) obtained by setting to zero those singular values of a less than some $e > 0$. We give a modified method that does this. Consider the iteration

$$(8) \qquad X_{k+1/2} = (2I - X_k A) X_k$$

$$(9) \qquad X_{k+1} = X_{k+1/2}\, A\, X_{k+1/2} .$$

The eigenvalues of $X_k A$ satisfy

$$t_j^{(k+1)} = \left[(2 - t_j^{(k)})\, t_j^{(k)}\right]^2$$

Figure 9 illustrates the effect of this mapping. The fixed point is $T = (3 - \sqrt{5})/2$; $2-T = (1 + \sqrt{5})/2 \doteq 1.618$, $T \doteq .3820$. Evidently, eigenvalues in $(0,T)$ and $(2-T,2)$ are sent toward zero, while those in $(T,2-T)$ are sent toward 1. To compute $A^+(e)$ we would

1. Choose $a_0 = \min (2/(s_1^2 + e^2), T/e^2)$. This insures that the eigenvalues t_j corresponding to singular values $s_j > e$ are all closer to 1 than the other eigenvalues. Set $\underline{t} = a_0 e^2$ and $\bar{t} = a_0 s_1^2$.

2. Use the accelerated method (5) until the eigenvalues corresponding to singular values greater than e lie in $(T,2-T)$, i.e., until $\underline{t}^{(k)} > T$.

3. Use the method (8) - (9) until the necessary accuracy is obtained.

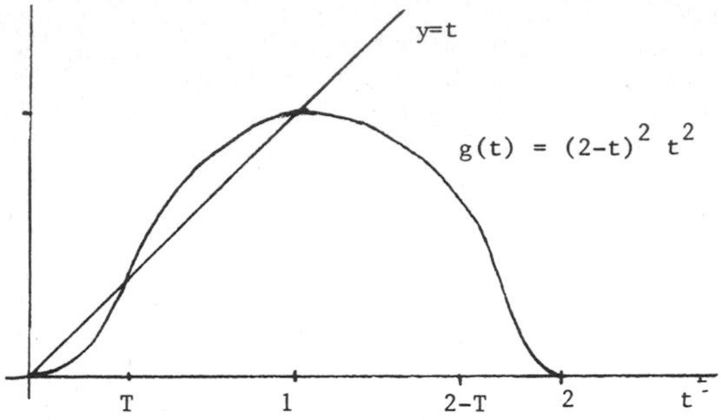

Figure 9
The Iteration Function for (8) - (9)

Further details and numerical experiments with the modified methods will be given in a later paper.

ACKNOWLEDGMENT

The author thanks Richard Brent for stimulating discussions on these topics.

REFERENCES

1. Ben-Israel, A. and Cohen, D., On iterative computation on generalized inverses and associated projections, SIAM J. Numer. Anal. 3 (1966) 410-419.

2. Brent, R.P., and Luk, F.T., A systolic architecture for almost linear-time solution of the symmetric eigenvalue problem, Report TR-CS-82-10, Dept. of Comp. Sci., Australian National Univ. (August 1982).

3. Brent, R.P, Kung, H.T., and Luk, F.T., Some linear-time algorithms for systolic arrays, Report TR 83-541, Dept. of Comp. Sci., Cornell Univ. (December 1982).

4. Owsley, N.L., High resolution spectral analysis by dominant mode enhancement, in: Kung, S.Y., Whitehouse, H.J., and Kailath, T. (eds.), VLSI and Modern Signal Processing (Prentice-Hall, Englewood Cliffs, New Jersey, 1984).

5. Parlett, B.N., The Symmetric Eigenvalue Problem (Prentice-Hall, Englewood Cliffs, New Jersey, 1980).

6. Schreiber, R., Systolic arrays for eigenvalue computation, in: Trimble, J. (ed.), Real Time Signal Processing V (SPIE Vol. 341, Bellingham, Washington, 1982).

7. Schreiber, R., Systolic arrays: high performance parallel machines for matrix computation, in: Birkhoff, G. and Schoenstadt, A. (eds.), Elliptic Problem Solvers (Academic Press, to appear).

8. Schulz, G., Iterative berechnung der reziproken Matrix, Z. Angew. Math. Mech. 13 (1933) 57-59.

9. Söderström, T. and Stewart, G.W., On the numerical properties of an iterative

method for computing the Moore-Penrose generalized inverse, SIAM J. Numer. Anal. 11 (1974) 61-74.

10. Whitehouse, H.J., Speiser, J.M., and Bromley, K., Signal processing applications of systolic array technology, in: Kung, S.Y., Whitehouse, H.J., and Kailath, T. (eds.), VLSI and Modern Signal Processing (Prentice-Hall, Englewood Cliffs, New Jersey, 1984).

ASYMPTOTIC EXPANSION AND HOMOGENEIZATION

EXPANSION ASYMPTOTIQUE ET HOMOGENEISATION

Computing Methods in Applied Sciences and Engineering, VI
R. Glowinski and J.-L. Lions (Editors)
Elsevier Science Publishers B.V. (North-Holland)

REMARQUES SUR L'HOMOGENEISATION

Jacques-Louis Lions

Collège de France
et
I.N.R.I.A.
Institut National de Recherche en Informatique et en Automatique
Domaine de Voluceau
B.P. 105 - 78153 Le Chesnay
France

ABSTRACT

Brief survey of some methods for asymptotic expansions for various perforated materials, with a periodic structure.
Asymptotic expansions for spectral problems are also indicated ans some open questions are mentionned.

INTRODUCTION

La théorie de l'homogénéisation présente des aspects variés :

(i) opérateurs différentiels à *coefficients rapidement oscillants* et à structure périodique, ou presque périodique, ou aléatoire,

(ii) opérateurs différentiels dans des *domaines perforés* avec une structure périodique, ou bien dans des domaines avec des écrans perforés (et structure périodique), ou bien encore dans des domaines avec obstacles arrangés non périodiquement ou de manière aléatoire.

Nous allons présenter ici quelques résultats, dûs à divers auteurs mentionnés dans la Bibliographie, concernant la situation *(ii)* ci-dessus.

Nous rappellerons également quelques problèmes ouverts.

Le plan est le suivant :

1. - Position du problème. Cas du problème de Dirichlet.

2. - Calcul des variations asymptotique .

3. - Quelques problèmes spectraux.

4. - Questions diverses.

Bibliographie.

POSITION DU PROBLEME. CAS DU PROBLEME DE DIRICHLET

1.1. Définition des ouverts perforés.

Soit $Y \subset \mathbb{R}^n$ le cube unité $Y =]0,1[^n$ et soit $\mathcal{O}$ un ouvert avec $\overline{\mathcal{O}} \subset Y$.

Soit (cf. Figure 1) $S = \partial\mathcal{O}$ et soit $\mathcal{Y} = Y \backslash \mathcal{O}$.

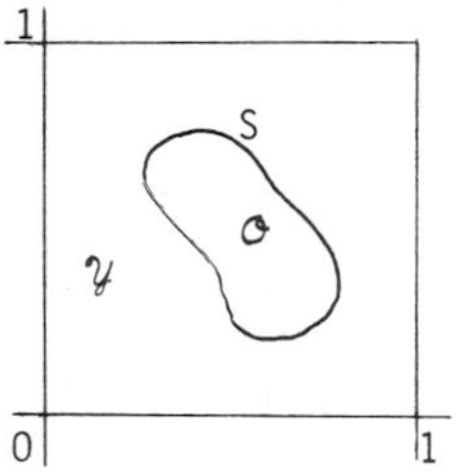

Figure 1.

On considère la fonction m donnée par

(1.1) $\quad m(y) = 1$ si $y \in \mathcal{Y}$; 0 si $y \in \mathcal{O}$

et prolongée périodiquement à $\mathbb{R}^n$ avec une période 1 en toutes les variables.

Si Ω est un ouvert borné de $\mathbb{R}^n$ de frontière Γ, on introduit

(1.2) $$\Omega_\varepsilon = \{x \mid x \in \Omega,\ m(\tfrac{x}{\varepsilon}) = 1\}.$$

Géométriquement Ω_ε est ce qui reste de Ω après avoir enlevé les ensembles $\varepsilon\mathcal{O}$ et tous leurs translatés de multiples entiers de ε parallèlement aux axes de coordonnées et qui rencontrent Ω.

La frontière de Ω_ε comprend deux parties jouant des rôles distincts :

(1.3) $$\begin{cases} \partial\Omega_\varepsilon = \Gamma_\varepsilon \cup S_\varepsilon, \\ \Gamma_\varepsilon \subset \Gamma, \quad S_\varepsilon = \text{l'ensemble union des bords des "trous" ou} \\ \text{"obstacles" } \varepsilon\mathcal{O} \text{ (et leurs translatés).} \end{cases}$$

Remarque 1.1.

Il peut arriver qu'une composante de S_ε consiste en seulement une *partie* de la frontière du "trou" (cf. Figure 2)■

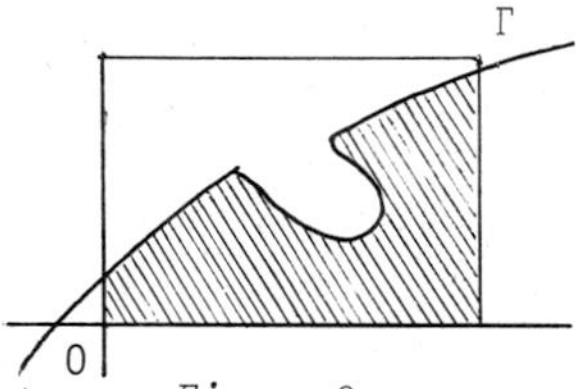

Figure 2.

Remarque 1.2.

Lorsque les $\varepsilon\mathcal{O}$ (et leurs translatés) sont considérés comme des "trous", on a affaire à des *ouverts perforés à structure périodique*. Dans des modélisations de *milieux poreux*, on peut considérer les $\varepsilon\mathcal{O}$ (et leurs translatés) comme des obstacles.

On va considérer le paramètre ε *comme "petit" au regard des dimensions de* Ω *et on va chercher les développements asymptotiques* (en ε) *des solutions de certains problèmes aux limites dans* Ω_ε.

1.2. Problème de Dirichlet.

Soit f donnée dans $L^2(\Omega)$ (pour fixer les idées). Soit u_ε la solution du problème de Dirichlet

(1.4) $$-\varepsilon^2 \Delta u_\varepsilon = f \quad \text{dans } \Omega_\varepsilon,$$

(1.5) $u_\varepsilon = 0$ sur $\Gamma_\varepsilon \cup S_\varepsilon$.

Remarque 1.3.

Dans (1.4) il s'agit en fait au 2ème membre de la restriction de f à Ω_ε. ∎

Remarque 1.4.

Le facteur ε^2 dans (1.4) est évidemment une simple question d'écriture ; on peut toujours poser $\varepsilon^2 u_\varepsilon = w_\varepsilon$. ∎

Naturellement, pour tout $\varepsilon > 0$ fixé, (1.4)(1.5) définit u_ε de manière unique. *On cherche un développement asymptotique* de u_ε.

1.3. Ansatz.

On cherche u_ε sous la forme

$$(1.6) \qquad u_\varepsilon = u_0 + \varepsilon u_1 + \varepsilon^2 u_2 + \ldots$$

où

$$(1.7) \qquad u_j = u_j(x, x/\varepsilon) ,$$

avec

(1.8) $\left| \begin{array}{l} u_j(x,y) \text{ défini dans } \Omega \times \mathcal{Y} , \\ u_j(x,y) \text{ périodique en } y \text{ (de période 1 en toutes les variables),} \end{array} \right.$

$$(1.9) \qquad u_j(x,y) = 0 \quad \text{si} \quad x \in \Omega, \quad y \in S.$$

Remarque 1.5.

1) Les calculs qui vont suivre sont *formels*. Il faut donc, dans chaque cas, justifier le procédé.

2) La structure des fonctions $u_j(x,y)$ tient compte de la "double structure" géométrique de Ω_ε.

3) Les conditions (1.9) correspondant (sous réserve de convergence de la série (1.6)) à $u_\varepsilon = 0$ sur S_ε. Il serait naturel d'imposer également (pour tenir compte de $u_\varepsilon = 0$ sur Γ_ε)

$$(1.10) \qquad u_j(x,y) = 0 \quad \text{si} \quad x \in \Gamma , \ y \in \mathcal{O} .$$

On va voir que cela conduit à des difficultés.

4) Comme on le verra dans les exemples ci-après, la structure de l'Ansatz ci-dessus est *générale*. ∎

On utilise maintenant (1.6) (en supposant la série convergente de manière à pouvoir dériver terme à terme - sans préciser les espaces topologiques où cette convergence est supposée), et on porte (1.6) dans (1.4), puis on identifie les diverses puissances de ε dans le développement.

On note que, pour calculer $\Delta\phi(x,x/\varepsilon)$, on applique

$$(1.11)\quad \left|\begin{array}{l} \varepsilon^{-2}\Delta_y + 2\varepsilon^{-1}\Delta_{xy} + \Delta_x \ , \\ \Delta_{xy} = \dfrac{\partial^2}{\partial x_i \partial y_i} \ ^{(1)} \end{array}\right.$$

à $\phi(x,y)$ puis on remplace y par x/ε.

L'identification donne alors :

$$(1.12)\quad \left|\begin{array}{l} -\Delta_y u_0 = f \ , \\ -\Delta_y u_1 - 2\Delta_{xy} u_0 = 0 \ , \\ \text{----------------} \end{array}\right.$$

Dans $(1.12)_1$, x joue *le rôle de paramètre*. Si donc l'on introduit $w_0(y)$ par

$$(1.13)\quad \left|\begin{array}{l} -\Delta_y w_0 = 1 \quad \text{dans } \mathcal{Y}, \\ w_0 = 0 \text{ sur } S \ , \quad w_0 \text{ périodique en } y^{(2)} \end{array}\right.$$

alors

$$(1.14)\quad u_0(x,y) = w_0(y) f(x).$$

Alors $(1.12)_2$ donne

$$(1.15)\quad -\Delta_y u_1 = 2 \frac{\partial w_0}{\partial y_i} \frac{\partial f}{\partial x_i} \ ;$$

on introduit $w_i(y)$ par

$$(1.16)\quad \left|\begin{array}{l} -\Delta_y w_i = 2\dfrac{\partial w_0}{\partial y_i} \quad \text{dans } \mathcal{Y} \ , \\ w_i = 0 \text{ sur } S \ , \quad w_i \text{ périodique ;} \end{array}\right.$$

alors

$$(1.17)\quad u_1 = w_i(y) \frac{\partial f}{\partial x_i}(x)$$

et ainsi de suite. ■

---

(1) On adopte la convention de sommation des indices répétés.

(2) I.e. w_0 et ses dérivées prennent des valeurs égales sur les faces opposées de Y.

Remarque 1.6.

Les u_0, u_1, ... étant *définis de manière unique*, les conditions (1.10) *ne peuvent pas être satisfaites, au moins avec l'Ansatz* (1.6), *sauf si* f *vérifie des conditions aux limites*, i.e. si f et un certain nombre de (ou toutes) ses dérivées sont nulles sur Γ.

Il n'est pas difficile d'obtenir *dans ce cas* des estimations d'erreur du type

$$(1.18)\qquad \| u_\varepsilon - (u_0+\varepsilon u_1+ \dots +\varepsilon^m u_m \|_{H^1(\Omega_\varepsilon)} \leq C\, \varepsilon^m$$

(où $H^1(\Omega_\varepsilon)$ désigne l'espace de Sobolev d'ordre 1 sur Ω_ε). Cf. J.L.LIONS [1][2]..

Remarque 1.7.

Si $f \neq 0$ sur Γ, alors $u_0(x,y)$ n'est pas nul sur Γ et il faut donc introduire des termes supplémentaires *de couche limite* dans le développement (1.6). On trouvera quelques indications dans ce sens dans J.L. LIONS [3] et la Bibliographie de ce travail. ∎

2. CALCUL DES VARIATIONS ASYMPTOTIQUE

2.1. Principe de Dirichlet.

Le problème (1.4)(1.5) équivaut à la recherche de

$$(2.1)\qquad \inf J_\varepsilon(v)\ ,\quad v \in H_0^1(\Omega_\varepsilon)$$

où

$$(2.2)\qquad \left| \begin{array}{l} J_\varepsilon(v) = \frac{\varepsilon^2}{2}\, a_\varepsilon(v,v) - (f,v) \\ a_\varepsilon(u,v) = \displaystyle\int_{\Omega_\varepsilon} \frac{\partial u}{\partial x_j}\frac{\partial v}{\partial x_j}\, dx, \quad (f,v) = \int_{\Omega_\varepsilon} fv dx, \end{array} \right.$$

$$(2.3)\qquad H_0^1(\Omega_\varepsilon) = \{\varphi \mid \varphi \in H^1(\Omega_\varepsilon),\quad \varphi = 0 \text{ sur } \partial\Omega_\varepsilon\}\ .$$

On note que, si $\phi(x,y)$ est donnée dans $\Omega\times\mathcal{Y}$, continue (par exemple) sur $\bar\Omega\times\bar{\mathcal{Y}}$, périodique en y, alors

$$(2.4)\qquad \left| \begin{array}{l} \displaystyle\int_{\Omega_\varepsilon} \phi(x,x/\varepsilon)dx \longrightarrow \iint_{\Omega\times\mathcal{Y}} \phi(x,y)dxdy \\ \quad\text{lorsque } \varepsilon\to 0. \end{array} \right.$$

On prend dans $J_\varepsilon(v)$ des fonctions v données par

$$(2.5)\qquad v = v_\varepsilon(x) = v_0(x,x/\varepsilon) + \varepsilon v_1(x,x/\varepsilon) + \dots$$

où l'on suppose la somme *finie* , et où

$$(2.6)\qquad \left| \begin{array}{l} v_j(x,y) \text{ est définie dans } \Omega\times\mathcal{Y}, \text{ périodique en } y, \text{ et} \\ v_j(x,y) = 0 \quad \text{si} \quad y\in S. \end{array} \right.$$

Tenant compte de (2.4), on voit que $J_\varepsilon(v)$ est "voisin" de

$$\mathcal{J}_\varepsilon(v_0+\varepsilon v_1+\ldots) = \frac{\varepsilon^2}{2}\iint_{\Omega\times\mathcal{Y}} |(\varepsilon^{-1}\nabla_y + \nabla_x)(v_0+\varepsilon v_1+\ldots)|^2\,dxdy - \iint_{\Omega\times\mathcal{Y}} f(v_0+\varepsilon v_1+\ldots)\,dxdy. \tag{2.7}$$

C'est la *fonctionnelle homogénéisée* de $J_\varepsilon(v)$. ■

Développant $\mathcal{J}_\varepsilon(v_0+\varepsilon v_1+\ldots)$, il vient

$$\mathcal{J}_\varepsilon(v_0+\varepsilon v_1+\ldots) = \mathcal{J}_0(v_0)+\varepsilon\mathcal{J}_1(v_0,v_1) + \varepsilon^2\mathcal{J}_2(v_0,v_1,v_2) + \ldots\ , \tag{2.8}$$

$$\left|\begin{array}{l} \mathcal{J}_0(v_0) = \frac{1}{2}\iint_{\Omega\times\mathcal{Y}} |\nabla_y v_0|^2\,dxdy - \iint_{\Omega\times\mathcal{Y}} f\,v_0\,dxdy, \\ \mathcal{J}_1(v_0,v_1) = \iint_{\Omega\times\mathcal{Y}} \nabla_y v_0\,(\nabla_y v_1 + \nabla_x v_0)dxdy - \iint_{\Omega\times\mathcal{Y}} f\,v_1, \\ \mathcal{J}_2(v_0,v_1,v_2) = \frac{1}{2}\iint_{\Omega\times\mathcal{Y}} |\nabla_y v_1 + \nabla_x v_0|^2 + \iint_{\Omega\times\mathcal{Y}} (\nabla_y v_2 + \nabla_x v_1)\,\nabla_y v_0\,dxdy - \\ \qquad - \iint_{\Omega\times\mathcal{Y}} f v_2\,dxdy\ , \\ \ldots\ldots\ldots\ldots \end{array}\right. \tag{2.9}$$

On *prend ensuite pour* v_0 *l'élément qui minimise* $\mathcal{J}_0(v_0)$, soit u_0 donné par

$$\iint_{\Omega\times\mathcal{Y}} \nabla_y u_0\,\nabla_y v_0\,dxdy = \iint_{\Omega\times\mathcal{Y}} f\,v_0\,dxdy \tag{2.10}$$

où v_0 vérifie (2.6) avec v_0, $\nabla_y v_0 \in L^2(\Omega\times\mathcal{Y})$.

On retrouve ainsi u_0 *donnée par* (1.13)(1.14).

On note ensuite que (en utilisant (2.10))

$$\mathcal{J}_1(u_0,v_1) = \iint_{\Omega\times\mathcal{Y}} \nabla_y\,u_0\,\nabla_X u_0\,dxdy \tag{2.11}$$

est connu et indépendant de v_1.

Utilisant (2.10), on voit que

$$\mathcal{J}_2(u_0,v_1,v_2) = \frac{1}{2}\iint_{\Omega\times\mathcal{Y}} |\nabla_y v_1 + \nabla_x u_0|^2\,dxdy + \iint_{\Omega\times\mathcal{Y}} \nabla_y u_0\,\nabla_x v_1\,dxdy, \tag{2.12}$$

donc *indépendant de* v_2.

On minimise cette fonction en v_1, ce qui donne u_1 caractérisée par

$$\iint_{\Omega\times\mathcal{Y}} \nabla_y u_1\,\nabla_y v_1\,dxdy + \iint_{\Omega\times\mathcal{Y}} (\nabla_x u_0\,\nabla_y v_1 + \nabla_Y u_0\,\nabla_x v_1)\,dxdy = 0 \tag{2.13}$$

ce qui *redonne l'équation* $(1.12)_2$, *et ainsi de suite.* ■

Remarque 2.1.

Le procédé que l'on vient de donner est *formel*, et doit être justifié dans chaque cas particulier. Il a été introduit et utilisé dans J.L. LIONS [2], Chap. 1, § 5.

Dans le cas du problème de Dirichlet, le procédé précédent est plus compliqué que le calcul direct du N° 1. Mais nous allons maintenant donner des exemples où le procédé précédent, dit du *Calcul des Variations Asymptotique*, conduit assez rapidement à des remarques utiles.

2.2. Conditions aux limites du type de Neumann sur S_ε.

Nous considérons maintenant l'équation

(2.14) $$-\Delta u_\varepsilon = f \quad \text{dans} \quad \Omega_\varepsilon,$$

avec

(2.15) $$u_\varepsilon = 0 \quad \text{sur} \quad \Gamma_\varepsilon$$

et

(2.16) $$\frac{\partial u_\varepsilon}{\partial \nu} + \varepsilon \alpha u_\varepsilon = 0 \quad \text{sur} \quad S_\varepsilon,$$

où $\alpha > 0$.

La solution u_ε est *la* fonction qui minimise

(2.17) $$J_\varepsilon(v) = \frac{1}{2}\int_{\Omega_\varepsilon} |\nabla v|^2 dx + \frac{1}{2}\varepsilon\, \alpha \int_{S_\varepsilon} v^2 dS_\varepsilon - \int_{\Omega_\varepsilon} fv\, dx,$$

dans l'espace

(2.18) $$V_\varepsilon = \{v | \quad v \in H^1(\Omega_\varepsilon), \quad v = 0 \text{ sur } \Gamma_\varepsilon\}.$$

Le *calcul des variations asymptotique* conduit à introduire

(2.19) $$\begin{aligned}\mathcal{J}_\varepsilon(v_0+\varepsilon v_1+ \dots) = \frac{1}{2}\iint_{\Omega\times\mathcal{Y}} & |(\varepsilon^{-1}\nabla_y+\nabla_x)(v_0+\varepsilon v_1+ \dots)|^2 \, dxdy + \\ + \frac{\alpha}{2}\iint_{\Omega\times S} & |v_0+\varepsilon v_1+ \dots|^2 \, dxdy - \iint_{\Omega\times\mathcal{Y}} f(v_0+\varepsilon v_1+\dots)dxdy,\end{aligned}$$

où

(2.20) $$v_j \text{ est défini dans } \Omega\times\mathcal{Y}, \text{ périodique en } y$$
(*sans* la condition $v_j(x,y) = 0$ si $y \in S$).

La fonction $\mathcal{J}_\varepsilon(v_0+\varepsilon v_1+ \dots)$ est *finie* lorsque $\varepsilon \to 0$ si, et seulement si,

(2.21) $$\nabla_y v_0 = 0 \quad , \text{ i.e.} \quad v_0 = v_0(x).$$

Alors

(2.22) $$\mathcal{J}_\varepsilon(v_0+\varepsilon v_1+ \dots) = \mathcal{J}_0(v_0;v_1) + \varepsilon\, \mathcal{J}_1(v_0,v_1,v_2) + \dots \quad ,$$

$$(2.23)\quad \mathcal{J}_0(v_0,v_1) = \frac{1}{2}\iint_{\Omega\times\mathcal{Y}} |\nabla_y v_1+\nabla_x v_0|^2\, dxdy + \frac{\alpha}{2}\iint_{\Omega\times S} v_0^2\, dxdy - \iint_{\Omega\times\mathcal{Y}} fv_0 dxdy,$$

On considère alors le problème

$$(2.24)\quad \inf \mathcal{J}_0(v_0,v_1)$$

sur l'espace des fonctions v_0, v_1 telles que

$$(2.25)\quad \begin{cases} \nabla_x v_0 + \nabla_y v_1 \in L^2(\Omega\times\mathcal{Y}), \\ v_0 = 0 \text{ sur } \Gamma\ , \quad v_1 \text{ périodique en } y\ ^{(1)}. \end{cases}$$

Il faut en fait se placer dans *l'espace quotient* par la relation d'équivalence

$$\{v_0,v_1\} \sim \{\hat{v}_0,\hat{v}_1\} \text{ si } v_1 - \hat{v}_1 \text{ est indépendant de } y.$$

En effet la quantité

$$\left(\iint_{\Omega\times\mathcal{Y}} |\nabla_y v_1+\nabla_2 v_0|^2\, dx\, dy\right)^{\frac{1}{2}} = q(v_0,v_1)$$

est une semi norme : $q(v_0,v_1) = 0$ équivaut à $v_0 = 0$ et v_1 indépendant de y.
On trouve donc comme solution de (2.24) le couple $\{u_0,u_1\}$, avec

$$(2.26)\quad \iint_{\Omega\times\mathcal{Y}} (\nabla_y u_1+\nabla_2 u_0)(\nabla_y v_1+\nabla_2 v_0)dxdy + \alpha\iint_{\Omega\times S} u_0 v_0\, dxdy = \iint_{\Omega\times\mathcal{Y}} fv_0\, dxdy. \quad \blacksquare$$

Remarque 2.2.

Le calcul asymptotique direct pour (2.14)(2.15)(2.16) conduit à

$$\begin{cases} -\Delta_y u_0 = 0, \\ \dfrac{\partial u_0}{\partial\nu_{(y)}} = 0 \text{ sur } S\ , \quad u_0 \text{ périodique en } y\ , \end{cases}$$

donc $u_0 = u_0(x)$. Ensuite

$$(2.27)\quad \begin{cases} -\Delta_y u_1 = 0\ , \\ \dfrac{\partial u_1}{\partial\nu_{(y)}} + \nu_j \dfrac{\partial u_0}{\partial x_j} = 0 \text{ sur } S\ , \end{cases}$$

puis

---

(1) Ces conditions ont un sens.

$$(2.28)\quad \left|\begin{array}{ll} -\Delta_y u_2 - 2\Delta_{xy} u_1 - \Delta_x u_0 = f & \text{dans } \mathcal{Y}, \\ \dfrac{\partial u_2}{\partial \nu_{(y)}} + \nu_j(y)\dfrac{\partial u_1}{\partial x_j} + \alpha u_0 = 0 & \text{sur } S . \end{array}\right.$$

Le problème (2.18) admet une solution en u_2 si et seulement si

$$(2.29)\quad \int_S (\nu_j(y)\frac{\partial u_1}{\partial x_j} + \alpha u_0)dS - 2\int_{\mathcal{Y}} \Delta_{xy} u_1 dy - \int_{\mathcal{Y}} \Delta_x u_0\, dy = \int_{\mathcal{Y}} f\, dy.$$

On vérifie que (2.26) équivaut à (2.28)(2.29).

2.3. Un problème en dynamique des fluides.

On se donne Ω_ε comme précédemment, avec $\Omega_\varepsilon \subset \mathbb{R}^n$, n=2 ou n=3.
Des problèmes de *condensation* (cf. C. CONCA [1][2]) conduisent à la recherche de u_ε qui minimise

$$(2.30)\quad J_\varepsilon(v) = \nu\int_{\Omega_\varepsilon} e_{ij}(v)\, e_{ij}(v)dx + \frac{\varepsilon\alpha}{2}\int_{S_\varepsilon} v_n^2 dS_\varepsilon - \int_{\Omega_\varepsilon} fv dx$$

où

$$e_{ij}(v) = \frac{1}{2}\left(\frac{\partial v_i}{\partial x_j} + \frac{\partial v_j}{\partial x_i}\right) ,$$

$$v_n = v_i\, \nu_i \quad , \quad \nu = \{\nu_i\} = \text{normale à } S_\varepsilon ,$$

$$\alpha > 0.$$

On minimise $J_\varepsilon(v)$ dans l'ensemble des v telles que

$$(2.31)\quad \operatorname{div} v = 0 \quad \text{dans } \Omega_\varepsilon$$

et

$$(2.32)\quad v = 0 \text{ sur } \Gamma_\varepsilon .$$

Le *calcul des variations asymptotique* conduit à l'introduction de la fonctionnelle $\mathcal{J}_\varepsilon(v_0 + \varepsilon v_1 + \ldots)$:

$$(2.33)\quad \left|\begin{array}{l} \mathcal{J}_\varepsilon(v_0+\varepsilon v_1+ \ldots) = \nu\sum\displaystyle\iint_{\Omega\times\mathcal{Y}} |(\varepsilon^{-1} e_{ijy} + e_{ijx})(v_0+\varepsilon v_1+ \ldots)|^2 dxdy + \\ + \dfrac{\alpha}{2}\displaystyle\iint_{\Omega\times S} |(v_0+\varepsilon v_1+ \ldots)_n|^2 dxdS - \iint_{\Omega\times\mathcal{Y}} f(v_0+\varepsilon v_1+ \ldots)dxdy. \blacksquare \end{array}\right.$$

Comme au N° 2.2. précédent, $\mathcal{J}_\varepsilon(v_0+\varepsilon v_1+ \ldots)$ est *finie* lorsque $\varepsilon \to 0$ si, et seulement si,

$$(2.34)\quad e_{ijy}(v_0) = 0.$$

Alors $\quad v_0 = v_0(x)$ et

(2.35) $\mathcal{J}_\varepsilon(v_0+\varepsilon v_1+\ \ldots) = \mathcal{J}_0(v_0,v_1) + \varepsilon \mathcal{J}_1(v_0,v_1,v_2) + \ldots$,

avec

(2.36) $$\mathcal{J}_0(v_0 v_1) = \nu\sum\iint_{\Omega\times\mathcal{Y}} |e_{ijy}(v_1)+e_{ijx}(v_0)|^2 dxdy + \frac{\alpha}{2}\iint_{\Omega\times S} v_{0_n}^2 dxdS - \iint_{\Omega\times\mathcal{Y}} fv_0 dxdy.$$

La condition $\operatorname{div} v_\varepsilon = 0$ donne

$$(\varepsilon^{-1}\operatorname{div}_y + \operatorname{div}_x)(v_0+\varepsilon v_1+\ \ldots) = 0$$

soit

(2.37) $$\begin{cases} \operatorname{div}_y v_1 + \operatorname{div}_x v_0 = 0 \ , \\ \operatorname{div}_y v_2 + \operatorname{div}_x v_1 = 0 \ , \quad \ldots . \end{cases}$$

On doit donc minimiser $\mathcal{J}_0(v_0,v_1)$ pour les v_0,v_1 tels que

(2.38) $\operatorname{div}_x v_0 + \operatorname{div}_y v_1 = 0$,

ce qui conduit à u_0,u_1 avec

(2.39) $$\left|\begin{aligned} &2\nu\sum\iint_{\Omega\times\mathcal{Y}} (e_{ijy}(u_1)+e_{ijx}(u_0))(e_{ijy}(v_1)+e_{ijx}(v_0))dxdy + \\ &+ \alpha\iint_{\Omega\times S} u_{0_n} v_{0_n} dxdS = \iint_{\Omega\times\mathcal{Y}} fv_0 dxdy, \end{aligned}\right.$$

pour tout v_0, v_1 avec (2.38) et $v_o = 0$ sur Γ.

Cela conduit à une équation elliptique en u_0, assez compliquée, dûe à C. CONCA [1][2] où sont effectués les calculs directs. ■

Remarque 2.3.

Le cas où $u_\varepsilon = 0$ sur S_ε conduit à la loi de Darcy ; cf. J.L. LIONS [2][4], E. SANCHEZ-PALENCIA [1] . ■

3. QUELQUES PROBLEMES SPECTRAUX

3.1. Opérateur de Dirichlet.

On considère encore Ω_ε comme précédemment et on cherche *le spectre de l'opérateur* $-\Delta$ *pour les conditions aux limites de Dirichlet sur* $\partial\Omega_\varepsilon$.

Soit donc u_ε avec

(3.1) $-\Delta u_\varepsilon = \lambda(\varepsilon)u_\varepsilon$, $u_\varepsilon \in H_0^1(\Omega_\varepsilon)$, $u_\varepsilon \neq 0$.

On introduit ϕ par

(3.2) $$\begin{cases} -\Delta_y\phi = \alpha\phi \quad \text{dans } \mathcal{Y}, \\ \phi = 0 \ \text{sur } S \ , \ \phi \text{ périodique}, \\ \alpha = \textit{1ère valeur propre}, \end{cases}$$

ϕ *étant choisie par*

(3.3) $\phi(y) > 0$ dans $\mathcal{Y}$

(et normalisée par exemple par $\int_{\mathcal{Y}} \phi^2 \, dy = 1$). Si l'on définit ϕ_ε par

(3.4) $\phi_\varepsilon(x) = \phi(\frac{x}{\varepsilon})$, ϕ étant prolongée périodiquement, on définit (*sans difficulté, grâce à* (3.3)) w_ε par

(3.5) $u_\varepsilon = \phi_\varepsilon w_\varepsilon$.

Alors, toujours grâce à (3.3), (3.1) *équivaut* à

(3.6) $-\phi_\varepsilon \Delta(\phi_\varepsilon w_\varepsilon) = \lambda(\varepsilon) \phi_\varepsilon^2 w_\varepsilon$

avec

(3.7) $w_\varepsilon = 0$ sur Γ_ε.

et *sans conditions aux limites sur* S_ε (puisque ϕ_ε s'annule sur S_ε comme la distance à S_ε).

Mais utilisant $(3.2)_1$ on voit que

(3.8) $-\phi_\varepsilon \Delta (\phi_\varepsilon w) = B_\varepsilon w + \varepsilon^{-2} \alpha \phi_\varepsilon^2 w$

où

(3.9) $B_\varepsilon w = -\frac{\partial}{\partial x_i} (\phi_\varepsilon^2 \frac{\partial w}{\partial x_i})$.

Alors (3.6) équivaut à

(3.10) $B_\varepsilon w_\varepsilon = \mu(\varepsilon) w_\varepsilon$, avec (3.7),

(3.11) $\mu(\varepsilon) = \lambda(\varepsilon) - \varepsilon^{-2}\alpha$.

Il reste donc à obtenir le spectre de B_ε. On utilise les ansatz

(3.12) $w_\varepsilon = w_0 + \varepsilon w_1 + \dots$, $w_j = w_j(x, \frac{x}{\varepsilon})$,
$\mu(\varepsilon) = \mu_0 + \varepsilon \mu_1 + \dots$

où

(3.13) $w_j(x,y)$ est définie dans $\Omega \times \mathcal{Y}$, périodique en y (et sans condition aux limites imposée sur S).

On a

(3.14)
$$
\begin{aligned}
B_\varepsilon &= \varepsilon^{-2} B_0 + \varepsilon^{-1} B_1 + B_2 \\
B_0 &= -\frac{\partial}{\partial y_i} (\phi(y)^2 \frac{\partial}{\partial y_i}), \\
B_1 &= -\frac{\partial}{\partial y_i} (\phi(y)^2 \frac{\partial}{\partial x_i}) - \frac{\partial}{\partial x_i}(\phi(y)^2 \frac{\partial}{\partial y_i}) , \\
B_2 &= -\frac{\partial}{\partial y_i} (\phi(y)^2 \frac{\partial}{\partial x_i}) .
\end{aligned}
$$

L'équation (3.10) devient :

$$(3.15)\quad \left|\begin{array}{l} B_0 w_0 = 0 \ , \\ B_0 w_1 + B_1 w_0 = 0 \ , \\ B_0 w_2 + B_1 w_1 + B_2 w_0 = \mu_0 \phi^2 w_0 \ , \\ \dots\dots\dots\dots\dots\dots \end{array}\right.$$

On déduit de $(3.15)_1$ et de la périodicité de w_0 en y que $w_0 = w_0(x)$ ne dépend pas de y . Alors $(3.15)_2$ se réduit à

$$(3.16)\quad B_0 w_1 = \left(\frac{\partial}{\partial y_i}\phi(y)^2\right) \frac{\partial w_0}{\partial x_i} .$$

Si donc l'on introduit $\psi_j = \psi_j(y)$ - définie à une fonction de x additive près - par

$$(3.17)\quad \left|\begin{array}{l} B_0 \psi_j = \frac{\partial}{\partial y_j} (\phi(y))^2 \ \text{dans} \ \mathcal{Y}, \\ \psi_j \ \text{périodique en} \ y \ , \end{array}\right.$$

on a

$$(3.18)\quad w_1 = \psi_j \frac{\partial w_0}{\partial x_j} + \tilde{w}_1(x) .$$

L'équation $B_0 w_2 = F$, w_2 périodique, admet une solution si et seulement si

$$\int_{\mathcal{Y}} F \, dy = 0 \ ,$$

de sorte que $(3.15)_3$ donne, pour que w_2 existe :

$$(3.19)\quad \int_{\mathcal{Y}} (B_1 w_1 + B_2 w_0) dy = \mu_0 \int_{\mathcal{Y}} \phi^2 w_0 dy = \mu_0 w_0 \ ,$$

ce qui donne :

$$(3.20)\quad \mathcal{B} w_0 = \mu_0 w_0,$$

où

$$(3.21)\quad \mathcal{B} = -b_{ij} \frac{\partial^2}{\partial x_i \partial x_j} \ ,$$

$$(3.22)\quad b_{ij} = \int_{\mathcal{Y}} \phi(y)^2 \left[\delta_i^j + \frac{\partial \psi_j}{\partial y_i}\right] dy.$$

On ajoute les conditions aux limites

$$(3.23)\quad w_0 = 0 \ \text{sur} \ \Gamma.$$

On peut vérifier (cf. M. VANNINATHAN [1]) que l'opérateur $\mathcal{B}$ ainsi obtenu est *elliptique* de sorte que *le problème* (3.20)(3.23) *admet un spectre discret*

$$(3.24)\quad 0 < \mu_{01} \le \mu_{02} \le \cdots \le \mu_{0n} \le \cdots \ .$$

On montre (M. VANNINATHAN, loc. cit.) que *les valeurs propres de* $-\Delta$ *dans* Ω_ε *(pour les conditions de Dirichlet) sont données par*

$$(3.25)\quad \lambda_n = \varepsilon^{-2}\alpha + \mu_{0n} + O(\varepsilon).$$

3.2. Quelques problèmes ouverts.

Les résultats du N° 3.1. reposent, *de manière apparemment essentielle,* sur le fait que *la première fonction propre définie dans* (3.2) *est* > 0 *dans* $\mathcal{Y}$. Faute (notamment) d'une extension de cette propriété pour Δ^2, *le problème suivant est ouvert :* trouver le développement asymptotique du spectre $\lambda(\varepsilon)$ pour

$$(3.26)\quad \left| \begin{array}{l} \Delta^2 u_\varepsilon = \lambda(\varepsilon)\, u \ \text{ dans } \Omega_\varepsilon, \quad u_\varepsilon \neq 0, \\ u_\varepsilon = \dfrac{\partial u_\varepsilon}{\partial \nu} = 0 \text{ sur } \Gamma_\varepsilon \cup S_\varepsilon. \end{array} \right.$$

Remarque 3.1.

L'exemple (3.26) est mentionné à titre d'illustration. *Toutes* les questions de ce type pour les opérateurs elliptiques d'ordre ≥ 4 et pour les systèmes elliptiques sont *ouvertes*.

3.3. Conditions aux limites non locales sur les trous.

Le problème des vibrations propres d'un faisceau de tubes élastiques immergés dans un liquide est étudié, par les méthodes de l'homogénéisation, par J. PLANCHARD [1] [2] . Il s'agit, avec les notations précédentes, de trouver u_ε et $\lambda(\varepsilon)$ solutions de

$$(3.27)\quad \left| \begin{array}{l} \Delta u_\varepsilon = 0 \ \text{ dans } \Omega_\varepsilon, \\ \dfrac{\partial u_\varepsilon}{\partial \nu} = 0 \text{ sur } \Gamma_\varepsilon, \\ \dfrac{\partial u_\varepsilon}{\partial \nu} = \lambda(\varepsilon)\, \sigma_\varepsilon(u_\varepsilon) \text{ sur } S_\varepsilon \end{array} \right.$$

où

$$(3.28)\quad \sigma_\varepsilon(\varphi) = \nu_i\left(\frac{x}{\varepsilon}\right) \int_{\gamma_\varepsilon} \nu_i\left(\frac{x}{\varepsilon}\right) \varphi(x)\, d\gamma_\varepsilon(x)$$

où γ_ε désigne le bord du trou sur lequel on calcule dans $(3.27)_3$ la dérivée normale. Dans (3.28) $\{\nu_i(y)\}$ désigne la normale à S dirigée vers l'intérieur de $\mathcal{O}$.

Remarque 3.2.

On suppose que Ω est un rectangle dans $\mathbb{R}^2$ et que les trous *ne rencontrent pas* Γ.■

On cherche u_ε, $\lambda(\varepsilon)$, sous la forme

$$(3.29)\quad \left| \begin{array}{l} u_\varepsilon = u_0 + \varepsilon u_1 + \ldots \ , \quad u_j = u_j\left(x, \dfrac{x}{\varepsilon}\right), \\ \lambda(\varepsilon) = \varepsilon^{-2}\mu + \ldots \ , \end{array} \right.$$

$u_j(x,y)$ ayant la structure habituelle.

On trouve ainsi (cf. J. PLANCHARD, loc. cit. et J. PLANCHARD, F.N. REMY et P. SONNEVILLE [1] .

$$(3.30)\quad \begin{cases} -\Delta_y u_0 = 0 \ , \\ \dfrac{\partial u_0}{\partial \nu} = \mu\sigma(u_0) \ \text{ sur } \ S \ , \\ u_0 \ \text{ périodique} \end{cases}$$

où

$$(3.31)\qquad \sigma(u_0) = \nu_i(y) \int_S \nu_i(y) u_0 \, dS(y).$$

On obtient un spectre discret pour μ , en travaillant sur *l'espace quotient* $H^1_{per}(\mathcal{Y})/\mathbb{R}$, $H^1_{per}(\mathcal{Y})$ étant l'espace des fonctions de $H^1(\mathcal{Y})$ et *périodiques* (i.e. prenant des valeurs égales sur les faces opposées de Y).

4. QUESTIONS DIVERSES

4.1. Autres Ansatz.

Considérons dans Ω_ε le problème

$$(4.1)\quad \begin{cases} -\varepsilon^3 \, \Delta u_\varepsilon + u_\varepsilon = f \ \text{ dans } \ \Omega_\varepsilon , \\ u_\varepsilon = 0 \ \text{ sur } \Gamma_\varepsilon \cup S_\varepsilon . \end{cases}$$

On se convaincra facilement que les ansatz des types qui précèdent ne conviennent pas.

On est alors conduit à introduire

$$(4.2)\qquad u_\varepsilon = u_{0\varepsilon} + \varepsilon u_{1\varepsilon} + \ldots$$

où les $u_{j\varepsilon}(x,y)$ dépendent de ε.

On obtient ainsi (formellement)

$$(4.3)\quad \begin{cases} -\varepsilon\Delta_y u_{0\varepsilon} + u_{0\varepsilon} = f \ , \\ -\varepsilon\Delta_y u_{1\varepsilon} + u_{1\varepsilon} - 2\varepsilon \, \Delta_{xy} \, u_{0\varepsilon} = 0 \ , \\ -\varepsilon\Delta_y u_{2\varepsilon} + u_{2\varepsilon} - 2\varepsilon \, \Delta_{xy} \, u_{1\varepsilon} - \varepsilon\Delta_x \, u_{0\varepsilon} = 0 \ , \\ \ldots\ldots\ldots\ldots\ldots\ldots\ldots \end{cases}$$

$$(4.4)\qquad u_{j\varepsilon} = 0 \ \text{si} \ \ y \in S \ .$$

On introduit alors $w_{0\varepsilon}$ par

$$(4.5)\quad \begin{cases} -\varepsilon \, \Delta_y \, w_{0\varepsilon} + w_{0\varepsilon} = 1 \ \text{ dans } \mathcal{Y}, \\ w_{0\varepsilon} = 0 \ \text{ sur } \ S \ , \ \ w_{0\varepsilon} \ \text{ périodique} \end{cases}$$

d'où

(4.6) $u_{0\varepsilon} = w_{0\varepsilon}(y) f(x)$,

et ainsi de suite. Le développement correspondant peut être justifié. ■

Remarque 4.1.

Dans le travail de B. DESGRAUPES [1] on trouvera de nombreux développements du type précédent. ■

Remarque 4.2.

Dans les cas où la structure "périodique" (approximative !) dépend de la solution- comme dans des modèles de turbulence ; cf. la conférence de C. BEGUE, D.Mc. LAUGHIN G. PAPANICOLAOU et O. PIRONNEAU [1] - on introduit des ansatz où $y = x/\varepsilon$ est remplacé par $y = \frac{\theta(x,t)}{\varepsilon}$, $\theta(x,t)$ étant lui-même donné par la solution d'un système d'Equations aux Dérivées Partielles faisant intervenir les termes du développement. Tout cela est lié aux idées de V. MASLOV [1] . ■

4.2. Problèmes raides et homogénéisation.

Dans plusieurs questions, on a affaire à des problèmes à structure périodique *et* avec des coefficients d'ordres de grandeur très différents selon les régions. Cela *peut* entrer dans un "mélange" de problèmes "raides" (tels que présentés dans J.L. LIONS [6]) et de problèmes d'homogénéisation.
Cf. J.L. LIONS [5], G.P. PANASENKO [1][2] et également les notes [3] de l'Auteur.

Pour les aspects numériques récents de la théorie de l'homogénéisation, on renvoie à la conférence de D. BEGIS, S. DINARI, G. DUVAUT, A. HASSIM, F. PISTRE [1] dans ces Proceedings. ■

4.3. Problèmes de convergence.

Dans le cas des *matériaux composites*, la méthode la plus puissante pour justifier, au moins le premier terme du développement, est celle de L. TARTAR [1], basée sur des estimations de l'énergie et sur des idées de la *compacité par compensation*, (F. MURAT [1], L. TARTAR [2]).

Pour les milieux perforés ou à obstacle, on se reportera à la Bibliographie déjà indiquée et à D. CIORANESCU et J. SAINT JEAN PAULIN [1][2].

4.4. Obstacles irréguliers.

Dans le cas d'obstacles (ou de trous) irréguliers, on se reportera à D.CIORANESCU et F. MURAT [1] et à la Bibliographie de ces travaux.

4.5. On trouvera d'autres aspects de ces questions dans les travaux de O.A.OLEINIK et ses collaborateurs [1] et dans ceux de N.C. BAHBALOV [1]. On consultera également l'intéressant exposé de E. SANCHEZ-PALENCIA [2] et la Bibliographie de ce travail.

Un point de vue, peut-être moins constructif, mais plus général, est développé par E. DE GIORGI et son Ecole. Nous renvoyons à E. DE GIORGI [1] et à l'importante Bibliographie de ce travail.

-=-

BIBLIOGRAPHIE

N.C. BAHVALOV [1] Livre à paraître.

D. BEGIS, S. DINARI, G. DUVAUT, A. HASSIM, F. PISTRE [1] MODULEF et les matériaux composites. Ces Proceedings.

C. BEGUE, C.Mc LAUGHIN, G. PAPANICOLAOU, O. PIRONNEAU [1] Turbulence modelling by homogenization of microstructures. Ces Proceedings.

A. BENSOUSSAN, J.L. LIONS et G. PAPANICOLAOU [1] *Asymptotic Analysis for Periodic Structures*. North-Holland. 1978.

D. CIORANESCU et F. MURAT [1] Un terme étrange venu d'ailleurs. Parties I et II. dans "Non linear partial differential equations and their applications". Collège de France Seminar, Vol. II et III, Ed. H. BREZIS et J.L. LIONS, Research Notes in Mathematics, 60 et 70, Pitman, (1982), pp. 98-138 et 154-178.

D. CIORANESCU et J. SAINT JEAN PAULIN [1] Homogenization in open sets with holes. J.M.A.A. 71 (1979), pp. 590-607.

[2] Homogénéisation de problèmes d'évolution dans des ouverts à cavités. C.R. Acad. Sci. Paris, A. tome 286 (1978), pp. 899-902.

C. CONCA [1] On the application of the homogenization theory to a class of problems arising in fluid mechanics. I. Theoretical results. A paraître.

[2] II. Numerical results. A paraître.

E. DE GIORGI [1] Generalized Limits in Calculus of Variations, dans "Topics in Functional Analysis", ed. par F. STROCCHI, E. ZARANTONELLO, E. DE GIORGI, G. DALMASO et L. MODICA, Scuola Normale Superiore Pisa, 1980/81, pp. 117-148.

B. DESGRAUPES [1] Thèse de 3ème cycle, Paris, 1980.

J.L. LIONS [1] Asymptotic expansions in perforated media with a periodic structure. Rocky Mountains Journal (1980), 10, pp. 125-140.

[2] *Some Methods in the Mathematical Analysis of Systems and their Control*. Science Press, Beijing et Gordon et Breach, New York, 1981.

[3] Remarques sur les problèmes d'homogénéisation dans des milieux à structure périodique et sur quelques problèmes raides. Dans le Cours CEA-INRIA-EDF, Juillet 1983.

[4] Some problems connected with Navier Stokes equations. Lectures at ELAM-IV, Lima 1978, Actas, Lima, 1979, pp. 222-286.

[5] Remarques sur les aspects numériques de la méthode d'homogénéisation dans les matériaux composites, dans *Etude Numérique des Grands Systèmes*, ed. par G.I. MARCHOUK et J.L. LIONS, Dunod, 1978, pp. 15-38.

[6] *Perturbations singulières dans les problèmes aux limites et en Contrôle optimal*. Lecture Notes, Springer 323, 1973.

V. MASLOV [1] Conférence au Congrès International des Mathématiciens, Varsovie, Août 1983.

F. MURAT [1] Compacité par compensation. Annali Scuola Normale Superiore di Pisa, Sc. Fis. et Mat. (IV), 3 (1978), pp.489-507.

O.A. OLEINIK, G.A. IOSIFIAN et G.P. PANASENKO [1] Développement asymptotique de la solution dy système de l'élasticité dans des milieux perforés. Mat. Sbornik, 1983, 120 (162), N° 1, pp. 22-41.

G.P. PANASENKO [1] Solutions algébriques et valeurs propres d'équations elliptiques avec coefficients fortement variables. Doklady Akad. Nauk, 1980, t. 252, N° 6, pp. 1320-1325.

[2] The principle of splitting an averaged operator for a nonlinear system of equations in periodic and random skeletal structures. Soviet Mat. Doklady, Vol. 25 (1982), N° 2, pp. 290-295.

J. PLANCHARD [1] Développements asymptotiques des fréquences propres d'un faisceau de tubes élastiques plongés dans un fluide incompressible. E.D.F. Bulletin Direction des Etudes et Recherches. Série C, N° 2 (1982), pp. 65-76.

[2] A paraître.

J. PLANCHARD, F.N. REMY, P. SONNEVILLE [1] A simplified method for determining acoustic and tube eigenfrequencies in that exchangers. Journal Pressure Vessel Technology. 104 (1982), pp. 175-179.

E. SANCHEZ-PALENCIA [1] *Non homogeneous media and vibration theory*. Lecture Notes in Physics. Springer Verlag 127, 1980.

[2] Homogenization method for the study of composite media. Dans *Asymptotic Analysis* II, Lecture Notes in Mathematics, 1983, pp. 192-214.

L. TARTAR [1] Oeuvres complètes non publiées.

[2] Compensated compactness and applications to Partial Differential Equations, dans *Non linear Analysis and Mechanics* : Heriot-Watt Symposium, Vol. IV, Research Notes in Mathematics, 39, Pitman (1979), pp. 136-212.

-=-

Computing Methods in Applied Sciences and Engineering, VI
R. Glowinski and J.-L. Lions (Editors)
Elsevier Science Publishers B.V. (North-Holland)

ASYMPTOTIC EXPANSIONS FOR FINITE ELEMENT APPROXIMATION OF ELLIPTIC PROBLEM ON POLYGONAL DOMAINS

by Lin Qun (1) and Lu Tao (2)

(1) Institute of Systems Science, Academia Sinica, Beijing, China and Bonn University, Bonn.

(2) Chengdu Branch of Academia Sinica, Chengdu, China.

Abstract : Consider the elliptic problem (1) on a convex polygonal domain Ω in R^2. This paper concerns with finite element method on a special partition of Ω, in which Ω is divided into several triangles Ω_i $(i=1,\ldots,m)$ with all vertices at the boundary and each Ω_i is divided into a uniform triangulation (with triangles of size h_i). It is shown that there exists an asymptotic expansion (2) (3) for the linear finite element approximation defined on the special partition as above. As a by-product we can see that the superconvergence for the gradient can be obtained by using the means of the gradients for two adjacent elements as the approximations to the gradient at the mid-points of common sides for all elements within each Ω_i.

1.

Let us consider the simple model problem : Find $u \in H_0^{1,2}$ such that

$$- \Delta u = f \quad \text{in } \Omega. \tag{1}$$

We approximate (1) by linear finite element method. First we divide Ω in several triangles Ω_i $(i=1,\ldots,m)$ with all vertices at the boundary and then divide each Ω_i in a uniform triangulation $I_i = \{K_i\}$ with triangles K_i of size h_i. Let u^h be the linear finite element solution of (1) with respect to the partition $I = \bigcup_{i=1}^{m} I_i$ of Ω and u^I the linear interpolation of u on I. We have

Theorem. If $u \in C^4$ then there exists ϕ_i $(i=1,\ldots,m)$ independent on h such that

$$u^h - u^I = \sum_{i=1}^{m} \phi_i \, h_i^2 + r, \tag{2}$$

$$\| r \|_C \le o\,\Big(\sum_{i=1}^{m} h_i^2 \Big) \, \| u \|_{C^4}. \tag{3}$$

Proof : Follow Frehse [4] [5] and scott [10] we define the Green function $G_0 = G(z_0, z)$ at point $z_0 \in \Omega$ and the corresponding finite element approximation G_0^h. Then

$$u^h(z_0) - u^I(z_0) = (\nabla(u^h - u^I), \nabla G_0^h) = (\nabla(u - u^I), \nabla G_0^h)$$

$$= \sum_{i=1}^{m} \sum_{\kappa_i} \int_{\partial\kappa_i} (u - u^I) \, \frac{\partial}{\partial n} \, G_0^h \, ds, \tag{4}$$

Where $\partial\kappa_i$ extend all over the sides of all elements $\kappa_i \subset \Omega_i$. We now consider a typical triangle Ω_i = ABC and calculate the line-integrals along direction 23 (see Fig.1)

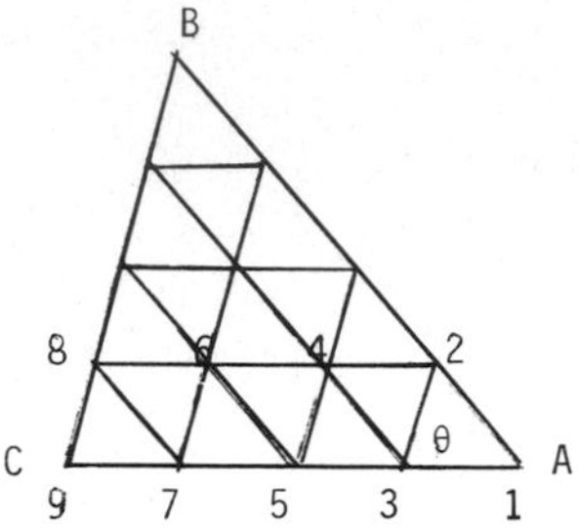

Fig. 1 (Ω_i)

$$\Sigma \int_{23} (u-u^I) \frac{\partial}{\partial n} G_o^h \, ds$$
$$= \Sigma -\sin\theta \int_{23} (u-u^I) \frac{\partial}{\partial x} G_o^h \, ds$$
$$+ \Sigma \cos\theta \int_{23} (u-u^I) \frac{\partial}{\partial y} G_o^h \, ds$$
$$= \Sigma_i^s + \Sigma_i^c .$$

Since G_o^h is a peciewise linear function on T_i one has

$$\frac{\partial}{\partial x} G_o^h = \frac{1}{2\alpha} ((y_3-y_1)(G_o^h(2) - G_o^h(1)) + (y_1-y_2)(G_o^h(3) - G_o^h(1))) \text{ in } 123,$$
$$\frac{\partial}{\partial x} G_o^h = \frac{1}{2\alpha} (-(y_3-y_1)(G_o^h(3) - G_o^h(4)) - (y_1-y_2)(G_o^h(2) - G_o^h(4)) \text{ in } 234,$$

..., where α is the area of triangle 123 and (x_j,y_j) the coordinate of point j. Then

$$\Sigma_i^s = \frac{\sin\theta}{2\alpha} (y_2-y_3)((G_o^h(2)- G_o^h(1) + G_o^h(3) - G_o^h(4)) \int_{23} (u-u^I) \, ds$$
$$+ (G_o^h(4) - G_o^h(3) + G_o^h(5) - G_o^h(6)) \int_{45} (u-u^I) \, ds$$
$$+ (G_o^h(6) - G_o^h(5) + G_o^h(7) - G_o^h(8)) \int_{67} (u-u^I) \, ds)$$
$$- \frac{\sin\theta}{2\alpha} ((y_9-y_7)\,(G_o^h(8)-G_o^h(7)) + (y_7-y_8)\,(G_o^h(9)-G_o^h(7))) \int_{89} (u-u^I) \, ds$$
$$+ \ldots$$
$$= \frac{\sin\theta}{2\alpha} (y_2-y_3)((G_o^h(2)-G_o^h(1)) \int_{23} (u-u^I) \, ds + (G_o^h(4)-G_o^h(3))(\int_{45} - \int_{23})(u-u^I) \, ds$$
$$+ (G_o^h(6) - G_o^h(5)) (\int_{67} - \int_{45})(u-u^I) \, ds + (G_o^h(8)-G_o^h(7))(\int_{89} - \int_{67})(u-u^I) \, ds)$$

$$- \frac{\sin\theta}{2\alpha} (y_7 - y_8) (G_o^h(9) - G_o^h(8)) \int_{89} (u - u^I)\, ds.$$

Since $u \in C^4$, by the Euler-Maclaurin formula,

$$\int_{23} (u-u^I)\, ds = \frac{1}{12} l_{23}^3 (D_{23}^2 u(M_{23}) - D_{23}^2 u(M_{21}) + D_{23}^2 u(M_{21})) + O(h_i^5) \| u \|_{C^4}$$

$$= \frac{1}{12} l_{23}^3 \frac{1}{2} l_{24} \frac{1}{\alpha} \int_{123} D_{24} D_{23}^2 u(z) dz + \frac{1}{12} l_{23}^3 \frac{1}{l_{12}} \int_{12} D_{23}^2 u\, ds + O(h_i^5) \| u \|_{C^4},$$

$$\left(\int_{45} - \int_{23}\right)(u-u^I)\, ds - \frac{1}{12} l_{23}^3 (D_{23}^2 u(M_{45}) - D_{23}^2 u(M_{23})) + O(h_i^5) \| u \|_{C^4}$$

$$= \frac{1}{12} l_{23}^3 l_{24} \frac{1}{2\alpha} \int_{2345} D_{24} D_{23}^2 u(z) dz + O(h_i^5) \| u \|_{C^4},$$

$$\int_{89} (u-u^I)\, ds = \frac{1}{12} l_{23}^3 \frac{1}{l_{89}} \int_{89} D_{23}^2 u\, ds + O(h_i^5) \| u \|_{C^4},$$

..., where $l_{jk} = |j-k|$ and M_{jk} is the midpoint between j and k.

Nothing that

$$G_o^h(2) - G_o^h(1) = l_{12} D_{12} G_o^h,$$

$$G_o^h(4) - G_o^h(3) = l_{34} D_{34} G_o^h = l_{34} \frac{1}{2\alpha} \int_{2345} D_{34} G_o^h(z) dz,$$

$$G_o^h(9) - G_o^h(8) = l_{23} D_{23} G_o^h,$$

... and setting

$$c_i h_i^2 = \frac{\sin\theta}{48\alpha^2} (y_2 - y_3) l_{23}^3 l_{24} l_{12},$$

$$b_i h_i^2 = \frac{\sin\theta}{24\alpha} (y_2 - y_3) l_{23}^3, \quad a_i h_i^2 = - \frac{\sin\theta}{24\alpha} (y_7 - y_8) l_{23}^3 ,$$

we have

$$\Sigma_i^s = c_i h_i^2 \int_{\Omega_i} D_{23}^2 D_{24} u(z) D_{34} G_o^h(z)\, dz + b_i h_i^2 \int_{AB} D_{23}^2 u D_{AB} G_o^h\, ds$$

$$+ a_i h_i^2 \int_{BC} D_{BC}^2 u D_{BC} G_o^h\, ds + O(h_i^3) \|u\|_{C^4} \int_{\Omega_i} |\nabla G_o^h(z)|\, dz.$$

Integrating by parts and noting that

$$G_o^h(A) = G_o^h(B) = G_o^h(C) = o$$

we obtain

$$\Sigma_i^s = C_i\, h_i^2 \int_{\Omega_i} D_{23}^2\, D_{24}\, u\,(z)\, D_{34}\, G_o^h(z)dz - b_i\, h_i^2 \int_{AB} D_{AB}\, D_{23}^2\, u.\, G_o^h\, ds$$

$$- a_i\, h_i^2 \int_{BC} D_{BC}^3\, u.\, G_o^h\, ds + O(h_i^3)\, \|u\|_{C^4} \int_{\Omega_i} |\nabla G_o^h(z)|\, dz. \qquad (5)$$

Instead of use G_o^h in (5) one may use G_o and obtains ([4] [5])

$$\Sigma_i^s = C_i\, h_i^2 \int_{\Omega_i} D_{23}^2\, D_{24}\, u(z)\, D_{34}\, G_o\,(z)\, dz - b_i\, h_i^2 \int_{AB} D_{AB}\, D_{23}^2\, u.\, G_o\, ds$$

$$- a_i\, h_i^2 \int_{BC} D_{BC}^3\, u.\, G_o\, ds + (\sum_i h_i^3\, |\ln h_i|)\, \|u\|_{C^4}.$$

We also have the similar expansions for Σ_i^c and for directions 24 and 34. Then (2) (3) follows from the summation of these estimates.

Remark 1. A great deal of attention has been focused on the asymptotic expansions for finite difference method on the rectangular domains and the smooth domains. See Böhmer [1], Marchuk and Shaidurov [8], Munz [9], Stetter [11]. Recently the authors [6][7] have studied the asymptotic expansions for finite element method on a triangular domain. Our method in proving (2) (3) is based on the Green function techniques developed in Frehse [4][5] and Scott [10] and differ entirely from those considered in [1] [8] [9] [11].

2.

Let us observe now the superconvergence for the gradient. We consider the directional derivatives along the edge 23 and 14 respectively (see Fig. 1) :

$$D_{23}\, u\,(M_{23}) = \frac{u(2) - u(3)}{l_{23}} + O(h^2),\quad D_{23}\, u^h\,(M_{23}) = \frac{u^h(2) - u^h(3)}{l_{23}},$$

$$D_{14}\, u\,(M_{23}) = \frac{u(1) - u(4)}{l_{14}} + O(h^2),\quad D_{14}\, u^h\,(M_{23}) = \frac{u^h(1) - u^h(4)}{l_{14}},$$

Then by (4), if $u \in C^3$,

$$|D_{23}(u(M_{23}) - u^h(M_{23}))| = \frac{1}{l_{23}}\, |u(2) - u^h(2) - ((u(3) - u^h(3))| + O(h^2)$$

$$= \frac{1}{l_{23}}\, |(\nabla(G_2^h - G_3^h),\, \nabla(u-u^I))| + O(h^2)$$

$$\le Ch\, \|u\|_{C^3} \int_{\Omega} |\nabla(G_2^h - G_3^h)|\, dz + O(h^2) \le O(h^2|\ln h|)$$

$$|D_{14}\,(u(M_{23}) - u^h\,(M_{23}))| \le O(h^2|\ln h|)$$

Thus we have, if $u \in C^3$,

$$\nabla(u(M_{23}) - u^h\,(M_{23})) = O(h^2|\ln h|). \qquad (6)$$

Remark 2. Bramble, Nitsche and Schatz [2] have proved the estimate (6) in the totally regular case. And Zlamal [12] Chen [3] have proved the estimate (6) in the strong regular case for some M_{23}; see also [13].

Acknowledgment. The authors wish to express their sincere tanks to professor J. Frehse whose talk make this paper to be possible. We also wish to thank professors F. Brezzi, R. Glowinski, H. Helfrich, K. Hoffmann, R. Scott and L. Wahlbin for their discussions and comments.

BIBLIOGRAPHY

[1] Böhmer, K., Math. Z. 177, 235-255 (1981)

[2] Bramble, J., Nitsche, J. and Schatz, A., Math. Comp. 677-688 (1975)

[3] Chen, C., Numer. Math. (A Journal of Chinese University) 2, 12-20 (1980)

[4] Frehse, J., Proc. of China-France Symposium on FEM, Science Press, Beijing, 1983

[5] Frehse, J. and Rannacher R., Bonn. Math. Schrift, No 89 (1976)

[6] Lin, Q., Lu, T. and Shen, S., J. Comp. Math. 4 (1983

[7] Lin, Q., Lu, T. and Shen, S., Research Report IMS - 11, Chengdu Branch of Acad. Sinica, Chengdu, China (1983)

[8] Marchuk, G. and Shaidurov, V., Difference methods and their extrapolations, Springer-Verlag, 1983

[9] Munz, H., Math. Comp. 36, 155-170 (1981)

[10] Scott, R., Math. Comp. 30 (1975)

[11] Stetter, H., Numer. Math. 7, 18-31 (1965)

[12] Zlamal, M., Lecture Notes in Math. 606,351-262 (1977)

[13] Křížek, M. and Neittaanmäki, P., Univ. Jyväshylä, Dept. Math. Preprint 23 (1983)

PARTICLE AND SPECTRAL METHODS

METHODES PARTICULAIRES ET SPECTRALES

Computing Methods in Applied Sciences and Engineering, VI
R. Glowinski and J.-L. Lions (Editors)
Elsevier Science Publishers B.V. (North-Holland)

ON SOME APPLICATIONS OF THE PARTICLE METHOD

P.-A. RAVIART
Analyse Numérique
Université P. & M. Curie
4, place Jussieu
75230 Paris Cédex 05

We describe the particle method of approximation of linear hyperbolic equations of the first order and we give some typical convergence results. Next, we consider a kinetic equation involving transport and scattering terms and we propose a numerical scheme based on a coupled "particle method-moment method".

1. Introduction.

Among the numerical methods used for solving partial differential equations, particle methods are far less popular than conventional methods such as finite difference methods, finite element methods or spectral methods. However, these particle methods seem to be fairly well adapted to the numerical simulation of transport models which arise in a variety of physical situations (transport of neutral particles or charged particles) especially when collision processes are neglected. Moreover, particle methods have always a clear physical interpretation which makes them very attractive. For a discussion of particle models in various physical problems, we refer to Hockney & Eastwood [13]. For an application of the particle method to the numerical simulation of incompressible fluid flows at large Reynolds numbers, see the recent review of Leonard [15] on the vortex method.

Mathematically, particle methods have to be first considered as numerical methods of approximation of linear hyperbolic equations of the first order where the solution is approximated by a set of Dirac measures (the particles). In fact, the corresponding numerical analysis is now well understood at least in the case of the pure Cauchy problem (see [17], [18]). Difficulties arise however when we need to consider more general situations where the model takes into account collision phenomena via diffusion or scattering terms. The most common way for overcoming this difficulty consists in using a Monte Carlo methodology such as random-walk techniques similar to those used by Chorin [7] [8]. The drawback of this method is its low rate of convergence and one can be tempted by a deterministic method. In that direction, see [11].

In this paper, we shall give a brief discussion of the particle approximation of linear hyperbolic equations of the first order and we shall describe an application to the numerical solution of a kinetic equation involving scattering terms.

2. The particle approximation of linear hyperbolic equations of the first order.

Let Ω be an open subset of $\mathbb{R}^d$ with boundary Γ. We denote by n the unit outward normal to Γ. We consider the linear differential operator of the first order written in conservation form :

$$Lu \equiv \frac{\partial u}{\partial t} + \sum_{i=1}^{d} \frac{\partial}{\partial x_i} (a_i\, u) + a_0\, u$$

where the coefficients $a_i = a_i(x)$, $0 \leqslant i \leqslant d$, are "sufficiently" smooth functions defined in $\overline{\Omega}$. We set

$$a = (a_1,\ldots,a_d)$$

and

$$\Gamma^- = \{x \in \Gamma \ ; \ (a\cdot n)(x) < 0\} \ , \ \Gamma^+ = \{x \in \Gamma \ ; \ (a\cdot n)(x) > 0\} \ ,$$

$$Q_T = \Omega \times]0,T[\quad , \quad \Sigma_T^{\pm} = \Gamma^{\pm} \times]0,T[\quad , \quad 0 < T < +\infty \ .$$

Then, given three functions $u_0 : \Omega \to \mathbb{R}$, $f : Q_T \to \mathbb{R}$ and $g : \Sigma_T^- \to \mathbb{R}$, we consider the following mixed boundary value problem for the operator L

$$\text{(2.1)} \qquad \begin{cases} Lu = f \ \text{ in } \ Q_T \ , \\ u \ = g \ \text{ on } \ \Sigma_T^- \ , \\ u(\cdot,0) = u_0 \ \text{ in } \ \Omega \ . \end{cases}$$

A measure u on $\overline{Q}_T$ is called a measure solution of (2.1) if

$$\text{(2.2)} \qquad \langle u, L^\star\varphi \rangle_{Q_T} = \langle f,\varphi\rangle_{Q_T} + \langle u_0, \varphi(\cdot,0)\rangle_{\Omega} - \langle a\cdot n\, g, \varphi \rangle_{\Sigma_T^-}$$

holds for any function $\varphi \in C_0^1(\overline{Q}_T)$ which vanishes on Σ_T^+ and for $t = T$. In (2.2), $L^\star$ is the formal adjoint of L , i.e.

$$L^\star \varphi = - \frac{\partial \varphi}{\partial t} - \sum_{i=1}^{d} a_i \frac{\partial \varphi}{\partial x_i} + a_0\, \varphi \ ,$$

and $\langle \cdot,\cdot \rangle$ is the duality pairing between the measures and the functions with compact support. Clearly, a classical solution of (2.1) is a measure solution of (2.1).

In order to introduce the fundamental examples of particle solutions of

Problem (2.1), we first consider the characteristic curves associated with the operator

$$\frac{\partial}{\partial t} + \sum_{i=1}^{d} a_i \frac{\partial}{\partial x_i} \quad .$$

They are defined as the solutions of the differential system

$$\frac{dX}{dt} = a(X,t) \quad , \quad X = (X_1,\ldots,X_d) \ . \tag{2.3}$$

Given $x \in \Omega \cup \Gamma^-$ and $s \in [0,T]$, we denote by $t \to X(t;x,s)$ the solution of (2.3) which satisfies the initial condition

$$X(s) = x \quad , \tag{2.4}$$

and we set

$$\tau(x,s) = \sup\{t; t \in [s,T] \ , \ X(t;x,s) \in \Omega\} \ . \tag{2.5}$$

Let us next consider the following examples of measure solutions of Problem (2.1).

Example 1. Denote by $\delta(x-x_0)$ the Dirac measure located at the point $x_0 \in \mathbb{R}^d$. Then, if

$$u_0 = \delta(x-x_0) \quad , \quad x_0 \in \Omega \quad , \quad f = 0 \quad , \quad g = 0 \quad ,$$

the unique measure solution of (2.1) is given by

$$u(x,t) = \begin{cases} \alpha_0(t)\ \delta(x-X(t;x_0,0)) & , \quad 0 \leqslant t \leqslant \tau(x_0,0) \quad , \\ 0 & , \quad \tau(x_0,0) < t \leqslant T \quad , \end{cases}$$

where

$$\alpha_0(t) = \exp(-\int_0^t a_0(X(s;x_0,0),s)ds) \ .$$

Example 2. Choose

$$u_0 = 0 \ , \ f = \delta(x-x_0) \otimes \delta(t-t_0) \ , \ x_0 \in \Omega \ , \ t_0 \in [0,T] \ , \ g = 0 \ .$$

Then, the corresponding measure solution of (2.1) is given by

$$u(x,t) = \begin{cases} \beta_0(t)\delta(x-X(t;x_0,t_0)) & , \quad 0 \leqslant t \leqslant \tau(x_0,t_0) \quad , \\ 0 & , \quad \tau(x_0,t_0) < t \leqslant T \quad , \end{cases}$$

where

$$\beta_0(t) = H(t-t_0)\ \exp(-\int_{t_0}^t a_0(X(s;x_0,t_0),s)ds)$$

and

$$H(t) = \begin{cases} 1 & t \geq 0 \ , \\ 0 & t < 0 \ . \end{cases}$$

Example 3. Finally, we take

$$u_o = 0 \ , \ f = 0 \ , \ g = \delta(x-x_o) \otimes \delta(t-t_o) \ , \ x_o \in \Gamma^- \ , \ t_o \in [\,0,T\,] \ .$$

Again, one can easily check that the corresponding measure solution is given by

$$u(x,t) = \begin{cases} \gamma_o(t)\ \delta(x-X(t;x_o,t_o)) & , \quad 0 \leq t \leq \tau(x_o,t_o) \ , \\ 0 & , \quad \tau(x_o,t_o) < t \leq T \ , \end{cases}$$

where

$$\gamma_o(t) = -\ (a\cdot n)(x_o)\ H(t-t_o)\ \exp(-\int_{t_o}^{t} a(X(s;x_o,t_o),s)ds) \ .$$

In each above example, $u(\cdot,t)$ is proportional to a Dirac measure whose trajectory in the (x,t)-space is a characteristic curve.

Now, the particle method of approximation of the general problem (2.1) consists first in approximating the functions u_o , f and g by a linear combination of Dirac measures :

$$u_o \simeq u_h^o = \sum_{j\in J} \alpha_j\ \delta(x-x_j) \quad , \quad x_j \in \Omega \quad ,$$

$$f \simeq f_h = \sum_{k\in K} \beta_k\ \delta(x-x_k) \otimes \delta(t-t_k) \quad , \quad x_k \in \Omega \ , \ t_k \in [\,0,T\,] \ ,$$

$$g \simeq g_h = \sum_{\ell\in K} \gamma_\ell\ \delta(x-x_\ell) \otimes \delta(t-t_\ell) \quad , \quad x_\ell \in \Gamma^- \ , \ t_\ell \in [\,0,T\,] \ .$$

Next, we solve exactly the problem :

$$(2.6) \quad \begin{cases} L\, u_h = f_h & \text{in } Q_T \ , \\ u_h = g_h & \text{on } \Sigma_T^- \ , \\ u_h(\cdot,0) = u_h^o & \text{in } \Omega \ . \end{cases}$$

Using the linearity of Problem (2.6) and the results of the above examples, the unique measure solution u_h of (2.6) is easily constructed. Note that the particle method is conservative variant of the classical method of characteristics.

Concerning the computational cost of the particle method, one has to determine the trajectory and the weight of each particle. Except in very simple cases, this has to be done numerically. In practice, one solves a differential system of dimension d for the trajectory and a coupled differential equation for the weight of each particle. For instance, in the case of Example 1, $t \to X_o(t) = X(t;x_o,0)$ is the solution of

$$\frac{dX_o}{dt}(t) = a(X_o(t),t) \quad , \quad X_o(0) = x_o \quad ,$$

while $t \to \alpha_o(t)$ is the solution of

$$\frac{d\alpha_o}{dt}(t) + a_o(X_o(t),t)\,\alpha_o(t) = 0 \quad , \quad \alpha_o(0) = 1 \ .$$

3. Approximation results.

It remains to compare the exact solution u of Problem (2.1) and its particle approximation u_h solution of Problem (2.6). For simplicity, we restrict ourselves to the case $\Omega \equiv \mathbb{R}^d$ so that

(3.1) $$u_h(x,t) = \sum_{j\in J} \alpha_j(t)\delta(x-X(t;x_j,0)) + \sum_{k\in K} \beta_k(t)\delta(x-X(t;x_k,t_k))$$

where

$$\alpha_j(t) = \alpha_j \exp\left(\int_0^t a_o(X(s;x_j,0),s)ds\right) \quad ,$$

$$\beta_k(t) = \beta_k \, H(t-t_k) \exp\left(-\int_{t_k}^t a_o(X(s;x_k,t_k),s)ds\right) \ .$$

In order to obtain for all $t \in [0,T]$ a continuous approximation of $u(\cdot,t)$, we introduce a cut-off function $\zeta \in C^o(\mathbb{R}^d) \cap L^1(\mathbb{R}^d)$ such that

$$\int_{\mathbb{R}^d} \zeta \, dx = 1 \ .$$

Then, we set for all $\varepsilon > 0$

$$\zeta_\varepsilon(x) = \frac{1}{\varepsilon^d}\,\zeta\left(\frac{x}{\varepsilon}\right)$$

and

(3.2) $$u_h^\varepsilon(\cdot,t) = u_h(\cdot,t) \star \zeta_\varepsilon \ ,$$

i.e.,

(3.3) $$\left\{ \begin{aligned} u_h^\varepsilon(x,t) = & \sum_{j\in J} \alpha_j(t)\ \zeta_\varepsilon(x-X(t;x_j,0)) + \\ & + \sum_{k\in K} \beta_k(t)\ \zeta_\varepsilon(x-X(t;x_k,t_k)) \ . \end{aligned} \right.$$

We can now compare the functions $u(\cdot,t)$ and $u_h^\varepsilon(\cdot,t)$. Let us then describe a typical convergence result. We first introduce a uniform grid in $\mathbb{R}^d$ with meshwidth Δx : for all $j = (j_1,\ldots,j_d) \in \mathbb{Z}^d$, we set

$$x_j = (j_i\ \Delta x)_{1 \leq i \leq d} \ .$$

Assuming that the initial condition u_o is continuous, we take

$$(3.4) \qquad u_h^o = \sum_{j \in \mathbb{Z}^d} \alpha_j\ \delta(x-x_j) \quad , \quad \alpha_j = \Delta x^d\ u_o(x_j) \ .$$

Next, we introduce a time-step Δt and we set :

$$t_n = n\ \Delta t \quad , \quad t_{n+1/2} = (n+\tfrac{1}{2})\ \Delta t \quad , \quad n \in \mathbb{N} \ .$$

Assuming again that the function f is continuous, we choose

$$(3.5) \qquad f_h = \sum_{j \in \mathbb{Z}^d} \sum_{n \in \mathbb{N}} \beta_j^{n+1/2}\ \delta(x-x_j) \otimes \delta(t-t_{n+1/2})$$

with

$$(3.6) \qquad \beta_j^{n+1/2} = \Delta x^d\ \Delta t\ f(x_j, t_{n+1/2}) \ .$$

Now, an error estimate for $u(\cdot,t) - u_h^\varepsilon(\cdot,t)$ depends clearly on the smoothness of the exact solution u but also on the properties of the cut-off function ζ. For specificity, we consider only the most popular choice of the function ζ. Denote by χ the characteristic function of the unit hypercube $[-\frac{1}{2}, \frac{1}{2}]^d$ and set

$$(3.7) \qquad \zeta = \chi \star \chi$$

so that ζ is the usual d-dimensional hat function. Then one can prove the following result

<u>Theorem</u>. <u>Assume that the data</u> u_o <u>and</u> f <u>are sufficiently smooth and</u>

$$(3.8) \qquad \frac{\Delta x}{\varepsilon} \leq C \ .$$

<u>There exists a constant</u> $C = C(T, u_o, f)$ <u>independent of</u> Δx, Δt <u>and</u> ε <u>such that</u>

$$(3.9) \qquad \| u(\cdot,t_n) - u_h^\varepsilon(\cdot,t_n)\|_{L^\infty(\mathbb{R}^d)} \leq C(\varepsilon^2 + (\frac{\Delta x}{\varepsilon})^2 + \Delta t^2) \ , \ t_n \leq T \ .$$

Note that the convergence of the method occurs if and only if $\frac{\Delta x}{\varepsilon} \to 0$ as Δx and ε tend to zero.

For various extensions of the above theorem and the corresponding proofs, we refer to [17], [18].

The case $\Omega \neq \mathbb{R}^d$, i.e. $\Gamma = \phi$, is more technical to handle. For some first results in that direction, see [6].

When the source term f is $\neq 0$, we introduce at each time $t_{n+1/2}$ a new set of particles. After some time-steps, the total number of particles can become too large. Thus, for computational economy, we need to reinitialize the whole set of particles. This can be done without deteriorating the rate of convergence of the particle method. For the details, see [18].

Concerning the convergence of the vortex method of approximation of the Euler equations for an ideal incompressible fluid, we refer to the papers of Beale & Majda [1] [2] [3] and the work of Cottet [9] [10]. See also [17]. For an application to the numerical solution of the Vlasov-Poisson equations, cf. [12]. For a proof of convergence of Chorin's method for reaction-advection-diffusion equations, see Brenier [5].

4. Application to a kinetic problem.

Let us now show how the above ideas can be applied in order to solve a model kinetic equation involving transport and scattering terms which arises in radiative transfer problems. We first consider the one-speed particle transport problem : Find a function $I = I(x,\mu,t)$ solution of the transport equation

$$\frac{1}{v}\frac{\partial I}{\partial t} + \mu \frac{\partial I}{\partial x} + \sigma I = S \quad , \quad 0 < x < R \quad , \quad -1 < \mu < 1 \quad , \quad t > 0 \tag{4.1}$$

with the boundary conditions

$$\begin{cases} I(0,\mu,t) = g_0(\mu,t) & , \quad 0 < \mu < 1 \quad , \\ I(R,\mu,t) = g_R(\mu,t) & , \quad -1 < \mu < 0 \quad , \end{cases} \tag{4.2}$$

and the initial condition

$$I(x,\mu,0) = I_0(x,\mu) \ . \tag{4.3}$$

Problem (4.1) - (4.3) is clearly a special case of Problem (2.1). The characteristic curves in the (x,μ,t)-space are the straight lines

$$\mu = c_1 \quad , \quad x - \mu v t = c_2 \ .$$

Hence, the particle method may be conveniently used for numerically solving Problem (4.1) - (4.3).

If we include scattering terms, a reasonable model consists in replacing (4.1) by

$$\frac{1}{v}\frac{\partial I}{\partial t} + \mu \frac{\partial I}{\partial x} + \sigma I = S + \sum_{\ell=0}^{2} b_\ell \Phi_\ell \tag{4.4}$$

where

(4.5) $$\Phi_\ell = \Phi_\ell(x,t) = \frac{1}{2}\int_{-1}^{+1} \mu^\ell \, I(x,\mu,t)d\mu \quad , \quad \ell = 0,1,\ldots$$

and $b_\ell = b_\ell(x,\mu,t)$. This is the case of Thomson scattering for which $b_1 = 0$.

We want to generalize the particle method to the more realistic Problem (4.2) - (4.4). One way is to use a Monte-Carlo technique for dealing with the scattering terms. Another solution consists in using a coupled particle method - moment method that we describe below.

Assume for simplicity that the source term S and the coefficient σ are independent of μ . Then, taking the moments of order 0 and 1 of Equation (4.4) with respect to μ gives :

(4.6) $$\begin{cases} \dfrac{1}{v}\dfrac{\partial\Phi_0}{\partial t} + \dfrac{\partial\Phi_1}{\partial x} + \sigma\,\Phi_0 = S + \dfrac{1}{2}\displaystyle\sum_{\ell=0}^{2}\left(\int_{-1}^{+1} b_\ell \, d\mu\right)\Phi_\ell \quad , \\ \dfrac{1}{v}\dfrac{\partial\Phi_1}{\partial t} + \dfrac{\partial\Phi_2}{\partial x} + \sigma\,\Phi_1 = \dfrac{1}{2}\displaystyle\sum_{\ell=0}^{2}\left(\int_{-1}^{+1} \mu\, b_\ell \, d\mu\right)\Phi_\ell \quad . \end{cases}$$

Since the variable μ has been eliminated, we have reduced by one the dimension of the problem. However, the system (4.6) is not closed because Φ_2 cannot be expressed in general in terms of the first two moments Φ_0 and Φ_1 . We set

(4.7) $$\Phi_2 = \gamma\,\Phi_0$$

where $\gamma = \gamma(x,t)$ is called the variable Eddington factor and is an unknown of the problem.

The numerical method of solution of Problem (4.2) - (4.4) is a predictor - corrector method : the predictor phase uses the moment equations (4.6) while the corrector phase is based on the full kinetic equations. More precisely, the algorithm is as follows :

(i) Assume that I^n is a particle approximation of $I(\cdot,\cdot,t_n)$ at time t_n . Compute Φ_ℓ^n , $0 \leqslant \ell \leqslant 2$, approximation of $\Phi_\ell(\cdot,t_n)$ and $\gamma^n = \Phi_2^n(\Phi_0^n)^{-1}$.

(ii) Using γ^n as an approximation of γ in (t_n,t_{n+1}) , solve the moment equations (4.6) by a classical finite-difference method. We thus obtain a prediction $\tilde{\Phi}_\ell^{n+1}$ of $\Phi_\ell(\cdot,t_{n+1})$, $0 \leqslant \ell \leqslant 2$.

(iii) Using the predicted values $\tilde{\Phi}_\ell^{n+1}$ in the right-hand side of (4.4), solve the equations (4.2) - (4.4) in (t_n,t_{n+1}) by means of the particle method. This provides the desired particle approximation I^{n+1} of $I(\cdot,\cdot,t_{n+1})$ at time t_{n+1}.

In order to make clear the point (i) of the algorithm, it remains to show how to compute from I^n grid values of Φ_ℓ^n. In fact, we use the cut-off technique. Setting

$$\varepsilon = \frac{R}{I+1} ,$$

we introduce the grid

$$x_i = i\,\varepsilon \quad , \quad 0 \leqslant i \leqslant I+1$$

and we denote again by ζ the one-dimensional hat function defined by (3.6). Then, we take

$$\Phi_\ell^n(x_i) = \left(\left(\frac{1}{2}\int_{-1}^{+1} \mu^\ell \, I^n \, d\mu\right) \star \zeta_\varepsilon\right)(x_i) \quad , \quad 1 \leqslant i \leqslant I$$

where

$$\zeta_\varepsilon(x) - \frac{1}{\varepsilon}\,\zeta\left(\frac{x}{\varepsilon}\right) .$$

The case of the boundary points corresponding to the indices $i = 0$ and $i = I + 1$ deserves a special treatment.

This type of coupled particle method - moment method seems to have been first introduced by Mason [16] and extended by Brackbill & Forslund [4] for plasma calculations.

Note that one could have also used a standard finite difference method or finite element method instead of the particle method in order to solve the one-speed transport problem (4.1) - (4.3). In this respect, the coupled method has obvious connections with the synthetic diffusion acceleration method of solution of the stationary neutron transport equation. See [14] for instance for a discussion of this method.

References.

[1] Beale, J.T., and Majda, A., "Vortex methods I : Convergence in three dimensions", Math. Comp., 32, 1-27 (1982).

[2] Beale, J.T., and Majda, A., "Vortex methods II : Higher order accuracy in two and three dimensions", Math. Comp., 32, 29-56 (1982).

[3] Beale, J.T., and Majda, A., "Vortex methods for fluid flow in two or three dimensions" (preprint).

[4] Brackbill, J.U., and Forslund, D.W., "An implicit method for electromagnetic plasma simulation in two dimensions", J. Comput. Phys., 46, 271-308 (1982).

[5] Brenier, Y., "A particle method for one dimensional nonlinear reaction advection diffusion equations", Math. Comp. (to appear).

[6] Cherfils, C., (in preparation).

[7] Chorin, A.J., "Numerical study of slightly viscous flow", J. Fluid. Mech., 57, 785-796 (1973).

[8] Chorin, A.J., "Vortex sheet approximation of boundary layers", J. Comput. Phys., 27, 428-442 (1978).

[9] Cottet, G.H., "Méthodes particulaires pour l'équation d'Euler dans le plan", Thèse de 3ème cycle, Paris (1982).

[10] Cottet, G.H., "Méthodes particulaires pour l'équation d'Euler incompressible : Formulation faible et estimations d'erreurs en deux et trois dimensions" (in preparation).

[11] Cottet, G.H., and Gallic, S., "Une méthode de décomposition pour une équation de type convection - diffusion combinant résolution explicite et méthode particulaire", C.R. Acad. Sc. Paris, t. 297, Série I, 133-136 (1983).

[12] Cottet, G.H., and Raviart, P.A., "Particle methods for the one-dimensional Vlasov-Poisson equations", SIAM J. Numer. Anal. (1984).

[13] Hockney, R.W., and Eastwood, J.W., "Computer Simulation Using Particles", Mac Graw Hill, New-York (1981).

[14] Larsen, E.W., "Unconditionally stable diffusion - acceleration of the transport equation", Transport Theory and Stat. Physics, 11, 29-52 (1982).

[15] Leonard, A., "Vortex methods for flow simulations", J. Comput. Phys., 37, 283-335 (1980).

[16] Mason, R.J., "Implicit moment particle simulation of plasmas", J. Comput. Phys., 41, 233-244 (1981).

[17] Raviart, P.A., "An analysis of particle methods" CIME Course in Numerical Methods in Fluid Dynamics, Como, July 1983 (to be published in Lecture Notes in Mathematics, Springer Verlag).

[18] Raviart, P.A., "Particle approximation of linear hyperbolic equations of the first order", Conference on Numerical Analysis, Dundee 1983 (to be published in Lecture Notes in Mathematics, Springer Verlag).

Computing Methods in Applied Sciences and Engineering, VI
R. Glowinski and J.-L. Lions (Editors)
Elsevier Science Publishers B.V. (North-Holland)

RECENT ADVANCES IN THE ANALYSIS OF CHEBYSHEV AND LEGENDRE SPECTRAL METHODS FOR BOUNDARY VALUE PROBLEMS

CLAUDIO CANUTO

Istituto di Analisi Numerica del C.N.R.
C.so C. Alberto, 5 - 27100 Pavia (Italy)

Different treatments of boundary conditions for Chebyshev and Legendre approximations of elliptic problems are analyzed. Moreover, an easily implementable radiation condition of global type is proposed, as a part of an accurate Chebyshev-Fourier method for exterior elliptic problems in the plane.

INTRODUCTION

This paper addresses some questions about the efficient and accurate treatment of boundary conditions with spectral methods, in the approximation of problems of elliptic type.

Direct (explicit) and indirect (implicit) ways of imposing boundary conditions are considered in Chapter 1, for some one dimensional model problems. The analysis ensures the stability and convergence of these methods. The behavior of the spectrum of the corresponding operators is also studied, since this is relevant when iterative algorithms of solution are used.

Some of the methods discussed in Chapter 1 have been applied in building up a spectral approximation to elliptic problems in exterior domains of the plane (see [LHC]). The foremost results of this investigation are anticipated in Chapter 2. The precision produced by spectral methods is lost if a poorly accurate radiation condition on the artificial boundary is imposed. Therefore, a far field condition of integral type is introduced, which appears to be very natural in a spectral context: it can be exactly satisfied by the spectral solution, and it can be imposed through a "fast", efficient procedure. Numerical evidence shows that this radiation condition guarantees infinite-order accuracy on the spectral solution.

Part of the research reported here was done while the author was in residence at ICASE, NASA Langley Research Center, in Hampton (Virginia).

This paper was typed by Mrs. O. Spada using *troff* under UNIX on VAX 11/780 at the University of Pavia.

1. BOUNDARY CONDITION IN CHEBYSHEV AND LEGENDRE METHODS

Boundary value problems can be discretized by spectral methods in a variety of ways, depending on how the differential equation is approximately satisfied and on how the boundary conditions are taken into account. Before going into the details of the boundary conditions treatment, we briefly recall some typical aspects of Chebyshev and Legendre methods (complete discussions of spectral methods can be found, e.g., in [GO],[PT],[GHV]).

The approximate solution is a global algebraic polynomial of degree N, and the differential equation is satisfied after projection onto a finite dimensional subspace of global polynomials. The projection operator (i.e., the subspace and the inner product which define the projection) characterize the type of spectral approximation (pure Galerkin, tau, collocation). In all cases, one has to *evaluate* the differential operator on polynomial functions: what is typically needed are the values of the operator at $N+1$ selected points in the domain, or the values of the first $N+1$ "Fourier" coefficients of the operator in the expansion in the Chebyshev or Legendre orthogonal basis. If the operator has variable coefficients or non-linear terms, this evaluation can be accurately and efficiently done by the *pseudospectral* technique, which we now describe.

Denote by $w = w(x)$ either the Chebyshev or the Legendre weight in the interval $(-1,1) \subset \mathbf{R}$, i.e.,

$$w(x) = 1/\sqrt{1-x^2} \qquad (\textit{Chebyshev weight})$$

or

$$w(x) \equiv 1 \qquad (\textit{Legendre weight}),$$

and consider the $N+1$ nodes

$$-1 = x_N < x_{N-1} < \cdots < x_1 < x_0 = 1 \tag{1.1}$$

of the Gauss-Lobatto quadrature formula for w in $[-1,1]$. The x_j's are the relative extrema of ψ_N, the N-th Chebyshev or Legendre orthogonal polynomial, in the interval $[-1,1]$. For the Chebyshev weight, one has the analytical expression $x_j = \cos j\pi/N$, $j = 0,\dots,N$. The Gauss-Lobatto quadrature formula is exact for polynomials of degree up to $2N-1$.
The nodes (1.1) uniquely define a global interpolation operator I_N by polynomials of degree N

$$I_N v \in \mathbf{P}_N : \ (I_N v)(x_j) = v(x_j), \qquad j = 0,\dots,N. \tag{1.2}$$

Assume we are given a second order uniformly elliptic differential operator in $(-1,1)$

$$Lu = -(au_x)_x + bu \quad , \quad -1 < x < 1, \tag{1.3}$$

where a and b are smooth functions satisfying $a(x) \ge a_0 > 0$ and $b(x) \ge b_0 > 0$. If $u \in \mathbf{P}_N$ is a polynomial of degree N, the *pseudospectral* realization of Lu is given by

$$L_{sp} u = -\frac{d}{dx}[I_N(a\frac{du}{dx})] + I_N(bu). \tag{1.4}$$

Any space derivative is replaced by the space derivative of the global interpolant polynomial. Only the values of u and the coefficients at the nodes (1.1) are required to evaluate the operator L pseudospectrally.

Since u and $L_{sp}u$ are both polynomials of degree N, they can be equivalently represented either by their values at the $N+1$ nodes (1.1), or by their "Fourier" coefficients with respect to the Chebyshev or Legendre basis. The choice between the two systems of representation (in the *physical space* or in the *frequency space*) depends on the type of spectral approximation chosen for the full boundary value problem. Here, we will systematically use the physical representation, i.e., we will identify a N-th degree polynomial v with the set of its values $\{v(x_j) \mid j = 0,\dots,N\}$. The reader should keep in mind that the dual approach is possible.

The mapping

$$\{v(x_j)\}_{0 \le j \le N} \to \{\frac{dv}{dx}(x_i)\}_{0 \le i \le N} \quad , \quad v \in \mathbf{P}_N \tag{1.5}$$

is the core of any pseudospectral realization of a differential operator. The matrix D_x corresponding to (1.5) has entries D_{ij} given by

$$D_{ij} = \frac{c_i}{c_j} \frac{(-1)^{i+j}}{x_i - x_j} \qquad (i \neq j)$$

$$D_{ii} = -\frac{x_i}{2(1-x_i^2)} \quad , \quad D_{00} = -D_{NN} = \frac{2N^2+1}{6}$$

(with $c_0 = c_N = 2$, $c_i = 1$ for $0 < i < N$) for the Chebyshev nodes, and by

$$D_{ij} = \frac{\psi_N(x_i)}{\psi_N(x_j)} \frac{1}{x_i - x_j} \qquad (i \neq j)$$

$$D_{ii} = 0 \quad (1 \le i \le N-1) \ , \ D_{00} = -D_N = \frac{N(N+1)}{4}$$

(ψ_N denotes the N-th Legendre polynomial) for the Legendre nodes. The product $D_x v$ requires $0(N^2)$ operations. For the Chebyshev nodes, (1.5) can be performed in $0(N \log_2 N)$ operations by a fast transform method, consisting of differentiating in the frequency space and going back and forth in the physical space by a real cosine transform (see [CO] for more details). The existence of this fast transform is the main reason why Chebyshev points have been used so far instead of Legendre points. The choice between matrix multiplication or fast transform method depends on how "fast" is the available transform.

Usually the transform method is competitive for N large enough only (the fastest transform the present author is aware of is a fully vectorized Chebyshev transform on the Cray-1, which is competitive for N larger than 8 (Morchoisne, private communication)). It follows that Legendre points may be used whenever few unknowns per space dimension are enough to attain accuracy.

1.1. A Dirichlet Boundary Value Problem

Assume that we want to solve the following Dirichlet boundary value problem:

$$Lu_{exact} = f \quad in \quad -1 < x < 1 \quad , \quad u_{exact}(\pm 1) = 0 \, ,$$

where L is given by (1.3) and f is continuous in $[-1,1]$. A *pseudospectral-collocation* method of approximation consists of matching the values of $L_{sp} u$ and f at the interior Gauss-Lobatto nodes, satisfying exactly the Dirichlet conditions at the boundary nodes:

$$\begin{cases} u \in \mathbf{P}_N \\ (L_{sp}u)(x_j) = f(x_j) \qquad j = 1,\ldots,N-1 \\ u(x_0) = u(x_N) = 0 \end{cases} \tag{1.6}$$

(A *pseudospectral-tau* method would rather consist of matching the Chebyshev or Legendre coefficients of $L_{sp}u$ and f up to the order $N-2$).
Stability and convergence analyses of scheme (1.6) have been given in [GO],[CQ_1], [CQ_2]. For the Chebyshev nodes and a non-constant coefficient $a(x)$ in the elliptic operator L, the analysis relies upon the fulfillment of an "inf-sup" condition of the Nečas-Babuška type (see [CQ_2] for more details). In general the resulting error estimate is

$$\|u - u_{exact}\|_{1,w} \leq C_1 N^{1-r}\, \|u_{exact}\|_{r,w} + C_2 N^{-s}\, \|f\|_{s,w}$$

where

$$\|v\|^2_{m,w} = \sum_{k=0}^{m} \int_{-1}^{1} |D_x^k v|^2\, w(x)dx\ .$$

Computationally, the matrix corresponding to (1.6), which we still denote by L_{sp}, is full and severely ill-conditioned (the condition number is $O(N^4)$). Moreover, when a transform method is used, its entries are not computed explicitly. For these reasons, iterative methods have been proposed for solving systems arising from spectral approximations like (1.6), often together with a preconditioning strategy. Perhaps the simplest example of such a technique is the Richardson iteration with preconditioning, first suggested for spectral methods in [Mo] and [O]:

$$u^{k+1} = u^k - \alpha A^{-1}(L_{sp}u^k - f)\,, \qquad k \geq 0 \tag{1.7}$$

(for the Dirichlet problem (1.6), the equality is satisfied at the interior nodes). A necessary condition for convergence is that the real parts of the eigenvalues of $A^{-1}L_{sp}$ be of one sign. The preconditioning matrix A is usually an easily invertible approximation of the operator L, typically arising from a low order finite difference or finite element discretization at the collocation nodes. In 2D or 3D, A is an incomplete LU decomposition of such a matrix. In this way the condition number of $A^{-1}L_{sp}$ is reduced to $O(1)$ or $O(N)$ with little computational effort (see [O], [WH],[CQ_3] for the details).

The efficiency of the Richardson method relies on the choice of the acceleration parameter α. The value $\alpha = 2/(m+M)$, where m and M are respectively the smallest and the largest modulus of the eigenvalues of $A^{-1}L_{sp}$, minimizes the spectral radius of the iteration matrix. m can be easily estimated but M is generally unknown. Hence a dynamical choice of α is preferred. In [HLAD] the choice $\alpha^{k+1} = 2/(m + M^k)$ is suggested, where M^k is computed by the assumption $\|\delta u^k\| / \|\delta u^{k-1}\| = |1 - \alpha^k M^k|$. In [WH] α^k is chosen to minimize the residual $L_{sp}u^{k+1} - f$. A priori, this seems to be the best strategy. Unlikely, it requires that the symmetric part of the matrix $L_{sp}A^{-1}$ be definite, which is not always true for spectral operators as we shall see later. Recently, the Richardson method has been successfully coupled with a multigrid strategy for Chebyshev approximations ([ZWH], [PZH]).

Of course, any other stable time advancing scheme can be used as an iterative scheme for solving spectral systems. In particular the DuFort-Fraenkel method has received considerable attention (see, e.g., [CQ_3] and the references therein). An alternative approach is the use of conjugate gradient methods for non-symmetric matrices, like Axelsson's methods (Métivet-Morchoisne, private communication). However the definiteness of the symmetric part of the spectral matrices is required in this case.

1.2. Neumann and Robin Boundary Conditions

Consider now the boundary value problem:

$$\begin{cases} -u_{xx} + u = f \quad , \quad -1 < x < 1 \\ B_+ u = u_x + \alpha u = 0 \quad at \ x = 1 \\ B_- u = u_x - \beta u = 0 \quad at \ x = -1 \end{cases} \tag{1.8}$$

This problem is well-posed if we assume $\alpha \geq 0$ and $\beta \geq 0$. We have chosen $a(x) \equiv b(x) \equiv 1$ in (1.3) for simplicity of notations only, since the boundary treatments described here apply to general variable coefficient operators.

Neumann or third-type boundary conditions can be treated similarly to Dirichlet boundary conditions, by imposing them *explicitly* at the boundary nodes:

$$\begin{cases} u \in {}_{-N} \\ (-u_{xx} + u)(x_j) = f(x_j) \quad j = 1,\dots,N-1 \\ B_+ u(x_0) = 0 \\ B_- u(x_N) = 0 \end{cases} \tag{1.9}$$

However, another way of putting boundary conditions, proposed by D. Gottlieb for time-dependent problems, is possible. In this approach boundary conditions are imposed *implicitly*, by using them in evaluating the overall spectral operator. The scheme is

$$\begin{cases} u \in \mathrm{P}_N \\ -\dfrac{d}{dx}[I_N(B(u).u_x)](x_j) + u(x_j) = f(x_j) \\ \qquad\qquad j = 0,1,\dots,N-1,N \end{cases} \tag{1.10}$$

where $B(u)$ is a linear operator (matrix) acting on u_x , which modifies the values of u_x at the boundary nodes according to (1.8), i.e.,

$$B(u) = diag(-\alpha u(x_0), 1, 1, \dots 1, \beta u(x_N)).$$

Note that with the explicit treatment the numerical solution satisfies the boundary conditions exactly, while the differential equation is collocated at the interior nodes only. Conversely, in the implicit treatment of the boundary conditions the equation is collocated at all the nodes, but the solution does *not* satisfy the boundary condition exactly. Let us briefly analyze each method separately.

a - ***Explicit imposition of the boundary conditions.***

The explicit method (1.9) can be written in matrix form, with obvious notations, as follows:

$$L_E u \equiv \begin{bmatrix} -B_+ \\ \cdots \\ -D_{xx} + I \\ \cdots \\ B_- \end{bmatrix} \cdot \begin{bmatrix} u_i \end{bmatrix} = \begin{bmatrix} 0 \\ \cdots \\ f_k \\ \cdots \\ 0 \end{bmatrix}, \tag{1.11}$$

where $D_{xx} = D_x . D_x$ is the matrix of the second derivative at the Gauss-Lobatto nodes.

Concerning the analysis of this scheme, Gottlieb and Lustman [GL] proved that the eigenvalues of the associated problem

$$\begin{cases} \psi \in \mathbf{P}_N \\ \psi_{xx}(x_j) = \lambda \psi(x_j) \quad j = 1,\dots,N-1 \\ B_+\psi(x_0) = B_-\psi(x_N) = 0 \end{cases} \tag{1.12}$$

are real, positive and distinct. On the other hand, A. Quarteroni and the author [CQ$_2$] proved stability and convergence, by using the equivalent variational formulation of (1.9)

$$\begin{cases} u \in \mathbf{P}_N \;, \; B_+u\,(x_0) = B_-u\,(x_N) = 0 \\ a_N(u,v) = (f\,,v)_N \quad \forall v \in \mathbf{P}_{N-2}\,, \end{cases}$$

where

$$a_N(u,v) = \sum_{j=1}^{N-1} (-u_{xx} + u)(x_j)\, v(x_j)\, (1 - x_j^2)\, w_j$$

and

$$(f\,,v)_N = \sum_{j=1}^{N-1} f(x_j)\, v(x_j)\, (1 - x_j^2)\, w_j \;;$$

$\{w_j\}$ are the weights of the Gauss-Lobatto quadrature formula. Set $\eta(x) = (1 - x^2)\, w(x)$ and for $m \geq 0$ define $\|v\|^2_{m,\eta} = \sum_{k=0}^{m} \int_{-1}^{1} |D_x^k v|^2 \eta(x)\, dx$, the natural norm of the weighted Sobolev space H^m_η . In [CQ$_2$] it is shown that a_N satisfies the stability condition

$$\sup_{v \in \mathbf{P}_{N-2}} \frac{a_N(u,v)}{\|v\|_{0,\eta}} \geq C \|u\|_{2,\eta}$$

($C > 0$ independent of N), and the following error estimate holds

$$\|u - u_{exact}\|_{2,\eta} \leq C N^{2-s} \{\|u\|_{s,w} + \|f\|_{s-2,w}\} .$$

From the computational point of view, system (1.11) may be solved by an iterative procedure like (1.7). Convergence requires that the matrix L_E , or possibly a preconditioned form $A^{-1} L_E$, has eigenvalues with real parts of one sign. We are not aware of any theoretical result in this sense (note that the result does not follow from Gottlieb-Lustman's analysis, since the matrix corresponding to (1.12) has order $N-1$, while L_E has order $N+1$). However, the numerical computation of the eigenvalues of L_E up to the order 129 for different values of α and β shows that they all are real, positive and distinct, for both the Chebyshev and the Legendre points. This property is preserved for the eigenvalues of the preconditioned matrix $A^{-1}L_E$, where A corresponds to the second order centered finite difference approximation of problem (1.8). It follows that the use of Richardson's method, or similar methods, is heuristically justified. On the contrary, the computer shows that the symmetric part of L_E , or $A^{-1}L_E$, is indefinite: the minimal residual strategy may fail for problem (1.11).

b - *Implicit imposition of the boundary conditions.*

The matrix representation of the implicit method (1.10) is given by

$$L_I u \equiv \left(-D_x \right) \cdot \begin{pmatrix} -\alpha \cdots 0 \cdots \\ \cdots \\ D_x \\ \cdots \\ \cdots 0 \cdots \beta \end{pmatrix} \cdot \left(u_i \right) + \left(u_k \right) = \begin{pmatrix} f_0 \\ \cdots \\ f_k \\ \cdots \\ f_N \end{pmatrix} . \tag{1.13}$$

(Of course, the reader should not forget that both (1.9) and (1.10) can be implemented by fast transform methods when the Chebyshev nodes are used).

A somewhat complete theory can be provided for the Neumann boundary conditions ($\alpha = \beta = 0$) . Let us consider the case of Legendre points first. Multiply the j-th equation in (1.10) by $u(x_j)w_j$ and sum; recall that the w_j's are the weights of the Gauss Lobatto quadrature formula for the Legendre weight. The higher order term can be transformed as follows:

$$- \sum_{k=0}^{N} \frac{d}{dx}[I_N(B(u)u_x)](x_j)u(x_j)\, w_j = \int_{-1}^{1} I_N(B(u)u_x)\, u_x\, dx$$
$$= \int_{j=1}^{N-1} u_x^2(x_j) w_j .$$

We have used the exactness of the Gauss-Lobatto formula for polynomials of degree $2N-1$, and integration by parts. Therefore, the following discrete energy estimate holds:

$$\sum_{j=1}^{N-1} u_x^2(x_j)w_j + \sum_{j=0}^{N} u^2(x_j)w_j = \sum_{j=0}^{N} f(x_j)u(x_j)w_j . \tag{1.14}$$

This proves the stability (with respect to the discretization parameter N) of the implicit treatment of the boundary conditions.

Note that the values of u_x at the boundary do not appear in (1.14). Since the homogeneous Neumann condition has been exploited in the evaluation of the spectral operator only, the normal derivative does not necessarily vanish at $x = \pm 1$. Of course, we expect that u_x be small there. This is precisely what happens. It can be shown (see [C]) that the following estimate holds:

$$|u_x(\pm 1)| \leq CN^{2-s} \|u_{exact}\|_{s,w} .$$

This proves that the boundary conditions are satisfied *up to a spectral error* when they are imposed implicitly in a pseudo-spectral scheme.

Finally, one can obtain the following convergence estimate

$$\|u - u_{exact}\|_{1,w} \leq CN^{1-s} (\|u_{exact}\|_{s,w} + \|f\|_{s-1,w}) . \tag{1.15}$$

The details are contained in [C].

Estimate (1.14) easily provides useful information on the spectrum of the matrix L_I . Replacing f with λu yields that the eigenvalues of L_I are real positive and uniformly bounded from below. Hence the iterative methods described in Section 1.1 can be used to solve (1.10). Moreover (1.14) shows that L_I is positive definite with respect to the discrete inner product

$$(u,v)_{N,w} = \sum_{j=0}^{N} u_j v_j w_j ,$$

i.e.,

$$(L_I v, v)_{N,w} > 0 \quad \text{if } v \neq 0 .$$

We conclude that the minimal residual strategy can be successfully applied.

The eigenvalues of L_I remain positive even when third type conditions are imposed (i.e., when $\alpha \neq 0$ or $\beta \neq 0$). However L_I fails to be positive definite for α or β large enough.

Assume now that the Chebyshev points are used. Equation (1.10) can be equivalently written as

$$-v_x + u = I_N f ,$$

with $v = I_N[(B(u)u_x] \in \mathbf{P}_N$. This is an identity between polynomials of degree N . Following D. Gottlieb, we can differentiate termwise, multiply by vw and integrate over (-1,1), getting

$$\int_{-1}^{1} -v_{xx} vw dx + \sum_{j=1}^{N-1} u_x^2(x_j) w_j = \int_{-1}^{1} (I_N f)_x vw dx ; \tag{1.16}$$

the w_j's are now the weights of the Gauss-Lobatto formula with respect to the Chebyshev weight w . The integral on the left-hand side is positive and dominates the H_w^1-norm of v , since v vanishes at the boundary (see $[CQ_1]$). The integral on the right-hand side can be bounded by $\|I_N f\|_{0,w} \cdot \|v\|_{1,w}$. It follows that the approximation is stable with respect to the parameter N. Moreover, an error estimate formally similar to (1.15) can be proved (see [C]).

Replacing f with λu in (1.16) proves the reality and positiveness of the eigenvalues of L_I even in the Chebyshev case. However (1.16) does not ensure the definiteness of L_I. Actually the computer shows that the symmetric part of L_I is indefinite. Hence we are facing a situation where the use of Legendre's points instead of Chebyshev's produces better properties in the matrix corresponding to the same numerical scheme.

c - *Conclusions.*

Two different ways of imposing Neumann or third type boundary conditions - one explicit, the other implicit - have been analyzed. Both of them produce stable and convergent pseudospectral-collocation schemes, whose solutions can be efficiently computed by the same iterative procedures. Moreover, the numerical computation of the eigenvalues of the matrices generated by the two methods shows that the spectra are very close to each other. Under this respect, the explicit and the implicit treatment of boundary conditions appear essentially equivalent.

However, there are some considerations supporting a larger versatility of the implicit treatment. Most of the commonest finite difference or finite element methods discretize the differential equation ***up to the boundary***, making use of the boundary conditions in this process. They are the exact counterpart of a pseudospectral scheme with implicit imposition of the boundary conditions. It follows that such schemes may be preconditioned in the most natural way.
On the other hand, the implicit treatment of boundary conditions is the most efficient strategy in parabolic problems, when explicit time-advancing schemes are used. The equation is simultaneously advanced at all the grid points in the domain, since the boundary conditions intervene in the evaluation of the space operator only. Instead, the explicit treatment would require a projection step after every time advancing step, in order to satisfy the boundary conditions exactly.

2 AN EXTERIOR ELLIPTIC PROBLEM IN THE PLANE

In this chapter, some of the results of a joint work with S. Hariharan and L. Lustman are reported. Full details and further results are presented in [LHC].

We are concerned with the approximation by spectral methods of elliptic equations in exterior domains. Let D be a body, represented by a bounded, simply connected domain in the plane, with smooth boundary Γ. Define $\Omega = \mathbb{R}^2 - \overline{D}$ and denote by $r = \sqrt{x^2+y^2}$ the distance from the origin. Consider the following exterior problem:

$$\begin{cases} \Delta u = 0 & \text{in } \Omega \\ u = g & \text{on } \Gamma \\ B_\infty(u) \to 0 & \text{as } r \to \infty , \end{cases} \tag{2.1}$$

where g is a given, smooth function on Γ. (2.1.3) is a radiation condition at infinity, which prescribes the behavior of the solution far from the body and makes the problem mathematically well-posed.

Many physical situations enter into this formulation. For instance, in electrostatics D is a conductor in the field of a line source, u is the electrostatic potential and $B_\infty(u) = u - \log r$. In fluid-dynamics, u is the velocity potential of an incompressible irrotational flow around a body, and $B_\infty(u) = u - U_0 r \cos\vartheta - \frac{\Gamma}{2\pi}\log r$, where U_0 is the main stream velocity in the x-direction and Γ is the circulation. In the electromagnetic or acoustic scattering by an obstacle the Poisson equation is replaced by the Helmoltz equation $\Delta u + k^2 u = 0$, and the radiation condition is of the Sommerfeld type. We will not consider Helmoltz problems although the treatment of the radiation condition presented here can be easily extended to these problems.

Problem (2.1) can be efficiently solved by a boundary integral approach. The explicit knowledge of Green's function, satisfying the radiation condition at infinity, allows to restate the problem through an integral equation on Γ . A number of different techniques have been proposed to solve the boundary equation numerically, and an abundant and growing literature is available. However, there are situations of interest where this approach is not efficient, or even fails because the kernel is not known. This occurs, for instance, when the differential operator has variable coefficients, which tend to a constant state at infinity.

Therefore, an alternative approach has been followed, consisting of the direct discretization of the elliptic equation, so far by finite difference or finite element methods. The computational domain is bounded by an artificial surface Γ_∞ surrounding the body at a finite distance, and the radiation condition at infinity is replaced by a suitable far field condition

$$B(u) = 0 \quad on \ \Gamma_\infty .$$

The operator B may be local, i.e., $B(u)$ at each point on Γ_∞ depends on the values of u and certain derivatives of u at the same point; or global, if $B(u)$ is determined by the whole behavior of u on Γ_∞ . Local radiation conditions have been considered, e.g., in [BGT],[KM],[G]. In particular Bayliss, Gunzburger and Turkel derive a sequence of radiation conditions of increasing order, which will be considered in Section 2.2. Global radiation conditions have been derived, e.g., in [FM], [M],[MCM] by separation of variables.

The cut-off of the domain by the artificial boundary, and the radiation boundary condition on this surface, are responsible for an error on the approximate solution. It is desirable that this error be comparable with the discretization error introduced by the numerical procedure. If spectral methods are used, this requirement is even more stringent, since a careless treatment of the far field condition may result in the complete waste of the infinite-order accuracy guaranteed by such methods. We will now present a spectral procedure to approximate problem (2.1), in which the treatment of the radiation condition is both computationally efficient and accurate, producing overall spectral accuracy on the solution.

2.1. A Chebyshev-Fourier Method

We assume that the origin lies within the body and that the boundary Γ may be described in polar coordinates by a smooth function $r = R(\varphi)$, $0 \le \varphi < 2\pi$. The artificial boundary Γ_∞ is chosen to be the circumference $r = R_\infty$, for a suitable $R_\infty > 0$.

The reduced domain $\Omega' = \{(r\cos\varphi, r\sin\varphi) \mid 0 \le \varphi < 2\pi \ , \ R(\varphi) < r < R_\infty\}$ is mapped into the computational domain $\hat{\Omega} = \{(s,\vartheta) \in \mathbb{R}^2 \mid -1 < s < 1\ ,\ 0 \le \vartheta < 2\pi\}$ via a smooth (non-conformal) transformation of the type

$$\begin{cases} \varphi = \vartheta \\ r = r(s,\vartheta)\ , \end{cases} \tag{2.2}$$

with $r(-1,\vartheta) = R(\vartheta)$ and $r(1,\vartheta) \equiv R_\infty$, $0 \le s < 2\pi$. Different expressions for $r(s,\vartheta)$ have been tested. The best results were given by exponential forms of the stretching such as $r(s,\vartheta) = \alpha(\vartheta)\exp(\beta(\vartheta)s)$. Actually, the exact solution depends logarithmically on r at infinity, hence linearly on s if this transform is used, so that it is correctly approximated by Chebyshev polynomials in s . Moreover, the exponential stretching prevents the unnecessary crowding of the Chebyshev points near the artificial boundary.

Under the mapping (2.2) the Laplace equation is transformed into a variable coefficient elliptic equation in the (s,ϑ) coordinates, which takes the form

$$Lu = au_{ss} + bu_{s\varphi} + cu_{\varphi\varphi} + du_s = 0$$

for suitable coefficients $a > 0$, $c > 0$, b and d .

The solution u is periodic in ϑ , satisfies a Dirichlet condition at $s = -1$ and a radiation condition, to be specified, at $s = 1$. Therefore, we look for an approximate solution u^N which is a trigonometric polynomial of degree N in ϑ and an algebraic polynomial of degree M in s ,

$$u^N = \sum_{m=0}^{M} \ \sum_{|k| \le N}{}^{*}\ \hat{u}^N_{mk}\ T_m(s)\,e^{ik\vartheta}$$

The asterisk means that $\hat{u}^N_{mN} = \hat{u}^N_{m(-N)}$ for any m ; $T_m(s) = \cos(m\arccos s)$ is the m-th Chebyshev polynomial of first kind.

Rather than solving for the coefficients $\hat{u}^N_{mk}$ in this orthogonal expansion, our primary unknowns are the values u^N_{ij} of u^N at the Chebyshev-Fourier nodes

$$(s_i\,,\vartheta_j) = (\cos\frac{i\pi}{M},\ \frac{j\pi}{N})\ , \qquad i = 0,\ldots,M \quad j = 0,\ldots,2N-1\ .$$

Derivatives are computed spectrally at these nodes, which are also the collocation points for the spectral operator L_{sp} . L_{sp} takes into account the radiation condition on Γ_∞ , as indicated in the next section.

2.2. Construction of a global radiation condition on Γ_∞ .

We describe our method assuming that the radiation condition at infinity is

$$B_\infty(u) = u - \log r \to 0 \quad as \quad r \to \infty \tag{2.4}$$

(Other radiation conditions can be treated similarly). Under this assumption, the solution of (2.1) can be expanded in the convergent series

$$u(r,\varphi) = \log r + \sum_{k \ne 0} \frac{a_k}{r^{|k|}}\,e^{ik\varphi}\ . \tag{2.5}$$

First of all, let us briefly recall that a family of local radiation conditions can be derived following Bayliss, Gunzburger and Turkel [BGT]. The differential operators

$$B_m = \prod_{j=1}^{m} \left(\frac{\partial}{\partial r} + \frac{2j-1}{r}\right) \qquad m = 1,2,\ldots$$

exactly annihilate the terms of order up to $2m$ in the series (2.5). Hence

$$B_m(u - \log r) = 0\left(\frac{1}{r^{2m+1}}\right) \qquad m = 1,2,\ldots$$

Neglecting the right-hand side, one gets a family of boundary conditions of increasing order to be satisfied by the approximate solution on Γ_∞. For instance, the first two radiation conditions are

$$\left(\frac{\partial}{\partial r} + \frac{1}{r}\right)(u^N - \log r)_{|r=R_\infty} = 0 \quad (\textit{first order radiation condition}) \tag{2.6}$$

and

$$\left(\frac{\partial^2}{\partial r^2} + \frac{4}{r}\frac{\partial}{\partial r} + \frac{2}{r^2}\right)(u^N - \log r)_{|r=R_\infty} = 0 \tag{2.7}$$

$$(\textit{second order radiation condition}).$$

Condition (2.6) is a Robin-type condition, which can be imposed spectrally either explicitly or implicitly, according to the methods described in Chapter 1 . Conditions like (2.7) involve higher order derivatives in the normal direction, which could be accurately evaluated spectrally. However, the direct imposition of these conditions results in the non-definiteness of the matrix corresponding to the numerical scheme, hence in the non-convergence of most of the iterative methods of solution. Rather, higher order derivatives must be eliminated via the partial differential equation, yielding a boundary operator of first order in the normal direction. The new operator may be of intricate implementation and expensive evaluation, although it produces definite matrices. However, the main drawback of these local radiation conditions, when applied in conjunction with spectral methods, is that they introduce errors of order $1/R_\infty^3$, $1/R_\infty^5$ and so on, as shown in [BGT]. At moderate R_∞, they may destroy the spectral accuracy of the numerical method. An example of such a behavior will be given in the next Section.

A different radiation condition may be introduced, which does not suffer from such disadvantages. It is a global condition, similar to the one considered in [FM] for finite element approximations of Helmoltz problems. Unlike [FM], the implementation of our condition in a spectral context does not require any further approximation, being computationally extremely efficient.

In order to derive this radiation condition, let us differentiate term by term (2.5) in the r direction

$$u_r = \frac{1}{r}\left[1 - \sum_k |k| \frac{a_k}{r^{|k|}} e^{ik\varphi}\right].$$

Note that for any fixed r , $a_k/r^{|k|}$ is the k-th Fourier coefficient of the periodic function $\vartheta \to u(r,\vartheta)$ (see (2.5))

$$\frac{a_k}{r^{|k|}} = \frac{1}{2\pi}\int_0^{2\pi} u(r,\vartheta)\, e^{-ik\vartheta}\, d\vartheta = \hat{u}_k(r,.),$$

whence

$$u_r = \frac{1}{r}[1 - \frac{1}{2\pi}\int_0^{2\pi} \sum_k |k| e^{ik(\varphi-\vartheta)} u(r,\vartheta)\, d\vartheta] .$$

It follows that the solution of (2.1) satisfies on every circumference of radius r the integro-differential boundary condition

$$u_r + \frac{1}{r} K * u = \frac{1}{r} \tag{2.8}$$

where $K(\vartheta) = \sum_k |k| e^{ik\vartheta}$ and $*$ denotes periodic convolution.

It is precisely this condition that we impose on the approximate solution u^N at $r = R_\infty$, for instance through an implicit procedure. Then $K * u^N$ must be evaluated at the nodes on Γ_∞ . But $u^N(R_\infty, .)$ is a trigonometric polynomial of degree N , i.e., $\hat{u}_k^N(R_\infty, .) = 0$ if $|k| > N$. Thus,

$$K * u^N = \sum_{|k| \le N} |k| \hat{u}_k^N (R_\infty, .) e^{ik\varphi} .$$

The values of $K * u^N$ at the nodes on Γ_∞ may be computed from the values of u^N at the same nodes by one FFT, 2N complex multiplications and one inverse FFT. Hence the radiation condition can be imposed *exactly* and *efficiently*.

Let us stress that, unlike the local boundary conditions (2.6)-(2.7), no error derives from approximating the true radiation condition: the exact solution and the numerical one satisfy the *same* condition on the artificial boundary. The error only depends on R_∞ through the measure of the finite domain Ω' on which the problem is numerically solved.

2.3. A Pseudospectral Algorithm. Numerical Results

The global radiation condition proposed in the previous section was imposed implicitly, during the pseudospectral realization of the operator L . At each iteration, u_s is computed. The values at $s = 1$ are corrected in order to satisfy eq. (2.8) (note that the transformation (2.2) is such that u_s is proportional to u_r on the exterior boundary). The modified u_s is used to compute u_{ss} and $u_{s\vartheta}$.

The differential equation was collocated at the nodes (s_i, ϑ_j) for $0 \le i < M$ and $0 \le j \le 2N-1$ (at the nodes corresponding to $i = M$ the Dirichlet condition was imposed directly). The residual was driven to zero by preconditioned Richardson iterations, according to (1.7). The preconditioning matrix A was built up starting from a finite difference approximation of L at the Chebyshev-Fourier nodes, which used a 7-point molecule (the standard five points plus the North-East and South-West points to represent the mixed term $u_{s\vartheta}$) . In order to preserve the band structure of the matrix, the global radiation condition (2.8) was approximated by a local operator $u_r + K_{fd} u = 1/r$, with

$$(K_{fd}\, u)_{1j} = k_+ u_{1,j+1} + k_0 u_{1j} + k_- u_{1,j-1} .$$

k_+, k_0 and k_- are determined in such a way that the approximation is exact for $u = 1$, $\sin\vartheta$, $\cos\vartheta$ on the exterior boundary. This process yields a seven diagonal matrix, with extra non-zero entries due to the periodicity of u in the ϑ direction. The matrix A was chosen as the Meijerink-van der Vorst seven-diagonal incomplete LU decomposition of this matrix (modified to take into account the periodicity) (see [MV]). The acceleration parameter α in (1.7) was determined by the minimal residual strategy. Although convergence is not guaranteed for this choice of α , no failure of the algorithm was observed for the tested problems.

In order to compare the behavior of our global radiation condition, the first two Bayliss-Gunzburger-Turkel radiation conditions were implemented too. The same pseudospectral algorithm was used. Here we present some typical results. We assume that the body surface is described by the function $R(\varphi) = 1 + .4\cos 2\vartheta$ (see Fig.1), and that the exact solution of problem (2.1) is given by $u(r,\varphi) = \log r + \dfrac{\cos 2\vartheta}{r^2} + 3\,\dfrac{\cos 3\varphi}{r^3}$.

Figure 2 represents the convergence histories of the iterative method (1.7) for different mesh sizes, when the global radiation condition (2.8) is imposed. RES denotes the relative residual in the l^2 -norm. It is seen that the preconditioning is effective, since the rate of reduction of the residual is essentially independent of the number of unknows.

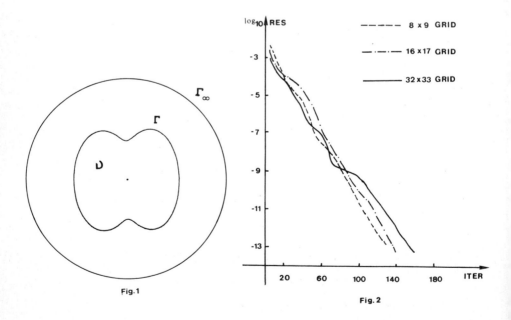

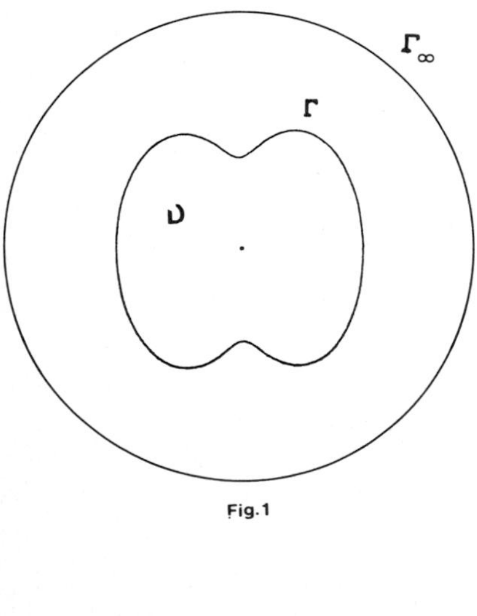

Fig.1

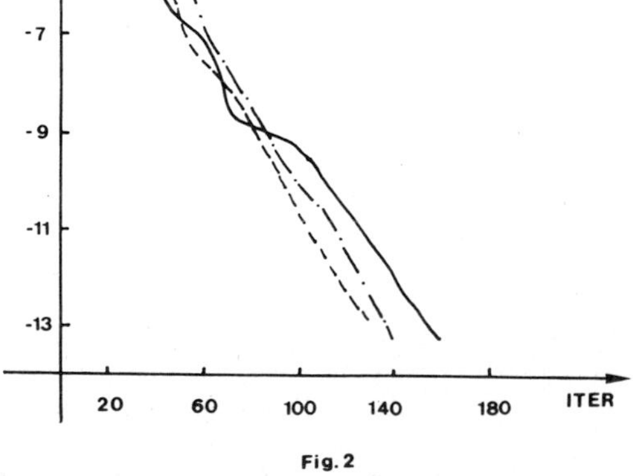

Fig. 2

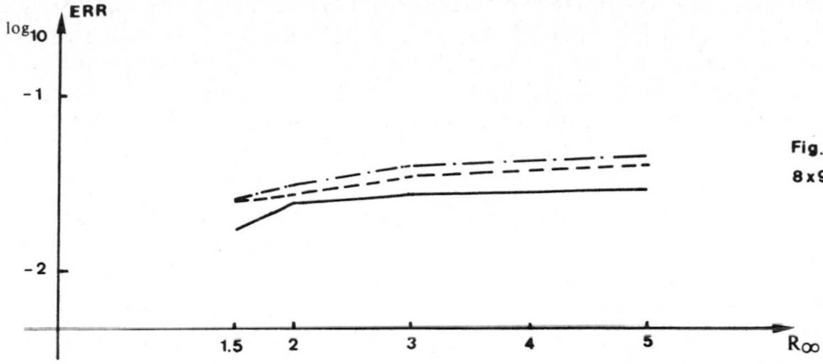

Fig. 3
8x9 GRID

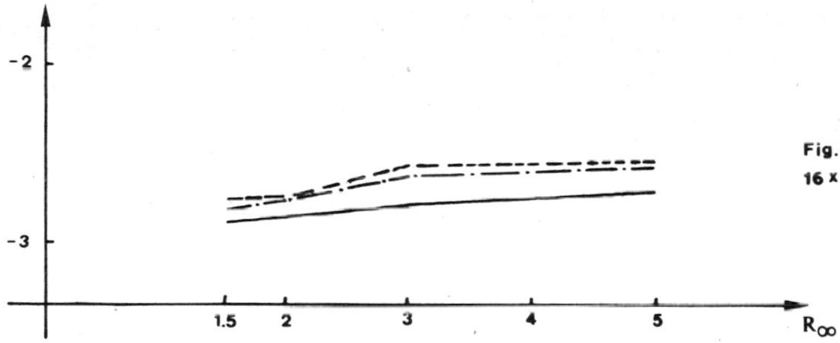

Fig. 4
16 x 17 GRID

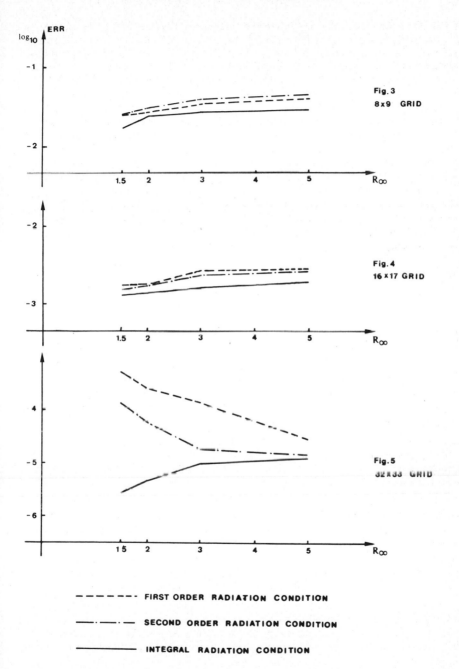

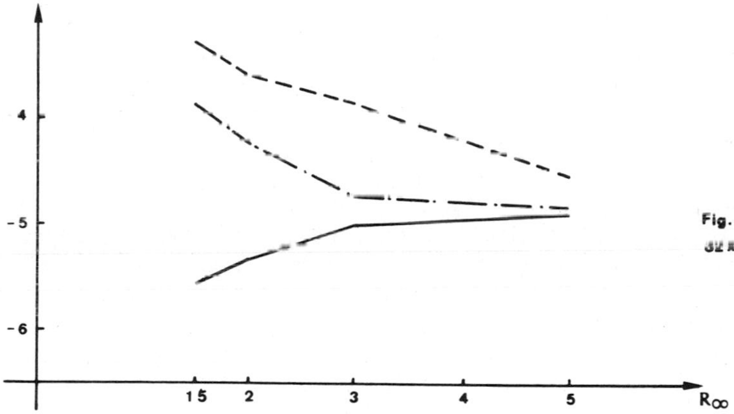

Fig. 5
32 x 33 GRID

- - - - - - FIRST ORDER RADIATION CONDITION

—·—·— SECOND ORDER RADIATION CONDITION

———— INTEGRAL RADIATION CONDITION

The error behavior, as a function of the distance of the artificial boundary, is reported in Figures 3, 4 and 5 for different mesh sizes. ERR denotes the relative error in the maximum norm. The accuracy deriving from the use of the global radiation condition may be compared with the one produced by the boundary conditions (2.6) and (2.7). The global radiation condition always gives the better results, and is the sole to guarantee the spectral accuracy at any value of R_∞. More precisely , the three methods behave equivalently when 8x9 or 16x17 grid points are used: the error is dominated by the discretization error (which decays spectrally) being almost insensitive to the position of the artificial boundary. Things are radically different on the 32x33 mesh. The local radiation conditions introduce errors prevailing over the errors produced by the spectral method: the global error increases when the artificial boundary is brought close to the body, where the spectral accuracy is completely lost. On the contrary, the error decreases if the integral condition id used, since the mesh size becomes finer and finer and no error comes from the artificial boundary. Exceedingly good results are produced at small R_∞. This is relevant since in many applications only the information on the solution near the body is required.

REFERENCES

[BGT] - A. BAYLISS, M.D. GUNZBURGER and E. TURKEL - Boundary Conditions for the Numerical Solutions of Elliptic Equations in Exterior Regions - SIAM J. Appl. Math. , 42(1982), 430-451.

[C] - C. CANUTO - Boundary Conditions in Chebyshev and Legendre Methods - ICASE Report (1984), to appear.

[CQ_1] - C. CANUTO and A. QUARTERONI - Spectral and Pseudo-Spectral Methods for Parabolic Problems with Non-Periodic Boundary Conditions - Calcolo, 18(1981), 197-218.

[CQ_2] - C. CANUTO and A. QUARTERONI - Variational Methods in the Theoretical Analysis of Spectral Approximations - in [GHV].

[CQ_3] - C. CANUTO and A. QUARTERONI - Preconditioned Minimal Residual Methods for Chebyshev Spectral Calculations - ICASE Report 83-28(1983).

[FM] - G.J. FIX and S.P. MARIN - Variational Methods in Underwater Acoustic Problems - J. Comput. Phys., 28(1978), 253-270.

[G] - C.I. GOLDSTEIN - The Finite Element Method with Non-Uniform Mesh Sizes Applied to the Exterior Helmoltz Problem - Numer. Math., 38(1982), 61-82.

[GHV] - D. GOTTLIEB, M.Y. HUSSAINI and R.G. VOIGT (editors) - *Proceedings of the Workshop on Spectral Methods (Hampton, Virginia; August 1982)* - SIAM (Philadelphia), 1984.

[GL] - D. GOTTLIEB and L. LUSTMAN - The Spectrum of the Chebyshev Collocation Operator for the Heat Equation - SIAM J. Numer. Anal., 20(1983), 909-921.

[GO] - D. GOTTLIEB and S.A. ORSZAG - *Numerical Analysis of Spectral Methods: Theory and Applications*, SIAM (Philadelphia), 1977.

[HLAD] - P. HALDENWANG, G. LABROSSE, S. ABBOUDI and M. DEVILLE - Chebyshev 3-D Spectral and 2-D Pseudo-Spectral Solvers for the Helmoltz Equation - (to appear).

[KM] - G.A. KRIEGSMANN and C. S. MORAWETZ - Solving the Helmoltz Equation for Exterior Problem with Variable Index of Refraction: I - SIAM J. Sci. Statist. Comput., 1(1980), 371-385.

[LHC] - L. LUSTMAN, S.I. HARIHARAN and C. CANUTO - Spectral Methods for Exterior Elliptic Problems - ICASE Report (1984), to appear.

[M] - S.P. MARIN - A Finite Element Method for Problems involving the Helmoltz Equation in Two-dimensional Exterior Regions - Ph.D. Thesis, Carnegie-Mellon University (Pittsburg), 1978.

[Mo] - Y. MORCHOISNE - Resolution of Navier-Stokes Equations by a Space-Time Spectral Method - La Rech. Aerosp., 5(1979), 293-306.

[MCM] - R.C. MacCAMY and S.P. MARIN - A Finite Element Method for Exterior Interface Problems - Int. J. Math. Math. Sci.,3(1980), 311-350.

[MV] - J.A. MEIJERINK and H.A. van der VORST - An Iterative Solution Method for Linear Systems of Which the Coefficient Matrix is a Symmetric M-Matrix - Math. Comput., 31(1977), 148-162.

[O] - S.A. ORSZAG - Spectral Methods for Problems in Complex Geometries - J. Comput. Phys., 37(1980), 70-92.

[PT] - R. PEYRET and T.D. TAYLOR - *Computational Methods for Fluid Flow* - Springer (New York), 1982.

[PZH] - T.N. PHILLIPS, T.A. ZANG and M.Y. HUSSAINI - Preconditioners for, the Spectral Multigrid Method - ICASE Report 83-48 (1983).

[ZWH] - T.A. ZANG, Y.S. WONG and M.Y. HUSSAINI - Spectral Multigrid Methods for Elliptic Equations II - ICASE Report 3-12(1983).

[WH] - Y.S. WONG and M.M. HAFEZ - A Minimal Residual Method for Transonic Potential Flow - ICASE Report 82-15(1982).

NONLINEAR ELASTICITY AND STRUCTURAL MECHANICS

ELASTICITE NON LINEAIRE ET MECANIQUE DES STRUCTURES

Computing Methods in Applied Sciences and Engineering, VI
R. Glowinski and J.-L. Lions (Editors)
Elsevier Science Publishers B.V. (North-Holland)

ALGORITHMES EFFICACES POUR LES CALCULS EN POST-FLAMBEMENT

EFFICIENT ALGORITHMS FOR POST BUCKLING COMPUTATION

C. PETIAU & C. CORNUAULT

Avions Marcel Dassault-Bréguet Aviation
SAINT-CLOUD

ABSTRACT

We present a set of algorithms for nonlinear elastic structural analysis involving large displacements, which are particularly efficient for post buckling of stiffened panels.

We took advantage of two fundamental remarks :

- Within the frame of finite element discrete models and the Lagrangian formulation, the elastic total potential energy is a fourth degree algebraic expression, and more precisely a sum of squares of trinomials. So its minimization in a given direction is reduced exactly to the resolution of a third degree scalar equation, which leads to a quite cheap line search technique,

- Practically, geometrical nonlinearities of thin panels mainly lie in membrane effects ; therefore we can separate linear flexural terms from nonlinear membrane terms and compute them only once at the beginning of the procedure.

Following these remarks, we have developed and incorporated in our general Finite Element computer program ELFINI a branch specially adapted to fast computation of geometrically nonlinear problems. In this program, the solution procedure is based on the previously outlined minimization of the total potential energy, implemented in conjunction with an iterative method, which can be chosen among a range of classical techniques : Newton Raphson, Conjugate Newton, B.F.G.S.

Some of the many comparisons we achieved between these algorithms are described. These comparisons were performed :

- on conventional cases frequently quoted in Literature, proving the efficiency of our tool in taking the most severe nonlinearities into account,
- on cases typical of most of our applications, i.e. involving many simultaneous local bucklings, while keeping some appreciable global stiffness,
- in computing a complex test simulating the buckling of a wing upper panel of Super Mirage 4000.

1 - INTRODUCTION

Nous présentons une synthèse des travaux menés par AMD-BA avec le soutien de la DRET sur le développement d'un logiciel efficace pour l'analyse des structures dans le domaine non linéaire géométrique, en visant particulièrement les problèmes d'équilibre en post-flambement des panneaux minces raidis utilisés dans l'industrie aéronautique ; le détail de cette étude est exposé dans la référence 1 .

Nous avons posé le problème de la recherche des équilibres stables comme celui de la minimisation du potentiel total ; cette hypothèse couvre les cas de non linéarité de grand déplacement, d'élasticité non linéaire et de contact ; elle ne s'applique pas directement dans les cas de plasticité, de frottement et de force appliquée non conservative.

2 - PRINCIPES THEORIQUES

Les algorithmes que nous présentons s'appuient sur les 3 remarques indépendantes suivantes :

2.1 - Le potentiel élastique en grand déplacement peut être discrétisé par une fonction algébrique du 4ème degré.

X étant le vecteur des déplacements des noeuds, en tout point M de la structure, le champ de déplacement $u(M)$ peut être obtenu par les opérateurs linéaires d'interpolation des éléments finis, soit :

$$u(M) = [N(M)]X$$

Cette relation linéaire ne s'applique pas si on utilise comme degré de liberté des grandes rotations.

La formulation du tenseur de déformation de Green ε_{ij} est :

1 $$\varepsilon_{ij} = \frac{1}{2}\left(\left[\frac{\partial u_i}{\partial x_j}\right] + \left[\frac{\partial u_j}{\partial x_i}\right] + \left[\frac{\partial u_j}{\partial x_i}\right]\left[\frac{\partial u_i}{\partial x_j}\right] \right)$$

U étant une fonction linéaire de X , il en résulte que ε_{ij} est une fonction du 2ème degré des composantes de X .

Pour un matériau à loi d'élasticité linéaire, le tenseur de contrainte $\sigma_{ij} = [H]\ \varepsilon_{ij}$ est aussi une fonction du 2ème degré des déplacements.

L'énergie élastique interne s'écrit :

2 $$W_{int}(X) = \int_V [\varepsilon_{ij}]_t\ [H]\,[\varepsilon_{ij}]\ dv$$

Il résulte de la relation 1 que l'énergie élastique est fonction algébrique du 4ème degré des composantes de .

L'opérateur de la loi de comportement $[H]$ étant symétrique, cette fonction est discrétisable par une somme de carrés de trinômes.

Dans nos applications habituelles, les forces extérieures dérivent elles aussi d'un potentiel $W_{ex}(X)$ qui est linéaire (forces constantes) ou représentable par une fonction polynomiale du 3ème degré (pression constante).

Le potentiel total $W_{TOT}(X) = W_{int}(X) - W_{ex}(X)$ peut être donc lui aussi représenté par une fonction algébrique du 4ème degré.

Nous exploitons cette remarque dans la phase de minimisation du potentiel total dans une direction de descente V (ou "Line Search"), soit :

$W_{TOT}(X + \rho V)$ minimum par rapport à ρ .

$W_{TOT}(X + \rho V)$ étant un polynôme du 4ème degré, la recherche du minimum se ramène à la résolution d'une équation algébrique du 3ème degré dont on choisit la racine correspondant au minimum absolu ; les coefficients du polynôme en ρ sont calculés <u>explicitement</u> à partir de X et de V par une procédure détaillée en annexe, dont le coût est du même ordre que celui d'un seul calcul de $W_{TOT}(X)$.

Cette méthode de "Line Search" exacte est la pierre d'achoppement de nos travaux ; c'est elle qui garantit la convergence inconditionnelle de tous les algorithmes et cela particulièrement dans les méthodes de gradient conjugué qui sont très exigeantes quant à la précision de la phase de "Line Search".

Son avantage est considérable vis-à-vis des procédures de "Line Search" itératives qui outre un volume de calcul 10 à 20 fois plus important ne garantissent pas l'obtention du minimum absolu.

2.2 - <u>Linéarisation et découplage des effets de flexion, Super Elément linéaire</u>

Pour les éléments de flexion simple (poutre, plaque), on pose que la non linéarité n'intervient que par les effets de membrane.

Cette hypothèse simplificatrice nécessite deux conditions :

- que les éléments finis soit suffisamment petits pour que les variations de longueur d'arc sur un élément puissent être assimilées à celle des cordes, ce qui est cohérent en pratique avec la finesse des maillages exigée pour l'emploi des éléments à faible degré d'interpolation dont nous avons l'habitude,
- que les rotations restent modérées (inférieures en valeur absolue à $\pi/4$), ce qui est le cas pratique dans nos problèmes.

Sur le plan du coût des calculs, l'intérêt est considérable ; l'énergie élastique s'écrit sous la forme :

$$W_{int}(X) = W_{int\,membrane}(X) + \frac{1}{2} X_t \left[K_{Flexion}\right] X$$

La matrice de rigidité partielle des termes de flexion $[K_{Flexion}]$ est calculée initialement ; dans les itérations, les forces internes correspondantes sont obtenues par simple multiplication de matrice creuse.

En plus des termes de flexion, nous nous donnons la possibilité de garder linéaire une partie de la structure dont la matrice de rigidité est elle aussi calculée préalablement aux itérations non linéaires, cette zone linéaire peut avoir tout ou partie de ses degrés de liberté condensés par une technique de super élément.

2.3 - <u>Préconditionnement, métrique optimale</u>

L'idée directrice est de se placer implicitement dans une métrique optimale pour un problème linéaire tangent, cette métrique est celle d'un espace Y

relié à l'espace des degrés de liberté par la relation $Y = [L]\,X$.
$[L]$ étant une décomposition triangulaire d'une matrice de rigidité tangente $[K_{tg}(X)] = [L]_t\,[L]$.

Il est aisé de démontrer qu'un déplacement égal au gradient du potentiel total dans l'espace Y et restitué dans l'espace X s'écrit :

$$D(x) = [K_{tg}(X)]^{-1} R(x)$$

avec $R(x) = \partial W_{tot}(X)/\partial X$, résidu de l'équation d'équilibre dans l'état X.

L'utilisation implicite du changement de métrique affecte les produits scalaires, soit :

$$(X_1 \cdot X_2)_{\text{métrique } Y} = Y_{1t}\; Y_2 = X_1 [L]_t [L] X_2 = X_1 [K_{tg}(X)] X_2$$

2.4 - Algorithmes de calcul

A partir des remarques précédentes, nous avons étudié 4 types d'algorithmes pour la résolution des équilibres, ils diffèrent par la technique de choix de la direction de descente ; pour un problème à forces imposées variant proportionnellement par palier, l'organigramme général est le suivant :

Palier $k = 0$

- Déplacement X = X initial (en général X = 0)

Palier $k = k + 1$

- Charges appliquées au palier k $F_{ex} = \lambda_K\, F_{ex}$.
- Matrice de rigidité tangente $[K_{tg_0}(X)] = [\partial^2 W_{tot}(X)/\partial X_i\, \partial X_j]$
- Calcul implicite de $[K_{tg_0}(X)]^{-1}$ (factorisation de Gauss)
- Résidu des équations d'équilibre $R(x) = \partial W_{tot}(X)/\partial X$
  → Si $|R(x)| \simeq 0$ (retour au palier $k = k + 1$)
- Direction de descente V : NEWTON-CONJUGUE / BFGS / NEWTON MODIFIE / NEWTON RAPHSON
- Coefficient du polynôme en ρ : $W_{tot}(X + \rho V)$
- Résolution équation du 3ème degré $\partial W_{tot}(X + \rho V)/\partial \rho = 0$
  ρ_{opt} = racine correspondant à $W(X + \rho V)$ minimal
- Incrément des déplacements $X = X + \rho_{opt}\, V$ (retour au calcul du résidu)

Les procédures de calcul de direction de descente V que nous utilisons sont maintenant classiques, elles ont été présentées par de nombreux auteurs (référence 2, 3, 4), soit :

- Méthode de Newton Conjugué

C'est la méthode du gradient conjugué non linéaire des variantes "Fletcher Reeves" et "Polak Ribiere" (Référence 2) dans la métrique optimale du problème tangent de la fin de chaque palier ; la direction de descente V à chaque itération s'obtient par la relation

- $D = [K\,tg_0\,(X)]^{-1}\;R\,(X)$ gradient conditionné dans la métrique de $[Ktg\,(X)]$ du début de palier
- 1ère itération $V = -D$
- autre itération $V = -D + \gamma V$ précédent

$\gamma = (D - D_{\text{précédent}}).R(X)/D_{\text{précédent}}.R(X)_{\text{précédent}}$ (Polak Ribiere)

$\gamma = D\,.\,R(X)/D_{\text{précédent}}.R(X)_{\text{précédent}}$ (Fletcher Reeves)

L'efficacité du conditionnement est liée à la variation de forme géométrique entre chaque palier ; en pratique, dans les cas de Snap Through important il est intéressant de reconditionner après quelques dizaines d'itérations (10 à 20) en cas de non convergence ; le fait que $[Ktg\,(X)]$ ne soit plus définie positive n'est pas rédhibitoire.

- Méthode de quasi Newton (BFGS)

L'idée directrice de cette méthode classique réf. 3, 4, 5) est de se placer dans la métrique d'une matrice Hessienne approximée à chaque itération par un opérateur "sécant", une des formes de cet opérateur les plus pratiques en éléments finis est :

$$[K_i]^{-1} = \prod_{j=i-1}^{1}\left[I - \delta_j \varphi_j^t\right]\left[K_0\,tg\,(X)\right]^{-1}\prod_{j=1}^{i-1}\left(I - \varphi_j \delta_j^t\right)$$

Les vecteurs δ_i et φ_i sont obtenues par les relations :

$$2 \quad \delta_i = D_i/D_i^t\,(R_i - R_{i-1}) \qquad 3 \quad \varphi_i = R_{i+1} - R_i\left(1 + \left(\frac{D_i^t(R_i - R_{i-1})}{D_i^t\,R_i}\right)^{\frac{1}{2}}\right)$$

avec :

$$R_i = R\,(X_i) \qquad D_i = [K_i]^{-1} R_i \qquad X_{i+1} = X_i - \rho_{opt}\,D_i$$

Pour ne pas expliciter des opérateurs non creux, les relations précédentes s'appliquent par récurrence ; leur inconvénient est de nécessiter le stockage de 2 vecteurs supplémentaires à chaque itération et d'effectuer un nombre croissant de produits scalaires ou équivalent avec les itérations ; en pratique si la convergence n'est pas acquise en 10 ou 20 itérations on redémarre l'algorithme en refactorisant dans l'état d'arrivée, la non-positivité de $[K_i]$ qui entraine l'obtention du nombre négatif sous le radical de la relation 3 , se contourne en utilisant une formule de récurrence plus coûteuse présentée planche 1.

Pour ces nombres d'itérations faibles, le volume de mémoire nécessaire au stockage des vecteurs δ_i et φ_i est nettement moins important que celui de la forme factorisée de $K_0\,tg\,(X)$.

- Newton Raphson

C'est la méthode du gradient simple dans la métrique la plus adaptée, celle de $[K_{tg}(X)]$ à chaque itération, soit la direction de descente :

$$V_i = \left[K_{tg}\ (X)\right]^{-1} R\ (X)$$

Avec Line Search, c'est la méthode qui converge le plus rapidement mais avec des itérations beaucoup plus coûteuses car elles incluent une factorisation de $[K_{tg}\ (X)]$

Comme dans le cas précédent, la non-positivité de $[K_{tg}(X)]$ n'est pas un obstacle.

- Newton modifié

Nous citons pour mémoire cette méthode qui est un gradient simple dans la métrique de $[K_{tg_0}(X)]$ en début de palier ; avec Line Search elle converge beaucoup plus lentement que Newton conjugué qui ne diffère que par 1 produit scalaire supplémentaire par itération.

2.5 - Déplacements imposés, déplacements contrôlés

D'une façon générale, il est intéressant de se ramener chaque fois qu'on le peut à des conditions aux limites de déplacements imposés car les variations de géométrie, nuisibles à l'efficacité du préconditionnement, y sont plus limitées.

Nous donnons planche 1 l'algorithme général correspondant à des conditions aux limites hybrides de forces et de déplacements imposés, son principe est le même que précédemment, il faut prendre la précaution de démarrer les itérations sur une forme X solution linéarisée de l'incrément de déplacement ; on évite ainsi le calcul des résidus $R(X)$ dans un état comportant des discontinuités importantes.

Dans les problèmes comportant des conditions aux limites de forces imposées réparties, variant proportionnellement à un facteur de charges λ on ne peut pas se ramener directement à un problème de déplacement imposé, pour éviter de trop grands sauts entre les déformées des points calculés, il peut être intéressant d'échanger une des inconnues de déplacement imposé contre le paramètre de charges λ ; on est ainsi amené à chercher les facteurs de charges conduisant à ce qu'un déplacement prenne une succession de valeurs données ; on procède par itération sur un problème à déplacements et forces imposées, avec des facteurs de charges variables, de façon à annuler la composante de réaction $R_i(X)$ au point de déplacement contrôlé ; dans cette itération extérieure, l'équation scalaire de réaction nulle au point contrôlé est résolue par une méthode de Newton qui est présentée planche 1.

La procédure du déplacement contrôlé trouve son intérêt dans les cas où le passage du flambage entraine un changement de forme considérable ; il est alors intéressant de contrôler un point dont on intuite que le déplacement sera monotone avec la charge et dont l'appui évite l'effondrement de la structure.

2.6 - Méthode de Gradient Projeté, Newton Conjugué projeté

Au départ ces méthodes, surtout dans leur version Newton Conjugué projeté, ont été conçues pour traiter les problèmes de minimisation de fonction avec contrainte d'inégalité linéaire ; nous les utilisons largement dans nos processus d'optimisation structurale.

Dans notre problème l'adjonction de contraintes d'inégalité sur les déplacements trouve 2 applications :

- prise en compte des contacts,
- contrôle de déplacements pour le suivi des branches instables dans les Snap through.

Nous détaillons dans la référence 1 l'algorithme de Newton Conjugué Projeté ; il diffère de Newton Conjugué par les points suivants :

- direction de descente projetée (dans la métrique optimale) sur les contraintes d'inégalité touchées,
- Line Search limité éventuellement à la première contrainte touchée,
- Abandon des contraintes touchées en fonction du signe des multiplicateurs de Lagrange des contraintes correspondantes.

Quand on utilise cet algorithme pour le suivi des branches instables, on se donne des contraintes sur la valeur absolue de l'amplitude d'un certain nombre de déplacements significatifs ; on procède alors en itérant sur le facteur de charges de façon à annuler le multiplicateur de Lagrange (la réaction) de la dernière contrainte touchée ; on contourne ainsi les principales difficultés d'application de la méthode du déplacement contrôlé du § 2.5.

3 - CONSIDERATIONS INFORMATIQUES

Les algorithmes présentés ont été programmés dans la branche NLIRAP de notre logiciel général ELFINI ce qui permet de bénéficier de toutes les facilités de maillage intéractif, visualisation Super Elément, gestion de banque de données, etc., particulièrement performantes de ELFINI.

L'organisation de NLIRAP est la suivante :

INITIE

Conditionnement de tableaux élaborés par ELFINI
- Géométrie et caractéristique des éléments, conditions aux limites, charges,
- Correspondance Noeud-degré de liberté inconnue,
- Matrice de rigidité constante, profil matrice de rigidité,
- Définition des paliers de charges de déplacement imposé et contrôlé.

CANEVAS

Compactage matrice de rigidité constante, partition en 3 (couplage DDL libre, imposée-libre, imposée-imposée)

GRACØN | **BFGS** | **NEWTØN**

Algorithme de résolution avec variantes de déplacement imposé et contrôlé :

- GRACØN : Méthode de Newton Conjugué
- BFGS : Méthode BFGS avec variante pour $[K_{tg}(X)]$ non défini positif
- NEWTØN : Méthode de Newton Raphson

Modules de services

MATTG : matrice de rigidité tangente, $K_{tg}(X)$
FØRTG : forces de précontrainte et potentiel total, $F_{int}(X)$, $W_{TOT}(X)$
COEFPO : coefficients du polynôme en ρ de l'énergie interne $W_{TOT}(X+\rho V)$
MINI 4 : minimum potentiel total (résolution équation du 3e degré et tri des racines)
FACGAU : factorisation de Gauss par méthode "profil"
RESGAU : résolution système linéaire profil

Pour gagner en performance, nous avons choisi de garder en mémoire centrale tous les tableaux utilisés dans les itérations de NLIRAP, en particulier le plus important qui est la matrice de rigidité factorisée, il n'y a donc pas d'entrée sortie programmée à l'intérieur des itérations.

Aujourd'hui, cette conception entraine des limitations sur la taille des problèmes traitables de l'ordre de 5000 à 8000 degrés de liberté pour une occupation mémoire maximale de 7000 Koctets sur IBM 3081 ; cette limite va être prochainement levée par l'utilisation d'un système d'exploitation à adressage virtuel étendu et nous pensons que des problèmes de l'ordre de 15000 à 25000 DDL pourront être traités raisonnablement sur une machine disposant d'une mémoire réelle de 32000 Koctets.

Les opérations sont effectuées en double précision (64 Bits) dans NLIRAP à l'exception de la factorisation qui peut être effectuée en simple précision (32 Bits) pour limiter l'encombrement de la matrice factorisée ; nous n'avons pas vu dans nos tests de différence avec une factorisation en double précision, cela tant sur la précision des résultats que sur la vitesse de convergence ; cette tolérance résulte de ce que la factorisation n'intervient que dans le préconditionnement et non dans les calculs de résidu.

4 - TESTS

4.1 - Tests élémentaires

Ces tests nous démontrent le bon fonctionnement des algorithmes et permettent de comparer nos résultats à ceux publiés dans la littérature ; à ce niveau, nous ne nous attardons pas sur les comparaisons de performances entre algorithmes car elles ne sont pas représentatives des applications envisagées, en effet :

- le faible nombre de degrés de liberté et le caractère très "bande" des matrices de rigidité réduisent d'un ordre de grandeur les coûts relatifs des factorisations vis-à-vis de ceux des résolutions et des tabulations de forces internes ; ce qui avantage la méthode de Newton,
- les cas académiques de post-flambage et de "Snap through" peuvent conduire à des sauts de forme considérables dans les méthodes à forces imposées, ce qui rend moins efficace le préconditionnement ; ce phénomène est moins accentué dans les problèmes de flambage de panneaux quand on espère une résistance post critique,
- les cas académiques se résolvent quelquefois aisément par les méthodes de déplacement contrôlé parce que le choix du point de contrôle est évident.

- Poutre sur 2 appuis en compression (voir planche 2)

Pour ces calculs menés en déplacement imposé aux extrémités, nous avons utilisé 2 types de maillage :

- treillis de barres (120 éléments),
- poutres de flexion (20 éléments),

les rigidités linéaires en compression et flexion sont identiques.

Ces modélisations ont été choisies pour déceler une éventuelle influence des approximations entrainés par l'hypothèse de linéarisation de la flexion.

Les résultats des deux modélisations présentées planche 2 sont quasiment identiques jusqu'au moment où sont atteintes des rotations de l'ordre de $\pi/4$; le calcul en élément de flexion linéarisée devient franchement aberrant quand les rotations maximales atteignent $\pi/2$.

A titre de curiosité, nous avons conduit les calculs de la poutre en treillis jusqu'au bouclage (voir planche 2) au moment du croisement les méthodes avec "Line Search" procèdent à un retournement complet de la poutre alors que la méthode de Newton Raphson simple continue sur les solutions à boucle, ce phénomène s'explique par le fait que les méthodes avec "Line Search", détectent dans ce cas la présence des solutions retournées dont l'énergie est beaucoup plus faible que celle des solutions à boucle.

- Plaque carrée articulée sur 4 bords (planche 3)

Le maillage est composé de 200 éléments triangle de plaque ; outre les appuis simples en $\mathfrak{z}$ sur les 4 bords, les conditions aux limites sont des déplacements imposés en $\mathfrak{x}$ sur AB et CD, de plus pour se placer dans les cas de validité des calculs de Timoshenko on contraint les déplacements en sur les côtés AD et BC à être égaux (rigidité à plat de bordure infinie).

Les charges critiques théoriques de flambage sont retouvées.

Nous présentons sur la planche 3 la comparaison de l'évolution de la largeur travaillante, nos résultats sont confrontés à ceux des théories de Karmann et de Marguerre ainsi qu'à des points expérimentaux réunis par Vallat (Réf. 6).

- "Snap through" d'une coque cylindrique encastrée chargée verticalement en son centre (planche 4)

Ce sont deux exemples classiques auxquels se réfèrent souvent les auteurs ; leur schématisation est représentée planche 4 , ils se différencient uniquement par leur épaisseur : 12,7 mm (1er cas) et 6,35 mm (2e cas).

Sur la planche 4, nous présentons nos résultats comparés à ceux obtenus par Sabir et Lock et ceux de Crisfield avec sa variante de la méthode de Riks (Réf. 7).

Nous présentons sur les tableaux ci-dessous les nombres d'itérations nécessaires à la convergence pour un résidu maximum de 1 daN.

Déplacement au centre imposé e = 12,7 mm

Palier											
	D en mm	3,85	7,40	10,5	13	15,5	16,3	17,5	19,5	22,5	26,5
	F en daN	121	189	208	185	110	85	61	54	94	219
Newton Raphson		1	1	2	2	2	2	2	2	2	
Newton Conjugué		3	3	2	3	7	4	8	4	3	4
B.F.G.S.		3	3	2	3	5	4	4	4	3	4

Forces au centre imposé e = 12,7 mm

Palier					
	F en daN	123.	195.	220.	280.
	D en mm	3,92	7,95	26,51	27,86
Newton Raphson		2	3	6	2
Newton Conjugué		5	5	32	3
B.F.G.S.		4	5	22	3

Déplacement au centre imposé, e = 6,35 mm

Palier													
Palier	D en mm	4	7	10	12,5	14,5	15,5	16,3	16,8	16,9	17,8	20,5	23
	F en daN	29.	43.	53.	57.	54.	45.	20.	36.	35.	34.	25.	10.
Newton Raphson		1	1	1	1	2	2	3	5	1	1	1	1
Polak Ribiere		3	3	3	3	3	3	12	50	1	1	3	2
B.F.G.S.		3	3	3	3	3	3	6	20	1	1	3	2

Force au centre imposée, e = 6,35 mm

Palier	D en mm	3,94	6,76	9,63	28,81	29,50	
	F en daN	29	42,5	52,5	58	60	70
Newton Raphson		3	2	3	7	1	2
Polak Ribiere		5	4	4	50	1	2
B.F.G.S.		4	4	4	34	1	2

Pour pouvoir effectuer la comparaison au résultat de la référence présentée planche 4 pour le calcul des branches instables, nous avons mélangé des paliers de forces imposées et de déplacement imposé ; ceci dit l'intérêt de notre approche par rapport à la méthode de continuation présentée dans la référence 6 est de ne pas avoir besoin de suivre les branches instables.

- Post-flambage de panneaux galbés en compression (planches 5,6,7)

Il s'agit de 3 plaques cylindriques appuyées sur les deux génératrices latérales et comprimées longitudinalement par déplacement imposé des extrémités. Outre les formes de flambage, nous présentons sur les planches 5,6 et 7 l'évolution de la contrainte de compression moyenne en fonction des déplacements imposés ; nous situons sur ces courbes les charges de flambage semi-empiriques recommandées par la référence 8 .

Le premier panneau (planche 5) du type "long" avec une relativement faible courbure est très représentatif des mailles de revêtement du fuselage d'un avion ; il présente un Snap Through notable au moment du flambage conduisant à des formes classiques dites "Damier".

Le passage du Snap Through a demandé 38 itérations par Newton conjugué et 23 itérations par BFGS, une refactorisation au bout de 10 itérations amène Newton Conjugué à 2 x 10 + 1 itérations et BFGS à 10 + 9 itérations.

Les autres points demandent environ 6 itérations par palier par l'une ou l'autre méthode.

Les points inférieurs de la branche post-flambée ont été obtenus par décharge après le flambage.

Le second panneau planche 6 ne diffère du précédent que par une courbure plus importante, les formes d'équilibre post-flambées sont beaucoup plus tourmentées (flambage en pointe de diamant), il présente des solutions multiples ; le palier de Snap Through principal est relativement dispendieux.

Les résultats présentés ont été obtenus avec la méthode de Newton Raphson avec Line Search (35 itérations pour le Snap Through).

Le troisième panneau, planche 7, est du type cylindre court, il ne présente pas de saut de branche au moment du flambage, son comportement se rapproche de celui des plaques planes, la convergence en post-flambage est acquise en quelques itérations, la comparaison à la référence 8 est alors excellente.

4.2 - Tests de performance : flambage d'une âme de longeron

Nous avons choisi cet exemple, planches 8 et 9, pour présenter la comparaison des performances des algorithmes étudiés, car ce cas est bien représentatif de nos applications : flambage quasi-simultané de nombreux modes locaux et résistance réelle en post-flambage. La structure étudiée représente 3 mailles d'une âme de longeron séparée par des raidisseurs et dont les semelles sont supposées stabilisés latéralement : le modèle est constitué d'éléments finis de plaque triangulaire et de poutre, il comporte 1705 degrés de liberté.

Nous avons envisagé 2 types de sollicitations :

- Effort vertical en extrémité, introduit par un déplacement imposé, qui sollicite la poutre en cisaillement et en flexion.

Nous avons calculé une série de 11 paliers allant très au-delà du flambage initial, les formes de cloques évoluent et sont relativement complexes (planche 8) ; la non linéarité globale représentée par la courbe effort fonction du déplacement du point chargé reste faible.

Nous présentons sur le tableau ci-dessous les performances comparées des différentes méthodes, en nombre d'itérations nécessaires à la convergence pour un résidu maximum de 1 daN.

Flèche	9 mm	11 mm	14 mm	15 mm	20 mm	25 mm	30 mm	35 mm	40 mm	45 mm	50 mm
N.R. simple	2	1	X								
N.R.M. simple	3	1	X								
N.R. + Line Search	2	1	4	2	3	2	7	2	4	4	3
N.R.M. + Line Search	2	1	58	7	39	41	54	20	100	100	47
N.C. Fletcher Reeves	2	1	15	4	13	18	12	9	40	34	11
N.C. Polak Ribiere	2	1	9	4	14 10+1	10	8	9	49 10+7	42 10+3	11
B.F.G.S.	2	1	8	4	9	8	7	9	16 10+4	15 10+2	8

* en deuxième ligne, les résultats obtenus par N.C. et B.F.G.S. avec une refactorisation au bout de 10 itérations.

X pas de convergence.

Ces résultats appellent les remarques suivantes :

- sans Line Search, les méthodes classiques de Newton Raphson et Newton Raphson modifié ne passe pas le flambage,
- c'est la méthode de Newton Raphson avec Line Search qui demande le moins d'itérations ; mais elle est plus coûteuse que Newton Conjugué et BFGS car elle nécessite des factorisations à chaque itération,
- la méthode BFGS converge la plus vite dans les cas difficiles si on ne limite pas le nombre des itérations,
- si on refactorise au bout de 10 itérations, Newton Conjugué converge aussi vite et devient plus rentable du fait de sa simplicité.

Pour permettre au lecteur de mieux cerner les coûts relatifs de ces méthodes, nous donnons les temps d'exécution de chacun des modules de base utilisé pour un ordinateur IBM 3081.

coefficient du polynôme dérivé de l'énergie	0,185 s
énergie et forces internes	0,174 s
matrice tangente	0,650 s
résolution simple	0,251 s
supplément BFGS à la 10e itération	0,463 s
factorisation	3,594 s

- <u>Chargement à plusieurs composantes</u>

On a ajouté au chargement précédent un effort de compression réparti sur la section d'extrémité ; comme on ne peut plus dans ce cas procéder par déplacement imposé, nous avons utilisé avec l'algorithme de Newton Conjugué les procédures suivantes :

. Forces imposées,
. Forces imposées et déplacements contrôlés en z à l'extrémité.

Les formes de post-flambage obtenues sont présentées planche 9 ; les convergences comparées des 2 méthodes sont les suivantes pour un résidu maximal de 1 daN.

forces imposées	λ (chargement)	0.4	0.5	0.6	0.62	0.66	0.70	0.85
	W (mm)	8.69	10.94	13.42	13.94	14.98	16.06	20.26
	nbre d'itérations	3	10	15	3	5	5	15
déplacement contrôlé 1 composante d'un point de chargement)	λ (chargement)	0.46	0.502	0.584	0.623	0.661	0.716	0.841
	W (mm)	10.	11.	13.	14.	15.	16.5	20.
	nbre d'itérations	2+1	5+1	9+1	3+1	3+1	5+1	10+1

4.3 - Calcul d'un essai partiel de compression multiaxiale et de cisaillement de l'extrados de la voilure du Mirage 4000 (planche 10)

Cet essai avait pour but de restituer sur un élément de panneau raidi intégral les champs de contraintes complexes calculées sur l'extrados de la voilure du Mirage 4000 pour en évaluer la résistance en post-flambage : l'éprouvette a été chargée en flexion différentielle sur deux axes de façon à reconstituer les champs de contraintes visés dans sa partie centrale.

Le calcul a été effectué en deux niveaux :

- calcul linéaire de l'ensemble de l'éprouvette comportant 8900 degrés de liberté avec essentiellement des éléments de flexion plaque et poutre à décalage de fibre neutre pour les raidisseurs intégraux ; le maillage, visualisé planche 10, représente la moitié supérieure de l'éprouvette, qui est muni de condition aux limites anti-symétrique,
- calcul non linéaire de la partie centrale représenté par un modèle comportant 8 mailles séparées par des raidisseurs ; à la frontière on prend des conditions aux limites mixtes dans le calcul précédent :
  . dans le plan des mailles, des forces imposées (réaction du calcul linéaire),
  . perpendiculairement à ce plan, on impose les déplacements du calcul linéaire.

Cette procédure de calcul en 2 étapes a été déterminée par la limitation à 5000 degrés de liberté du module NLIRAP et par le fait que la non linéarité reste très locale, la structure étant globalement linéaire.

Nous présentons planche 10 les déformées obtenues après flambage ainsi que la comparaison calcul essais de l'évolution des jauges de contraintes les plus caractéristiques jusqu'à la rupture.

Cette confrontation a amené les remarques suivantes :

- la simulation numérique nous a donné une bonne prévision globale ; l'effondrement de la structure par le flambage, l'apparition et l'évolution des non-linéarités étant semblables,
- le sens dans lequel apparaissent certaines cloques est inversé entre le calcul et l'essai, ce qui se traduit par une évolution opposée des contraintes mesurées et calculées sur certaines jauges ; ce type de phénomène, que nous avons souvent rencontré par ailleurs, est attribuable à l'existence latente de plusieurs solutions.

5 - CONCLUSIONS-DEVELOPPEMENTS

Nous disposons d'un outil efficace pour le calcul des non linéarités géométriques particulièrement adapté au post-flambage des panneaux minces raidis.

C'est notre procédure de "Line Search" exact qui permet de franchir les Snap Through sans être obligé de suivre les branches instables par continuation comme dans certaines méthodes précédemment préconisées (réf. 6) cette minimisation exacte permet de pratiquer la méthode de Newton Conjugué qui obtient ainsi des performances comparables à la méthode BFGS.

Nous appliquons aujourd'hui cet outil intensivement pour les calculs de notre nouvel avion de combat en matériaux composites ; nous en confrontons les résultats à ceux de nombreux essais partiels.

Nous poursuivons le développement dans plusieurs voies :

- adaptation aux nouvelles architectures de calculateur parallèle ; ce qui ne pose pas de problèmes insurmontables, si ce n'est qu'il faut disposer d'une taille mémoire homogène à la rapidité du calculateur,
- enrichissement de la bibliothèque d'éléments non linéaires ; en particulier la recherche d'éléments coque adaptés à l'expression algébrique du potentiel total,
- prise en compte des phénomènes de plasticité et de frottement ; l'idée est de procéder par une méthode de point fixe entre des phases de calcul d'équilibre en grand déplacement pour une déformation plastique donnée et des phases de calcul d'écoulement plastique en fonction des contraintes,
- amélioration de l'algorithme de minimisation proprement dit ; notre conviction profonde est qu'un progrès significatif peut encore être obtenu en exploitant plus à fond le fait que le potentiel total se discrétise sous la forme d'une somme de carrés de trinômes.

REFERENCES

1 C. Petiau & C. Cornuault, Document AMD-BA 22363 - Marché DRET 8034-262 Lot N° 3 Recherche d'algorithmes performants pour le calcul des équilibres en post-flambement.

2 Polak, Computational method in optimization : a unified approach. Academic Press, New York, 1971.

3 M.A. Crisfield, Incremental/iterative solution procedure for nonlinear analysis/numerical methods for nonlinear problems, Numerical methods for nonlinear problems - Vol. 1, Proceedings of the International Conference held at University College Swansea, 1980.

4 A. Pica & E. Hinton, The quasi Newton BFGS method in large deflection analysis plates. Numerical methods for nonlinear problems - Vol. 1. Proceedings of the International Conference held at University College Swansea, 1980.

5 M. Geradin, S. Idelsohn, M. Hogge, Computational strategies for the solution of large nonlinear problems via quasi-Newton methods. Computers and structures - June 1981 - Volume 13.

6 P. Vallat, Résistance des matériaux appliquée à l'aviation. Paris et Liège - Béranger - 1950.

7 M.A. Crisfield, A fast incremental/iterative soluton procedure that handles "snap through" Computers and structures - juin 1981 - Volume 13.

8 Bruhn, Analysis and design of flight vehicle structures.

BLOCK DIAGRAM

— LOAD STEP $\quad k = 0 \qquad X = X_0$

— LOAD STEP $\quad k = k + 1$

. LOAD LEVEL PARAMETER $\qquad \lambda = \lambda_k$

. PRESCRIBED DISPLACEMENT COMPONENTS $\quad \lambda \, X_{imp} \, k$

. CONTROLLED DISPLACEMENT COMPONENT $\quad \mathcal{X}_{cont} \, k$

— PARTITIONED TANGENT STIFFNESS MATRIX

$$k t g_0(X) = \begin{bmatrix} K\ell\ell & K\ell i \\ K\ell i_t & K i i \end{bmatrix} \quad \begin{array}{l} \longleftarrow \text{UNKNOWN DEGREES OF FREEDOM} \\ \longleftarrow \text{PRESCRIBED AND CONTROLLED DEGREES OF FREEDOM} \end{array}$$

— FACTORIZATION $\qquad \Longrightarrow \qquad [K\ell\ell]_0^{-1} = [L]_t \, [L]$

— INITIALIZATION OF DISPLACEMENTS AT THE BEGINNING OF THE LOAD STEP

$$X = \begin{vmatrix} X\ell + D\ell \\ X_{imp} \, k \end{vmatrix} \quad \text{avec } D\ell = - [K\ell\ell]^{-1} [K\ell i] \, \lambda \, \Delta X_{imp} \, k$$

$$\underline{i = i + 1}$$

DISPLACEMENT CONTROL POSSIBILITY

REACTION AT THE CONTROLLED D.O.F.

$r = $ COMPONENT OF $R(X)$

$r \simeq 0$

$\partial r / \partial \lambda = [K\ell c]_t \, [K\ell\ell]^{-1} \, Fex_\ell - \int ex_c$

$\lambda k = \lambda k - r / (\partial r / \partial \lambda)$

— COMPUTATION OF THE TOTAL POTENTIAL ENERGY $\quad W \, tot \, (X) \quad$ AND GENERALIZED INTERNAL FORCES $\quad Fint \, (X)$

— GRADIENT OF THE TOTAL POTENTIAL ENERGY $\qquad R(X) = \partial W \, tot \, (X) / \partial X$

$$R(X) = Fint \, (X) - \lambda k \, f_{EXT}$$

$| R_\ell(X) \simeq 0 |$

— COMPUTATION OF THE DESCENT DIRECTION V

— MINIMIZATION OF $W \, tot \, (X + \rho V)$ WITH RESPECT TO ρ

. COEFFICIENTS OF THE 4TH DEGREE POLYNOMIAL $\qquad W \, tot \, (\rho)$

. SOLVING THE 3RD DEGREE EQUATION $\qquad \partial W \, tot \, (\rho) / \partial \rho = 0$

. $\rho \, opt = \quad$ THE ROOT CORRESPONDING TO $\quad W \, tot \, (\rho) \, mini$

— UPDATE OF DISPLACEMENTS

$$X_\ell = X_\ell + \rho \, opt \, V$$

COMPUTATION OF THE DESCENT DIRECTION V :

NEWTON-RAPHSON	BFGS	CONJUGATE-NEWTON
. COMPUTATION OF $\quad ktg(X)$	$. \varphi = R_i(X) - R_{i-1}(X)$	SCALED GRADIENT :
. FACTORIZATION $\quad [K\ell\ell]^{-1} = [L]_t \, [L]$	$. \delta = \ell_{i-1} \, V_{i-1}$	$D_i = [K_{\ell\ell}]_0^{-1} \, R_i \, (X)$
$V_i = - [K\ell\ell]^{-1} \, R_i(X)$	$V_i = - \left[(1 - \frac{\delta \varphi^t}{\delta^t \varphi}) K_{i-1}^{-1} (1 - \frac{\varphi \delta^t}{\delta^t \varphi}) + \frac{\delta \delta^t}{\delta^t \varphi} \right] R_i(X)$	. FIRST ITERATION : $\qquad V_1 = - D_1$
	$(V_1 = - [K_{\ell\ell}]_0^{-1} R(X)$	. ITERATION i :
	AT FIRST ITERATION)	$V_i = - D_i + \gamma \, V_{i-1}$
		$\begin{cases} \gamma = (D_i - D_{i-1}) . R_i \, (X) / D_{i-1} . R_{i-1} \, (X) \\ \qquad \text{(POLAK-RIBIERE)} \\ \gamma = D_i . R_i \, (X) / D_{i-1} . R_{i-1} \, (X) \\ \qquad \text{(FLETCHER-REEVES)} \end{cases}$

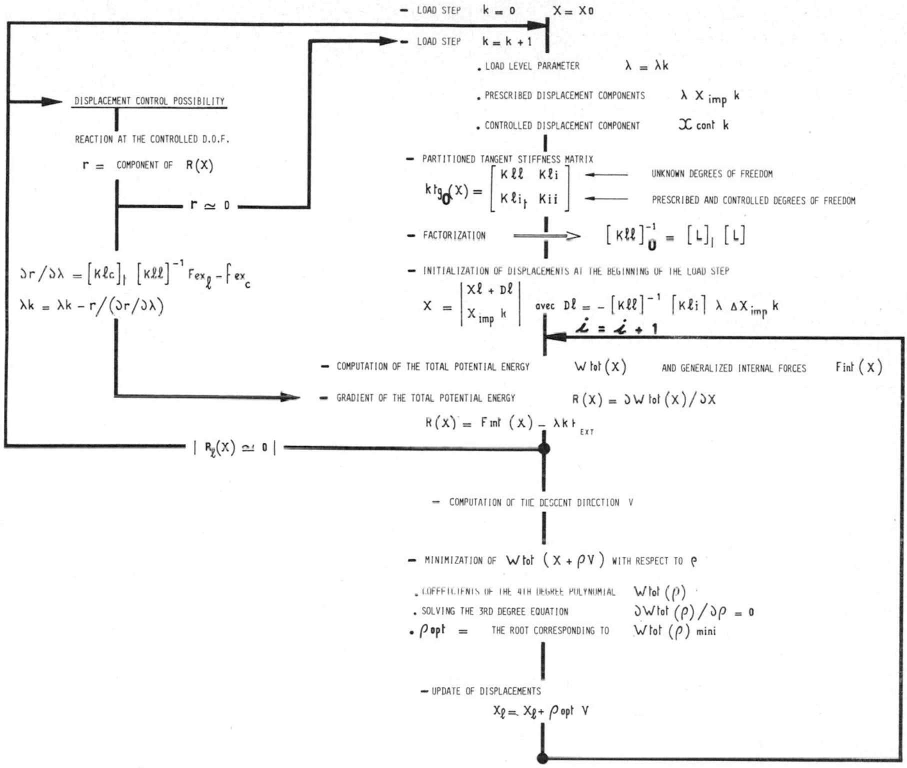

PL 2

SIMPLY SUPPORTED BEAM UNDER COMPRESSION

COMPARISON BETWEEN BEAM ELEMENTS AND TRUSS FRAMES

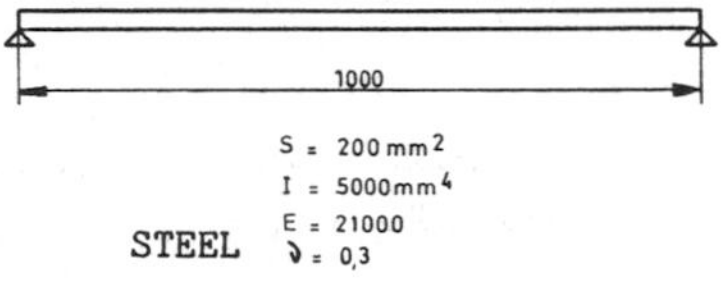

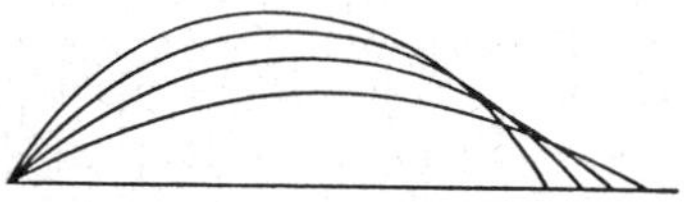

FLEXURAL BEAM ELEMENTS

TRUSS FRAME

STABILITY OF THE SOLUTION DERIVED FROM LINE SEARCH

NEWTON RAPHSON WITHOUT LINE SEARCH

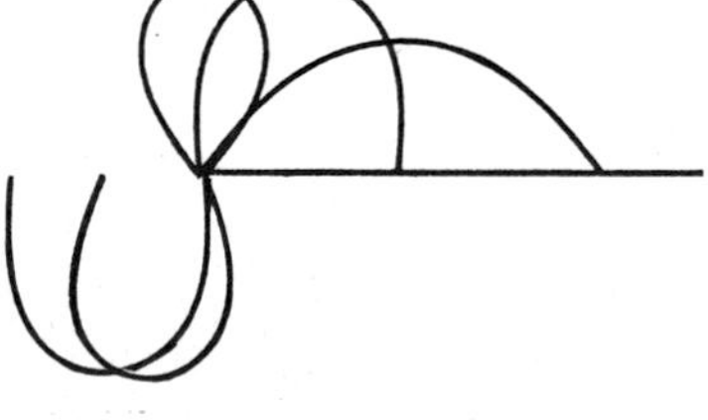

NEWTON RAPHSON WITH LINE SEARCH

PL 3

HINGED SQUARE PLATE UNDER COMPRESSION - EFFECTIVE WIDTH

OUT-OF-PLANE DEFLECTION CONTOUR LINES - NORMAL STRESSES

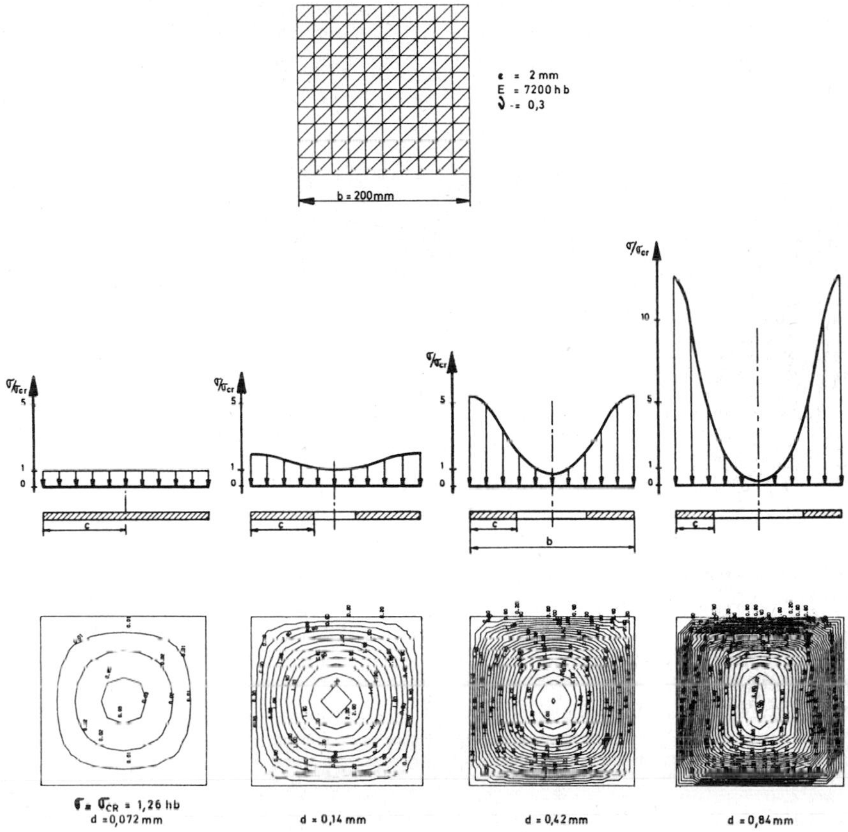

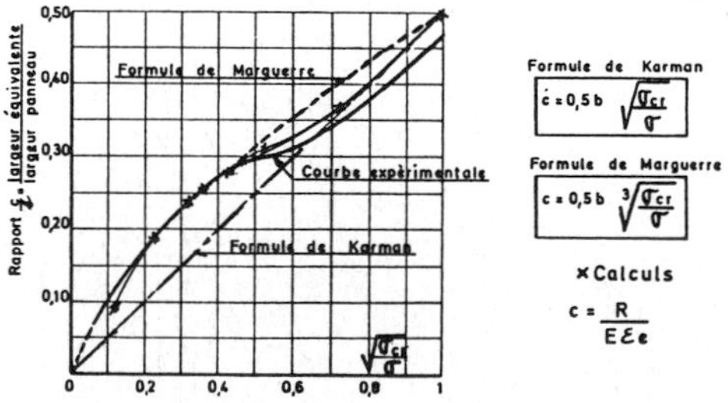

COMPARISON WITH DESIGN FORMULAS

PL 4

HINGED CYLINDRICAL SHELL WITH A CENTRAL POINT LOAD

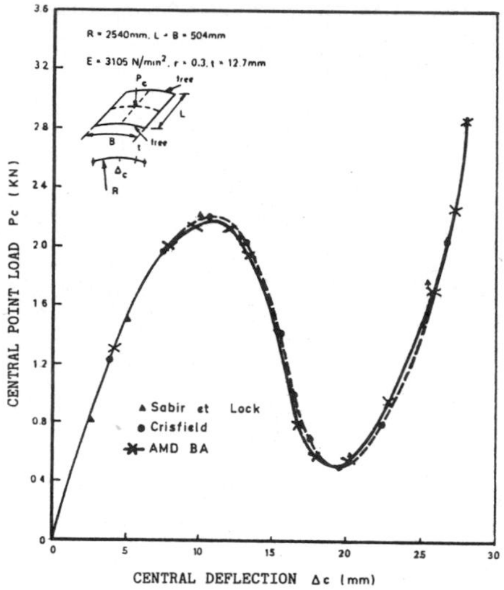

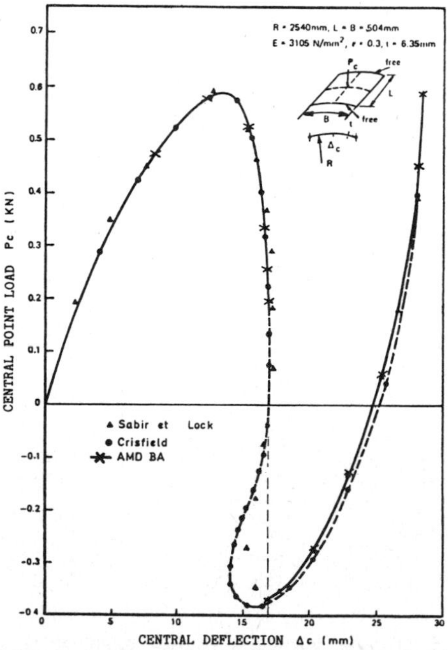

PL 5

POST-BUCKLING OF A CURVED SHEET PANEL UNDER AXIAL COMPRESSION

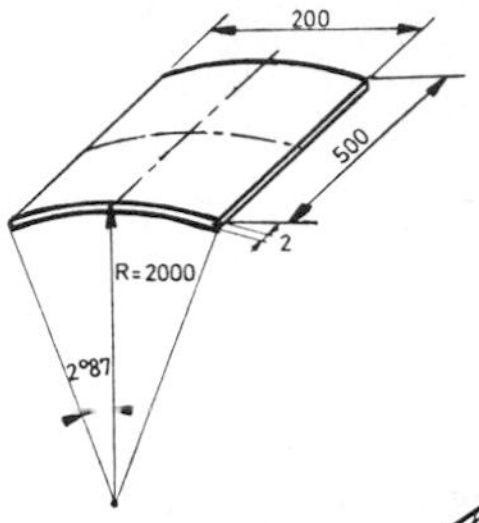

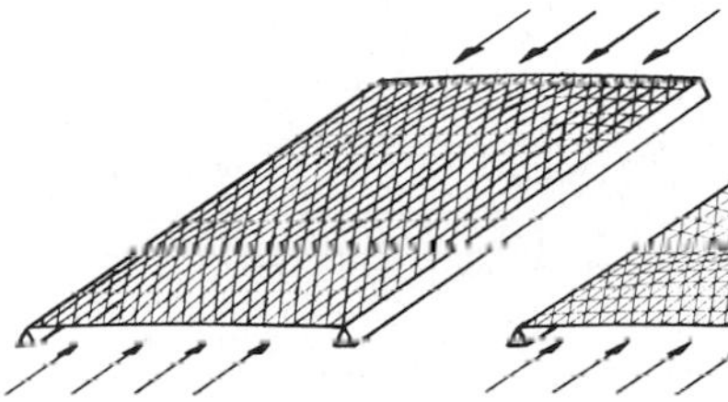

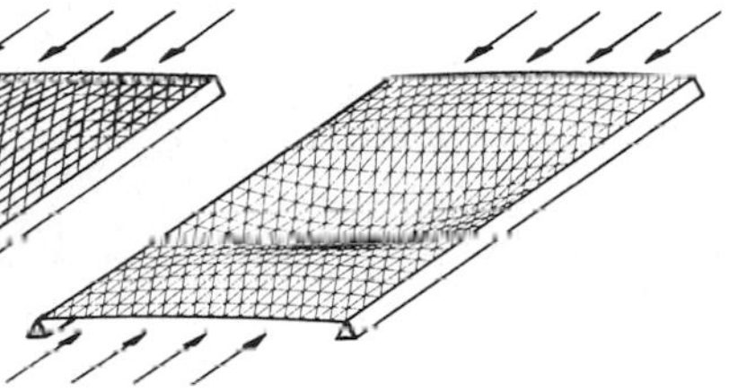

① σ_{moy} = 7 hb, d = 0,5 mm

BEFORE BUCKLING

② σ_{moy} = 7 hb, d = 0,8 mm

POST-BUCKLING

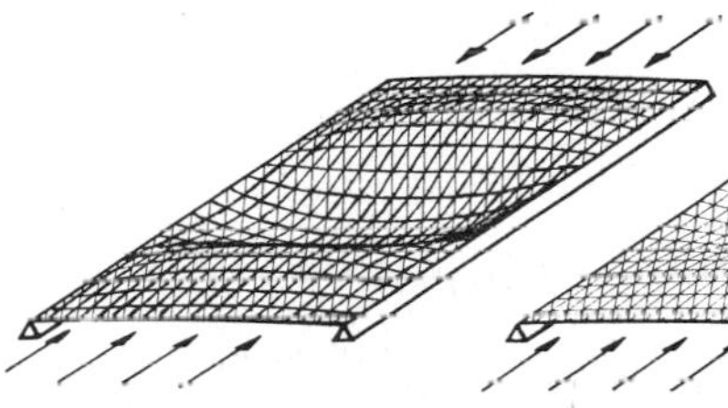

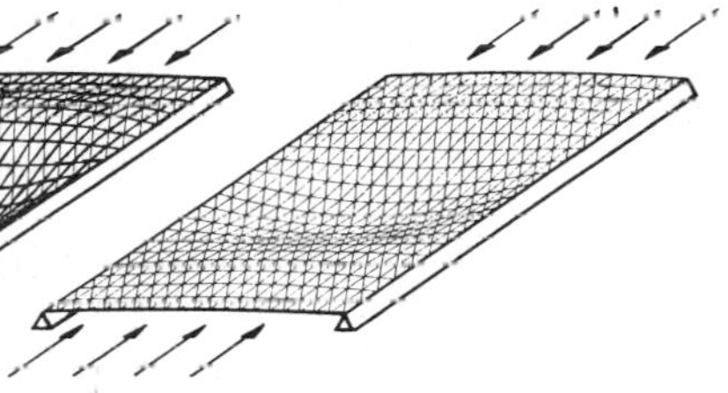

③ σ_{moy} = 12 hb, d = 1,5 mm

POST-BUCKLING

④ σ_{moy} = 3,5 hb, d = 0.35 mm

POST-BUCKLING

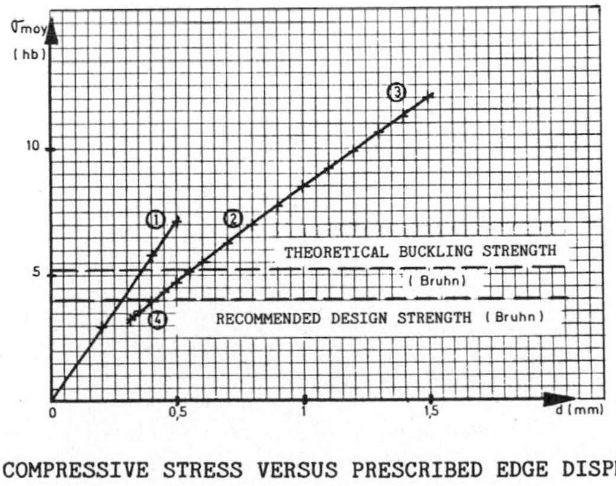

AVERAGE COMPRESSIVE STRESS VERSUS PRESCRIBED EDGE DISPLACEMENT

PL 6

POST-BUCKLING OF A CURVED SHEET PANEL UNDER AXIAL COMPRESSION

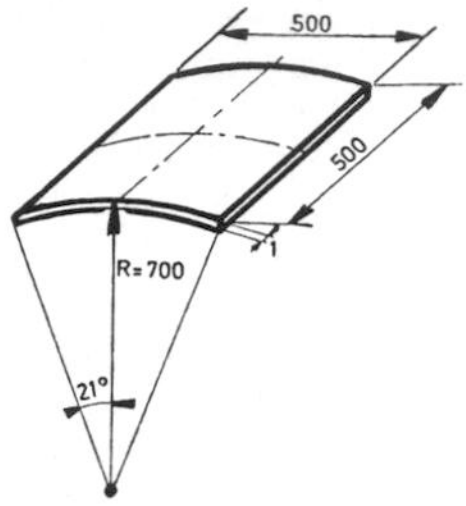

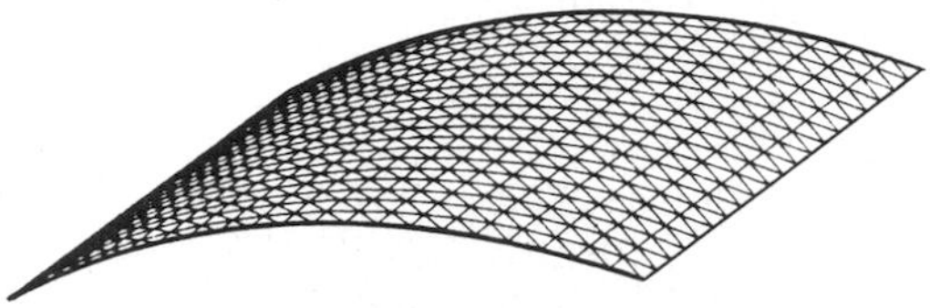

$\sigma_m = 4{,}05$ hb d = 0,42 mm

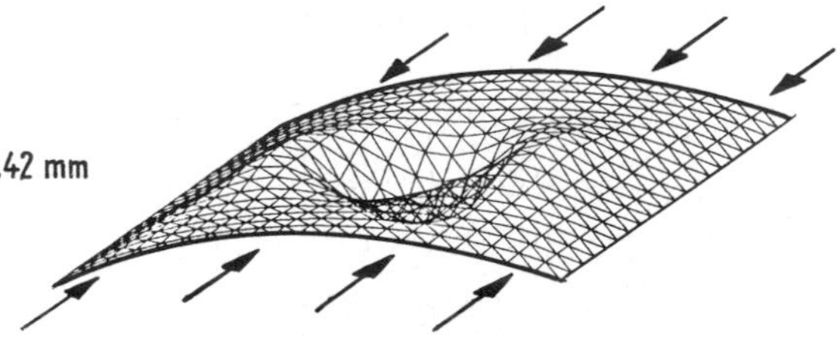

$\sigma_m = 3{,}9$ hb d = 0,73 mm

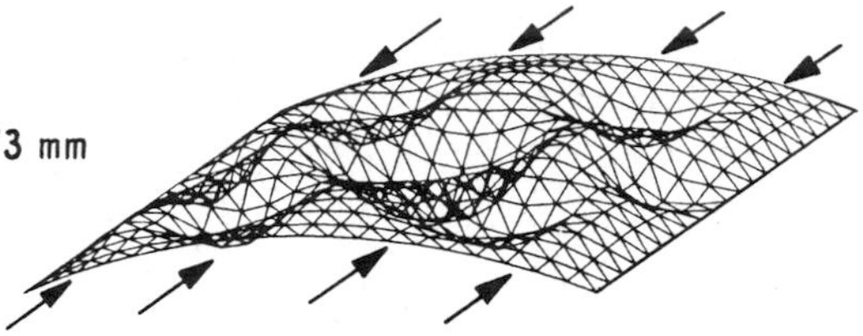

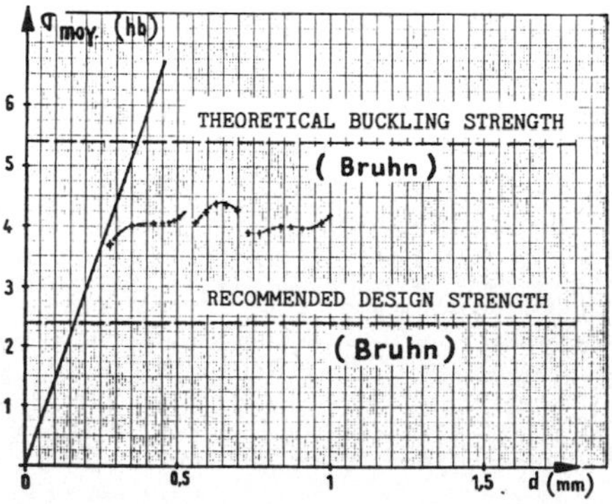

AVERAGE COMPRESSIVE STRESS VERSUS PRESCRIBED EDGE DISPLACEMENT

PL 7

POST-BUCKLING OF A CURVED SHEET PANEL

UNDER AXIAL COMPRESSION

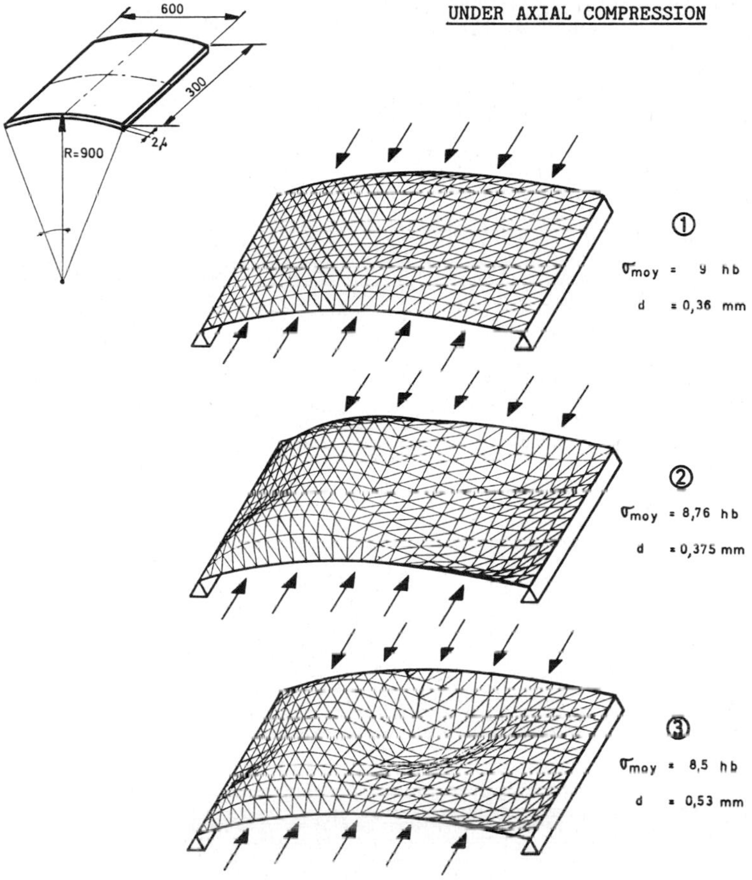

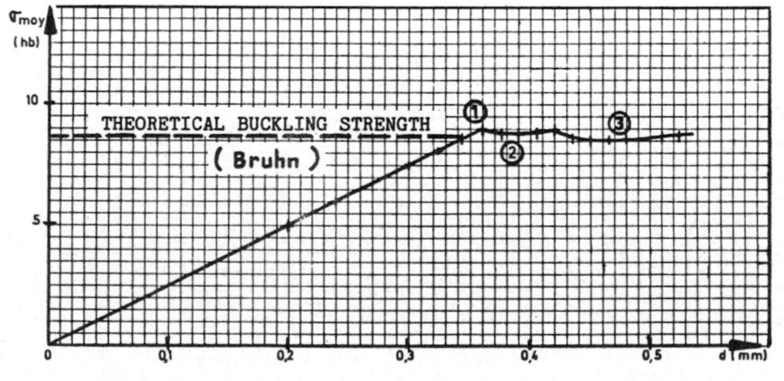

AVERAGE COMPRESSIVE STRESS VERSUS PRESCRIBED EDGE DISPLACEMENT

PL 8

POST-BUCKLING OF A SPAR WEB

UNDER SHEAR AND BENDING

OUT-OF-PLANE DEFLECTION CONTOUR LINES

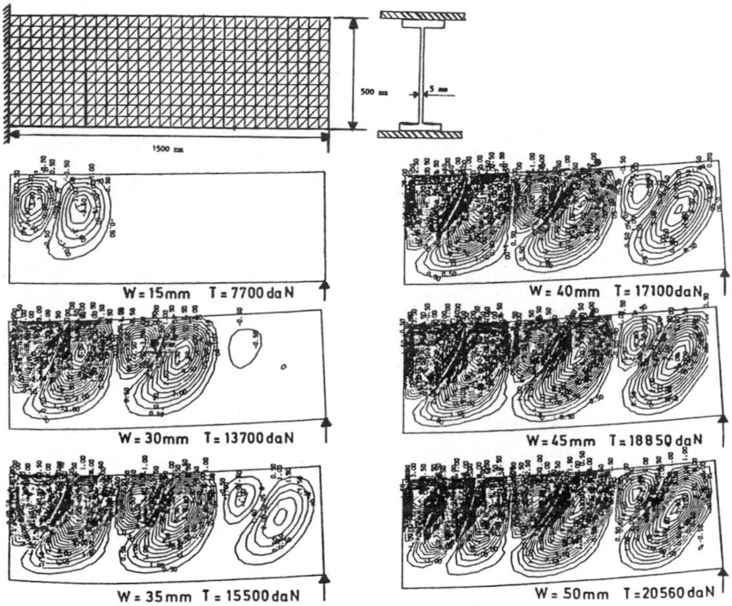

SPAR WEB UNDER SHEAR AND BENDING - SHEARING LOAD VERSUS DEFLECTION

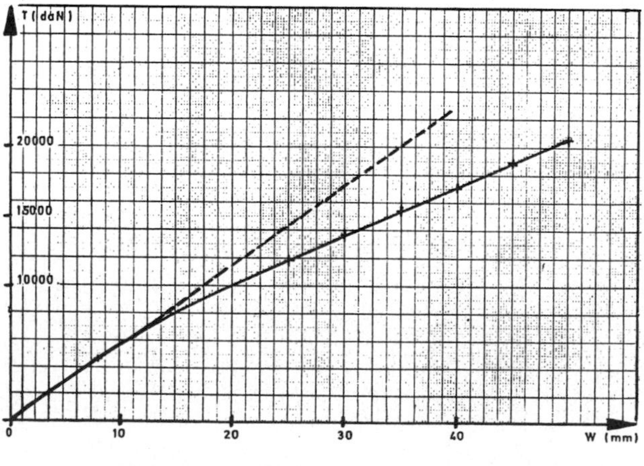

PL 9

POST-BUCKLING OF A SPAR WEB

UNDER SHEAR BENDING AND COMPRESSION

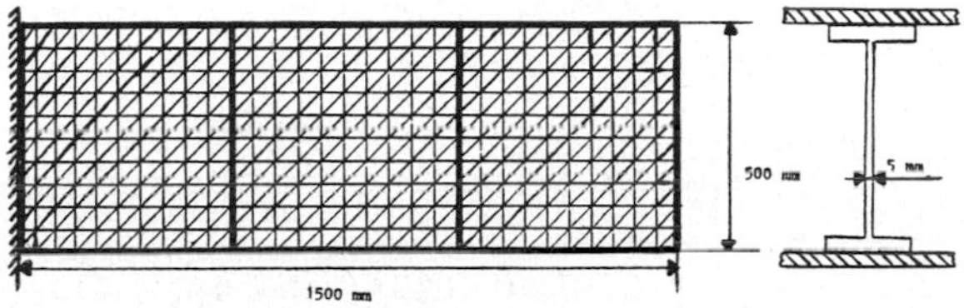

OUT-OF-PLANE DEFLECTION

CONTOUR LINES

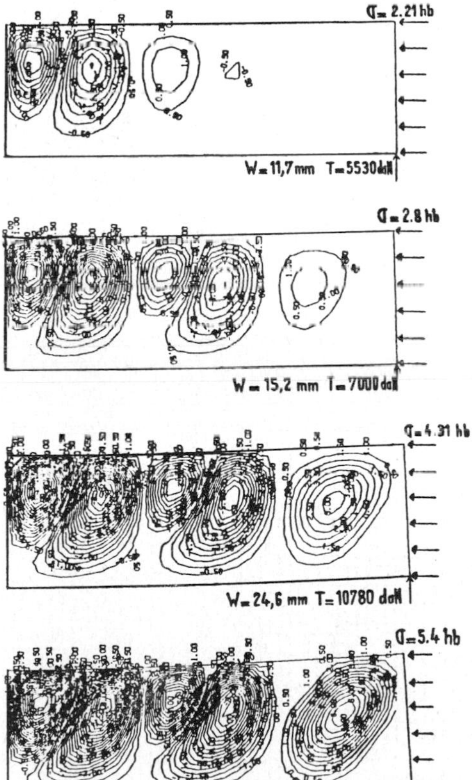

PL 10

BUCKLING OF A WING BOX OF THE SUPER MIRAGE 4000

MESH OF HALF THE SPECIMEN TEST

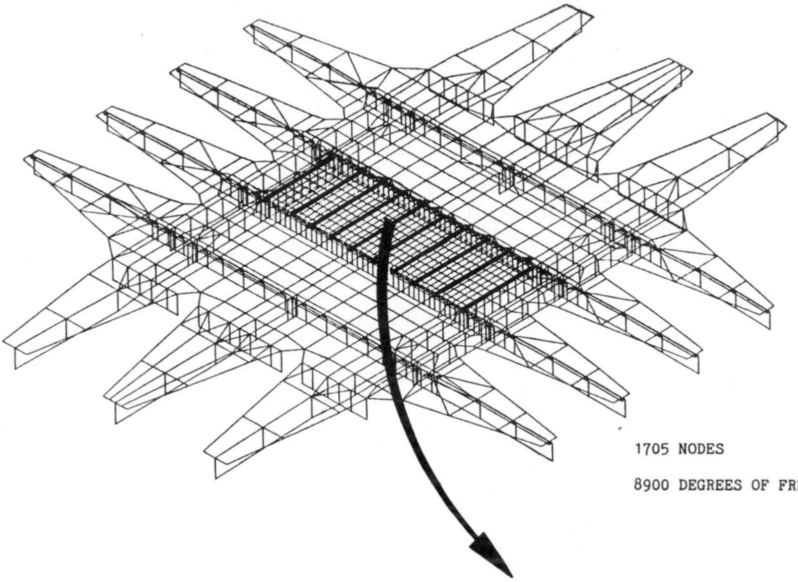

1705 NODES

8900 DEGREES OF FREEDOM

WAVES IN THE MIDDLE PANEL DERIVED FROM COMPUTATION, AT LOAD LEVEL 1.6

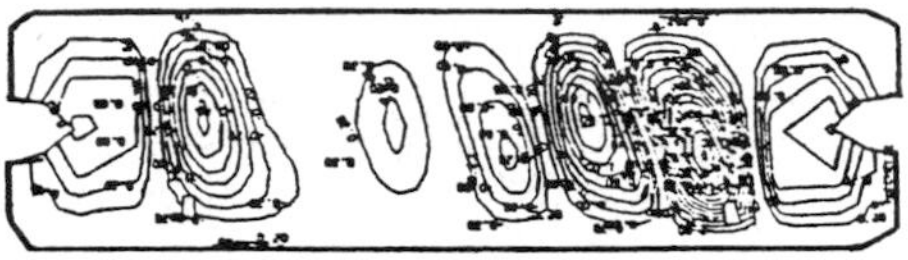

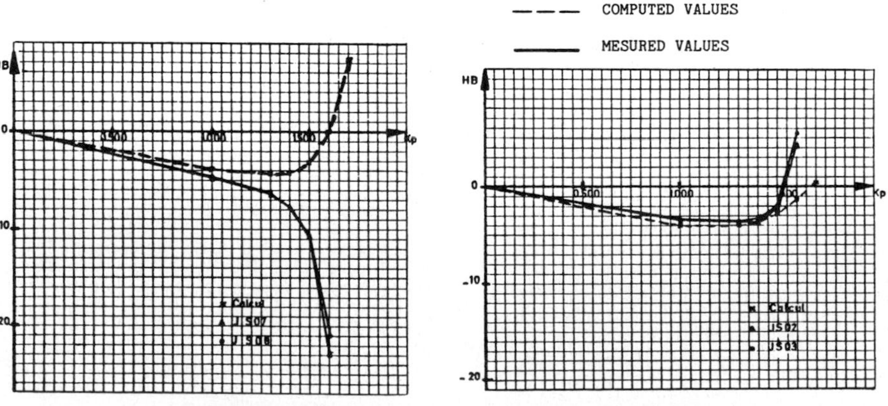

COMPARISON BETWEEN THE TEST AND THE COMPUTATIONS

STRAIN-GAGES LOCATED AT THE TOP AND THE BOTTOM OF THE STIFFENERS

ANNEXE

CALCUL DES COEFFICIENTS DU POLYNOME DERIVEE DE L'ENERGIE $\dfrac{\partial W_{TOT}(X+\rho V)}{\partial \rho}$

- Le tenseur des déformations de Green pour des déplacements $X+\rho V$ en un point M d'un élément se met sous la forme d'un polynôme du second degré :

$$[\varepsilon] = \frac{1}{2}\left(\left[\frac{\partial(X+\rho V)}{\partial M}\right] + \left[\frac{\partial(X+\rho V)}{\partial M}\right]^t + \left[\frac{\partial(X+\rho V)}{\partial M}\right]^t \left[\frac{\partial(X+\rho V)}{\partial M}\right]\right)$$
$$= [\varepsilon_0] + \rho[\varepsilon_1] + \rho^2[\varepsilon_2]$$

avec:

$$\begin{cases} [\varepsilon_0] = \dfrac{1}{2}\left(\left[\dfrac{\partial X}{\partial M}\right] + \left[\dfrac{\partial X}{\partial M}\right]^t + \left[\dfrac{\partial X}{\partial M}\right]^t\left[\dfrac{\partial X}{\partial M}\right]\right) \\ [\varepsilon_1] = \dfrac{1}{2}\left(\left[\dfrac{\partial V}{\partial M}\right] + \left[\dfrac{\partial V}{\partial M}\right]^t + \left[\dfrac{\partial X}{\partial M}\right]^t\left[\dfrac{\partial V}{\partial M}\right] + \left[\dfrac{\partial V}{\partial M}\right]^t\left[\dfrac{\partial X}{\partial M}\right]\right) \\ [\varepsilon_2] = \dfrac{1}{2}\left[\dfrac{\partial V}{\partial M}\right]^t\left[\dfrac{\partial V}{\partial M}\right] \end{cases}$$

où pour un vecteur U, $\left[\dfrac{\partial U}{\partial M}\right]$ désigne la matrice $\left[\dfrac{\partial u_i}{\partial x_{oj}}\right]$.

- Pour chaque élément, la dérivée par rapport à ρ de l'énergie de déformation élastique $W = \int_{\mathcal{V}} \frac{1}{2}[\varepsilon]^t[H][\varepsilon]\,d\mathcal{V}$ est un polynôme du 3ème degré :

$$\frac{\partial W}{\partial \rho} = \int_{\mathcal{V}} \left([\varepsilon_0] + \rho[\varepsilon_1] + \rho^2[\varepsilon_2]\right)^t [H]\left([\varepsilon_1] + 2\rho[\varepsilon_2]\right) d\mathcal{V}$$
$$= a_0 + a_1\rho + a_2\rho^2 + a_3\rho^3$$

avec : $a_0 = \int_{\mathcal{V}}\left([\varepsilon_0]^t[H][\varepsilon_1]\right)d\mathcal{V}$, $a_1 = \int_{\mathcal{V}}\left(2[\varepsilon_0]^t[H][\varepsilon_2] + [\varepsilon_1]^t[H][\varepsilon_1]\right)d\mathcal{V}$

$a_2 = \int_{\mathcal{V}}\left(3[\varepsilon_1]^t[H][\varepsilon_2]\right)d\mathcal{V}$, $a_3 = \int_{\mathcal{V}}\left(2[\varepsilon_2]^t[H][\varepsilon_2]\right)d\mathcal{V}$

- Pour la structure toute entière et un chargement extérieur constant :

$$W_{TOT} = W_{INT} - F_{ex}^t\,(X+\rho V)$$

$$\frac{\partial W_{TOT}}{\partial \rho} = A_0 + A_1\rho + A_2\rho^2 + A_3\rho^3$$

avec :

$$A_0 = \sum_{\text{éléments}} a_0 - F_{ex}^t V\ ,\quad A_i = \sum_{\text{éléments}} a_i \ \text{pour}\ i = 1,2,3$$

Computing Methods in Applied Sciences and Engineering, VI
R. Glowinski and J.-L. Lions (Editors)
Elsevier Science Publishers B.V. (North-Holland)

SIMULATION NUMERIQUE DES CONTRAINTES AUX INTERFACES PROTHESE DE HANCHE-OS

M. BERNADOU (1) ; P. CHRISTEL (2) ; J.M. CROLET (1)

La pose de prothèses de hanches est une opération classique de nos jours. Cependant, à moyen et long terme, il est observé des descellements qui affectent principalement l'implant cotyloïdien et nécessitent généralement une réintervention d'urgence. Dans ce travail, nous proposons une méthode de simulation numérique permettant de calculer les contraintes principales dans les différents composants ainsi que les vecteurs de contraintes aux interfaces cupule-ciment et ciment-os. Les résultats obtenus permettent d'expliquer les phénomènes de descellements de manière très cohérente avec les observations cliniques.

INTRODUCTION

L'utilisation de la méthode des elements finis dans la résolution numérique de problèmes de biomécanique orthopédique a débuté voici une dizaine d'années. Parmi ces problèmes figurent la déformation des os, les fractures et leurs réductions, les articulations (hanche, genou) et les prothèses correspondantes.
Un bilan des résultats obtenus est dressé par HUISKES et CHAO [1983].

L'utilisation de ces méthodes de simulation numérique pour la résolution de tels problèmes est très attractive étant donné la grande difficulté, sinon la quasi-impossibilité, d'effectuer des mesures expérimentales "in situ".

Par la suite, nous allons nous intéresser aux problèmes liés à la pose de prothèses de hanches. Ces problèmes ont déjà fait l'objet d'un certain nombre de travaux, en particulier ceux de JACOB, HUGGLER, DIETSCH et SCHRIEBER [1976], GOEL, VALLIAPPAN et SVENSSON [1978], VASU, CARTER et HARRIS [1982], PEDERSEN, CROWNINSHIELD, BRAND et JOHNSTON [1982].

Dans ce travail nous proposons une méthode de simulation numérique bi-dimensionnelle par éléments finis permettant d'approcher :

---

(1) INRIA, Domaine de Voluceau, B.P. 105, Rocquencourt 78153 LE CHESNAY Cedex

(2) Laboratoire de Recherches Orthopédiques, Faculté de Médecine Lariboisière St Louis, 10 avenue de Verdun, 75010 PARIS

i) le champ des contraintes principales dans les différentes parties d'une coupe plane d'une hanche appareillée (cupule et os iliaque) ;

ii) les vecteurs de contraintes aux interfaces des différents matériaux.

L'originalité de cette étude repose surtout sur cette seconde approche ii) et sur l'utilisation que l'on fait de ces résultats pour expliquer la plupart des phénomènes de descellements de prothèses de hanche (partie cotyloïdienne).

En dépit des approximations assez grossières qui ont été faites (modèle bidimensionnel, os élastiques homogènes et isotropes par morceaux, encastrement de l'os iliaque...) et de la relative simplicité de la méthode proposée, il est assez remarquable que les résultats numériques trouvés reflètent harmonieusement les observations cliniques.

Dans cette rédaction nous rappelons tout d'abord au paragraphe II l'essentiel du problème médical. La méthode de simulation numérique est détaillée dans le paragraphe III, puis nous analysons les résultats obtenus dans le paragraphe IV. Enfin, nous énonçons quelques problèmes ouverts dans le paragraphe V.

Remerciements : Cette étude a largement bénéficié du soutien apporté par les chercheurs des projets MENUSIN et MODULEF de l'INRIA, d'une part, du Laboratoire de Recherches Orthopédiques de l'Hôpital Saint-Louis, d'autre part. En particulier, nous tenons à remercier G. DUVAUT pour les conseils qu'il nous a donnés durant l'avancement de ce travail.

II - LE PROBLEME MEDICAL

II.1. Présentation sommaire de la hanche humaine

La hanche (Fig. II.Ia) est l'articulation proximale du membre inférieur. Ses trois axes de rotation lui permettent d'orienter ce membre dans toutes les directions. Les surfaces articulaires comportent :

* la tête fémorale constituée par les deux tiers d'une sphère de 40 à 50 mm de diamètre ;
* la cavité cotyloïde de forme semi-sphérique qui reçoit la tête fémorale.

II.2. Prothèses de hanche

Suite à diverses déficiences, il est aujourd'hui classique de remplacer l'articulation de la hanche par une prothèse totale de hanche composée (Fig. II.1b)

* d'un implant fémoral : la tête de l'implant remplace la tête du fémur et la queue sert à l'ancrage dans l'os.
* d'un implant cotyloïdien (appelé cupule) : il est attaché au bassin et il est destiné à recevoir la tête de l'implant fémoral.

Il existe essentiellement deux techniques de poses :

i) Prothèses "cimentées" : Le ciment acrylique - polymétacrylate de méthyl - polymérise "in situ". Il est introduit en phase pâteuse, poussé en force dans les infractuosités de l'os spongieux et dans les trous spécialement creusés par le chirurgien dans le cotyle osseux. Les composants prothétiques sont alors positionnés dans le ciment pâteux. Le durcissement demande 5 à 10 mn et s'accompagne d'un dégagement de chaleur : de 45 à 92° C selon les conditions locales, la masse de ciment et la température ambiante.

ii) Prothèses sans ciment (ancrage biologique) : Dans le cas où le but est d'obtenir une repousse osseuse au contact de la prothèse, la fixant ainsi sans l'intermédiaire de ciment, il faut réunir plusieurs conditions :

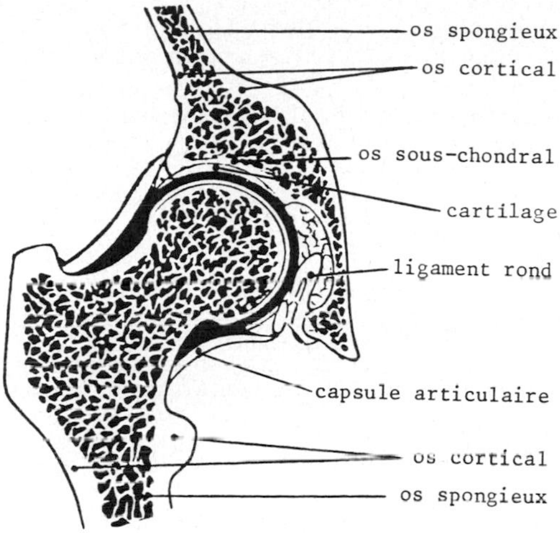

1a : hanche normale

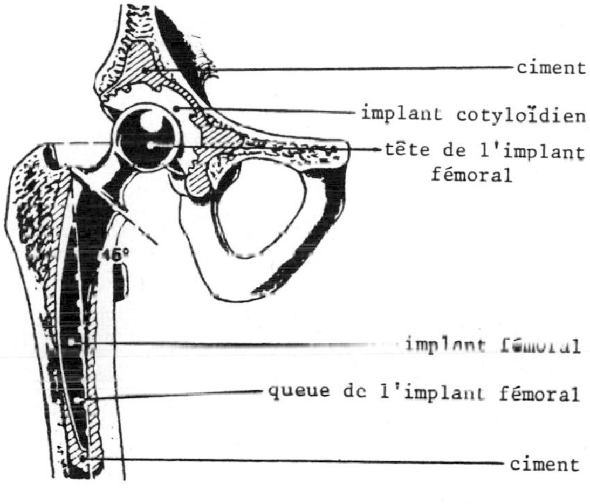

1b : hanche prothésée

Fig. II.1 : Coupes verticales d'une hanche humaine, normale et prothésée

. créer sur la prothèse des irrégularités, des pores ou des dessins permettant un meilleur ancrage direct.

. réaliser un ajustage serré de la prothèse et maintenir un os de surface vivant.

. ne pas avoir de mouvements entre l'implant et l'os au début de la période post-opératoire afin que l'os puisse "envahir par repousse" les pores de l'implant.

La pose sans ciment nécessite pour réussir une bonne repousse osseuse et une longue période (2 à 3 mois) de non-appui du pied au sol. C'est pourquoi il reste de nombreuses indications de prothèses totales cimentées : âge avancé, morphologie articulaire et/ou mauvaise qualité du support osseux du patient.

Par la suite nous étudions la prothèse OSTEAL-CERAVER qui se pose avec ciment. Il s'agit d'un système modulaire composé de plusieurs pièces interchangeables :

*) un implant fémoral sans tête en alliage de titane ;

**)deux types de têtes fémorales s'emboitant sur l'implant fémoral, l'une en alumine, l'autre en acier ;

***) deux types d'implants cotyloïdiens, l'un en alumine, l'autre en polyéthylène.

II.3. Accidents observés sur les implants cotyloïdiens

Après quelques années il apparait parfois un certain nombre de complications : usure, descellement avec ou sans migration de la cupule. La nature de ces complications est liée au type de matériau utilisé :

i) Migration et descellement des cupules en polyéthylène :
CHARNLEY [1979] a montré que les prothèses totales de hanche qui avaient été posées 12 à 15 ans auparavant présentaient dans 65 % des cas un liseré radio-transparent partiel, dans 14 % des cas un liseré radio-transparent total de 2 mm d'épaisseur à l'interface os-ciment et s'accompagnaient dans 11 % des cas d'une migration progressive de la pièce cotyloïdienne.

ii) Descellement des cupules en alumine :
Les descellements se produisent à l'interface alumine-ciment. Ils ont des conséquences graves puisqu'ils nécessitent une réintervention immédiate.

III - LA SIMULATION NUMERIQUE

III.1. Orientation

Dans ce paragraphe nous proposons une méthode de simulation numérique permettant de déterminer les contraintes principales dans la cupule (implant cotyloïdien), dans le ciment et dans l'os spongieux entourant la cupule, ainsi que les vecteurs de contraintes aux interfaces implant-ciment et ciment-os.

La connaissance de ces diverses contraintes nous permettra dans le paragraphe IV d'analyser, au moins du point de vue mécanique, certains des phénomènes observés lors de descellements d'implants cotyloïdiens ainsi que certains aléas de la pose de la cupule (imprécision dans l'orientation, épaisseur de ciment trop importante...).

III.2. Formulations continues et discrètes du problème de l'élasticité linéaire

Le problème continu (voir DUVAUT-LIONS [1972]) :

Considérons un problème d'élasticité posé dans un ouvert Ω. Les champs de déplacements $\underset{\sim}{u} = (u_i)$ et de contraintes $\underset{\sim}{\sigma} = (\sigma_{ij})$ satisfont (les indices latins prennent ici les valeurs 1 ou 2 et l'on utilise la convention de sommation sur les indices répétés) :

(2.1) $$\sigma_{ij,j} + f_i = 0 \text{ dans } \Omega$$

(2.2) $$\sigma_{ij}(\underset{\sim}{u}) = a_{ijkl}\,\varepsilon_{kl}(\underset{\sim}{u}) \quad , \quad \varepsilon_{ij}(\underset{\sim}{u}) = \frac{1}{2}(u_{i,j} + u_{j,i})$$

+ conditions aux limites (C.L.).

Ce problème admet la formulation variationnelle suivante : Trouver $\underset{\sim}{u} \in \underset{\sim}{V}$ tel que

(2.3) $$a(\underset{\sim}{u},\underset{\sim}{v}) = g(\underset{\sim}{v}) \quad , \quad \forall \underset{\sim}{v} \in \underset{\sim}{V} \, ,$$

(2.4) $a(\underset{\sim}{u},\underset{\sim}{v}) = \int_\Omega a_{ijkl}\ \varepsilon_{kl}(\underset{\sim}{u})\ \varepsilon_{ij}(\underset{\sim}{v})\ dx$

(2.5) $g(\underset{\sim}{v})$ = énergie potentielle des forces extérieures

où $\underset{\sim}{V}$ est l'espace des déplacements admissibles, i.e.,

(2.6) $\underset{\sim}{V} = \{\underset{\sim}{v} = (v_1,v_2)\ ;\ v_i \in H^1(\Omega) + C.L.\}$.

Pour des données assez régulières le problème (2.3) admet une solution et une seule.

Le problème approché

Pour simplifier, nous supposons i) que le domaine Ω est polygonal ce qui permet de le trianguler exactement, puis de construire un espace de dimension finie $\underset{\sim}{X}_h$ associé à des éléments finis de type P_1 de Lagrange ; ii) que la prise en compte des conditions aux limites permet de définir un sous-espace $\underset{\sim}{V}_h \subset \underset{\sim}{X}_h$ tel que $\underset{\sim}{V}_h \in \underset{\sim}{V}$ (approximation interne).

Dès lors, le problème discret s'énonce : Trouver $\underset{\sim}{u}_h \in \underset{\sim}{V}_h$ tel que

(2.7) $a(\underset{\sim}{u}_h,\underset{\sim}{v}_h) = g(\underset{\sim}{v}_h)\quad ,\quad \forall \underset{\sim}{v}_h \in \underset{\sim}{V}_h$.

Comme l'approximation est interne, ce problème admet une solution unique $\underset{\sim}{u}_h$ vérifiant (CIARLET [1978]) :

(2.8) $\|\underset{\sim}{u} - \underset{\sim}{u}_h\|_{(H^1(\Omega))^3} = O(h)$.

III.3. Calcul des contraintes et des vecteurs de contraintes aux interfaces

L'approximation du champ de déplacement n'est qu'un intermédiaire commode. Pour comprendre les phénomènes de descellements, il importe de connaitre les valeurs des contraintes "internes" et, surtout, de connaitre les surcontraintes - forces de tractions ou de cisaillement - apparaissant à des interfaces de matériaux de caractéristiques mécaniques parfois très différentes. Citons par exemple les interfaces "alumine-ciment" et "alumine-os".

Calcul des contraintes "internes" :
Connaissant le champ des déplacements approchés $\underset{\sim}{u}_h$ les relations (2.2) nous permettent de trouver une approximation du champ des contraintes dans tout le domaine Ω.

Calcul des vecteurs de contraintes aux interfaces :
Le champ approché des contraintes "internes" est en général discontinu d'un triangle au suivant ce qui constitue un grave défaut sur le plan de la formulation et de l'interprétation mécanique. D'où l'idée de calculer des vecteurs de contraintes aux interfaces qui respectent le principe mécanique de l'équilibre.

Principe de la méthode (DUVAUT et PISTRE [1982]) :
Considérons une partition de l'ouvert Ω en deux sous-ensembles connexes ω_1 et ω_2 suffisamment réguliers. Nous supposons que la solution $\underset{\sim}{u}$ du problème (2.3) est assez régulière de telle sorte qu'il est loisible de multiplier l'équation (2.1) par une fonction $\underset{\sim}{v} = (v_i)$ assez régulière, puis d'intégrer par parties sur le domaine ω_1. Il vient

(3.1) $\int_{\partial\omega_1\cap\partial\omega_2} F_{i1} v_i d\gamma = \int_{\omega_1} (\sigma_{ij}\ \varepsilon_{ij}(\underset{\sim}{v}) - f_i v_i) dx - \int_{\partial\omega_1\cap\partial\Omega} \sigma_{ij} v_i n_{j1} d\gamma$

où

(3.2) $\underset{\sim}{F}_1 = (F_{i1}) = (\sigma_{ij} n_{j1})$

désigne l'action exercée par le milieu ω_2 sur le milieu ω_1 et où $\underset{\sim}{n}_1 = (n_{j1})$ désigne la normale unitaire extérieure au domaine ω_1.

On définit alors l'approximation discrète F^h_{i1} des composantes F_{i1} comme suit :

i) on suppose que la frontière $\partial\omega_1 \cap \partial\omega_2$ est une ligne polygonale, elle-même réunion de côtés de la triangulation de l'ouvert Ω ;

ii) on désigne par $\partial \underset{\sim}{X}_h$ l'espace discret de dimension un tel que $\partial \underset{\sim}{X}_h$ = trace de $\underset{\sim}{X}_h$ sur $\partial\omega_1 \cap \partial\omega_2$;

iii) par analogie avec la relation (3.1), les approximations F^h_{i1} des composantes F_{i1} sont alors définies par : Trouver $(F^h_{i1}) \in \partial \underset{\sim}{X}_h$ tel que, pour tout $\underset{\sim}{v} \in \underset{\sim}{X}_h$, on ait :

$$(3.3) \quad \left\{ \begin{aligned} \int_{\partial\omega_1 \cap \partial\omega_2} F^h_{i1} v_i \, d\gamma = & \int_{\omega_1} (\sigma_{ij}(\underset{\sim}{u}_h)\, \varepsilon_{ij}(\underset{\sim}{v}) - f_i v_i)\, dx \\ & - \int_{\partial\omega_1 \cap \partial\Omega} \sigma_{ij}(\underset{\sim}{u}_h)\, v_i n_{j1}\, d\gamma , \end{aligned} \right.$$

où $\underset{\sim}{u}_h$ est la solution unique du problème (2.7).

Théorème III.1. : L'équation (3.3) a une solution unique

Démonstration : Soit a_m, m = 1,...,M, les noeuds de la triangulation situés sur la ligne polygonale $\partial\omega_1 \cap \partial\omega_2$. L'équation (3.3) est vérifiée pour tout $\underset{\sim}{v} \in \underset{\sim}{X}_h$, donc en particulier pour toute fonction $\underset{\sim}{v}$ dont les seuls degrés de liberté non nuls se trouvent sur $\partial\omega_1 \cap \partial\omega_2$. D'où

$$(3.4) \quad \left\{ \begin{aligned} & [F^h_{11}(a_1)\; F^h_{21}(a_1)\; F^h_{11}(a_2)\; F^h_{21}(a_2) \ldots F^h_{21}(a_M)]^t [v_1(a_1)\; v_2(a_1) \ldots v_2(a_M)] \\ & \quad = [G^h_{11}(a_1) \ldots G^h_{21}(a_M)]^t [v_1(a_1) \ldots v_2(a_M)] \end{aligned} \right.$$

où les différents coefficients $F^h_{i1}(a_m), v_i(a_m), i=1,2$; $m=1,\ldots,M$ désignent les degrés de liberté des champs de vecteurs $\underset{\sim}{F}^h_1$ et $\underset{\sim}{v}$. Par ailleurs les coefficients $G^h_{i1}(a_m), i=1,2$; $m=1,\ldots,M$ sont obtenus immédiatement à partir des matrices de rigidité et des seconds membres élémentaires calculés dans la phase d'assemblage, d'une part, et à l'aide du champ approché $\underset{\sim}{u}_h$ que l'on vient de déterminer, d'autre part.
Comme la relation (3.4) doit être vérifiée pour toute fonction $\underset{\sim}{v} \in \underset{\sim}{X}_h$, il vient directement $F^h_{i1}(a_m) = G^h_{i1}(a_m)$, i = 1,2 et m = 1,...M.

□

La Fig. III.1 présente une triangulation de Ω s'appuyant sur la ligne polygonale $\partial\omega_1 \cap \partial\omega_2$. On y indique les degrés de liberté des composantes F^h_{i1}.

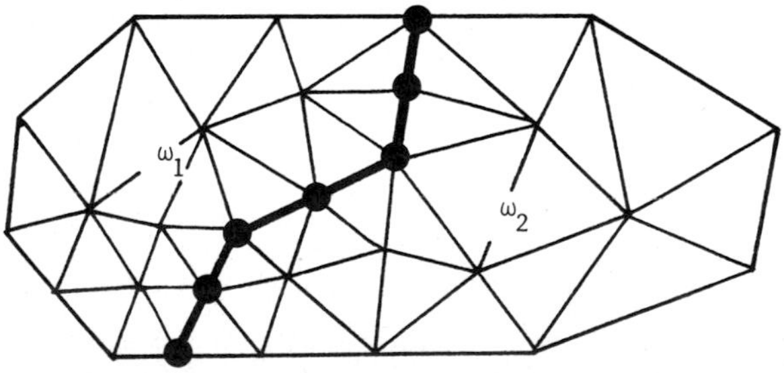

Fig. III.1 :
Les degrés de liberté F^h_{i1} pour une approximation P_1

D'une manière entièrement analogue à (3.2), on définirait l'action $\underset{\sim}{F}_2 = (F_{i2}) = (\sigma_{ij} n_{j2})$ exercée par le milieu ω_1 sur le milieu ω_2, puis, par analogie avec (3.3), les approximations F^h_{i2} des composantes F_{i2} : Trouver $(F^h_{i2}) \in \partial \underset{\sim}{X}_h$ tel que pour tout $\underset{\sim}{v} \in \underset{\sim}{X}_h$, on ait :

$$(3.5)\quad \begin{cases} \displaystyle\int_{\partial\omega_1\cap\partial\omega_2} F^h_{i2} v_i \, d\gamma = \int_{\omega_2} [\sigma_{ij}(\underset{\sim}{u}_h)\, \varepsilon_{ij}(\underset{\sim}{v}) - f_i v_i]\, dx \\ \displaystyle\qquad - \int_{\partial\omega_2\cap\partial\Omega} \sigma_{ij}(\underset{\sim}{u}_h)\, v_i n_{j2}\, d\gamma \,. \end{cases}$$

En ajoutant les relations (3.3) et (3.5) il vient grâce à (2.1) et à $\underset{\sim}{n}_1 + \underset{\sim}{n}_2 = \underset{\sim}{0}$:

$$\int_{\partial\omega_1\cap\partial\omega_2} (F^h_{i1} + F^h_{i2}) v_i \, d\gamma = 0, \forall \underset{\sim}{v} \in \underset{\sim}{X}_h$$

soit

$$F^h_{i1} + F^h_{i2} = 0$$

ce qui est le résultat cherché : équilibre des actions de ω_2 sur ω_1 et de ω_1 sur ω_2, ceci pour les approximations discrètes.

Il convient de souligner que

(i) les champs de vecteurs de contraintes approchés sont continus le long de $\partial\omega_1\cap\partial\omega_2$ (par construction) ;

(ii) ce type d'approche peut être étendu à d'autres types d'espaces d'éléments finis et au cas tridimensionnel.

Interprètation mécanique :
Chacun des vecteurs $\underset{\sim}{F}_1$ peut être décomposé en une composante normale et une composante tangentielle

$$\underset{\sim}{F}_1 = \underset{\sim}{F}_{N1} + \underset{\sim}{F}_{T1}$$

où $\underset{\sim}{F}_{T1}$ est une force de cisaillement et $\underset{\sim}{F}_{N1}$ est $\begin{cases} \text{une pression si } \underset{\sim}{F}_{N1}\cdot\underset{\sim}{n}_1 < 0 \,, \\ \text{une traction si } \underset{\sim}{F}_{N1}\cdot\underset{\sim}{n}_1 > 0 \,. \end{cases}$

Cette analyse, menée en parallèle avec des résultats expérimentaux devrait permettre de déterminer les seuils à partir desquels une traction entraine le décollement et un cisaillement entraîne un glissement.

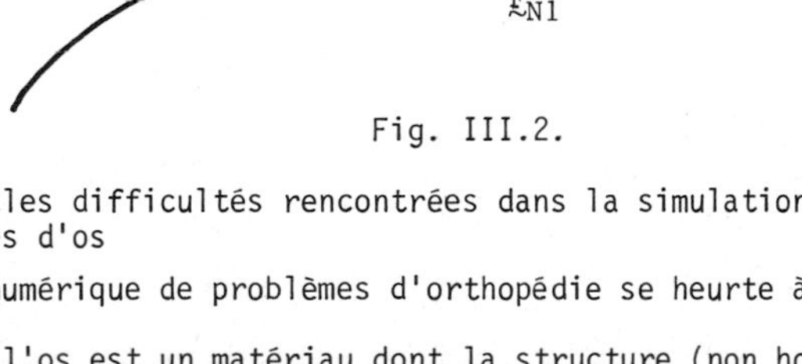

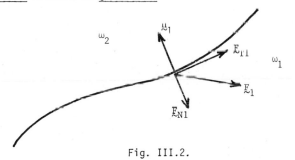

Fig. III.2.

III.4. Principales difficultés rencontrées dans la simulation numérique de problèmes d'os

La simulation numérique de problèmes d'orthopédie se heurte à de nombreuses difficultés :

i) l'os est un matériau dont la structure (non homogène, anisotrope) et le comportement ne sont pas encore totalement connus. Par la suite, en toute première approximation, on considèrera que l'os est élastique, homogène, isotrope par morceaux.

ii) l'os est un matériau vivant qui se remodèle suivant un processus encore mal identifié. Ceci rend très mal aisée la simulation numérique du comportement d'un os à long terme.

iii) on ne dispose pas de données expérimentales sur les conditions aux limites appliquées à la frontière de l'os iliaque. Il y a plus d'une vingtaine de muscles qui entourent la hanche en s'attachant sur le bassin et qui assurent la mobilité, en première approximation, on supposera que l'os iliaque est encastré à ses extrêmités et on négligera l'action des muscles.

iv) la forme particulièrement complexe de l'os iliaque et la nature des charges appliquées à la hanche inciteraient à utiliser un modèle tridimensionnel. Mais alors on serait conduit à des systèmes linéaires de très grandes dimensions ce qui ne permettrait pas d'utiliser des maillages fins dans les zones critiques.

Pour ces différentes raisons nous nous limitons dans cette étude à la considération d'un modèle bidimensionnel, ce qui parait être suffisant au vu des résultats obtenus.

III.5. Les problèmes considérés

Dans ce travail, nous proposons une méthode de simulation numérique permettant de calculer les contraintes principales dans la partie cotyloïdienne de la hanche normale et dans les cupules prothétiques correspondantes, ainsi que les vecteurs de contraintes aux interfaces. Les données des problèmes considérés sont les suivantes :

III.5.1. Description géométrique

* Forme générale : La forme du domaine osseux a été obtenue d'après une radiographie d'une coupe vertico-frontale du bassin d'un homme de 72 ans. Bien que le plan vertical de la coupe ne corresponde pas au plan dans lequel se trouvent les charges maximales, il en est proche et il a en outre le mérite de contenir les extrêmités proximales et distales de l'aile iliaque les plus éloignées de la région cotyloïdienne. Ces régions extrêmales de l'os iliaque sont considérées comme encastrées ; on peut ainsi espérer que l'effet de cette condition sur la zone d'étude qui nous intéresse (la partie cotyloïdienne de la hanche, ou la cupule, et leur support osseux immédiat) est minimum.

* Domaine osseux : Le domaine osseux a été décomposé en plusieurs sous domaines (Fig. III.3) qui ont pu être observés sur la radiographie de la coupe osseuse. On distingue l'os cortical (en pointillé) et cinq domaines d'os spongieux de caractéristiques mécaniques différentes, chaque domaine étant considéré comme homogène et isotrope.

* Implant et ciment : Tous les implants cotyloïdiens simulés ont un diamètre intérieur de 32 mm et un diamètre extérieur de 50 mm ; ils ont un positionnement bien déterminé : la surface de base doit se trouver dans un plan orienté à 45° par rapport au plan horizontal d'un patient en position debout.

Pour positionner correctement l'implant, le chirurgien est fréquemment contraint d'enlever (par fraisage) la partie d'os sous-chondral située au niveau du toit du cotyle. Nous supposons dans le cas standard que cette région a été retirée. L'autre cas sera examiné comme variante.

Chaque implant est scellé dans l'aile iliaque grâce à du ciment acrylique. Dans le cas standard l'épaisseur de ciment est supposée constante (3 mm). Le cas d'épaisseurs variables sera également examiné.

III.5.2. Types d'implants retenus

Nous allons étudier la cupule en alumine, la cupule en polyéthylène, et la cupule en polyéthylène entourée de métal (chrome/cobalt ou alliage de titane).

III.5.3. Les charges et les conditions aux limites

Nous supposons que le modèle considéré est soumis à deux types de conditions. Ce

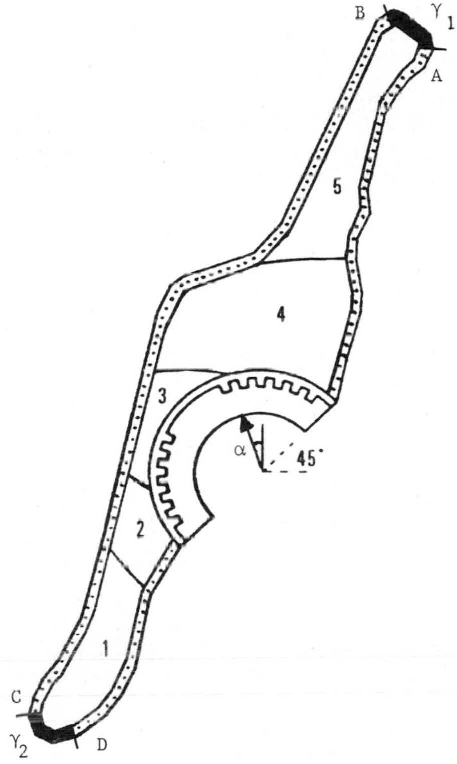

Figure III.3
Décomposition du domaine osseux prothésé

sont avec les notations de la Figure III.3 :

* déplacement nul sur les frontière $\gamma_1 = \widehat{AB}$ et $\gamma_2 = \widehat{CD}$

** une répartition de pressions appliquée à la surface interne de l'implant et dont la résultante est orientée de $\alpha = 16°$ par rapport à la verticale. Cette distribution de pressions, représentant l'action de l'implant fémoral sur l'implant cotyloïdien a été étudiée dans CROLET [1979]. Dans ce travail, il est montré que le contact entre les pièces prothétiques peut être approché par le contact d'une sphère de caractéristiques mécaniques (E,σ) et d'une demi-sphère de caractéristiques mécaniques (E',σ') ayant pour rayons respectifs R_1 et R_2 $(R_1 < R_2)$.

La coupole de contact (Fig. III.4) a un rayon de base r qui est solution de $f(x) = 0$ avec

$$f(x) = x\,R_2 - (R_2^2 - x^2)\log\left(\sqrt{\frac{R_2+x}{R_2-x}}\right) - \frac{2\,D\,F\,R_1}{3(R_2-R_1)}$$

$$* \; D = \frac{3}{4}\left(\frac{1-\sigma^2}{E} + \frac{1-\sigma'^2}{E'}\right)$$

* F est la charge appliquée ; nous prendrons la charge F égale à 3 fois la masse d'un corps moyen c'est à dire F = 1800 N.

La pression maximale est donnée par

$$P_{max} = \frac{3rQ}{\Pi D} \quad \text{où} \quad Q = \frac{1}{2}\,\frac{R_2 - R_1}{R_2 R_1} \; .$$

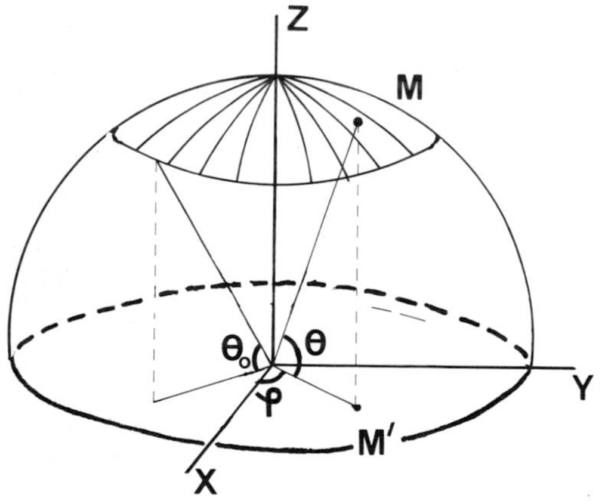

$M(r,\phi,\theta)$

$I = [\theta_o, \frac{\Pi}{2}]$

$J = [0,\theta_o]$

Fig. III.4 :
La coupole de contact

On a alors la distribution de pressions suivante qui dépend de la nature des matériaux en contact par le biais du terme D :

$$\begin{cases} \theta \in I & P(\theta) = P_{max}\sqrt{1 - \dfrac{R_2^2\cos^2\theta}{r^2}} \\ \theta \in J & P(\theta) = 0 \\ \theta = \theta_o & P(\theta_o) = 0 \end{cases}$$

Nous donnons dans le Tableau III.1 les valeurs numériques des caractéristiques mécaniques des matériaux étudiés.

	E(MPa) (Young)	σ (Poisson)
os spongieux I (dom. 1 et 5 Fig. III.3)	40	0,18
os spongieux II (dom. 2 et 4 Fig. III.3)	350	0,22
os spongieux III (dom. 3 Fig. III.3)	1000	0,247
os cortical	6200	0,326
os sous-chondral	6200	0,3

	E(MPa)	σ
cartilage	14	0,3
ciment	2 200	0,4
alumine	350 000	0,4
polyéthylène	600	0,4
chrome/cobalt ou titane/aluminium/ vanadium	11 000	0,4

Tableau III.1 :

Caractéristiques mécaniques des matériaux simulés

En pratique, nous étudions trois cas de charges directement liés à la nature du contact entre les composants de l'articulation :

* une distribution de pressions "naturelle" (hanche normale) : voir CROLET [1979].

** une distribution de pressions de type AA (Alumine - Alumine) correspondant au contact d'une cupule en alumine et d'un implant fémoral en alumine.

*** une distribution de pressions de type PM (Polyéthylène - Métal) correspondant à un contact cupule en polyéthylène et implant fémoral en métal (titane ou chrome-cobalt).

III.5.4. Techniques et aléas de poses

Pour une même cupule, les techniques et aléas de pose varient selon l'état du malade et le chirurgien qui opère. Pour chaque type de cupule envisagé nous avons donc considéré plusieurs configurations d'implantation :

i) Cas de référence : c'est le cas où il n'y a pas de lame dense sous chondrale et où la cupule est entourée d'une couche concentrique de ciment ;

ii) Changement d'orientation de la cupule : il peut se faire qu'à la pose, la cupule ne soit pas très bien orientée. Nous étudions ici les effets d'une telle variation pour des rotations (dans le sens positif) variant de 5° en 5° depuis la position de référence, i.e., 45° par rapport à l'horizontale, jusqu'à 65°.

iii) Présence de tout ou partie de la lame dense sous-chondrale : les cas où le chirurgien a pu garder la lame dense sous-chondrale sont simulés en présence

*) soit d'une couche concentrique de ciment ;

**) soit d'une couche de ciment au niveau du ligament rond (Fig. II.1) seulement et d'un ancrage direct cupule-os sous-chondral ailleurs.

iv) Surépaisseurs de ciment : deux cas sont examinés
*) couche épaisse de ciment au toit du cotyle et mince derrière le ligament rond ;

**) couche mince de ciment au toit du cotyle et épaisse derrière le ligament rond.

III.6. La mise en oeuvre numérique

La mise en oeuvre numérique des différents problèmes étudiés utilise le code MODULEF [1983].

III.6.1. La triangulation a été réalisée à l'aide d'un mailleur automatique à partir d'une digitalisation de la forme osseuse et de ses diverses régions telles qu'elles apparaissent sur la radiographie d'une coupe antéro-postérieure de l'os iliaque initial.

Ainsi le modèle représentant le cotyle normal comprend 799 triangles et 463 noeuds tandis que le maillage du cas appareillé comporte 1361 triangles et 748 noeuds (Figure III.5). Pour ce dernier cas nous avons considéré 184 triangles pour la couche concentrique de ciment (3 mm d'épaisseur) et 496 triangles pour la cupule.

Dans le cas "appareillé", nous avons retriangulé le domaine mais cela n'a entrainé que des variations faibles (3 à 4 %) des résultats numériques enregistrés dans la cupule et la région péricotyloïdienne.

III.6.2. L'espace d'éléments finis. A la triangulation définie en III.6.1, nous associons l'espace d'éléments finis classique construit à l'aide de triangles de type (1).

IV - ANALYSE DES RESULTATS OBTENUS

IV.1. Orientation

Dans ce paragraphe nous analysons les résultats obtenus pour les problèmes formulés dans la section III.5. Pour plus de détails, voir BERNADOU-CHRISTEL et CROLET [1983].

IV.2. Champs des contraintes principales dans les différentes zones de la hanche normale ou appareillée

IV.2.1. Cas du cotyle normal. On trouvera à la Fig. IV.1. ci- après un tracé du champ des contraintes principales obtenu dans le cas d'un cotyle normal. On constate que le cartilage transmet les efforts qui lui sont imposés et que l'os sous-chondral est partout fortement contraint. Le champ de contraintes obtenu dans l'os spongieux sert de référence. Il convient de le rapprocher de la distribution des travées osseuses obtenue par radiographie. Les travées osseuses et les contraintes principales ont, à quelques exceptions près, les mêmes orientations.

IV.2.2. Cas des implants en alumine et en polyéthylène. Nous avons rassemblé, sur la Figure IV.2, les tracés des champs de contraintes principales existant dans l'os spongieux obtenus pour des cupules en alumine (chargement du type AA) et en polyéthylène (chargement de type PM) dans le cas de référence explicité en III.5.4(i). Ces résultats montrent que :

i) dans les cas d'un implant cotyloïdien en alumine
*) la cupule a un comportement de poutre avec, aux abords du ciment, des valeurs de contraintes très élevées (130 à 240 MPa).

**) le ciment est faiblement contraint (2 à 5 MPa).

***) l'os spongieux connait une répartition de contraintes similaire à celle du cas normal en grandeur (3 à 5 MPa) et direction.

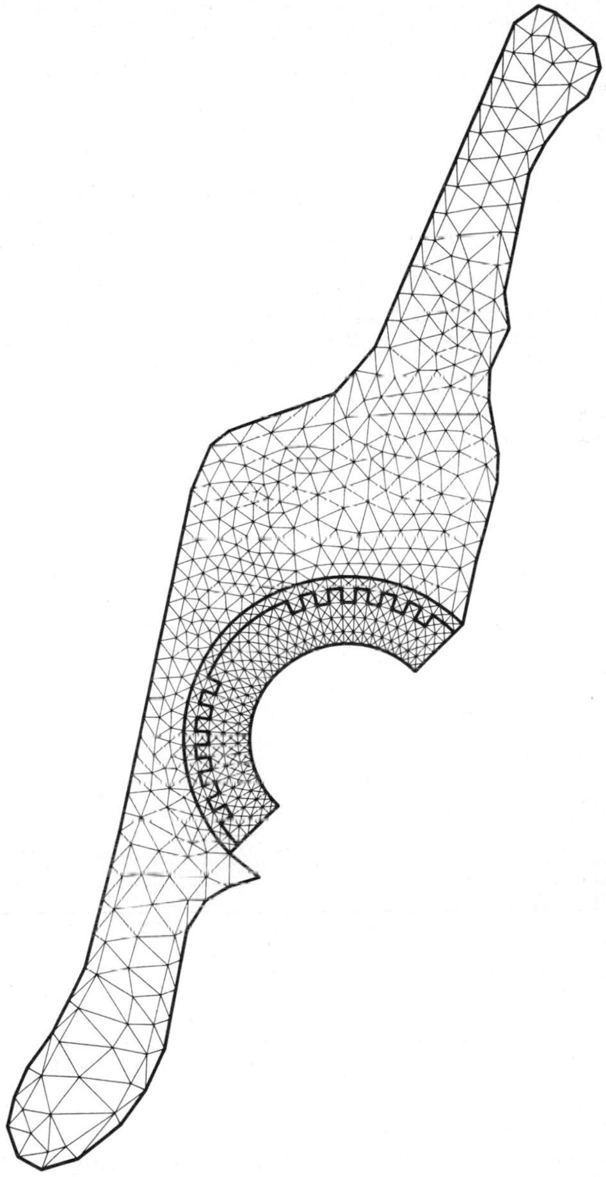

Figure III.5
Triangulation d'une coupe d'os iliaque appareillé

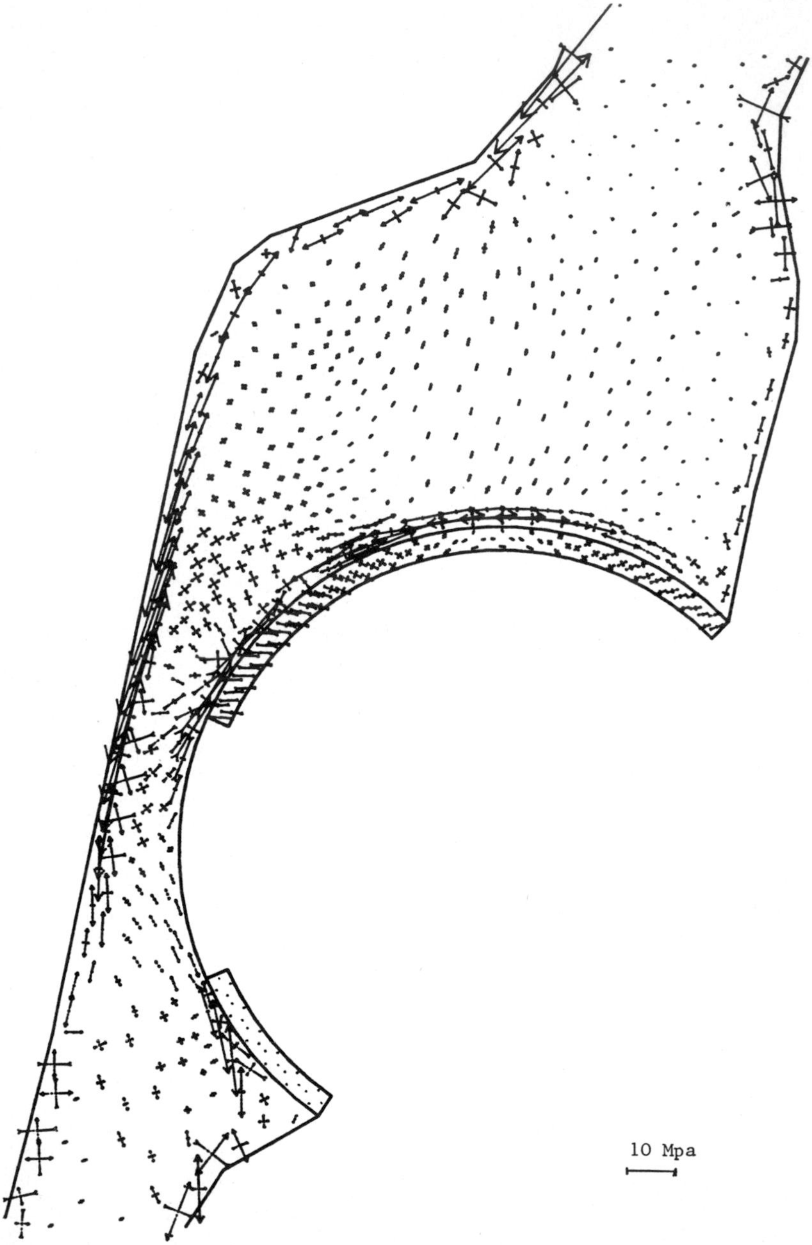

Figure IV.1
Contraintes principales dans le cas d'un cotyle normal

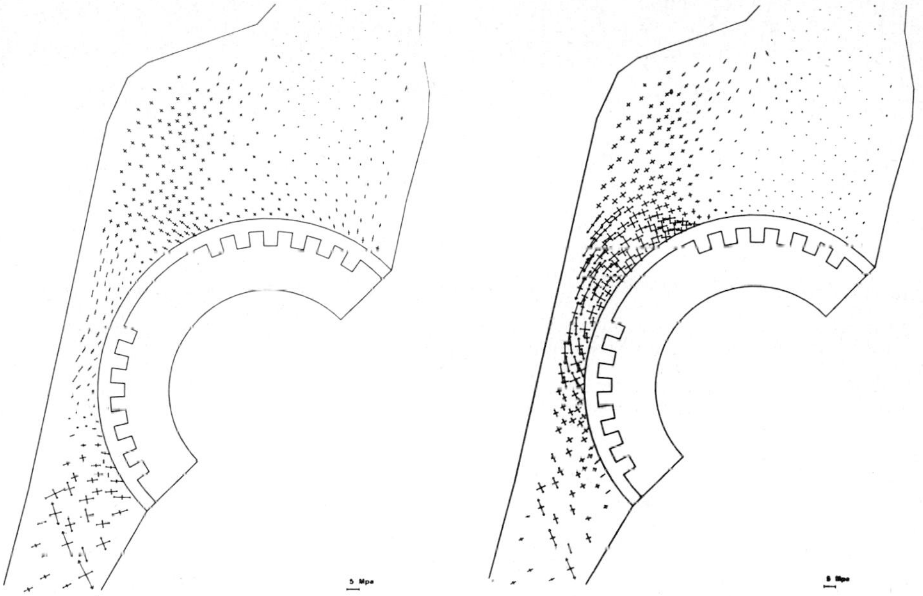

2a - Cupule en alumine

2b - Cupule en polyéthylène

Figure IV.2

Champs de contraintes principales dans l'os spongieux d'une hanche appareillée.

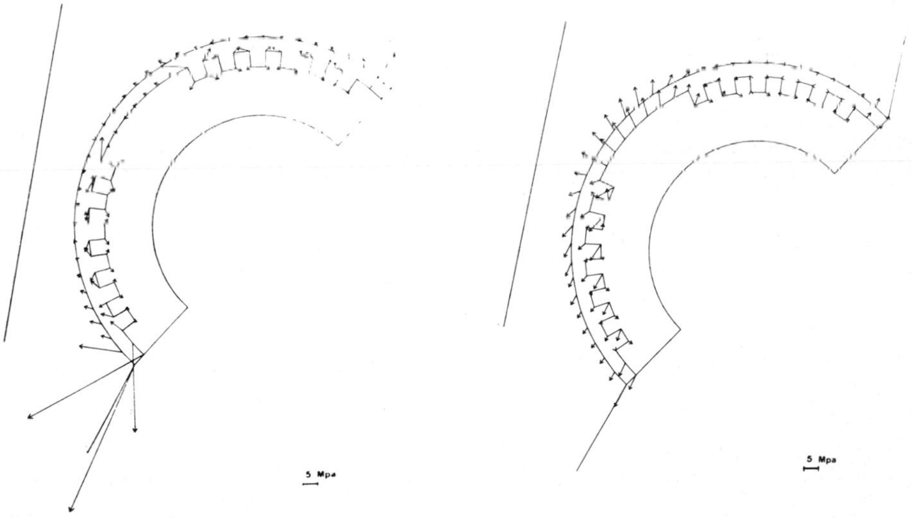

3a - Cupule en alumine

3b - Cupule en polyéthylène

Figure IV.3

Vecteurs contraintes aux interfaces pour des cupules en alumine et en polyéthylène.

ii) dans le cas d'un implant cotyloïdien en polyéthylène
*) la cupule est moyennement contrainte (30 MPa) sur la face chargée, 5 à 10 MPa sur la face cimentée.

**) le ciment est moyennement contraint (5 à 10 MPa).

***) l'os spongieux est beaucoup plus contraint (7 à 12 MPa) que dans le cas "normal". Les directions de contraintes se répartissent de manière concentrique autour de l'implant, les tractions étant supérieures aux compressions. Cette disposition favorise l'apparition de liserés à l'interface os-ciment et peut-être la migration de l'implant dans l'os spongieux (ce qui a été observé par J. CHARNLEY [1979]).

Un des palliatifs préconisés par des chirurgiens américains à ce problème de migration consiste à utiliser une cupule en polyéthylène sur sa face interne et en métal sur sa face externe. Le métal joue le rôle de poutre (en 2D) ce qui diminue les contraintes, en particulier les tractions, dans l'os spongieux. Le calcul montre également que les contraintes sont diminuées de moitié dans la partie en polyéthylène.

Ajoutons que dans tous les cas examinés, les résultats obtenus montrent que *) les déplacements sont faibles ; **) l'os cortical joue le rôle de "coquille" ; ***) l'ancrage de la cupule sur l'os cortical est un gage essentiel de stabilité.

IV.3. Analyse des vecteurs de contraintes aux interfaces

On aborde ici la partie la plus originale de ce travail. Le calcul des vecteurs contraintes aux interfaces des différents matériaux va nous permettre de donner une explication de nature mécanique aux différents modes de descellements observés cliniquement. Tous les résultats de cette section sont relatifs au cas de référence explicité en III.5.4(i).

IV.3.1. Les vecteurs de contraintes aux interfaces. Sur la Figure IV.3 nous donnons les tracés de ces vecteurs de contraintes aux interfaces cupule-ciment, ciment-os pour les cupules en alumine et en polyéthylène. Il convient de souligner que :

i) dans le cas d'un implant cotyloïdien en alumine
*) le ciment a tendance à se décoller de la cupule à chaque cran ;

**) au pôle géométrique de la cupule il n'apparait que des forces de cisaillement ;

***) à aucun endroit l'alumine n'exerce une "poussée franche" sur le ciment ;

****) de même les efforts exercés par le ciment sur l'os sont des efforts de cisaillement pour l'essentiel, avec une petite composante de pression par endroit. Une exception cependant : une forte pression sur l'os cortical favorisant fort heureusement l'ancrage de la cupule.

ii) dans le cas d'un implant cotyloïdien en polyéthylène
les forces exercées par la cupule en polyéthylène sur le ciment sont essentiellement des forces de compression ; la composante tangentielle de cisaillement est faible ; l'intensité de ces forces prend la valeur minima sur une zone importante couvrant largement le toit du cotyle.

iii) dans le cas d'un implant en polyéthylène serti de métal (non illustré), on observe la même répartition de contraintes d'interfaces pour chacune des trois interfaces polyéthylène-métal, métal-ciment, ciment-os. Il s'agit essentiellement de compression sauf sur une zone située au toit du cotyle où il y a de la traction pure (cinq fois plus faible que la compression maximale).

Ajoutons que pour ces deux types de cupules en polyéthylène, les pressions exercées sur l'os cortical sont faibles (7 à 10 fois moins importantes que pour l'alumine).

IV.3.2. Moments des vecteurs de contraintes aux interfaces. Nous avons d'autre part calculé le moment de ces vecteurs de contraintes aux interfaces par rapport au centre "déplacé" du cotyle. On peut comparer, sur la Fig. IV.4, les répartitions de ces moments le long de l'interface os spongieux-ciment. On constate que cet

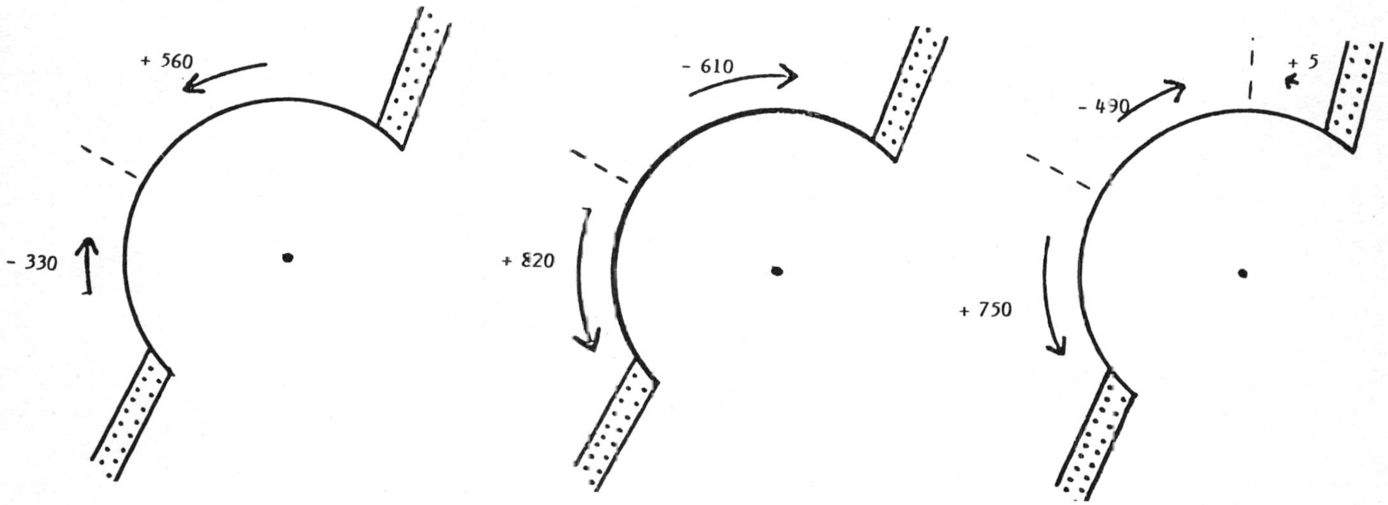

4a - Cupule en alumine

4b - Cupule en polyéthylène

4c - Cupule en polyéthylène/métal

Figure IV.4

Moments des charges appliquées à l'interface os-ciment (N × mm) interface se divise en deux parties : l'une couvrant largement le toit du cotyle et l'autre s'étendant sur l'arrière fond cotyloïdien. Chacune de ces régions a été déterminée comme étant le lieu des points d'application des forces d'interface qui ont le même effet de rotation : les moments de ces forces ont donc tous le même signe pour chacune de ces régions. Les valeurs indiquées sur la Fig. IV.4 sont les sommes des moments élémentaires.

Ajoutons que la somme algébrique des moments représentés sur la Figure IV.4 donne un moment s'étageant de + 210 à + 265 Nxmm. Ce résultat numérique confirme les observations expérimentales suivant lesquelles une cupule descellée se met à tourner dans le sens positif (dans le plan de coupe).

IV.3.3. Interprètation de ces résultats en terme de modes de descellements.

Il apparait déjà que les modes de descellements vont dépendre de la nature de l'implant.

i) Cas d'un implant en alumine. Deux facteurs vont intervenir dans le descellement :
*) les contraintes dans le ciment sont faibles ;
**) les forces de cisaillement ou de décollement sont prépondérantes à l'interface alumine-ciment et ont un moment résultant positif.

Ces deux facteurs correspondent très bien au processus très rapide de descellement observé en clinique : fracture du ciment et simultanément décollement de l'implant puis rotation de ce dernier.

ii) Cas d'un implant en polyéthylène. Deux facteurs, différents des précédents, vont intervenir dans le descellement :
*) le ciment exerce sur l'os cortical une très faible poussée alors que ce dernier devrait servir de point d'ancrage ;
**) la distribution de charges exercée par le ciment sur l'os spongieux a pour effet d'écarter le toit du cotyle et l'arrière fond cotyloïdien.

Dans le cas du polyéthylène il n'a a pas rupture brutale comme dans le cas de l'alumine, mais une phase d'évolution physique de l'implant et d'évolution de structure de l'os, ce qui permet de comprendre l'apparition des liserés radio-transparents décrits par CHARNLEY [1979]. Dans ce cas, pour étudier le phénomène du descellement, il conviendrait de prendre en compte le remodelage osseux, ce qui dépasse le cadre de cette étude.

iii) Enfin, pour un implant en polyéthylène serti de métal, les résultats sont tout à fait comparables à ceux trouvés dans le cas d'un implant en polyéthylène pur avec cependant des valeurs numériques plus faibles laissant présager une évolution de la structure osseuse plus lente que dans le cas de la cupule en polyéthylène.

IV.4. Techniques et aléas de pose

L'analyse détaillée des cas de références a été menée dans les sections IV.2 et IV.3. Nous examinons très brièvement les résultats numériques relatifs aux autres techniques de poses et aux différents aléas signalés dans la section III.5.4. Pour tous ces cas il y a une très bonne concordance entre les résultats numériques et les observations cliniques.

IV.4.1. Changement d'orientation de la cupule. Pour les deux types de cupules en polyéthylène pur ou en polyéthylène/métal, la simulation numérique montre que de légères variations dans l'orientation de ces cupules sont sans effet significatif sur les champs de contraintes et sur les vecteurs contraints aux interfaces excepté au voisinage des extrémités des ancrages.

Par contre, de légères variations dans l'angle de pose pour la cupule en alumine, peuvent s'avérer très dommageables, et ceci pour des variations d'angles d'à peine 5°. Il est à noter que ces effets sont peu perceptibles sur les tracés de contraintes principales mais extrêmement visibles sur les tracés de vecteurs de contraintes aux interfaces. D'où l'intérêt de notre approche.

IV.4.2. Conservation de la lame dense sous-chondrale. Les tracés des différentes contraintes ne sont pas fondamentalement changés par la conservation de la lame dense sous-chondrale dans le cas d'une cupule en polyéthylène ou en polyéthylène-métal.

Dans le cas d'une cupule en alumine, la conservation de la lame dense sous-chondrale a un effet stabilisateur, en particulier lorsque la cupule n'est pas cimentée au toit du cotyle. Cette dernière technique conduit à d'excellents résultats cliniques.

IV.4.3. Surépaisseurs de ciment. Ici encore l'approche numérique confirme les observations cliniques :
*) une surépaisseur de ciment au toit du cotyle a toujours des effets néfastes sur la tenue de la cupule ;
**) par contre, une surépaisseur de ciment derrière le ligament rond affecte peu la tenue de la cupule, et ceci quel que soit le matériau qui la constitue.

IV.5. Vers le "contrôle" de deux paramètres

A titre purement exploratoire nous avons fait varier les deux paramètres suivants :

IV.5.1. Géométrie de la cupule. Nous avons effectué quelques essais numériques avec d'autres formes de crantage de la partie externe de la cupule en alumine. Les résultats obtenus ne conduisent pas à une amélioration sensible de la configuration des vecteurs contraintes aux interfaces, ce qui nous conduit à penser que le réel problème de l'alumine est son extrême dureté.

IV.5.2. Matériaux utilisés pour fabriquer la cupule. L'alumine est excessivement dure (E = 350 000 MPa), le polyéthylène est excessivement mou (E = 600 MPa). Nous avons donc fait des essais avec des matériaux dont le module de Young prendrait des valeurs intermédiaires.

Pour des modules de Young inférieur à 7000 MPa les résultats obtenus sont similaires à ceux du cas polyéthylène.

Nous avons également fait des essais numériques sur une prothèse dont la paroi interne serait en alumine et la paroi externe en polyéthylène, ceci afin de combiner les qualités des deux matériaux et d'atténuer leurs défauts. Les résultats des essais numériques s'avèrent très encourageants... mais la technologie d'un tel assemblage parait être fort délicate... au moins pour le moment.

V - QUELQUES PROBLEMES OUVERTS

La méthode de simulation numérique que nous venons de proposer est assez simple et elle repose sur un certain nombre d'approximations assez grossières de la réalité. Malgré tout, les résultats obtenus sont en bon accord avec les observations cliniques. Cela incite à poursuivre l'étude dans quelques-unes des directions suivantes

i) Etude tridimensionnelle : Dans cette voie on pourrait reprendre la modélisation de OONISHI, ISHA et HASEGAWA [1983] en affinant certaines parties du maillage (notamment la partie cotyloïdienne) et en introduisant le calcul des vecteurs contraintes aux interfaces.

ii) Meilleure prise en compte des conditions aux limites : Il conviendrait de s'affranchir de l'hypothèse d'encastrement de l'os iliaque et de prendre en compte l'effet des muscles du bassin.

iii) Prise en compte des interactions implant fémoral-cupule.

iv) Mieux cerner les propriétés mécaniques des os impliqués.

v) Prise en compte du remodelage osseux.

vi) Prise en compte des fissures.

vii) Conception optimale de la forme des cupules et autres problèmes de contrôle optimal.

BIBLIOGRAPHIE

BERNADOU, M. ; CHRISTEL, P. et CROLET, J.M. [1983], Calcul des vecteurs contraintes aux interfaces de prothèses de hanche. Rapports de Recherche INRIA (à paraître).

CHARNLEY, J. [1979], Low friction arthrosplasty of the hip. Springer-Verlag, New York.

CIARLET, P.G. [1978], The Finite Element Method for Elliptic Problems. North-Holland, Amsterdam.

CROLET, J.M. [1979], Modélisation mathématique d'une hanche humaine appareillée et applications. Extensions à la hanche normale. Thèse de 3ème cycle. Université de Paris VI, Juin 1979.

DUVAUT, G. et LIONS, J.L. [1972], Les inéquations en mécanique et en physique. Dunod, Paris.

DUVAUT, G. et PISTRE, F. [1982], Calcul des vecteurs contraintes en approximation P_1 et P_2. C.R. Acad. Sc. Paris, 295, Série II, pp 827-830.

GOEL, V.K. ; VALLIAPPAN, S. and SVENSSON, N.L. [1978], Stresses in the normal pelvis. Comput. Biol. Med., 8, pp. 91-104.

HUISKES, R. and CHAO, E.Y.S. [1983], A survey of finite element analysis in orthopedic biomechanics : the first decade. J. Biomechanics, 16, n° 6, pp. 385-409.

JACOB, H. ; HUGGLER, A.H. ; DIETSCH, C. and SCHREIBER, A. [1976], Mechanical function of subchondral bone as experimentally determined on the acetabulum of the human pelvis. J. Biomechanics, 9, pp. 625-627.

MODULEF [1983], Présentation du Club Modulef, Notice 85, Version 3.4, INRIA, Juin 1983. Also available in english : "The Club Modulef - A library of computer procedures for finite element analysis", Notice 73, INRIA, Juin 1983.

OONISHI, H. ; ISHA, H. and HASEGAWA, T. [1983], Mechanical analysis of the human pelvis and its application to the artificial hip joint - by means of the three dimensional finite element method, J. Biomechanics, 16, n° 6, pp. 427-444.

PEDERSEN, D.R. ; CROWNINSHIELD, R.D. ; BRAND, R.A. and JOHNSTON, R.C. [1982], An axisymmetric model of acetabular components in total hip arthroplasty. J. Biomechanics, 15, pp. 305-315.

VASU, R. ; CARTER, D.R. and HARRIS, W.H. [1982], Stress distributions in the acetabular region. I. Before and after total joint replacement ; II. Effects of cement thickness and metal backing of the total hip acetabular component. J. Biomechanics, 15, pp. 155-170.

Computing Methods in Applied Sciences and Engineering, VI
R. Glowinski and J.-L. Lions (Editors)
Elsevier Science Publishers B.V. (North-Holland)

VIBRATIONS WITH UNILATERAL CONSTRAINTS

Michelle Schatzman*

Michel Bercovier**

To Mirta, thankfully.

1. INTRODUCTION

We would like to be able to compute the motion of an elastic system subject to unilateral constraints, i.e. material obstacles restrict the displacement of this system.

The mechanical systems we have in mind could be, for instance, a tyre rolling on the ground, a voluntarily or involuntarily loosened piece of machinery, a beam guided with ease.

We shall consider here only linear elasticity with small deformations, and shall concentrate on the numerical treatment of the unilateral constraints.

The most famous problem of static linear elasticity is Signorini's problem: given an elastic body, with reference configuration Ω, a partition of its boundary $\partial\Omega = \Gamma_F \cup \Gamma_u \cup \Gamma_{UL}$, a perfectly rigid obstacle filling the region $\mathcal{O}$ of space, we assume that Γ_{UL} is contained in $\partial\mathcal{O}$. The displacements $\bar{u}$ are given on Γ_u, the surface forces $\bar{f}$ on Γ_f, the volume forces F in Ω; the problem is to find a sufficient condition for the existence of an admissible static field of displacement u; the admissibility condition is that the body stay in $\overline{\mathcal{O}^c}$, the closure of the complementary region to the obstacle.

A moment of thought shows that the above problem does not possess an equilibrium solution for all given $\bar{f}$ and F, if Γ_u is empty: consider the problem of an obstacle which is the ceiling (in the frame of reference where the earth gravity is directed downwards), and the body is some parallepipedic box, which touches the ceiling in its reference configuration!

G. Fichera [15], J. L. Lions and G. Stampacchia [19] have given a sufficient condition which ensures the existence of a solution to Signorini's problem. Before elaborating on this condition, let us comment on the free boundary feature which is present in this problem: the part of Γ_{UL} where the deformed body has a contact with the obstacle is an unknown of the problem, as well as the normal strain on this part of Γ_{UL}. Similarly, the displacement is unknown on the part of Γ_{UL} where the deformed body does not have a contact with the obstacle.

* Centre de Mathematiques Appliquees, Ecole Polytechnique, 91128 Palaiseau Cedex, France.

** School of Applied Science and Technology, The Hebrew University, Jerusalem, Israel

The mathematical formulation of Signorini's problem is as follows: let $u(x)$ be the displacement of the point of Ω of coordinate x, and let the stress tensor be (under the assumption of small deformations)

$$(1.1) \qquad \varepsilon_{ij}(u) = \frac{1}{2}(u_{i,j} + u_{j,i}),$$

where $_{,i}$ denotes the partial differentiation with respect to the i-th spatial variable. As the material is assumed to be linear, its constitutive relation, which relates the strain tensor ε_{ij} and the stress tensor σ_{ij} is Hooke's law, written for a general material as

$$(1.2) \qquad \sigma_{ij} = a_{ijkl}\ \varepsilon_{kl} \quad \text{(with the convention of repeated indices)}$$

where a_{ijkl} is a tensor which has the following symmetry properties:

$$(1.3) \qquad a_{ijkl} = a_{klij} = a_{jikl}, \quad \text{for all } i, j, k \text{ and } l.$$

Newton's law is written, with the density of mass,

$$(1.4) \qquad \sigma_{ij,j} + \rho F_i = 0.$$

The elastic potential energy is

$$(1.5) \qquad a(u,u) = \int_\Omega a_{ijkl}\ \varepsilon_{ij}(u)\ \varepsilon_{kl}(u)\ dx,$$

and if the convex of constraints is

$$(1.6) \qquad K = \{u \in (H^1(\Omega))^3 \ / \ u_{|\Gamma_u} = \bar{u},\ u.n_{|\Gamma_{UL}} \leq 0\}$$

where $n = (n_i)_{i=1,2,3}$ is the exterior normal to Ω, Signorini's problem can be stated as follows:

$$(1.7) \qquad \text{To minimize over } K \quad \frac{1}{2}\, a(u,u) = \int_\Omega \rho F.u\ dx - \int_{\Gamma_f} f.u\ d\Gamma.$$

If we define the normal strain by

$$(1.8) \qquad \sigma_N = \sigma_{ij}\ n_i\ n_j$$

and the tangential strain by

$$(1.9) \qquad \sigma_T = (\sigma_{Ti})_{i=1,2,3} = (\sigma_{ij}\ n_j - \sigma_N\ n_i)_{i=1,2,3}$$

it can be shown that u solves (1.7) if and only if

$$(1.10) \qquad \sigma_{ij,j} + \rho F_i = 0$$

$$(1.11) \qquad u_{|\Gamma_u} = \bar{u}$$

(1.11) $\quad \sigma_{N}|_{\Gamma_f} = f_i$

(1.12) $\quad \sigma_N \le 0, \quad u_N \equiv u.n \le 0, \quad u_N\sigma_N \le 0, \quad \sigma_T = 0 \quad \text{on} \quad \Gamma_{UL}.$

A statement equivalent to (1.7) is

To find u in K such that

(1.13) $\quad a(u,v-u) - \int_\Omega F.(v-u)\,dx - \int_{\Gamma_f} f.(v-u)\,dx \ge 0,$ for all

for all v in K.

In the dynamical case, one has simply to replace (1.4) by a relation involving the acceleration

(1.15) $\quad \sigma_{ij,j} + \rho F_i = \ddot{u}_i,$

where, of course, ρ is the density of the material. In the dynamical case, we must make sure that no solid movement is involved; if this is the case, one has first to solve for the solid motion involved, including eventually reflexion on the obstacle(s), and then work in a frame of reference attached to the body, taking into account the inertia forces in the volume forces.

To avoid this problem, we may assume that

(1.16) $\quad \Gamma_u$ is of positive measure.

If this is not satisfied, we can assume that the existence condition for the static equilibrium is satisfied, and we work around such an equilibrium position; this condition can be written

(1.17) $\quad \int_\Omega F.r\,dx + \int_{\Gamma_f} f.r\,d\Gamma < 0,$

for all rigid displacement $r \neq 0$ belonging to K; we recall that r is a rigid displacement if it can be written as

(1.18) $\quad r(x) = a + b \wedge x,$

with $\wedge$ denoting the vector product in $\mathbb{R}^3$.

Then, formally, the variational formulation of the dynamical problem we consider is

(1.19)
$$u \in L^\infty(0,T;(H^1(\Omega))^3)$$
$$u_t \in L^\infty(0,T;(L^2(\Omega))^3)$$

which means that u has bounded energy for all time;

(1.20)
$$u(t) \in K, \quad \text{for all } t,$$
$$(\ddot{u},v-u) + a(u,v-u) - \int_\Omega F.(v-u)\,dx - \int_{\Gamma_f} f.(v-u)\,d \ge 0$$
for all v in K.

where (,) designates the scalar product of $(L^2(\Omega))^3$, with the weight ρ. Unfortunately, there is no nice existence theorem for (1.19)-(1.20), because of several theoretical difficulties, the main two of which are:

* the problem is not convex, or a perturbation of a convex problem;

* one has to treat the boundary condition properly.

There is one theorem for the special case of a vibrating rod with an obstacle at one end, which is given, together with a theory of its numerical approximation in [28]. Nevertheless, the approach of the dynamical Signorini's problem by penalizing on the boundary constraint seems to be promising.

Some related problem have been treated more or less completely:

* material points with obstacle. and more generally, systems with infinitely many degrees of freedom and obstacles: see[25] which contains an existence theory, and examples of non-uniqueness, [7], [8], [11], which contains a generalization to Riemannian situations, [22] which contains sufficient conditions for uniqueness, and [20] for a different approach.

* strings with obstacles; this subject, initiated by Amerio and Prouse [3], was studied afterwards in [1], [2], [26] which contains a uniqueness theorem for the problem considered in [1], [27], [12] and [13] which contain the partially elastic or inelastic bounce case of [3], [14], [6] which considers the inelastic reflexion of a string against a convex obstacle, and [5], which contains an explicit formula for the solution of the problem with a plane obstacle.

* wave equation with an unilateral constraint at the boundary; this problem was considered in [18], as a simplified model for Signorini's dynamical problem.

The problem of main concern in most of the above papers is to prove that the energy is conserved in a convenient sense; in some of the cases, the energy is conserved as a consequence of the variational formulation; in others, the conservation of the energy, or a definite loss of energy due to the contact has to be imposed on the solution. We shall see in the course of this paper the implications of this situation on the computational problems.

At least formally, one has to expect the automatic conservation of the energy when the constraint takes place on a set of zero measure, such as a portion of a boundary.

2. VARIATIONAL ALGORITHMS, AND THEIR WANTED AND UNWANTED PROPERTIES.

We would like to approximate numerically the solution of a problem which can be somewhat sloppily formulated as

$$\text{(2.1)} \quad \begin{cases} (\ddot{u},v-u) + a(u,v-u) \geq (f,v-u), & \text{for all } v \text{ in } K, \\ u(t) \in K, & \text{for all } t, \end{cases}$$

where u takes its values in $V = (H^1(\Omega))^N$, K is a closed convex subset of V, (,) is the scalar product of $H = (L^2(\Omega))^N$, a is a bilinear continuous symmetric form on V

(2.2) $$|a(u,v)| \leq M \|u\| \|v\|, \ \forall\ u,v \text{ in } V,$$
$$a(u,u) \geq \|u\|^2, \ \forall\ u \text{ in } V,$$

where α is some positive number.

Whenever necessary, the scalar product on H, and the duality product between V and V' can be identified.

The first and simplest idea to approximate (2.1) is to stay in the frame of variational inequalities, using a good linear scheme to take care of the linear part of the problem, and introducing the constraint naturally in the variational inequality. We start by reviewing a class of linear schemes.

Let A be the operator from V to V' defined by

(2.3) $$a(u,v) = (Au,v), \ \forall\ u,v \in V.$$

If V_h is a sequence of finite dimensional subspaces of V, an operator A_h is defined by

(2.4) $$(A_h u,v) = a(u,v), \ \forall\ u,v \in V_h.$$

Consider the linear problem

(2.4) $$\frac{d^2u}{dt^2} + Au = f.$$

Then, a class of approximations, centered in time, and which involve only the time levels n and $n-1$ in the computation of the time level $n+1$ is given by

(2.5) $$\frac{u^{n+1} - 2u^n + u^{n-1}}{\Delta t^2} + A_h[\theta u^{n+1} + (1-2\theta)u^n + \theta u^{n-1}] = f^n,$$

with θ a real number in the interval $[0,\frac{1}{2}]$. This scheme is termed "θ-method"; a different θ-method, for evolution equations of first order in time has been introduced by P. A. Raviart in [23]; it is not very interesting for the first order formulation of (2.4), because it is generally not centered in time, and the conditional stability theory cannot be made. A θ-method adapted to (2.4) is introduced in [4], this is the scheme (2.5).

The stability of the scheme is proved as follows: (2.5) is equivalent to the following variational equality

(2.6) $$\left(\frac{u^{n+1} - 2u^n + u^{n-1}}{\Delta t^2},v\right) + a(\theta u^{n+1} + (1-2\theta)u^n + \theta u^{n-1},v) =$$

$$= (f^n,v), \ \forall\ v \text{ in } V_h.$$

Substitute $v = u^{n+1} - u^{n-1}$ in (2.6); then, the first term of the equality becomes

(2.7) $$\left|\frac{u^{n+1} - u^n}{\Delta t}\right|^2 - \left|\frac{u^n - u^{n-1}}{\Delta t}\right|^2$$

the second term can be written as

$$a(u^{n+1} + 2\theta u^n + u^{n-1}, u^{n+1} - u^{n-1}) +$$
$$+ (1 - 4\theta)\, a(u^n, u^{n+1} - u^{n-1}).$$

From the identity $a(u,v) = (a(u+v,u+v) - a(u-v,u-v))/4$, we obtain for the second term

$$(a(u^{n+1} + u^n, u^{n+1} + u^n) + (4\theta - 1)a(u^{n+1} - u^n, u^{n+1} - u^n))/4 -$$
$$- (a(u^n + u^{n-1}, u^n + u^{n-1}) + (4\theta - 1)a(u^n - u^{n-1}, u^n - u^{n-1}))/4.$$

Let

$$E(\theta,\Delta t,u,v) = |v|^2 + a(u+(\Delta t\ v/2), u + (\Delta t\ v/2)) + \tag{2.8}$$
$$+ (4\theta - 1)\, a(v,v)\, \Delta t^2/4.$$

With notation (2.8), we have an "energy conservation" for the scheme

(2.5):

$$E(\theta,\Delta t,u^n, \frac{u^{n+1}-u^n}{\Delta t}) - E(\theta,\Delta t,u^{n-1}, \frac{u^n-u^{n-1}}{\Delta t}) = (f^n, \frac{u^{n+1}-u^{n-1}}{\Delta t})\Delta t. \tag{2.9}$$

To make sure that (2.9) implies stability, we must check that there is a constant K independent of h such that

$$E(\theta,\Delta t,u,v) \geq K|v|^2, \quad \forall\ v \in V_h. \tag{2.10}$$

Under (2.10), a discrete Gronwall inequality on (2.9) yields stability. If

$$\theta \geq 1/4 \quad \text{(unconditional stability)}, \tag{2.11}$$

(2.10) is always satisfied, because the part of E which contains a is always non-negative. If (2.11) is not satisfied, let

$$L_h = \max\ \{a(u,u)\ |u|^{-2}\ /\ u \in V_h\}. \tag{2.12}$$

Then,

$$E(\theta,\Delta t,u,v) \geq |v|^2\ [1 - (4\theta-1)L_h\ \Delta t^2/4],$$

and thus (2.10) is satisfied if there exists $\delta > 0$ such that

$$\Delta t \leq 2(1 - \delta)/(L_h(4\theta-1))^{1/2} \quad \text{(conditional stability)}. \tag{2.13}$$

The variational version of (2.6) with convex constraints can be written as

$$u^{n+1} \in K_h \tag{2.14}$$

(2.15)
$$\left(\frac{u^{n+1} - 2u^n + u^{n-1}}{t^2}, v - u^{n+1}\right) + a(\theta u^{n+1} + (1-2\theta)u^n + \theta u^{n-1}, v - u^{n+1}) \geq (f^n, v - u^{n+1}), \quad \forall v \in K_h,$$

where K_h is a closed convex subset of V_h, which approximates K in a suitable sense. Then, not very surprisingly, upon substitution of v by u^{n-1} in (2.15), we obtain the inequality

$$E\left(\theta, \Delta t, u^n, \frac{u^{n+1} - u^n}{t}\right) - E\left(\theta, \Delta t, u^{n-1}, \frac{u^n - u^{n-1}}{\Delta t}\right) \leq$$
$$\leq \left(f^n, \frac{u^{n+1} - u^n}{\Delta t}\right)\Delta t,$$

which is similar to (2.9). Therefore, the conditions of linear stability are the same as the conditions of non-linear stability.

This apparently promising scheme was tried by H. Tayari in his these de 3eme cycle [29], on the case of a vibrating string with an obstacle; the mathematical formulation of this problem is (see [26])

(2.17)
$$\Box u \geq 0$$
$$u \geq -K \qquad \text{in } (0,L) \times (0,T)$$
$$\operatorname{supp} \Box u \subset \{(x,t)/u(x,t) = -K\}$$

(2.18)
$$u(x,0) = u_0(x) \geq -K \qquad \text{on } (0,L)$$
$$u_t(x,0) = u_1(x)$$

(2.19)
$$u(0,t) = u(L,t) = 0 \qquad \text{on } (0,T)$$

To ensure uniqueness, one has to add an energy condition, which can be written as

(2.20)
$$\partial(u_x^2 + u_t^2)/\partial t - \partial(2u_x u_t)/\partial x = 0, \quad \text{in the sense of distributions.}$$

Then, this problem possesses a unique solution [26] in the space of functions of bounded energy, provided that K is strictly positive.

An exact solution is known for $K = 1/2$, $L = 1$, and initial data

$$u_0(x) = 0$$

$$u_1(x) = -\pi \sin \pi x.$$

In fact, many exact solutions of this problem are known, due to the work of C. Reder [24], and subsequently to the work of A. Haraux and H. Cabannes [9,10,17]; see the work of these authors for subtle properties of periodicity and almost periodicity of the solutions of (2.17)-(2.20). The solutions of (2.17)-(2.20) can be constructed explicitly, thanks to an explicit formula, proved in [5], together with the convergence of a numerical scheme adapted to this formula; therefore (2.17)-(2.20) is a good test problem.

The computations of Tayari yielded the following remarkable result: the scheme (2.15), applied to the case of the vibrating string with an obstacle converges experimentally, but the limit displayers an enormous loss of energy, so that (2.15) is not consistent with (2.20). On Fig. 1, which is extracted from [29], one can see that the computed string "sticks" to the obstacle, immediately after touching it, instead of bouncing back, and thus, the kinetic energy is lost through the shock.

To understand better this phenomenon, let us consider a simplified problem; to find the trajectory of a material point whose motion is unidimensional, in the absence of exterior forces, knowing that this point bounces back without loss of energy when it hits a given perfectly rigid obstacle. If $u(t)$ denotes the abscissa of the material point at time t, and if the obstacle occupies the region $\{u \leq 0\}$, the solution of this problem with initial data

$$(2.21) \qquad u(0) = 1, \quad \dot{u}(0) = (du/dt)(0) = -1$$

is

$$(2.22) \qquad u(t) = \begin{cases} 1 - t, & \text{if } t \leq 1, \\ t - 1, & \text{if } t \geq 1. \end{cases}$$

It is immediate that u satisfies the following family of inequalities

$$(2.23) \qquad \int_0^T \ddot{u}(v - u)\, dt \geq 0,$$

for all v such that

$$(2.24) \qquad v \text{ is continuous}, \quad v \geq 0 \text{ on } [0,T].$$

In the present case, $\ddot{u}$ is a Dirac mass, located at $t = 1$. But, for any other function u is given by

$$(2.25) \qquad u(t) = \begin{cases} 1-t, & \text{if } t \leq 1, \\ 0, & \text{if } 1 \leq t \leq t_0, \\ a(t-t_0), & \text{if } t_0 \leq t \leq T, \end{cases}$$

where a is an arbitrary positive number, and t_0 is arbitrary in the interval $[1,T]$, satisfies (2.23), for all v satisfying (2.24); it can be shown indeed that all solutions of (2.23) - (2.24) with the initial condition (2.21) are of the form (2.25). It is the energy condition

$$|\dot{u}| = \text{constant}$$

which allows us to select the solution (2.22) from the large family (2.25) of functions satisfying (2.21) and (2.23). We can see thus, that the variational

Let us see what the scheme (2.15) would give for (2.21)-(2.23):

$$(2.26)\qquad \left(\frac{u^{n+1} - 2u^n + u^{n-1}}{t^2}, v - u^{n+1}\right) \geq 0, \ \forall\ v \geq 0,$$

$$u^{n+1} \geq 0.$$

This is equivalent to

$$u^{n+1} = \max(2u^n - u^{n-1}, 0).$$

Starting with initial conditions

$$u^0 = 1,$$

$$u^1 = 1 - \Delta t,$$

we obtain

$$u^n = 1 - n\Delta t, \quad \text{as long as} \quad n\Delta t \leq 1.$$

If N is the smallest integer such that $N\Delta t > 1$,

$$u^N = \max(2(1 - (N-1)\Delta t) - (1 - (N-2)\Delta t), 0) =$$

$$= \max\ (1 - N\Delta t, 0) = 0;$$

$$U^{N+1} = \max\ (-u^{N-1}, 0) = 0,$$

and, subsequently,

$$u^n = 0, \quad \forall\ n \geq N.$$

Therefore, the limit of this approximation is

$$(2.27)\qquad u(t) = \begin{cases} 1-t & \text{if } t \leq 1, \\ 0 & \text{if } t \geq 1. \end{cases}$$

Therefore, the scheme (2.26) is consistant with the assumption of a total loss of energy in the shocks.

If we want the energy to be conserved, a different scheme has to be used. Before seeing which ideas could be employed for taking care of this problem, it is important to mention that (2.15) is not always a bad scheme. In the case of a vibrating rod with an obstacle at one end, we have shown in [28] that (2.15), with $\theta = 0$ is a convergent scheme; the main reason for this convergence is that, in this case, the conservation of the energy for the continuous problem is a consequence of the other equations, i.e. it is contained implicitly in the variational formulation of the problem. We conjecture that, whenever the conservation of the energy is a consequence of the other equations, (2.15) should give reasonable results. Nevertheless, numerically, some energy can be lost, because, it is very clear that in no finite dimensional case the conservation of the energy is a consequence of the other equatons. The larger the proportion of the degrees of freedom where the constraint must hold, the larger the loss of energy is to be expected.

3. NON-VARIATIONAL ALGORITHMS.

A first idea to take into account the conservation of the energy is the following: consider a finite dimensional problem with constraints, and assume that the constraint has to be satisfied for M degrees of freedom; then, one could try to compute (at least approximately) the instants at which each of these degrees of freedom hits the obstacle, and then, to bounce back these degrees, with the corresponding reflexion of the velocity. This is the approach which has been taken by P. Lotstedt in [20]. A similar approach has been taken by D. Osmont in [21], but for the following continuum problem: two elastic beams, with different mechanical constants hit one another during their motion. To take care of the unilateral constraint, D. Osmont modifies it into a penalty method, assuming that the obstacle is very stiff, instead of perfectly rigid, and considers the following equation

(3.1) $$\frac{d^2u^\lambda}{dt^2} + Au^\lambda + \lambda^{-1}(u^\lambda - P_K u) = f,$$

where $P_K u$ is the projection of u on the convex set K, with respect to the L^2 norm. In this case, one can see that the time step must be smaller than $\pi\sqrt{\lambda}$, by considering the penalized version of (2.21), (2.23) and (2.24):

(3.2) $$\ddot{u}^\lambda + \lambda^{-1}\min(u^\lambda, 0) = 0$$

(3.3) $$u^\lambda(0) = 1, \quad \dot{u}^\lambda(0) = -1.$$

The solution of (3.2)-(3.3) can be explicitly computed and is given by

$$u^\lambda(t) = \begin{cases} 1 - t, & t \le 1 \\ -\sqrt{\lambda}\,\sin(t/\sqrt{\lambda}), & 1 \le t \le 1 + \pi\sqrt{\lambda}, \\ t - 1 - \pi\sqrt{\lambda}, & 1 + \pi\sqrt{\lambda} \le t, \end{cases}$$

which clearly converges to the desired energy-conserving solution, given by (2.22). The differential equation can be approximated by an implicit method, which has to be chosen carefully; in the case computed by Osmont, there is a large number of bouncing instants, which are not all significative; Osmont chooses between them with a lot of ingenuity, and little justification. No exact solution being available, in this case, he proceeds to an experimental comparison, and gets excellent agreement.

We take a different path here: we believe that the computational complexity of the "bounce tracking" methods is probably too high for any realistic problem, and we resort to a different method, in the guise of a product formula, similar to Trotter-Kato's formula.

Assume that we work in a finite-dimensional space V_h and that the convex set K is a product

(3.5) $$K = \prod_{i \in I} K^i$$

of closed convex sets K^i, each of which spans a vector space V^i, and V^i is orthogonal to V^j in H for $i \neq j$. Assume that the boundary of K^i is of class C^1, and piecewise C^3, with each of the pieces either of strictly positive Gaussian curvature, or affine. Then, according to [25], there is uniqueness for the billiard ball flow, which is described as follows:

* when the billiard ball does not hit the boundary, its velocity is constant;

* when the billiard ball hits the boundary, the tangential component of the velocity is conserved, and the normal component is reversed.

This definition allows for motion on the boundary of K^i, along the geodesics.

The billiard ball flow will be denoted by $S_i(t)$, so that if u is given in K^i, and v in the half space tangent to K^i and containing K^i, $S_i(t)\binom{u}{v}$ has for components the position and the velocity at the instant $t + 0$ of the billiard ball which started with position u and velocity v at the instant 0.

We consider for simplicity a convex set which contains zero, and a problem with no forcing term. We may decompose u as

$$u = \sum_{i \in I} u_i + u_* \tag{3.6}$$

where u_i is the orthogonal projection of u_i on V_i, and u_* the projection on the orthogonal of the sum of the V_i's.

Consider the linear evolution problem

$$u_{tt} + Au = f \tag{3.7}$$

We want to define a scheme of the form

$$\frac{u^{n+1} - u^n}{t} = v^{n+1} \tag{3.8}$$

$$\frac{v^{n+1} - v^n}{t} = G(h,\Delta t,u^n,v^n),$$

which is stable, and conserves the energy in the following sense: there exists a quadratic form $E(h,\Delta t,u,v)$ and a positive constant K such that

$$E(h,\Delta t,u,v) \geq K\ |v|^2\ , \tag{3.9}$$

for all pairs $(h,\Delta t)$ in the considered sequence of approximations, and moreover, the following numerical energy conservation law holds:

$$E(h,\Delta t,u+v\Delta t,v+G(h,\Delta t,u,v)\ t) = E(h,\Delta t,u,v), \tag{3.10}$$

so that, if a sequence (u^n,v^n) is defined inductively by (3.8), $E(h,\Delta t,u,v)$ is independent of n.

The θ-method (2.6) satisfies the requirements (3.8)-(3.10): by definition of the θ-method,

$$G(h,\Delta t,u,v) = -(I+\theta\Delta t A_h)^{-1}A_h u;$$

(3.9) is satisfied if the method is stable, either with condition (2.11) or with condition (2.13), and (3.10) holds with

$$E(h,\Delta t,u,v) = |v|^2 + a(u+(\Delta t v)/2, u+(\Delta t v/2)) + \tag{3.11}$$
$$+ (4\theta - 1)a(v,v)\Delta t^2/4,$$

according to (2.5).

The first proposed scheme is the following: given $\begin{matrix} u^n \\ v^n \end{matrix}$ such that

$$\begin{aligned} & u_i^n \in K^i \\ & v_i^n \in T_{K^i}(u_i^n) = \text{the tangent cone at } u_i^n \text{ to } K^i, \end{aligned} \tag{3.12}$$

let

$$\begin{aligned} v^{n+1/2} &= v^n + \Delta t \; G(h,t,u^n,v^n) \\ u^{n+1/2} &= u^n + \Delta t \; v^{n+1/2} \end{aligned} \tag{3.13}$$

Then, to take care of the reflexion process, we let

$$\begin{matrix} u_i^{n+1} \\ v_i^{n+1} \end{matrix} = S_i(\Delta t) \begin{matrix} u_i^{n+1/2} \\ v_i^{n+1/2} \end{matrix} \tag{3.14}$$

$$\begin{matrix} u_*^{n+1} \\ v_*^{n+1} \end{matrix} = \begin{matrix} u_*^{n+1/2} \\ v_*^{n+1/2} \end{matrix} \tag{3.15}$$

If K^i is one dimensional, the evaluation of S_i is easy; if not, one will have to approximate S_i suitably.

Observe that, if $u_i^{n+1/2}$ belongs to K^i, then

$$\begin{matrix} u_i^{n+1} \\ v_i^{n+1} \end{matrix} = \begin{matrix} u_i^{n+1/2} \\ v_i^{n+1/2} \end{matrix}$$

as, in this case, the billiards motion from $\begin{matrix} u_i^n \\ v_i^{n+1/2} \end{matrix}$ to $\begin{matrix} u_i^{n+1} \\ v_i^{n+1} \end{matrix}$ is simply uniform motion. Thus, if $u_i^{n+1/2}$ is in K_i for each i, the energy of $\begin{matrix} u^{n+1} \\ v^{n+1} \end{matrix}$ is equal to the energy of $\begin{matrix} u^n \\ v^n \end{matrix}$.

From its very definition, the scheme (3.13) - (3.15) furnishes a pair

$\begin{pmatrix} u^{n+1} \\ v^{n+1} \end{pmatrix}$ which satisfies (3.12). Numerical experiences show that

* <u>for finite dimensional problems, there is numerical convergence</u>;

* <u>for continuum problems, the convergence is poor, because the scheme generates energy</u>.

This phenomenon can be understood as follows: in the case of the θ -method, for instance, the numerical energy (3.11) has a kinetic energy part, $|v|^2$, and the remainder is the potential energy part. The step (3.14)-(3.15) of the scheme does not modify the kinetic energy, but it increases the rest of the energy, by increasing the discrete spatial derivatives between points at which there is a reflexion, and points at which there is none.

This can be taken care of by the following refinement of (3.13)-(3.15). Assuming that $\begin{pmatrix} u^n \\ v^n \end{pmatrix}$ satisfies (3.12), we define

(3.16) $$v^{n+1/3} = v^n + \Delta t\ G(h,\Delta t,u^n,v^n)$$
$$u^{n+1/3} = u^n + \Delta t\ v^{n+1/3}$$

(3.17) $$\begin{pmatrix} u_i^{n+2/3} \\ v_i^{n+2/3} \end{pmatrix} = S_i(\Delta t) \begin{pmatrix} u_i^{n} \\ v_i^{n+1/3} \end{pmatrix}$$

(3.18) $$\begin{pmatrix} u_*^{n+2/3} \\ v_*^{n+2/3} \end{pmatrix} = \begin{pmatrix} u_*^{n+1/3} \\ v_*^{n+1/3} \end{pmatrix}$$

and an additional scaling step to ensure the conservation of the energy:

$$\begin{pmatrix} u^{n+1} \\ v^{n+1} \end{pmatrix} = \begin{pmatrix} u^{n+2/3} \\ v^{n+2/3} \end{pmatrix} E(h,\Delta t,u^n,v^n)^{1/2}\ E(h,\Delta t,u^{n+2/3},v^{n+2/3})^{-1/2}.$$

In finite dimensional situations, it can be proved that this scheme converges as Δt goes to zero to an energy conserving solution; there is no proof, even in the best understood continuum situations, but numerical experiments in the case of the string with an obstacle show good agreement between the known solution and the calculated solution; see Fig. 2. We display the results of another numerical experiment, in the case of an elastic plane beam, of finite thickness, using one rank of standard bilinear elements; the mass matrix is lumped; see Fig. 3 - The beam's movement is activated by a unit speed upward at time t=0. The beam falls back on the obstacle and rebounds. Explicit time step was bounded by $1/\lambda max$ and was here $\Delta t = 0.005$. Drawn figures from bottom to top are given every 1000 time steps. Deformations are amplified by factor 100 on the drawing.

4. CONCLUSION.

We have shown in this article some difficulties which appear in the understanding, and in the computation of vibration problems with an obstacle. We expect that some of the above considersations should be applicable to other nonlinear vibration problems, involving different constraints, such as inextensibility, or different nonlinearities in the hyperbolic operator itself.

The main conclusion of our work, is that one should be very cautious, when using the variational approach in this setting: sometimes, it is very good, but it can yield also highly inconsistent results.

REFERENCES

[1] L. Amerio, Su un problema di vincoli unilaterali per l'equazione non omogenea della corda vibrante, I.A.C. (istito per le applicazioni del calcolo "Mauro Picone") Publ., Ser. D. 109 (1976), 3-11.

[2] L. Amerio, On the motion of a vibrating string through a moving ring with a continuously variable diameter, Rend. Acc. Naz. Lincei 62(1977), 134-142.

[3] L. Amerio and G. Prouse, Study of the motion of a string vibrating against an obstacle, Rend. Mat. (2) 8 Serie VI (1975), 563-585.

[4] A. Bamberger, Approximation numerique des systemes hyperboliques lineaires, Cours de troisieme cycle, Laboratoire d'Analyse Numerique, Universite Pierre et Marie Curie, Paris, 1981-1982.

[5] A. Bamberger, and M. Schatzman, New results on the vibrating string with a continuous obstacle, S.I.A.M. J. Math. Anal. 14 (1983), 560-595.

[6] R. Burridge, J. Kapprass and C. Morshedi, The Sitar string, a vibrating string with a one-sided inelastic constraint, S.I.A.M. J. Appl. Math. 42(1982), 1231-1251.

[7] G. Buttazzo and D. Percivale, The bounce problem on n-dimensional Riemannian manifolds, Atti. Accad. Naz. Lincei, Rend. Cl. Sci. Fis. Mat. Natur., (8) 70 (1981), 246-250.

[8] G. Buttazzo and D. Percivale, On the approximation of the elastic bounce problem on Riemannian manifolds, to appear, J. Differential Equations.

[9] H. Cabannes, Mouvements periodiques d'une corde vibrante en presence d'un obstacle rectiligne, Z. Angew. Math. Phys. 31 (1980), 473-482.

[10] H. Cabannes, and A. Haraux, Mouvements presque periodiques d'une corde vibrante en presence d'un obstacle fixe, rectiligne ou ponctuel, International J. Non-linear Mech. 16 (1981), 449-458.

[11] M. Carriero and E. Pascali, Il problema del rimbalzo unidimensionale, Rend. Mat. 13 (1980), 541-553.

[12] C. Citrini, Sull'urto parzialmente elastico o anelastico di una corda vibrante contra un ostacolo I, Atti Accad. Naz. Lincei, Rend. Cl. Sci. Fis. Mat. Natur. (8) 59 (1975), 368-376 (1976).

[13] C. Citrini. Sull'urto parzialmente elastico o anelastico di una corda vibrante contra un ostacolo II, Atti Accad. Naz. Lincei, Rend. Cl. Sci. Fis. Mat. Natur. (8) 59 (1975), 667-676 (1976).

[14] C. Citrini, The energy theorem in the impact of a string vibrating against a point-shaped obstacle, Rend. Accad. Naz. 62 (1977), 143-149.

[15] G. Fichera, Problemi elastostatici con vincoli unilaterali: il problema di Signorini con ambigue condizioni al contorno, Atti Accad. Naz. Lincei 8 (1963-1964), 91-140.

[16] R. Glowinski, J. L. Lions, and R. Tremolieres, Numerical analysis of variational inequalities, North Holland, Amsterdam, New-York, 1981.

[17] A. Haraux and H. Cabannes, Almost periodic motion of a string vibrating against a straight fixed obstacle, Nonlinear Anal. 7 (1983), 129-141.

[18] G. Lebeau and M. Schatzman, A wave problem in a half-space, with a unilateral constraint at the boundary, to appear in J. Differential equations.

[19] J. L. Lions and G. Stampacchia, Variational inequalities, Comm. Pure. Appl. Math. 20 (1967), 493-519.

[20] P. Lotstedt, Mechanical systems of rigid bodies subject to unilateral constraints, S.I.A.M. J. Math. Anal. 42 (1982), 281-296.

[21] D. Osmont, Computation of the dynamic response of structures with unilateral constraints (contact) - Comparison with experimental results, Comput. Meth. Appl. Mech. Eng. 34 (1982), 847-859.

[22] D. Percivale, Uniqueness in the elastic bounce problem, to appear in J. Differential Equations.

[23] P. A. Raviart, Sur l'approximation de certaines equations d'evolution lineaires et non-lineaires, J. de Math. Pures Appl. 46 (1967), 11 107 and 109-183.

[24] C. Reder, Etude qualitative d'un probleme hyperbolique avec constrainte unilaterale, these de 3eme cycle, Universite Bordeaux 1, 1979.

[25] M. Schatzman, A class of nonlinear differential equations of second order in time, Nonlinear Analysis, 2 (1983), 560-595.

[26] M. Schatzman, A hyperbolic problem of second order with unilateral constraints: the vibrating string with a concave obstacle, J. Math. Annal. Appl. 73 (1980), 138-101.

[27] M. Schatzman, Un probleme hyperbolique du 2eme ordre avec constraintes unilatérales: la corde vibrante avec obstacle ponctuel, J. Differential Equations 36 (1980), 295-334.

[28] M. Schatzman and M. Bercovier, On the numerical approximation of a vibration problem with unilateral constraints, rapport interne, Centre de Mathematiques Appliquees, Ecole Polytechnique, Palaiseau, to appear in 1984.

[29] H. Tayari, These de troisieme cycle, Laboratoire d'Analyse Numerique, Universite Pierre et Marie Curie, Paris, 1983.

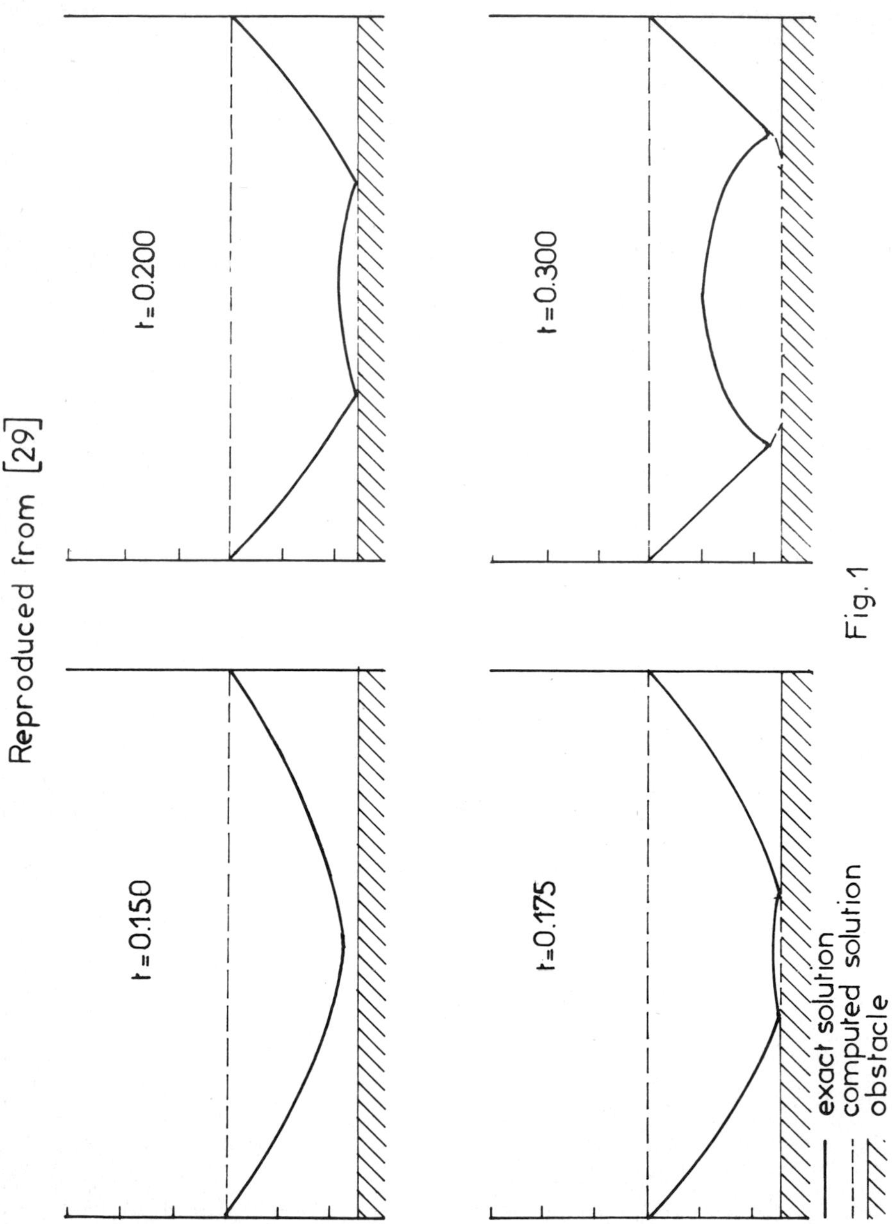
Reproduced from [29]
t = 0.150
t = 0.200
t = 0.175
t = 0.300
exact solution
computed solution
obstacle

Fig. 1

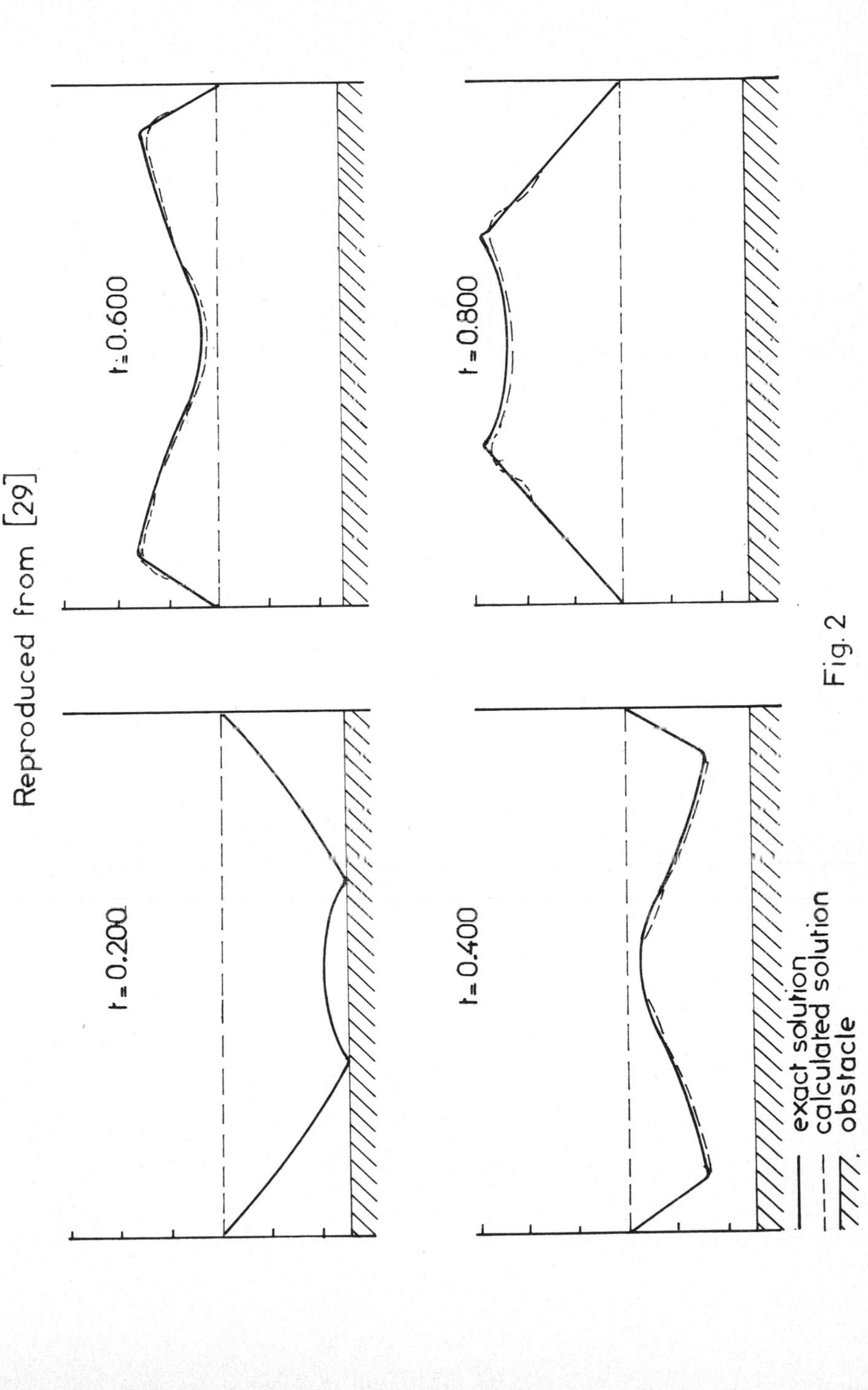
Reproduced from [29]
t=0.200
t=0.400
t=0.600
t=0.800
exact solution
calculated solution
obstacle

Fig. 2

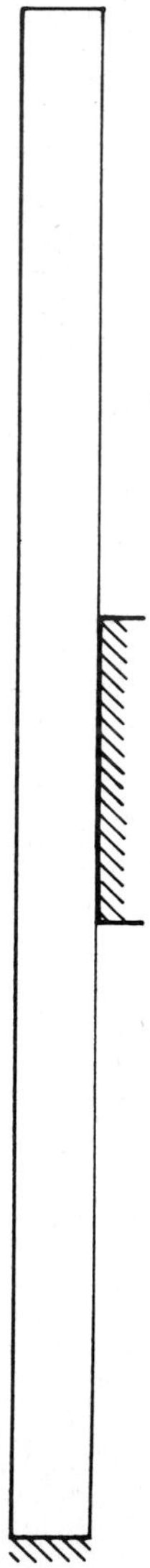

Fig. 3a

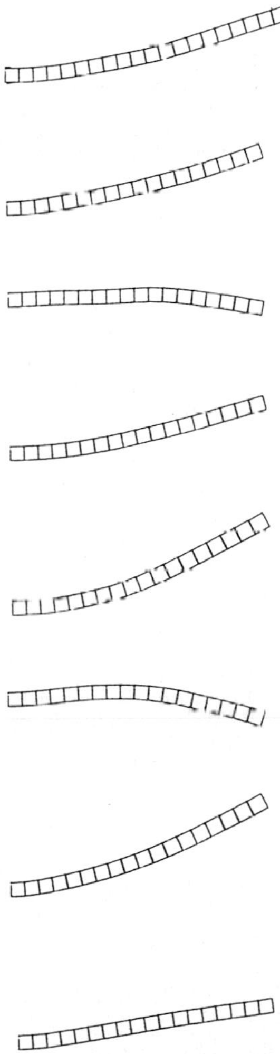

Fig. 3b

Computing Methods in Applied Sciences and Engineering, VI
R. Glowinski and J.-L. Lions (Editors)
Elsevier Science Publishers B.V. (North-Holland)

NUMERICAL SOLUTION OF A MODEL PROBLEM FROM COLLAPSE LOAD ANALYSIS

Michael L. Overton

Courant Institute of Mathematical Sciences
New York University
New York, N. Y. 10012
U.S.A.

We consider a model problem from collapse load analysis discussed recently by Strang. The analytic solution is a characteristic function with a jump discontinuity. We develop a method for solving a discretized version of the model problem, which requires the minimization of a convex piecewise differentiable function. Numerical results are presented.

INTRODUCTION

In collapse load analysis the following problem arises (see Strang [16]):

$$\min_u \int_\Omega \|\nabla u\| \tag{1.1a}$$

subject to the Dirichlet boundary conditions

$$u(x,y) = f(x,y) \quad \text{on} \quad \partial\Omega \tag{1.1b}$$

and the constraint

$$\int_\Omega c(x,y)\, u(x,y) = 1 \tag{1.1c}$$

Here $\|\nabla u\|$ denotes the Euclidean norm of the gradient of $u(x,y)$, i.e. $\sqrt{u_x^2+u_y^2}$. We shall restrict Ω to be a square with side two. The functions f and c are usually smooth functions, defined on $\partial\Omega$ and Ω respectively. Sometimes the constraint (1.1c) does not appear, but this is required to obtain a nontrivial solution if, for example, $f(x,y) \equiv 0$. In order for a minimum to be achieved, the appropriate space of functions to consider is BV, the functions of bounded variation. Strang shows, using a result of Fleming [9], that when $c(x,y) \equiv 1$, $f(x,y) \equiv 0$, the solution $u(x,y)$ is a characteristic function with a jump discontinuity along a curve Γ in Ω. Specifically, $u(x,y)$ has constant positive value inside Γ and zero value outside Γ in the following figure, where Γ is the solid curve and the broken lines represent $\partial\Omega$:

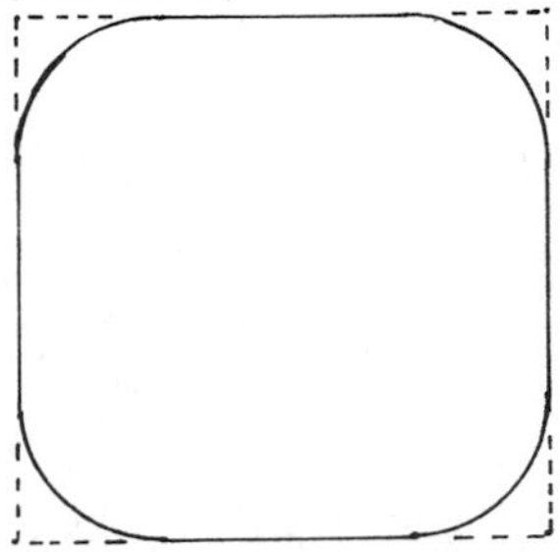

(1.2)

The curve Γ consists of four circular arcs of radius $\rho = 1/(1+\sqrt{\pi}/2) \simeq 0.530$, plus parts of the sides of the square. Once it is recognized that the solution is a characteristic function, (1.1) reduces to a classical isoperimetric problem, since the quantity to be minimized reduces to measuring the length of Γ while the constraint fixes the area inside Γ. The solution therefore defines the region in the square with minimal perimeter given fixed area (maximal area given fixed perimeter).

In this paper we are concerned with obtaining the numerical solution of a discrete approximation to (1.1). Let us triangulate Ω as follows:

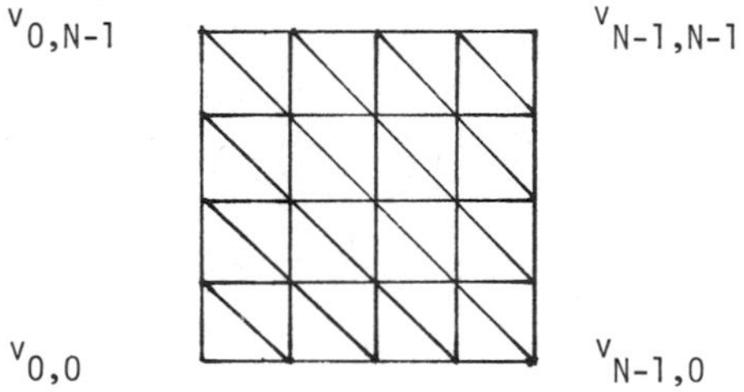

Let $h = 1/(N-1)$ be the mesh size, where there are N mesh points in each direction. We replace u in (1.1) by a piecewise linear finite element approximation v, obtaining a finite dimensional optimization problem. The variables are the unknown function values $v(x_i,y_j)$ at the given mesh points, $i = 0,\ldots,N-1$, $j = 0,\ldots,N-1$. Because v is piecewise linear, the norm of its gradient $v = (v_x,v_y)^T$ on a "lower" triangle:

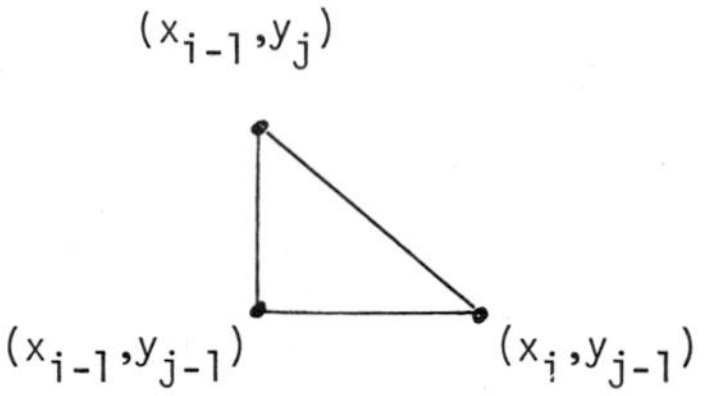

multiplied by the area of the triangle, is given by

$$\|r_{ij1}\| = \left\| \begin{bmatrix} \frac{h}{2}\,(v(x_i,y_{j-1}) - v(x_{i-1},y_{j-1})) \\ \frac{h}{2}\,(v(x_{i-1},y_j) - v(x_{i-1},y_{j-1})) \end{bmatrix} \right\| .$$

Let us write

$$r_{ij1} = A_{ij1}{}^T v \in \mathbb{R}^2$$

where v is the vector $\in \mathbb{R}^{N^2}$ of unknowns. The matrix A_{ij1} has dimension $N^2 \times 2$ and has only two nonzero components in each column. We use r_{ij2} and A_{ij2} for the corresponding notation for "upper" triangles. Thus the discretized problem is:

$$\min_v \sum_{i,j,k} \|r_{ijk}\| \qquad (i=1,\ldots,N-1,\ j=1,\ldots,N-1,\ k=1,2) \tag{1.3a}$$

subject to the boundary conditions

$$v(x_i,y_j) = f(x_i,y_j) \quad \text{for} \quad (x_i,y_j) \in \partial\Omega \tag{1.3b}$$

and the constraint

$$\frac{1}{N^2} \sum_{i,j} c(x_i,y_j)\, v(x_i,y_j) = 1 \,. \tag{1.3c}$$

Although (1.3) is a convex finite-dimensional optimization problem, it is not easily solved, because the objective function to be minimized is only piecewise differentiable with respect to v if any of the r_{ijk} equal zero. Now the residual r_{ijk} is zero precisely if the solution is constant on the corresponding triangle. The solution of (1.1) (with $c \equiv 1$, $f \equiv 0$) is constant everywhere except across Γ where it jumps. We expect therefore that the solution to the discretized problem will be constant over most of Ω but varying in some band around Γ. In this case (1.3) is indeed a nondifferentiable optimization problem.

AN ALGORITHM FOR MINIMIZING A SUM OF EUCLIDEAN NORMS

We have recently proposed [14] an algorithm for solving nondifferentiable optimization problems of this form. Somewhat more generally, it solves the convex problem:

$$\min_{v\in\mathbb{R}^n} F(v) = \sum_{i=1}^{m} \|r_i(v)\| \tag{2.1}$$

where

$$r_i(v) = A_i^T v - b_i \in \mathbb{R}^\ell .$$

We are changing the notation slightly, writing i where before we wrote the triple (i,j,k), and now allowing the residual r_i to be affine rather than linear in v. As before, $\|\cdot\|$ denotes the Euclidean vector norm. Linear constraints may be included without difficulty, but we omit them for convenience. The problem (2.1) dates back to Fermat and has an interesting history. In its simplest form, with $n = \ell = 2$, it asks for the point in the plane which minimizes the sum of distances between it and m given points. A more interesting version involves a weighted sum of the distances. With this variation, the solution coincides with one of the given points if the corresponding weight is large enough. This problem is also associated with the names of Steiner and Weber and is often known as the single facility location problem. The multifacility location problem arises when more than one point in the plane is to be determined, i.e. $n > \ell = 2$. The weighted sum of norms to be minimized then involves the distances between each pair of unknowns as well as the distances between the unknown and fixed points. This particular problem is discussed in more detail by Calamai and Conn [2,3,4], who give an algorithm closely related to ours.

The algorithm as described in [14] is intended to solve (2.1) when the matrices $\{A_i\}$ are dense and n is not too large; a full rank condition is also required. None of these conditions holds in (1.3). The purpose of this paper is to explain how to adapt the algorithm to solve (1.3), and to report on the advantages and difficulties of such an approach.

<u>Notation</u>. Using ∇ to denote gradient with respect to v, we have:

$$g_i(v) \equiv \nabla\|r_i(v)\| = \frac{1}{\|r_i(v)\|} A_i r_i(v)$$

$$G_i(v) \equiv \nabla^2\|r_i(v)\| = \frac{1}{\|r_i(v)\|} A_iA_i^T - \frac{1}{\|r_i(v)\|^3} A_ir_i(v)r_i(v)^TA_i^T$$

provided that $\|r_i(v)\| \neq 0$. Note that the Hessian term $G_i(v)$ is unbounded as $\|r_i(v)\| \to 0$, but that the gradient term $g_i(v)$ remains bounded, although it is of course discontinuous at v if $r_i(v) = 0$. Now let us define

$$g(v) = \sum_{\|r_i(v)\|\neq 0} g_i(v)$$

and

$$G(v) = \sum_{\|r_i(v)\|\neq 0} G_i(v) .$$

If the gradient and Hessian of F(v) are defined it is clear that they are given by g(v) and G(v) respectively. The discontinuities in the gradient and consequent unboundedness in the Hessian are what cause the difficulty in solving (2.1).

In order to be able to handle the discontinuities consider the following idea. Define the active set at a point v as

$$J(v) = \{i \mid \|r_i(v)\| = 0\}, \tag{2.2}$$

i.e. those indices associated with zero residuals. We refer to r_i as an active residual or active term at v if $i \in J(v)$. The idea of the algorithm is to project the objective function F(v) into the linear manifold where zero residuals remain unchanged. In this space we see that F(v) is locally continuously differentiable, since discontinuities occur only in directions crossing the manifold. To make this precise let

$$\hat{A} = \hat{A}(v) = [A_{i_1} A_{i_2} \ldots] \quad \text{where} \quad J = \{i_1, i_2, \ldots\}$$

i.e. the matrix whose columns are the coefficient matrices of the active residuals. Let $Z = Z(v)$ be a matrix with maximal column rank such that $\hat{A}^TZ = 0$, i.e. such that the columns of Z span the null space of $\hat{A}^T$. The significance of the matrix Z is that

$$r_i(v+p) = r_i(v) = 0 \quad \text{if} \quad i \in J(v) \quad \text{and} \quad p \in R(Z) ,$$

the range space of Z. Now for any v consider the matrix Z and define F restricted to the space $v \oplus R(Z)$ as

$$F_Z(p_Z) = F(v + Zp_Z).$$

Consider the gradient and Hessian of F_Z, which will be called respectively the projected gradient and projected Hessian. Because the active terms are fixed for all p_Z, we have

$$\nabla F_Z(p_Z) = Z^Tg(v + Zp_Z)$$

$$\nabla^2F_Z(p_Z) = Z^TG(v + Zp_Z)Z .$$

Thus the projected gradient and projected Hessian are defined and differentiable in a neighborhood of v regardless of whether any residuals are zero.

Optimality conditions. Let us consider conditions for a point v to be a solution to (2.1), making use of the above notation. Clearly a necessary condition for v to be a solution is that the projected gradient is zero, since otherwise F could be decreased along a direction in $R(Z(v))$. Thus for optimality we require

$$Z^T g(v) = 0 . \tag{2.3}$$

By the definition of Z an equivalent condition is

$$g(x) = \sum_{i \in J(v)} A_i t_i \tag{2.4}$$

for some vectors $t_i \in R^\ell$ for each $i \in J(v)$. We call the $\{t_i\}$ Lagrange vectors. By considering the change in F along directions not in $R(Z)$, it can be shown that a necessary and sufficient condition for v to solve (2.1) is that (2.4) hold for some vectors $\{t_i\}$ satisfying

$$\|t_i\| \leq 1 \quad \text{for all} \quad i \in J(v) \tag{2.5}$$

(see [12],[14], [19]).

The algorithm. The basic idea of the algorithm is as follows. Given a point $v^{(0)}$ with an active set $J(v^{(0)})$, we proceed to minimize the function restricted to $v^{(0)} \oplus R(Z(v^{(0)}))$, where the zero residuals remain unchanged. At the kth step of the iteration, a direction of search is computed by solving the projected Newton system of equations, utilizing the projected gradient $Z^T g$ and projected Hessian $Z^T GZ$. A special line search is used to obtain the next iterate $v^{(k+1)}$. During the course of this iteration we may reduce other residuals to zero, in which case they are added to the active set and the restricting manifold is reduced in dimension. Once the projected gradient $Z^T g$ is sufficiently small, the Lagrange vectors $\{t_i\}$ should be computed. These are uniquely defined if $\hat{A}$ has full column rank. If $\|t_i\| \leq 1$ for all $i \in J(v)$, the procedure terminates. If $\|t_j\| > 1$ for some j, the j is deleted from the active set J, the dimension of the manifold is increased, and the process is continued.

The details of the algorithm may be found in [14]. Here we outline the main steps which must be performed to obtain a new iterate $v^{(k+1)}$ from $v^{(k)}$. It is assumed that $\hat{A}$ has full column rank and that matrix factorizations are practical; such is not the case for (1.3).

Choose active set, compute Z, and make almost active terms exactly active if necessary. (2.6)

Find the active set J and the matrix $\hat{A}$ at $v^{(k)}$. Compute Z from the QR factorization of $\hat{A}$. If some residuals are almost active, compute a point $\tilde{v}$ which makes these terms exactly active. If $F(\tilde{v}) < F(v^{(k)})$, then replace $v^{(k)}$ by $\tilde{v}$ and restart this step; otherwise reject these terms as inactive.

Check optimality (2.7)

If the projected gradient norm $\|Z^T g\|$ is small, compute the Lagrange vectors $\{t_i\}$ using the QR factorization of $\hat{A}$. If $\|t_i\| \leq 1$ for all $i \in J$ then stop; solution has been approximated. Otherwise delete an appropriate term from J and set p to a projected steepest descent direction.

Solve projected Newton system (2.8)

If $\|Z^T g\|$ is not small, solve

$$(Z^T GZ) p_Z = -Z^T g$$

using a Choleski factorization and set the direction of search $p = Zp_Z$.

Line search (2.9)

Use a special line search, which takes account of the form of F, to obtain a point $v^{(k+1)} = v^{(k)} + \alpha p$, with $F(v^{(k+1)}) < F(v^{(k)})$.

One interesting aspect of the algorithm is illustrated by considering the weighted Fermat problem:

$$\min_{v \in \mathbb{R}^2} \| v - \begin{bmatrix} -1 \\ 0 \end{bmatrix} \| + \omega \| v - \begin{bmatrix} 0 \\ 1 \end{bmatrix} \| + \| v - \begin{bmatrix} 1 \\ 0 \end{bmatrix} \| . \qquad (2.10)$$

Let the initial iterate be given by, say, $v^{(0)} = [3,2]^T$. We have $J(v_0) = \emptyset$ and $Z(v_0) = I$. If $\omega = 1$, the solution v^* is a point with $J(v^*) = \emptyset$, and the iterates generated by the algorithm converge quadratically to v^* because of standard properties of Newton's method. Now suppose that $\omega = 2$. It is easily verified that the solution v^* is $[0,1]^T$, with J containing one index. Now there is no obvious reason to suspect that the convergence of $v^{(k)}$ to v^* will be rapid, since F is not differentiable at v^* and the usual quadratic convergence property does not apply. In fact the Hessian $\nabla^2 F$ is unbounded in any neighborhood of v^* and undefined at v^*. It is the case, however, that this ill-conditioning of $\nabla^2 F$, combined with the effect of the special line search algorithm, actually causes quadratic convergence to v^*. Once the convergence is recognized, the exact step to v^* may be taken by (2.6). Since the new restricting manifold has zero dimension, the Lagrange vector is then computed by (2.7), and the algorithm terminates. This quadratic convergence result is similar in flavor to the well known cubic convergence of the Rayleigh quotient iteration to find an eigenvector of a symmetric matrix. In both cases the superlinear convergence is caused by the fact that the ill-conditioning of the matrices is in precisely the desired direction. For further details see [14].

ADAPTING THE ALGORITHM TO SOLVE THE MODEL PROBLEM

Let us now consider how to adapt the algorithm of Section 2 to solve (1.3). It is convenient to continue using the notation of Section 2, i.e. writing A_i for A_{ijk}, n for N^2, etc. The first question is how to represent the active set and the matrix $\hat{A}$. First note that $\hat{A}$ is certain to be rank deficient. This is easily seen by considering the following example:

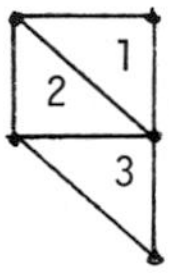

(3.1)

If the solution v is constant across triangles 1 and 3, then it must also be constant across triangle 2. Thus the three residuals being zero are not independent properties. In general, the independent columns of $\hat{A}$ can be written in the form

$$\begin{bmatrix} 1&1&1&1&&&&&\\ -1&&&&&&&&\\ &-1&&&&&&&\\ &&-1&&&&&&\\ &&&-1&&&&&\\ &&&&1&1&&&\\ &&&&-1&&&&\\ &&&&&-1&&&\\ &&&&&&1&1&\\ &&&&&&-1&&\\ &&&&&&&-1&\\ &&&&&&&&\ddots \end{bmatrix} \tag{3.2}$$

Recall that $\hat{A}^T p = 0$ is the equation which defines directions along which the active set does not change (zero residuals remain zero). Thus the first diagonal block of (3.2) corresponds to a region such as (3.1), where five function values $v(x_i, y_j)$ are constrained to retain the same common value. This common value, is, however, free to change. The other blocks of (3.2) correspond to other regions in Ω with constant v.

Now let us add columns to (3.2) to account for the constraints, since any direction of search p must also retain feasibility. Some of the regions with constant solution may include one or more boundary points, and hence the common function value must be fixed. Thus one column should be added to the corresponding block, making it square. Other blocks corresponding to regions away from the boundary remain unchanged. Boundary points not in any of the constant regions form their own 1×1 blocks. Let us re-order the blocks so that those corresponding to regions including boundary points appear first. The last rows, which contain no diagonal blocks, correspond to free interior points. Finally, one column corresponding to the constraint (1.3c) must be added. The matrix thus has the form:

$$\tilde{A} = \begin{bmatrix} 1&1&1&&&&&&&&&&&c_1\\ &-1&&&&&&&&&&&&c_2\\ &&-1&&&&&&&&&&&\\ &&&1&&&&&&&&&&\\ &&&&1&&&&&&&&&\\ &&&&&\ddots&&&&&&&&\vdots\\ &&&&&&1&1&1&1&&&&c_{\alpha_1}\\ &&&&&&-1&&&&&&&\vdots\\ &&&&&&&-1&&&&&&\\ &&&&&&&&-1&&&&&\\ &&&&&&&&&-1&&&&c_{\beta_1}\\ &&&&&&&&&&1&1&&c_{\alpha_2}\\ &&&&&&&&&&-1&&&\\ &&&&&&&&&&&-1&&c_{\beta_2}\\ &&&&&&&&&&&&\ddots&\vdots\\ &&&&&&&&&&&&&c_\gamma\\ &&&&&&&&&&&&&\vdots\\ &&&&&&&&&&&&&c_n \end{bmatrix}$$

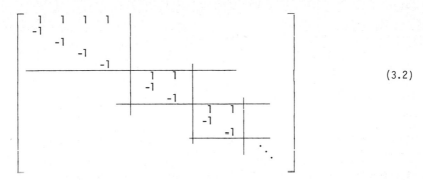

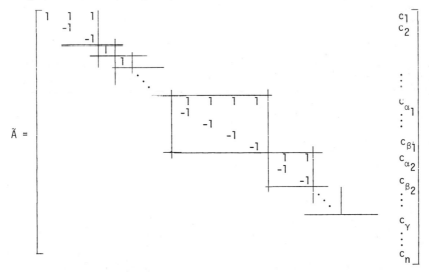

Storing and factoring $\tilde{A}$ would be prohibitively expensive. Fortunately, this is not necessary. A full rank matrix Z which spans $N(\tilde{A}^T)$ can be written in the following form:

$$Z = \left[\begin{array}{c|c|c|c} & 0 & & \\ \hline 1 & & & \\ 1 & & & \\ 1 & & & \\ 1 & & & \\ 1 & & & \\ \hline & 1 & & \\ & 1 & & \\ & 1 & & \\ \hline & & \ddots & \\ \hline & 0 & & \\ d_1 & d_2 & \cdots & \end{array}\right] \begin{array}{l} \\ (\text{row } \alpha_1) \\ \\ \\ \\ \\ \\ (\text{row } \alpha_2) \\ \\ \\ \\ \\ (\text{row } \gamma) \\ \\ \end{array}$$

$$\text{where } \; d_i = -\,(c_{\alpha_i} + \ldots + c_{\beta_i}) / c_n \,. \tag{3.3}$$

Note that Z is not orthogonal, which would be desirable for numerical stability. However, this is not critical. We must however ensure that we do not divide by a small number in (3.3). To avoid this, the largest of the values $|c_\gamma|,\ldots,|c_n|$ is used, with the rows interchanged accordingly. (Strictly speaking, there might be no single free interior points, i.e. $\gamma > n$, but in this unlikely case the definition of Z can be changed to have two or more dense rows instead of one.)

The best way to represent the active set and the matrices $\tilde{A}$ and Z seems to be as follows, using a linked list structure. One linked list is maintained for every block in $\tilde{A}$, including the null blocks corresponding to free interior points. These lists are of two kinds: <u>fixed</u> lists, corresponding to square blocks in $\tilde{A}$, and <u>variable</u> lists corresponding to other blocks. Thus each fixed list contains points whose common function value is fixed by the boundary conditions, and each variable list contains points whose common function value may vary. Initially, if $v^{(0)}$ has no constant regions, each boundary point is put in its own fixed list and each interior point is put in its own variable list. (Neumann conditions could also be handled by putting two adjacent points from each segment orthogonal to the boundary in their own variable list.) We also maintain a table which specifies the list containing each point.

We now focus on the best ways to carry out steps (2.6) through (2.9) of the algorithm, using the data structure just described. Let us first consider (2.6). At the beginning of each iteration, the list structure has a form which defines the values of J (and Z) at the previous iteration. It is now first necessary to add to J any indices corresponding to residuals which became "exactly" zero during the line search. This is done by merging the appropriate lists. For example, consider

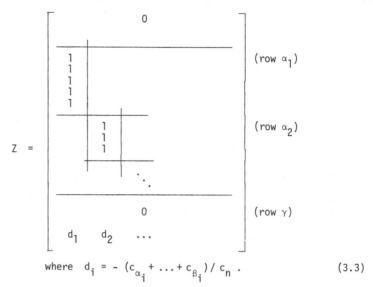

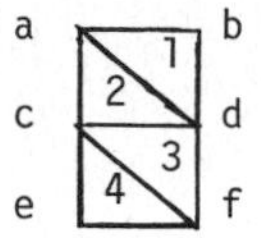

(3.4)

supposing that at the previous iteration there was a zero residual across triangle 1 only, and that the values of v at points b, d, f are fixed by the Dirichlet boundary conditions. Reflecting this, there would be one fixed list specifying the values of v at (a,b,d), a second fixed list specifying the value of v at f, and two variable lists reflecting the free values at c and e. Now suppose that in the subsequent line search, residuals were reduced to zero across triangles 2 and 3 (this assumes that the Dirichlet data at d and f are the same, and involves a change only to the value at c). The two fixed lists and the variable list for c would all be merged, creating one fixed list and reflecting the new definitions of J and Z. The merge operation is trivial as it simply requires appending one linked list to another.

Step (2.6) must also consider forcing residuals which are "almost" zero to become "exactly" zero. This is a more complicated operation for two reasons. First, the current list structure must be saved, since the new vector $\tilde{v}$, for which the relevant residuals are exactly zero, may have a higher value of F than the current iterate, in which case it must be discarded. Second, the vector $\tilde{v}$ must be chosen to satisfy the constraint (1.3c). When two variable lists are merged, with two slightly different common values of v, the new variable common value can be chosen to ensure that (1.3c) is satisfied, provided that $c(x_i,y_j)$ is a strictly positive function on the whole mesh. This is the case for Strang's model problem where $c \equiv 1$. When a variable and a fixed list are merged, the variable common value must be changed to the fixed common value, potentially violating (1.3c). The simplest way to deal with this situation is to perform all the merge operations, and then restore feasibility by scaling all the variable values by a constant factor. Again, it is easy to see that this is possible provided c is a positive function. The merging of two fixed lists is not permitted when the residual is not "exactly" zero, i.e. within machine accuracy.

It is of some interest to note that in practice, many residuals are reduced exactly to zero by the line search. This contrasts with the situation for the single and multi-facility location problems, e.g. (2.10) with $\omega = 2$, where the convergence of the residual to zero is an iterative process with quadratic convergence. The reason for the different behavior here is not the rank deficiency of $\hat{A}$ per se, since this is also present in a different form for the multifacility location problem. Consider again the example (3.4). Given any $p \in R(Z)$, the only component relevant to the proposition of incorporating triangle 2 into the constant region of triangle 1 is the change in the value of v at c, or, if (a,b,d) are in a variable list, the changes of the value at c and the value at (a,b,d). Thus it is normally possible to pick a steplength α which will make $v + \alpha p$ constant across triangles 1 and 2 (when (a,b,d) are in a variable list, it is just a matter of weighting the two values correctly). In other words, for almost any $p \in R(Z)$, the extra residual can be reduced to zero along p. In the weighted Fermat problem (2.10), with $\omega = 2$, almost no direction points from a given v in the plane to the solution v^*, the only point where the relevant residual is zero.

Let us now consider the question of solving the projected Newton system (2.8). It is clearly best to use an iterative method to solve this linear system, so that it is not necessary to form or store Z^TGZ. Instead, a subroutine is required to multiply Z^TGZ onto any vector y. This can be accomplished by first multiplying by Z, then by G, and then by Z^T. Thus subroutines are required to multiply any vector y by G, Z and Z^T respectively. The first uses:

$$Gy = \sum_{i \notin J} \frac{1}{\|r_i\|} A_i (A_i^T y - \frac{(r_i^T A_i^T y)}{\|r_i\|^2} r_i)$$

where the terms corresponding to active residuals are omitted as explained in Section 2. The subroutines which multiply a vector y by Z and Z^T use the list structure, which represents Z, directly. These are also required to compute the vectors $Z^T g$ and Zp_Z. Each of the three subroutines requires O(n) operations for the matrix-vector multiplication.

Following [7], we do not usually carry out the iterative solution of the linear system to full accuracy. Let us use the conjugate gradient method to solve the system, denoting its iterates by w_k; the relevant formulas may be found in [7], with $w_0 = 0$. We terminate the iteration, setting $p_Z = w_k$, if $\|Z^T G Z w_k + Z^T g\| \leq \text{CGTOL}\|Z^T g\|$, but with a maximum of $k \leq$ CGMAX imposed unless $\|Z^T g\| \leq$ CGTHRESH. This rule is rather arbitrary, but it allows avoiding the expense of too many conjugate gradient iterates when an accurate solution of (2.8) is not required, namely away from the minimum on the current manifold. The overall work to approximately solve (2.8) is thus of the order of n times the number of conjugate gradient iterates used. Of course, it would be desirable to precondition the conjugate gradient iteration, but it is not clear how to do this since $Z^T GZ$ is not explicitly available.

Let us now consider the line search (2.9). There is no difficulty in using the line search of [14] unchanged, since all the necessary information is available. Usually one step of this line search obtains a sufficient decrease in F. This step requires estimating the zeros of all the inactive residuals in the direction along the line. Let us say that there are q inactive residuals, and let $\{z_j\}$, j = 1,...,q, be the estimates of the zeros. It is explained in [14] how these values are compared with an estimate of the minimum of F along the line, supposing that F is smooth. If F does not appear to be smooth, because of the fact that the latter estimate lies beyond one or more of the $\{z_j\}$, it is necessary to compute the minimum of a piecewise linear function defined by the $\{z_j\}$. This computation amounts to finding a weighted median of q points, and is most easily done by a partial "Heapsort" of the values, which requires O(q log q) operations. However, it is possible to do this computation in O(q) time, although this is not worthwhile unless q is very large. See [1,15] for further details. In any case, this is not a significant factor at least in the latter stages of minimizing (1.3), since then q, the number of inactive residuals, is much less than m, the number of triangles.

Let us turn our attention to (2.7), by far the most difficult step. As already mentioned, the matrix $\hat{A}$, when all columns are included, is highly rank deficient, and hence the Lagrange vectors of (2.4) are not uniquely defined. Consequently it is very difficult to tell whether a given point v, which is optimal on the current restricting manifold, is in fact optimal in the whole space. Furthermore, even if it were known that v is not optimal, it would be very difficult to decide which terms to delete from the active set to obtain a descent direction. This type of difficulty is called <u>degeneracy</u> in the context of linear programming and linearly constrained optimization. The only known methods guaranteed to overcome degeneracy require time exponential in the number of active terms, which is out of the question. We therefore have the unsatisfactory situation that we must terminate the algorithm with a "solution" that we know to be optimal on the final restricting manifold, but for which we are unable to verify optimality in the whole space.

This difficulty seems to be inherent in problem (1.3), because of the extreme built-in degeneracy. However, there are a number of positive remarks which can be made. First, assuming that the tolerances of [14] are chosen small enough so that an accurate solution is demanded, note that there is no possibility of the algorithm terminating with too few residuals set to zero. The only danger is that too many of them may have been set to zero. Second, the situation is comparable to the well known problem of finding the global minimum of a nonconvex function. Several runs may be made, using different starting values or different choices of the parameters, and the lowest "minimum" found may be taken as the best candidate for the solution. Third, and most important, the numerical results of the next section indicate that, at least for a certain initial guess $v^{(0)}$, the "solutions" to which the algorithm converges in practice are very nearly optimal. Fourth, we mention a suggestion of Dax [6] at the end of the next section.

We conclude this section by describing an alternative approach to solving (1.3), based on ideas of [6,8] and others. This requires consideration of the hyperbolic approximating (HAP) function of [8], given by

$$\hat{F}(v) = \sum_{i=1}^{m} (\|r_i(v)\|^2 + \varepsilon_H^2)^{1/2} . \qquad (3.5)$$

The function $\hat{F}$ is differentiable everywhere provided that $\varepsilon_H > 0$. The smaller the value of ε_H, the better $\hat{F}$ approximates F, but a consequence of this, of course, is that the smaller the value of ε_H, the worse is the conditioning of $\hat{F}$. The alternative method, then, simply requires minimizing $\hat{F}$ directly, using a straightforward Newton method. This has the advantage that the degeneracy is avoided and optimality of a point minimizing $\hat{F}$ can be verified. However, there are four serious difficulties with this approach. First, a minimum of $\hat{F}$, not F, is obtained. This may not be important, since (1.3) is derived from (1.1) by a discretization, but it means that a solution to (1.3) is not obtained. Second, if ε_H is small, the extreme ill-conditioning of $\hat{F}$ makes the convergence of a Newton method very slow. Third, each Newton step requires the solution of a linear system with much larger dimension than that of (2.8), since there are no zero residuals to reduce the dimension of the space. Especially in the last stages of the iteration, the dimension of (2.8) is much less than n, and consequently (2.8) can be approximately solved using fewer conjugate gradient iterates than a full Newton system requires. Fourth, if ε_H is small, the conditioning of the full Newton system is much worse than the conditioning of (2.8), and hence the convergence of the "inner" iteration can be expected to be much slower than that for (2.8). This consideration is separate from the convergence of the "outer" nonlinear iteration (the second point) and the dimension of the linear system (the third point).

This alternative algorithm, minimizing $\hat{F}$, is easily implemented in the same program that was prepared for the direct minimization of F. In particular, although the alternative algorithm does not allow zero residuals, the same linked list data structure is used to impose the boundary conditions and the constraint (1.3c). Thus the "full Newton" system is actually implemented in the form (2.8), but with the matrix Z having rank equal to the number of variables minus the number of constraints. This corresponds to a matrix $\tilde{A}$ containing only the columns which describe the constraints, i.e. the 1×1 blocks for the Dirichlet conditions, and the last column for (1.3c).

NUMERICAL RESULTS

The results were obtained using Fortran on a VAX 11/780 at the Courant Mathematics and Computing Laboratory. Double precision arithmetic, i.e. approximately 16 decimal digits of accuracy, was used throughout. We restricted the experiments to Strang's model problem, i.e. (1.1) with $f \equiv 0$ and $c \equiv 1$, whose solution is shown in (1.2). Because of the symmetry, we actually solved the problem on the lower left quarter of the square shown in (1.2). Thus the domain was a unit square, with N mesh points in each direction, homogeneous Dirichlet boundary conditions on the lower and left sides, and with the points on the other two sides constrained to match pairwise. This is conveniently implemented using the linked list data structure, with an initial configuration of one fixed list for each point on the lower and left sides, one variable list for each pair of points on the upper and right sides, and one variable list for the top right corner point and each interior point.

We chose the initial vector $v^{(0)}$ as follows, where i and j range from 0 (lower or left side) to N-1 (upper or right side):

$$v^{(0)}(x_i,y_j) = \begin{cases} 0 & \text{if } i=0 \text{ or } j=0 \text{ (Dirichlet boundary conditions)} \\ N-1 & \text{if } i+j \geq N \\ i+j-1 & \text{otherwise} \end{cases}$$

This vector was then scaled to satisfy (1.3c). The corresponding piecewise linear function is constant, i.e. has zero residuals, in the shaded part of:

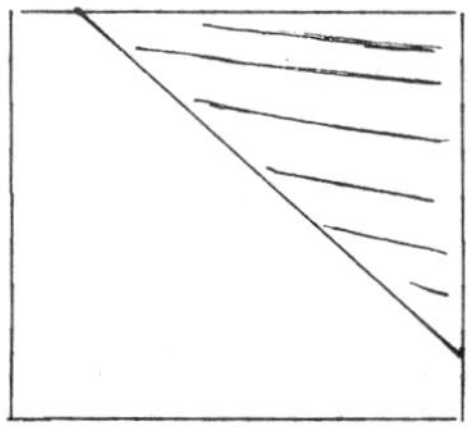

and is linear across the remainder of the grid except at the boundary. This choice of $v^{(0)}$ is a compromise between $v^{(0)}$ = constant in the interior, which has too many zero residuals, and v = linear, not constant, in the interior, which having no zero residuals causes Z^TGZ to be singular initially.

The results are summarized in Tables I and II. Table I gives the results of solving (1.3) when high accuracy is demanded. In this case the parameters of [14] were set as follows: $\varepsilon_{MCH} = 10^{-16}$, $\varepsilon_{ACT} = 10^{-4}$, $\varepsilon_{PG} = 10^{-6}$ (except 10^{-5} for N = 33), $\varepsilon_{LINE} = 10^{-12}$, $\eta = 0.9$. Of these, the most important is ε_{PG}, which specifies how small the norm of the projected gradient, $\|Z^Tg\|$, is to be made. The parameter ε_{ACT} determines when an attempt is made in (2.6) to make "almost" active residuals "exactly" active. This parameter may be reduced dynamically by the program; see [14] for details. The conjugate gradient iteration parameters were set as follows: CGTOL = 10^{-1}, CGTHRESH = 10^{-2}, and CGMAX as shown in the table. The other columns in the table specify N (the number of mesh points in each direction), the column rank of Z at the computed solution (i.e. the dimension of the final restricting manifold), the final value of F, ITER (the number of major iterates required), and CGITER (the total number of conjugate gradient iterates). The most interesting thing to note in Table I is the consistency of the final values of F determined by the various choices of CGMAX. The dependence of CGITER, the best measure of the total amount of work, on CGMAX, is also interesting although somewhat inconclusive.

The minimizing solution v is shown for the case N = 13 in Table III. Note that there are two regions where v is constant, the large one in the top right and the small one in the bottom left. The band of varying values between the two regions is, of course, the result of the discretization. One can see that the circular arc Γ of (1.2) is reflected in this band of varying values since, for example, $v(x_2,y_2) = 0.209$ is greater than $v(x_1,y_3) = v(x_3,y_1) = 0.149$.

In order to show how well the solution of (1.3) approximates the solution of (1.1) as N is increased, we display in Figure 1 graphs of the solution v plotted along the diagonal of the square. The results from Table I are shown for N = 9, 17 and 33. The step function shown is the analytic solution of (1.1). The distance from the mesh point (0,0) to the discontinuous jump in the solution, divided by the length of the diagonal, is $\rho(1-1/\sqrt{2}) \approx 0.155$, where ρ is given following (1.2). The analytic solution has a value $u = 1/(1-\rho^2+\pi\rho^2/4) \simeq 1.064$ in the constant positive region, with a corresponding minimal F value of $2-2\rho+\pi\rho/2 \simeq 1.772$.

Table II summarizes results when a much less accurate solution of (1.3) is required. Here we show results both for the direct minimization of F and for the minimization of the HAP function $\hat{F}$, described at the end of Section 3. In the latter case, we set $\varepsilon_H = 10^{-1}\cdot(h^2/2)$, where $h = 1/(N-1)$, since the residual sizes are proportional to the areas of the triangles. If a much larger value of ε_H is used, the function $\hat{F}$ approximates F very poorly; if a much smaller value is used, the computation time becomes excessive. We set ε_{PG}, the tolerance on the projected gradient, to 10^{-1}. Again, a much smaller value of ε_{PG} leads to excessive computation time, since $\hat{F}$ is ill-conditioned. We therefore also used $\varepsilon_{PG} = 10^{-1}$ for the runs which minimize F directly. We also set $\varepsilon_{ACT} = 10^{-2}$ for these runs, except $\varepsilon_{ACT} = 10^{-4}$ for N = 33; this is actually relevant only for $N \leq 9$, since for larger values of N the residual sizes are smaller. The other parameters had the same values as for Table I. Note that it is F, not $\hat{F}$, which is shown in the column for the HAP results, as well as for the direct results. The interesting observation to make from Table II is that even when an inaccurate solution is required, the method which directly minimizes F has substantial advantages over the HAP method. In order to show the difference in the accuracy of the solutions from Tables I and II, the solution obtained from minimizing the HAP function when N = 17 is plotted in Figure 1.

Dax [6] has made an interesting suggestion regarding the use of the HAP function in the context of multifacility location problems. He suggests switching to the minimization of $\hat{F}$ only after reaching the minimum of F on the current manifold, as a way of potentially escaping from a nonoptimal point. If a lower value of F is obtained in the process, a switch can be made back to the minimization of F. This seems a useful idea, especially if $v^{(0)}$ is poorly chosen, e.g. $v^{(0)}$ = constant in the interior. In this case the direct algorithm would terminate immediately, but a switch to minimizing $\hat{F}$ produces a lower value of F without difficulty. However, this idea does not produce any benefits for the runs given in Table I, apparently because the direct minimization produced very nearly optimal results.

CONCLUDING REMARKS

In this paper we have confined our attention to the description of a method which obtains accurate solutions to (1.3), a discretization of (1.1). Some alternatives to (1.1) which we have not considered include allowing discontinuous elements (Johnson [11]) and solving the primal problem for which (1.1) is the dual (Strang [16]). Other relevant papers include [5,10,13,17,18]. We conclude with the remark that if a method such as the one described in this paper is to be used to obtain very accurate solutions to (1.1) or related problems, some sort of mesh refinement near the curve of jump discontinuity would probably be desirable.

ACKNOWLEDGMENTS

The author would like to thank Gene Golub for introducing him to the topic of this paper. Thanks also go to Robert Kohn and Olof Widlund for many helpful conversations, and to Gilbert Strang for his interest and comments.
This work was supported in part by NSF Grant MCS-8302021 and in part by the U.S. Department of Energy under Contract DE-AC02-76-ER03077-V.

REFERENCES

[1] Bleich, C. and Overton, M. L., A linear-time algorithm for the weighted median problem, Computer Science Dept. Report No. 75, Courant Institute of Mathematical Sciences, New York (April 1983).

[2] Calamai, P. H., On numerical methods for continuous location problems, Ph.D. Thesis, Dept. of Systems Design, University of Waterloo, Waterloo, Ontario (1983).

[3] Calamai, P. H. and Conn, A. R., A stable algorithm for solving the multi-facility location problem involving Euclidean distances, SIAM J. Scient. and Stat. Comp. 1 (1980) 512-526.

[4] Calamai, P. H. and Conn, A. R., A second-order method for solving the continuous multifacility location problem, in: G. A. Watson (ed.), Numerical Analysis (Proc. 9th Biennial Conf., Dundee, Scotland), Lecture Notes in Mathematics 912 (Springer Verlag, 1982), 1-25.

[5] Christiansen, E., Computation of limit loads, Report 2, Matematisk Institut, Odense University (1979).

[6] Dax, A., The use of Newton's method for solving Euclidean multifacility location problems, Hydrological Service, P. O. Box 6381, Jerusalem (1983).

[7] Dembo, R. S. and Steihaug, T., Truncated Newton algorithms for large scale unconstrained optimization, Math. Programming 26 (1983) 190-212.

[8] Eyster, J. W., White, J. A. and Wierwille, W. W., On solving multifacility location problems using a hyperboloid approximation procedure, AIIE Transactions 5 (1973) 1-6.

[9] Fleming, W. H., Functions with generalized gradients and generalized surfaces, Annali di Matematica 44 (1957) 93-103.

[10] Grierson, D. E., Collapse load analysis, NATO-ASI Lecture Notes on engineering plasticity by mathematical programming, University of Waterloo, Waterloo, Ontario (August 1977).

[11] Johnson, C., Error estimates for some finite element methods for a model problem in perfect plasticity, Report, Chalmers University of Technology, Gothenburg, Sweden (1981).

[12] Kuhn, H. W., On a pair of dual nonlinear programs in: J. Abadie (ed.), Nonlinear Programming (North-Holland, Amsterdam, 1967) 38-54.

[13] Matthies, H., Problems in plasticity and their finite element approximation, Ph.D. Thesis, Mass. Inst. of Tech. (1978).

[14] Overton, M. L., A quadratically convergent method for minimizing a sum of Euclidean norms, Math. Programming 27 (1983) 34-63.

[15] Reiser, A., A linear selection algorithm for sets of elements with weights, Information Processing Letters 7 (1978) 159-162.

[16] Strang, G., A minimax problem in plasticity theory, in: Nashed, M.Z. (ed.), Functional Analysis Methods in Numerical Analysis, Lecture Notes in Mathematics 701 (Springer Verlag, 1979).

[17] Strang, G. and Matthies, H., Mathematical and computational methods in plasticity, presented at Sept. 1978 IUTAM Conf. on Variational Methods in the Mechanics of Solids, Evanston, Illinois (1979).

[18] Yang, W. H., A practical method for limit torsion problems, Computer Meth. in Appl. Mech. and Eng. 19 (1979) 151-158.

[19] Rockafeller, R.T., Convex Analysis (Princeton University Press, 1970).

TABLE I

Results of Minimizing F Accurately ($\varepsilon_{PG} = 10^{-6}$)

N	CGMAX	RANK(Z)	F	ITER	CGITER
5	5	3	2.5377626	7	14
	10	3	2.5377626	7	14
9	5	9	2.2305853	24	92
	10	9	2.2305853	22	91
13	5	19	2.1199518	59	231
	10	20	2.1198360	57	383
	15	20	2.1198360	40	286
	20	21	2.1198359	41	400
17	5	35	2.0629985	122	623
	10	36	2.0629969	78	496
	15	41	2.0629921	57	533
	20	41	2.0629921	85	919
33	5 ($\varepsilon_{PG}=10^{-5}$)	140	1.9759079	562	5133

TABLE II

Results of Minimizing Both F and the HAP Function $\hat{F}$ Inaccurately ($\varepsilon_{PG} = 10^{-1}$)

		Results of Minimizing F Directly				Results of Minimizing the HAP Function $\hat{F}$			
N	CGMAX	RANK(Z)	F	ITER	CGITER	RANK(Z)	F	ITER	CGITER
5	5	3	2.542	4	5	12	2.553	7	30
	10	3	2.542	4	5	12	2.549	5	30
9	5	9	2.236	10	18	56	2.260	25	120
	10	9	2.239	10	29	56	2.252	12	103
13	5	54	2.184	8	12	132	2.169	44	215
	10	27	2.128	17	57	132	2.155	18	163
	15	25	2.122	26	126	132	2.146	16	159
	20	18	2.121	20	110	132	2.142	11	200
17	5	79	2.099	20	43	240	2.131	48	231
	10	66	2.083	22	95	240	2.102	30	278
	15	55	2.077	22	134	240	2.104	20	207
	20	62	2.082	15	118	240	2.100	16	226
33	5	438	2.060	24	79	992	2.101	46	197
	15	370	2.031	92	874	992	2.051	33	385

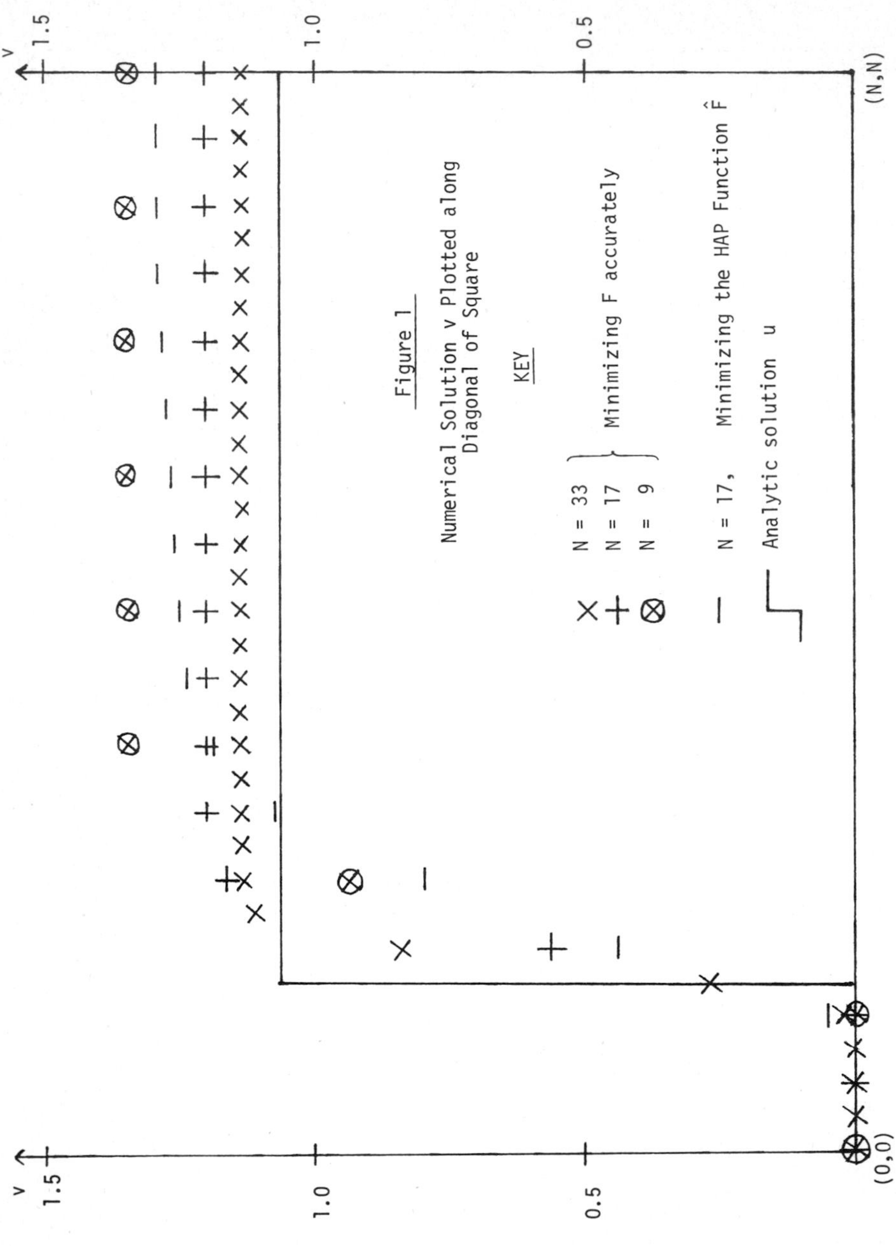

Figure 1
Numerical Solution v Plotted along Diagonal of Square

TABLE III
Numerical Solution $v(x_i,y_j)$ for N = 13 ($\varepsilon_{PG} = 10^{-6}$, CGMAX = 10,15,20)

j \ i	0	1	2	3	4	5	6	7	8	9	10	11	12
12	0	1.252	1.252	...								...	1.252
11	0	1.252	1.252	...									⋮
10	0	1.252	1.252	...									
9	0	1.252	1.252	...									
8	0	1.252	1.252	...									
7	0	1.232	1.252	1.252	...								
6	0	1.123	1.252	1.252	...								
5	0	0.857	1.229	1.252	1.252	...							
4	0	0.482	1.077	1.242	1.252	1.252	...						
3	0	0.149	0.686	1.120	1.242	1.252	1.252	1.252	...				
2	0	0	0.209	0.686	1.077	1.229	1.252	1.252	1.252	1.252	1.252	1.252	1.252
1	0	0	0	0.149	0.482	0.857	1.123	1.232	1.252	1.252	1.252	1.252	1.252
0	0	0	0	0	0	0	0	0	0	0	0	0	0

INCOMPRESSIBLE FLOWS

ECOULEMENTS INCOMPRESSIBLES

Computing Methods in Applied Sciences and Engineering, VI
R. Glowinski and J.-L. Lions (Editors)
Elsevier Science Publishers B.V. (North-Holland)

Global existence theorems on nonstationary stratified flow of incompressible viscous fluid

Hisashi OKAMOTO

Department of Math., Faculty of Science
University of Tokyo
113 Japan

Stratified flow means here that the fluid is nonhomogeneous (i.e., the density is not constant) but the density is invariant along the stream-line of the fluid particles. We deal with the equations describing unsteady motion of nonhomogeneous incompressible viscous fluid. This system of equations is a generalization of the Navier-Stokes system. We obtain existence and uniqueness of a strong solution. Global solution is obtained under conditions very similar to those in the Navier-Stokes system, i.e., we always have a global solution in $\Omega \subset \mathbb{R}^2$, while, in $\Omega \subset \mathbb{R}^3$, global existence is ensured if the initial values are small.

§1. Introduction.

We consider nonhomogeneous incompressible viscous fluid. Incompressibility and nonhomogeneity can occur at the same time, from which follows the nullity of the Lagrange differentiation of the density. Thus the system of the equations which we shall deal with is written as follows.

$$(1.1)\qquad \frac{\partial\rho}{\partial t} + u\cdot\nabla\rho = 0 \qquad (0 < t , \; x \in \Omega),$$

$$(1.2)\qquad \rho\{ \frac{\partial u}{\partial t} + (u\cdot\nabla)u \} = \Delta u - \nabla p \qquad (0 < t , \; x \in \Omega),$$

$$(1.3)\qquad \operatorname{div} u = 0 \qquad (0 < t , \; x \in \Omega).$$

Here ρ, u, p are mass density, velocity vector and pressure of the fluid, respectively. Ω is a flow region, which is assumed to be a bounded domain in $\mathbb{R}^2$ or $\mathbb{R}^3$ with a smooth boundary $\partial\Omega$. For simplicity we assume that the viscosity is equal to the unity. We solve the system (1.1-3) under the initial-boundary conditions below.

$$(1.4)\qquad u(0,x) = a(x) , \qquad \rho(0,x) = \rho_0(x) \qquad (x \in \Omega),$$

$$(1.5)\qquad u|_{\partial\Omega} = 0 \qquad (0 < t).$$

If $\rho_0(x) \equiv$ constant, then our system reduces to the Navier-Stokes system. Therefore it is just natural that we can prove the exitence theorems for (1.1-5) which are almost of the same form as those for the Navier-Stokes system.

Mathematical study for the system (1.1-5) was initiated by Kazhikhov [3]. There he proved existence of a weak solution of Hopf-type and also a classical solution. However, he did not show the uniqueness of the solution. On the other hand, Lions [5] constructed a weak solution of another kind under most general assumptions on a and ρ_o. Unique solution was obtained in Ladyzhenskaya and Solonnikov [4]. They used Sobolev space of L^p-type and obtained a unique strong solution, requiring that $p > n$ (= the dimension of Ω). Their solution is global for $\Omega \subset \mathbb{R}^2$ but local in time for $\Omega \subset \mathbb{R}^3$.

Our aim is to show the unique existence of a strong solution and to obtain a global solution for $\Omega \subset \mathbb{R}^2$ and also for $\Omega \subset \mathbb{R}^3$ under reasonable conditions on a and ρ_o. Roughly speaking, our results are as follows.

I) If $\Omega \subset \mathbb{R}^2$ and if a and ρ_o are sufficiently smooth, then the solution exists globally and uniquely.

II) If $\Omega \subset \mathbb{R}^3$, a is sufficiently smooth and small and if ρ_o is sufficiently smooth and sufficiently close to a constant, then the solution exists globally and uniquely.

We first construct a solution local in time, using the Kato-Tanabe theory of evolution equations. Then global solution is obtained by connecting local solutions successively, of which the two-dimensional case is dealt with by making use of a certain estimate in [4], while our new result for the three-dimensional case is a consequence of an argument inspired by Matsumura and Nishida [7]. We use Sobolev spaces of L^2-type, which makes it possible to base our methods on the theory of evolution equations in Hilbert spaces. Moreover, the idea for obtaining a local solution is basically the same as that in Fujita and Kato [1]. Complete proofs of our theorems are in Okamoto [9].

§2. Mathematical formulation.

We solve (1.1-5) by the method employed in Fujita and Kato [1]. To this end we use the following notation (n is the dimension of Ω, $n = 2$ or 3).

$C_{o,\sigma}^{\infty}(\Omega) = \{ v = (v_1,\cdots,v_n) \in C_o^{\infty}(\Omega)^n \; ; \; \mathrm{div}\, v = 0 \}$,

V ; the closure of $C_{o,\sigma}^{\infty}(\Omega)$ in $H^1(\Omega)^n$,

H ; the closure of $C_{o,\sigma}^{\infty}(\Omega)$ in $L^2(\Omega)^n$.

We denote by P the orthogonal projection from $L^2(\Omega)^n$ onto H. The L^2-norm is denoted by $\| \;\; \|$ and L^{∞}-norm by $\| \;\; \|_{\infty}$. Then we define what is called the Stokes operator A by

$$A = -P\Delta , \qquad D(A) \text{ (= the domain of } A \text{) } = H^2(\Omega)^n \cap V.$$

We remark that A is a positive definite self-adjoint operator in H.

Now we formulate (1.1-5) as a quasilinear evolution equation in $W^{1,\infty}(\Omega)\times H$:

$$\frac{\partial \rho}{\partial t} + u\cdot\nabla\rho = 0 \qquad \text{a.e. } (t,x) \in [0,\infty[\, \times\Omega , \tag{2.1}$$

$$P\rho(t)\frac{du}{dt} + Au(t) + F_{\rho}u(t) = 0 \qquad (0 < t), \tag{2.2}$$

$$u(0) = a , \quad \rho(0,x) = \rho_o(x). \tag{2.3}$$

Here we have put $\rho(t) = \rho(t,\cdot)$, $F_{\rho}w = P\{\rho(t)(w\cdot\nabla)w\}$. Note that $r \in W^{1,\infty}(\Omega)$ is equivalent to say that r is uniformly Lipschitz continuous. Our goal is to show the following three theorems.

THEOREM 1. Suppose that $a \in D(A^{\eta})$ for some $\eta > n/4$ and that $\rho_0 \in W^{1,\infty}(\Omega)$. Suppose also that, for some positive constants m and ℓ, the function ρ_0 satisfies

$$m < \rho_0(x) < \ell \qquad (x \in \Omega).$$

Then there exists a positive constant T_0 such that we have a unique solution

$$u \in C([0,T_0];D(A^{\eta})) \cap C(]0,T_0];D(A)) \cap C^1(]0,T_0];H)$$

and

$$\rho \in W^{1,\infty}(]0,T_0[\times\Omega).$$

Here and hereafter $D(A^{\eta})$ is the domain of A^{η} and $D(A^{\eta}) = H^{2\eta}(\Omega)\cap V$ in our case.

THEOREM 2. If Ω is a two-dimensional domain, then the solution always exists in $[0,\infty[$.

THEOREM 3. If Ω is a three dimensional domain, then there exists a global solution so long as $\| A^{\eta}a\|$ and $\| \nabla\rho_0\|_{\infty}$ are sufficiently small.

§3. Outline of the proof of THEOREM 1.

We consider only the three dimensional case. As in [1] we adopt the successive approximation method. The scheme is as follows. Firstly put

$$u_1(t) = e^{-tA}a\,, \qquad \rho_1(t,x) = \rho_0(x).$$

Then, for n = 1,2,···, we solve

$$\text{(3.1)} \qquad \frac{\partial\rho_{n+1}}{\partial t} + u_n\cdot\nabla\rho_{n+1} = 0\,, \qquad \rho_{n+1}(0,x) = \rho_0(x)$$

and

$$\text{(3.2)} \qquad P\rho_n(t)\frac{du_{n+1}}{dt} + Au_{n+1} = -F_{\rho_n}u_n(t)\,, \qquad u_{n+1}(0) = a.$$

Since (3.2) is linear with respect to u_{n+1}, we can apply standard theory of linear equations of evolution (Tanabe [8]).

The initial value problem (3.1) is explicitly solved by the method of characteristics to yield

$$\rho_{n+1}(t,x) = \rho_0(\xi_{0,t}(x))$$

where $\xi_{s,t}(x)$ is a solution of

$$\text{(3.3)} \qquad \frac{d}{ds}\xi_{s,t}(x) = u_n(s,\xi_{s,t}(x))\,, \qquad \xi_{t,t}(x) = x.$$

To solve (3.2) we put $B_n(t) = P\rho_n(t)\cdot$. Then (3.2) is rewritten as

$$\frac{du_{n+1}}{dt} + B_n(t)^{-1}Au_{n+1}(t) = -B_n(t)^{-1}F_{\rho_n}u_n(t), \qquad u_{n+1}(0) = a.$$

We solve this equation by showing that $B_n(t)^{-1}A$ ($0 < t < T$) generates an evolution operator $U(t,s)$ ($0 \leqq s \leqq t \leqq T$) in H (see,e.g., Tanabe [8]). For any $T > 0$ fixed, it is shown that the approximate solutions $\{\rho_n, u_n\}_{n=1}^{\infty}$ exists in [0,T]. Then we choose $T_0 \leqq T$ such that u_n is convergent in $C([0,T];D(A^{\eta}))$.

We now examine technical details of the scheme above. We first consider (3.1). The following proposition ensures existence of ρ_{n+1} and several estimates needed for proving the convergence of $\{\rho_n, u_n\}$.

PROPOSITION 3.1 (Ladyzhenskaya and Solonnikov). Suppose that $\rho_0 \in W^{1,\infty}(\Omega)$ and that we are given a function $u \in C([0,T]\times\Omega)$ with

$$L \equiv \int_0^T \| \nabla u(t) \|_{\infty} dt < +\infty. \tag{3.4}$$

Then there exists a unique $\rho \in W^{1,\infty}(]0,T[\times\Omega)$ such that

$$\frac{\partial \rho}{\partial t} + u \cdot \nabla \rho = 0 \qquad (\text{ a.e. } (t,x) \in]0,T[\times \Omega),$$

$$\rho(0,x) = \rho_0(x).$$

Furthermore the solution ρ satisfies the following inequalities for all $t \in [0,T]$.

$$m \leqq \rho(t,x) \leqq \ell, \tag{3.5}$$

$$\| \nabla\rho(t) \|_{\infty} \leqq c \| \nabla\rho_0 \|_{\infty} e^L, \tag{3.6}$$

$$\| \frac{\partial\rho}{\partial t}(t) \|_{\infty} \leqq c \| \nabla\rho_0 \|_{\infty} \| u(t) \|_{\infty} e^L. \tag{3.7}$$

Here and in what follows c means a positive constant depending only on Ω, m and ℓ.

REMARK. Under the hypotheses of this proposition the characteristic curve $\xi_{s,t}(x)$ exists uniquely. The function $u_1(t) = e^{-tA}a$ clearly satisfies (3.4), hence we obtain $\rho_2(t,x)$.

We next solve (3.2). Let $\rho \in W^{1,\infty}(]0,T[\times\Omega)$ and $v \in C([0,T];D(A^{\eta}))$ be given. Suppose that

$$m \leqq \rho(t,x) \leqq \ell \qquad (0 < t < T, \ x \in \Omega), \tag{3.8}$$

$$M_0 \equiv \sup\{ |\frac{\partial\rho}{\partial t}(t,x)| ; 0 < t < T, x \in \Omega \} < +\infty, \tag{3.9}$$

$$\| A^{\alpha}v(t) \| \leqq N \qquad (0 < t < T, \ \alpha = 5/8, \eta), \tag{3.10}$$

(3.11) $$\| Av(t) \| \leq Nt^{-(1-\eta)} \qquad (0 < t \leq T),$$

(3.12) $$\| A^{5/8}\{v(t) - v(s)\} \| \leq N(t-s)^{\theta}s^{-\theta} \qquad (0 < s < t < T).$$

Here N and $\theta \in]1/4,3/8[$ are positive constants. Then we have

PROPOSITION 3.2. We choose γ and θ such that $1/4 < \gamma < \theta < 3/8$ and fix them. Under the conditions (3.8-12) there exists a unique

$$u \in C([0,T];D(A^{\eta}))\cap C(]0,T];D(A))\cap C^1(]0,T];H)$$

satisfying

$$B_{\rho}(t)\frac{du}{dt} + Au(t) + F_{\rho}v(t) = 0 \qquad (0 < t < T),$$

$$u(0) = a.$$

Furthermore we have the inequalities below.

(3.13) $$\| A^{\alpha}u(t) \| \leq c\| A^{\alpha}a \| + t\phi(M_0) \| A^{\alpha}a \| + \phi(M_0)N^2t^{1-\alpha} \qquad (0 < t < T)$$
$$(\alpha = 5/8 \text{ or } \eta),$$

(3.14) $$\| A^{5/8}\{u(t) - u(s)\} \| \leq (t-s)^{\theta}\left\{cs^{-\theta}\| A^{5/8}a \| + \phi(M_0) \| A^{5/8}a \| + \phi(M_0)N^2\right\}$$
$$(0 < s < t < T)$$

(3.15) $$\| Au(t) \| \leq ct^{-(1-\eta)} \| A^{\eta}a \| + \phi(M_0)t^{\eta} \| A^{\eta}a \| + \phi(M_0)N^2 \qquad (0 < t < T),$$

(3.16) $$\| \nabla u(t) \|_{\infty} \leq ct^{-(1+\gamma-\eta)} \| A^{\eta}a \| + \phi(M_0)t^{\eta-\gamma} \| A^{\eta}a \| + \phi(M_0)N^2t^{-\gamma}$$
$$(0 < t < T),$$

where we have put $\phi(M_0) = c(1+M_0)^3\exp(cM_0)$.

We can easily verify that $v(t) = u_1(t) = e^{-tA}a$ and $\rho(t,x) \equiv \rho_0(x)$ satisfy the hypotheses of this proposition. Consequently we obtain $u_2(t)$.

PROPOSITION 3.2 is proved in Okamoto [9]. Since $u(t)$ is given by

$$u(t) = U(t,0)a - \int_0^t U(t,s)B_{\rho}(s)^{-1}F_{\rho}v(s)ds ,$$

the proof is carried out by estimates for $A(t)^{\alpha}U(t,s)A(s)^{-\beta}$ ($A(t) \equiv B_{\rho}(t)^{-1}A$). For instance we use

$$\| A(t)^{\alpha}U(t,s)A(s)^{-\beta}\| \leqq c(t-s)^{-(\alpha-\beta)} + \phi(M_0)(t-s)^{-(\alpha-\beta)+1}$$

$$(0 \leqq \beta < \alpha < 2 , 0 \leqq \beta < 1),$$

where it should be noticed that c does not depend on M_0.

Using PROPOSITIONS 3.1 and 3.2 successively, we obtain $\{\rho_n,u_n\}_{n=1}^{\infty}$ in $[0,T]$. To study the convergence we put for $n = 1,2,\cdots$

$$N_n(t) \equiv \max\left(\sup_{0<s<t} \| A^{5/8}u_n(s)\| , \sup_{0<s<t} \| A^{\eta}u_n(t)\| , \sup_{0<s<t} s^{1-\eta}\| Au_n(s)\| , \right.$$

$$\left. \sup_{0<r<s<t} (s-r)^{-\theta}r^{\theta}\| A^{5/8}\{u_n(s) - u_n(r)\}\| \right)$$

and

$$M_n(t) \equiv \max\left(\sup_{0<s<t} \| \nabla\rho_n(s)\|_{\infty} , \sup_{0<s<t} \| \frac{\partial\rho_n}{\partial s}(s)\|_{\infty} \right) .$$

Then we can choose a $T_1 \leqq T$ such that $\{N_n(T_1)\}_n$ and $\{M_n(T_1)\}_n$ are bounded sequences. This fact, then, ensures the convergence of $\{\rho_n,u_n\}$ as follows. Putting $\sigma_n \equiv \rho_n - \rho_{n-1}$, we have

$$\frac{\partial\sigma_{n+1}}{\partial t} + u_n\cdot\nabla\sigma_{n+1} = - (u_n - u_{n-1})\cdot\nabla\rho_n$$

$$\sigma_{n+1}(0,x) = 0.$$

The following lemma is useful for estimating σ_n.

<u>LEMMA 3.1.</u> Suppose that $v \in C([0,T_1]\times\overline{\Omega})$ satisfies

$$\int_0^{T_1} \| \nabla v(t)\|_{\infty} dt < +\infty .$$

Let $r \in W^{1,\infty}(]0,T_1[\times \Omega)$ satisfy

$$\frac{\partial r}{\partial t} + v\cdot\nabla r \equiv g \in L^1(]0,T_1[;L^{\infty}(\Omega)).$$

Then we have

$$\| r(t)\|_{\infty} \leqq \| r(0)\|_{\infty} + \int_0^t \| g(s)\|_{\infty} ds \qquad (0 < t < T).$$

This lemma is proved in [9] but the proof is essentially due to Bardos and Frisch [10]. We now obtain

$$\| \sigma_{n+1}(t) \|_\infty \leq K\int_0^t \| u_n(s) - u_{n-1}(s) \|_\infty ds$$

by the lemma above and the boundedness of $N_n(T_1)$ and $M_n(T_1)$.

We next put $w_n(t) = u_n(t) - u_{n-1}(t)$. Then it satisfies

$$B_n(t)\frac{dw_{n+1}}{dt} + Aw_{n+1} = -\{ F_{\rho_n} u_n(t) - F_{\rho_{n-1}} u_{n-1}(t)\} - \{B_n(t) - B_{n-1}(t)\}\frac{du_n}{dt},$$

$$w_{n+1}(0) = 0.$$

Putting $h_n(t) = \sup_{0<s<t} \| A^\eta w_n(s) \|$, we finally obtain

$$\| A^\eta w_{n+1}(t) \| \leq c\int_0^t (t-s)^{-\eta}\{ h_n(s) + \| \sigma_n(s) \|_\infty \}ds$$

$$\leq c'\int_0^t (t-s)^{-\eta} h_n(s)ds.$$

Since $h_n(s)$ is increasing in s, the right hand side is increasing in t. Hence we have

$$h_{n+1}(t) \leq c'\int_0^t (t-s)^{-\eta} h_n(s)ds.$$

By the induction we have for $n = 3,4,\cdots$

$$h_n(t) \leq \frac{\{c' t^{1-\eta}\Gamma(1-\eta)\}^{n-2}}{\Gamma(1+(n-2)(1-\eta))} h_2(t) \qquad (0 < t < T_1)$$

where Γ is Euler's gamma function. Hence it holds that

$$\sum_{n=1}^{\infty} h_n(T_1) < +\infty ,$$

which implies that $\{u_n\}$ is convergent in $C([0,T_1];D(A^\eta))$ and that $\{\rho_n\}$ is convergent in $C([0,T_1] \times \bar{\Omega})$. It is proved in [9] that $\lim u_n$ and $\lim \rho_n$ are actually a strong solution.

§4. Outline of the proof of THEOREMS 2 and 3.

We first consider THEOREM 2, i.e., the case of $\Omega \subset \mathbb{R}^2$. In this case we have the following a priori estimates.

$$(4.1) \qquad m\| u(t)\|^2 + 2\int_0^t \| \nabla u(s)\|^2 ds \leqq \ell \| a\|^2 ,$$

$$(4.2) \qquad \| A^{1/2}u(t)\|^2 + \int_0^t \| Au(s)\|^2 ds \leqq c$$

$$\text{with } \; c = c(\Omega, m, \ell, \| A^{1/2}a\|).$$

These estimates are obtained if we multiply the equation (2.2) by $u(t)$ and $B_\rho(t)^{-1}Au(t)$, respectively, and if we integrate it with respect to t. Then the inequality (4.2) is used to show the following

LEMMA 4.1. For all $T > 0$ there are positive constants L and $\theta \in]0,1[$ depending only on T, Ω, m, ℓ, $\| A^{1/2}a\|$ and $\| \nabla\rho_0\|_\infty$ such that

$$\sup_{x \in \Omega} |\rho(t,x) - \rho(s,x)| \leqq L|t-s|^\theta \qquad (s,t \in [0,T]).$$

We finally get to the estimates below

$$\| A^\eta u(t)\| \leqq K \qquad (0 < t < T),$$

$$\int_0^t \| \nabla u(s)\|_\infty ds \leqq K \qquad (0 < t < T)$$

with $K = K(T,\Omega,m,\ell,\| A^\eta a\| ,\| \nabla\rho_0\|_\infty)$, so long as the solution exists in $[0,T]$. Making use of these estimates and PROPOSITION 3.1, we can show THEOREM 2. For the details, see [9].

The proof of THEOREM 3 is different in its nature from the proof for the case of the Navier-Stokes system in $\Omega \subset \mathbb{R}^3$. Our proof was inspired by Matsumura and Nishida [7]. The proof is easy if we admit the following a priori estimate.

LEMMA 4.2. There are positive constants $\delta = \delta(\Omega,\ell)$, $\varepsilon = \varepsilon(\Omega,m,\ell)$ and $c_j = c_j(\Omega,m,\ell)$ ($j = 1,2$) such that the following assertion holds true. Take an arbitrary $T > 0$ and suppose that

$$\sup_{0<t<T} \| \frac{\partial\rho}{\partial t}(t)\|_\infty \leqq \varepsilon , \quad \sup_{0<t<T} e^{\delta t}\| A^{5/8}u(t)\| \leqq \varepsilon$$

and $\| A^{\eta} a \| \leqq 1$. Then we have for any $t \in]0,T]$

$$\| A^{\alpha} u(t) \| \leqq c_1 \| A^{\alpha} a \| e^{-\delta t} \qquad (\alpha = 5/8 \text{ or } \eta),$$

$$\| \nabla\rho(t) \|_{\infty} \leqq c_2 \| \nabla\rho_0 \|_{\infty} \quad \text{and} \quad \| \frac{\partial\rho}{\partial t}(t) \|_{\infty} \leqq c_2 \| \nabla\rho_0 \|_{\infty}.$$

Using LEMMA 4.2 in the same way as in Matsumura and Nishida [7], we can show THEOREM 3. The details are in Okamoto [9].

† Partially supported by the Fûjukai.

References.

[1] H. Fujita and T. Kato : On the Navier-Stokes Initial Value Problem. I, Arch. Rational Mech. Anal., 16 (1964) 269-315.

[2] T. Kato : Quasilinear Equations of Evolution with Applications to Partial Differential Equations, Springer Lecture Notes in Math., 448 (1975) 25-70.

[3] A.V. Kazhikhov : Solvability of the initial and boundary value problem for the equations of motion of an inhomogeneous incompressible fluid (Russian), Doklady Akad. Nauk., 216 (1974) 1008-1010.

[4] O.A. Ladyzhenskaya and V.A. Solonnikov : Unique solbability of an initial and boundary-value problem for viscous incompressible nonhomogeneous fluids (Russian), Zap. Nauk. Sem. Leningrad Otd. Mat. Inst. Steklov SSSR., 52 (1975) 52-109.

[5] J.L. Lions : On Some Problems Connected with Navier-Stokes Equations, in Nonlinear Evolution Equations, Academic Press, New York, 1978.

[6] J.E. Marsden : Well-posedness of the equations of a nonhomogeneous perfect fluid, Comm. Partial Diff. Eqs., 1 (1976) 215-230.

[7] A. Matsumura and T. Nishida : The initial value problem for the equations of motion of viscous and heat-conductive gases, J. Math. Kyoto Univ. 20 (1980) 67-104.

[9] H. Okamoto : On the equation of nonstationary stratified fluid motion : Uniqueness and existence of the solution, to appear in J. Fac. Sci. Univ. Tokyo

[10] C. Bardos and U. Frisch : Finite-Time Regularity for Bounded and Unbounded Ideal Incompressible Fluids Using Hölder Estimates, Springer Lecture Notes in Math. 565 1-13.

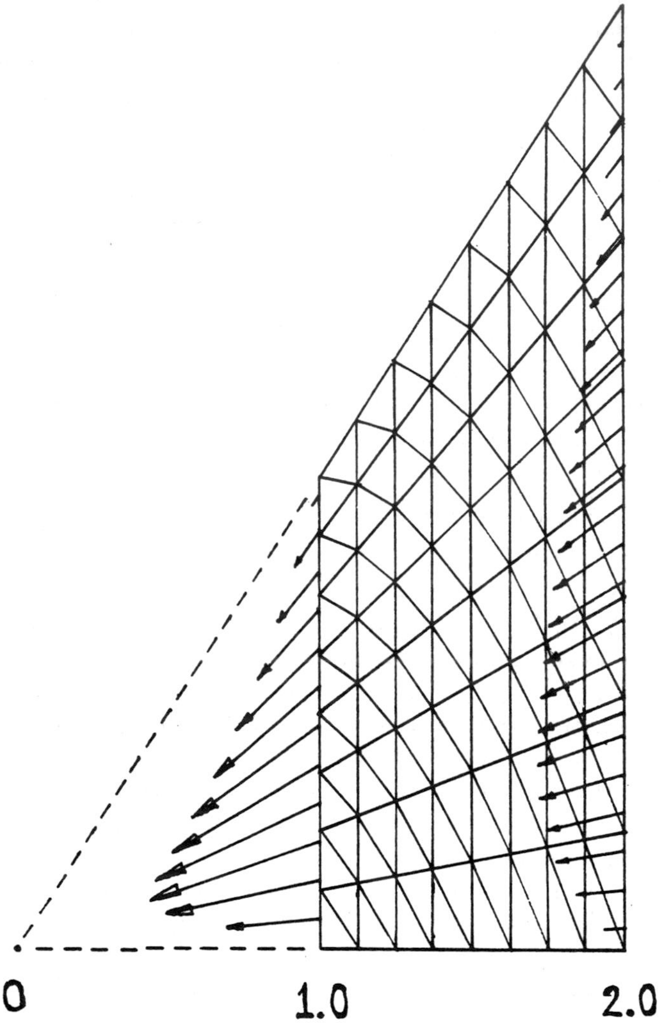

Figure 2.1
Domain and mesh used for Jeffrey-Hamel flow calculation with method (2.2). Arrows indicate boundary data (velocity is zero on walls parallel to dashed lines).

assessments were preferred over qualitative, or pictorial, comparisons with other numerical methods on test problems with unknown solutions. This demanding evaluation of the technique's performance led to both an uncovering of subtle bugs in the coding as well as to a succinct evaluation of the method's performance. Tests of this type are planned as well for the methods to be discussed in the next section.

In all cases, the behavior of the method (2.2), with respect to decreasing mesh size h, fit that theoretically predicted. For example, the singularity of the stick-slip flow limits the velocity convergence in L^2 to first-order in h, even though a better approximation rate is possible for the solution by piecewise-linear functions. This reflects the well known "pollution" effect of singular boundary-conditions (see [13] for details). For the flow between converging walls, the solution (and boundary conditions) are regular, and second-order convergence (in all norms) was obtained with respect to h for the velocity approximation. Typical behavior for the relative L^2 error in approximating the Jeffrey-Hamel flow is given in Figure 2.2; the problem being solved is exactly as depicted in Figure 2.1, and the solid dot in Figure 2.2 refers precisely to the triangulation in Figure 2.1. The other triangulations summarized in Figure 2.2 were obtained similarly by dividing the region into quadrilaterals by connecting evenly spaced points on the boundary and then adding the diagonal going from upper-left to lower-right in each quadrilateral.) For this problem, the pressure approximation converged at first-order in h with respect to the L^2-norm, again as theoretically predicted. However, without special treatment (e.g., mesh refinement near the walls) of the boundary layer, the constant multiplying the appropriate power of h in the error grows substantially with the Reynolds' number, as would be expected. For more details concerning the testing of the method (2.2), see [13].

3. HIGHER-ORDER METHODS FOR THE STOKES PROBLEM.

In the previous section, we described computational results for a particular low-order method for the Navier-Stokes equations. The method performed as predicted theoretically and is quite adequate for a wide range of applications. However, the method (2.2) and others like it have various drawbacks, in addition to low accuracy. Thus it was felt there was a need to understand more generally the possibilities for other, higher-order methods. As no general theory appeared to exist for problems with divergence constraints, it was

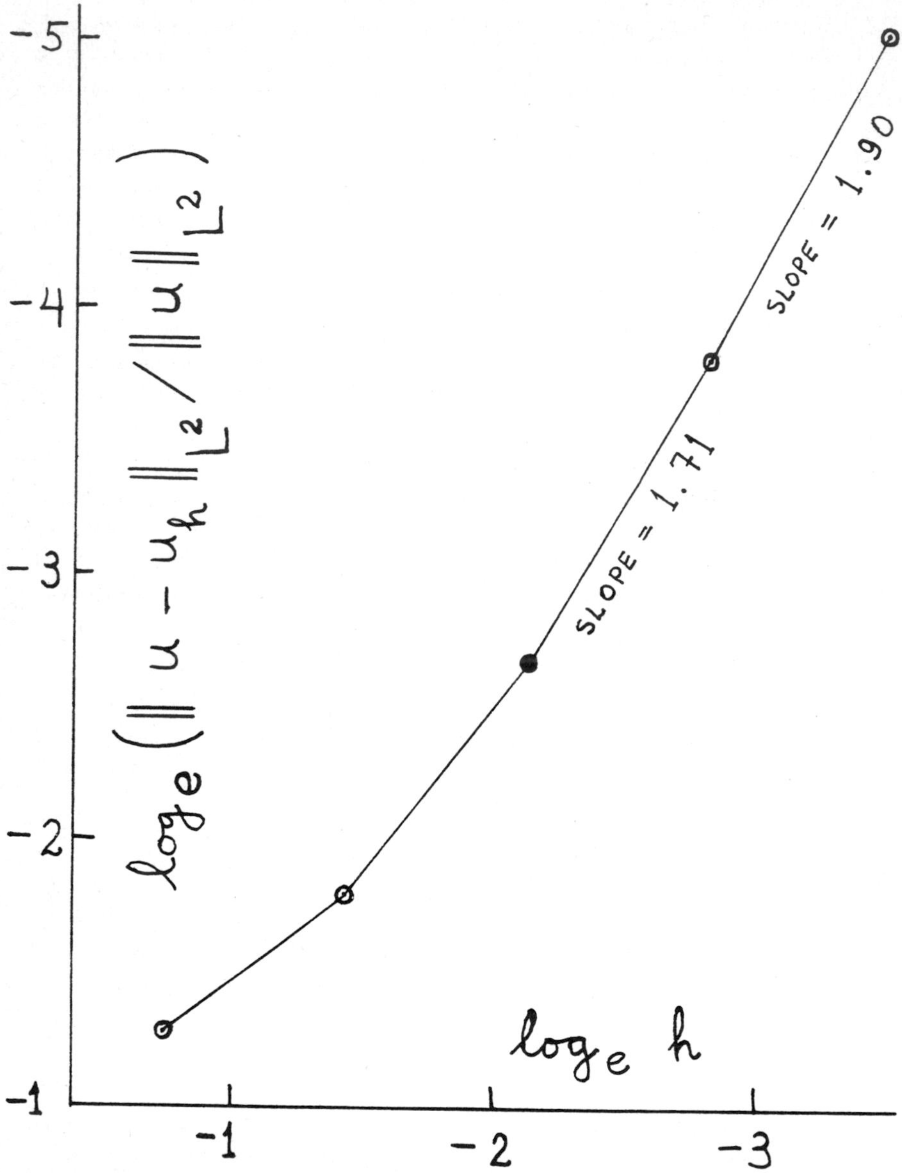

Figure 2.2

Relative L^2-error $\left(\| u - u_h \|_{L^2(\Omega)} / \| u \|_{L^2(\Omega)} \right)$ for the method (2.2) as a function of mesh size h (see 2.1) for Jeffrey-Hamel flow with boundary data as shown in Figure 2.1. The solid dot refers to the triangulation in Figure 2.1.

decided to try to develop such a theory. In this section and the following three sections we describe the results of this attempt, which have been obtained jointly with M. Vogelius [14,15], concerning families of methods for the Stokes problem having some highly desirable properties. In particular, we shall discuss the Lagrange family of elements and shall give an outline of the justification of the method, including verification of the "inf-sup" stability condition. (The latter is abstracted from [14] and is intended only to give the spirit of the proof, not full details of the proof; for full details of the argument, the reader is referred to [14].) This family of methods has the properties that

- the discrete velocity approximation is exactly divergence free
- the "inf-sup" condition is satisfied with a constant independent of the mesh size
- the velocity and pressure can be calculated via the penalty method (without reduced integration), cf. Glowinski [7], or they may be calculated separately in view of the fact that
- there is an explicit basis for the divergence-zero subspace of the discrete velocity space.

We now describe the method in question. For the sake of brevity only, we shall assume that Ω is simply connected. Let $\{T_h : 0 < h \leq 1\}$ denote a family of triangulations of Ω parametrized by mesh size, h (see 2.1). We shall assume that the family is quasi-uniform, i.e., that there is some constant $\rho_0 > 0$ such that

$$(3.1) \qquad \rho(\tau) \geq \rho_0 h \quad \forall \ \tau \in T_h \quad \forall \ 0 < h \leq 1 ,$$

where $\rho(\tau)$ denotes the supremum of diameters of discs contained in τ. From now on, we shall suppress the subscript h when referring to a particular triangulation in order to simplify notation.

If T' is an arbitrary subset of a triangulation T, let Ω' denote the corresponding polygonal domain:

$$(3.2) \qquad \Omega' = \text{interior} \left(\overline{\bigcup_{\tau \in T'} \tau} \cap \Omega \right).$$

(We shall use similar notation for T^s, T^m, etc., namely Ω^s, Ω^m, etc.) For i = 0 or 1 and p any integer not less than i + 1, let

$$P^{p,i}(T')$$

denote the set of C^i piecewise polynomials of degree at most p with respect to the mesh T'. That is, $\phi \in P^{p,i}(T')$ provided a) $\phi \in C^i(\Omega')$ and b) $\phi|_\tau$ is a polynomial whose degree does not exceed p for all $\tau \in T'$.

The standard Galerkin method for the Stokes problem (cf. Girault and Raviart [6]) utilizes two spaces, V_h to approximate the velocity and Π_h to approximate the pressure. The basic stability condition required (cf. Babuska [1] and Brezzi [2]) is that

$$(3.3) \qquad 0 < \beta = \inf_{q \in \Pi_h} \sup_{v \in V_h} (\mathrm{div}\, v, q) / \|v\|_{H^1(\Omega)} \|q\|_{L^2(\Omega)} ,$$

where "div" denotes the divergence operator on vector functions, $(\, , \,)$ denotes the $L^2(\Omega)$ inner-product and $H^1(\Omega)$ is the usual Sobolev space. We shall be interested here in the natural choices

$$V_h = P^{p,0}(T_h) \times P^{p,0}(T_h)$$

or a corresponding subspace satisfying essential boundary conditions, and

$$\Pi_h = \mathrm{div}\, V_h = \{q = \mathrm{div}\, v : v \in V_h\} .$$

Note that with this choice of Π_h, we always have $\beta > 0$ in (3.3). Moreover, the resulting Galerkin approximation is exactly divergence-free. However, it may be that β tends to zero as $h \to 0$, which corresponds to instability of the method, or that $\Pi_h = \mathrm{div}\, V_h$ is so complicated as to be computationally impractical. We shall see that neither holds provided $p \geq 4$. The derivation of this result proceeds in three steps. First, we identify some simple, local constraints that hold on the image of the divergence operator applied to continuous piecewise polynomials of any order when the triangulation takes a particular form at a vertex. Second, we show how, by counting dimensions, one can see that these are the only constraints for $p \geq 4$. Finally, we sketch how one shows that the constant β in (3.3) can be chosen independent of h in this case.

4. CONSTRAINTS ON THE IMAGE OF THE DIVERGENCE OPERATOR.

We shall say that a vertex, s, in T' is a singular interior vertex (cf. [11]) provided a) s does not lie on $\partial\Omega'$ and b) precisely four edges meet at s and c) these edges lie on two straight lines (i.e., considering the edges consecutively, each edge is parallel to its second-nearest neighbor)--see Figure 4.1.

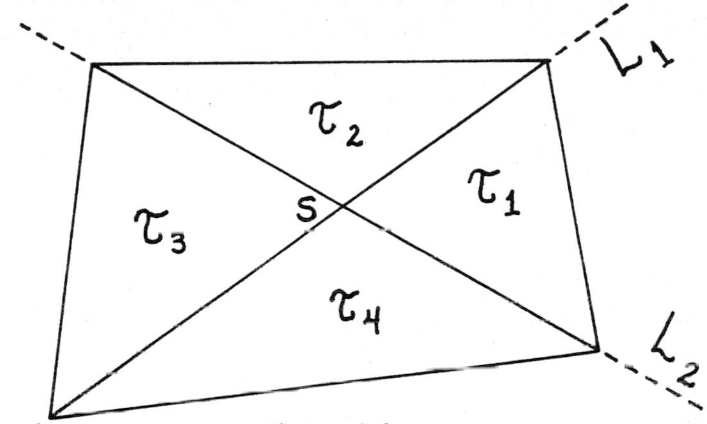

Figure 4.1
Singular interior vertex

Following [17], we introduce, for $p \geq 0$, the space

$$P^{p,-1}(T')$$

of (discontinuous) piecewise polynomials ϕ of degree p which satisfy the restriction

<u>R1</u>: at any singular interior vertex s in T'

$$\sum_{i=1}^{4} (-1)^i (\phi|_{\tau_i})(s) = 0 \tag{4.1}$$

where $\tau_1, \ldots, \tau_4$ are the four triangles meeting at s, numbered consecutively, as shown in Figure 4.1.

That is, $P^{p,-1}(T')$ may be viewed as a collection of independent polynomials, one per triangle, viz. $\{\phi|_\tau : \tau \in T'\}$, linked only by the constraint (4.1). The reason for introducing the restriction (4.1) is the following, which generalizes results of Nagtegaal, Parks and Rice [12], Mercier [9], [10] and, Fix, Gunzberger and Nicolaides [4].

<u>Proposition 4.1</u>

For any $T' \subset T$ and $p \geq 1$,

$$\operatorname{div}\left(P^{p,0}(T') \times P^{p,0}(T')\right) \subset P^{p-1,-1}(T') .$$

<u>Proof</u>. Obviously, if $v = (v_1, v_2) \in P^{p,0}(T')^2$, then $\operatorname{div} v$ is a (possibly discontinuous) piecewise polynomial of degree $p - 1$. Thus

we only have to verify that (4.1) holds for $\phi = \operatorname{div} v$, which we do following the argument of Mercier [10]. Let s be a singular vertex as shown in Figure 4.1, and let L_1 and L_2 be the lines containing the edges meeting at s. Let w_i, $i = 1,2$, be the polynomial of degree p given by

$$(4.2) \qquad w_i = \sum_{j=1}^{4} (-1)^j (v_i|_{\tau_j}) .$$

Note that $(v_i|_{\tau_1} - v_i|_{\tau_2})$ vanishes on the line L_1, as does $(v_i|_{\tau_3} - v_i|_{\tau_4})$, $i = 1,2$. Thus each w_i must vanish on L_1. Similarly each w_i must vanish on L_2 as well. Hence both w_i's vanish to second order at s, and in particular,

$$\frac{\partial w_1}{\partial x_1}(s) + \frac{\partial w_2}{\partial x_2}(s) = 0 .$$

This clearly implies (4.1) for $\phi = \operatorname{div} v$. Q.E.D.

When boundary conditions are imposed, a new set of restrictions at the boundary become important. Let

$$\mathring{P}^{p,i}(T')$$

denote the subpsace of $\mathring{P}^{p,i}(T')$ consisting of functions vanishing to $(i+1)$-st order on $\partial\Omega'$, $i = 0,1$ and $p \geq i+1$. That is, functions in $\mathring{P}^{p,i}(T')$ are required to be zero on $\partial\Omega'$, $i = 0,1$; and, in addition, functions in $\mathring{P}^{p,1}(T')$ are required to have a vanishing normal derivative.

We now introduce the notion of singular boundary vertex (cf. [17]). A vertex $s \in \partial\Omega'$ is said to be a singular boundary vertex in T' if all of the edges in T' meeting at s fall on two straight lines. There are thus four possible configurations, depending on the number, k, of triangles meeting at s, as shown in Figure 4.2. The reason that these configurations play a distinguished role is that the triangulation T' could be extended to a triangulation $\overline{T'}$ by adding triangles outside Ω' (shown by dotted lines in Figure 4.2) in such a way that s would be a singular interior vertex in $\overline{T'}$. Of course, the case $k = 4$ does not require the addition of triangles; it is essentially the same as a singular interior vertex. Again following [17], for $p \geq 0$ we let

$$\tilde{P}^{p,-1}(T')$$

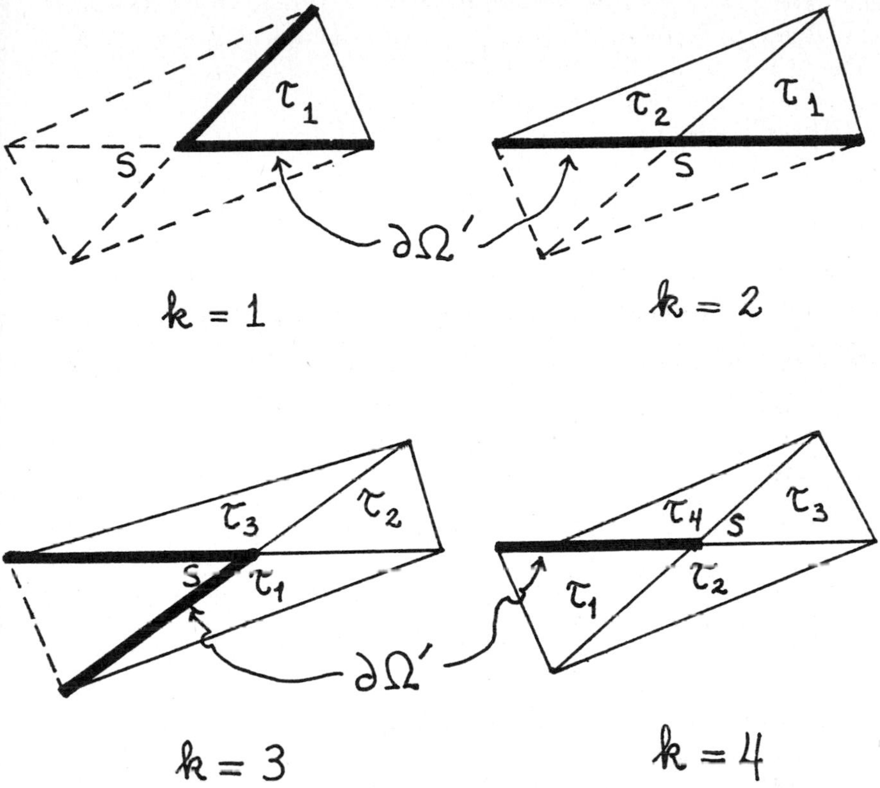

Figure 4.2
The four types of singular boundary vertices. The dark lines indicate edges on $\partial\Omega'$. The dashed lines represent an extension of the triangulation T' to form $\overline{T'}$.

denote the subspace of $P^{p,-1}(T')$ consisting of functions ϕ such that the following two restrictions hold:

R2: At any singular boundary vertex s in T'

$$\sum_{i=1}^{k} (-1)^{i} (\phi|_{\tau_i})(s) = 0 \tag{4.3}$$

where $\tau_1,\ldots,\tau_k$ are all the triangles meeting at s, numbered consecutively, as shown in Figure 4.2 (k may be any number from 1 to 4 depending on the particular vertex s).

R3: For any connected component Ω'' of Ω',

(4.4) $$\int_{\Omega''} \phi(x_1,x_2)\,dx_1dx_2 = 0 .$$

Proposition 4.2

For any $\mathcal{T}' \subset \mathcal{T}$ and any $p \geq 1$,

$$\operatorname{div}(\mathring{P}^{p,0}(\mathcal{T}') \times \mathring{P}^{p,0}(\mathcal{T}')) \subset \tilde{P}^{p-1,-1}(\mathcal{T}') .$$

Proof. Any function $v \in \mathring{P}^{p,0}(\mathcal{T}')^2$ can be extended by zero outside Ω' to a continuous function $\bar{v}$. Utilizing the extended triangulation $\overline{\mathcal{T}'}$ indicted in Figure 4.2 by dashed lines, we may view $\bar{v} \in P^{p,0}(\overline{\mathcal{T}'})^2$. Since each s is a singular interior vertex in $\overline{\mathcal{T}'}$, we have (4.1) for $\operatorname{div} \bar{v}$. But since $\bar{v} \equiv 0$ outside Ω', (4.1) for $\bar{v}$ implies (4.3) for v. (For the case of a singular boundary vertex with the case $k = 4$, the reasoning is the same, except that no extensions need be made at the slit-- v is already continuous as it vanishes there.) Condition (4.4) is an obvious consequence of the divergence theorem. Q.E.D.

5. CHARACTERIZING THE RANGE OF THE DIVERGENCE OPERATOR

In the previous section, we gave simple, local constraints on the image of the divergence operator on spaces of continuous, piecewise polynomial spaces. However, it is not clear that one would expect these to be the only constraints on the image, and indeed there are examples where this is known to be false (cf. [15]). In particular, additional global constraints occur in some cases that make Π_h impractical from a computational point of view. Thus the following result (first published by Vogelius [17] but also derived independently by the author) is quite remarkable.

Proposition 5.1

For $p \geq 4$ one has

$$\operatorname{div} P^{p,0}(\mathcal{T}_h)^2 = P^{p-1,-1}(\mathcal{T}_h) .$$

The proof of this result is repeated in [14], via two independent techniques. One proof, a combinatorial argument, proves that the divergence operator is surjective simply by calculating the dimension of the image and observing that it equals $\dim P^{p-1,-1}(\mathcal{T}_h)$, which is easily computed from the definition. To compute the dimension of the image, one simply needs to know the dimension of the kernel of the divergence operator, since the dimension of $P^{p,0}(T_h)^2$ is well known. To discover what the kernel is, recall that $P^{p,1}(T_h)$ denotes the space of all C^1 piecewise polynomials of degree not greater than p with respect to T_h. Also, recall that the curl operator in two dimensions takes the form

$$(5.1) \qquad (\nabla \times \phi =) \operatorname{curl} \phi = \left(\frac{\partial \phi}{\partial y}, -\frac{\partial \phi}{\partial x}\right)$$

for any, say, C^1 function ϕ. The following simple fact appears to be fundamental in the study of the divergence operator on spaces of piecewise polynomials.

Lemma 5.1

Suppose Ω is simply connected. Consider the divergence operator

$$\operatorname{div}: (P^{p,0}(T_h))^2 \to P^{p-1,-1}(T_h).$$

Then, for any $p \geq 1$, its kernel is given by

$$\ker \operatorname{div} = \operatorname{curl} P^{p+1,1}(T_h) .$$

Proof. If $v \in P^{p,0}(T_h)^2$ and $\operatorname{div} v \equiv 0$, then we can obviously find $\phi \in C^1(\Omega)$ such that

$$v = \operatorname{curl} \phi .$$

But since both $v_x = \phi_{,y}$ and $v_y = -\phi_{,x}$ are piecewise polynomials of degree p, all (weak) derivatives of order $p+2$ of ϕ must vanish in the interior of each $\tau \in T_h$. Thus ϕ must be a piecewise polynomial of degree $p+1$. Q.E.D.

The proof of Proposition 5.1 then follows easily from the results of [11], which also provides an explicit basis for the kernel of the divergence operator in this case in view of Lemma 5.1; for details, see [14]. Note that a key point is that the kernel of the curl operator consists of only constant functions in two dimensions. Similarly, the case of Dirichlet boundary conditions can be considered as follows.

Proposition 5.2

For $p \geq 4$, one has

$$\operatorname{div} \mathring{P}^{p,0}(T_h) \times \mathring{P}^{p,0}(T_h) = \tilde{P}^{p-1,-1}(T_h) .$$

The combinatorial proof of this proposition is similar and

follows from

Lemma 5.2

Suppose Ω is simply connected and $p \geq 1$. Then

$$\ker \operatorname{div}|_{\mathring{P}^{p,0}(T_h)^2} = \operatorname{curl} \mathring{P}^{p+1,1}(T_h)$$

where we recall that $\mathring{P}^{p+1,1}(T_h)$ is the subspace of $P^{p+1,1}(T_h)$ consisting of functions that vanish to second order on $\partial\Omega$.

Proof. Using the same ϕ as in the proof of Lemma 5.1, we see immediately from (5.1) that if $v = \operatorname{curl}\phi$ vanishes on $\partial\Omega$, then so does $\nabla\phi$. Thus ϕ is constant on $\partial\Omega$, say equal to $\phi_0 \in \mathbb{R}$. Then defining

$$\tilde{\phi} = \phi - \phi_0 ,$$

we still have $v = \operatorname{curl}\tilde{\phi}$, and both $\tilde{\phi}$ and $\frac{\partial}{\partial n}\tilde{\phi}$ vanish on $\partial\Omega$. Thus $\tilde{\phi} \in \mathring{P}^{p+1,1}(T_h)$. Q.E.D.

The dimension of $\mathring{P}^{p+1,1}(T_h)$ can be calculated as an extension of the results of [11], cf. [14], although it is not an obvious consequence of the former paper. Moreover, an explicit basis for $\mathring{P}^{p+1,1}(T_h)$ is derived in [14] which then, via Lemma 5.2, gives a basis for the divergence-free subspace of V_h which can be used in computations. The extenstion of these results to three dimensions poses a challenging problem. It seems that the first step would be to understand analogues to Lemmas 5.1 and 5.2. They are not immediately obvious in three dimensions due to the lack of "coercivity" of the curl operator in this case.

6. VERIFICATION OF THE INF-SUP CONDITION

It is obvious that the constant β in (3.3) must tend to zero for "nearly singular" vertices, e.g., for vertices where four triangles meet with opposing edges that are nearly parallel. This is simply because the dimension of the range of the divergence operator drops down by one when the edges become parallel. (Observe that β^{-1} is a bound for the norm of the inverse of the divergence operator.) However, we can define a parameter that allows us to quantify the nearness to singularity as follows.

For every vertex, s, in the triangulation, let r be the number of interior edges in T meeting at s, and let $\theta_1,\ldots,\theta_{r+1}$

be the (interior) angles formed by adjacent edges, numbered consecutively around s. We note that if s is an interior vertex, then $\theta_{r+1} = \theta_1$. Define

$$R(s)^{-1} = \min \{ |\theta_i + \theta_{i+1} - \pi| : 1 \leq i \leq r \}. \tag{6.1}$$

Obviously, $R(s) = \infty$ if and only if s is singular. Thus define

$$\begin{aligned} R(\mathcal{T}) &= \max\{R(s) : s \text{ is a non-singular interior vertex}\} \\ \mathring{R}(\mathcal{T}) &= \max\{R(s) : s \text{ is any nonsingular vertex}\}. \end{aligned} \tag{6.2}$$

That is, to define $R(\mathcal{T})$, we consider only interior vertices, and we exclude ones that are exactly singular. The definition of $\mathring{R}(\mathcal{T})$ involves all vertices, but excluding exactly singular vertices, either on the boundary or in the interior. We shall refer to $R(\mathcal{T})$ and $\mathring{R}(\mathcal{T})$ as the regularity constants for $\mathcal{T}$.

It is now clear that an inverse exists for $\mathrm{div}|_{P^{p,0}(\mathcal{T})^2}$ (resp. for $\mathrm{div}|_{\mathring{P}^{p,0}(\mathcal{T})^2}$) whose norm is bounded by a constant that depends only on $R(\mathcal{T})$ (resp. on $\mathring{R}(\mathcal{T})$), the quasi-uniformity constant ρ_0 in (3.1), and the number of triangles in $\mathcal{T}$, by finite dimensionality and compactness. This will be crucial in our derivation of uniform bounds for β in (3.3) independent of the mesh size h. In the following we shall assume that $\partial\Omega$ is Lipschitzian, i.e., there are no slits in Ω.

<u>Theorem 6.1</u>

For $p \geq 4$, $V_h = P^{p,0}(\mathcal{T}_h)^2$ and $\Pi_h = \mathrm{div}\, V_h = P^{p-1,-1}(\mathcal{T}_h)$, the constant β in (3.3) can be chosen independent of the mesh size h, depending only on the regularity constant $R(\mathcal{T}_h)$ in (6.2) and the quasi-uniformity constant ρ_0 in (3.1).

<u>Theorem 6.2</u>

For $p \geq 4$, $V_h = \mathring{P}^{p,0}(\mathcal{T}_h)^2$ and $\Pi_h = \mathrm{div}\, V_h = \tilde{P}^{p-1,-1}(\mathcal{T}_h)$, the constant β in (3.3) can be chosen independent of the mesh size h, depending only on the regularity constant $\mathring{R}(\mathcal{T}_h)$ in (6.2) and the quasi-uniformity constant ρ_0 in (3.1).

The dependence of β on the degree p is discussed in [14], where it is shown that β tends to zero at most algebraically in p^{-1}. We shall now sketch how, say, Theorem 6.2 is proved; see [14] for complete details. Given $\phi \in \tilde{P}^{p-1,-1}(\mathcal{T}_h)$, we wish to construct

$v \in \mathring{P}^{p,0}(T_h)^2$ such that

$$(\mathrm{div}\, v, \phi) \geq \alpha \|\phi\|^2_{L^2(\Omega)}$$

and $\|v\|_{H^1(\Omega)} \leq C\|\phi\|_{L^2(\Omega)}$. Then $\beta = \alpha/C$. The argument proceeds in several steps.

1) First construct $w^1 \in \mathring{P}^{4,0}(T)^2$ such that $(\mathrm{div}\, w^1 - \phi)$ vanishes at all vertices in T and $\|w^1\|_{H^1(\Omega)} \leq C\|\phi\|_{L^2(\Omega)}$. This is a local construction; at any vertex, s, let T^s be the star of the vertex, i.e., the union of all triangles meeting at s. Then Proposition 5.2 applied to T^s says that we can find $w^{1,s}$ supported in Ω^s (see 3.2) such that $(\mathrm{div}\, w^{1,s} - \phi)$ vanishes at s. Homogeneity guarantees the required bound on $\|w^{1,s}\|_{H^1(\Omega^s)}$, and summation over s defines w^1.

2) Next, construct a collection of "macro-elements" $T^m \subset T$ of mesh size $H = Kh$ and a function ϕ_H such that $\phi_H|_{\Omega^m} \in \mathring{P}^{1,0}(T^m)$ and

$$\int_{\Omega^m} \phi_H dxdy = \int_{\Omega^m} (\mathrm{div}\, w^1 - \phi)\, dxdy$$

for all Ω^m. Note that $\phi_H \in \tilde{P}^{1,-1}(T)$ because it is continuous and vanishes on the boundary. Moreover, ϕ_H can be chosen to be "smooth," i.e., such that $\|\phi_H\|_{L^2(\Omega)} + H\|\phi_H\|_{H^1(\Omega)} \leq C\|\phi\|_{L^2(\Omega)}$.

3) Consider

$$\psi = \phi - \mathrm{div}\, w^1 - \phi_H .$$

Then $\psi|_{\Omega^m} \in \tilde{P}^{p-1,-1}(T^m)$ for all m (it vanishes at the boundary vertices of T^m and has mean zero on Ω^m). Thus we may find, by Proposition 5.2, $w^2|_{\Omega^m} \in \mathring{P}^{p,0}(T^m)^2$ such that $\mathrm{div}\, w^2 = \psi$ and $\|w^2\|_{H^1(\Omega^m)} \leq C(K)\|\psi\|_{L^2(\Omega^m)}$ for all m. Here the constant $C(K)$ depends only on the number of triangles in T^m, which is measured by $K = H/h$, and the constants ρ_0 and R in (3.1) and (6.2), respectively, by the equivalence of norms on a finite dimensional space and a compactness argument. Since w^2 vanishes on each $\partial\Omega^m$, we see that w^2 can be viewed as being in $\mathring{P}^{p,0}(T)^2$, and furthermore, $\|w^2\|_{H^1(\Omega)} \leq C(K)\|\psi\|_{L^2(\Omega)} \leq C(K)\|\phi\|_{L^2(\Omega)}$.

4) Choose $u_H \in H^2(\Omega) \cap \overset{\circ}{H}{}^1(\Omega)$ such that $\operatorname{div} u_H = \phi_H$, $\|u_H\|_{H^1(\Omega)} \leq C\,\|\phi_H\|_{L^2(\Omega)}$ and $\|u_H\|_{H^2(\Omega)} \leq C\,\|\phi_H\|_{H^1(\Omega)}$ (see [14]). Let $u_h \in \overset{\circ}{P}{}^{p,0}(T_h)^2$ be the $H^1(\Omega)$ projection of u_H. Then

$$\begin{aligned}
\|\phi_H - \operatorname{div} u_h\|_{L^2(\Omega)} &\leq C\,\|u_H - u_h\|_{H^1(\Omega)} \\
&\leq C\,h\,\|u_H\|_{H^2(\Omega)} \leq C\,h\,\|\phi_H\|_{H^1(\Omega)} \\
&\leq C(h/H)\|\phi\|_{L^2(\Omega)} = C\,K^{-1}\|\phi\|_{L^2(\Omega)}
\end{aligned}$$

Now choose K so large that

$$\|\phi_H - \operatorname{div} u_h\|_{L^2(\Omega)} \leq \frac{1}{2}\,\|\phi\|_{L^2(\Omega)},$$

and let K be fixed from now on. Note also that

$$\|u_h\|_{H^1(\Omega)} \leq \|u_H\|_{H^1(\Omega)} \leq C\|\phi_H\|_{L^2(\Omega)} \leq C\|\phi\|_{L^2(\Omega)} .$$

5) Finally, define

$$v = w^1 + w^2 + u_h .$$

Then $\|v\|_{H^1(\Omega)} \leq C\,\|\phi\|_{L^2(\Omega)}$ and

$$\phi - \operatorname{div} v = \phi_H - \operatorname{div} u_h .$$

Therefore

$$\begin{aligned}
(\phi - \operatorname{div} v, \phi) &\leq \|\phi_H - \operatorname{div} u_h\|_{L^2(\Omega)}\ \|\phi\|_{L^2(\Omega)} \\
&\leq \frac{1}{2}\,\|\phi\|^2_{L^2(\Omega)} .
\end{aligned}$$

Thus $(\operatorname{div} v, \phi) \geq \frac{1}{2}\,\|\phi\|^2_{L^2(\Omega)}$ as required.

This completes the (sketch of the) proof of Theorem 6.2. The reader will note a difficulty in obtaining a constant in the bound for w^2 that depends only on $\overset{\circ}{R}(T)$ and not $\overset{\circ}{R}(T^m)$ for all m. This is avoided in [14] by separate considerations using the fact that ψ actually vanishes at all boundary vertices. Moreover, the dependence on K is also bounded explicitly via a constructive argument in [14].

7. COMPUTATIONAL RESULTS WITH PIECEWISE QUARTICS

In this section, we present some preliminary computational results for the method discussed in the previous four sections. They have been obtained using the IMSL code TWODEPEP which uses the standard penalty method (cf. Glowinski [7]) to enforce the divergence constraint. One would expect singular vertices to have an effect only on the pressure calculation, if at all, and this possibility has not been explored as yet. So far, we have tested the code only on the Jeffrey-Hamel flow; the relative $L^2(\Omega)$-error for the velocity approximation for a Reynolds' number of about twenty using the mesh in Figure 7.1 is 2.7% versus 7.1% for the Crouzeix-Raviart scheme (2.2) using the mesh shown in Figure 2.1. (These two calculations involved identical boundary conditions, as indicated by the arrows in both figures.) For such coarse triangulations as shown in Figure 7.1, the error does not behave like the asymptotically predicted $O(h^5)$ with respect to mesh refinement, except for very small Reynolds' numbers. However, refinement of the mesh in Figure 7.1 by subdividing each triangle into four similar triangles of half the size reduced the error by at least an order of magnitude in all cases tested so far. Moreover, the error on the mesh shown in Figure 7.1 is reasonable even for fairly large Reynolds' numbers, as indicated in Table 7.1.

Reynolds' number	6.9	18	65	222
Relative L^2-error	0.013	0.027	0.069	0.15

Table 7.1

Relative L^2-error ($\|u - u_h\|_{L^2(\Omega)} / \|u\|_{L^2(\Omega)}$) as a function of Reynolds' number. The case $Re = 18$ corresponds to the boundary data as shown in Figure 7.1. In each calculation, the mesh shown in Figure 7.1 was used.

More complete results concerning this method will be reported elsewhere. The results reported in this section were obtained with the assistance of Mr. S. Wineberg.

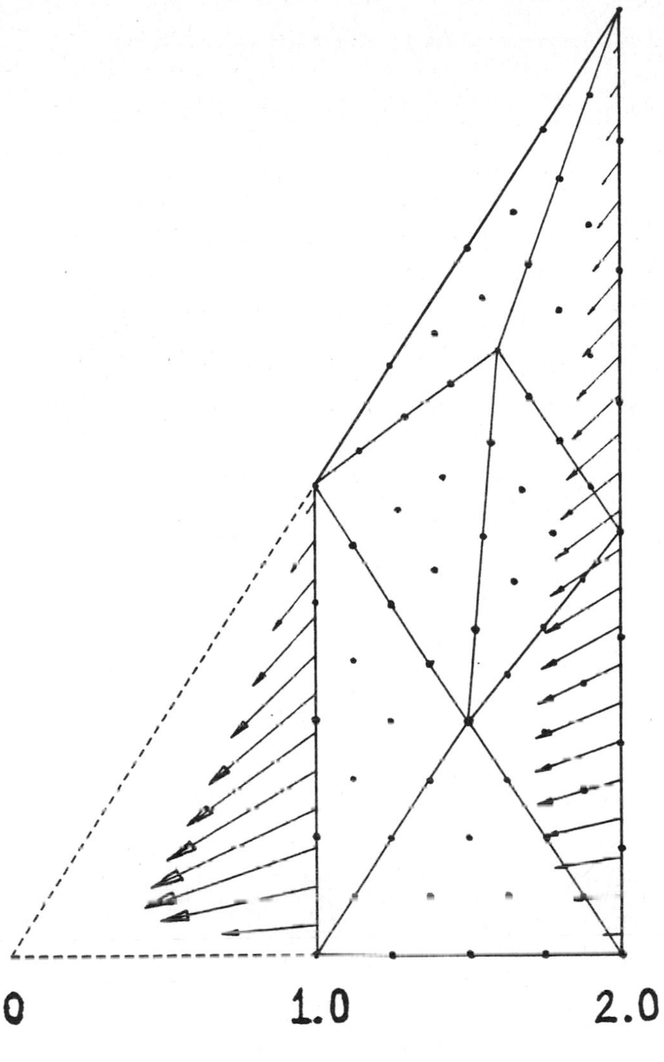

Figure 7.1

Domain, mesh and boundary conditions for Jeffrey-Hamel flow calculation using method discussed in Sections 3 through 6 with p = 4. (Boundary conditions are same as in Figure 2.1). The dots indicate location of nodal variables.

REFERENCES:

[1] I. Babuška, "Error-bounds for the finite element method," Numer. Math. 16 (1971) 322-333.

[2] F. Brezzi, "On the existence, uniqueness and approximation of saddle-point problems arising from Lagrangian multipliers," R.A.I.R.O. 8, R2 (1974) 129-151.

[3] M. Crouzeix and P.-A. Raviart, "Conforming and nonconforming finite element methods for solving the stationary Stokes equations," R.A.I.R.O. 7e année, R-3 (décember 1973), 33 à 76.

[4] G.J. Fix, M.D. Gunzberger and R.A. Nicolaides, "On mixed finite element methods for first-order elliptic systems," Numer. Math. 37 (1981) 29-48.

[5] S.M. Gallic, "Système de Stokes en dimension 3: formulation en ψ et formulation en u,p dans le cas axisymétrique," thèse, Univ. de Paris VI, 1982.

[6] V. Girault and P.-A. Raviart, Finite Element Approximation of the Navier-Stokes Equations, Springer Lecture Notes in Math. #749, Berlin: Springer-Verlag, 1979.

[7] R. Glowinski, Numerical Methods for Nonlinear Variational Problems, 2nd Ed., Springer-Verlag Series in Comp. Physics, to appear.

[8] M. Jean and W.G. Pritchard, "The flow of fluids from nozzles at small Reynolds numbers," Proc. Roy. Soc. London A370 (1980), 61-72.

[9] B. Mercier, Lectures on Topics in Finite Element Solution of Elliptic Problems, Springer-Verlag, 1979.

[10] B. Mercier, "A conforming finite element method for two-dimensional, incompressible elasticity," Int. J. Numer. Meths. Eng. 14 (1979) 942-945.

[11] J. Morgan and R. Scott, "A nodal basis for C^1 piecewise polynomials of degree $n \geq 5$," Math. Comp. 29 (1975) 736-740.

[12] J.C. Nagtegaal, D.M. Parks and J.R. Rice, "On numerically accurate finite element solutions in the fully plastic range," Comp. Meth. Appl. Mech. & Eng. 4 (1974) 153-177.

[13] W.G. Pritchard, Y. Renardy, and L.R. Scott, "Tests of a numerical method for viscous, incompressible flow. I: Fixed-domain problems," to appear.

[14] L.R. Scott and M. Vogelius, "Norm estimates for a maximal right inverse of the divergence operator in spaces of piecewise polynomials," Tech. Note BN-1013, Inst. Phys. Sci. & Tech., Univ. of Maryland, 1983.

[15] L.R. Scott and M. Vogelius, "Conforming finite element methods for incompressible and nearly incompressible continua," to appear in Large Scale Computations in Fluid Mechanics, S.Osher, ed., Lect. Appl. Math. 22, Providence: AMS.

[16] F. Thomasset, Implementation of Finite Element Methods for Navier-Stokes Equations, Berlin: Springer-Verlag, 1981.

[17] M. Vogelius, "A right-inverse for the divergence operator in spaces of piecewise polynomials. Applications to the p-version of the finite element method," Numer. Math. 41 (1983) 19-37.

Computing Methods in Applied Sciences and Engineering, VI
R. Glowinski and J.-L. Lions (Editors)
Elsevier Science Publishers B.V. (North-Holland)

APPLICATION DES METHODES DE DECOMPOSITION AUX CALCULS NUMERIQUES EN HYDRAULIQUE INDUSTRIELLE

J.P. BENQUE*, J.P. GREGOIRE**, A. HAUGUEL*, M. MAXANT**.

Electricité de France
Direction des Etudes et Recherches
* L.N.H. CHATOU ** M.M.N. CLAMART
FRANCE

Les applications numériques de l'hydraulique industrielle nécessitent le traitement d'un grand nombre de variables (problèmes tridimensionnels), régies par des équations complexes (Navier-Stokes), dans des géométries compliquées. Les ordinateurs actuels ne permettent pas le traitement direct de ces problèmes. On montre ici comment, en séparant les opérateurs qui interviennent, en découplant les différentes variables, et en décomposant le domaine de calcul, on peut espérer résoudre les gros problèmes industriels.

Différents tests sur la qualité et la rapidité des méthodes sont présentés, ainsi que les applications industrielles dans des domaines de l'hydraulique très différents.

INTRODUCTION

La plupart des modélisations en hydraulique, que cela soit dans le domaine de l'environnement ou de l'industriel conduisent à des équations du même type. En effet, pour les déterminer, on part généralement des équations de Navier-Stokes auxquelles on fait subir quelques modifications. La forme générale de ces équations est la suivante :

$$(1) \quad \frac{1}{C^2(P)}\frac{\partial P}{\partial t} \;\Big|\; \qquad \;\Big|\; + \nabla . U \;\Big|\; \qquad = 0$$

$$(2) \quad \frac{\partial U}{\partial t} \;\Big|\; + \nabla (u(U) \otimes U) \;\Big|\; + \nabla P \;\Big|\; - Diff(U) = F(U,T)$$

$$(3) \quad \frac{\partial T}{\partial t} \;\Big|\; + \nabla (u(U) . T) \;\Big|\; \qquad \;\Big|\; - Diff(T) = S(U,T)$$

Convection — continuité — Diffusion

L'équation (1) traduit la conservation de la masse, (2) celle de la quantité de mouvement, (3) celle d'un champ scalaire : température par exemple. U représente un flux lié à la quantité de mouvement, u(U) est la vitesse de transport par le courant, P est liée à la pression, C(P) est la célérité des ondes de pression (infinie pour un fluide incompressible).

Finalement, les mécanismes d'évolution de U et de T traduits par les équations (2) et (3) sont un transport par le courant (convection) et une diffusion, sous la contrainte de continuité traduite par l'équation (1) et par l'intermédiaire de ∇P.

Il existe des différences fondamentales de comportement, représentées par des opérateurs de nature mathématique différente, entre ces mécanismes. La convection hyperbolique polarise le comportement dans le sens du courant. La diffusion parabolique peut être non linéaire et représente le plus souvent les effets de la turbulence. La continuité hyperbolique (elliptique si $C = \infty$) est toujours isotrope et couple ainsi les différentes composantes de U.

La plupart des modèles numériques développés au L.N.H. ces dix dernières années exploite cette idée simple qu'un algorithme n'est performant que s'il n'a qu'un seul but à atteindre. On a donc élaboré différents algorithmes en découpant les équations en plusieurs parties et en appliquant une résolvante différente pour chacune d'elle. On est donc ainsi ramené à inverser des systèmes de petites tailles, symétriques et pas trop mal conditionnés. Au cours du temps, des progrès ont été faits sur la précision du raccord entre ces résolutions ; en partant de simples pas fractionnaires pour envisager maintenant des raccords pratiquement transparents. Les chapitres I et II présente les derniers développements sur la convection et son raccord naturel, cette première étape étant fondamentale pour la qualité des solutions obtenues.

Le développement considérable de la puissance des ordinateurs a également permis le traitement de problèmes de plus en plus complexes. Résolues d'abord en dimension 1 puis deux, on résout maintenant les problèmes tridimensionnels, mais on est dans ce contexte, et pour quelques années à la limite de la puissance. Aujourd'hui les résolutions de 40 points au cube ne sont possibles qu'avec un traitement particulier de l'étape de continuité qui découple les trois composantes de la vitesse et la pression. Les idées actuelles à E.D.F. pour traiter cette étape sont présentées au chapitre III.

Enfin, les progrès réalisés dans la modélisation de la turbulence qui se développe au voisinage des parois ont été réalisés de paire avec une meilleure description de la géométrie de celles-ci. On est ainsi passé des différences finies en maillage rectiligne, aux différences finies en mailles curvilignes pour travailler maintenant sur les éléments finis. Ceci s'est accompagné bien sûr d'un surcoût de traitement numérique à la limite des ordinateurs actuels en tridimensionnels. Pour les applications industrielles, il convient d'exploiter au maximum la régularité du maillage (comme on le fait en différences finies) tant en conservant une bonne description des frontières (comme on le fait en éléments finis). Se pose alors le problème du raccord de sous-structures géométriques traitées indépendamment éventuellement sur des calculateurs parallèles. Le premier pas dans cette voie est présenté au chapitre IV où est traité le raccord en éléments finis.

Différentes applications industrielles sont ensuite présentées en conclusion.

Nous avons volontairement choisi de placer ce papier dans le cadre éléments finis, mais il convient de noter que les différentes méthodes présentées sont applicables (et appliquées) en différences finies ou plus précisément sur des maillages à éléments rectangulaires réguliers. Pour cette raison, nous adopterons des notations matricielles, les matrices pouvant être construites par différences finies et nous n'aborderons pas les techniques informatiques qui diffèrent pour les deux méthodes.

I. PRINCIPES GENERAUX

On se place dans un contexte "éléments finis", c'est-à-dire que les champs U, P et T sont définis dans des espaces discrétisés à l'aide de fonctions de base :

$V_h \qquad U_j = \sum_{i=1}^{N} U_{ji} \, \varphi_i$: pour les composantes U_j $j = 1, 2, 3$ de la vitesse $\overline{U}$

$T_h \qquad T = \sum_{1}^{N} T_i \ \varphi_i$

$P_h \qquad P = \sum_{1}^{N} P_i \ \pi_i$

Une discrétisation par différences en temps des équations (1) (2) (3) entre les instants t^n et t^{n+1} nous donne les équations à résoudre au cours d'un pas de temps Δt ($\alpha = \frac{1}{\Delta t}$). Selon l'idée directrice des méthodes d'éléments finis, on peut écrire une formulation faible de ces équations en les multipliant par des fonctions "test" que nous appellerons et ψ . γ Cependant, contrairement aux méthodes classiques de Galerkin ou de Petrov - Galerkin dans lesquelles les fonctions test ne dépendent pas du temps, nous prendrons les fonctions ψ définissant vitesse et température telles que :

$$(4) \quad \begin{cases} \dfrac{\partial \psi}{\partial t} + U^n . \nabla \psi = 0 \\ \psi(t = t^{n+1}, x) = \psi^{n+1}(x) = \varphi(x) \end{cases}$$

Les fonctions test vérifient donc une équation de transport dans le champ U^n entre t^n et t^{n+1}. Cette propriété permet de simplifier la formulation faible que l'on obtient en multipliant les équations (2) et (3) par ψ et en intégrant sur le domaine Ω ainsi que sur l'intervalle de temps $]t^n, t^{n+1}[$. (cf. [1], [2]).

On obtient finalement :

$$(5) \int_\Omega \frac{\alpha}{C^2} P^{n+1} \pi \, d\Omega + \int_\Omega \pi \operatorname{div} \bar{U}^{n+1} \, d\Omega = \int_\Omega \frac{\alpha}{C^2} \bar{P}^n \pi \, d\Omega$$

$$(6) \int_\Omega \left(\alpha \, Ui^{n+1} \varphi + \nu \, \bar{\nabla} \, Ui^{n+1} . \bar{\nabla} \varphi - P \frac{\partial \varphi}{\partial xi}\right) d\Omega = \int_\Omega \left(\alpha Ui^n \psi^n + F \varphi\right) d\Omega + TBU$$

$$(7) \int_\Omega \left(\alpha \, T^{n+1} + K \nabla T^{n+1} . \nabla \varphi\right) d\Omega = \int_\Omega \left(\alpha T^n . \psi^n + S . \varphi\right) d\Omega + TBT$$

Dans ces équations TBU et TBT sont des termes de bord permettant de prendre en compte les conditions aux limites naturelles (et qui seront explicités au chapitre II). De plus, les hypothèses suivantes ont été utilisées :

- Implicitation des termes de diffusion et de pression.
- Champ transporteur constant pendant Δt : U^n.
- Fonctions test de l'espace des pressions constantes en temps.

Il est ensuite facile d'obtenir un système matriciel en remplaçant les champs U, T et P par leurs discrétisations sur les fonctions de base et en écrivant les équations pour toutes les fonctions test. On fait ainsi apparaître les matrices :

$a_{ij} = \int_\Omega \varphi_i . \varphi_j \, d\Omega$: Identité éléments finis.

$l_{ij} = \int_\Omega \nabla \varphi_i . \nabla \varphi_j \, d\Omega$: Laplacien et les intégrales élémentaires :

$$P_{ij} = \int_\Omega \varphi_i \psi_j^{\,n} \, d\Omega.$$

Les matrices (a_{ij}) et (l_{ij}) sont obtenues de manière classique, elles sont symétriques et indépendantes du pas de temps considéré.

Les termes P_{ij} qui apparaissent au second membre sont plus difficiles à évaluer car les fonctions $\psi_j^{\,n}$ ne coïncident pas avec le maillage. Ce sont ces termes et eux seuls qui permettent de prendre en compte la convection.

On obtient donc finalement le système :

$$(8) \begin{cases} AU - {}^{T}BP = M = F + S(T) \\ BU + CP \;\; = D \\ A'T \qquad\;\; = E \end{cases}$$

Dans ce système les matrices A, A' et C sont symétriques et les vecteurs F et E sont des seconds membres faisant intervenir la convection.

L'algorithme général de résolution est donc :

(i) Calcul des seconds membres E et F en calculant $\psi_j^{\,n}$ par transport des fonctions φ_j en utilisant la méthode des caractéristiques et en évaluant les termes

$$P_{ij} = \int_\Omega \varphi_i \, \psi_j^{\,n} \, d\Omega$$

(ii) Résolution de l'étape thermique A'T = E et calcul du terme source S(T).

(iii) Résolution du problème de STOKES généralisé.

$$\begin{aligned} AU - {}^{T}BP &= M \\ BU + CP &= D \end{aligned}$$

Ce type de traitement de la convection conduit à un schéma du deuxième ordre en temps. Il constitue un découplage naturel de la convection de type Lagrange-Euler.

Le calcul des fonctions de base suppose u constant entre les instants t^n et t^{n+1}. Ceci constitue une linéarisation des termes de convection. Pour un bon traitement de la non-linéarité de cet opérateur, on est amené à utiliser un pas de temps qui reste de l'ordre de 1 pour le nombre de Courant construit sur la vitesse.

II. CONVECTION

II.1. Principe

Comme nous l'avons vu précédemment l'étape de convection revient à calculer les fonctions $\psi_j^{\,n}$ "transportées" des fonctions de base φ_j en remontant le temps entre t^{n+1} et t^n, puis à calculer les termes $P_{ij} = \int_\Omega \varphi_i \, \psi_j^{\,n} \, d$. De plus, il intervient un terme de bord qui n'a pas été explicité au chapitre I.

Dans le but de simplifier la présentation, nous nous intéressons ici au cas particulier de la seule équation de transport d'un scalaire f dans un domaine monodimensionnel.

$$(9)\quad \begin{cases} \dfrac{\partial f}{\partial t} + \dfrac{\partial uf}{\partial x} = 0 & \text{sur } = [a,b] \\ f\big|_{\Gamma-} = f_0 & \text{entre } t^n \text{ et } t^{n+1} \end{cases}$$

$\Gamma-$ étant la partie de la frontière où la vitesse entre dans le domaine, par exemple a si u est positif sur Ω.

En imposant à Ψ_j de vérifier :

$$\begin{cases} \dfrac{\partial \Psi_j}{\partial t} + u \dfrac{\partial \Psi_j}{\partial x} = 0 \\ \Psi_j\,(t = t^{n+1}) = \varphi_j \end{cases}$$

le même calcul formel d'intégration conduit ici à :

$$(10)\ \forall\, j \quad \sum_i f_i^{n+1} < \varphi_i, \varphi_j > = \sum_i f_i^{nh} < \varphi_i, \Psi_j^n > + \int_{t^n}^{t^{n+1}} uf\Psi_j\Big|_a dt - \int_{t^n}^{t^{n+1}} uf\Psi_j\Big|_b dt$$

Où les crochets <, > signifient $\int_\Omega d\Omega$ ce qui est une notation classique pour $< \varphi_i, \varphi_j >$ car les deux fonctions appartiennent au même espace de discrétisation : il s'agit d'un produit interne. Par contre la notation utilisée ne correspond plus à un produit interne pour les termes $< \varphi_j, \Psi_i^n >$ car Ψ_i^n, résultat du transport de φ_j, n'est pas connu sur les fonctions de base du domaine. Par exemple avec des fonctions de base linéaires par éléments :

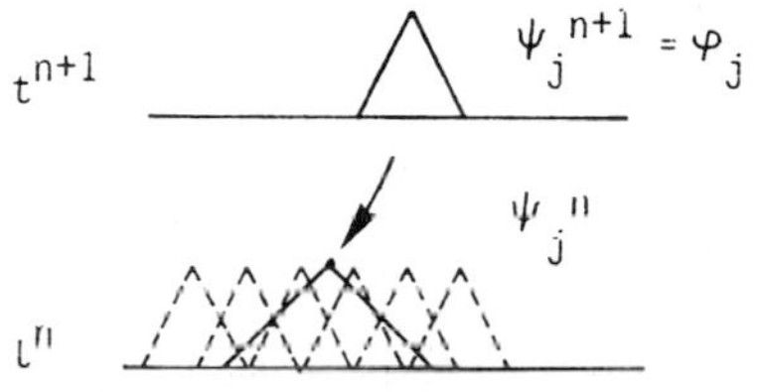

Le calcul des produits $< \varphi_i, \Psi_j^n >$ est fondamental. A une dimension d'espace, l'intégration exacte est possible, et le schéma possède alors des propriétés remarquables ([1] , [2]) : amortissement et déphasage négligeables pour des longueurs d'onde supérieures à 5 pas d'espace. A deux dimensions, il faut avoir recours à l'intégration numérique. La méthode utilisée au L.N.H. a été présentée dans [2]. A titre d'exemple et pour en illustrer les performances, la figure 1 présente en haut le test sévère du cône tournant dans un champ de rotation solide. L'amortissement du cône, qui s'étend sur 4 éléments seulement, après 1 tour (48 pas de temps), est de 12 % avec les éléments considérés de type triangles P2 et se stabilise ensuite (12,2 % après 2 tours).

II.2. Conditions aux limites

Les travaux récents ont porté sur le calcul des termes de bord, où interviennent les conditions aux limites naturelles du problème. Dans notre exemple, il s'agit de :

$$TB = \int_{t^n}^{t^{n+1}} uf\Psi_j\Big|_a dt - \int_{t^n}^{t^{n+1}} uf\Psi_j\Big|_b dt$$

On remarque immédiatement que seule la frontière où la vitesse U est entrante amènera une contribution de bord car si u = 0 le terme s'annule trivialement et si la vitesse est sortante la seule fonction Ψ_j non nulle au bord n'est différente de

0 qu'à l'instant t^{n+1} : son intégrale sur $[t^n, t^{n+1}]$ est donc nulle. Supposons que : $u > 0$ en a.

u étant constant pendant l'intervalle Δt on peut poser le changement de variable h = ut et écrire

$$\int_{\Delta t} u_a f_a \psi_{j_a} dt = \int_{\Delta h} f_a \psi_{j_a} dh$$

en supposant que f_a varie entre f_a^n et f_a^{n+1} : valeur de la condition limite donnée au système (9) il est possible d'introduire une fonction de base supplémentaire φ_0 extérieure au domaine et de prolonger la fonction φ_1 à l'extérieur pour ramener le calcul de l'intégrale de bord à celui d'une intégrale d'espace similaire aux termes $<\varphi_i \psi_j^n>$

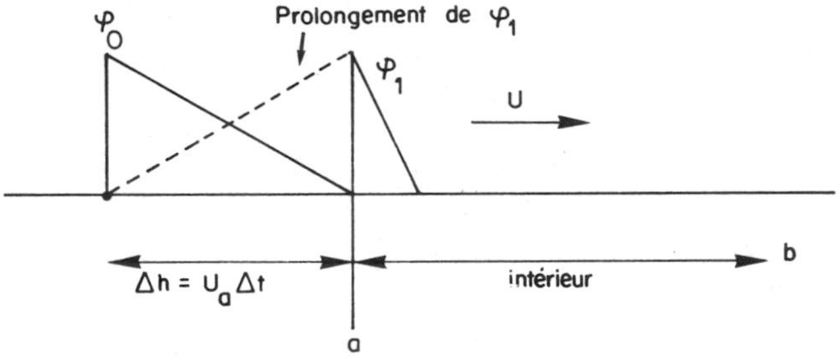

ainsi la fonction f est définie sur [a b] uΔh par :

$$f^n = \sum_{i=0}^{N} f_i^n \varphi_i \quad \text{avec} \quad f_0^n = f_a^{n+1}$$

avec $\Delta h = u\Delta t$ toutes les fonctions ψ_j^n $1 \leqslant j \leqslant N$ resteront intérieures au domaine $[a, b]$ u Δ h, et le calcul du second membre se fait alors de façon identique pour toutes les fonctions de base. Il s'agit donc d'un traitement naturel de la condition à la limite.

En dimension deux, le même principe est généralisable ; les frontières à flux entrants sont bordées d'une couche d'éléments Q1 rectilignes construits sur les vitesses de bord comme l'indique le schéma ci-dessous :

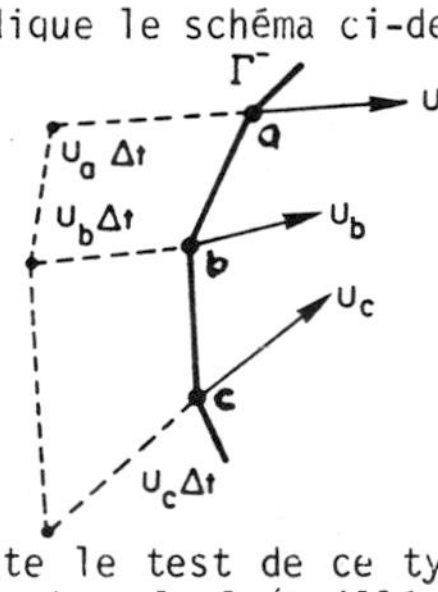

La figure 1, en bas, présente le test de ce type de formulation sur le cas d'un cône entrant dans le domaine de calcul (maillé en triangles P2) soumis à un champ de transport uniforme.

III. PROBLEME DE STOKES GENERALISE

Rappelons ici que l'objectif est de découpler les différentes variables en vue de résolutions tridimensionnelles. Il existe de tels solveurs opérationnels et efficaces en dimension 2 (Glowinski-Pironneau par exemple), mais leur extension au cas 3D devient prohibitive. C'est pourquoi, nous sommes revenus à un algorithme plus classique (Uzawa), mais en recherchant un préconditionnement efficace pour en accentuer la convergence.

III.1. Algorithme d'Uzawa préconditionné ([4], [5])

Après l'étape de convection et après le calcul éventuel de la température, il reste à résoudre un problème de Stokes généralisé :

$$(11)\quad \begin{cases} AU - {}^{T}BP = M \\ BU + CP = D \end{cases}$$

L'algorithme d'Uzawa est un algorithme de gradient sur le système symétrique associé à (11).

$$(12)\quad (B\,A^{-1}\,{}^{T}B + C)\,P = -\,B\,A^{-1}\,M + D$$

Lorsque la célérité est infinie (fluide incompressible, C = 0, D = 0) le système (12) est relativement mal conditionné et l'algorithme d'Uzawa converge lentement. Dans ce cas, qui est celui du problème de Stokes, on accélère la convergence en utilisant un préconditionnement basé sur le fait, qu'aux conditions aux limites près, (12) s'interprète comme la discrétisation de :

$$(13)\quad \begin{aligned} &\Delta p = \operatorname{div}(s) \qquad \text{(s résultat de la convection)} \\ &\frac{\partial p}{\partial n} = 0 \end{aligned}$$

qui fournit donc un champ de pression "très proche" de la solution du problème de Stokes. L'algorithme d'Uzawa préconditionné comporte alors deux étapes par itération :

(i) p^n étant l'estimation de p à l'itération n, calculer u^n solution de :

$$A\,U^n = {}^{T}B\,p^n + M$$

(ii) calculer p^{n+1} solution de

$$E\,p^{n+1} = E\,p^n + \rho B\,U^n$$

où la matrice E est la discrétisation élément fini du problème (13) :

$$C_{ij} = \int_{\Omega} \operatorname{grad}\pi i \,\operatorname{grad}\pi j \,d\Omega$$

Compte tenu de la définition du préconditionnement, celui-ci est d'autant meilleur que la diffusion est faible (que le nombre de Reynolds est grand) comme le montre les tests ci-après. Cette propriété constitue un avantage indéniable pour les problèmes industriels.

Cet algorithme découple non seulement la pression des vitesses, mais également les composantes de la vitesse entre elles (puisque la pression est rejetée au second membre) lorsque les conditions aux limites sont de type vitesses imposées. De plus, les matrices intervenant pour chaque composante sont les mêmes.

Une difficulté peut apparaître pour certaines conditions aux limites couplant les composantes de la vitesse entre elles (condition de type "paroi" par exemple). Dans ce cas, la solution étudiée en collaboration avec l'I.N.R.I.A. consiste à itérer par une méthode de gradient sur l'opérateur de bord qui associe aux conditions de Dirichlet les conditions aux limites recherchées (cf. [7]).

III.2. Tests numériques

On compare ici l'efficacité des algorithmes simple et préconditionné sur l'exemple de la cavité carrée, maillée avec 200 triangles (121 noeuds P1, 441 noeuds P2), à différents nombres de Reynolds : 100 - 400 - 1000 ([5]).

L'algorithme d'Uzawa consiste à résoudre à chaque itération 2 problèmes P2 (l'un pour u, l'autre pour v) et, en cas de préconditionnement, 1 problème P1 en plus. Chaque résolution P2 ou P1 s'effectue par gradient conjugué préconditionné par la matrice Cholesky incomplet, dans ces conditions la résolution du problème P2 nécessite en moyenne 7 itérations, celle du problème P1 : 19 itérations. Le tableau suivant compare pour chaque Reynolds étudié l'évolution du résidu d'une itération à l'autre dans l'algorithme d'Uzawa simple et préconditionné. L'erreur de divergence initiale provient sur ce test d'un changement de pas de temps qui accentue donc l'effet de la convection (second membre) volontairement plus grand (le pas de temps choisi correspond à un nombre de Courant de 10). En phase d'exploitation, le nombre de Courant reste plus faible (~1) pour le traitement des non-linéarités.

Dans ces conditions, il semble possible de limiter le nombre d'itération à 2 au plus pour les applications pratiques (cf. exemple ci-après). On constate sur ce test l'efficacité du préconditionnement ; on note également que l'adjonction d'un problème P1 reste d'un effet marginal sur le temps calcul par itération ce qui montre tout l'intérêt du préconditionnement. On vérifie enfin que le préconditionnement est d'autant plus efficace que le nombre de Reynolds est plus élevé.

	Re 100		Re 400		Re 1000	
	Uzawa simple $\|\|div\ u\|\|$	Uzawa précondi-tionné $\|\|div\ u\|\|$	Uzawa simple $\|\|div\ u\|\|$	Uzawa précondi-tionné $\|\|div\ u\|\|$	Uzawa simple $\|\|div\ u\|\|$	Uzawa précondi-tionné $\|\|div\ u\|\|$
Valeur initiale	$.133\times10^{-1}$	$.133\times10^{-1}$	$.156\times10^{-1}$	$.156\times10^{-1}$	$.176\times10^{-1}$	$.176\times10^{-1}$
Après un gradient simple	$.124\times10^{-1}$	$.627\times10^{-2}$	$.137\times10^{-1}$	$.695\times10^{-2}$	$.138\times10^{-1}$	$.672\times10^{-2}$
Après un gradient conjugué	$.115\times10^{-1}$	$.415\times10^{-2}$	$.104\times10^{-1}$	$.427\times10^{-2}$	$.111\times10^{-1}$	$.360\times10^{-2}$
" " "	$.941\times10^{-2}$	$.297\times10^{-2}$	$.888\times10^{-2}$	$.269\times10^{-2}$	$.710\times10^{-2}$	$.250\times10^{-2}$
" " "	$.799\times10^{-2}$	$.220\times10^{-2}$	$.678\times10^{-2}$	$.221\times10^{-2}$	$.529\times10^{-2}$	$.160\times10^{-2}$
Temps calcul moyen/ itération IBM 3081	3,9 s	4,1 s	3,8 s	4,0 s	3,8 s	3,9 s

La figure 2 présente le profil médian de vitesse dans le cas Re = 400 (convection faible + Uzawa préconditionné) comparé à la référence de Burggraff (cf. [3]). La figure présente également le résultat plus ancien du même calcul pour une formulation classique par caractéristiques de la convection (transport de l'inconnue et non des fonctions de base).

La figure 2 présente enfin le résultat du calcul de l'écoulement tourbillonnaire derrière un cylindre (cf. aussi [4]). Dans ce dernier cas, le nombre d'itérations de la méthode d'Uzawa préconditionné a été limité à 2. Le Reynolds de l'écoulement est de 400.

On trouve par calcul un nombre de Strouhal de 0,195 en accord total avec les résultats expérimentaux. Le coefficient de trainée de pression est 1,2 (valeur surestimée d'environ 20 % par rapport aux données expérimentales).

IV. SOUS-STRUCTURATION

Le principe de la méthode de sous-structuration est de partitionner le domaine de calcul en sous-structures afin d'obtenir, sur chacune de celles-ci, des problèmes indépendants. La difficulté est alors de trouver les conditions aux limites sur les frontières intérieures qui coordonnent le système, c'est-à-dire qui sont telles que la solution formée des solutions de chaque sous-structure vérifie le système global.

Les deux avantages majeurs de cette méthode sont :

- d'une part, elle permet de prendre en compte la notion de sous-structuration "régulière" et donc d'avoir des algorithmes beaucoup plus rapide du fait de la régularité du maillage (dans un cadre élément fini).
- d'autre part c'est une méthode qui, du point de vue algorithmique, est bien adaptée aux architectures parallèles des nouveaux et futurs calculateurs.

IV.1. Principe adopté (cf. [6])

Nous allons présenter la méthode dans le cadre des équations de Stokes (célérité infinie). De plus, nous n'aborderons pas le problème de la détermination des seconds membres encore à l'étude (raccord en convection).

Comme nous l'avons vu plus haut les équations de Stokes une fois discrétisées peuvent s'écrire :

$$(14)\qquad \begin{aligned} Au - {}^{T}Bp &= S \quad \text{dans}\ \Omega \\ Bu &= 0 \quad \text{dans}\ \Omega \\ u &= u_D \quad \text{sur}\ \Gamma \end{aligned}$$

partitionnons le domaine Ω en deux sous-structures $(\Omega^e)^2_{e=1}$. Soit $(\mathcal{T}h^e)^2_{e=1}$ les triangulations correspondantes qui sont elles-mêmes un partitionnement de $\mathcal{T}h$ triangulation de Ω.

Notons π^e, V^e les espaces d'approximations des pressions et vitesses, W^e l'espace engendré par les fonctions de base associées aux noeuds de $\partial\Omega^e$.

On définit alors le sous-espace $\widetilde{M}^e$ de We par :

$$\widetilde{M}^e = \left\{ \lambda^e \in W^e \ ;\ \lambda^e = 0 \text{ sur } \Gamma^e \text{ et} \int \lambda^e . n^e = 0 \right\}$$

où $\Gamma^e = \partial\Omega^e \cap \partial\Omega$ et $\mathcal{E} = \partial\Omega_1 \cap \partial\Omega_2$

on note alors $\widetilde{M} = \prod_{e=1}^{2} \widetilde{M}^e$

On définit alors la forme bilinéaire k sur $\widetilde{M} \times \widetilde{M}$ par :

$$k(\lambda, \mu) = \sum_{e=1}^{2} \left(\int_{\Omega^e} \alpha\, u^e_\lambda\, \mu^e + \int_{\Omega^e} \nu \bar{\bar{\nabla}} u^e_\lambda\ \bar{\bar{\nabla}} \mu^e - \int_{\Omega^e} p^e_\lambda \operatorname{div} \mu^e \right)$$

ainsi que le second membre b :

$$(b,\mu) = \sum_{e=1}^{2} \left(\int_{\Omega^e} \alpha u_0^e \mu^e + \int_{\Omega^e} \nu \bar{\bar{\nabla}} u_0^e \; \bar{\bar{\nabla}} \mu^e - \int_{\Omega^e} P_0^e \operatorname{div} \mu^e \right)$$

où $(u_\lambda^e, P_\lambda^e)$ et (u_0^e, μ_0^e) sont solutions de :

$$\begin{cases} A^e u_\lambda^e - {}^T B^e P_\lambda^e = 0 \text{ dans } \Omega^e \\ B^e u_\lambda^e = 0 \text{ dans } \Omega^e \\ u_\lambda^e = \lambda^e \text{ sur } \mathcal{E} \\ u_\lambda^e = 0 \text{ sur } \Gamma^e \end{cases} \qquad \begin{cases} A^e u_0^e - {}^T B^e P_0^e = S^e \\ B^e u_0^e = 0 \\ u_0^e = w^e \\ u_0^e = u_D^e \end{cases}$$

avec w^e tel que $\int_{\partial\Omega^e} u_0^e . n = 0$

Alors (14) est équivalent à :

$$(15) \qquad \begin{cases} K\tilde{\lambda} = b & \\ A^e u^e - {}^T B^e u^e = S^e & \text{dans } \Omega^e \\ B^e u^e = 0 & \text{dans } \Omega^e \\ u^e = \tilde{\lambda}^e + w^e & \text{sur } \mathcal{E} \\ u^e = u_D^e & \text{sur } \Gamma^e \end{cases} \quad \text{pour } e = 1,2$$

Cette méthode dite "primale" consiste donc à chercher une condition de type Dirichlet telle que les contraintes se raccordent. La méthode "duale" consiste à chercher une condition de type contrainte qui raccorde les vitesses.

L'opérateur dual s'écrit alors :

$$n(\varphi,\psi) = \sum_{e=1}^{2} \left(\int_{\mathcal{E}} (-1)^{e+1} u^e \varphi \psi \, d\gamma \right)$$

et les sous problèmes de Stokes sont alors résolus avec des conditions de type contrainte sur $\mathcal{E}$.

Dans les deux cas, la méthode consiste donc à construire au préalable l'opérateur raccord ; chaque pas de temps nécessitant ensuite la construction des seconds membres, la résolution du problème du raccord, puis la résolution de chaque sous-structure. L'ensemble de ces opérations est indépendant de l'algorithme de Stokes choisi ; celui-ci étant en fait utilisé en boîte noire.

IV.2. Exemples

Nous avons testé ces deux méthodes dans le cas du problème de Stokes sur la cavité carrée à paroi mobile déjà présentée. Nous présentons les résultats (conf. figure 4) de la méthode primale, la méthode duale donnant les mêmes résultats et étant plus facile à mettre en oeuvre. Il apparaît que ces deux méthodes conduisent à de légères différences de pression sur les frontières raccord. On ne raccorde en effet exactement que la contrainte et la vitesse. Cela dit, le gradient de pression en x est bien symétrique par rapport à la médiane verticale de la cavité carrée, et le gradient de pression en y est bien anti-symétrique.

Nous avons aussi testé ces deux méthodes sur les équations de Saint-Venant (célérité finie cette fois) dans le cas d'une onde amortie se propageant dans un bassin rectangulaire. Les résultats obtenus sont très bons aussi bien pour le débit que pour la hauteur d'eau, comme le montre la figure 3, même dans le cas d'une frontière raccord quelconque.

La méthodologie du raccord étant maintenant au point, il convient d'exploiter la régularité du maillage dans certaines sous-structures, l'idée étant pour les gros calculs industriels de mailler la plus grande partie du domaine de façon régulière et de le raccorder à un maillage quelconque décrivant mieux la géométrie des parois.

CONCLUSION

Les applications numériques de l'hydraulique industrielle nécessitent le traitement d'un grand nombre de variables dans des géométries complexes.

Les figures 5 et 6 en fournissent deux exemples très différents. La figure 5 présente des calculs en cours de surcotes de tempête en mer : en haut, la Manche sous l'action d'un vent d'ouest ; en bas le maillage du plateau continental de la Mer du Nord qui servira aux simulations de tempêtes réelles. La figure 6 présente un calcul plus ancien en différences finies d'un transitoire thermique dans le collecteur chaud du surgénérateur Super Phénix II. Dans ce dernier cas et malgré les 40 000 points de discrétisation, la géométrie pourtant relativement simple pour une structure industrielle, n'était qu'approximée.

Les solutions retenues consistent d'abord à séparer les opérateurs, la convection étant en particulier traitée à part (ce qui conduit alors à des matières symétriques) ; elles consistent ensuite à découpler les variables, composantes de la vitesse et pression, afin de diminuer de façon notable la taille des systèmes à inverser. Elles consistent enfin à décomposer le domaine de calcul en sous-structures traitées de façon indépendante.

Les méthodes développées à E.D.F. dans cette optique ont été présentées. Elles sont bien adaptées aux équations instationnaires : l'évolutif en temps étant un bon moyen de traiter les non-linéarités des écoulements turbulents. Les différents tests ont montré leur efficacité. On entre maintenant dans la phase laborieuse de la mise en forme industrielle.

REFERENCES

1 J.P. BENQUE, J. RONAT : "Quelques difficultés des modèles numériques en hydraulique". 5th International Symposium on Computing Methods in Applied Sciences and Engineering. Versailles (FRANCE) décembre 1981.

2 J.P. BENQUE, G. LABADIE, J. RONAT : "A new-finite element method for the Navier-Stokes equations coupled with a temperature equation". To be published in "International Journal for Numerical Method in Fluids".

3 J.P. BENQUE, B. IBLER, A. KERAMSI, G. LABADIE : "A finite element method for the Navier-Stokes Equations". 3rd International Conference on F.E.M. in Flow Problems, Bannf. Alberta (CANADA) June 81.

4 J.P. BENQUE, P. ESPOSITO, G. LABADIE : "New decomposition finite element methods for the Stokes problems and the Navier-Stokes equations". Numerical Method in Laminar and Turbulent Flows, Seattle (U.S.A.) Août 83.

5 J. GOUSSEBAILE, J.P. GREGOIRE, A. HAUGUEL : "Iterative Stokes Solvers and Splitting Techniques for Industrial Flows". 5th International Conference on F.E.M. in Flow Problems, Austin (U.S.A.) January 84.

6 J.P. MARTINAUD : "Résolution des équations de Saint-Venant et de Stokes par une méthode de décomposition de domaine". 16ème Colloque d'Analyse Numérique, Lorient (FRANCE) Mai 83.

7 J. GOUSSEBAILE, G. LABADIE, L. REIHNART : Publication commune INRIA-LNH à paraître dans "International Journal for Numerical Methods in Fluids".

Figure 1 : Tests schématiques de la convection faible cone dans un champ de vitesse tournant

Etat initial

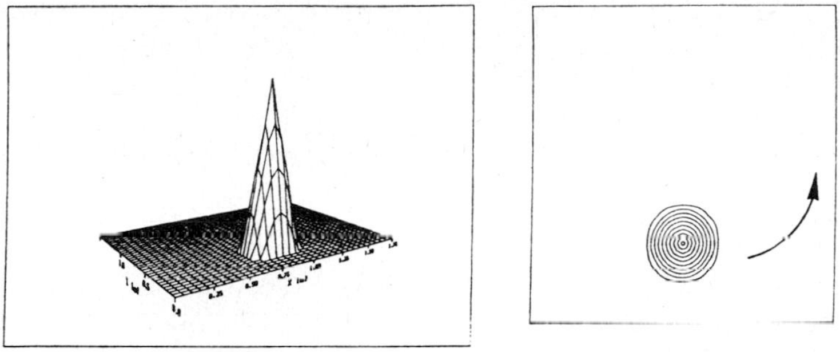

1/2 tour
24 pas de temps

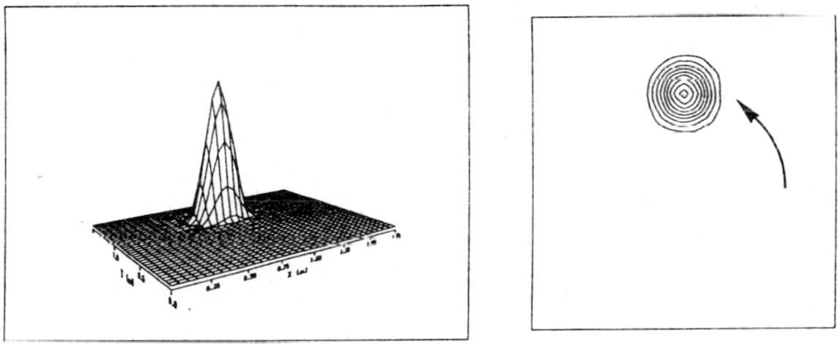

1 tour
48 pas de temps

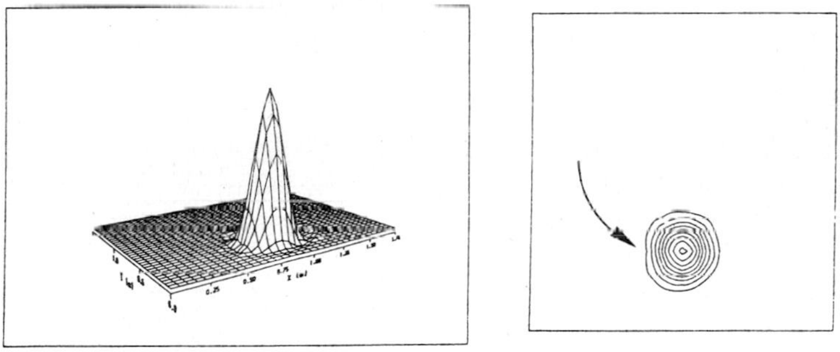

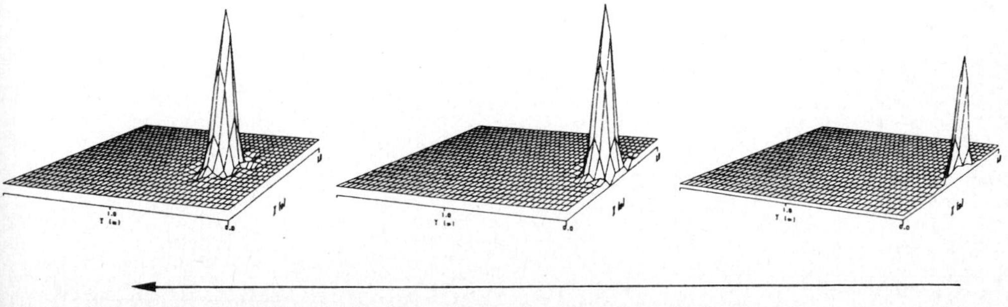

Figure 2 : Méthode d'UZAWA préconditionnée

Profil médian de vitesse dans une cavité à paroi mobile à Reynolds = 400

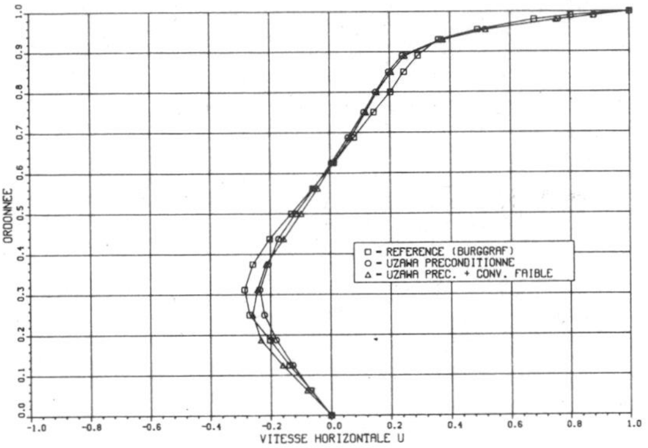

Ecoulement autour d'un cyclindre **Reynolds = 400**

Maillage

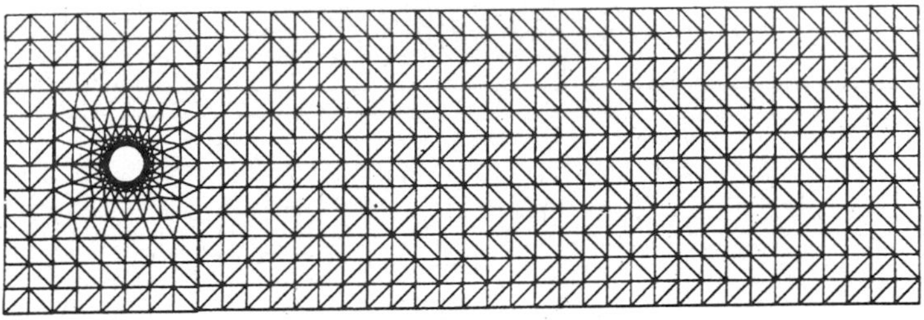

Champ de vitesse

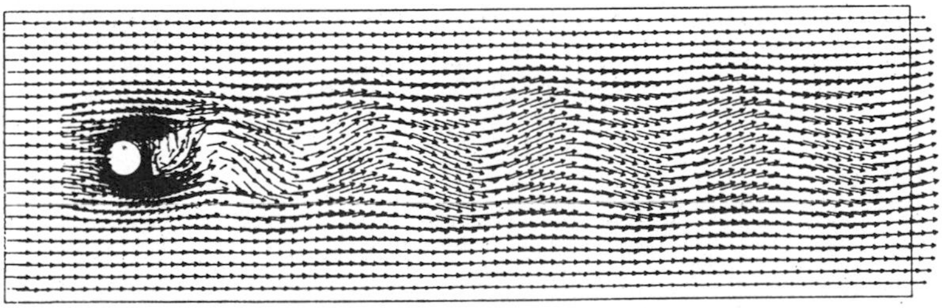

Champ de pression

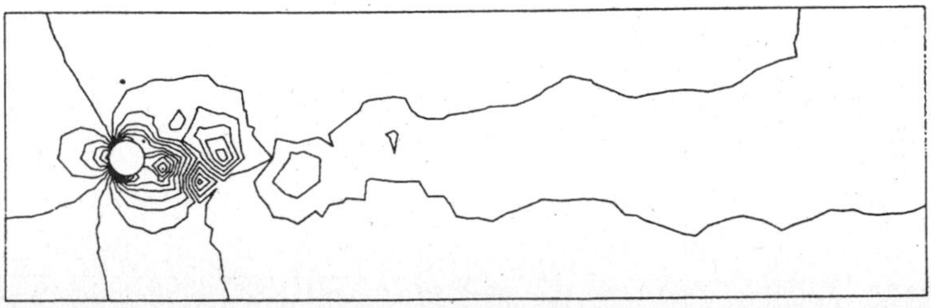

Figure 3 : Résolution des équations de Saint-Venant par sous-structuration
Onde amortie se propageant dans un bassin

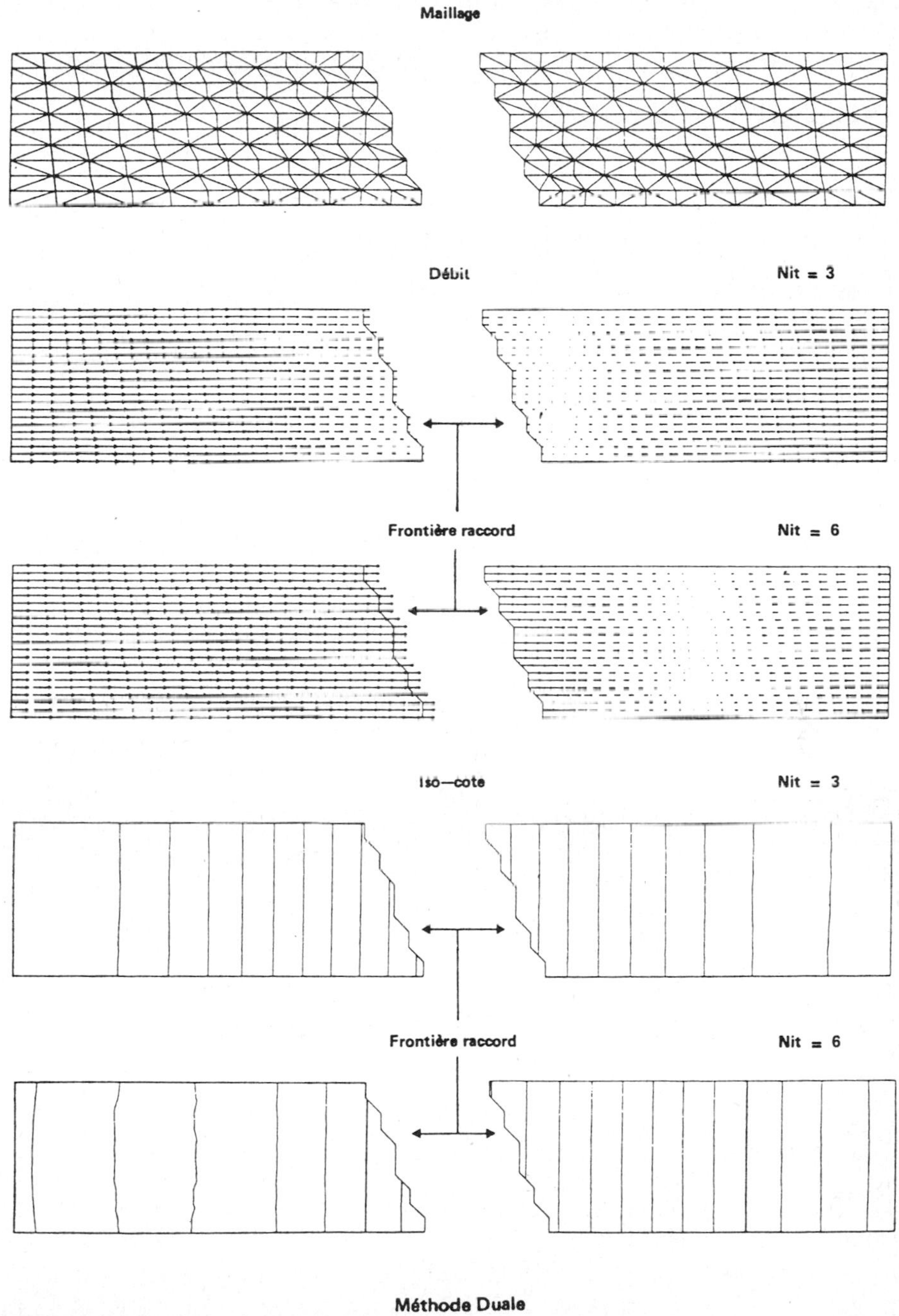

Méthode Duale

Figure 4 : Résolution des équations de Stokes par sous-structuration
Test de la cavitée carrée

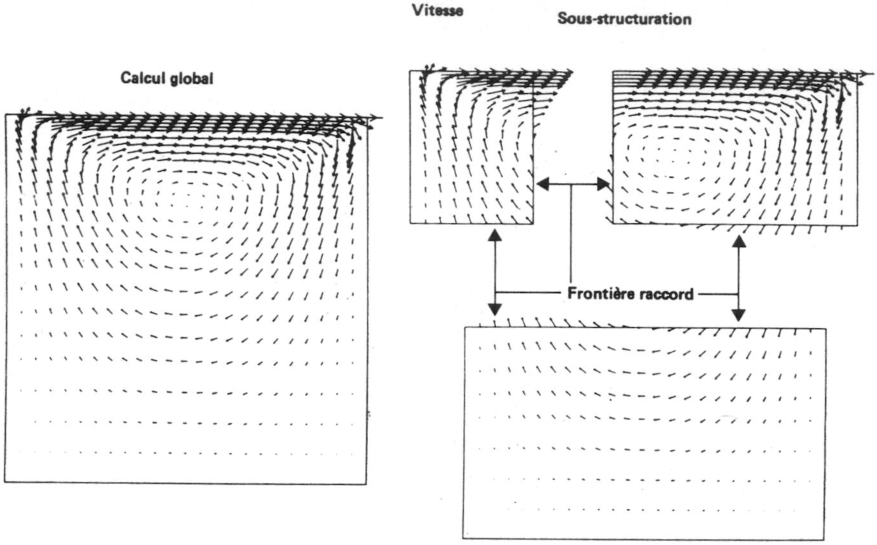

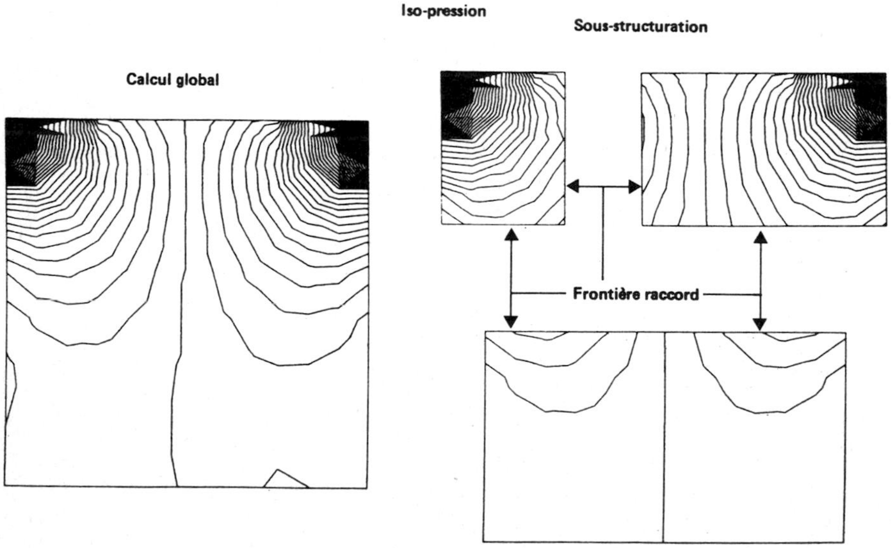

Méthode Primale

Figure 5 : Equations de St VENANT

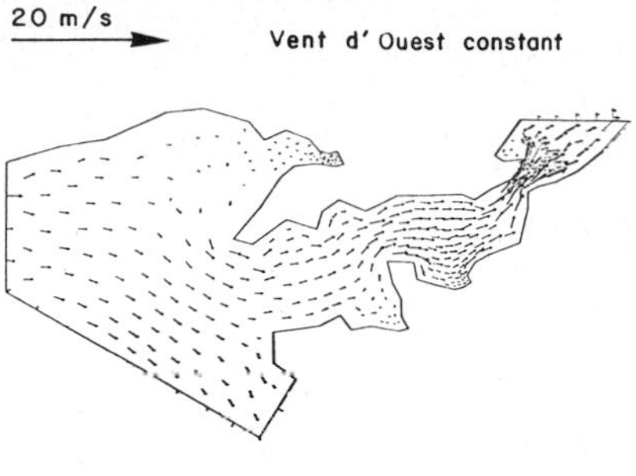

Courants engendrés par un vent d'ouest en Manche

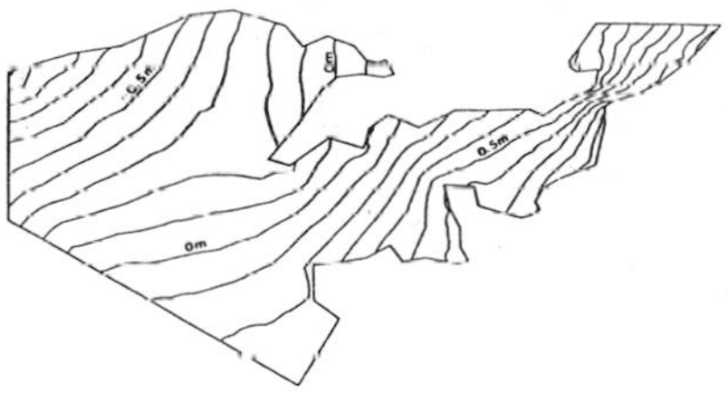

Surcote engendrée par un vent d'ouest en Manche

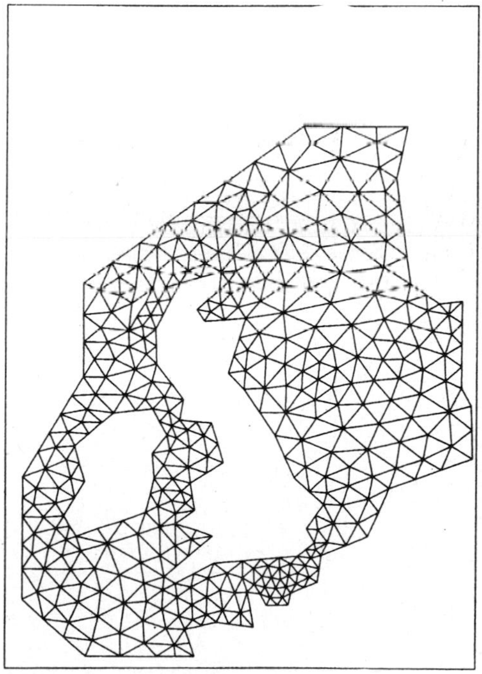

Maillage du plateau continental de l'Europe du Nord Ouest pour l'étude des surcotes de tempête

Figure 6 : Calcul d'un transitoire thermohydraulique dans
le collecteur chaud de Super Phenix 2

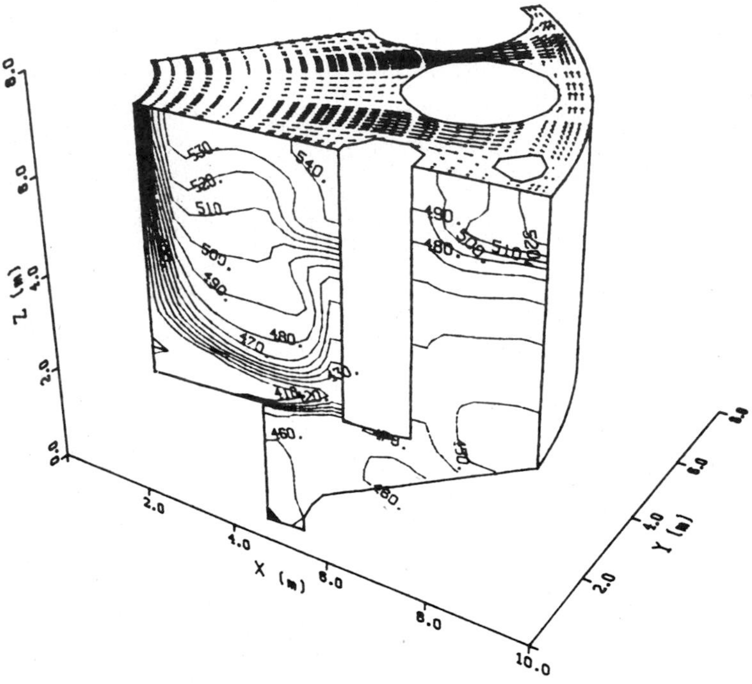

COMPRESSIBLE FLOWS

ECOULEMENTS COMPRESSIBLES

Computing Methods in Applied Sciences and Engineering, VI
R. Glowinski and J.-L. Lions (Editors)
Elsevier Science Publishers B.V. (North-Holland)

On a class of TVD Schemes for Gas Dynamic Calculations

H.C. YEE† AND R.F. WARMING†
NASA Ames Research Center, Moffett Field, California

AMI HARTEN‡
Tel-Aviv University, Tel Aviv and New York University, New York

Several Techniques for the construction of nonlinear, explicit, second-order accurate, high-resolution schemes for hyperbolic conservation laws have been developed in recent years [1-5]. The goal of constructing these highly nonlinear schemes is to simulate complex flow fields more accurately. They are algorithms based in part on either an exact or approximate Riemann solver. These schemes are constructed under a common theme; i.e., to achieved second order accuracy without introducing spurious oscillations near discontinuities by employing some kind of feed back mechanism. Some of these schemes were based on physical and geometric arguments; and some were based on more formal mathematical approaches. Most of these schemes were independently derived, and thus they are very much different in form, methodology and design principle. However, from the standpoint of numerical analysis, these schemes are total variation diminishing (TVD) for nonlinear scalar hyperbolic conservation laws and for constant coefficient hyperbolic systems. The notion of TVD schemes was introduced by Harten [3,6].

The purpose of this paper is to review a subset of the class of TVD schemes. We start with a discussion of first-order TVD schemes, and then follow with a brief description of the construction of second-order TVD schemes. The recently developed implicit TVD schemes will also be included [6,7]. We will conclude with some transient and steady-state calculations to illustrate the applicability of these schemes to the Euler equations.

---

†Research Scientist, Computational Fluid Dynamics Branch.
‡Associate Professor, School of Mathematical Sciences.

References

[1] B. van Leer, "Towards the Ultimate Conservative Difference Scheme. V. A Second-Order Sequel to Godunov's Method," J. Comp. Phys., Vol. 32, 1979, pp. 101-136.

[2] P. Colella and P.R. Woodward, "The Piecewise-Parabolic Method (PPM) for Gas-Dynamical Simulations," LBL report no. 14661, July 1982.

[3] A. Harten, "A High Resolution Scheme for the Computation of Weak Solutions of Hyperbolic Conservation Laws," NYU Report, oct. 1981 and J. Comp. Phys., Vol. 49, 1983, pp. 357-393.

[4] P.L. Roe, "Some Contributions to the Modelling of Discontinuous Flows," to appear in Proc. AMS-SIAM Summer Seminar on Large Scale Computations in Fluid Mechanics, Univ. of Calif. at San Diego, June 27-July 8, 1983.

[5] S. Osher, "Shock Modeling in Transonic and Supersonic Flow," to appear in Recent Advances in Numerical Methods in Fluids, Vol. 4, Advances in Computational Transonics, W.G. Habashi Ed., Pineridge Press, 1984.

[6] A. Harten, "On a Class of High Resolution Total-Variation-Stable Finite-Difference Schemes," to appear in SIAM J. Num. Anal.

[7] H.C. Yee, R.F. Warming and A. Harten, "Implicit Total Variation Diminishing (TVD) Schemes for Steady-State Calculations," AIAA Paper No. 83-1902, Proc. of the AIAA 6th Computational Fluid Dynamics Conference, Danvers, Mass., July, 1983.

Computing Methods in Applied Sciences and Engineering, VI
R. Glowinski and J.-L. Lions (Editors)
Elsevier Science Publishers B.V. (North-Holland)

MULTIDIMENSIONAL EXPLICIT DIFFERENCE SCHEMES FOR HYPERBOLIC CONSERVATION LAWS

Bram van Leer
Department of Mathematics and Computer Science
Delft University of Technology
Delft, The Netherlands

First- and second-order explicit difference schemes are derived for a three-dimensional hyperbolic system of conservation laws, without recourse to dimensional factorization. All schemes are upwind biased and optimally stable.

1. INTRODUCTION

For the solution of initial-value problems governed by hyperbolic conservation equations some fine numerical techniques are available. Most methods are based on a one-dimensional scheme

$$u^{n+1} = L_x(\Delta t)\ u^n \ , \tag{1}$$

integrating the system of conservation laws

$$u_t + [f(u)]_x = 0 \tag{2}$$

from time t^n to $t^{n+1} = t^n + \Delta t$. A multi-dimensional system like

$$u_t + [f(u)]_x + [g(u)]_y + [h(u)]_z = 0 \tag{3}$$

may always be approximated by a sequence of one-dimensional steps; most commonly used is

$$u^{n+1} = L_z(\Delta t)\ L_y(\Delta t)\ L_x(\Delta t)\ u^n \ , \tag{4.1}$$

$$u^{n+2} = L_x(\Delta t)\ L_y(\Delta t)\ L_z(\Delta t)\ u^{n+1} \ , \tag{4.2}$$

with second-order accuracy in time at every other time-level [1].

The convenience of such a factorization is twofold. Firstly, in developing multi-dimensional methods one may concentrate on one-dimensional operators; secondly, multi-dimensional codes reduce to a sequence of one-dimensional sweeps.

A display of the power of the above approach is found in the review paper by Woodward and Colella [2] on the numerical simulation of two-dimensional, strongly compressible flow.

Considering the success of dimensional factorization we may ask ourselves if there

---

The research reported here was partially supported under NASA Contract No. NAS1-15810 while the author was in residence at ICASE, NASA Langley Research Center, Hampton, VA 23665, during the summer of 1982.

is any point in designing genuinely multi-dimensional methods, i.e. methods that can not be implemented as a sequence of one-dimensional operators. The answer still is "yes", but may very well tend to "no" if computers will continue to grow, in speed and capacity, at the current pace. The same, however, may have been said, more than 20 years ago, on the matter of developing second-order methods. In both cases the increase in complexity of the methods is meant to pay off via an increase in efficiency.

Genuinely multi-dimensional schemes are most efficient in modeling essentially multi-dimensional phenomena in a relatively small region, such as push-pull flow or flow around a sharp corner. If such regions have a strong influence on the overall solution, then the use of a multi-dimensional scheme is recommendable.

In the further sections of this paper I shall indicate how to construct second-order difference schemes in two or three space dimensions. First-order schemes were published earlier [3]; their construction is briefly reviewed.

2. A FIRST-ORDER SCHEME

In order to derive a first-order scheme for Eq.(3), consider the scalar linear equation

$$u_t + au_x + bu_y + cu_z = 0, \quad a > 0 \ , \ b > 0 \ , \ c > 0 \ , \tag{5}$$

in combination with a piecewise uniform initial-value distribution

$$u^n(x,y,z) = u^n_{ijk}, \quad x_{i-\frac{1}{2}} < x < x_{i+\frac{1}{2}}, \ y_{j-\frac{1}{2}} < y < y_{j+\frac{1}{2}}, \ z_{k-\frac{1}{2}} < z < z_{k+\frac{1}{2}}, \tag{6}$$

with $x_{i\pm\frac{1}{2}} = x_i \pm \frac{1}{2}\Delta x$, $y_{j\pm\frac{1}{2}} = y_j \pm \frac{1}{2}\Delta y$, $z_{k\pm\frac{1}{2}} = z_k \pm \frac{1}{2}\Delta z$.

The exact solution of (5) at time t^{n+1} in terms of the initial values at t^n is

$$u^{n+1}(x,y,z) = u^n(x-a\Delta t, y-b\Delta t, z-c\Delta t). \tag{7}$$

Inserting the initial values (6) into (7) yields, for sufficiently small Δt, the following difference scheme:

$$\begin{aligned} u^{n+1} = \Big[1 &- \sigma_x\Delta_x\{1 - \tfrac{1}{2}\sigma_y\Delta_y(1 - \tfrac{1}{3}\sigma_z\Delta_z) - \tfrac{1}{2}\sigma_z\Delta_z(1 - \tfrac{1}{3}\sigma_y\Delta_y)\} \\ &- \sigma_y\Delta_y\{1 - \tfrac{1}{2}\sigma_z\Delta_z(1 - \tfrac{1}{3}\sigma_x\Delta_x) - \tfrac{1}{2}\sigma_x\Delta_x(1 - \tfrac{1}{3}\sigma_z\Delta_z)\} \\ &- \sigma_z\Delta_z\{1 - \tfrac{1}{2}\sigma_x\Delta_x(1 - \tfrac{1}{3}\sigma_y\Delta_y) - \tfrac{1}{2}\sigma_y\Delta_y(1 - \tfrac{1}{3}\sigma_x\Delta_x)\}\Big]u^n, \end{aligned}$$

$$\sigma_x = a\Delta t/\Delta x < 1 \ , \quad \sigma_y = b\Delta t/\Delta y < 1 \ , \quad \sigma_z = c\Delta t/\Delta z < 1 \ ; \tag{8}$$

here Δ_x, Δ_y, Δ_z denote backward (upwind) differencing in the x-, y-, z-direction.

The terms in the above formula have been especially arranged in order to facilitate their interpretation. The terms in parentheses are one-dimensional updates over a time-step 1/3 Δt, with the first-order upwind scheme. The braced terms are two-dimensional updates over a time-step ½ Δt, and represent the values of u at cell interfaces, averaged both in space and in time. The full operator, between brackets, shows the three-dimensional upwind nature of the scheme.

Godunov [4] has indicated how to transform upwind differencing for a scalar linear convection equation into a method for a nonlinear hyperbolic system of the form

(2). For scheme (8) his recipe implies replacement of $\sigma_x\Delta_x u$ by

$$\frac{\Delta t}{\Delta x}\left[(\Delta_x f)^+ + (\nabla_x f)^-\right] \tag{9.1}$$

or

$$\frac{\Delta t}{\Delta x}\left[\Delta_x(f^+) + \nabla_x(f^-)\right] , \tag{9.2}$$

where ∇ denotes forward differencing and the superscripts + and − indicate the splitting of a flux-difference or a flux in parts associated with forward and backward signals, respectively (see [5] and [6]). The expressions $\sigma_y\Delta_y u$ and $\sigma_x\Delta_x u$ are replaced analogously.

Note that (8) can be factorized exactly as indicated in (4):

$$u^{n+1}_{ijk} = (1-\sigma_z\Delta_z)(1-\sigma_y\Delta_y)(1-\sigma_x\Delta_x)\, u^n_{ijk} \ . \tag{10}$$

While this property in general is lost when extending the scheme to a linear or nonlinear hyperbolic system, it nevertheless shows that dimensional splitting basically is a sound idea.

3. SECOND-ORDER SCHEMES

Second-order schemes for Eq. (3) may be derived, again, by extending schemes for Eq. (5). In order to achieve second-order accuracy in time and space we must use a piecewise linear initial-value distribution (see [7]), i.e.,

$$u^n(x,y,z) = u^n_{ijk} + \frac{x-x_i}{\Delta x}\,\delta_x u^n_{ijk} + \frac{y-y_i}{\Delta y}\,\delta_y u^n_{ijk} + \frac{z-z_k}{\Delta z}\,\delta_z u^n_{ijk},$$

$$x_{i-\frac12} < x < x_{i+\frac12}\ ,\quad y_{j-\frac12} < y < y_{j+\frac12}\ ,\quad z_{k-\frac12} < z < z_{k+\frac12}\ ; \tag{11}$$

here $\delta_x u$, $\delta_y u$ and $\delta_z u$ represent locally averaged differences of u, e.g. $\delta_x = \frac12(\Delta_x + \nabla_x)$.

Combining (11) with the exact solution (7) of (5) reveals that the second-order scheme contains terms up to $O[(\Delta t)^2]$ when restricted to one dimension, up to $O[(\Delta t)^3]$ in two dimensions, and up to $O[(\Delta t)^4]$ in three dimensions. The two- and three-dimensional schemes therefore can not be equivalent to products of steps with the one-dimensional scheme.

For simplicity, consider first the two-dimensional scheme; with the terms arranged as in (8) it reads

$$\begin{aligned} u^{n+1}_{ijk} = \Big[1 &- \sigma_x\Delta_x\big\{1 + \tfrac12(1-\sigma_x)\delta_x - \tfrac12\sigma_y\Delta_y\big[1 + (\tfrac12 - \tfrac23\sigma_x)\delta_x + (\tfrac12 - \tfrac13\sigma_y)\delta_y\big]\big\} \\ &- \sigma_y\Delta_y\big\{1 + \tfrac12(1-\sigma_y)\delta_y - \tfrac12\sigma_x\Delta_x\big[1 + (\tfrac12 - \tfrac23\sigma_y)\delta_y + (\tfrac12 - \tfrac13\sigma_x)\delta_x\big]\big\}\Big]u^n_{ijk}, \end{aligned}$$

$$\sigma_x < 1\ ,\quad \sigma_y < 1\ ,\quad \sigma_z < 1. \tag{12}$$

When changing this scalar scheme into a scheme for a nonlinear system, we must replace $\sigma_x\delta_x u^n_{ijk}$ by

$$\frac{\Delta t}{\Delta x}\left\{ f\left[(1+\tfrac12\delta_x)u^n_{ijk}\right] - f\left[(1-\tfrac12\delta_x)u^n_{ijk}\right] \right\} , \tag{13}$$

a central difference of fluxes inside of volume (ijk). The resulting method is unattractive because of the many intermediate steps needed for a full update; moreover, it is not clear that all steps are relevant for a non-diagonalizable hyperbolic system.

A slight simplification results when the third-order terms on the right-hand side of (12) are taken together, yielding

$$\sigma_x\sigma_y\Delta_x\Delta_y\left[\tfrac{1}{2}(1-\sigma_x)\delta_x + \tfrac{1}{2}(1-\sigma_y)\delta_y\right], \tag{14}$$

and then redistributed in a more convenient fashion over the expressions in braces, e.g. with weight ½ in each expression. The most drastic simplification results by omitting <u>all</u> third-order terms. This does not appear to affect the stability condition of the scheme. Its short-wave stability is still dictated by the first-order scheme (8), while its long-wave stability is slightly improved (dropping (14) causes the long waves to slow down).

The three-dimensional scalar linear version of this simplemost scheme is

$$\begin{aligned} u^{n+1}_{ijk} = \Big[1 &- \sigma_x\Delta_x\{1 + \tfrac{1}{2}(1-\sigma_x)\delta_x - \tfrac{1}{2}\sigma_y\Delta_y(1-\tfrac{1}{3}\sigma_z\Delta_z) - \tfrac{1}{2}\sigma_z\Delta_z(1-\tfrac{1}{3}\sigma_y\Delta_y)\} \\ &- \sigma_y\Delta_y\{1 + \tfrac{1}{2}(1-\sigma_y)\delta_y - \tfrac{1}{2}\sigma_z\Delta_z(1-\tfrac{1}{3}\sigma_x\Delta_x) - \tfrac{1}{2}\sigma_x\Delta_x(1-\tfrac{1}{3}\sigma_z\Delta_z)\} \\ &- \sigma_z\Delta_z\{1 + \tfrac{1}{2}(1-\sigma_z)\delta_z - \tfrac{1}{2}\sigma_x\Delta_x(1-\tfrac{1}{3}\sigma_y\Delta_y) - \tfrac{1}{2}\sigma_y\Delta_y(1-\tfrac{1}{3}\sigma_x\Delta_x)\}\Big]u^n_{ijk}, \\ &\sigma_x < 1\,,\quad \sigma_y < 1\,,\quad \sigma_z < 1. \end{aligned} \tag{15}$$

It turns out that a two-dimensional version of this scheme for the Euler equations was derived independently by P. Collela (private communication, 1983) and has already been applied successfully to an aerodynamics flow problem in a curvilinear grid [8].

4. CONCLUSIONS

In the preceding sections it has been shown that fully three-dimensional difference schemes for hyperbolic systems of conservation laws can be constructed on the basis of a simple convection principle.

The resulting schemes are explicit, upwind biased and stable under a combination of one-dimensional Courant-Friedrichs-Lewy conditions. The latter property they share with dimensionally factorized schemes.

A two-dimensional second-order scheme of this kind has been put into practice by Eidelman, Colella and Shreeve [8]. A numerical comparison between factorized and non-factorized schemes remains to be made.

References

1. A.R. Gourlay and J.Ll. Morris, J. Comp. Phys. <u>5</u> (1970), 229.
2. P.R. Woodward and P. Colella, " The numerical simulation of two-dimensional fluid flow with strong shocks", Lawrence Livermore Lab. Report UCRL 86952 (1982), to appear in J. Comp. Phys.
3. B. van Leer, "Computational methods for ideal compressible flow". NASA Contractor Report, ICASE, July 1983.
4. S.K. Godunov, Matem. Sb. <u>47</u> (1959), 271.
5. P.L. Roe, "Fluctuations and signals, a framework for numerical evolution problems", in <u>Numerical Methods for Fluid Dynamics</u>, eds. K.W. Morton and M.J. Baines, Academic Press, New York, 1982.

6. A. Harten, P.D. Lax and B. van Leer, SIAM Review 25 (1983), 35.
7. B. van Leer, J. Comp. Phys. 23 (1977), 276.
8. S. Eidelman, P. Colella and R.P.S. Shreeve, "Application of the Godunov method and higher order extensions of the Godunov method to cascade flow modeling", AIAA paper 83-1941-CP, AIAA 6th Computational FLuid Dynamics Conference, July 1983, Danvers, MA.

Computing Methods in Applied Sciences and Engineering, VI
R. Glowinski and J.-L. Lions (Editors)
Elsevier Science Publishers B.V. (North-Holland)
INRIA, 1984

EFFICIENT CONSTRUCTION AND UTILISATION OF APPROXIMATE RIEMANN SOLUTIONS

P.L. Roe
J.Pike

Royal Aircraft Establishment
Bedford UK

Flux-difference-splitting (FDS) methods for the unsteady Euler equations are reviewed, and it is shown that any of the proposed methods can be regarded as a prescription for finding the entries in a "Riemann table", which records the effect on each flow quantity of the wave system generated by the differences between neighbouring pairs of states. For an associated finite-difference scheme to be efficient, it is first of all necessary that the table is constructed economically, and then that the information contained in it should be fully used. We describe a scheme which is efficient in both respects.

INTRODUCTION

There is at present a lot of interest in the development of numerical schemes for hyperbolic systems of partial differential equations; particularly in schemes whose numerical processes attempt to model fairly closely the physical events taking place in the flow. For a long time the equal importance of conservation laws and wave phenomena have been realised, but until quite recently it had not been apparent how to devise efficient numerical methods which give the same degree of emphasis to both. The pioneering work of Godunov [1] remained of largely academic interest, since it led to an algorithm which was only first order accurate, and was distinctly expensive to use. Certain aspects of Godunov's method, however, have very strongly influenced much of the more recent research; this is especially true of the use to which he puts exact solutions of the Riemann problem. To make our meaning clear we must recapitulate the basic Godunov scheme for a set of one-dimensional conservation laws

$$\underline{u}_t + \underline{F}(\underline{u})_x = 0 \qquad (1.1)$$

We assume that at time t^n data $\underline{u}^n(x)$ is available, where $\underline{u}$ is a vector-valued function of the <u>continuous</u> real variable x. The first step is to replace this data by a piecewise constant approximation $\underline{v}^n(x)$, such that within a given interval $(x_{i-\frac12}, x_{i+\frac12})$ $\underline{v}^n(x)$ assumes the constant state

$$\underline{v}_i^n = \frac{1}{x_{i+\frac12} - x_{i-\frac12}} \int_{x_{i-\frac12}}^{x+\frac12} \underline{u}^n(x)dx \qquad (1.2)$$

In the language of finite elements, this stage would be called a <u>projection</u>. In this way we obtain an approximate problem which can be solved exactly, because in a portion of x, t space $(x_{i+\frac12} - h, x_{i+\frac12} + h)$ and $(t^n, t^n + \Delta t)$ defined by sufficiently small values of h and Δt, we can make use of the known algebraic solution to the Riemann problem [2]. The Riemann problem is, by definition, the interaction between two adjacent, and initially uniform, states. When this exact solution U(x, t) to the approximate problem has been worked out, we take

$$\underline{u}^{n+1}(x) = U(x, t^{n+1})$$

and repeat the process of alternating projections and solutions.

The theoretical foundations of Godunov's method are very sound. The projection step (1.2), and the subsequent set of Riemann solutions, all respect the conservation laws. The Riemann solutions serve to propagate information in the correct directions, and to maintain a proper distinction between compression and expansion waves. This latter feature ensures that Godunov's scheme obeys an appropriate entropy law [3]. Despite these virtues, however, the two halves of Godunov's scheme are, from the viewpoint of computational efficiency, ill-matched. Solving the Riemann problem exactly is an iterative process, which in the case of gasdynamics involves at each iteration the taking of logical decisions and the evaluation of fractional powers. It is a computational extravagence to employ this expensive machinery in the presence of the large truncation errors induced by (1.2).

In recognition of this fact, several authors [4-8] have devised approximate Riemann solvers, which provide, with varying degrees of accuracy, some or all of the information regarding wave speeds and intensities which could have been got from the exact solution. When these approximate solvers are substituted into the Godunov scheme, there is usually very little deterioration in the numerical results (see [5] and [9] for examples). At the same time, in order to remove the incompatibility from the other half of the partnership, there have been devised more elaborate numerical schemes within which the Riemann solvers play a role. Common to most of these, however, is the adoption of some assumed form for $\underline{u}^n(x)$, the continuously defined data whose projection is represented by $\underline{u}_i^n$. Harten, in the first-order version of his high resolution scheme [9, 10], and also in his collaborative work with Hyman on moving grids [11], assumes piecewise constant data within each interval, whereas van Leer [12] assumes piecewise linear data. In each case, the data is assumed to be discontinuous at almost all the interval edges. Woodward and Colella [13] assume piecewise parabolic data, which may be continuous if the average values within the cells vary sufficiently slowly. If the data varies rapidly, however, discontinuities at the edges of intervals are again allowed.

In all these methods [9-13], it is assumed that the purpose of the Riemann solver is to resolve the discontinuous parts of the data. The other parts of the algorithms then have to be carefully matched to the assumed behaviour of the smooth parts of the data. In contrast, we wish to adopt a point of view which is, intentionally, less clear cut in its interpretation. We still regard the numerical data as representing the average state inside a cell (eqn (1.2)), partly because this provides a natural way to deal with irregular grids [14], but we do not attempt to be precise about the distribution of states within the cell. Instead, we argue loosely that if most of cell i is occupied by fluid more-or-less in state $\underline{u}_i$, and most of cell (i+1) by fluid more-or-less in state $\underline{u}_{i+1}$, then whatever is happening within these cells, it will be a set of wave interactions closely resembling the wave interactions which take place in the Riemann problem defined by $(\underline{u}_i, \underline{u}_{i+1})$. To solve the Riemann problem approximately is then a perfectly natural procedure, because the error in the solution does no more than reflect our ignorance about what is really happening. We admit, however, one exception to this line of argument. If the two states are such that they can be connected by a shock transition, then it is in the nature of hyperbolic systems that either they are so connected or that they will be soon. We therefore attach especial importance to Riemann solvers which are able to recognise shock transitions.

We will use the word 'explanation' to denote whatever items of information about wave speeds, amplitudes, etc, will prove most helpful to incorporate into a numerical scheme. The best matching between a numerical scheme and an approximate Riemann solver will occur when the kind of explanation we wish for is generated as economically as possible.

In Section 2 we set out with some supporting arguments, the particular form of explanation that will in our judgement be the most generally useful. We show, for the particular case of the unsteady Euler equations in one space dimension, that it can be completely generated from a knowledge of three scalar quantities. The non-linear algebraic problem of finding these quantities turns out to have a unique solution, which is precisely the 'square-root-averaging' procedure of Ref [5]. The present paper therefore supplements and continues the analysis of [5] in the following ways.

(a) It puts forward in more detail the computational motives which underlay the somewhat abstract presentation given in [5].

(b) Certain properties of the averaging procedure which were not apparent from [5] are easily demonstrated in the course of its present derivation.

(c) We then demonstrate how we can generate, along the lines suggested in [15] and [16], a very simple numerical scheme which advances the solution in time, using almost no information, apart from that contained in the 'explanations'. This scheme can be third-order accurate in the case of linear advection, second-order accurate in smooth regions of non-linear flows, and free from spurious oscillations around discontinuities.

(d) To demonstrate the robustness and practicality of the method, we apply it to a rather severe test problem proposed by Woodward and Colella, [17], involving the collision, in a non-uniform environment, between two blast waves having pressure ratios of 10^4 and 10^5.

(e) We present a simple modification of the method which ensures that it meets an evolutionary form of the entropy condition.

THE CHOICE OF A DESCRIPTION

We consider two adjacent states, which we shall refer to as $\underline{u}_L$, $\underline{u}_R$, and propose these as left-hand and right-hand input states for the Riemann problem. In general we expect the solution to consist of m waves (which may be shocks, contacts or expansions) separating a total of (m+1) piecewise constant states, including the two given input states (Fig 1). Here m is the number of independent conservation laws contained in (1.1). There are many possible ways to describe the solution once it has been obtained. We could give the constant states between the waves, or we could give the jumps across the waves. In either event we could use the primitive variables (usually thought of as ρ, u, p for Eulerian gasdynamics) or the conserved variables (more correctly described perhaps as densities or concentrations of conserved variables, ρ, $m = \rho u$, $e = p/(\gamma-1) + \frac{1}{2}\rho u^2$). We could also use the fluxes of the conserved quantities (ρu, $p+\rho u^2$, $u(e+p)$), or any set of derived quantities, like temperature, entropy, or kinetic energy. For conservation laws such that the characteristic equations can be integrated to produce Riemann invariants, we could use those.

It hardly seems possible to point to any set of variables as being for all purposes inherently superior to any other; indeed, an intelligent human calculator would change variables many times in the course of building up a picture of some complex flow. Lombard and his co-workers make some interesting observations on this topic [8, 18, 19], and in the present formulation of our work we have found these very stimulating. Of course a solution in one form can always be translated into another form, but the effort of translation may not be negligible, and information which is immediately apparent from one form may not be obvious in another. Our own views concerning the best presentation of a solution, although based on long reflection and debate, are still cautious and provisional. We are inclined to suppose that the best policy for the Euler equations is to work primarily with the conserved variables, since we have found no other way of ensuring that the conservation laws are obeyed. However, as we shall show shortly, it turns out to be both possible and simple to enforce those laws by making use, not of the flux variables, but of the primitive variables. At least for one-dimensional problems, therefore, we propose to carry the primitive variables as auxiliary data. In multi-dimensional problems, we would save storage by computing the primitive variables as they are needed.

The foregoing may help to explain our choice of a format for solutions to the Riemann problem. It comprises a table with m columns, corresponding to the m waves, and (m+2) rows. The first m rows make up a square matrix of which the entry in the jth row and k^{th} column is

$$\Delta u_j^{(k)} = (\text{change in } u_j \text{ across the } k^{th} \text{ wave}) \tag{2.1}$$

The entries in the $(m+1)^{th}$ row are

$$a^{(k)} = (\text{average velocity of the } k^{th} \text{ wave}) \tag{2.2}$$

The entries in the $(m+2)^{th}$ row are

$$\delta^{(k)} = (\text{spreading rate of the } k^{th} \text{ wave}) \tag{2.3}$$

The quantities $\delta^{(k)}$ are needed to ensure a proper distinction between shocks and rarefaction fans. We define $\delta^{(k)}$ as the positive difference between the speeds of the fastest and slowest characteristics of the kth family. This is a positive quantity for expansion waves, but zero for contacts and shocks.

We could fill in this table using any of the Riemann solvers referenced above, and we could exploit other methods also. The wave path analysis of Osher and Solomon [20] can be regarded in this way, as discussed by Roe [21]. We shall, however, attach particular significance to tables having the following conservation property; for j=1,..., m we require

$$(F_j)_R - (F_j)_L = \textstyle\sum_k a^{(k)} \Delta u_j^{(k)} \tag{2.4}$$

Note that we could use this formula in many different ways. Harten, Lax, and van Leer [4] begin by estimating the $a^{(k)}$ and then arrive at a set of jumps $\Delta u_j^{(k)}$ consistent with (2.4) by integrating around various control volumes. Colella [6] estimates the $\Delta u_j^{(k)}$ and then, in effect, solves the m simultaneous equations comprising (2.4) to find the $a^{(k)}$. Colella's process could also be used to define average wave speeds, if the $\Delta u^{(k)}$ were taken from the exact Riemann solution or from Osher and Solomon's analysis.

A further possibility is to define the $a^{(k)}$ and the $\Delta u_j^{(k)}$ in such a way that (2.4) is an algebraic identity. This is the approach we take here. For the special case of ideal compressible flow it leads to rather simple algebra, and generates a first-order algorithm equivalent to that given in [5]. That algorithm has the property that if $\underline{u}_i$, $\underline{u}_{i+1}$ are states which can be connected by a single shockwave or contact discontinuity, then no other wave system is activated. Colella's method shares that property, and has the advantage of being applicable to gases with non-ideal equations of state. Osher and Solomon's method does not have the "shock-recognising" property, but other advantages are claimed for it. Regardless of the actual method used to obtain the approximate (or, indeed, exact) solution to the Riemann problem, we will refer to the information specified in (2.1)-(2.3) as a Riemann table. In this way we impose some unity on the various first-order methods. More importantly, however, we will show that once the information has been assembled as a Riemann table, it can be processed without regard to its origin so as to yield higher-order "monotone" solutions. We observe that the second-order method of Harten [9] could be formulated in much the same way.

CONSTRUCTING THE RIEMANN TABLE

In this section we are concerned with algebraic detail, and restrict attention to the unsteady Euler equations for ideal compressible gasdynamics in one dimension. The Eulerian conservation form is

$$\frac{\partial}{\partial t}\underline{u} + \frac{\partial}{\partial x}\underline{F} = 0 \tag{3.1}$$

where

$$\underline{u} = \begin{vmatrix} \rho \\ \rho\, u \\ e \end{vmatrix}, \quad \underline{F} = \begin{vmatrix} \rho u \\ p + \rho u^2 \\ u(e + p) \end{vmatrix} \tag{3.2}$$

The presure p and total energy e are related by

$$e = \frac{p}{\gamma - 1} + \frac{1}{2}\rho u^2 \tag{3.3}$$

We also use the notations

$$a = (\gamma p/\rho)^{\frac{1}{2}} \tag{3.4}$$

for the sound speed, and

$$H = (e+p)/\rho \tag{3.5}$$

for the enthalpy.

Now consider the Riemann problem with initial data (ρ_L, u_L, p_L), (ρ_R, u_R, p_R). If both these states are close to some reference state $(\hat{\rho}, \hat{u}, \hat{p})$, the problem can be linearised about these reference conditions, and then tabulated in the form proposed above, with $\hat{a}$ having its natural meaning

TABLE I

	Wave number (k)		
	1	2	3
$\Delta u_1^{(k)}$	$\alpha^{(1)}$	$\alpha^{(2)}$	$\alpha^{(3)}$
$\Delta u_2^{(k)}$	$\alpha^{(1)}(\hat{u} - \hat{a})$	$\alpha^{(2)}\hat{u}$	$\alpha^{(3)}(\hat{u} + \hat{a})$
$\Delta u_3^{(k)}$	$\alpha^{(1)}(\frac{\hat{a}^2}{\gamma-1} - \hat{u}\hat{a} + \frac{\hat{u}^2}{2})$	$\alpha^{(2)}\frac{\hat{u}^2}{2}$	$\alpha^{(3)}(\frac{\hat{a}^2}{\gamma-1} + \hat{u}\hat{a} + \frac{\hat{u}^2}{2})$
$a^{(k)}$	$\hat{u} - \hat{a}$	$\hat{u}$	$\hat{u} + \hat{a}$

where

$$\left.\begin{aligned} \alpha^{(1)} &= \frac{1}{2\hat{a}^2}(\Delta p - \hat{\rho}\hat{a}\Delta u) \\ \alpha^{(2)} &= \frac{1}{\hat{a}^2}(\hat{a}^2\Delta\rho - \Delta p) \\ \alpha^{(3)} &= \frac{1}{2\hat{a}^2}(\Delta p + \hat{\rho}\hat{a}\Delta u) \end{aligned}\right] \quad (3.6)$$

and

$$\Delta(\bullet) = (\bullet)_R - (\bullet)_L$$

For the moment we omit mention of the spreading rate, since all characteristic slopes are constant in the linearised problem. It is easily checked that

$$\sum_k \Delta u_j^{(k)} = \Delta u_j \quad (3.7)$$

and that

$$\sum_k a^{(k)} \Delta u_j^{(k)} = \Delta F_j \quad (3.8)$$

with error terms which are $O(\Delta^2)$.

If the left and right states are widely different, we propose the following algebraic problem. Find some average state $(\tilde{\rho}, \tilde{u}, \tilde{p})$ such that (3.7), (3.8) are identically satisfied, when this average state is used to construct Table I, including the computation of the coefficients (3.6). At first sight, the proposal is unpromising; six simultaneous non-linear equations are to be satisfied by three scalar quantities $(\tilde{\rho}, \tilde{u}, \tilde{a})$. However, it can quickly be seen that the first equation of (3.7) is satisfied by any average state, and that the second equation of (3.7) is the same as the first of (3.8). It also turns out that the third of (3.7) is implied by the first and second of (3.8), so that (3.8) are the only equations we need to solve.

By substituting the coefficients $\alpha^{(k)}$ from (3.6),and also the contents of Table I,into the first two of (3.8), we find

$$\Delta(\rho u) = \tilde{\rho}\Delta u + \tilde{u}\Delta\rho \quad (3.9)$$

and

$$\Delta(\rho u^2) = 2\tilde{\rho}\tilde{u}\Delta u + \tilde{u}^2\Delta\rho \quad (3.10)$$

If we then eliminate $\tilde{\rho}$ from these equations we obtain for $\tilde{u}$ the quadratic equation

$$\tilde{u}^2\,\Delta\rho - 2\tilde{u}\,\Delta(\rho u) + \Delta(\rho u^2) = 0 \quad (3.11)$$

whose solutions are

$$\begin{aligned} \tilde{u} &= \frac{\Delta(\rho u) \pm [\{\Delta(\rho u)\}^2 - \Delta\rho\Delta(\rho u^2)]^{\frac{1}{2}}}{\Delta\rho} \\ &= \frac{\rho_R u_R - \rho_L u_L \pm (\rho_L\rho_R)^{\frac{1}{2}}(u_R - u_L)}{\rho_R - \rho_L} \\ &= \frac{[\rho_R \pm(\rho_L\rho_R)^{\frac{1}{2}}]u_R - [\rho_L \pm (\rho_L\rho_R)^{\frac{1}{2}}]u_L}{\rho_R - \rho_L} \end{aligned}$$

If we take the positive roots, we do not arrive at a useful answer, but the negative roots lead to

$$\tilde{u} = \frac{\rho_L^{\frac{1}{2}} u_L + \rho_R^{\frac{1}{2}} u_R}{\rho_L^{\frac{1}{2}} + \rho_R^{\frac{1}{2}}} \tag{3.12}$$

This is the 'square root averaging' which was derived from apparently very different considerations in [5]. Substituting this value of u back into (3.9) leads quickly to

$$\tilde{\rho} = (\rho_L \rho_R)^{\frac{1}{2}} \tag{3.13}$$

The third equation of (3.11) is now, in effect, a direct expression for $\tilde{a}$. Again we substitute the trial solution, and this time we find

$$\tilde{\rho}\tilde{a}^2 \Delta u = \gamma[\Delta(up) - \tilde{u}\Delta p] + \frac{\gamma-1}{2}[\Delta(\rho u^3) - \tilde{u}^3 \Delta\rho - 3\tilde{\rho}\tilde{u}^2 \Delta u] \tag{3.14}$$

After this equation has been expanded on the right-hand side, a factor Δu cancels throughout, leaving

$$r_L r_R \tilde{a}^2 = \gamma \frac{r_L p_R + r_R p_L}{r_L + r_R} + \frac{\gamma-1}{2} \frac{r_L^2 r_R^2 (u_R - u_L)^2}{(r_L + r_R)^2} \tag{3.15}$$

where for typographical clarity we have introduced

$$r_L = (\rho_L)^{\frac{1}{2}}, \quad r_R = (\rho_R)^{\frac{1}{2}} \tag{3.16}$$

The expression (3.15) is not quite as simple as (3.12) and (3.13), but it does demonstrate two nice properties. First, that if we have physically realistic data, the square of $\tilde{a}$ must come out positive, and secondly, that our solution is invariant with respect to all uniformly moving observers because only velocity differences appear in (3.6) or (3.15). Actually, we have not quite finished with (3.15). Define a mean enthalpy

$$\tilde{H} = \frac{\tilde{a}^2}{\gamma-1} + \frac{\tilde{u}^2}{2} \tag{3.17}$$

Substituting (3.15) into this gives

$$\tilde{H} = \frac{r_L H_L + r_R H_R}{r_L + r_R} \tag{3.18}$$

which is, again, the same result as in [5]. In practice, $\tilde{a}$ can be found more quickly from (3.17), (3.18) than from (3.15). Finally, it may be checked that the coefficients (or wave strengths) given by (3.6) are identically equal to those derived from the matrix analysis in [5]; in particular, they share the property that whenever $\underline{u}_L$, $\underline{u}_R$ can be connected by a single shock or contact discontinuity, then every other wave is assigned zero strength.

From a mathematical point of view, it is quite striking to find such simple results emerging; one would like to know whether similar results hold for some special class of conservation laws. We were intrigued to learn that a very closely related result has been found by M. Brio of the University of California at Los Angeles. He sought the average state $\underline{u}^*(u_L, u_R)$, such that

$$\underline{\Delta F} = A(\underline{u}^*)\Delta\underline{u} \tag{3.19}$$

This is a weaker condition than (3.8), but its solution is $u^* = \tilde{u}$, $a^* = \tilde{a}$, ρ^* arbitrary.

From a philosophical point of view, it is interesting to see that the conservation condition (3.8) is met without any direct mention of the flux variables appearing in Table I or in equations (3.6). This also has practical consequences. Let us consider three-dimensional finite-volume methods in which the state of the flow is described by the average values of conserved variables within each cell. To balance the fluxes around the cell boundaries we might construct flux differences from combinations of the seven quantities (p, ρu^2, ρv^2, ρw^2, ρuv, ρvw, ρuw). Our formulae are constructed from differences of the five quantities (ρ, p, u, v, w). Of course, this apparent saving is more than offset by having to construct the five special averages ($\tilde{p}$, $\tilde{u}$, $\tilde{v}$, $\tilde{w}$, $\tilde{a}$). However, using these averages not only restores the conservation properties but also provides a full description of the flux difference splitting.

We will conclude this section by stating, in algorithmic form, a one-dimensional first order, conservative, FDS scheme algebraically identical to that in [5]. Data consisting of ρ, ρu, e is supposed to be given in cells of a uniform length Δx.

(a) Compute $R_{i+\frac{1}{2}} = (\rho_{i+1}/\rho_i)^{\frac{1}{2}}$

(b) Compute, in this order,

$$\tilde{\rho}_{i+\frac{1}{2}} = R_{i+\frac{1}{2}}\,\rho_i$$

$$\tilde{u}_{i+\frac{1}{2}} = (R_{i+\frac{1}{2}}\,u_{i+1} + u_i)/(R_{i+\frac{1}{2}}+1)$$

$$\tilde{H}_{i+\frac{1}{2}} = (R_{i+\frac{1}{2}}\,H_{i+1} + H_i)/(R_{i+\frac{1}{2}}+1)$$

$$\tilde{a}_{i+\frac{1}{2}} = [(\gamma-1)(\tilde{H}-\tfrac{1}{2}\tilde{u}^2)]^{\frac{1}{2}}$$

(c) Compute

$$p_i = (\gamma-1)[e_i - \tfrac{1}{2}\rho_i u_i^{\,2}]$$

$$u_i = (\rho u)_i/\rho_i$$

(d) Compute

$$\Delta p_{i+\frac{1}{2}} = p_{i+1} - p_i$$

$$\Delta u_{i+\frac{1}{2}} = u_{i+1} - u_i$$

$$\Delta \rho_{i+\frac{1}{2}} = \rho_{i+1} - \rho_i$$

(e) Compute $\alpha^{(1)}$, $\alpha^{(2)}$, $\alpha^{(3}$ from (3.6) and hence fill in Table I.

Note that all these stages, although they aim at providing the information needed to create an upwind scheme, do not yet involve any logic, and are therefore suitable for parallel or vector processing. Note also that the only abnormal expense is the computation of an additional square root at stage (a). For flows involving relatively small density fluctuations this can be evaluated cheaply by means of simple approximations, valid if x is close to unity, such as

$$x^{\frac{1}{2}} \doteq \frac{1}{4.5}\left[6 + x - 5/(1+x)\right] \quad (3.20)$$

The final stage (f) is the only one to involve logic.

(f) Subtract $(\Delta t/\Delta x)a^{(k)}\Delta u_j^{(k)}$ from $(u_j)_{i+1}$ if $a^{(k)} > 0$, or from $(u_j)_i$ if $a^{(k)} < 0$

ADMISSIBILITY CONDITIONS

The algorithm (a)-(f) above is identically equivalent to the first-order algorithm described in [5]. It therefore repeats the defects of that method; in particular the defect of admitting non-physical solutions. Let us recall how these come about. Consider systems of conservation laws such that the mapping $\underline{u} \to \underline{F}$ is many-one; for exmple, in the Euler equations, if $\underline{F}$ is given, a quadratic equation must be solved to find $\underline{u}$ giving two mappings of $\underline{u}$ onto $\underline{F}$. Let $\underline{u}_A$, $\underline{u}_B$ be two states which map to the same flux vector $\underline{F}_{AB}$, and consider two alternative sets of data, D_1, D_2, whose elements are

$$D_1 = \{\underline{u}_A, \underline{u}_A, \ldots, \underline{u}_A, \underline{u}_B, \ldots, \underline{u}_B\} \quad (4.1a)$$

$$D_2 = \{\underline{u}_B, \underline{u}_B, \ldots, \underline{u}_B, \underline{u}_A, \ldots, \underline{u}_A\} \quad (4.1b)$$

In general, one of D_1, D_2 will represent a physically admissible shockwave (in the case of the Euler equations, one such that pressure and density increase in the flow direction) and the other of D_1, D_2 will be inadmissible [2, 3, 4, 9]. However, either set of data will give rise to $\underline{F} \equiv \underline{F}_{AB}$, and the FDS algorithm accepts either set as a stable solution which need not be altered. To construct an algorithm which accepts only the admissible data it is necessary to break the symmetry of the scheme with respect to D_1, D_2. This is the purpose of the final entry in the Riemann table; that is, the spreading rate associated with each wave.

Note that there are several ways of attempting to ensure physical correctness, and that we make use of geometrical ideas rather then ideas based upon the concept of entropy, even though these latter are quite feasible for the Euler equations [3, 4,9]. One of our reasons is that there are important physical systems for which the concept of entropy is not available, but where the geometry of characteristics does reliably discriminate between admissible and inadmissible solutions. One such system involves the flow of a combustible mixture [22]. A second reason, deriving from our theme of numerical efficiency, is that the information needed to implement the geometrical arguments has already been generated (in hidden form) in the previous section.

The geometrical condition is, of course, that an admissible discontinuity must be associated with a wave family whose characteristics are non-divergent. That is, if the shock speed is S, and if it associated with waves of the kth family whose speeds are $a_L^{(k)}$, $a_R^{(k)}$ in the states to the left and right of the shock, then

$$a_L^{(k)} \geq S \geq a_R^{(k)} \quad (4.2)$$

Normally in gasdynamics either both inequalities are strict inequalities (shocks), or else both are in fact equalities (contacts), but in other systems equality on just one side of (4.2) is possible. Clearly, if one of (4.1a, b) satisfies (4.2) the other does not. [See Jeffrey [23] for a proof that if (4.2) is not satisfied in a continuum solution, the discontinuity is unstable].

The task is to introduce a similar instability into the numerical scheme. Suppose we encounter $\underline{u}_A$, $\underline{u}_B$ as consecutive states i, i+1. If we meet them in their admissible order we need take no action, but if the order is reversed the algorithm must respond. Conservation still requires the total change to be zero, so the only appropriate response is to make equal and opposite changes to the cells i, i+1. We now modify the algorithm described in the previous section so as to achieve this property whilst retaining continuous dependence on the data.

Let $a^{(k)}_{i+\frac{1}{2}}$ be the average wavespeed, and $\delta^{(k)}_{i+\frac{1}{2}}$ the spreading rate (whose estimation we have yet to describe) for the kth wave across the $i+\frac{1}{2}$ interface. We will ensure that $\delta^{(k)}_{i+\frac{1}{2}}$ is zero for compression waves or contacts. For rarefaction waves there will be respectively least and greatest wavespeeds equal to $a - \frac{1}{2}\delta$, $a + \frac{1}{2}\delta$. We test to see if both of these are of the same sign; if so it will still be appropriate to use the unmodified algorithm. Therefore, almost the whole cost of the modification comes from making this simple test for each wave system and each pair of consecutive cells. If the least and greatest speeds are of different sign, we will send to the ith cell a signal

$$\frac{\Delta t}{2\Delta x}(a - \tfrac{1}{2}\delta)\Delta u \tag{4.3a}$$

and to the $(i+1)^{th}$ cell a signal

$$\frac{\Delta t}{2\Delta x}(a + \tfrac{1}{2}\delta)\Delta u \tag{4.3b}$$

the indices $i+\frac{1}{2}$ and k being understood. The total signal is, as before, $\nu\Delta u$, where $\nu = a\,\Delta t/\Delta x$ is the Courant number, and this ensures conservation. If the test is only just met, one of the signals is zero.

It remains to give an inexpensive prescription for evaluating δ in the case of the Euler equations. Across simple waves, with k=1 or 3, there is no change of the generalised Riemann invariants [23]

$$J^{(1)} = u + \frac{2}{\gamma-1}\,a \tag{4.4a}$$

$$J^{(3)} = u - \frac{2}{\gamma-1}\,a \tag{4.4b}$$

Hence, the change in u–a across a wave of the first family, or u+a across a wave of the third family is, in either case

$$\delta^{(k)} = \frac{\gamma+1}{2}\,\Delta u^{(k)}_{+} \tag{4.5}$$

where $\Delta u^{(k)}$ is the change in fluid velocity across the kth wave and the notation x_+ denotes the positive part $x_+ = \frac{1}{2}(x + |x|)$. Later we employ the analogous notation $x_- = \frac{1}{2}(x - |x|)$.

The expression (4.5) is exact; it is more in the spirit of approximate Riemann solvers to replace it by an approximation which uses information already available. From the facts that for an ideal gas $a^2 = \gamma p/\rho = \partial p/\partial\rho$ we readily obtain across either the first or third wave

$$\Delta a \simeq \frac{\gamma-1}{2\rho a}\,\Delta p \tag{4.6}$$

From this approximate equation, and the constancy of $J^{(1)}$ across the first wave, we obtain

$$\delta^{(1)} \simeq -\frac{(\gamma+1)\tilde{a}}{2\tilde{\rho}}\,\alpha_{-}^{(1)} \tag{4.7a}$$

and, by a similar argument

$$\delta^{(3)} = \frac{(\gamma+1)\tilde{a}}{2\tilde{\rho}}\,\alpha_{+}^{(3)} \tag{4.7b}$$

For the Euler equations, of course

$$\delta^{(2)} = 0 \tag{4.7c}$$

For other systems of equations, alternatives to (4.7) would have to be sought. It might be necessary to take into account any distinctive type of non-linearity associated with a particular system.

HIGHER-ORDER ALGORITHMS

In the previous sections we described, in effect, a first-order FDS algorithm for solving initial-value-problems. The required information was assembled as a set of Riemann tables; in computing terms this set defines an array having dimensions $m \times (m+2) \times (n-1)$, where m is the number of unknowns (or, equivalently, the number of independent wave systems) and n is the number of cells in the computing domain. We now employ this information to obtain higher-order solutions which give clear resolution of any discontinuities which may appear in the flow. Our procedure is independent of the method used to construct the Riemann tables. Indeed, it is independent of the particular system of conservation laws for which the Riemann problems have been solved.

The technique relies on the still somewhat heuristic concept that by calculating separately the changes in each flow variable due to each wave system we are effectively computing the solution to m^2 independent scalar problems. It is therefore sufficient to describe the scalar case, with the understanding that u can be interpreted as u_j, a as $a^{(k)}$, ν as $\nu^{(k)}$, and Δu as $\Delta u_j^{(k)}$. Frequently the subscript $i+\frac{1}{2}$ is also to be understood. Any other spatial index is always explicitly stated. We introduce $\sigma^{(k)} = \text{sgn}\,\nu^{(k)}$ and note that the first-order algorithm involves sending a <u>signal</u> of strength $\nu_{i+\frac{1}{2}}\,\Delta u_{i+\frac{1}{2}}$ to be <u>subtracted</u> from u at $i+\frac{1}{2}+\frac{1}{2}\sigma$. An algorithm equivalent to the Lax-Wendroff scheme results from sending $\frac{1}{2}\nu(1-\nu)\Delta u$ to i, and $\frac{1}{2}\nu(1+\nu)\Delta u$ to i+1. Let the weaker of these two signals, that is $\frac{1}{2}\nu(1-|\nu|)\Delta u$, be written as Δu^*. Then a reformulation of Lax-Wendroff is that after completing the first-order algorithm we subtract Δu^* from the physically unrealistic target $i+\frac{1}{2}-\frac{1}{2}\sigma$, but maintain conservation by adding Δu^* at $i+\frac{1}{2}+\frac{1}{2}\sigma$. This transfer can be regarded as analogous in its purpose to the antidiffusion stage of Flux-Corrected Transport [24, 25]. In the scheme being described here, the low-order component of an FCT method is represented by the first-order FDS algorithm, and the higher-order component will be either Lax-Wendroff, or a small perturbation of it. As with FCT, the key element which allows us to avoid the 'wiggles' associated with classical high-order methods is to limit the magnitude of the antidiffusive term. The particular recipe which we adopt for this limiting process has been described elsewhere [16, 26] but for completeness we repeat it, with some discussion, below.

Define

$$b_1 = \Delta u^*_{i+\frac{1}{2}} \tag{5.1}$$

$$b_2 = \Delta u^*_{i+\frac{1}{2}-\sigma} \tag{5.2}$$

In the Lax-Wendroff scheme, transferring the effect of b_1 from the physically realistic target $i+\frac{1}{2}+\frac{1}{2}\sigma$ to the unrealistic target $i+\frac{1}{2}-\frac{1}{2}\sigma$ is something which contravenes the proper direction for information flow, but is justifiable on formal mathematical grounds whenever the flow is sufficiently smooth. In fact, for smooth flows this non-physical transfer term could be estimated with sufficient accuracy by examining the interaction between any other nearby pair of adjacent cells. Thus, b_2, the estimate taken from the upwind cell, is equally valid from the viewpoint of accuracy. If we do decide to use b_2, then the scalar version of the algorithm is "wholly upwinded", and the systems version is wholly upwinded with respect to each wave system.

Experience shows that wholly upwinded algorithms are very effective indeed at capturing shockwaves which are strong and either stationary or nearly so, but are much less effective if the shocks are weak or rapidly moving. The practical criterion seems to be whether the shockwave is sufficiently strong to reverse the sign of $a^{(k)}$. When this is not the case, the wholly upwinded schemes cannot claim to be any more physically realistic than Lax-Wendroff. Information, although sent in the right direction, is being propagated too fast, and wiggles build up ahead of the shockwave rather than behind it [26, 27]. A recipe which deals equally well with both strong and weak shockwaves is to transfer neither b_1 nor b_2, but some "average" value of the two. The essential features of the averaging process are that if b_1, b_2 are close to each other in value, then the average $B(b_1, b_2)$ shall be close to both of them, and if b_1, b_2 are very different then the average shall be close to the smaller of the two. The first of these properties ensures that second-order accuracy is retained for smooth flows; the second property reduces the disruptive effect of the anti-diffusion in non-smooth flows. Note that the averaging function B must have the homogeneous property $B(kb_1, kb_2) = kB(b_1, b_2)$; otherwise the result will depend on the units of measurement.

Before specifying the averaging procedure more precisely, we note that many authors have by now published schemes which are akin to the above, and are often identical to it in the simple special case of linear advection. All the methods are supported by heuristic arguments. At first, these arguments seem different, but closer examination reveals strong similarities between the arguments used to support the methods, as well as between the methods themselves. Historical precedence goes to van Leer, whose work is mostly based on the idea of preprocessing data which is assumed to be piecewise linear within cells. No cell is allowed, however, to have a variation much stronger than that found in its neighbours. Instead, discontinuities are allowed to develop at cell interfaces, and this strengthens the first-order terms in the algorithm at the expense of certain second-order terms. The culmination of this approach is the somewhat complex MUSCL code described in [12], which has since been considerably simplified, see for example [28]. Harten [9] supports his scheme by arguments which associate certain second-order terms with a spurious wavespeed, which must be limited to avoid the spread of wiggles. Since the effect of a wave is proportional to the product of its speed and amplitude, it probably makes no real difference which of the two we choose to limit. Harten has shown [10] that his scheme is identical to MUSCL in the case of linear advection, and in fact both schemes are then also identical to the present scheme.

An averaging function which appears widely in these references is the function, christened by Sweby the "minmod" function, which is zero if b_1, b_2 have opposite sign, and otherwise equals whichever of b_1, b_2 is closer to zero. If this function is inserted into each scheme at the appropriate place, then the results are very similar indeed [29]. Something which has much more effect on the quality of the results is to change the averaging function, and van Leer [30] has recommended the harmonic mean, defined by

$$B(b_1, b_2) = \frac{2b_1 b_2}{b_1 + b_2} \qquad \text{if } b_1 b_2 \geqslant 0$$
$$= 0 \qquad \text{if } b_1 b_2 \leqslant 0 \qquad (5.3)$$

as well as other choices [31], all designed to produce a value biased toward the smaller input. Numerical experiments reveal that all such choices work well with moving shocks as well as with stationary shocks, but still find difficulty if the shocks are very weak, in the sense of producing only small changes in $a^{(k)}$. This will not be serious if the shocks are also weak in the sense of having small amplitude. The case which causes practical difficulty is the so-called linearly degenerate case [23, 26] which manifests itself in the Euler equations as the contact discontinuity, across which pressure is continuous, but density is not. The relevant "wavespeed" is the fluid velocity (or, in multidimensional flows, its normal component) and this is also a continuous quantity. Experiments on this case reveal [32] a slow spreading of discontinuous data, such that the transition regions will exceed any stated width within finite time.

Woodward and Colella have described [13] an elaborate MUSCL-type code, containing a mechanism which appears to treat contact discontinuities rather well, by actually augmenting the "anti-diffusive" terms under some circumstances. We can incorporate this effect very cleanly into the present scheme, by adopting an averaging function such that if b_1, b_2 are not very different, then the average is biassed toward the larger of the two. The following rule has been found remarkably effective.

If b_1, b_2 are of different sign choose
$$B(b_1, b_2) = 0.$$

If b_1, b_2 are of the same sign, and differ by a factor greater than 2.0, choose
$$B(b_1, b_2) = 2 \text{ minmod } (b_1, b_2) \qquad (5.4)$$

If b_1, b_2 are of the same sign, and differ by a factor less than 2.0, choose
$$B(b_1, b_2) = \text{maxmod } (b_1, b_2)$$

where maxmod is the function which selects the argument furthest from zero. The rule (5.4) defines a continuous function of two variables, which has been christened "Superbee". It has the remarkable property, established by extensive numerical experiments [32] that when it is used as the averaging function in the present scheme, and the scheme is specialised to linear advection, then discontinuities in the initial data no longer diffuse without limit. By this we mean that a given fraction of the amplitude is contained within a zone of finite width for all time. Further information regarding this, and other, averaging functions is available in [32, 33, 34]. A note of caution is required by Sweby's observation [34] that the Superbee function can give rise to instability in non-linear problems, unless the CFL number is kept below ½. He recommends that the use of Superbee should be reserved for the degenerate fields (ie contact discontinuities) and a more conventional average, such as (5.3), employed for the remaining waves.

We shall now set out the above scheme in a more explicitly algorithmic form; supposing that steps (a) - (f) as described in Section 3 have already been carried out.

(g) To ensure admissible solutions, test each pair of cells for sonic expansion waves. Flag any interface found with an integer indicating the wave number k involved. For each flagged interface where $a^{(k)}_{i+\frac{1}{2}} \geq 0$, transfer a signal

$$\frac{\Delta t}{2\Delta x}\left(a^{(k)}_{i+\frac{1}{2}} - \tfrac{1}{2}\,\delta^{(\kappa)}_{i+\frac{1}{2}}\right)\,\Delta\underline{u}^{(k)}_{i+\frac{1}{2}} \tag{5.5a}$$

from i+1 to i. For each flagged interface where $a^{(k)}_{i+\frac{1}{2}} \leq 0$, transfer a signal

$$\frac{\Delta t}{2\Delta x}\left(a^{(k)}_{i+\frac{1}{2}} + \tfrac{1}{2}\,\delta^{(k)}_{i+\frac{1}{2}}\right)\,\Delta\underline{u}^{(k)}_{i+\frac{1}{2}} \tag{5.5b}$$

from i to i+1.

(h) To compute a second-order correction, begin by finding the non-physical signals needed to implement the present version of Lax-Wendroff, ie

$$\Delta\underline{u}^{*(k)}_{i+\frac{1}{2}} = \tfrac{1}{2}\,\nu^{(k)}_{i+\frac{1}{2}}\left(1 - |\,\nu^{(k)}_{i+\frac{1}{2}}\,|\right)\Delta\underline{u}^{(k)}_{i+\frac{1}{2}} \tag{5.6}$$

For any flagged interface, it seems satisfactory to set $\Delta\underline{u}^{*(k)}_{i+\frac{1}{2}} = 0$.

(i) Let $b_1 = b^{(k)}_{i+\frac{1}{2}}$ be a scalar which acts as a norm for the vector $\Delta\underline{u}^{*(k)}_{i+\frac{1}{2}}$. In practice, we use merely the first component, ie the change in density.

Let $b_2 = b^{(k)}_{i+\frac{1}{2}-\sigma}$, where $\sigma = \operatorname{sgn} a^{(k)}_{i+\frac{1}{2}}$, be the norm associated with the same wave family in the upwind cell. Let us define a limiting factor

$$\beta = B(b_1,\, b_2)/b_1 \tag{5.7}$$

where B is one of the averaging functions discussed above. Because B must be homogeneous, this may be written

$$\beta = B(1,\, b_2/b_1) \tag{5.8}$$

(j) Transfer $\beta\Delta u^{*(k)}_{i+\frac{1}{2}}$ from $i+\frac{1}{2}+\frac{1}{2}\sigma$ to $i+\frac{1}{2}-\frac{1}{2}\sigma$.

Steps (g) to (j) complete a second-order, "monotone", FDS algorithm for one-dimensional initial-value problems. These steps are difficult to vectorise, and their efficient implementation depends on the machine being used. On a CRAY 1, it has proved possible to update about 60,000 data points in each second of CPU time.

Boundary procedures for one-dimensional flow are usually very simple. Closed ends are dealt with by using image cells. Open ends merely require a statement that there are no incoming waves. Again, image cells can be used, with a Riemann table describing the interaction between the image cells and the last real cell. The only time this table is ever consulted is for information about the strength of waves which may be entering the domain. Therefore, this table is filled with zeroes for all the waves, including the unused outgoing waves. Boundary procedures which model more complex events usually need to be supplied from ad hoc arguments.

NUMERICAL RESULTS

To illustrate the robust nature of our scheme, we have applied it to a test problem proposed by Colella and Woodward [17]. Since this problem involves very strong wave interactions, we feel that it gives convincing support to our strategy of splitting the flux differences into "scalar" components. We also take satisfaction that the code used to produce these results contains no "adjustable constants" or "problem-specific" features. The most empirical aspect of the code is the choice of an averaging function, and even this is closely based on the scalar analysis [32, 33, 34]. In fact, we also choose this test problem because it allows us to display the improvements brought about by a more sophisticated averaging function.

The problem involves flow in a parallel pipe extending over $0 \leq x \leq 1$ and closed at each end. At t=0, the tube contains an ideal gas ($\gamma = 1.4$), of unit density everywhere, and at rest. However, two diaphragms at $x = 0.1$ and $x = 0.9$ separate the gas into three regions with different pressures. In the central region $p = 10^{-2}$, and at the left and right $p = 10^3$ and $p = 10^2$ respectively. The diaphragms burst simultaneously and strong shockwaves rush into the central region, followed by contact discontinuities, whilst expansion fans move out to each end and are then reflected. When the shockwaves meet, a brief spike of very high pressure and density is created, and a third contact discontinuity is created. A more detailed description is to be found in [17].

Our graphs (see Fig 3) show density distributions at t = 0.010, 0.016, 0.026, 0.028, 0.030, 0.032, 0.034 and 0.038. The left-hand member of each pair shows results computed using the "minmod" averaging function, and the right-hand member shows results from the Superbee function. The computing grid comprises 1200 equal intervals. In every plot we can identify two shockwaves, each usually marked by two points occupying the transition region. The other regions of strong variation are the contact discontinuities, which are not so cleanly represented. In fact, they extend, in the left hand plots, over 25 to 30 mesh intervals, but in the right-hand plots the extent is typically 6 or 7 intervals. This fourfold increase in resolution implies an order of magnitude reduction in the computing times required to achieve results of similar quality at the same elapsed times.

ACKNOWLEDGEMENTS

We are grateful to Dr K-H Winkler, of the Max Planck Institute for Astrophysics in Munich, for advice on how to structure our methods for the CRAY computer. The numerical results in Fig 3 were produced by P Glaister at the University of Reading.

REFERENCES

1. GODUNOV, S.K., A difference method for the numerical computation of discontinuous solutions of hydrodynamic equations. *Mat Sbornik*, 47, 3, pp 271-306, 1959. Translated as JPRS 7225 by US Dept of Commerce, 1960.

2. LAX, P.D., Hyperbolic systems of conservation laws and the mathematical theory of shock waves. SIAM, Philadelphia, 1972.

3. HARTEN, A., On the symmetric form of systems of conservation laws with entropy. *J. Comp. Phys*, 49, Jan 1983.

4. HARTEN, A., LAX, P.D., van LEER, B., On upstream differencing and Godunov-type schemes for hyperbolic conservation laws. *SIAM Review*, 25, 1, p 35, Jan 1983.

5. ROE, P.L., Approximate Riemann solvers, parameter vectors, and difference schemes. *J. Comp. Phys*, 43, 2 pp 357-372, 1981.

6. COLELLA, P., Approximate solution of the Riemann problem for real gases. Preprint, Lawrence Berkeley Laboratory, Univ. of Calif, May 1982.

7. PANDOLFI, M., A contribution to the numerical prediction of unsteady flows. AIAA Paper 83-0121, 1983.

8. LOMBARD, C.K., Conservative supra-characteristics method for splitting hyperbolic systems of gasdynamics for real and perfect gases. NASA CR 166307, Jan 1982.

9. HARTEN, A., High resolution schemes for hyperbolic conservation laws. *J. Comp. Phys*, 49, March 1983.

10. HARTEN, A., On second-order accurate Godunov-type schemes. NYU Report, 1982.

11. HARTEN, A., HYMAN, J.M., A self-adjusting grid for the computation of weak solutions of hyperbolic conservation laws. *J. Comp. Phys*, 50, May 1983.

12. van LEER, B., Towards the ultimate conservative differencing schemes, Part V, *J. Comp. Phys*, 32, p 101, 1979.

13. COLELLA, P., WOODWARD, P.R., The piecewise parabolic method (PPM) for gas dynamical simulations. Report LBL 14661, Lawrence Berkeley Laboratory, Univ. of Calif., July 1982.

14. PIKE, J., Euler equation solutions on irregular grids. RAE Report in preparation.

15. ROE, P.L., The use of the Riemann problem in finite difference schemes, in *Proc. 7th Int. Conf. Num. Meth. Fl. Dyn.*, ed. W.C. Reynolds, R.W. MacCormack, Lecture Notes in Physics, 141, Springer 1981.

16. ROE, P.L., BAINES, M.J., Algorithms for advection and shock problems, in *Proc. 4th GAMM Comf. Num. Meth. Fl. Mech.*, ed H. Viviand, Vieweg 1982.

17. WOODWARD, P.M., COLELLA, P., The numerical simulation of two-dimensional fluid flow with strong shocks. Report UCRL 86952, Lawrence Livermore National Laboratory, Univ. of Calif., 1982.

18. LOMBARD, C.K., OLIGER, J., YANG, J.Y., A natural conservative flux-difference splitting for the hyperbolic systems of gasdynamics. AIAA Paper 82-0976, 1982.

19. YANG, J.Y., LOMBARD, C.K., BERSHADER, D., A characteristic flux-difference splitting for the hyperbolic conservation laws of inviscid gasdynamics. AIAA Paper 83-0040, 1983.

20. OSHER, S., SOLOMAN, F., Upwind schemes for hyperbolic systems of conservation laws. *Math. Comp.*, 38, 1982.

21. ROE, P.L., Numerical modelling of shockwaves and other discontinuities in *Numerical Methods in Aeronautical Fluid Dynamics*, ed. P.L. Roe, Academic Press, 1982.

22. CHORIN, A.J., MARSDEN, J.E., *A mathematical introduction to fluid mechanics.*, Springer, 1979.

23. JEFFREY, A., *Quasilinear hyperbolic systems and waves*, Research Notes in Mathematics No 5, Pitman 1976.

24. ZALESAK, S., Fully multidimensional flux-corrected transport algorithms for fluids, *J. Comp. Phys.*, 31, p 335, 1979.

25. LE ROUX, A.Y., Convergence of an accurate scheme for first order quasi linear equations. *RAIRO Num. Anal.*, 15, 2, pp 151-170, 1981.

26. ROE, P.L., Fluctuations and Signals, a framework for numerical evolution problems, in *Numerical Methods for Fluid Dynamics*, ed. K.W. Morton and M.J. Baines, Academic Press, 1982.

27. ROE, P.L., Numerical algorithms for the linear wave equation, RAE TR 81047, 1981.

28. LEVEQUE, R.J., GOODMAN, J.B., TVD Schemes in one and two space dimensions. *Proc. 15th AMS-SIAM Summer Seminar in Applied Mathematics*, to appear.

29. HARTEN, A., *Proc. 15th AMS-SIAM Summer Seminar in Applied Mathematics*, to appear.

30. van LEER, B., Towards the ultimate conservative differencing scheme, Part III, *J. Comp. Phys.*, 23, 1977.

31. van ALBADA, G.D., van LEER, B., ROBERTS, W.W., A comparative study of computational methods in cosmic gas dynamics. ICASE Report 81-24, August 1981.

32. ROE, P.L., BAINES, M.J., Asymptotic behaviour of some non-linear schemes for linear advection, *Proc. 6th GAMM Conference Num. Meth. Fl. Mech.* ed R. Piva, M. Pandolfi, Vieweg, to appear.

33. ROE, P.L., Some contributions to the modelling of discontinuous flows. *Proc. 15th AMS-SIAM Summer Seminar in Applied Mathematics*, to appear.

34. SWEBY, P.K., High resolution TVD schemes using flux limiters, *Proc. 15th AMS-SIAM Summer Seminar in Applied Mathematics*, to appear.

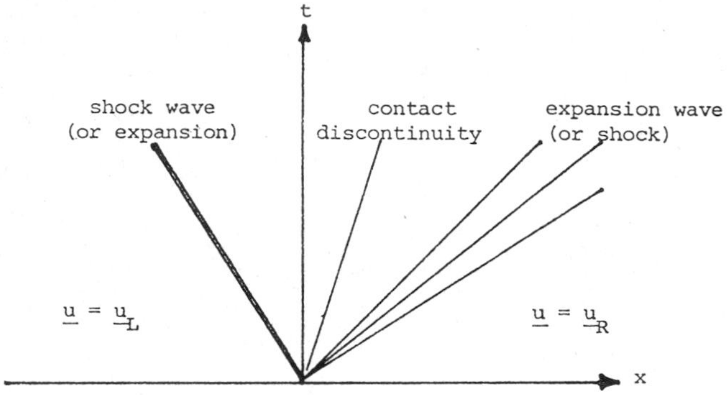

Fig 1 - General solution to the Riemann problem for the Euler equations.

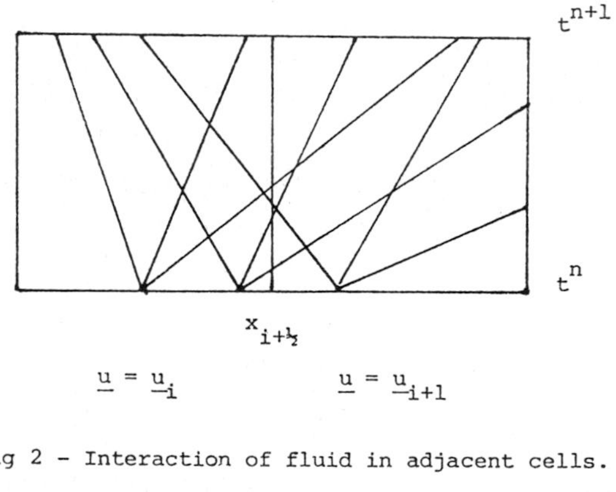

Fig 2 - Interaction of fluid in adjacent cells.

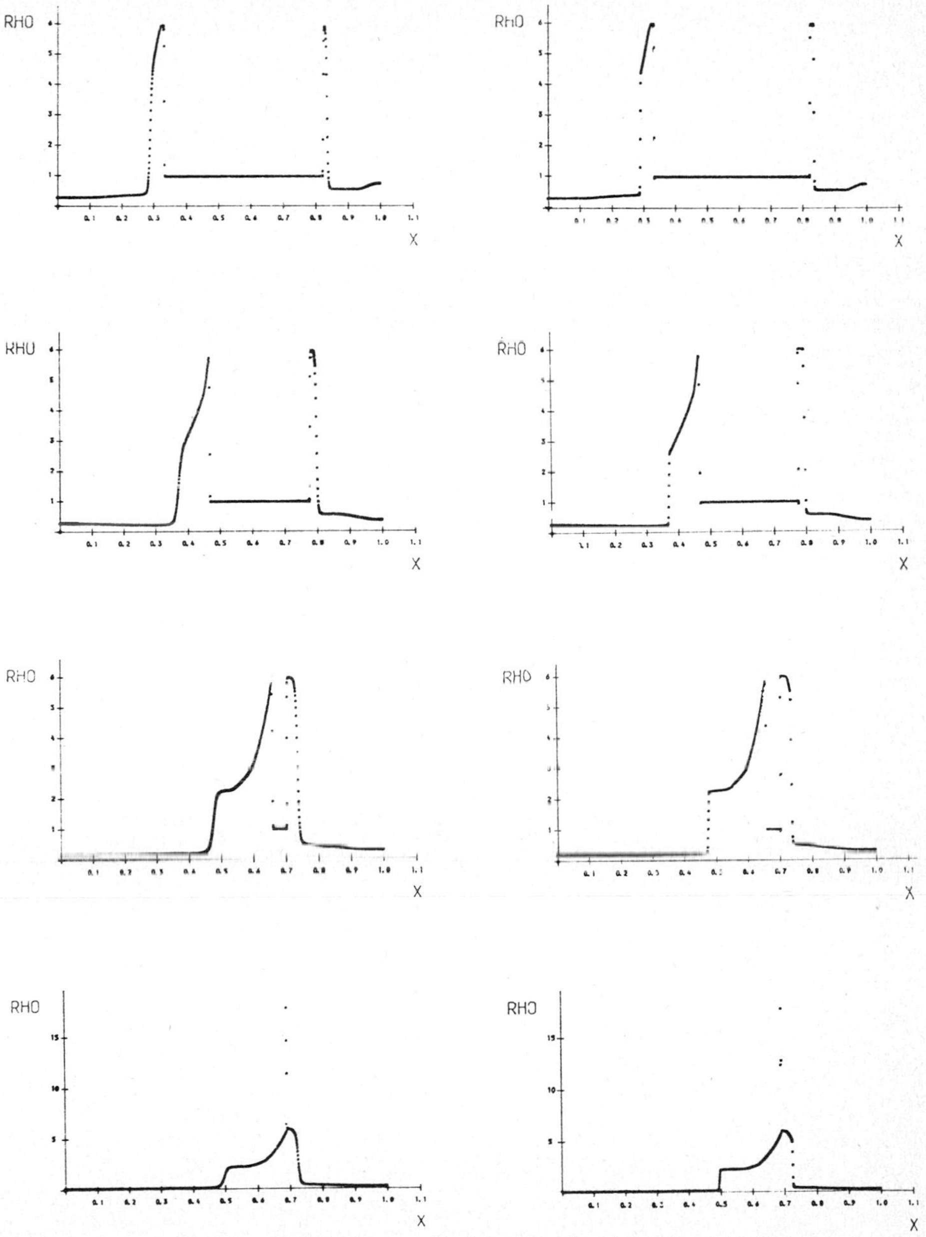

Fig 3 - Distributions of density for the interacting blast wave problem at times t = 0.010 (at top), 0.016, 0.026, and 0.028 (at bottom). Results on left employ minmod, on right Superbee. Note the changes of vertical scale.

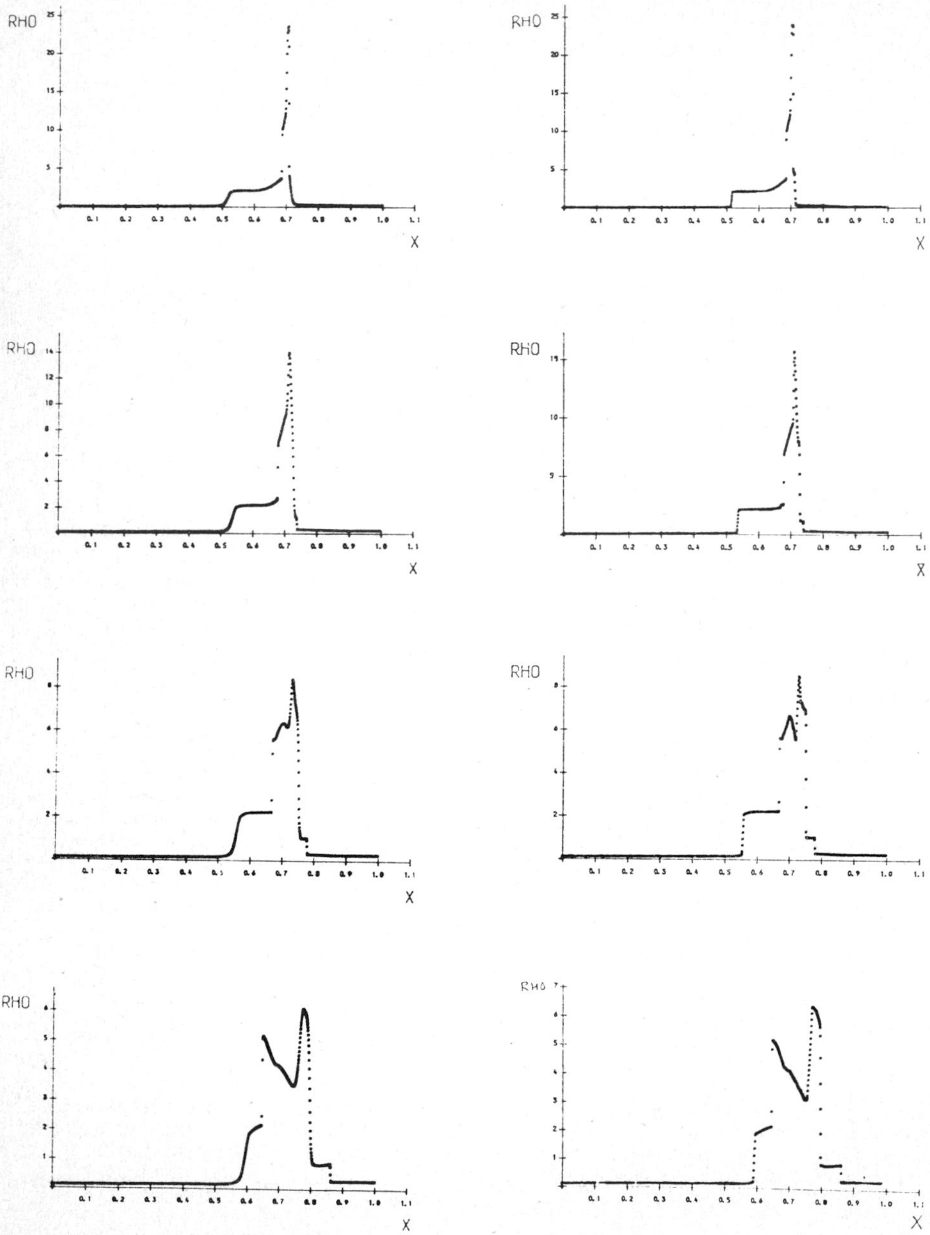

Fig 3 (Contd) - Results at times t = 0.030 (at top), 0.032, 0.034, and 0.038 (at bottom).

Computing Methods in Applied Sciences and Engineering, VI
R. Glowinski and J.-L. Lions (Editors)
Elsevier Science Publishers B.V. (North-Holland)

A SHOCK-CAPTURING FINITE ELEMENT METHOD

Thomas J.R. Hughes

Division of Applied Mechanics
Durand Building
Stanford University
Stanford, CA 94305
U.S.A.

Finite element methods are developed for a one-dimensional singular perturbation problem involving a nonlinear flux function. In the limit of vanishing dissipation, the numerical flux is found to be almost identical to that due to Engquist and Osher, the only difference being in the treatment of transonic compression. This facilitates sharper resolution of stationary shocks, but is accompanied by loss of smoothness and monotonicity.

1. INTRODUCTION

Shock capturing finite difference schemes and recently developed finite element methods for convective-diffusive equations seem, at first glance, to be totally disparate objects. The finite difference schemes make heavy use of nonlinear wave propagation theory, the mathematical theory of hyperbolic conservation laws, and, in particular, solutions of the Riemann problem. The well-known schemes have been interpreted as various types of Riemann solvers (van Leer [20]). On the other hand, finite element methods have been recently developed (see, e.g., Dendy [3], Brooks-Hughes [1], Hughes-Brooks [10], Morton [12], and Wahlbin [21]) which, for linear problems, yield optimal error estimates in integral norms (see Johnson [11] and Nävert [15]). The paper of Morton [13] exhibits a fundamental link in that similar qualitative features, particularly "switching" aspects, possessed by certain finite difference methods, also emanate from finite element methods of characteristic type. In this paper we develop this theme further. We employ as model problems one-dimensional, singular perturbation cases which are intimately related to the problem of hyperbolic conservation laws. In the limit of vanishing dissipation, the finite element method developed yields a numerical flux function which is almost identical to that of Engquist and Osher [4]. The only difference occurs in the treatment of "transonic compression". The present method results in sharper stationary shock resolution than both the Engquist-Osher and Godunov [5] schemes. It may be recalled that the Engquist-Osher scheme resolves shocks with two interior points and Godunov's resolves shocks with one interior point. Ours requires none. Unfortunately, the sharper resolution of stationary shocks is accompanied by some loss of smoothness of the flux function (ours, like Godunov's, is C^0, whereas Engquist-Osher's is C^1) and loss of monotonicity when coupled with the standard explicit one-step operator.

An outline of the remainder of the paper follows: In Section 2 we introduce scalar singular perturbation problem involving a nonlinear flux function. Its variational formulation is given in Section 3 and its finite element approximation is treated in Section 4. Because we anticipate that most readers of this paper will be more familiar with finite differences than finite elements, we go into greater detail in presenting basics than we normally would. The Petrov-Galerkin framework is employed and weighting functions are developed based upon the nodal Green's function problem for the adjoint operator. In the linear, constant-coefficient case the result is identical to Hemker's [8]. At this point we present arguments which justify the use of discontinuous approximations of the ideal (and continous) weighting functions.

In the limit of vanishing viscosity, the weighting functions for the variable-coefficient and nonlinear cases take on a particularly simple form. Some preliminary thoughts on the time-dependent case close out Section 4. A generalization to one-dimensional systems is presented in Section 5 and conclusions are drawn in Section 6.

2. SINGULAR PERTURBATION PROBLEM

Consider the following two-point boundary-value problem (BVP): Find $u = u(x)$, $x \in [x_L, x_R]$, such that

$$f_{,x} = \varepsilon u_{,xx} + \ell \tag{1}$$

$$u(x_L) = g \qquad \text{(Dirichlet b.c.)} \tag{2}$$

$$\varepsilon u_{,x}(x_R) = h \qquad \text{(Neumann b.c.)} \tag{3}$$

where f is the flux function; ε is a given positive constant; ℓ, the source, is a prescribed function of x; g and h are given constants; and an inferior comma denotes differentiation. The flux, in the most general case, is a function of u and x, that is, $f = f(u, x)$. Thus, in (1), $f_{,x} = D_1 f\, u_{,x} + D_2 f$. We shall assume f is a smooth function of its arguments. The notation $a = D_1 f$ is used in the sequel.

Remark There is nothing special about the boundary conditions chosen. We could have equally-well employed a Neumann condition on the left and a Dirichlet condition on the right, or Dirichlet conditions on both ends, without alteration of any of the results. Choosing one Dirichlet condition and one Neumann condition enables us to illustrate the use of "essential" and "natural" boundary conditions, respectively, in the variational formulation.

3. VARIATIONAL FORMULATION

To pose an equivalent weak, or variational, form of the BVP, we need to introduce two sets of functions: The set of trial solutions, $\mathcal{S}$, consists of all functions satisfying the Dirichlet boundary condition, (2), and certain technical differentiability conditions; $\mathcal{S}$ is a linear manifold. The set of weighting functions, $\tilde{\mathcal{V}}$, consists of all functions which satisfy the homogeneous version of the Dirichlet boundary condition, namely,

$$\tilde{w}(x_L) = 0 \quad , \quad \tilde{w} \in \tilde{\mathcal{V}} \tag{4}$$

and certain technical differentiability conditions; $\tilde{\mathcal{V}}$ is a linear space. Functions in $\tilde{\mathcal{V}}$ are sometimes referred to as "variations".

The weak form of the BVP is given as follows: Find $u \in \mathcal{S}$ such that for all $\tilde{w} \in \tilde{\mathcal{V}}$,

$$\int_{x_L}^{x_R} \left(\tilde{w}\, f(u, \cdot)_{,x} + \tilde{w}_{,x}\, \varepsilon u_{,x}\right) dx = \int_{x_L}^{x_R} \tilde{w}\, \ell\, dx + \tilde{w}(x_R)\, h \tag{5}$$

Integration by parts,

$$\int_{x_L}^{x_R} \tilde{w}\left(f(u,\cdot)_{,x} - \varepsilon\, u_{,xx} - \ell\right)dx + \tilde{w}(x_R)\left(\varepsilon\, u_{,x}(x_R) - h\right) = 0\ , \tag{6}$$

and standard arguments of the calculus of variations can be used to establish the equivalence of the weak solution with that of the strong solution, that is, the solution of (1)-(3). The Neumann boundary condition, (3), which is implied by (6), is referred to as a "natural" boundary condition. Because the Dirichlet conditions are built into the definitions of $\mathcal{S}$ and $\mathcal{V}$, they are referred to as "essential" boundary conditions.

4. FINITE ELEMENT FORMULATION

Finite element methods based upon (5) are developed by replacing $\mathcal{S}$ and $\tilde{\mathcal{V}}$ by finite-dimensional approximations, say $\mathcal{S}^h$ and $\tilde{\mathcal{V}}^h$, respectively. The superscript h refers to the mesh parameter which is defined to be the length of the largest element. It is assumed that $\mathcal{S}^h \subset \mathcal{S}$ and $\tilde{\mathcal{V}}^h \subset \tilde{\mathcal{V}}$. Thus functions in $\mathcal{S}^h$ and $\tilde{\mathcal{V}}^h$ inherit the boundary conditions and regularity of functions in $\mathcal{S}$ and $\tilde{\mathcal{V}}$, respectively. It is typical to represent $\mathcal{S}^h$ as the sum of a fixed function which satisfies the Dirichlet boundary condition, (2), and a space $\mathcal{V}^h$ whose members satisfy the homogeneous Dirichlet condition. Thus we write

$$\mathcal{S}^h = \{g^h\} + \mathcal{V}^h \tag{7}$$

where

$$g^h : [x_L\ , x_R] \to \mathbb{R} \tag{8}$$

$$g^h(x_L) = g \tag{9}$$

Equivalently, if $u^h \in \mathcal{S}^h$, then $u^h = g^h + v^h$ where $v^h \in \mathcal{V}^h$, $v^h(x_L) = 0$. Generally, $\mathcal{V}^h$ is constructed from simple piecewise polynomials which are globally C^0 (e.g. the piecewise linear hat functions). In the (Bubnov-) Galerkin formulation $\tilde{\mathcal{V}}^h = \mathcal{V}^h$. In this case the discrete equations have central-difference character and thus are unsuitable for the problem at hand. The general case in which $\tilde{\mathcal{V}}^h \neq \mathcal{V}^h$ can be used as a framework for developing more successful numerical methods. The idea is to construct $\tilde{\mathcal{V}}^h$ so that some measure of accuracy is optimized. This approach is currently referred to as a Petrov-Galerkin, or generalized Galerkin, formulation. The finite-dimensional counterpart of (5) is: Find $u^h \in \mathcal{S}^h$ such that for all $\tilde{w}^h \in \tilde{\mathcal{V}}^h$,

$$\int_{x_L}^{x_R}\left(\tilde{w}^h f(u^h,\cdot)_{,x} + \tilde{w}^h_{,x}\, \varepsilon\, u^h_{,x}\right)dx = \int_{x_L}^{x_R} \tilde{w}^h \ell\, dx + \tilde{w}^h(x_R)h \tag{10}$$

Let x_i, $i = 1, 2, \ldots, n_{np}$, represent the nodes of the finite element mesh. The nodes are assumed to be in ascending order from left to right with $x_1 = x_L$ and $x_{n_{np}} = x_R$. The error in the finite element solution at x_i is $e_i = e(x_i) = u^h(x_i) - u(x_i)$. We will attempt to develop finite element methods which minimize the e_i's. It is more typical in finite element methods to attempt to minimize error in integral norms such as L_2. Schemes of this type often exhibit oscillatory behavior about steep gradients in the exact solution. Morton [12] has emphasized that overshoot can be used to recover accurate structure of the steep gradient in L_2-optimal schemes. However, we are concerned that overshoots will have a debilitating effect in nonlinear problems and thus prefer to use the nodal measure of accuracy because there seems greater potential for creating methods with "monotone" character.

We will usually assume that $\mathscr{V}^h$ is constructed from linear combinations of the C^0 piecewise linear hat functions. This will facilitate direct comparison with commonly used finite difference methods. (Generalization to higher-order basis functions is straightforward.) Let $\tilde{N}_i$ be the basis function of $\tilde{\mathscr{V}}^h$ associated with node x_i , $i = 2, 3, \ldots, n_{np}$. (Likewise, the ith nodal basis function of $\mathscr{V}^h$ is denoted N_i .) The $\tilde{N}_i$'s are defined by the following conditions:

$$\tilde{N}_i(x_j) = \delta_{ij}, \quad i = 2, 3, \ldots, n_{np}, \quad j = 1, 2, \ldots, n_{np} \tag{11}$$

$$-\tilde{a}\,\tilde{N}_{i,x} = \varepsilon\,\tilde{N}_{i,xx} \tag{12}$$

where

$$\tilde{a} = \tilde{a}(u^h, x) = f(u^h, x)/u^h \tag{13}$$

and δ_{ij} is the Kronecker delta. We assume $\tilde{a}$ is a smooth function of its arguments. The linear differential operator of (12) is the formal adjoint of the nonlinear differential operator of (1). Equations (11) and (12) amount to two non-trivial Dirichlet problems for each $\tilde{N}_i$ associated with an interior node, and one for $\tilde{N}_{n_{np}}$. Clearly, the $\tilde{N}_i$'s will depend on u^h through $\tilde{a}$. The general solution for $i = 2, 3, \ldots, n_{np} - 1$ is given by

$$\tilde{N}_i = \int_{x_{i-1}}^{x} E_{i-1}(y)\,dy \Bigg/ \int_{x_{i-1}}^{x_i} E_{i-1}(y)\,dy, \quad x \in [x_{i-1}, x_i] \tag{14}$$

$$= 1 - \int_{x_i}^{x} E_i(y)\,dy \Bigg/ \int_{x_i}^{x_{i+1}} E_i(y)\,dy, \quad x \in [x_i, x_{i+1}] \tag{15}$$

$$= 0, \quad x \notin [x_{i-1}, x_{i+1}] \tag{16}$$

where

$$E_i(y) = \exp\left(- \int_{x_i}^{y} \tilde{a}(u^h(z), z)\,dz/\varepsilon \right) \tag{17}$$

The non-zero part of the solution for $i = n_{np}$ is given by (14).

To analyze the quality of the approximation, consider the following nodal Green's function problem for the adjoint operator

$$-\tilde{a}(u, \cdot)g_{i,x} = \varepsilon\, g_{i,xx} + \delta_i, \quad x \in\,]x_L, x_R[\tag{18}$$

$$g_i(x_L) = 0 \tag{19}$$

$$\left(g_i\, f(u, \cdot) + g_{i,x}\, \varepsilon\, u\right)\Big|_{x_R} = 0 \tag{20}$$

where $g_i \in \tilde{\mathcal{V}}$ is the Green's function corresponding to a Dirac delta function, $\delta_i(x) = \delta(x - x_i)$, located at node i . The discrete counterpart of (18)-(20) is

$$-\tilde{a}(u^h, \cdot)g^h_{i,x} = \varepsilon\, g^h_{i,xx} + \delta_i \quad , \qquad x \in \,]x_L , x_R[\tag{21}$$

$$g^h_i(x_L) = 0 \tag{22}$$

$$\left. \left(g^h_i \, f(u, \cdot) + g^h_{i,x} \, \varepsilon \, u^h \right) \right|_{x_R} = 0 \tag{23}$$

Note that $\tilde{a}(u^h(x), x)$ will be a C^0 function of x due to the dependence upon u^h . Consequently, g^h_i will be C^2 on $[x_L , x_R] - \{x_i\}$. By construction of the $\tilde{N}_i$'s , there exist constants, c_{ji}'s , such that

$$g^h_i = \sum_{i=2}^{n_{np}} c_{ji} \, \tilde{N}_i \quad , \tag{24}$$

that is, $g^h_j \in \tilde{\mathcal{V}}^h$.

To obtain an expression for the nodal error, first integrate (5) and (10) by parts,

$$\int_{x_L}^{x_R} \left(\tilde{a}(u, \cdot)\tilde{w}_{,x} - \varepsilon\, \tilde{w}_{,xx} \right) u \, dx + \left. \left(\tilde{w} \, f(u, \cdot) + \tilde{w}_{,x} \, \varepsilon \, u \right) \right|_{x_L}^{x_R} \tag{25}$$

$$= \int_{x_L}^{x_R} \tilde{w} \, \mathcal{f} \, dx + \tilde{w}(x_R) \mathcal{h} \tag{25}$$

$$\int_{x_L}^{x_R} \left(- \tilde{a}(u^h, \cdot)\tilde{w}^h_{,x} - \varepsilon\, \tilde{w}^h_{,xx} \right) u^h \, dx + \left. \left(\tilde{w}^h \, f(u^h, \cdot) + \tilde{w}^h_{,x} \, \varepsilon \, u^h \right) \right|_{x_L}^{x_R}$$

$$= \int_{x_L}^{x_R} \tilde{w}^h \mathcal{f} \, dx + \tilde{w}^h(x_R) \mathcal{h} \tag{26}$$

Equation (26) can be put in the following alternative form which explicitly acknowledges the fact $\tilde{w}^h$ is C^0 , but not C^1 :

$$\sum_{i=1}^{n_{np}-1} \int_{x_i}^{x_{i+1}} \left(- \tilde{a}(u^h, \cdot)\tilde{w}^h_{,x} - \varepsilon \, \tilde{w}^h_{,xx} \right) u^h \, dx$$

$$+ \left. \left(\tilde{w}^h \, f(u^h, \cdot) + \tilde{w}^h_{,x} \, \varepsilon \, u^h \right) \right|_{x_L}^{x_R}$$

$$= \int_{x_L}^{x_R} \tilde{w}^h \mathcal{f} \, dx + \tilde{w}^h(x_R) \mathcal{h} + \sum_{i=1}^{n_{np}-1} \left(\tilde{w}^h_{,x}(x_i^+) - \tilde{w}^h_{,x}(x_i^-) \right) \varepsilon \, u^h(x_i) \tag{26$'$}$$

The first integral in (26)′ vanishes identically by the definition of $\tilde{w}^h$. Substituting g_i for $\tilde{w}$ in (25) and g_i^h for $\tilde{w}^h$ in (26) results in, respectively,

$$\int_{x_L}^{x_R} \left(- \tilde{a}(u, \cdot)g_{i,x} - \varepsilon\, g_{i,xx}\right)u\, dx + \left(g_i\, f(u, \cdot) + g_{i,x}\, \varepsilon\, u\right)\Big|_{x_L}^{x_R} = \int_{x_L}^{x_R} g_i\, \mathscr{f}\, dx + g_i(x_R)\mathscr{h} \tag{27}$$

$$\int_{x_L}^{x_R} \left(- \tilde{a}(u^h, \cdot)g^h_{i,x} - \varepsilon\, g^h_{i,xx}\right)u^h\, dx + \left(g_i^h\, f(u^h, \cdot) + g^h_{i,x}\, \varepsilon\, u^h\right)\Big|_{x_L}^{x_R} = \int_{x_L}^{x_R} g_i^h\, \mathscr{f}\, dx + g_i^h(x_R)\mathscr{h} \tag{28}$$

Equations (27) and (28) may be simplified by using (18)-(20) and (21)-(23), respectively,

$$u(x_i) = \int_{x_L}^{x_R} g_i\, \mathscr{f}\, dx + g_{i,x}(x_L)\varepsilon \mathscr{g} + g_i(x_R)\mathscr{h} \tag{29}$$

$$u^h(x_i) = \int_{x_L}^{x_R} g_i^h\, \mathscr{f}\, dx + g^h_{i,x}(x_L)\varepsilon \mathscr{g} + g_i^h(x_R)\mathscr{h} \tag{30}$$

The expression for the nodal error immediately follows from (29) and (30):

$$e(x_i) = \int_{x_L}^{x_R} (g_i^h - g_i)\, \mathscr{f}\, dx + \left(g^h_{i,x}(x_L) - g_{i,x}(x_L)\right)\varepsilon \mathscr{g} + \left(g_i^h(x_R) - g_i(x_R)\right)\mathscr{h} \tag{31}$$

From (31) it is clear that if g_i^h is a good approximation of g_i then the nodal error will be small.

Remark 1. In the linear case, in which $f(u, x) = a(x)u$ and $\tilde{a} = a$, $g_i^h = g_i$, that is, the solution is exact at the nodes. Furthermore, (29) and (30) are explicit formulae, rather than identities, satisfied by u and u^h, respectively. The discrete equations are independent of the choice of basis functions used to define $\mathscr{S}^h$. Thus, the basis functions are used solely to recover the interelement behavior of u^h. By using C^0 piecewise polymials of degree k, the order of accuracy in L_2 is $k + 1$ and the order of accuracy in H^1 is k.

Remark 2. In the constant-coefficient case, in which $\tilde{a} = a = \text{const.}$, the weighting functions may be explicitly calculated:

$$\tilde{N}_i(x) = \left(e^{-\beta_{i-\frac{1}{2}}(x-x_{i-1})/h_{i-\frac{1}{2}}} - 1\right)/\left(e^{-\beta_{i-\frac{1}{2}}} - 1\right) , \qquad x \in [x_{i-1}, x_i] \tag{32}$$

$$= \left(e^{-\beta_{i+\frac{1}{2}}} - e^{-\beta_{i+\frac{1}{2}}(x-x_i)/h_{i+\frac{1}{2}}}\right)/\left(e^{-\beta_{i+\frac{1}{2}}} - 1\right) , \quad x \in [x_i, x_{i+1}] \tag{33}$$

where $\beta_{i+\frac{1}{2}} = a\, h_{i+\frac{1}{2}}/\varepsilon$ and $h_{i+\frac{1}{2}} = x_{i+1} - x_i$. These functions were first proposed by Hemker [8] and are illustrated in Figure 1. In the variable-coefficient and nonlinear cases, the ideal weighting functions would generally have to be evaluated by numerical quadrature or asymptotic approximations. Furthermore, the generalization to multi-dimensional cases appears very difficult. For these reasons not too much progress has been made along these lines and thus alternative approaches have been investigated (Morton [12]). However, we are mainly concerned with the inviscid limit $\varepsilon \to 0$ and it turns out that considerable simplicity is achieved thereby. In fact, very close relationships will be shown to exist with the numerical flux functions of common finite difference shock-capturing methods.

<u>Interlude: Variational Framework for Petrov-Galerkin Methods with Discontinuous Weighting Functions</u>

We wish to consider <u>approximations</u> of the ideal weighting function, possibly discontinuous. We note that the ideal weighting functions are smooth on element interiors and globally continuous. The form of (10), involving derivatives of $\tilde{w}^h$, is not suggestive of how discontinuous approximations of $\tilde{w}^h$ may be correctly employed. To this end let $w^h \in \mathcal{V}^h$ be the piecewise continuous polynomial interpolate of $\tilde{w}^h \in \tilde{\mathcal{V}}^h$. Let $p^h = \tilde{w}^h - w^h$. Clearly p^h is smooth on element interiors and zero at the nodes. With this notation, (7) can be put in the following form:

$$\int_{x_L}^{x_R} \left(w^h f(u^h, \cdot)_{,x} + w^h_{,x}\, \varepsilon\, u^h_{,x}\right)dx + \int_{x_L}^{x_R} p^h\left(f(u^h, \cdot)_{,x} - \varepsilon\, u^h_{,xx} - \ell\right)dx$$

$$= \int^{x_R} w^h \ell\, dx + w^h(x_D) \tag{34}$$

$$\int_{x_L} \left(w^h f(u^h, \cdot)_{,x} + w^h_{,x}\, \varepsilon\, u^h_{,x}\right)dx$$

$$+ \sum_{i=1}^{n_{np}-1} \int_{x_i}^{x_{i+1}} \left(p^h f(u^h, \cdot)_{,x} - \varepsilon\, u^h_{,xx} - \ell\right)dx$$

$$= \int_{x_L}^{x_R} w^h \ell\, dx + w^h(x_R)\, \boldsymbol{h} \tag{35}$$

Equation (35) is identical to (34) because p^h vanishes at the nodes and thus the nodal Dirac delta functions emanating from $u^h_{,xx}$ make no contribution. Equation (35) is a suitable starting point for discontinuous p^h. The idea is to employ convenient approximations which accurately reproduce the integral behavior of the ideal functions. This type of formulation was first presented in Hughes and Brooks [10] for the advection-diffusion equation and generalized to the incompressible Navier-Stokes equations by Brooks and Hughes [1]. Equation (35) enables justification of methods which initially seemed ad hoc (cf. [1, 10] with earlier presentations of the streamline upwind method, e.g. [9]). ■

Variable-coefficient Case

Next we wish to consider the variable coefficient case in which $f(u, x) = a(x)u$ and $\tilde{a} = a$. For simplicity we consider the case in which there is at most one zero of a in each element. In the limit $\varepsilon \to 0$, it can be shown that the ideal weighting functions exhibit only a limited dependence on a in that they are functions of sgn a. In this case, (14) and (15) may be combined as follows:

$$\left.\begin{aligned} \tilde{N}_i &= \frac{1}{2}(1 + b \,\mathrm{sgn}\, N_{i,x}) , \quad x \in [x_{i-1}, x_{i+1}] , \\ & \qquad i = 2, 3, \ldots, n_{np} - 1 \end{aligned}\right\} \tag{36}$$

where

$$\left.\begin{aligned} b(x) &= \int_{x_i}^{x_{i+1}} \mathrm{sgn}\, a(y) dy / h_{i+\frac{1}{2}} , \quad \text{if} \quad a_i \geq 0 \geq a_{i+1} , \\ & \qquad x \in [x_i, x_{i+1}] , \quad i = 1, 2, \ldots, n_{np} - 1 \end{aligned}\right\} \tag{37}$$

$$= \mathrm{sgn}\, a(x) , \quad \text{otherwise} \tag{38}$$

The nonzero part of weighting function $\tilde{N}_{n_{np}}$ is given by (36) with domain of definition restricted to $x \in [x_{n_{np}-1}, x_{n_{np}}]$. (The degenerate case of a identically zero in an element is covered by setting $\tilde{N}_i = N_i$.) If a has no zeros in an element then the weighting function is either zero or one in that element. If a has a zero then there are two cases: transonic compression (i.e., $a_i \geq 0 \geq a_{i+1}$) and transonic expansion (i.e., $a_i \leq 0 \leq a_{i+1}$). In transonic compression the weighting function is constant with amplitude given by (37). This corresponds to an internal discontinuity in the exact solution located at the zero of a. In transonic expansion the weighting function has an internal discontinuity at the

assume that a is approximated by its piecewise linear interpolate, say a^h. Then, in the absense of a source term, the discrete solution will still be exact because, as $\varepsilon \to 0$, g^h and $g^h_{,x}$ converge to g and $g_{,x}$ at the boundaries; see (31). The source term in this case contributes error terms of order the distance between the zeros of a and a^h (i.e. $O(h^2)$ for smooth a).

Nonlinear Case

Now let us consider the nonlinear case. For simplicity assume $f(x, u) = f(u)$ and f is a smooth convex function of u. This guarantees that if u^h is defined by C^0 piecewise linear interpolation, then $a(u^h)$ will have at most one zero per element. The secant function $\tilde{a} = f/u$ is assumed smooth, however, it will not equal the tangent, a, as in the linear case; see Figure 3. Nevertheless, $\mathrm{sgn}\, \tilde{a} = \mathrm{sgn}\, a$ ($= \mathrm{sgn}\, u$) and thus, in the limit $\varepsilon \to 0$, the definition of weighting functions is essentially the same as before, that is, replace

$a(\cdot)$ by $a(u^h(\cdot))$ in (37) and (38) and let $a_i = a(u^h(x_i))$ in (37). Figure 2 again depicts the weighting functions. It is instructive at this point to calculate the numerical flux functions which emanate from (36)-(38). Assuming $\mathscr{f} = 0$, the difference equation at an internal node i takes on the following form (we will temporarily drop the h superscript on u^h to simplify the notation):

$$\tilde{f}_{i+\frac{1}{2}} - \tilde{f}_{i-\frac{1}{2}} = 0 \tag{39}$$

where the <u>numerical flux</u> is defined by

$$\tilde{f}_{i+\frac{1}{2}} = \tilde{f}(u_{i+1}, u_i) = \frac{f_{i+1} + f_i}{2} - \frac{1}{2}\int_{u_i}^{u_{i+1}} b(u)a(u)du \tag{40}$$

in which

$$b(u) = \int_{u_i}^{u_{i+1}} \operatorname{sgn} a(v)dv/(u_{i+1} - u_i) \quad , \quad a_i \geq 0 \geq a_{i+1} \tag{41}$$

$$= \operatorname{sgn} a(u) \quad , \quad \text{otherwise} \tag{42}$$

and $f_i = f(u_i)$, $u_i = u(x_i)$. Change of variables has been used in (40) and (41). Combining (40)-(42) and recalling that $\operatorname{sgn} a = \operatorname{sgn} u$ leads to the simplified expressions:

$$\tilde{f}_{i+\frac{1}{2}} = \frac{f_{i+1} + f_i}{2} + \frac{u_{i+1} + u_i}{2}\,\frac{(f_{i+1} - f_i)}{(u_{i+1} - u_i)} \quad , \quad u_i \geq 0 \geq u_{i+1} \quad , \tag{43}$$

(i.e., transonic compression)

$$= \frac{f_{i+1} + f_i}{2} - \frac{1}{2}\int_{u_i}^{u_{i+1}} |a(u)|du \quad , \quad \text{otherwise} \tag{44}$$

<u>Remark</u> <u>Equation (44) is exactly the Engquist-Osher numerical flux function</u> (see [4, 16]). Thus the present formulation is the same as Engquist-Osher with regards to flux definition except for the treatment of transonic compression. (Part of the finite element "package deal" is a variationally consistent recipe for treating source and boundary terms.) It may also be recalled that the classical Godunov's flux function [5] is in agreement with Engquist-Osher except for transonic compression. An immediate conclusion is that <u>the present scheme is entropy satisfying</u> because it is identical to Godunov and Engquist-Osher in transonic expansion. All three schemes treat transonic compression differently and this directly translates into their effectiveness in resolving steady shocks. It is well known that the Engquist-Osher scheme produces steady shock profiles with two interior mesh points. The Godunov scheme is less dissipative in the sense that is resolves stationary shocks with one interior point. <u>The present scheme is the least dissipative of the three in that it resolves stationary shocks with no interior points and thus, in this sense, it is the ultimate shock capturing scheme</u>. A schematic illustration is presented in Figure 4. (There is one possible exception: If the shock front coincides with a mesh point then the value of u_i at that point <u>may</u> be zero.)

Shutting off the transonic compression switch, that is (43), and using (44) exclusively results in Engquist-Osher flux function treatment. This amounts to using discontinuous weighting functions (see Figure 2, Cases (a) and (e)-(h)), the duals of smooth solutions, for all transonic cases and is analogous to using rarefaction and overturned compression waves in the Riemann-solver context (van Leer [20]). The mirror-image, "all shocks" approach may be developed by employing only the constant weighting functions (see Figure 2, Cases (a)-(e)), which are the duals of discontinuous solutions. This is analogous to the interpretation of the Courant-Isaacson-Rees [2], Murman-Cole [14] and Roe [18] schemes given by van Leer [20]. However, this approach leads to the classical pathology of numerically stable, entropy-violating shocks; consequently, it is not pursued further herein.

Example. To gain further insight into the present formulation we consider Burgers' flux function: $f(u) = u^2/2$. In this case the numerical flux functions can be written as follows:

Godunov:
$$\tilde{f}^{G}_{i+\frac{1}{2}} = \frac{1}{2} \max \left\{ (u_i^+)^2 , (u_{i+1}^-)^2 \right\} \tag{45}$$

Engquist-Osher:
$$\tilde{f}^{EO}_{i+\frac{1}{2}} = \frac{1}{2} \left((u_i^+)^2 + (u_{i+1}^-)^2 \right) \tag{46}$$

Present method:
$$\tilde{f}_{i+\frac{1}{2}} = \tilde{f}^{EO}_{i+\frac{1}{2}} + \frac{1}{2} u_i^+ u_{i+1}^- \tag{47}$$

where $u^+ = \max \{u, 0\}$ and $u^- = \min \{u, 0\}$.

It may be noted that max and min are C^0 functions. However, $(u_i^+)^2$ and $(u_{i+1}^-)^2$ are C^1 . Thus $\tilde{f}^G_{i+\frac{1}{2}}$ and $\tilde{f}_{i+\frac{1}{2}}$ are C^0 functions whereas $\tilde{f}^{EO}_{i+\frac{1}{2}}$ is C^1 . Osher [16] has argued that this additional smoothness is a potentially valuable property which may be exploited in implicit algorithms.

A simple calculation reveals that the explicit, forward-in-time scheme,

$$u_i^{n+1} = u_i^n - \frac{\Delta t}{h_i} (\tilde{f}^n_{i+\frac{1}{2}} - \tilde{f}^n_{i-\frac{1}{2}}) , \tag{48}$$

in which $h_i = (h_{i+\frac{1}{2}} + h_{i-\frac{1}{2}})/2$, Δt is the time step, and the superscripts refer to the time level, is not monotonicity preserving [7]. To see this, consider the monotone profile:

$$\left.\begin{aligned} u_i^n &= +1 , & i &\le k \\ &= -\frac{1}{2} , & i &= k+1 \\ &= -1 , & i &\ge k+2 \end{aligned}\right\} \tag{49}$$

Then $u_k^{n+1} = 1 + \Delta t/(8h_k) > 1$. This lack of monotonicity is attributable to the decreased dissipation in transonic compression. Both the Godunov and Engquist-Osher schemes are monotone [7], consequently monotonicity preserving. Important convergence properties are possessed by monotone schemes [7]. ■

Thus the added resolution of the present method in capturing shocks is gained at the expense of a loss of smoothness compared with Engquist-Osher and a loss of monotonicity compared with Godunov and Engquist-Osher.

Further Remarks on the Time-dependent Case

If we wish to maintain the present definition of numerical flux in the time-dependent case then the most straightforward generalization is to simply replace

ℓ in (10) by $\ell - \dot{u}^h$, where $\dot{u}^h$ is the time derivative of u^h, and assume all quantities depend on time. In this case the nodal values of u^h carry the time dependence, viz.

$$u^h(x, t) = g^h(x, t) + \sum_{i=2}^{n_{np}} N_i(x) u_i^h(t) \tag{50}$$

However, in the nonlinear case, the weighting functions change with time due to their dependence upon u^h. In the absence of source terms, the semidiscrete equation at an internal nodes takes the form

$$\frac{1}{2}(\dot{\tilde{u}}_{i+\frac{1}{2}} + \dot{\tilde{u}}_{i-\frac{1}{2}}) + \frac{1}{h_i}(\tilde{f}_{i+\frac{1}{2}} - \tilde{f}_{i-\frac{1}{2}}) = 0 \tag{51}$$

The numerical rate terms can be calculated explicitly with the aid of (36)-(38). Schemes of this type tend to accentuate formal spatial accuracy rather than, say, monotonicity. For example, assume the linear constant-coefficient case in which a is positive. Then (51) can be shown to take the following form:

$$\frac{1}{2}(\dot{u}_i + \dot{u}_{i-1}) + \frac{a}{h_{i-\frac{1}{2}}}(u_i - u_{i-1}) = 0 \tag{52}$$

This is clearly an $O(h^2)$ semidiscrete scheme centered about $x_{i-\frac{1}{2}} = (x_i + x_{i-1})/2$. Furthermore, (52) implies the following energy identity:

$$\frac{d}{dt}\left(\sum_{i=1}^{n_{np}-1} h_{i+\frac{1}{2}}(u_i + u_{i+1})^2/8\right) + \frac{a}{2}u_{n_{np}}^2 = \frac{a}{2}g^2 \tag{53}$$

To achieve accurate and robust time integration schemes we are presently experimenting with formulations outside the weighted residual framework in which the $\dot{u}^h$-term is introduced into the variational equation muliplied by an <u>independent weighting function</u>. This weighting function is selected so as to produce schemes like those described in Harten [6] and Roe [19].

5. ONE-DIMENSIONAL SYSTEMS

Consider the following one-dimensional-system counterpart of (1)-(3):

$$\underset{\sim}{F}_{,x} = \varepsilon \underset{\sim}{U}_{,xx} + \underset{\sim}{\mathcal{F}} \tag{54}$$

$$\underset{\sim}{U}(x_L) = \underset{\sim}{\mathcal{G}} \tag{55}$$

$$\varepsilon \underset{\sim}{U}_{,x}(x_R) = \underset{\sim}{\mathcal{H}} \tag{56}$$

For the moment, assume the linear, constant-coefficient case in which $\underset{\sim}{F} = \underset{\sim}{A}\,\underset{\sim}{U}$ and $\underset{\sim}{A}$ is an $m \times m$ constant matrix. We assume

$$\underset{\sim}{A} = \underset{\sim}{S}\,\underset{\sim}{\Lambda}\,\underset{\sim}{S}^{-1} \tag{57}$$

where $\underset{\sim}{\Lambda}$ is a real, diagonal matrix. Consideration of the adjoint problem for the weighting function <u>matrix</u> leads to the following definition in the limit $\varepsilon \to 0$:

$$\tilde{N}_i = \frac{1}{2}(I + \text{sgn}\, N_{i,x}\, \text{sgn}\, A) \tag{58}$$

where I is the identity matrix, and

$$\text{sgn}\, A = S\, \text{sgn}\, \Lambda\, S^{-1} \tag{59}$$

In the variable-coefficient and nonlinear cases we propose replacing $\text{sgn}\, A$ in (58) by a matrix C which is defined as follows:

$$C = S\, B\, S^{-1} \tag{60}$$

where B is a diagonal matrix,

$$B = \begin{bmatrix} b_1 & & & \\ & b_2 & & \\ & & \ddots & \\ & & & b_m \end{bmatrix} \tag{61}$$

and for $\ell = 1, 2, \ldots, m$,

$$\left.\begin{aligned} b_\ell(x) &= \int_{x_i}^{x_{i+1}} \text{sgn}\, \Lambda_\ell(y)\, dy/h_{i+\frac{1}{2}}, \quad \text{if } \Lambda_\ell(x_i) \geq 0 \geq \Lambda_\ell(x_{i+1}), \\ & x \in [x_i, x_{i+1}], \quad i = 1, 2, \ldots, n_{np} - 1 \end{aligned}\right\} \tag{62}$$

$$= \text{sgn}\, \Lambda_\ell(x), \quad \text{otherwise} \tag{63}$$

This reduces to $\text{sgn}\, A$ in the linear constant-coefficient case.

The numerical flux vector can be written as

$$\tilde{F}_{i+\frac{1}{2}} = \tilde{F}(U_{i+1}, U_i) = \frac{1}{2}(F_{i+1} + F_i) - \frac{1}{2}\int_{U_i}^{U_{i+1}} C(U)A(U)\, dU \tag{64}$$

If exclusive use is made of (63), that is, if $C(U) = \text{sgn}\, A(U)$, then (64) becomes

$$\tilde{F}_{i+\frac{1}{2}} = \frac{1}{2}(F_{i+1} + F_i) - \frac{1}{2}\int_{U_i}^{U_{i+1}} |A(U)|\, dU \tag{65}$$

This is precisely Osher's numerical flux vector for systems. A simple and elegant procedure for exactly calculating the integral in (65) has been presented in [16, 17]. The transonic compression "switch", (62), would again seem to afford the potential for sharper resolution of shocks.

Finally, we note that the recipe above does not amount to exactly solving the adjoint Green's function problem except for the linear, constant-coefficient case. In fact, in the nonlinear case, the coefficient matrix of the adjoint operator, $-\tilde{A}^T$, is somewhat ambiguous in that it is only required to satisfy $\tilde{A}\, U = F$ and thus is non-unique. Fortunately, in the practically important case of the Euler equations, $\underset{\sim}{F}$ is a homogeneous function of degree 1 and thus we have the canonical choice $\underset{\sim}{\tilde{A}} = \underset{\sim}{A}$ which is consistent with the above formulation.

6. CONCLUSIONS

The present developments further illustrate the theme set forth in Morton [13], that is, that there is a close relationship between finite difference shock capturing methods and certain finite element methods. The new procedure developed herein yields a numerical flux function that is very close to Engquist-Osher [4]. The only difference occurs in transonic compression. The potential advantage of the present formulation is exact resolution of stationary shocks without interior points. The disadvantages are loss of smoothness of the flux function (ours is C^0) and loss of monotonicity of the standard explicit one-step scheme. By shutting off the transonic compression "switch" we attain exactly the Engquist-Osher flux function, along with a variationally consistent treatment of source and boundary-condition terms, and applicable to irregular meshing. Generalizations to the time-dependent case are briefly discussed. Further research is still needed in this area. A straightforward generalization is also presented for one-dimensional systems and, again, there is a similar, close relationship with the Osher numerical flux vector treatment. We plan on pursuing multidimensional aspects in future work and believe that there is considerable potential for generalizing the present ideas.

ACKNOWLEDGEMENT

This research was partially supported by the NASA Ames Research Center under Consortium Agreement NASA-NCA2-OR745-307 and by the NASA Langley Research Center under Grant NASA-NAG-1-361.

REFERENCES

1. A. N. Brooks and T.J.R. Hughes, "Streamline-Upwind/Petrov-Galerkin Formulations for Convection Dominated Flows with Particular Emphasis on the Incompressible Navier-Stokes Equations," Computer Methods in Applied Mechanics and Engineering, Vol. 32, 199-259 (1982).

2. R. Courant, E. Isaacson and M. Rees, "On the Solution of Nonlinear Hyperbolic Differential Equations by Finite Differences," Communications on Pure and Applied Mathematics, Vol. 5, 243-255 (1952).

3. J. E. Dendy, "Two Methods of Galerkin Type Achieving Optimum L^2 Rates of Convergence for First-order Hyperbolics," SIAM J. Numerical Analysis, Vol. 11, 637-653 (1974).

4. B. Engquist and S. Osher, "Stable and Entropy Satisfying Approximations for Transonic Flow Calculations," Mathematics of Computation, Vol. 34, 45-75(1980).

5. S. K. Godunov, "A Finite Difference Method for the Numerical Calculation of Discontinuous Solutions of the Equations of Fluid Dynamics," Mat. Sb., Vol. 47, 271-290 (1959).

6. A. Harten, "High Resolution Schemes for Hyperbolic Conservation Laws," Journal of Computational Physics, Vol. 49, 357-393 (1983).

7. A. Harten, J. M. Hyman and P. D. Lax, "On Finite Difference Approximations and Entropy Conditions for Shocks," Communications on Pure and Applied Mathematics, Vol. 29, 297-322 (1976).

8. P. W. Hemker, "A Numerical Study of Stiff Two-point Boundary Value Problems," Thesis, Mathematisch Centrum, Amsterdam, 1977.

9. T.J.R. Hughes and A. N. Brooks, "A Multidimensional Upwind Scheme With no Crosswind Diffusion," pp. 19-35 in Finite Element Methods for Convection Dominated Flows (ed. T.J.R. Hughes), AMD-Vol. 34, ASME, New York, 1979.

10. T.J.R. Hughes and A. N. Brooks, "A Theoretical Framework for Petrov-Galerkin Methods with Discontinuous Weighting Functions: Application to the Streamline Upwind Procedure" pp. 47-65 in Finite Elements in Fluids, Vol. 4, (eds. R. H. Gallagher et al.) John Wiley and Sons, London, 1982.

11. C. Johnson, "Finite Element Methods for Convection-diffusion Problems," Fifth International Symposium on Computing Methods in Engineering and Applied Sciences, INRIA, Versailles, 1981.

12. K. W. Morton, "Generalized Galerkin Methods for Steady and Unsteady Problems," pp. 1-32 in Numerical Methods for Fluid Dynamics (eds. K. W. Morton and M. J. Baines), Academic Press, London, 1982.

13. K. W. Morton, "Shock Capturing, Fitting and Recovery," Numerical Analysis Report 5/82, Department of Mathematics, University of Reading, 1982. Presented at the 9th International Conference on Numerical Methods in Fluid Dynamics, Aachen, June 28th-July 2nd, 1982.

14. E. Murman and J. D. Cole, "Calculations of Plane Steady Transonic Flow," AIAA J., Vol. 9, 114-121 (1971).

15. U. Nävert, "A Finite Element Method for Convection-diffusion Problems," Ph.D. Thesis, Department of Computer Sciences, Chalmers University of Technology, Göteborg, Sweden, 1982.

16. S. Osher, "Shock Modelling in Aeronautics," pp. 179-217 in Numerical Methods for Fluid Dynamics (eds. K. W. Morton and M. J. Baines), Academic Press, London, 1982.

17. S. Osher and F. Solomon, "Upwind Difference Schemes for Hyperbolic Systems of Conservation Laws," Mathematics of Computation, Vol. 38, 339-374 (1982).

18. P. L. Roe, "Approximate Riemann Solvers, Parameter Vectors, and Difference Schemes," Journal of Computational Physics, Vol. 43, 357-372 (1981).

19. P. L. Roe, "Some Contributions to the Modelling of Discontinuous Flows." Presented at the AMS/SIAM Summer Seminar on Large Scale Computation in Fluid Mechanics, Scripps Institution of Oceanography, University of California, San Diego, June 26th to July 8th, 1983.

20. B. van Leer "On the Relationship Between the Upwind-differencing Schemes of Godunov, Engquist-Osher and Roe," ICASE Report No. 81-11, NASA Langley Research Center, Hampton, Virginia, 1981.

21. L. B. Wahlbin, "A Dissipative Galerkin Method for the Numerical Solution of First-order Hyperbolic Equations," pp. 147-169 in Mathematical Aspects of Finite Elements in Partial Differential Equations, (ed. Carl de Boor), Academic Press, New York, 1974.

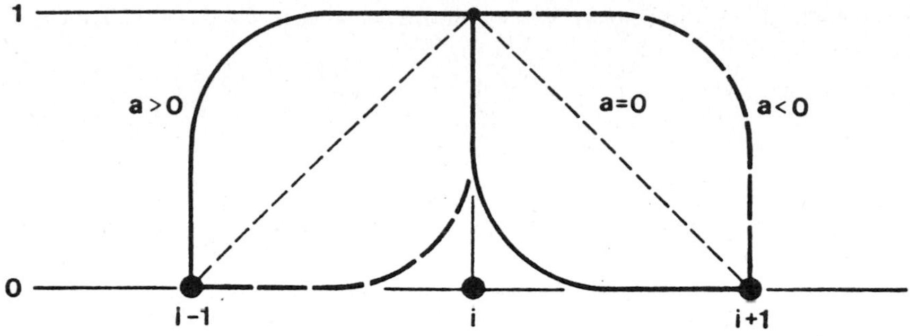

Figure 1. Hemker's weighting functions.

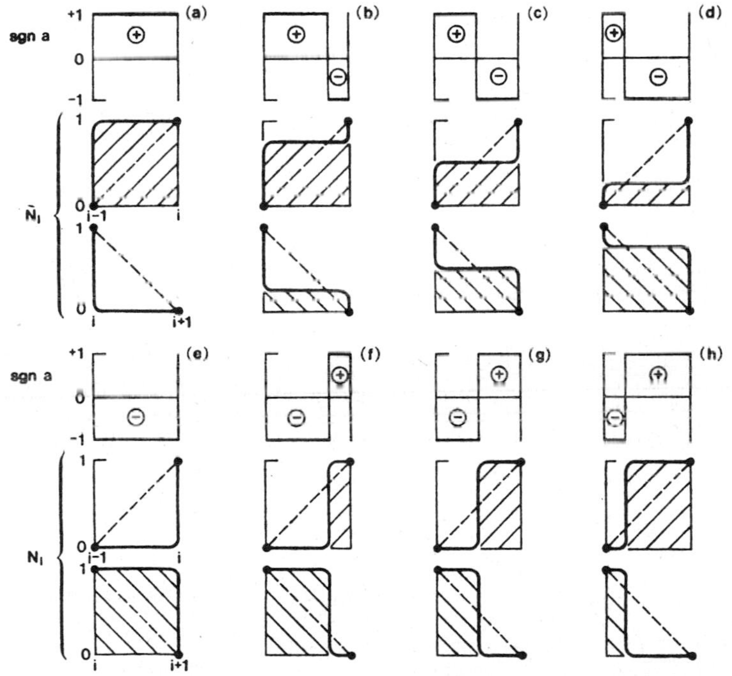

Figure 2. Weighting functions for the case $\varepsilon \to 0$. Case (a) is the supersonic case; Cases (b), (c) and (d) are transonic compressions; Case (e) is the subsonic case; and Cases (f), (g) and (h) are transonic expansion.

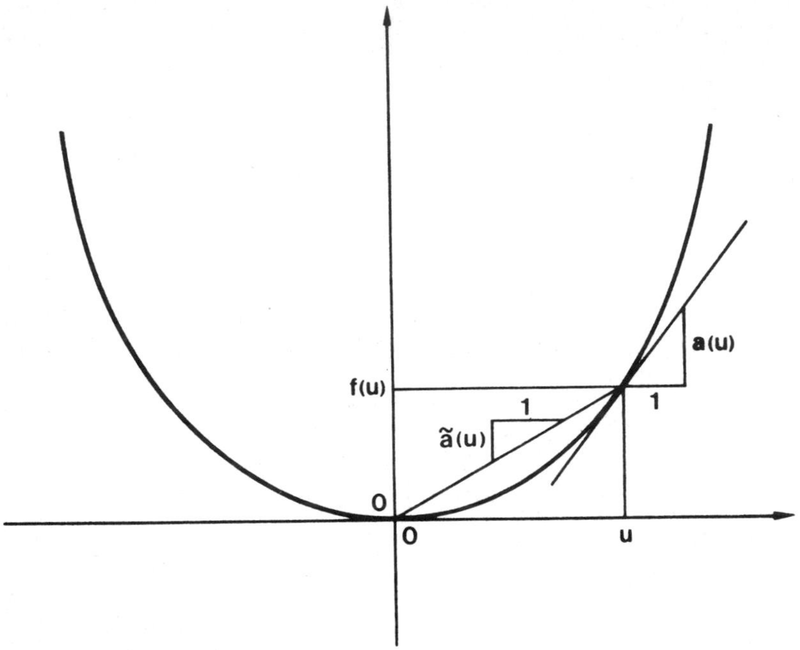

Figure 3. Convex flux function.

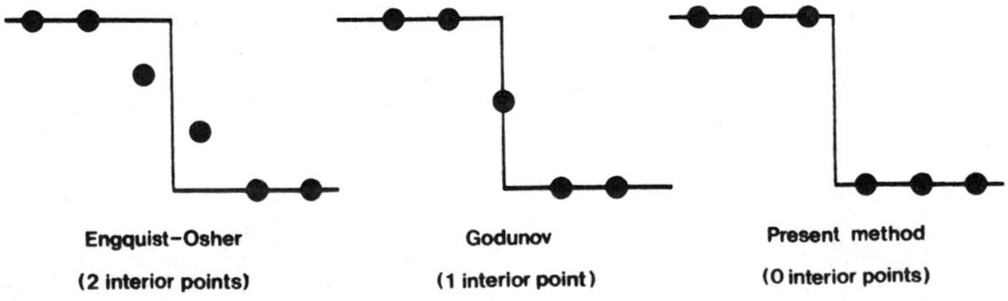

Figure 4. Resolution of stationary shocks.

Computing Methods in Applied Sciences and Engineering, VI
R. Glowinski and J.-L. Lions (Editors)
Elsevier Science Publishers B.V. (North-Holland)

TRIANGULAR FINITE ELEMENT METHODS FOR THE EULER EQUATIONS

Angrand F.*
Boulard V.**
Dervieux A.***
Périaux J.**
Vijayasundaram G.****

Triangular Finite Element Methods are developped for the unsteady Euler equations of compressible inviscid flows. A first-order Godunov-type scheme is presented and used for 3-D computations of internal flows. A second order Richtmyer-type implicite scheme is constructed ; an efficiency comparison with an earlier explicit version is developped with transonic 2-D test cases.

0. INTRODUCTION

During the last decade a large variety of transonic flow codes with both finite differences [1-3] and finite elements [4-6] appeared in the aerodynamic community. Although several fast and robust algorithms provided very efficient tools in an industrial context (aircraft design) the assumption of non vorticity (potential flow) appeared recently very restrictive and unadequate to more severe and realistic situation including flows with strong shocks vorticity and entropy production. Thus to the so called full potential scalar equation were favoured the Euler vector equation and an increasing interest is now devoted to the solution of the Euler equations for compressible inviscid fluids which are treated mainly via finite differences or finite volumes [7-13] and more recently via finite element methods [14-17]. The choice of a simplicial finite element approximation has proved very appropriate around a complete aircraft for solving the 3-D potential transonic equation [4] and therefore is used once more in the present computation.

This paper deals with simulations of transonic Euler flows using finite element type triangulations (respectively tetrahedrizations in 3-D).
With this choice, the mesh generation gets more manoeuvering abilities since non structured triangulations are allowed (the number of neighbors may differ from one node to another) and becomes more flexible and appropriate around local complicated configuration such as inlets/fuselage, wing/fuselage or fin/fuselage combinations.

Conversely the price to be paid consists of additional work both in approximation and method of solution.

In the following section 1, we shall review briefly an explicit Godunov type solver [Vijayasundaram][32] very interesting from an industrial viewpoint for its robust-

* INRIA-Rocquencourt, France
** AMD/BA, St Cloud, France
*** INRIA-Sophia-Antipolis, France
**** TIFR, Bangalore, India.

ness and refer to [B. Stoufflet] for efficient implicit extensions.

Section 2 focus on second order accuracy provided by a Richtmyer type scheme in which an implicit phase following Lerat [7], is introduced to increase the efficiency. The efficiency problem is particularly critical with finite elements involving expensive instructions of discrete equation and also non standard matrices (splitting into products of tridiagonal matrices is not possible) and an approach to deal with these difficulties is proposed.

1. A GODUNOV TYPE SCHEME

1.1. 3-D Extension

In this section we shall discuss the extension to 3-D case of the Godunov type introduced by G. Vijayasundaram ([32] thèse, for which some preliminary 2-D tests have been described in [19] .

The system to be solved is written

$$(1.1) \qquad \begin{aligned} & W_t + F(W)_x + G(W)_y = 0 && \text{2-D case} \\ & W_t + F(W)_x + G(W)_y + H(W)_z = 0 && \text{3-D case} \end{aligned}$$

We recall that the 2-D scheme can be presented as a Finite Volume scheme (for interior degrees of freedom) with cells constructed with medians of triangles (Fig. 1.1) ; then fluxes are computed with an upwind flux decomposition which can be written for two cells $\hat{i}$ and $\hat{j}$ around respectively vertices i and j as follows :

$$\Phi_{ij} = A^+_{\frac{i+j}{2}} W_i + A^-_{\frac{i+j}{2}} W_j$$

where W_i, W_j are the values of the unknown W at vertices i and j (Fig. 1.2)

$$A_{\frac{i+j}{2}} = (\eta_{x_{ij}} F' + \eta_{y_{ij}} G')\left(\frac{W_i+W_j}{2}\right)$$

$$\eta_{x_{ij}} = \int_{G^1_{ij} I_{ij} \cup I_{ij} G^2_{ij}} v_y \, d\sigma$$

$$\eta_{y_{ij}} = \int_{G^1_{ij} I_{ij} \cup I_{ij} G^2_{ij}} v_y \, d\sigma$$

$\vec{v} = (v_x, v_y)$ unitary vector normal to the boundary of the cell $\hat{i}$ pointing outward from $\hat{i}$.

The plus minus operators $A^+_{\frac{ij}{2}}$, $A^-_{\frac{ij}{2}}$ are defined as follows :

$$\Lambda_{\frac{i+j}{2}} = T^{-1}_{\frac{i+j}{2}} A_{\frac{i+j}{2}} T_{\frac{i+j}{2}} \quad \text{is diagonal}$$

$$A^+_{\frac{i+j}{2}} = T_{\frac{i+j}{2}} \Lambda^+_{\frac{i+j}{2}} T^{-1}_{\frac{i+j}{2}}$$

$$A^-_{\frac{i+j}{2}} = T_{\frac{i+j}{2}} \Lambda^+_{\frac{i+j}{2}} T^{-1}_{\frac{i+j}{2}}$$

$$\Lambda_{\frac{i+j}{2}} = \text{diag}\ (l_k)$$

$$\Lambda^+_{ij} = \text{diag}\ (\max\ (\lambda_k, 0))$$

For the 3-D case, the elementary cells î are constructed with median planes (see Fig. 1.3) and are decomposed in flux integration surfaces between neighbor vertices i,j (Fig. 1.4)

$$\Phi_{ij} = \frac{A^+_{i+j}}{2} W_i + \frac{A^-_{i+j}}{2} W_j$$

$$\frac{A_{i+l}}{2} = (n_{x_{ij}} F' + n_{y_{ij}} G' + n_{z_{ij}} H')\ (\frac{W_i + W_j}{2})$$

$$n_{x_{ij}} = \int_{\Sigma_{ij}} V_x\, d\sigma$$

(similar definition for $n_{y_{ij}}$ and $n_{z_{ij}}$).

The construction of the flux integration surface Σ_{ij} (a polyhedron) is described in Fig. 1.4. Indead the computation of the integrals $n_{x_{ij}}$, $n_{y_{ij}}$, $n_{z_{ij}}$ is done one for all for a given tetrahedrization.For a more details description, see V. Doulard [21] .

2. A RICHTMYER SCHEME

2.1. Theoretical statements for the advection model

2.1.1. Presentation of the scheme

We want to solve the following system

$$\text{(2.1)}\quad \begin{aligned} &\frac{\partial W}{\partial t} + V_1 \frac{\partial W}{\partial x} + V_2 \frac{\partial W}{\partial y} = 0 \quad \text{in } \mathbb{R}^2 \\ &W(x,0) = W_0(x) \end{aligned}$$

where $\vec{V} = (V_1, V_2)$ is a constant given velocity.

A classical explicit second-order two-level time discretization is the following

$$\text{(2.2)}\quad \begin{aligned} W^{n+1}(x) &= W^n(x) - \Delta t (V_1 W^n_x + V_2 W^n_y) \\ &+ \frac{\Delta t^2}{2} [V_1 (V_1 W^n_x + V_2 W^n_y)_x + V_2 (V_1 W^n_y)_y] . \end{aligned}$$

Let $\mathcal{H}$ be a set of strictly positive parameters with zero in its closure and let $(\mathcal{T}_h)_{h \in \mathcal{H}}$ be a family of triangulations such that h is the length of the largest segment of $\mathcal{T}_h$. We consider the following spaces

$$H_h = \{v \in L^2(\mathbb{R}^2)\ ;\ v \text{ is continuous }; v \text{ is linear on every triangle T of } \mathcal{T}_h\}\ ;$$

$$V_h = \{v \in H_h \cap H^1(\mathbb{R}^2)\} .$$

Then a Galerkin-type variational P_1 space discretization of (2.2) with consistent mass matrix is

$$\text{(2.3)}\quad \begin{cases} W_h^{n+1} \in V_h, \text{ and } \forall v \in V_h \\ \iint (W_h^{n+1} - W_h^n) v \, dxdy = \\ - \Delta t \iint v(V_1 \frac{\partial W_h^n}{\partial x} + V_2 \frac{\partial W_h^n}{\partial y}) \, dxdy \\ + \frac{\Delta t^2}{2} \iint (V_1 \frac{\partial W_h^n}{\partial x} + V_2 \frac{\partial W_h^n}{\partial y})(V_1 \frac{\partial v}{\partial x} + V_2 \frac{\partial v}{\partial y}) \, dxdy \end{cases}$$

where the sums $\iint$ are taken on $\mathbb{R}^2$.

We shall be also interested by the mass-lumped variant of (2.3) ; for the construction of this variant we need the following notations (h is fixed) :

- For any vertex A of $\mathcal{T}_h$, the integration zone $\hat{A}$ is defined as in Section 1 (Figure 1.1).

- S_o is the approximation space of functions which are constant on each $\hat{A}$:

$$S_o = \{v \in L^2(\mathbb{R}^2),\ v_{|\hat{A}} = \text{const.},\ \forall A \text{ vertex of } \mathcal{T}_h\}$$

- $\mathcal{S}_o$ is the trivial projection from H_h to S_o :

$$\text{(2.4)}\quad \begin{cases} \forall v \in H_h,\ \mathcal{S}_o v \in S_o \quad \text{and} \\ \mathcal{S}_o v_{|\hat{A}} = v(A), \quad \forall A \quad \text{vertex of } \mathcal{T}_h . \end{cases}$$

Then the mass-lumped variant of (2.3) is

$$\text{(2.5)}\quad \begin{cases} W_h^{n+1} \quad V_h, \text{ and } \forall v \quad V_h \\ \iint \mathcal{S}_o(W_h^{n+1} - W_h^n)\, \mathcal{S}_o v \, dxdy = \\ - \Delta t \iint v(V_1 \frac{W_h^n}{x} + V_2 \frac{W_h^n}{y}) \, dxdy \\ + \frac{\Delta t^2}{2} \iint (V_1 \frac{\partial W_h^n}{\partial x} + V_2 \frac{\partial W_h^n}{\partial y})(V_1 \frac{\partial v}{\partial x} + V_2 \frac{\partial v}{\partial y}) \, dx\, dy . \end{cases}$$

Let us study the stability of the two above schemes.

2.1.2. Energy analysis

We give an a priori L^2 estimate for the solution (W^n), uniformly with respect to the space increment h ; for this purpose, we need some uniform assumption concerning the triangulations :

The family $(\mathcal{T}_h)$ of triangulations, h taking all the values in $\mathcal{H}$ is assumed to satisfy

$\exists K' > 0$ such that $\forall h \in \mathcal{H}$, $\forall T_+$ and T_-

$$(2.6)\quad \begin{cases} \exists K' > 0 \text{ such that } \forall h \in \mathcal{H}, \ \forall T_+ \text{ and } T_- \\ \text{belonging to } \mathcal{T}_h \text{ and having exactly one common} \\ \text{side, we have} \\ \left| \dfrac{\text{area}(T_+) - \text{area}(T_-)}{\text{area}(T_+)} \right| \le K'h \end{cases}$$

Proposition 1 : Under assumption (2.6), for a discrete initial condition bounded as follows

$$\iint |W_h^o|^2 \, dx\, dy = N_1 < +\infty$$

$$\iint |\delta_o W_h^o|^2 \, dx\, dy = N_2 < +\infty$$

and for a bounded time interval $[0,T]$, there exists two real positive constants $K_1(K', N_1, T\, \|\vec{V}\|)$ and $K_2(K', N_2, T, \|\vec{V}\|)$ such that for any h in $\mathcal{H}$, any Δt satisfying the Courant-type condition

$$(2.7)\qquad \|\vec{V}\| \quad \Delta t < \frac{2}{3} \times \text{ smallest altitude of } \mathcal{T}_h ,$$

we have for scheme (2.3)

$$(2.8)\qquad \iint |W_h^n|^2 \, dx\, dy \le K_1$$

and for scheme (2.5)

$$(2.9)\qquad \iint |\delta_o \, W_h^n| \, dx\, dy \le K_2 .$$

For the mass-lumped variant, the proof can be found in [18] ; it is similar for the consistent mass matrix variant.

In practice, the limitation (2.7) is significant for the consistent mass matrix variant, for the mass-lumped variant we have generally used a less restrictive condition obtained by the Fourier analysis.

Proposition 2 : In the particular case of the periodic "seven point" triangulation $\hat{\mathcal{T}}_h$ defined by (see Figure [1.6])

$$\begin{cases} \hat{\mathcal{T}}_h = \hat{\mathcal{T}}_{1h} \cup \hat{\mathcal{T}}_{2h} \\ \hat{\mathcal{T}}_{1h} = \text{ set of triangles with some vertices} \\ \qquad \{[ih,jh] \ , \ [(i+1)h,(j+1)h] \ , \ [(i+1)h,jh]\} \quad (i,j)\in Z^2 \\ \hat{\mathcal{T}}_{2h} = \text{ set of triangles with some vertices} \\ \qquad \{[ih,jh] \ , \ [ih,(j+1)h] \ , \ [(i+1)h,(j+1)h]\} \ (i,j)\in Z^2 \end{cases}$$

The mass-lumped scheme (2.5) is L^2-stable for the Courant type condition

$$(2.10)\qquad \|V\| \ \ \Delta t < 1 \ \times \quad \text{smallest altitude of } \hat{\mathcal{T}}_h$$

The choice between the consistent mass matrix version (2.3) and the mass lumped one (2.5) needs the following comments.
We can be interested to use these schemes for two kinds of simulation

- transient simulations
- unsteady solutions of steady phenomena

The mass-lumped variant is cheaper for each time step because no non diagonal system has to be solved, the maximum allowed time step is greater (with (2.10) instead of (2.7)), and the scheme is more dissipative than the other one ; so since the two schemes give the same stationnary solution we think that the mass-lumped variant is clearly more efficient as an unsteady solver for steady simulation.

On the other hand, the consistent mass matrix scheme is theoretically and practically more accurate, with better dispersion properties (see Gresho [24], Donea-Giuliani [22] for discussions of this point). Then we think that the consistent mass matrix scheme should be prefered to obtain accurate transient simulations.

2.2. Lerat type implicit variant

We shall show that Lerat implicit scheme (see Lerat [26], Lerat-Sides Daru [28], Lerat AIAA [7]) can be extended in a natural way to Galerkin finite elements

2.2.1. Definition

We transform the explicit schemes (2.3), (2.5) as follows

$$(2.11)\quad \begin{cases} W^{n+1} \in V_h \quad \text{and} \quad \forall v \in V_h \\ \iint \mathcal{B}_0(W_h^{n+1} - W_h^n).\mathcal{B}_0 \; v \, dx\, dy \\ \frac{\Delta t^2}{2} \iint (\vec{V}.\vec{\nabla}(W_h^{n+1} - W_h^n)\} (\vec{V}.\vec{\nabla} v)\, dx\, dy \\ - \;\Delta t \iiint v\,(\vec{V}.\vec{\nabla}\, W_h^n) \\ - \frac{t^2}{2} \iint (\vec{V}.\vec{\nabla} W_h^n)(\vec{V}.\vec{\nabla} v)\, dx\, dy \end{cases}$$

where $\mathcal{B}_0$ holds for the identity operator (consistent matrix variant) or $\mathcal{S}_0$ (mass lumped variant).

We can remark that (2.11) is a delta-formulation with

$$W_h^n = W_h^{n+1} - W_h^n ,$$

and right hand side of (2.11) is the same than (2.3) and (2.5).
Let us study the stability of this implicit scheme.

2.1.2. Stability Analysis

Proposition 3 : The implicit scheme (2.11) is unconditionally L^2 stable for both cases $\mathcal{B}_0$ = Id and $\mathcal{B}_0 = \mathcal{S}_0$.

Proof :

We take $v = W_h^{n+1}$

We remark that

$$(b-a)b = \tfrac{1}{2}(b^2 - a^2 + (b-a)^2)$$

So we obtain

$$(2.12)\quad \begin{cases} 0 = \frac{1}{2}\iint|\mathcal{B}_o W_h^{n+1}|^2\,dxdy = \frac{1}{2}\iint|\mathcal{B}_o W_h^n|^2\,dxdy \\ +\frac{1}{2}\iint|\mathcal{B}_o W_h^{n+1} - \mathcal{B}_o W_h^n|^2\,dx\,dy \\ +\frac{\Delta t^2}{2}\iint(\vec{V}.\,\vec{\nabla}W_h^{n+1})(\vec{V}.\,\vec{\nabla}W_h^{n+1})dxdy \\ +\Delta t\iint W_h^{n+1}(\vec{V}.\,\vec{\nabla}W_h^n)dxdy \end{cases}$$

We use an integration by parts for the last term.

$$\iint W_h^{n+1}(\vec{V}.\,\vec{\nabla}W_h^n)dxdy = -\iint W_h^n(\vec{V}.\,\vec{\nabla}W_h^{n+1})dxdy$$

This formula is true because $\vec{V}$ is a constant vector.

We note also that

$$\iint W_h^{n+1}(\vec{V}.\vec{\nabla}W_h^{n+1})dxdy = \iint(W_h^{n+1})(\vec{V}.\vec{\nabla}W_h^{n+1})dxdy = \iint(W_h^{n+1}-W_h^n)(\vec{V}.\vec{\nabla}W_h^{n+1})dxdy$$

But $\vec{V}.\,\vec{\nabla}W_h^{n+1}$ is a constant on each triangle.
So we have also

$$\iint W_h^{n+1}(\vec{V}.\,\vec{\nabla}W_h^{n+1})dxdy = \iint \mathcal{B}_o(W_h^{n+1} - W_h^n)(\vec{V}.\,\vec{\nabla}W_h^{n+1})dxdy \quad ;$$

Let $\quad \mathcal{C} = \Delta t\,\vec{V}.\,\vec{\nabla}W_h^{n+1}$

(2.12) becomes

$$0 = \frac{1}{2}\iint\{|\mathcal{B}_o W_h^{n+1}|^2 - |\mathcal{B}_o W_h^n|^2 + |\mathcal{B}_o W_h^{n+1} - \mathcal{B}_o W_h^n|^2\}\,dxdy$$

$$(2.13)\qquad +\frac{1}{2}\iint \mathcal{C}^2\,dxdy + \iint .\,\mathcal{B}_o(W_h^{n+1} - W_h^n)\,\mathcal{C}\,dxdy$$

Let $\Theta = \mathcal{B}_o W_h^{n+1} - \mathcal{B}_o W_h^n$

From (2.13) we derive

$$\frac{1}{2}\iint|\mathcal{B}_o W_h^{n+1}|dxdy - \frac{1}{2}\iint|\mathcal{B}_o W_h^n|^2\,dxdy =$$

$$-\frac{1}{2}\iint\Theta^2\,dxdy - \frac{1}{2}\iint\mathcal{C}^2\,dxdy - \iint\Theta\,\mathcal{C}\,dxdy$$

$$= \frac{1}{2}\iint(-\Theta^2 - \mathcal{C}^2 + \Theta^2 + \mathcal{C}^2 - (\Theta+\mathcal{C})^2)dxdy$$

$$= -\frac{1}{2}\iint(\Theta+\mathcal{C})dxdy$$

So

$$\iint|\mathcal{B}_o W_h^{n+1}|dxdy - \iint|\mathcal{B}_o W_h^n|^2\,dxdy \le 0$$

which states a strong L^2-stability.

Remark 2.1

Let us assume that the velocity $\vec{V}$ has a vanishing vertical component then the Lax-Vendroff implicit term $\frac{\Delta t^2}{2}\iint [\vec{V}.\vec{\nabla}(W_h^{n+1}-W_h^n)][\vec{V}.\vec{\nabla}v]\,dxdy$ contains only a $\frac{\partial^2}{\partial x^2}$ derivative. For easy boundary conditions it could be interesting to add $\frac{\partial^2}{\partial y^2}$ derivative to get an elliptic fully two dimensional operator. A simple modification of the above proofs shows that the stability still holds.
More generally, with a time dependent velocity, we may replace the Lax-Wendroff implicit term by

$$\frac{\Delta t^2}{2}\iint \rho\,\vec{\nabla}(W_h^{n+1}-W_h^n).\vec{\nabla}\,v\,dxdy$$

with $\rho \geq \sup_t \|V\|$.

Remark 2.2.

The proof of proposition 3 still holds for the 3-D case and, with the consistent mass matrix for every Galerkin f.e.m. (with exact integration).

2.3. Euler solver

The Euler system for inviscid compressible flows (see Landau Lifchitz [33]) is written

$$\text{(2.14)}\quad \begin{aligned} &W_t + F(W)_x + G(W)_y = 0 \quad \text{in } \Omega \\ &+ \text{boundary conditions on } \partial\Omega \end{aligned}$$

where $W(x,y,t)$ is a vector of $\mathbb{R}^4$.

To solve (2.14), scheme (2.11) is extended in a deltaform splitted into a physical explicit phase and a mathematical implicit one.

2.3.1. Physical phase

The extension to systems is done by a two step process of Richtmyer type. The scheme is the following

Step 1 ; Predictor :

$$(2.15)_1 \quad \begin{cases} \forall T \in \mathcal{T}_h, \text{ and for } k = 1,2,3,4 \\ \tilde{W}_k(T) = \dfrac{1}{\text{area}(T)} \{\iint_T W_k^n\,dxdy \\ \qquad - \alpha\Delta t \int_{\partial T} [F_k(W^n)n_x + G_k(W^n)n_y]\,d\sigma\} \end{cases}$$

Step 2 ; Corrector :

$$\hat{W}^n \in (V_h)^4, \text{ and } \forall\phi \in (V_h)^4; \quad \forall k = 1,2,3,4,$$

(2.15)$_2$

$$\begin{cases} \iint_\Omega \mathcal{B}_o\left[\frac{W_k^n - W_k^n}{\Delta t}\right]\mathcal{B}_o\, \phi_k \, dxdy = \\ \quad \iint_\Omega \{\beta_1[F_k(W^n)\frac{\partial\phi k}{\partial x} + G_k(W^n)\frac{\partial\phi k}{\partial y}] \\ \quad + \beta_2\,[F_k(\tilde{W})\frac{\partial\phi k}{\partial x} + \tilde{G}_k(\tilde{W})\frac{\partial\phi k}{\partial y}]\}\, dxdy \\ + \iint_\Omega X\, f(W^n)\,\nabla(W_k^n).\nabla\phi \quad dxdy \\ - \int_{\partial\Omega} \phi_k[F_k(W^n)n_x + G_k(W^n)n_y] \quad d\sigma \end{cases}$$

With (following Lerat and Peyret [27])

$$\alpha = 1 + \frac{\sqrt{5}}{2}, \quad \beta_1 = \frac{2\alpha-1}{2\alpha}, \quad \beta_2 = \frac{1}{2\alpha} \qquad \square$$

The second integral of the right hand side is an artificial viscosity term constructed as in earlier papers (Angrand-Dervieux [18]).
Wall conditions are constructed by integrating the last term of (2.15)$_2$ with only the pressure terms, and for boundary conditions at infinity the integrand of this term is computed from both value at infinity (entropy, enthalpy and flow direction for inflow ; pressure for outflow) and interpolated values (pressure for inflow ; entropy, enthalpy and flow direction for outflow).
We used neither extrapolation nor upwinding.

2.3.2. Mathematical phase

Following Lerat-Sides-Daru [28] and also Remark 2.1, mathematical phase needs solutions of uncoupled second order elliptic systems :

(2.16)

$$\begin{cases} W^{n+1} \in (V_h)^4, \quad \forall\phi \in V_h^4 \quad \forall k = 1,2,3,4 \\ \iint \mathcal{B}_o(W_k^{n+1} - W_k^n)\mathcal{B}_o \quad \phi_k \, dxdy \\ + \beta\,\frac{\Delta t^2}{2}\iint \lambda\,\nabla(W_k^{n+1} - W_k^n)\,\nabla\phi_k \, dxdy \\ + \chi\iint f(W^n)\,\nabla(W_k^{n+1} - W_k^n)\,\nabla\phi_k \, dxdy = \\ \iint \mathcal{B}_o(W_k - W_k^n)\mathcal{B}_o \quad \phi_k \quad dxdy \end{cases}$$

λ is a positive real constant greater than the faster wave speed :

(2.17) $\qquad \lambda > \text{Max (velocity + sound speed)}$

β is a parameter of the scheme. From Lerat [7] $\beta = 1/2$ is good for accuracy but we have not established the strong stability in section 2.1.2. for $\beta < 1$. So we use $\beta = 1$.

For efficiency, we freeze the matrix for several time steps as long as (2.17) remains true together with a similar condition on the viscosity coefficient f(w) to insure the implicitness of the artificial viscosity term.

Remark 2.3.

The boundary conditions of this implicit phase are purely artificial. We found that the choice of Neumann conditions induced no damage to the boundary accuracy.

3. Numerical results

3.1. First 3-D experiments with the Godunov scheme

Because tetrahedra are fundamentally 3-D geometrical figures, we have the following property :

Even with regular meshes the 3-D code cannot give, for a 2-D problem the same result than the 2-D code. Then our first 3-D experiments consist of a comparison, for a given 2-D problem, of 2-D and 3-D results, in order to evaluate accuracy and efficiency of the 3-D version. The two codes are tested with a transonic flow in channel with a circular bump. For the definition of this problem, we refer to Viviand-Rizzi [29]. The 2-D triangulation (1512 vertices) is comparable with the finite difference mesh 72 × 21 proposed by the GAMM workshop [Ibid]. The 3-D triangulation is obtained from 3 planes of 2-D triangulation (Figure 3.1) ; then we have 3 × 1512 vertices.
For boundary conditions, wall conditions are applied on the z equal constant boundary plane.
Let us recall that, as in earlier works, conditions at infinity (inflow, outflow) are trivialy implemented by forcing all quantities (this is possible because the scheme is conveniently upwinded).

Accuracy evaluation

Although z derivatives are roughly discretized (the triangulation has only 3 ranges of vertices in that direction) it has been observed a good accordance between the 2-D and 3-D simulations. To illustrate this we present the $\text{Mach}_\infty = .85$ results. The only noticeable difference lies in some spiriously created entropy which modifies lightly the aspect of the isomach curves after the bump.Figures (3.2, 3.3).

Convergence properties

The Euler explicit (forward) scheme is used for time integration until obtention of steady solution. The comparison holds for four cases depending of the mach at infinity : $M_\infty = .5$ (completely subsonic), $M_\infty = .85$ (transonic ; subsonic at infinity) $M_\infty = 1.25$ and $M_\infty = 1.50$ (supersonic at infinity),Figures (3.4, 3.5).
The convergences of 2-D and 3-D cases are quite close to each others : for $M_\infty = .5$ we see the main difference, the convergence is quite fast, but a little slower for the 3-D case likely due to the reflections between the {z = constant} walls.
A relatively slow convergence is observed for both cases $M_\infty = .85$, $M_\infty = 1.25$ as well as for the 2-D as for the 3-D case.
When the supersonic domain is strongly dominant ($M_\infty = 2.5$) we get again a faster convergence.
To sum up, we have constructed a consistent 3-D generalization of 2-D Godunov type scheme ; a fair accordance is observed between the 2-D code and its 3-D extension for accuracy and efficiency.

3.2. Numerical results for the second order scheme

The shock tube simulation in the conditions of SOD [30] is used to test the transient accuracy of the scheme. The comparison of the temperature distributions are useful for the evaluation of contact discontinuity resolution. The consistent mass matrix scheme together with the implicit one with β equal 1/2 are clearly the best, with nearly constant values on each side of the contact discontinuity(Figure 3.6). For shock resolution we look at velocity distributions. The difference between the schemes are still more striking. The value β equal 1/2 appears again to be optimal, while symmetrical undershoots and overshoots are observed for the two other cases (β = 0, β = 1). The shock is nearly free of oscillation and free of smearing (only one point in the shock, compare with the explicit mass-lumped case)(Figure 3.7).

For steady computations (with the unsteady solver method) a realistic evaluation of the speed-up of the implicit version will be presented by comparing with the non-consistent explicit solver with local time step ; the local time step procedure is mainly efficient when the difference of size between the smallest and the largest triangles is important.
We present here two tests. The first test is the simulation of an internal flow in a channel with a quite little variation of the sizes of triangles. The second one is an external flow past a cylinder with high variation of the sizes between triangles at infinity and triangles near the body.

GAMM channel with circular bump

We use the same 2-D triangulation than in Section 1.2 with 1512 vertices (6048 unknowns). The memory core requirement for the matrix is about 35000 real variables. Figure 3.8 presents a comparison of the C_p distribution for the implicit and explicit simulations, with complete convergence.
In table 3.1, we present a comparison of implicit and explicit cases

	average CPU** cost: of 1 iteration	CFL	number of iterations to divide the residu* by 10^3	CPU cost to divide the residu by 10^3
explicit with local time step	0.21 s	.6	3000	11,02 mn
implicit	0.32 s	15	800	4,5 mn ratio = 2,4

Table 3.1

Convergences curves are presented in Figure 3.9. We have found the convergence with CFL between 10 and 20 are similar. We can use CFL greater than 40 but the solution is too smooth. Converged solutions for implicit and explicit scheme are similar. We can note that for CFL = .8, explicit scheme does not converge. The average CPU cost for the implicit iteration does not contains the several construction and factorization of matrices (less than 10 for one complete simulation). The division by 1000 of the residu is sufficient to obtain a converged solution.

Flow past a cylinder

The mach at infinity is .45, see Jameson [25] , we use a triangulation with 845 vertices (3380 unknowns). The memory core requirement for the matrix is about 17000 real variables. Figure 3.10 represents a comparison of the C_p distribution for the implicit and explicit simulations, with complete convergence.

** CRAY 1-S, Computational Costs supported by GCCVR.

* Residu equals to mean square root of density time derivative.

	average CPU cost of 1 iteration	CFL	number of iterations to divide the residu by 1000	CPU cost to divide the residu by 10^3
explicit with local time step	0,14	.6	3200	7,52 mn
implicit	0,31	10	500	2,6 mn , ratio ≃3

Table 3.2

From Table 3.2 we see that the external simulation exhibits a better ratio of efficiency. This is likely due to a less close far field troncature. Thus this second test provide a confirmation of the efficiency of the implicit solver.Figure 3.11 reprents several tests CFL =.6 is the explicit scheme. For CFL = 2, the implicit scheme is used as usually. To have a greater CFL, we have to increase continualy from 1 to 10 (or 5) in 50 iterations. When we compare the converged solutions, we can see that the implicit one is better than the explicit one.

CONCLUSION

Two schemes have been presented :

The Godunov-type scheme has proved his robustness and reliabily as well in 2-D as 3-D configurations, in sub., trans. and supersonic cases. With this scheme, no parameter or boundary condition adjustment is needed so that the computer program can be easily used as a "black box".

Concerning the second-order implicit scheme, we think that it represents a good trade-off between

- accuracy
- cost expense
- memory requirements
- facility of writing the computer program

Further experiments are now in progress for a precise evaluation of these two schemes within a wider range of problems.

BIBLIOGRAPHY

[1] JAMESON A.,"Numerical calculation of a transonic flow past a swept wing by a finite volume method", Computing Methods in Applied Sciences and Engineering, Part II (eds. R. Glowinski and J.L. Lions) (1977) ; Lecture Notes in Physics, Vol. 91, Springer, Berlin, pp. 125-148 (1979).

[2] HOLST T., "An implicit algorithm for the transonic fullpotential equation in conservative forme" ; in Computing Methods in Applied Sciences and Engineering (eds. R. Glowinski and J.L. Lions), North Holland, Amsterdam, pp. 157-174 (1980).

[3] B. WEDAN and J.C. SOUTH, "A Method for solving the transonic Full potentiel equation for general Confirmations", 6th AIAA Conference, DANVERS, Massachussetts, 1983.

[4] Glowinski R., Périaux J., "Finite element, least squares and domain decomposition methods for the numerical solution of nonlinear problems in fluid dynamics", Lectur Notes, CIME, July 1983, Como, Italy.

[5] AMARA M., "Analyse de methodes d'éléments finis pour des écoulements transsoniques" thèse. Paris 6 1983.

[6] AKAY H.U, ECER A., "Finite element analysis of transonic flows in highly staggered Cascades" AIAA paper 81-0210, 1981.

[7] LERAT A., "Implicit Methods of second-order accuracy for the Euler equations", AIAA paper n° 83-1925, 1983

[8] JAMESON A., Schmidt W., Turkel E., "Steady state solution of the Euler equations for transonic flow", AIAA Paper 81-1259, 1981.

[9] RIZZI A.W., "Damped Euler-equation Method to compute Transonic Flows around Wing-Body combinations", AIAA Journal, Vol. 20, No 10, 1982, pp. 1321-1328.

[10] OSHER S., SOLOMON F., Upwind difference schemes for hyperbolic systems of conservation laws, J. Math. Computation, (April 1982).

[11] YEE H.C., WARHING R.F., HARTEN A., Implicit Total Variation Diminishing (TVD) Schemes for Steady-state Calculations AIAA Paper 83-1902, 1983.

[12] STEGER J., WARMING R.F., Flux vector slitting for the inviscid gas dynamic equations with applications to finite difference methods, J. Comp. Phys., Vol. 40, No 2, pp. 263-293 (1981).

[13] VANLEER B., Towards the ultimate conservative difference scheme III, Upstream centered finite difference schemes for ideal compressible flow, J. Comp. Phys. 23, pp. 263-275 (1975).

[14] BAKER A.J., SOLIMAN M.O., "An accurate and efficient Finite Element Euler equation algorithm" in E. Krause(Ed.). Eight International Conference on Numerical Methods in Fluid Dynamics (Aachen, 1982), Lecture Notes in Physics, Vol. 170, Springer, 1982.

[15] BORREL M., MORICE Ph., "A Lagrangian-Eulerian approach to the computation of unsteady transonic flows", in H. Viviand (Ed.), Proceedings of the Fourth GAMM-Conference on Numerical Methods in Fluid Mechanics, Vieweg and Shon, Braunschweig/Wiesbaden, 1982.

[16] AKAY H.U., Ecer A., "Application of a Finite Element Algorithm for the solution of Steady Transonic Euler Equations", AIAA Paper 82-0970, 1982.

[17] HUGUES T.J.R., TEZDUYAR T.E., BROOKS A.N., "A Petrov-Calerkin finite elemente formulation for systems of conservation laws with special reference to the compressible Euler equations", Proceeding of the IMA Conference on Numerical Methods in Fluids Dynamics, March 29-31 1982, University of Reading, to appear, Academic Press.

[18] ANGRAND F., DERVIEUX A., Some explicit triangular Finite Element Schemes for the Euler equations, to appear in I.J. for Num. Methods in Fluids.

[19] ANGRAND F., BOULARD V., DERVIEUX A., PERIAUX J., VIJAYASUNDARAM G., Transonic Euler Simulations by means of Finite Element explicite Schemes, AIAA paper 83-1924, 1983.

[20] ANGRAND F., DERVIEUX A., LOTH L., VISAYASUNDARAM G., Simulation of Euler Transonic flows by means of explicit Finite Element-type schemes, INRIA Report no

[21] BOULARD V., Thesis, University of Paris VI (to appear)

[22] DONEA J, GUILIANI S. A simple method to generale high-order accurate convection operator for explicit schemes based on linear Finite elements, I.J. for Num. Math. in Fluids, 1, 63-79 (1981).

[23] GLOWINSKI R., OSHER (Ed.) workshop on Numerical Methods for the Euler Equations for compressible inviscid fluids Siam (to appear).

[24] GRESHO Ph. M, LEE R.L., SANI R.L., Advection-dominated flows with emphasis on the consequences of mass-lumping, in Finite Elements in Fluids, 3, Wiley, 1979-80.

[25] JAMESON A., SCHMIDT W., TURKEL E., Steady-State solution of the Euler equations for transonic flow, AIAA Paper 81-1259, 1981.

[26] LERAT A. Sur le calcul des solutions faibles des systèmes hyperboliques de loi de conservation à l'aide des schémas aux différences, Thesis, University of Paris VI, 1981.

[27] LERAT A., PEYRET R., Sur le choix de schémas aux différences du second ordre fournissant des profils de choc sans oscillations, Comptes Rendus Acad. SC. Paris, 277 A, 363-366 (1973).

[28] LERAT A., SIDES J., DARU V., An Implicit Finte-Volume Method for Solving the Euler equations, Lecture Notes in Physics, 170, 343-349 (1982).

[29] RIZZI A.W., VIVIAND H. (Eds.) Numerical methods for the computation of inviscid transonic flows with shock waves, Vieweg and Sohn, Braunschweig, Wiesbaden, 1981.

[30] SOD GA., "A survey of several finite difference methods for systems of nonlineaire hyperbolic conservation laws", J. Comp. Phys. 27, 1-31 (1978).

[31] STOUFFLET B. Thesis, University of Paris VI (to appear).
STOUFFLET B. Contribution to [23] .

[32] VIJAYASUNDARAM G., Résolution numérique des équations d'Euler pour des écoulements transsoniques avec un schéma de Godunov en Eléments Finis, Thesis, University of Paris VI, 1982-1983.

[33] LANDAU, LIFCHITZ, Mécanique des fluides ; éditions de Moscou (1971).

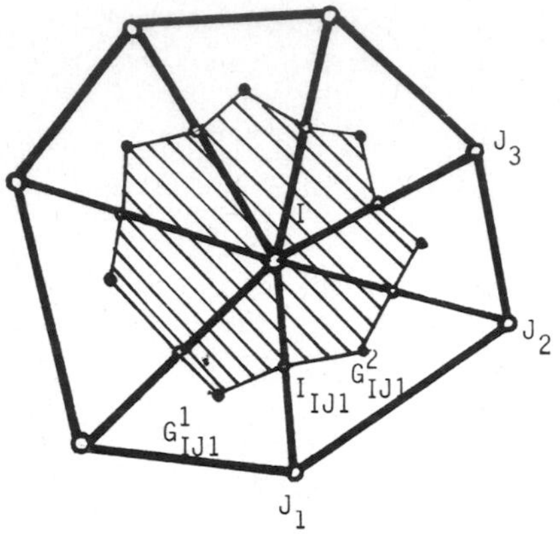

Fig. 1.1. Intégration zone around vertice I

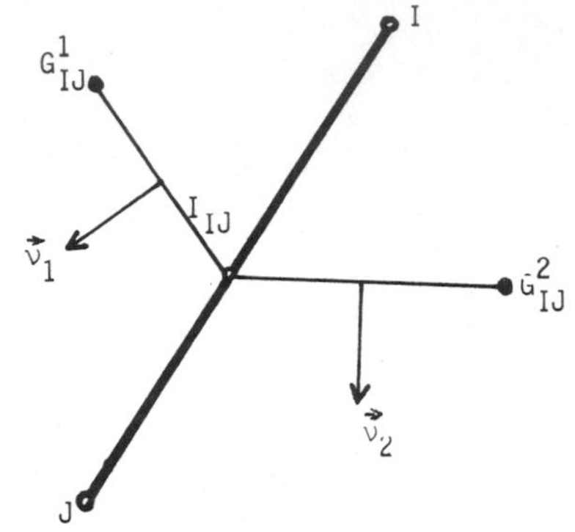

Fig. 1.2. Flux integration curve between I and J

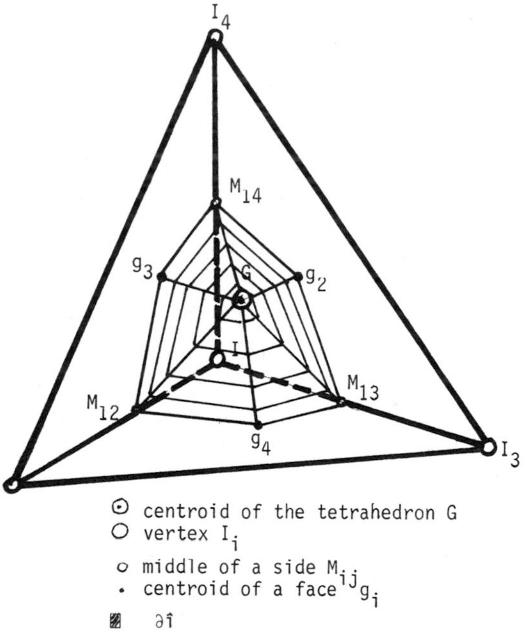

Fig.1.4. Definition of the integration zone $\hat{I}$ around vertex I

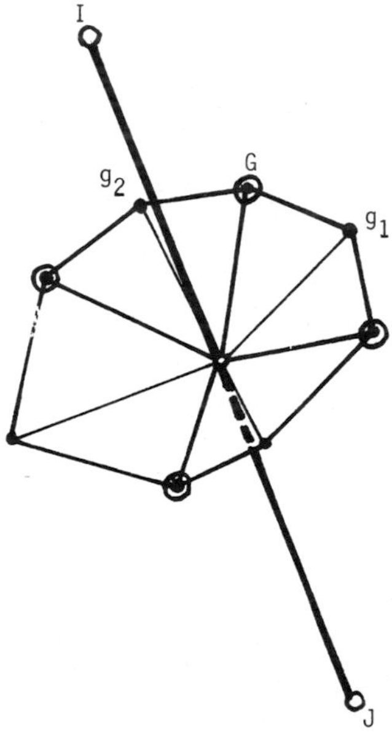

Fig. 1.5. Flux integration surface between I and J

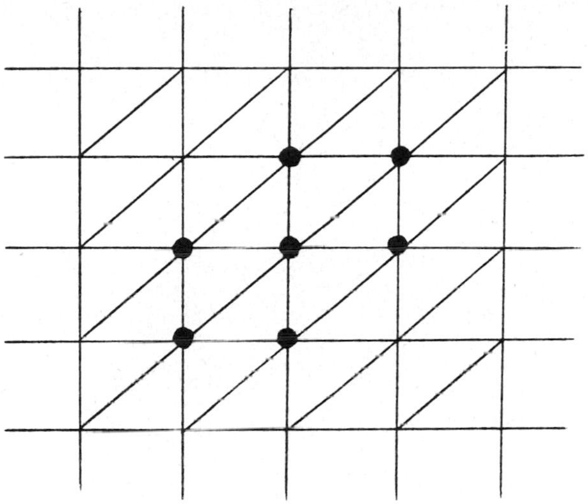

Fig. 1.6.

Periodic triangulation with
7 points cluster.

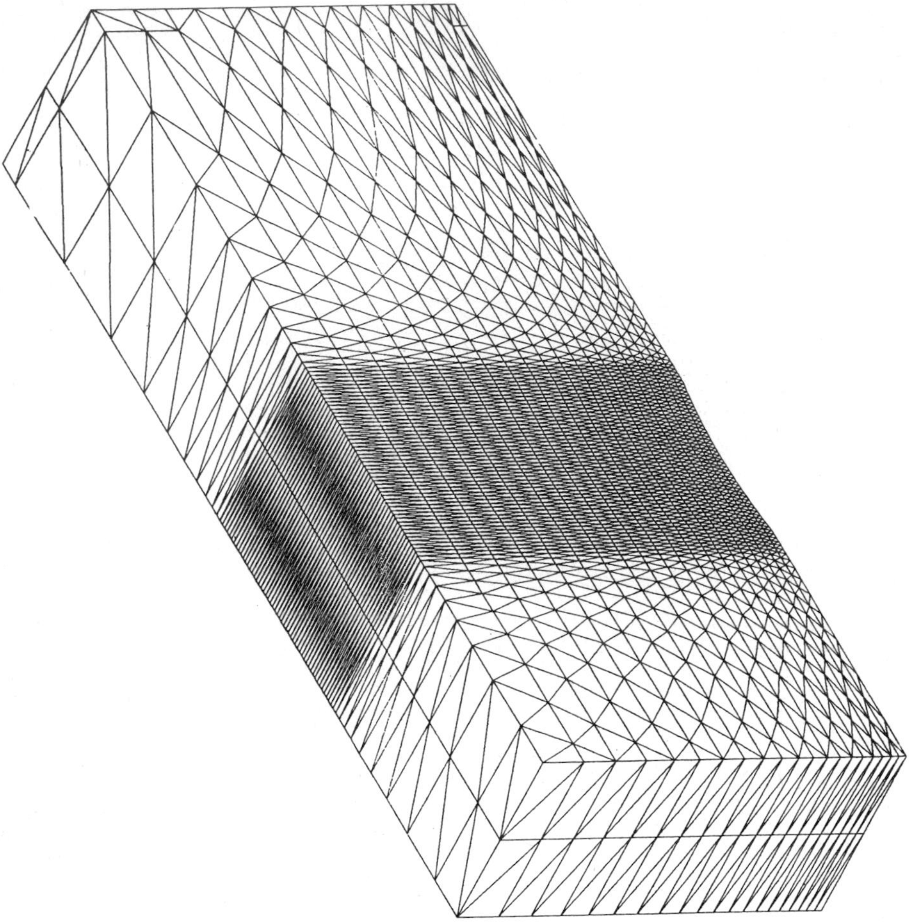

Fig. 3.1.

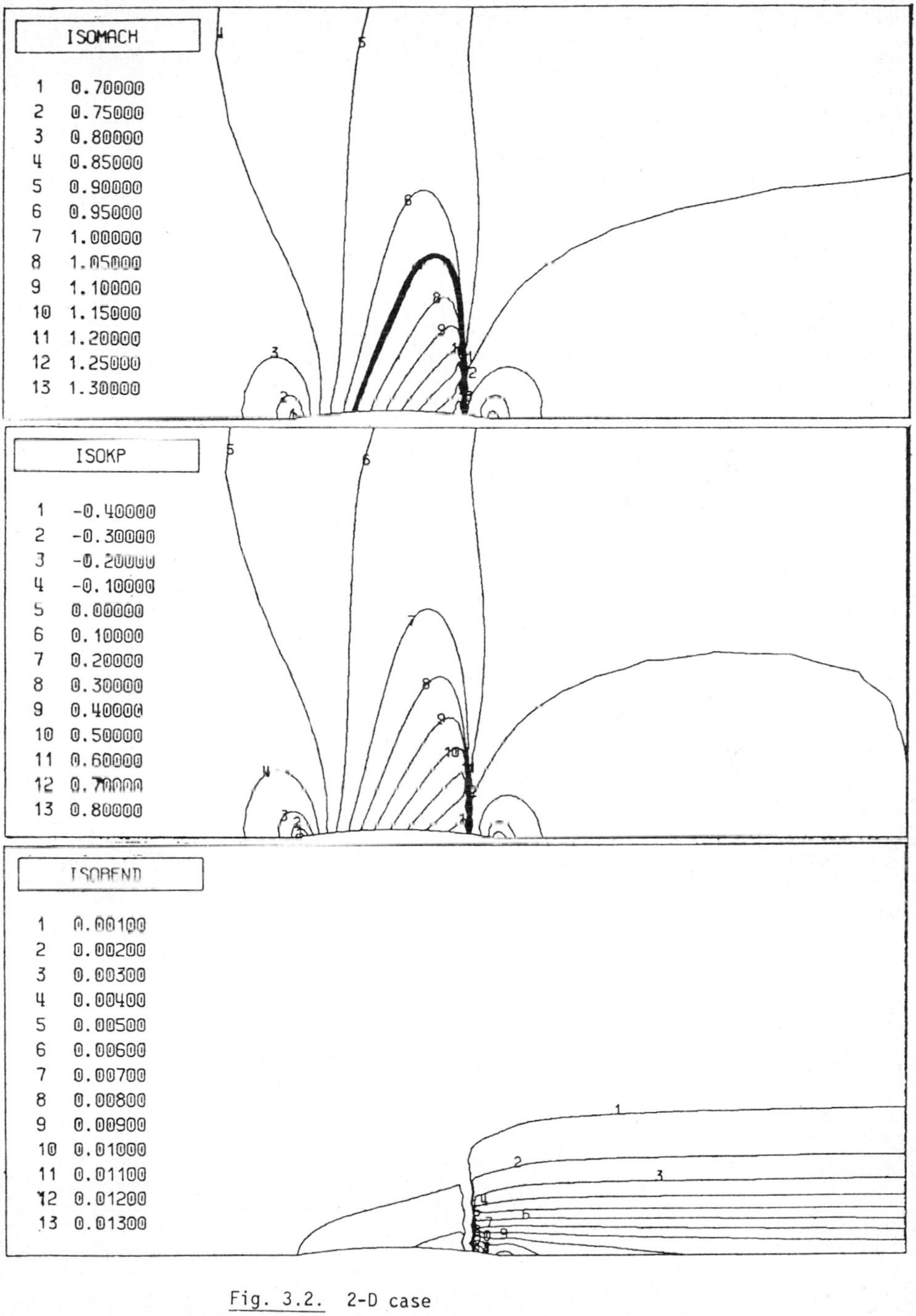

Fig. 3.2. 2-D case

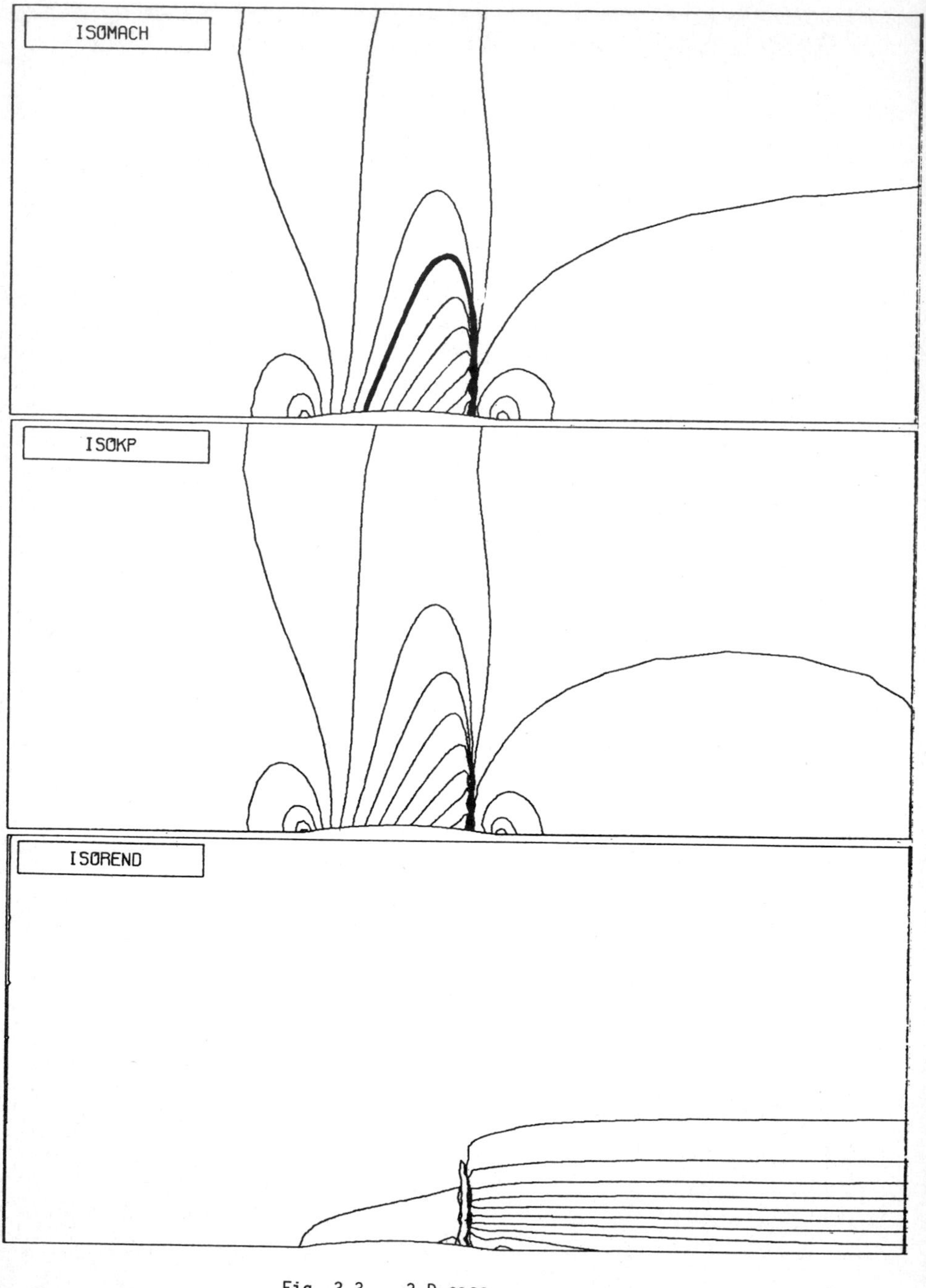

Fig. 3.3. 3-D case

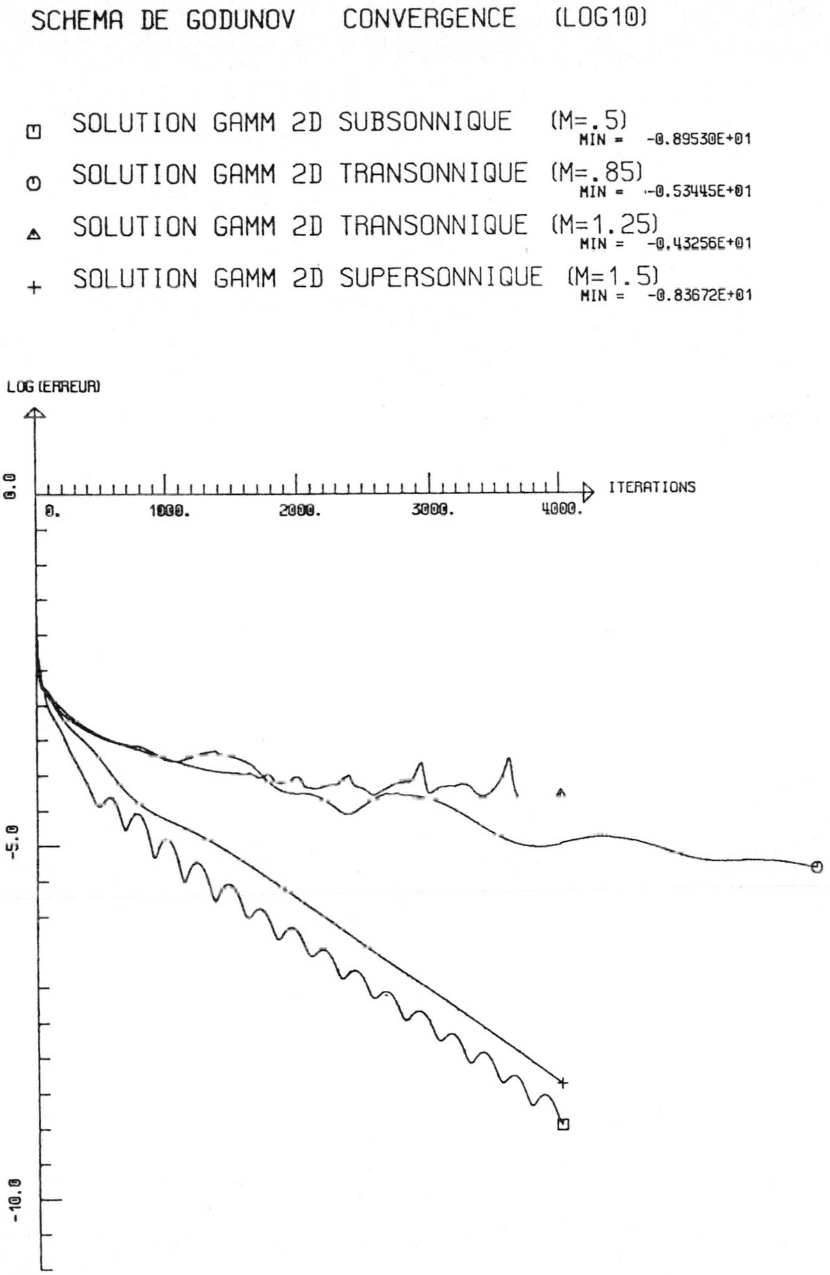
SCHEMA DE GODUNOV CONVERGENCE (LOG10)
SOLUTION GAMM 2D SUBSONNIQUE (M=.5) MIN = -0.89530E+01
SOLUTION GAMM 2D TRANSONNIQUE (M=.85) MIN = -0.53445E+01
SOLUTION GAMM 2D TRANSONNIQUE (M=1.25) MIN = -0.43256E+01
SOLUTION GAMM 2D SUPERSONNIQUE (M=1.5) MIN = -0.83672E+01
LOG (ERREUR)
0.0
-5.0
-10.0
ITERATIONS
0.
1000.
2000.
3000.
4000.

Fig. 3.4.

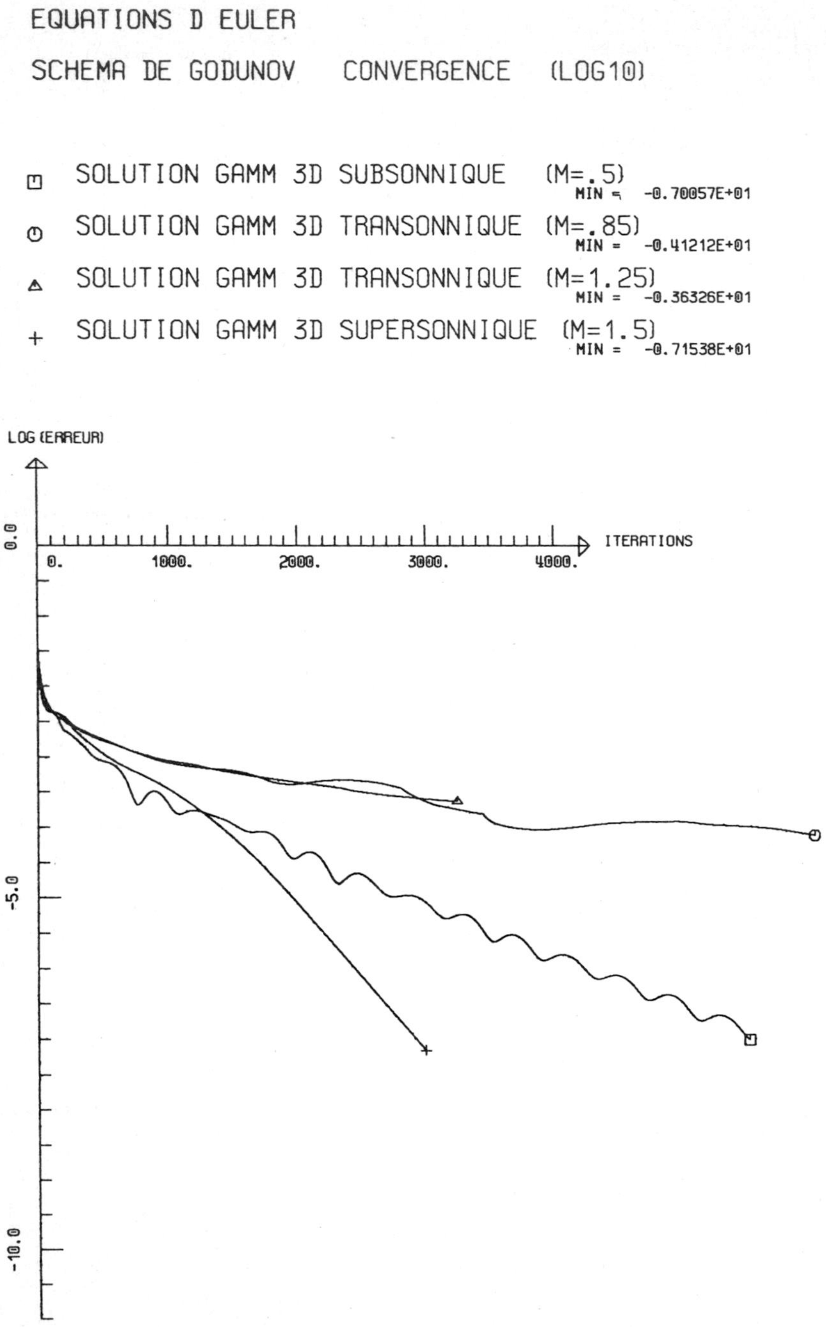
EQUATIONS D EULER
SCHEMA DE GODUNOV CONVERGENCE (LOG10)
SOLUTION GAMM 3D SUBSONNIQUE (M=.5)
MIN = -0.70057E+01
SOLUTION GAMM 3D TRANSONNIQUE (M=.85)
MIN = -0.41212E+01
SOLUTION GAMM 3D TRANSONNIQUE (M=1.25)
MIN = -0.36326E+01
SOLUTION GAMM 3D SUPERSONNIQUE (M=1.5)
MIN = -0.71538E+01
LOG (ERREUR)
0.0
-5.0
-10.0
ITERATIONS
0.
1000.
2000.
3000.
4000.

Fig. 3.5.

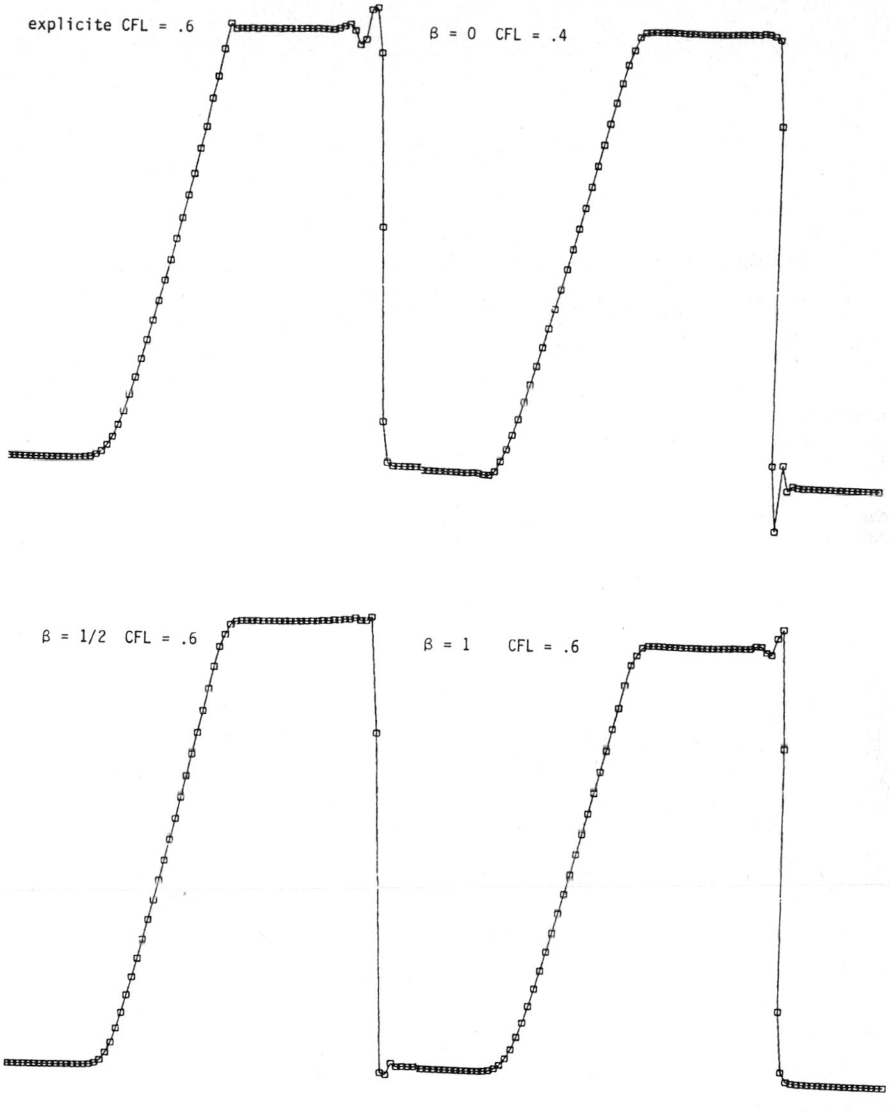

Fig. 3.6. Velocity

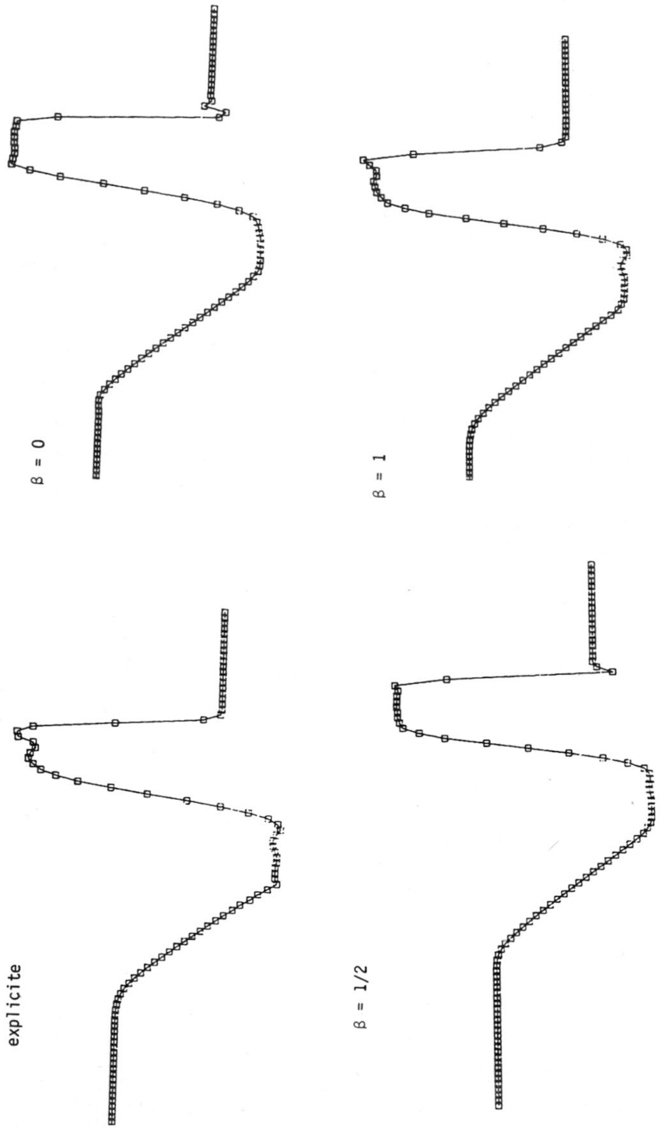

Fig. 3.7. Temperature

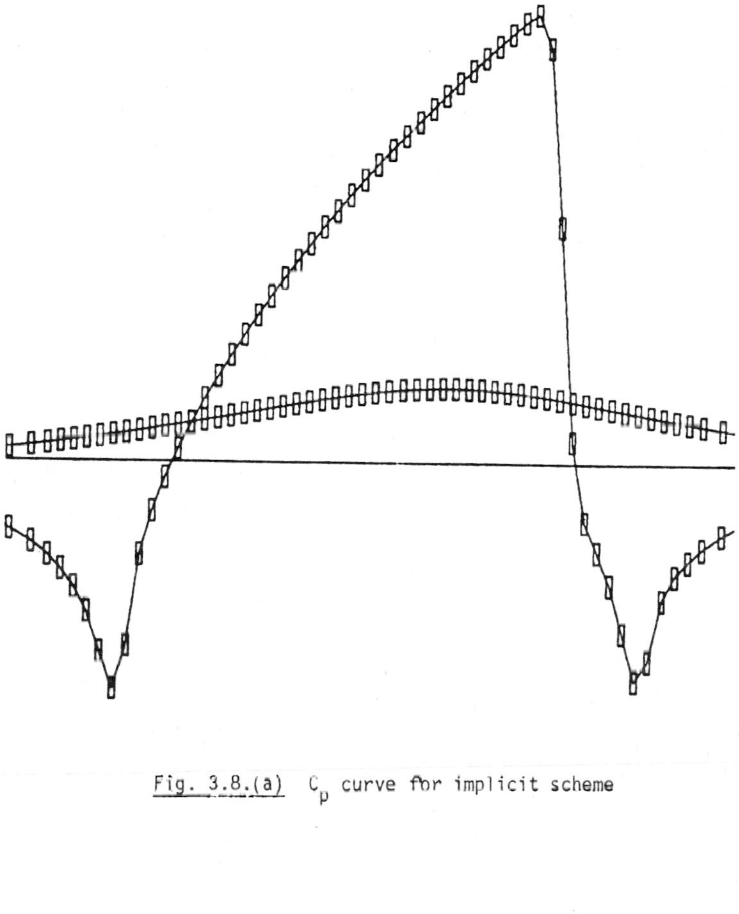

Fig. 3.8.(a) C_p curve for implicit scheme

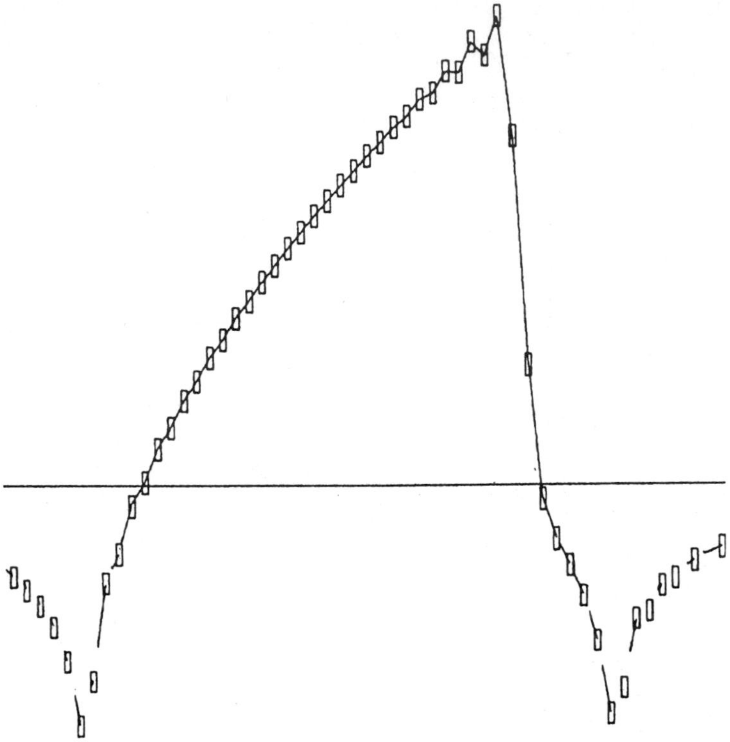

Fig.3.8.(b) C_p curve for explicit scheme

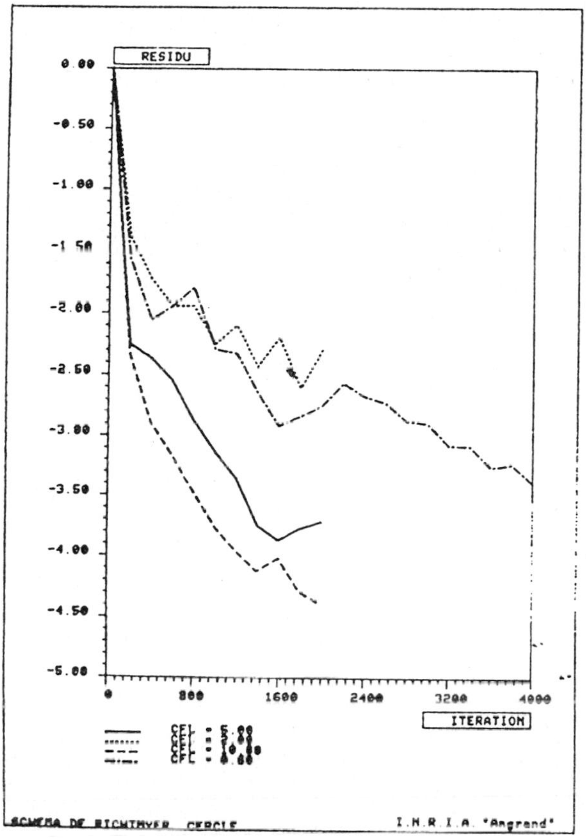
RESIDU
0.00
-0.50
-1.00
-1.50
-2.00
-2.50
-3.00
-3.50
-4.00
-4.50
-5.00
0
800
1600
2400
3200
4000
ITERATION
SCHEMA DE RICHTMYER CERCLE
I.N.R.I.A. 'Angrand'

Fig. 3.9.

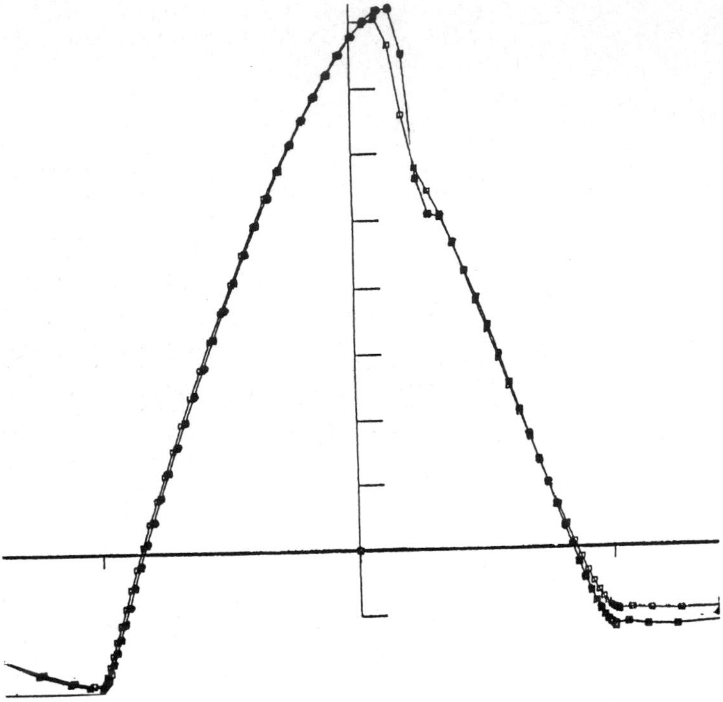

Comparison of C_p curves
□ explicit scheme
■ implicit scheme

Fig. 3.10.

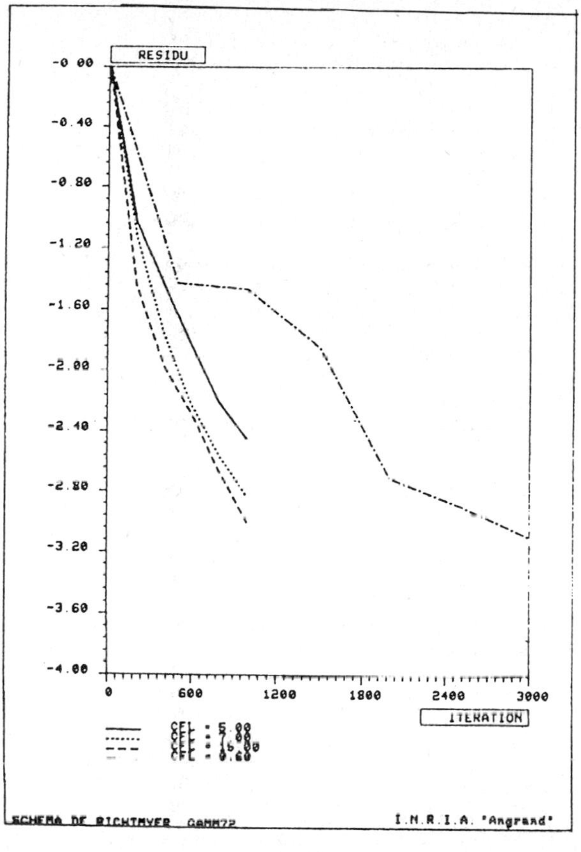
RESIDU
-0.00
-0.40
-0.80
-1.20
-1.60
-2.00
-2.40
-2.80
-3.20
-3.60
-4.00
0
600
1200
1800
2400
3000
ITERATION
CFL = 5.00
CFL = 7.00
SCHEMA DE RICHTMYER GAMM72
I.N.R.I.A. 'Angrand'

Fig. 3.11.

Computing Methods in Applied Sciences and Engineering, VI
R. Glowinski and J.-L. Lions (Editors)
Elsevier Science Publishers B.V. (North-Holland)

NUMERICAL METHODS FOR THE TIME DEPENDENT COMPRESSIBLE NAVIER-STOKES EQUATIONS

M.O. Bristeau*, R. Glowinski**, B. Mantel***, J. Périaux***, P. Perrier***

We discuss in this paper the numerical solution of the time dependent Navier-Stokes equations for compressible viscous fluids. Using a finite difference method for the time discretization we reduce the solution of the original problem to that of a sequence of nonlinear steady problems. These problems can be solved by least squares and conjugate gradient methods, once an appropriate space discretization by finite element has been defined. Numerical results which concern flows past airfoils are presented.

In this paper which is a sequel of [1],[2] we discuss the numerical solution of the *time dependent Navier-Stokes equations* modelling some class of compressible viscous flows. Most of the existing solution methods for the time dependent compressible Navier-Stokes equations are founded on *finite difference methods*, for both space and time discretizations (see [3]-[7] and the references therein) ; *spectral methods* of solution have been also considered (see [8],[9]). More recently finite element methods have been introduced to solve the above problem (see [10],[11] and also [1],[2]).

In this paper we would like to show that the methods used by the present authors for solving the time dependent Navier-Stokes equations for incompressible viscous fluids (see [12],[13],[14]) can be generalized to the compressible case. The general principle is to use an *implicit scheme* for the *time discretization* and to solve the resulting nonlinear problems by a method combining *finite elements* for the *space discretization* and *least squares* for the solution of the discrete problems. Numerical results are presented, they all concern flows past airfoils.

As a final comment we would like to mention the work of NISHIDA [15], where some theoretical results about the compressible Navier-Stokes equations have been proved, assuming that the data are "small".

2. THE GOVERNING EQUATIONS

Let $\Omega \subset \mathbb{R}^N$ (N=2,3 in practice) be the flow domain and Γ be its boundary ; we denote by $x = \{x_i\}_{i=1}^N$ the generic point of $\mathbb{R}^N$.

If we suppose that the fluid satisfies the *perfect gas law*

$$(2.1) \qquad p = (\gamma-1)\rho\, T$$

the flow is governed by the following set of (Navier-Stokes) equations, written in *non-conservative form* :

Continuity equation :

$$(2.2) \qquad \frac{\partial \rho}{\partial t} + \nabla . \rho \underset{\sim}{u} = 0,$$

Momentum equation :

$$(2.3) \qquad \rho \frac{\partial \underset{\sim}{u}}{\partial t} + \rho(\underset{\sim}{u}.\underset{\sim}{\nabla})\underset{\sim}{u} + (\gamma-1) \underset{\sim}{\nabla} \rho T = \frac{1}{Re}[\nabla^2 \underset{\sim}{u} + \frac{1}{3} \underset{\sim}{\nabla}(\underset{\sim}{\nabla}.\underset{\sim}{u})] ,$$

Energy equation (in two-dimensions) :

$$(2.4) \quad \begin{cases} \rho \frac{\partial T}{\partial t} + \rho \underset{\sim}{u}.\underset{\sim}{\nabla} T + (\gamma-1)\rho T \ \underset{\sim}{\nabla}.\underset{\sim}{u} = \frac{1}{Re} \{\frac{\gamma}{Pr} \ \underset{\sim}{\nabla}^2 T + \frac{4}{3} [(\frac{\partial u_1}{\partial x_1})^2 + \\ + (\frac{\partial u_2}{\partial x_2})^2 - \frac{\partial u_1}{\partial x_1}\frac{\partial u_2}{\partial x_2}] + (\frac{\partial u_1}{\partial x_2} + \frac{\partial u_2}{\partial x_1})^2\} . \end{cases}$$

In (2.1)-(2.4), ρ is the *density*, u is the *velocity*, p is the *pressure*, and T is the *temperature* of the fluid ; the constants Re, Pr and γ are the *Reynolds number*, the *Prandtl number* and the *ratio of specific heats*, respectively (γ = 1.4 in *air*).

A classical simplification is provided by the *constant total enthalpy* assumption

$$(2.5) \qquad \gamma T + \frac{|\underset{\sim}{u}|^2}{2} = H_o$$

where H_o is defined from the boundary conditions. If we suppose that (2.5) holds, (2.3) and (2.4) are replaced by

$$(2.6) \quad \begin{cases} \rho \frac{\partial \underset{\sim}{u}}{\partial t} + \rho(\underset{\sim}{u}.\underset{\sim}{\nabla})\underset{\sim}{u} - \frac{\gamma-1}{2\gamma} \underset{\sim}{\nabla} (\rho|\underset{\sim}{u}|^2) + \frac{\gamma-1}{\gamma} \ H_o \underset{\sim}{\nabla}\rho = \\ \frac{1}{Re}(\underset{\sim}{\nabla}^2 \underset{\sim}{u} + \frac{1}{3} \underset{\sim}{\nabla}(\underset{\sim}{\nabla}.\underset{\sim}{u})) . \end{cases}$$

The above equations are written in *dimensionless* form.

3. BOUNDARY AND INITIAL CONDITIONS

The right choice of boundary conditions for the compressible Navier-Stokes equations is a complicated theoretical problem by itself.
If we consider an *external flow*, like the one associated to Figure 3.1, we have used for our calculations the following boundary conditions (where

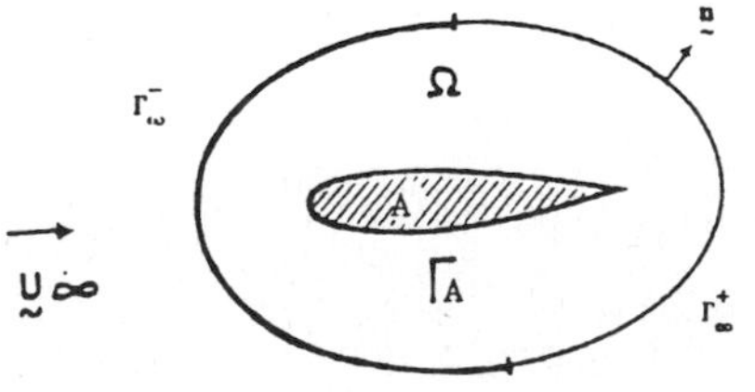

Figure 3.1.

(3.1) $\quad \Gamma_\infty^- = \{x \mid x \in \Gamma \; , \; \underset{\sim}{u}_\infty . \underset{\sim}{n}(x) < 0\}$,

(3.2) $\quad \Gamma_\infty^+ = \Gamma \setminus \Gamma_\infty^-)$

Inflow Boundary Conditions : We prescribe on Γ_∞^-

(3.3) $\quad \underset{\sim}{u} = \underset{\sim}{u}_\infty$

(3.4) $\quad \rho = 1$

(3.5) $\quad T = \dfrac{1}{\gamma(\gamma-1)M_\infty^2}$

where M_∞ denotes the *free stream* Mach number.

Outflow Boundary conditions : On Γ_∞^+ we prescribe *Neumann boundary conditions.*

Wall Boundary Conditions : On Γ_A we use the following conditions

(3.6) $\quad \underset{\sim}{u} = \underset{\sim}{0},$

(3.7) $\quad T = T_\infty \left(1 + \frac{\gamma-1}{2} M_\infty^2\right).$

We do not require a condition for ρ on Γ_A .

Remark 3.1. : In the fluid the *continuity, momentum* and *energy* equations are associated to the *density, velocity* and *temperature,* respectively. At the wall Γ_A, we have $\underset{\sim}{\nabla} . \underset{\sim}{u} = 0$, while the momentum equation reduces to

(3.8) $\quad (\gamma-1) \; \underset{\sim}{\nabla}\rho \; T = \frac{1}{Re}[\underset{\sim}{\nabla}^2 \underset{\sim}{u} + \frac{1}{3} \underset{\sim}{\nabla}(\underset{\sim}{\nabla} . \underset{\sim}{u})]$.

The temperature being deduced from the energy equation, we can use (3.8) to obtain ρ on Γ_A . □

Initial Conditions : At t = 0, we prescribe initial conditions for $\rho, \underset{\sim}{u}$, T, such as

(3.9) $\quad \rho(x,0) = \rho_0(x), \; \underset{\sim}{u}(x,0) = \underset{\sim}{u}_0(x), \quad T(x,0) = T_0(x).$

4. TRANSFORMATION OF THE EQUATIONS

In order to apply the methods previously used for the *incompressible probleme* (or the *shallow-water equations*) we introduce a new function σ defined by

(4.1) $\quad \sigma = \text{Log} \; \rho$.

With this new variable, the governing equations become

(4.2) $\quad \dfrac{\partial\sigma}{\partial t} + \underset{\sim}{\nabla} . \underset{\sim}{u} + \underset{\sim}{u} . \underset{\sim}{\nabla}\sigma = 0,$

(4.3) $\quad \dfrac{\partial \underset{\sim}{u}}{\partial t} + (\underset{\sim}{u} . \underset{\sim}{\nabla})\underset{\sim}{u} + (\gamma-1)(T\underset{\sim}{\nabla}\sigma + \underset{\sim}{\nabla} \; T) = \dfrac{e^{-\sigma}}{Re}(\underset{\sim}{\nabla}^2 \underset{\sim}{u} + \frac{1}{3} \underset{\sim}{\nabla}(\underset{\sim}{\nabla} . \underset{\sim}{u}))$,

(4.4) $\quad \dfrac{\partial T}{\partial t} + \underset{\sim}{u} . \; \underset{\sim}{\nabla} T + (\gamma-1)T\underset{\sim}{\nabla} . \underset{\sim}{u} = \dfrac{e^{-\sigma}}{Re}(\frac{\gamma}{Pr} \; \Delta T + F(\underset{\sim\sim}{\nabla u})),$

with

(4.5) $\quad F(\underset{\sim\sim}{\nabla u}) = \quad \frac{4}{3} \{(\frac{\partial u_1}{\partial x_1})^2 + (\frac{\partial u_2}{\partial x_2})^2 - \frac{\partial u_1}{\partial x_1} \frac{\partial u_2}{\partial x_2}\} \; + (\frac{\partial u_1}{\partial x_2} + \frac{\partial u_2}{\partial x_1})^2$

5. TIME DISCRETIZATION

We use for the *time discretization* an *implicit scheme* whose truncation error is $O(|\Delta t|^2)$; let define $\tilde{U}$ by

$$(5.1) \qquad \tilde{U} = \{\sigma, \tilde{u}, T\} \quad ;$$

(4.2)-(4.4) is clearly of the following general form

$$(5.2) \qquad \frac{\partial \tilde{U}}{\partial t} + G(\tilde{U}) = \tilde{0} \; .$$

Let $\Delta t > 0$ be a time step ; with $\tilde{U}^n$ approximating $\tilde{U}(n\Delta t)$ we discretize (5.2) by the following *two-step* scheme

$$(5.3) \qquad \frac{3\,\tilde{U}^{n+1} - 4\,\tilde{U}^n + \tilde{U}^{n-1}}{2\Delta t} + G(\tilde{U}^{n+1}) = \tilde{0} \quad ; n \geq 1 \; .$$

In (5.3), $\tilde{U}^0$ is obtained from the *initial data* (3.9), and $\tilde{U}^1$ by using two steps of an implicit backward Euler scheme with the time step $\frac{2\Delta t}{3}$ (to have the same matrices), and by interpolating. At step n+1 the system to be solved has the following formulation (we have dropped the indices)

$$(5.4) \qquad \alpha\sigma + \nabla.\tilde{u} + \tilde{u}.\nabla\,\sigma = g,$$

$$(5.5) \qquad \alpha\tilde{u} - \mu\,\Delta\tilde{u} + (\gamma-1)T_A\,\tilde{\nabla}\sigma - \Psi\,(\sigma,\tilde{u},\,T) = \tilde{f}\;,$$

$$(5.6) \qquad \alpha T - \Pi\Delta T - X(\sigma,\tilde{u},\,T) = h,$$

with :

(a) $\alpha = \frac{3}{2\Delta t}$,

(b) δ: a parameter of the order of the inverse of the maximum value of the density,

(c) $\nu = 1/Re$, $\mu = \nu\delta$, $\Pi = \gamma\nu\delta/Pr$,

(d) $\Psi(\sigma,\tilde{u},\,T) = -(\gamma-1)[\tilde{\nabla}T + (T-T_A)\tilde{\nabla}\sigma]$
$$- (\tilde{u}.\tilde{\nabla})\tilde{u} + \nu[e^{-\sigma}(\Delta\tilde{u} + \tfrac{1}{3}\,\tilde{\nabla}(\tilde{\nabla}.\tilde{u}) - \delta\,\Delta\tilde{u}\,] + \nu e^{-\sigma}F(\nabla\tilde{u})$$

(e) $X(\sigma,\tilde{u},\,T) = -(\gamma-1)T\tilde{\nabla}.\tilde{u} - \tilde{u}.\tilde{\nabla}T + \frac{\nu\gamma}{Pr}\,(e^{-\sigma}\Delta T - \delta\Delta T)$

(f) $g = \dfrac{4\,\sigma^n - \sigma^{n-1}}{2\Delta t}$, $\tilde{f} = \dfrac{4\tilde{u}^n - \tilde{u}^{n-1}}{2\Delta t}$, $h = \dfrac{4T^n - T^{n-1}}{2\Delta t}$

6. LEAST SQUARES SOLUTION OF (5.4)-(5.6).

6.1. Some functional spaces

Using the notation of Fig. 3.1, we introduce the following functional spaces

$$(6.1) \qquad R_r = \{\phi | \phi \in H^1(\Omega),\ \phi = r \ \ on \ \ \Gamma_\infty^-\} \ ,$$

$$(6.2) \qquad W_z = \{\tilde{v} | \tilde{v} \in (H^1(\Omega))^N,\ \tilde{v} = \tilde{z} \ \ on \ \Gamma_A \cup \Gamma_\infty^-\} \ ,$$

$$(6.3) \qquad V_s = \{\theta | \theta \in H^1(\Omega),\ \theta = s \ \ on \ \ \Gamma_A \cup \Gamma_\infty^-\} \ .$$

6.2. Least squares formulation of (5.4)-(5.6).

We use the *least squares-optimal control* approach discussed in [1],[12],[14],[16], i.e. we solve (5.4)-(5.6) via the *nonlinear least squares* problem

$$\text{(6.4)} \qquad \text{Min } J(\eta,\underset{\sim}{w},\tau), \qquad \{\eta,\underset{\sim}{w},\tau\} \in R_r \times W_z \times V_s ,$$

with

$$\text{(6.5)} \qquad \begin{aligned} J(\eta,\underset{\sim}{w},\tau) = \; & \frac{\alpha}{2}\int_\Omega |\sigma-\eta|^2\,dx \\ & + \frac{A}{2}\Big(\alpha\int_\Omega |\underset{\sim}{u}-\underset{\sim}{w}|^2 dx + \mu\int_\Omega |\nabla(\underset{\sim}{u}-\underset{\sim}{w})|^2\,dx\Big) \\ & + \frac{B}{2}\Big(\alpha\int_\Omega |T-\tau|^2 dx + \Pi\int_\Omega |\underset{\sim}{\nabla}(T-\tau)|^2\,dx\Big) , \end{aligned}$$

where, in (6.5), we have $A > 0$, $B > 0$ and where $\sigma, \underset{\sim}{u}, T$ are the solutions of the following state system

$$\text{(6.6)} \qquad \alpha\sigma + \underset{\sim}{\nabla}.\underset{\sim}{u} = g - \underset{\sim}{w}.\underset{\sim}{\nabla}\,\eta$$

$$\text{(6.7)} \qquad \alpha\underset{\sim}{u} - \mu\Delta\underset{\sim}{u} + (\gamma-1)\, T_A\underset{\sim}{\nabla}\sigma = \underset{\sim}{f} + \Psi(\eta,\underset{\sim}{w},\tau),$$

$$\text{(6.8)} \qquad \alpha T - \Pi\Delta T = h + X(\eta,\underset{\sim}{w},\tau),$$

$$\text{(6.9)} \qquad \{\sigma,\underset{\sim}{u},T\} \in R_r\times W_z\times V_s .$$

If we consider (6.6), (6.7), it reminds a Stokes problem, where the divergence free condition $\nabla.u = 0$ would have been replaced by (6.6). Solvers for such systems have been designed in [17] , for the solution of the shallow-water equations ; they generalize the Stokes solvers introduced by Glowinski and Pironneau in [18] (see also [19] and [12],[14]). These generalized Stokes solvers, applied to the compressible Navier-Stokes equations, are described in [1] ; they reduce the solution of systems like (6.6), (6.7) to the solution of a boundary integral equation and to simple problems of Poisson type (see [1] for the details, and also Appendix 1 for a brief description of such a method).

6.3. Iterative solution of the least squares problem (6.4)

It is quite convenient to solve the minimization problem (6.4) by a conjugate gradient algorithm operating in the space $R_r \times W_z \times V_s$. For a general description of such algorithms see [14], [16] ; in the particular case of problem (6.4) such an algorithm could be the following once $R_r \times W_z \times V_s$ are equipped with the product norm below

$$\{\phi,\underset{\sim}{v},\theta\} \to \{\alpha\int_\Omega \phi^2 dx + \alpha A\int_\Omega \underset{\sim}{v}^2 dx + \mu A\int_\Omega |\underset{\sim}{\nabla}\underset{\sim}{v}|^2 dx + \alpha B\int_\Omega \theta^2 dx + \Pi B\int_\Omega |\underset{\sim}{\nabla}\theta|^2 dx\}^{1/2} \; ;$$

<u>Step 0</u> : <u>Initialization</u>

$$\text{(6.10)} \qquad \{\eta^0,\underset{\sim}{w}^0,\tau^0\} \in R_r \times W_z \times V_s \;\; \textit{given}$$

solve then the following problems :

$$\text{(6.11)} \qquad \alpha\int_\Omega g^0_\eta\phi\,dx = \Big\langle\frac{\partial J}{\partial\eta}(\eta^0,\underset{\sim}{w}^0,\tau^0),\phi\Big\rangle \quad \forall\phi \in R_0 \; ; \; g^0_\eta \in R_0,$$

$$\text{(6.12)} \qquad \alpha\int_\Omega \underset{\sim}{g}^0_w.\underset{\sim}{v}\,dx + \mu\int_\Omega \underset{\sim}{\nabla}\underset{\sim}{g}^0_w.\underset{\sim}{\nabla}\underset{\sim}{v}\,dx = \Big\langle\frac{\partial J}{\partial\underset{\sim}{w}}(\eta^0,\underset{\sim}{w}^0,\tau^0),\underset{\sim}{v}\Big\rangle \quad \forall\underset{\sim}{v} \in W_0 ; \underset{\sim}{g}^0_w \in W_0 ,$$

$$\text{(6.13)} \qquad \alpha\int_\Omega g^0_\tau\,\theta\,dx + \Pi\int_\Omega \underset{\sim}{\nabla}g^0_\tau.\underset{\sim}{\nabla}\theta\,dx = \Big\langle\frac{\partial J}{\partial\tau}(\eta^0,\underset{\sim}{w}^0,\tau^0),\theta\Big\rangle \quad \forall\theta \in V_0 \; ; \; g^0_\tau \in V_0 ,$$

and set

(6.14) $z_\eta^0 = g_\eta^0 \,,\ \underset{\sim}{z}_w^0 = \underset{\sim}{g}_w^0 \,,\ z_\tau^0 = g_\tau^0 \,,$

Then for $n \geq 0$, *assuming* $\{\eta^n, \underset{\sim}{w}^n, \tau^n\}$, $\underset{\sim}{g}^n = \{g_\eta^n, \underset{\sim}{g}_w^n, g_\tau^n\}$, $\underset{\sim}{z}^n = \{z_\eta^n, \underset{\sim}{z}_w^n, z_\tau^n\}$ *known, define* $\{\eta^{n+1}, \underset{\sim}{w}^{n+1}, \tau^{n+1}\}$, $\underset{\sim}{g}^{n+1} = \{g_\eta^{n+1}, \underset{\sim}{g}_w^{n+1}, g_\tau^{n+1}\}$, $\underset{\sim}{z}^{n+1} = \{z_\eta^{n+1}, \underset{\sim}{z}_w^{n+1}, z_\tau^{n+1}\}$ *as follows*

Step 1 : Descent

Solve

$$(6.15)\quad \begin{cases} \lambda_n \in \mathbb{R} \ \text{ such that} \\ J(\{\eta^n, \underset{\sim}{w}^n, \tau^n\} - \lambda_n \underset{\sim}{z}^n) \leq J(\{\eta^n, \underset{\sim}{w}^n, \tau^n\} - \lambda \underset{\sim}{z}^n) \ \forall \lambda \in \mathbb{R} \,, \end{cases}$$

and set

$$(6.16)\quad \{\eta^{n+1}, \underset{\sim}{w}^{n+1}, \tau^{n+1}\} = \{\eta^n, \underset{\sim}{w}^n, \tau^n\} - \lambda_n \underset{\sim}{z}^n.$$

Step 2 : New descent direction

Solve

$$(6.17)\quad \alpha \int_\Omega g_\eta^{n+1} \phi \, dx = \langle \frac{\partial J}{\partial \eta}(\eta^{n+1}, \underset{\sim}{w}^{n+1}, \tau^{n+1}), \phi \rangle \quad \forall \phi \in R_0 \ ; \ g_\eta^{n+1} \in R_0,$$

$$(6.18)\quad \alpha \int_\Omega \underset{\sim}{g}_w^{n+1} . \underset{\sim}{v} \, dx + \mu \int_\Omega \underset{\sim}{\nabla} \underset{\sim}{g}_w^{n+1} . \underset{\sim\sim}{\nabla v} \, dx = \langle \frac{\partial J}{\partial \underset{\sim}{w}}(\eta^{n+1}, \underset{\sim}{w}^{n+1}, \tau^{n+1}), \underset{\sim}{v} \rangle \ \forall \underset{\sim}{v} \in W_0 ; \underset{\sim}{g}_w^{n+1} \in W_0,$$

$$(6.19)\quad \alpha \int_\Omega g_\tau^{n+1} \theta \, dx + \Pi \int_\Omega \underset{\sim}{\nabla} g_\tau^{n+1} . \underset{\sim}{\nabla} \theta \, dx = \langle \frac{\partial J}{\partial \tau}(\eta_\eta^{n+1}, \underset{\sim}{w}^{n+1}, \tau^{n+1}), \theta \rangle \ \forall \theta \in V_0 \ ; \ g_\tau^0 \in V_0 \,.$$

Compute then

$$(6.20)\quad \gamma_n = \frac{(\{g_\eta^{n+1}, \underset{\sim}{g}_w^{n+1}, g_\tau^{n+1}\}, \{g_\eta^{n+1} - g_\eta^n, \underset{\sim}{g}_w^{n+1} - \underset{\sim}{g}_w^n, g_\tau^{n+1} - g_\tau^n\})_1}{(\{g_\eta^n, \underset{\sim}{g}_w^n, g_\tau^n\} , \{g_\eta^n, \underset{\sim}{g}_w^n, g_\tau^n\})_1}$$

where, $\forall \ \{\phi, \underset{\sim}{v}, \theta\}, \{\phi', \underset{\sim}{v}', \theta'\}$,

$$(\{\phi, \underset{\sim}{v}, \theta\} , \{\phi', \underset{\sim}{v}', \theta'\})_1 = \alpha \int_\Omega \phi\phi' dx + \alpha A \int_\Omega \underset{\sim}{v}.\underset{\sim}{v}' dx + \mu A \int_\Omega \underset{\sim\sim}{\nabla v}.\underset{\sim\sim}{\nabla v}' \, dx + \alpha B \int_\Omega \theta\theta' \, dx + \Pi B \int_\Omega \underset{\sim}{\nabla}\theta.\underset{\sim}{\nabla}\theta' \, dx \,.$$

Finally define $\underset{\sim}{z}^{n+1}$ *by*

$$(6.22)\quad \underset{\sim}{z}^{n+1} = \underset{\sim}{g}^{n+1} + \gamma_n \underset{\sim}{z}^n \,. \qquad \square$$

Do $n = n+1$ *and go to* (6.15).

Remark 6.1. : From (6.1), it would have been more appropriate to use an H^1-norm over R_r ; numerical experiments show, however, that the choice of an L^2-norm yields satisfactory numerical results. □

A most important step, when applying algorithm (6.10)-(6.12) to solve (6.4), is the calculation of $J'(\eta^n, \underset{\sim}{w}^n, \tau^n)$; let us detail that calculation :

We have, by differentiation of (6.5),

$$(6.23)\quad \begin{cases} \delta J = \langle J'_\eta, \delta_\eta\rangle + \langle J'_{\underset{\sim}{w}}, \delta \underset{\sim}{w}\rangle + \langle J'_\tau, \delta\tau\rangle = \\ = \alpha \int_\Omega (\sigma-\eta)\,\delta\sigma\,dx + \alpha\int_\Omega (\eta-\sigma)\,\delta\eta\,dx \\ + \alpha A \int_\Omega (\underset{\sim}{u}-\underset{\sim}{w}).\delta\underset{\sim}{u}\,dx + \alpha A\int_\Omega (\underset{\sim}{w}-\underset{\sim}{u}).\delta\underset{\sim}{w}\,dx \\ + \mu A\int_\Omega \underset{\sim}{\nabla}(\underset{\sim}{u}-\underset{\sim}{w}).\underset{\sim}{\nabla}\,\delta\underset{\sim}{u}\,dx + \mu A\int_\Omega \underset{\sim}{\nabla}(\underset{\sim}{w}-\underset{\sim}{u}).\underset{\sim}{\nabla}\,\delta\underset{\sim}{w}\,dx \\ + \alpha B\int_\Omega (T-\tau)\delta T\,dx + \alpha B\int_\Omega (\tau-T)\delta\tau\,dx \\ + \Pi B\int_\Omega \nabla(T-\tau).\underset{\sim}{\nabla}\delta T\,dx + \Pi B\int_\Omega \underset{\sim}{\nabla}(\tau-T).\underset{\sim}{\nabla}\delta\tau\,dx \end{cases}$$

By differentiation of (6.8), we obtain

$$(6.24)\quad \begin{cases} \int_\Omega \delta T\,\theta\,dx + \Pi\int_\Omega \underset{\sim}{\nabla}\,\delta T.\underset{\sim}{\nabla}\,\theta\,dx = \\ = \int_\Omega \left(\frac{\partial X}{\partial\eta}\delta\eta + \frac{\partial X}{\partial \underset{\sim}{w}}\delta\underset{\sim}{w} + \frac{\partial X}{\partial\tau}\delta\tau\right)\theta\,dx \quad \forall\theta \in V_0\,; \end{cases}$$

taking $\theta = T-\tau$ in (6.24) we obtain that

$$(6.25)\quad \begin{cases} \alpha\int_\Omega (T-\tau)\delta T\,dx + \Pi\int_\Omega \underset{\sim}{\nabla}(T-\tau).\underset{\sim}{\nabla}\,\delta T\,dx = \\ \int_\Omega \left(\frac{\partial X}{\partial\eta}\delta\eta + \frac{\partial X}{\partial \underset{\sim}{w}}\delta\underset{\sim}{w} + \frac{\partial X}{\partial\tau}\delta\tau\right)(T-\tau)dx\,. \end{cases}$$

Setting $(\gamma-1)T_A = \beta$ in (6.7). We obtain from (6.6), (6.7)

$$(6.26)\quad \alpha\int_\Omega \delta\sigma\,\phi\,dx + \int_\Omega \underset{\sim}{\nabla}.\delta\underset{\sim}{u}\,\phi dx = -\int_\Omega \delta\underset{\sim}{w}.\underset{\sim}{\nabla}\eta\,\phi\,dx - \int_\Omega \underset{\sim}{w}.\underset{\sim}{\nabla}\delta\eta\,\phi\,dx \quad \forall\phi \in R_0,$$

$$(6.27)\quad \begin{aligned} &\alpha\int_\Omega \delta\underset{\sim}{u}.\underset{\sim}{v}\,dx + \mu\int_\Omega \underset{\sim}{\nabla}\delta\underset{\sim}{u}.\underset{\sim}{\nabla}\underset{\sim}{v}\,dx + \beta\int_\Omega \underset{\sim}{\nabla}\,\delta\sigma.\underset{\sim}{v}\,dx \\ &= \int_\Omega \left(\frac{\partial\psi}{\partial\eta}\delta\eta + \frac{\partial\psi}{\partial\underset{\sim}{w}}\delta\underset{\sim}{w} + \frac{\partial\psi}{\partial\tau}\delta\tau\right).\underset{\sim}{v}\,dx \quad \forall\underset{\sim}{v} \in W_0\,. \end{aligned}$$

We introduce now the pair $\{\chi,\underset{\sim}{\xi}\} \in R_0 \times W_0$, solution of the following *(linear)* system (the *adjoint* system) :

$$(6.28)\quad \alpha\int_\Omega \chi\,\phi\,dx + \beta\int_\Omega \underset{\sim}{\xi}.\underset{\sim}{\nabla}\phi\,dx = \alpha\int_\Omega (\sigma-\eta)\phi\,dx \quad \forall\phi \in R_0,$$

$$(6.29)\quad \begin{cases} \alpha\int_\Omega \underset{\sim}{\xi}.\underset{\sim}{v}\,dx + \mu\int_\Omega \underset{\sim}{\nabla}\underset{\sim}{\xi}.\underset{\sim}{\nabla}\underset{\sim}{v}\,dx + \int_\Omega \chi\,\underset{\sim}{\nabla}.\underset{\sim}{v}\,dx \\ = \alpha A\int_\Omega (\underset{\sim}{u}-\underset{\sim}{w}).\underset{\sim}{v}\,dx + \mu A\int_\Omega \underset{\sim}{\nabla}(\underset{\sim}{u}-\underset{\sim}{w}).\underset{\sim}{\nabla}\underset{\sim}{v}\,dx \quad \forall\underset{\sim}{v} \in W_0, \end{cases}$$

$$(6.30)\quad \{\chi,\underset{\sim}{\xi}\} \in R_0 \times W_0\,.$$

To obtain $\{\chi,\xi\}$ from σ,η,u,w we should use again a *generalized Stokes solver* (as the one discussed in Appendix 1).

Taking $\phi = \delta\sigma$ and $\underset{\sim}{v} = \delta\underset{\sim}{u}$ in (6.28), (6.29), respectively, we obtain

$$(6.31)\quad \alpha\int_\Omega (\sigma-\eta)\delta\sigma\,dx = \alpha\int_\Omega \chi\,\delta\sigma\,dx + \beta\int_\Omega \underset{\sim}{\xi}.\underset{\sim}{\nabla}\,\delta\sigma\,dx,$$

$$(6.32)\quad \begin{cases} \alpha A \int_\Omega (\underset{\sim}{u}-\underset{\sim}{w}).\delta\underset{\sim}{u}\, dx + \mu A \int_\Omega \nabla(\underset{\sim}{u}-\underset{\sim}{w}).\nabla\delta\underset{\sim}{u}\, dx \\ = \alpha \int_\Omega \underset{\sim}{\xi}.\delta\underset{\sim}{u}\, dx + \mu \int_\Omega \nabla\underset{\sim}{\xi}.\nabla\delta\underset{\sim}{u}\, dx + \int_\Omega \chi\, \underset{\sim}{\nabla}.\delta\underset{\sim}{u}\, dx \,, \end{cases}$$

respectively.

Taking now $\phi = \chi$ and $\underset{\sim}{v} = \underset{\sim}{\xi}$ in (6.26), (6.27) we obtain, by addition, from (6.31), (6.32),

$$(6.33)\quad \begin{cases} \alpha \int_\Omega (\sigma-\eta)\delta\sigma\, dx + \alpha A \int_\Omega (\underset{\sim}{u}-\underset{\sim}{w}).\delta\underset{\sim}{u}\, dx + \mu A \int_\Omega \nabla(\underset{\sim}{u}-\underset{\sim}{w}).\underset{\sim}{\nabla}\, \delta\underset{\sim}{u}\, dx \\ = \int_\Omega (\frac{\partial\psi}{\partial\eta}\, \delta\eta + \frac{\partial\psi}{\partial \underset{\sim}{w}}\, \delta\underset{\sim}{w} + \frac{\partial\psi}{\partial\tau}\, \delta\tau\,).\underset{\sim}{\xi}\, dx - \int_\Omega \delta\underset{\sim}{w}.\nabla\eta\, \chi\, dx - \int_\Omega \underset{\sim}{w}.\nabla\delta\eta\, \chi\, dx \,. \end{cases}$$

Combining (6.23), (6.25) and (6.33) we finally obtain :

$$(6.34)\quad \langle J'_\eta, \phi\rangle = \alpha \int_\Omega (\eta-\sigma)\phi\, dx + \int_\Omega [B(T-\tau)\frac{\partial X}{\partial\eta} + \frac{\partial\psi}{\partial\eta}.\underset{\sim}{\xi}]\phi\, dx - \int_\Omega \underset{\sim}{w}.\nabla\phi\, \chi\, dx, \ \forall\phi \in R_o,$$

$$(6.35)\quad \langle J'_w\, \underset{\sim}{v}\rangle = \alpha A \int_\Omega (\underset{\sim}{w}-\underset{\sim}{u}).\underset{\sim}{v}\, dx + \mu A \int_\Omega \nabla(\underset{\sim}{w}-\underset{\sim}{u}).\nabla\underset{\sim}{v}\, dx + \int_\Omega [B(T-\tau)\frac{\partial X}{\partial \underset{\sim}{w}}.\underset{\sim}{v} + (\frac{\partial\psi}{\partial \underset{\sim}{w}}\, \underset{\sim}{v}).\underset{\sim}{\xi}]\, dx - \int_\Omega \underset{\sim}{v}.\nabla\eta\, \chi\, dx, \ \forall \underset{\sim}{v} \in W_o$$

$$(6.36)\quad \langle J'_\tau, \theta\rangle = \alpha B \int_\Omega (\tau-T)\theta + \Pi B \int_\Omega \nabla(\tau-T).\nabla\theta\, dx + \int_\Omega [B(T-\tau)\frac{\partial X}{\partial\tau} + \frac{\partial\psi}{\partial\tau}.\underset{\sim}{\xi}]\, \theta\, dx, \ \forall\theta \in V_o.$$

7. SOME BRIEF COMMENTS ON THE SPACE DISCRETIZATION

The space discretization is achieved by a *finite element approximation* in which the spaces R_r, V_s (resp. W_z) are approximated by subspaces of

$$(7.1)\qquad H^1_h = \{\phi_h | \phi_h \in C^0(\bar{\Omega}),\ \ \phi_h|_T \in P_1\ \forall T \in \mathcal{T}_h\}$$

(resp. $H^1_h \times H^1_h$) defined from the boundary conditions satisfied by the functions belonging to R_r, W_z, V_s. In (7.1), $\mathcal{T}_h$ is a classical triangulation of Ω, and P_1 is the space of those polynomials in x_1, x_2 of degree ≤ 1.

We observe that unlike the incompressible Navier-Stokes equations, formulated with the primitive variables u and p, we have used similar approximations to approximate ρ, u and T ; the corresponding numerical results are quite reasonable, however (see [1],[2], and Sec. 8 of this paper).

8. NUMERICAL RESULTS

We present in this section the results of several numerical experiments in which is tested the solution method discussed in the above sections. We consider transonic flows around a standard NACA0012 airfoil. We use either a coarse triangulation with 800 nodes and 1514 triangles or a finer one with 3114 nodes and 6056 triangles ; on Figure 8.1., an enlargement of the fine triangulation near the NACA0012 airfoil is shown, while on Figure 8.2 the whole coarse triangulation is presented.

8.1. A comparison with an experimental result

On Figure 8.3, we compare the density lines obtained from the numerical algorithm and from experiments. The data for this case are the following :

(i) The *Mach number at infinity* is 0.8,
(ii) The *Reynolds number* is 73,
(iii) The *angle of attack* is 10°.

The measurements are due to Bernard and al. [20].

For this test, the code was run on the 3114 nodes grid. It can be assumed that the computational results would be still in better agreement with measurements if the adherence condition on velocity prescribed at the rigid boundary, was replaced by a slip condition simulating more closely the experimental conditions. The Mach lines are presented on Figure 8.4.

8.2. Simulation of a steady flow with a supersonic Mach number at infinity

For this test, we used only the 800 nodes triangulation. We prescribed the following data

(i) the *Mach number at infinity* is 1.1,

(ii) the *Reynolds number* is 100.

Figures 8.5 and 8.6 show density lines and Mach lines near the airfoil.

8.3. Simulation of an unsteady flow

Finally we consider an unsteady problem around the same NACA0012 airfoil at 30° of incidence. On Figure 8.7, velocity is plotted at different time steps. The initial solution was the corresponding steady incompressible flow.

8.4. Comments on the results of the numerical simulations.

The above numerical results agree quite well with the numerical experiments in [20] and also with those in [1], [2] obtained using the primitive variables, directly (we recall that in this paper we used the transformation $\sigma = \text{Log}\,\rho$) ; compared to the methods using primitive variables it seems that using σ instead of ρ , we do not need different approximations for the various functions occuring in the mathematical model. It will be interesting to check if this property still holds at higher Re's.

Acknowledgements :

This work was partly supported by DRET under contract n° 80/321.

References

[1] M.O. BRISTEAU, R. GLOWINSKI, B. DIMOYAT, J. PERIAUX, P. PERRIER, O. PIRONNEAU, Finite Element Methods for the Compressible Navier-Stokes equations, in *Proceedings of the AIAA Computation of Fluid Dynamics Conference, Danvers July 13-15, 1983*, AIAA Paper 83-1890.

[2] M.O. BRISTEAU, R. GLOWINSKI, B. MANTEL, J. PERIAUX, P. PERRIER, Finite Element Solution of the Navier-Stokes Equations for Compressible Viscous Fluids, in *Proceedings of the Fifth International Symposium on Finite Element and Flow Problems*, G.F. Carey, J.T. Oden eds., University of Texas at Austin, 1984, pp. 449-462.

[3] R.W. MAC CORMACK, A Numerical Method for Solving the Equations of Compressible Viscous Flow, *AIAA J., Vol. 20*, (1982), 9, pp. 1275-1281.

[4] R.M. BEAM, R.F. WARMING, An Implicit Factored Scheme for the Compressible Navier-Stokes Equations, *AIAA J., Vol. 16*, (1978), pp. 393-402.

[5] W.R. BRILEY, H. Mc DONALD, Solution of the Multidimensional Compressible Navier-Stokes Equations by a Generalized Implicit Method, *J. Comp. Physics., Vol. 24,* (1977), pp. 372-397.

[6] R. PEYRET, H. VIVIAND, Computation of Viscous Compressible Flows Based on the Navier-Stokes Equations, *AGARD-AG-212,* 1975.

[7] R. PEYRET, T.D. TAYLOR, *Computational Methods for Fluid Flow,* Springer, New-York, 1982.

[8] S. BOKMARI, M.Y. HUSSAINI, J.J. LAMBIOTTE, S.A. ORZAG, A Navier-Stokes Solution on the CYBER-203 by a Pseudo-Spectral Technique, *ICASE Report* N° 81-12, 1981.

[9] N. SATOFUKA, Modified differential quadrature method for numerical solution of multi-dimensional flow problems, in *Proceedings of the International Symposium on Applied Mathematics and Information Science,* Kyoto, March 1982, pp. 5.7-5.14.

[10] A.K. BAKER, Finite Element Analysis of Turbulent Flows, *Proceedings of the 1st International Conference on Numerical Methods in Laminar and Turbulent Flows,* Pentech Press, Swansea, 1978, pp. 203-229.

[11] C.H. COOKE, D.K. BLANCHARD, A shock capturing application of the finite element method, *Int. J. Num. Meth. Eng., Vol. 14,* (1979), pp. 271-286.

[12] M.O. BRISTEAU, R. GLOWINSKI, J. PERIAUX, P. PERRIER, O. PIRONNEAU, G. POIRIER, Application of Optimal Control and Finite Element Methods to the Calculation of Transonic Flows and Incompressible Flows, in *Numerical Methods in Applied Fluid Dynamics,* B. Hunt ed., Academic Press, London, 1980, pp. 203-312.

[13] R. GLOWINSKI, B. MANTEL, J. PERIAUX, Numerical Solution of the Time Dependent Navier-Stokes Equations for Incompressible Viscous Fluids by Finite Element and Alternating Direction Methods, in *Numerical Mathods in Aeronautical Fluid Dynamics,* P.L. Roe ed., Academic Press, London, 1982, pp. 309-336.

[14] R. GLOWINSKI, *Numerical Methods for Nonlinear Variational Problems,* Springer, New-York, 1984.

[15] A. MATSUMURA, T. NISHIDA, Initial boundary value problems for the equations of motion of general fluids, in *Computing Methods in Applied Sciences and Engineering, V,* R. Glowinski, J.L. Lions eds., North-Holland, Amsterdam, 1982, pp. 389-406.

[16] M.O. BRISTEAU, R. GLOWINSKI, J. PERIAUX, P. PERRIER, O. PIRONNEAU, G. POIRIER, On the numerical solution of nonlinear problems in fluid dynamics by least squares and finite element methods (I). Least squares formulations and conjugate gradient solution of the continuous problems, *Comp. Meth. Appl. Mech. Eng., Vol. 17/18,* (1979), pp. 619-657.

[17] J. GOUSSEBAILE, G. LABADIE, F. HECHT, L. REINHART, Finite Element Solution of the Shallow Waters Equations, by a quasi direct decomposition procedure, *Numerical Methods in Fluids* (to appear).

[18] R. GLOWINSKI, O. PIRONNEAU, Approximation par éléments finis mixtes du problème de Stokes en formulation vitesse-pression. Résolution des problèmes approchés. *C.R. Acad. Sci. Paris, 286A,* (1978), pp. 225-228.

[19] R. GLOWINSKI, O. PIRONNEAU, On numerical methods for the Stokes problem, in *Energy Methods in Finite Element Analysis,* R. Glowinski, E.Y. Rodin, O.C. Zienkiewicz ed., Wiley, Chichester, 1979, pp. 243-264.

[20] J.J. BERNARD, J. ALLEGRE, M. RAFFIN, Champs d'écoulements raréfiés compressibles autour d'un profil NACA 0012, *Contrat DRET 80/636, Rapport Technique.*

* INRIA Domaine de Voluceau - Rocquencourt
B.P. 105
78153 LE CHESNAY CEDEX FRANCE

** Université Pierre et Marie Curie
4, place Jussieu
75230 PARIS CEDEX 05 FRANCE

*** AMD / BA
78, quai Carnot
B.P. 300
92214 SAINT CLOUD FRANCE

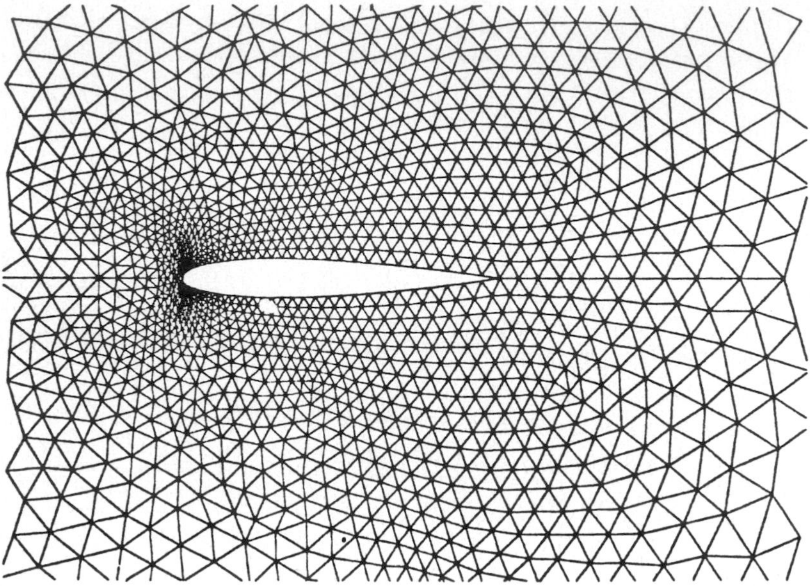

Figure 8.1. An enlargement of the fine triangulation around the NACA0012 airfoil.

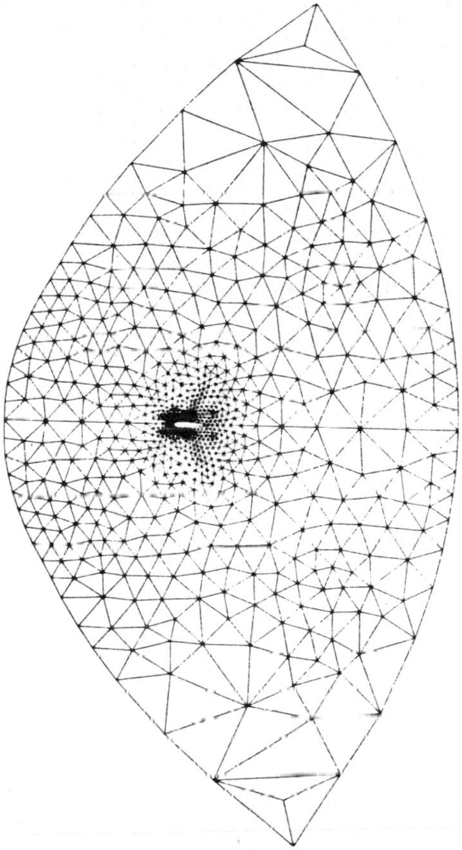

Figure 8.2. The coarse triangulation around the NACA0012 airfoil.

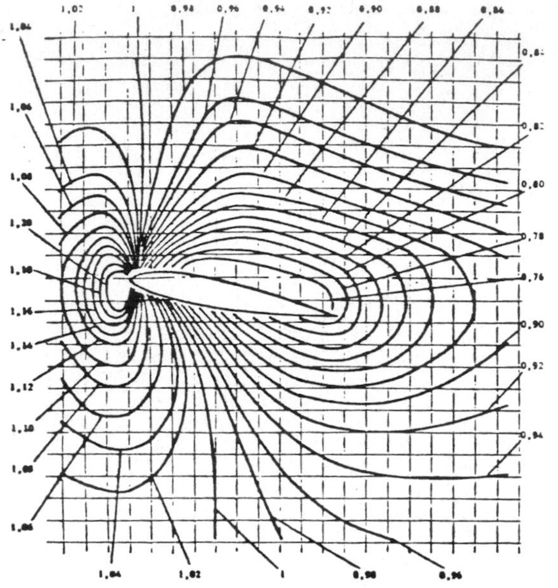

Experimental results

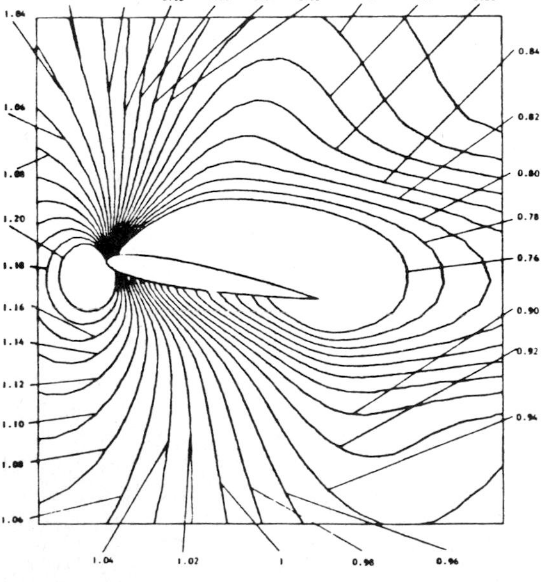

Computational results

Figure 8.3. Comparison of the density lines. $M_\infty = 0.8$, Re = 73., $\alpha = 10°$.

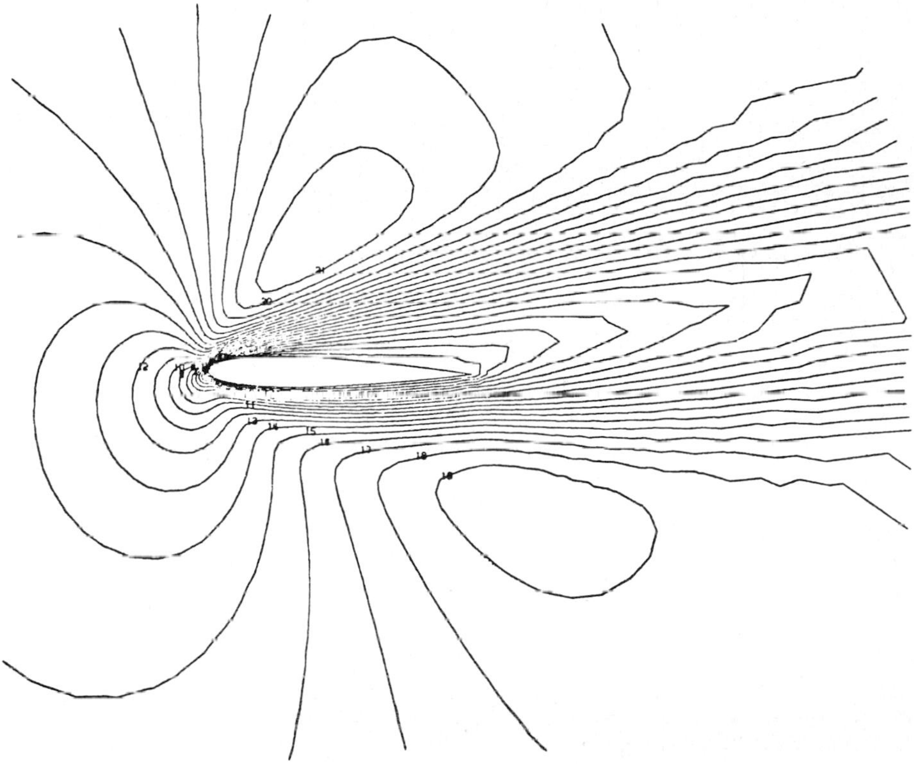

Figure 8.4. Mach lines
M_∞ = 0.8, Re = 73., α = 10°.

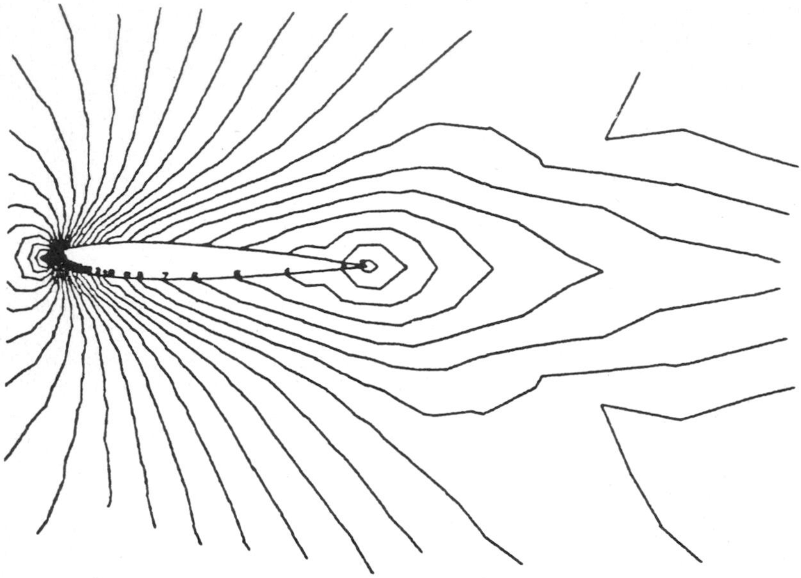

Figure 8.5. Density lines M = 1.1, Re = 100.

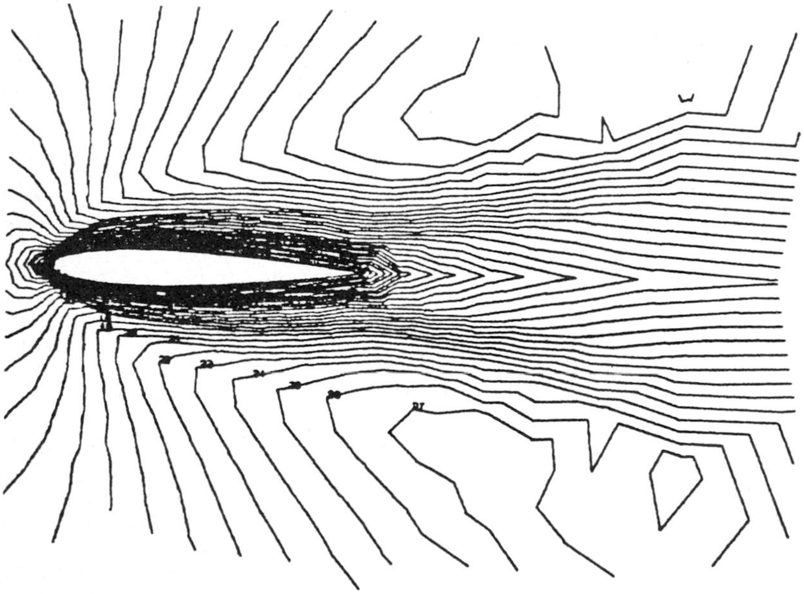

Figure 8.6. Mach lines M_∞ = 1.1, Re = 100.

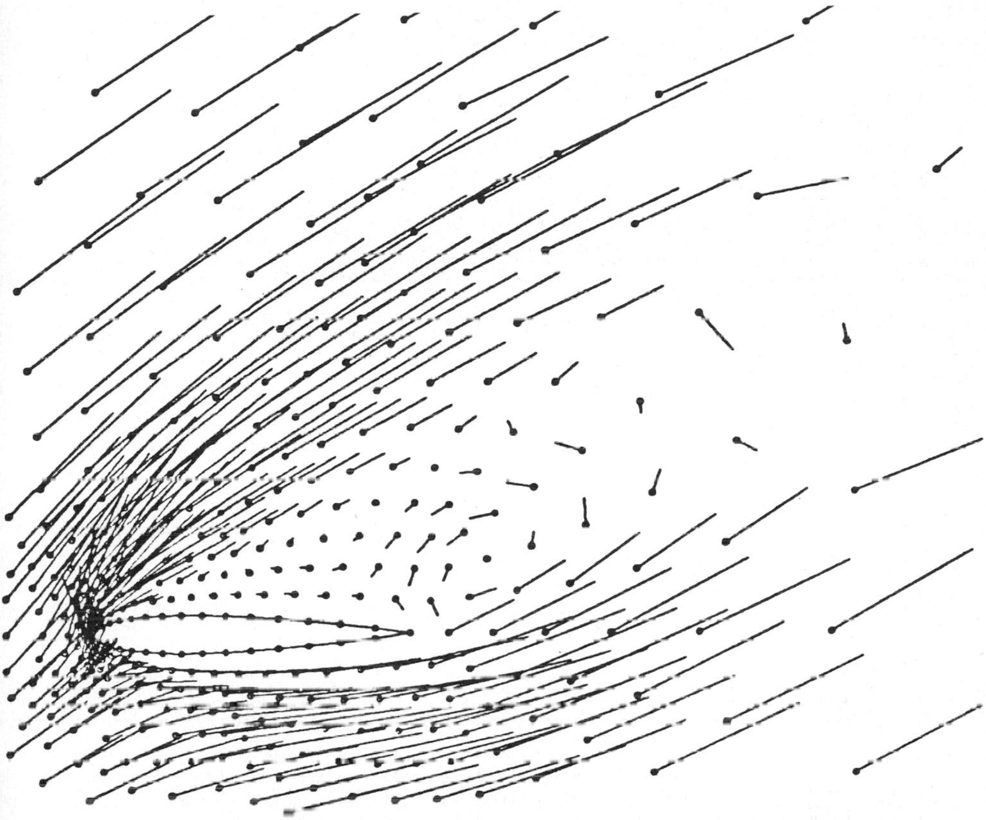

Number of time steps : 50

Figure 8.7 (a) : Velocity plottings

$M_\infty = 0.7$, $Re = 100$, $\alpha = 30°$, $\Delta t = 0.1$

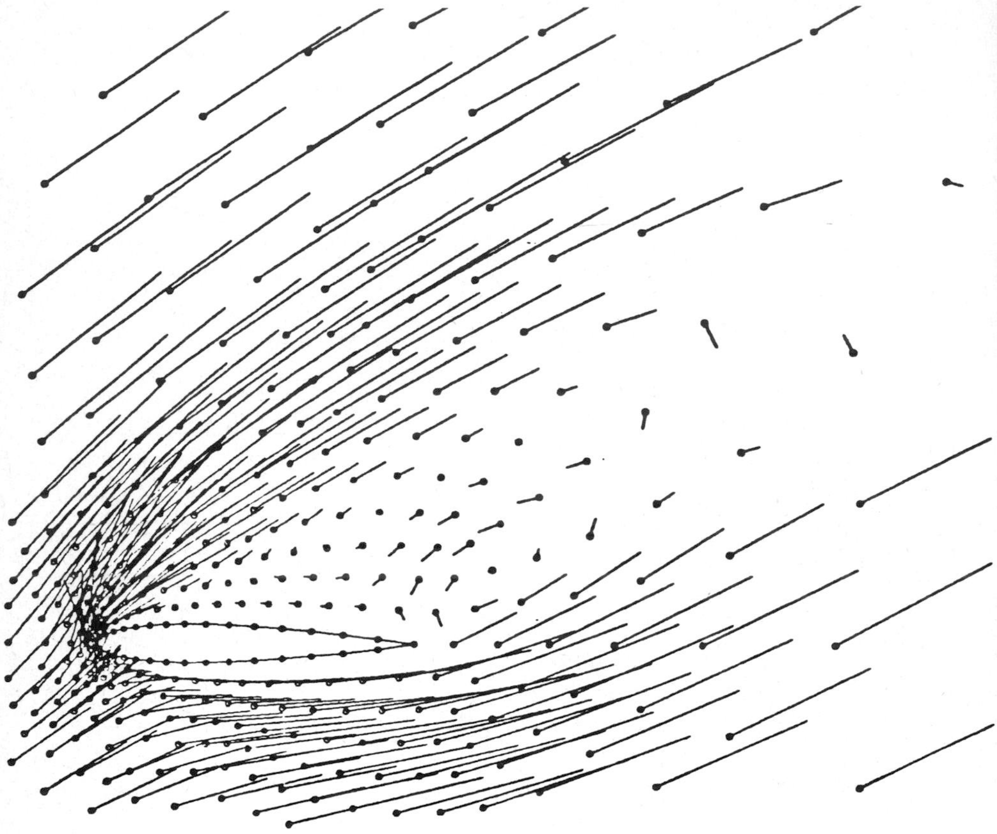

Number of time steps : 100

Figure 8.7 (b) : Velocity plottings

$M_\infty = 0.7$, Re = 100, $\alpha = 30°$, $\Delta t = 0.1$

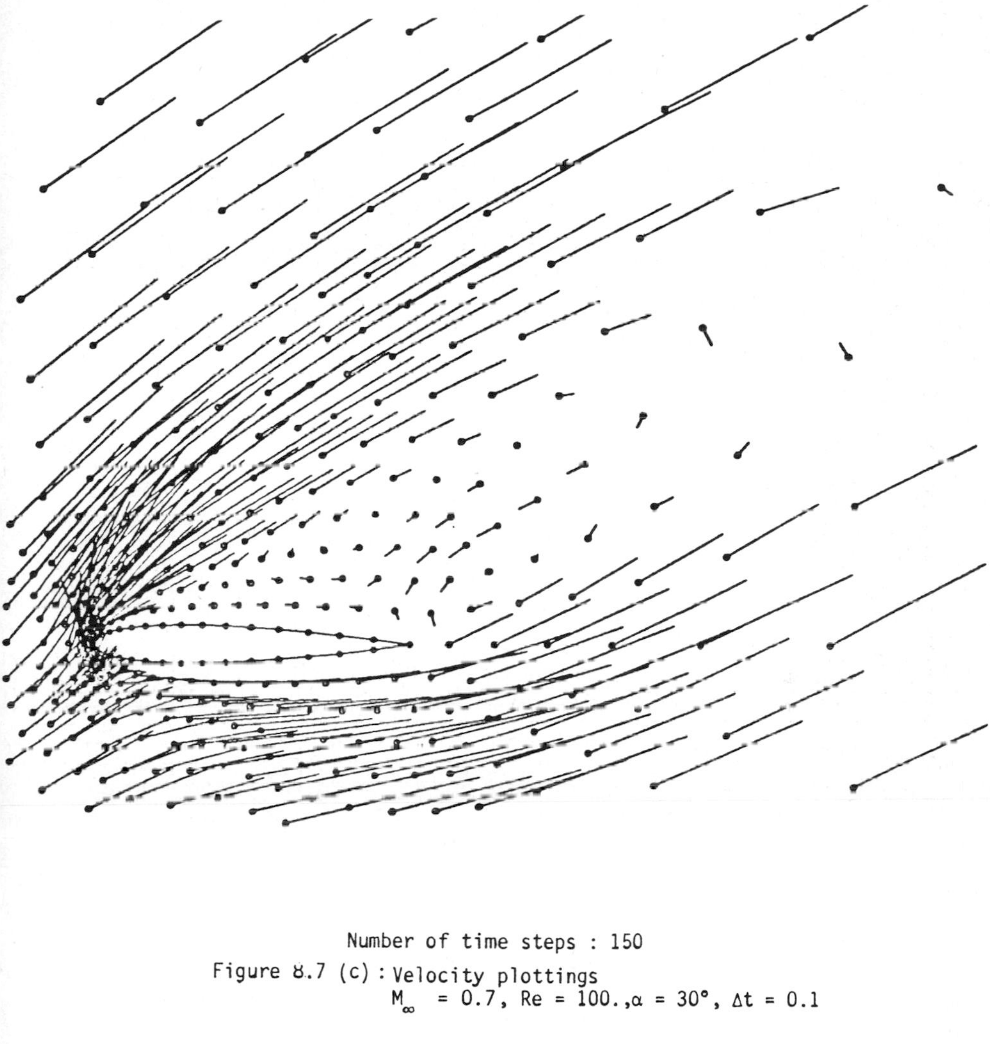

Number of time steps : 150

Figure 8.7 (c) : Velocity plottings
$M_\infty = 0.7$, $Re = 100.$, $\alpha = 30°$, $\Delta t = 0.1$

TURBULENT FLOWS

ECOULEMENTS TURBULENTS

Computing Methods in Applied Sciences and Engineering, VI
R. Glowinski and J.-L. Lions (Editors)
Elsevier Science Publishers B.V. (North-Holland)

LARGE EDDY SIMULATION: THE NEXT FIVE YEARS

Joel H. Ferziger

Department of Mechanical Engineering
Stanford University
Stanford, CA, U.S.A.

The prospect of major improvements in the performance of computers in the next five years means that large eddy simulation (LES), which has until now been strictly a research tool, may become a top-of-the-line engineering tool. In this paper, the historical development and past contributions of LES are reviewed. Then a discussion of the potential for applications of LES in new areas and of the developments needed to make LES a tool for the practicing engineer will be given.

I. HISTORICAL INTRODUCTION

The origins of what is now called large eddy simulation (LES) lie in the early, global, weather prediction models. Meteorologists quickly realized that computer resources permit use of only extremely coarse grids; in the early codes, the grid could hardly resolve the largest atmospheric structures. Modeling is required for the unresolved scales; considerable effort has been put into the development of such models.

It is hardly surprising that the first engineering application of LES was made by a meteorologist; Deardorff's pioneering paper of 1970 provided many of the foundations of the subject and influenced much of the later work. Although the practice of LES has been limited until recently to a small number of groups possessing the required resources, the increasing availability of large computers is allowing an increasing number of groups to participate in the development and application of the method.

The first work after Deardorff's was Schumann's thesis of 1973; following that, Schumann led a group at Karlsruhe that specialized in the simulation of heat transfer. Reynolds and the present author began work in 1972 and have concentrated on developing a fundamental formulation of the subject, systematic extension to more complex flows, and application to turbulence modeling. The NASA-Ames group, which began work in 1975, has specialized in state-of-the-art simulation of simple flows and on full simulation (see below). Leslie and his group in London began in 1976 to look at a number of issues, including the use of turbulence theories in developing subgrid scale models. In the past few years, several French groups have begun to apply LES; these include groups at Electricite de France, ONERA-Chatillon, Lyon, and Toulouse.

Full turbulence simulation (FTS), a close relative of LES, simulates turbulent flows without any modeling. The number of flows which can be simulated is more limited with this approach, and the Reynolds numbers are necessarily very small. The pioneering work in this field was done by Orszag and his group at MIT in 1970. The method has since been applied by a number of other groups, principally those which also employ LES.

With the exception of those in meteorology, the impact of LES on the simulation of flows of practical interest has been almost entirely indirect, due to the cost of the method. Recent advances in VLSI technology are producing dramatic reductions in the cost of computation, and large computers will become available to many more users in the near future. These improvements will make it possible for new groups to use LES, which may become the top-of-the-line engineering tool in the next decade. The coming supercomputers will also open up new directions for research in this field; this paper will explore some of these.

II. CONTRIBUTIONS OF LES TO DATE

A. Overview

As noted above, the practice of LES has been restricted to a few groups with access to large, fast computers. Furthermore, resource limitations required even those groups to limit the number of simulations they made. Consequently, runs have had to be selected with care; the choices generally reflected the goals of the particular research group. Most simulations made to date were aimed at demonstrating the potential of the approach and at exploring its capabilities and limitations and at turbulence model development, rather than at simulating flows of direct engineering interest.

This work has established that the conceptual basis of LES, i.e., that the important properties of turbulent flows are calculable by simulating the large eddies and modeling the small ones, is sound. At the lowest Reynolds numbers in simple geometries, it is possible to use FTS and no modeling is needed. At somewhat higher Reynolds numbers, most of the energy can be retained in the resolved part of the field, the results are not sensitive to the subgrid scale model, and satisfactory results are achieved with LES. Most applications require much higher Reynolds numbers, for which even the largest current LES programs can capture only a small fraction of the total turbulence energy. In these cases, one is demanding much more of the subgrid scale model, and a premium is placed on the ability of the model to accurately represent the unresolved scales of the turbulence.

In the remainder of this section, we shall review some of the accomplishments to date, with emphasis on those which seem most likely to be of use in the future. The author begs forgiveness if the emphasis appears to be on the work of his group. This reflects familiarity with that work rather than disregard for that of other groups.

B. Method Demonstration

Deardorff first demonstrated that the concepts behind LES are sound. He showed that many of the features of turbulent channel flow could be simulated on a relatively coarse grid. In this work, both the small-scale turbulence in the central region of the channel and everything occurring in the layers closest to the wall were treated by models.

The idea of using explicit filtering to define the large and small scales was introduced by Leonard. The Stanford group has continued to use this method. Others prefer Deardorff and Schumann's method, which combines filtering and numerical approximation. The principal advantagees to the two-step approach is that it clarifies thinking about what is computed and allows one to study subgrid-scale modeling more carefully.

The first flows considered by the Stanford group were homogeneous flows, the simplest turbulent flows. Using isotropic turbulence as the first test case, Kwak and Shaanan showed that LES could simulate this flow with a model that is independent of the grid size and the number of points used; this achievement provided faith that the subgrid-scale modeling concept had validity.

Use of FTS to test subgrid-scale models was first proposed by Clark et al. and was followed by McMillan and Ferziger. This work showed that Smagorinsky's model represents the energy transfer from the large to the small eddies well on the average, but poorly in detail; Moreover, the model becomes worse when applied to sheared or strained flows. This led to a search for new models, which eventually produced the scale-similarity model proposed by Bardina et al. and models based on two-point turbulence closures by Bertoglio et al. and Aupoix et al. These models are relatively new and untested but hold promise for the future.

At the completion of the work on homogeneous flows, it was decided to attack the next simplest flows--those which are homogeneous in two directions. The free shear flows in this category are the time-developing ones, while channel flows are the wall-bounded members of the set. Let us look at the results in this area.

At the outset, it appeared that the free shear flows would be the easier ones. However, they are the most energetic turbulent flows and grow very rapidly; these qualities make them difficult to follow for any length of time. One would like to use a grid which grows with the flow. Unfortunately, no one has yet found a method which accomplishes this without making severe approximations. Nonetheless, some progress has been made. Cain developed a method applicable to infinite domains, allowing him to simulate the transition of a mixing layer almost to the point of full development. Riley and Metcalfe produced interesting results on fully developed turbulent free-shear flows.

For wall-bounded flows, there has been considerable progress. Schumann, Grotzbach, and Kleiser have extended Deardorff's method to the computation of wall-bounded flows, including forced and natural convection heat transfer in planar and annular geometries. Comparisons with experimental data have been impressive.

The Deardorff-Schumann approach uses boundary conditions to represent the physics of the region closest to the wall. Since most of the interesting and important physics of wall-bounded flows occurs in these regions, the Stanford group felt it important to include this region in the simulation; no-slip wall conditions were used. Moin et al. demonstrated the feasibility of this approach, and Moin and Kim have refined the method and used it to discover new physics of turbulent flows near walls.

Recently, Baron has applied LES to flow of direct engineering interest in geometries more complicated than any considered previously. The grids used are coarse relative to the size of the eddies in the flow, but satisfactory results appear to have been achieved.

C. Impact on Turbulence Modeling

LES is currently too expensive for industrial use; the prediction of technological turbulent flows is done with simpler models. These include integral methods and, to an increasing degree, two-equation models. As the sophistication of these models increases, so does the need for a broad range of detailed experimental data. Such data are expensive to acquire; in some cases, the experimental techniques either do not exist or are not sufficiently accurate. This gap can be filled in part by LES--a goal of LES work for many years. Some success has been achieved.

Using either FTS or LES, it is possible to simultaneously compute the exact value of a quantity which is modeled by the simpler methods and the model prediction of it. By comparing these, it is possible to test the validity of the model and, assuming that the model is validated, to estimate the parameters of the model. For homogeneous flows, simulation methods have little to offer two-equation models, because the quantities have been measured accurately in experiments.

Useful information relating to Reynolds stress models has been obtained from simulations, cf. Rogallo, Feiereisen et al., and Shirani et al. The fluctuating pressure is particularly difficult to measure, and there are few data about the pressure-strain correlations that occur in the Reynolds stress equations. These terms are usually modeled in two parts, one associated with the mean velocity field (the rapid part), and a second deriving strictly from the turbulence (the slow part). Using FTS, the above-mentioned authors were able to show that the model commonly used for the slow terms is not very accurate. However, these models assume that the dissipation is isotropic, which is not the case. It turns out that, when the anisotropic component of the dissipation is added to the slow pressure-strain term, the model actually represents the combination well. Models for the rapid pressure-strain term did not fare well; attempts to find improvements were not very successful. This can be interpreted to mean that either different models for these terms or a new approach to modeling is needed.

Shirani et al. included a passive scalar in their work. They made the kinds of tests described above, with similar results. They also tested models of the turbulent Prandtl number. Several popular models were tested; all were found lacking in quality. They also presented a new model which their tests indicate to be better than the older ones, but it has not yet been tested in model simulations.

Less work on model testing has been done for inhomogeneous flows. In these flows, the LES results must be averaged over planes parallel to the main flow and, possibly, over time. This treatment yields a limited amount of data and prohibits detailed testing of models. The best one can do is to compute values of the model parameters as functions of the coordinate in the inhomogeneous direction and the tensor indices. Finding them not to be constant when the model assumes them to be constant is evidence of weakness in the model.

Schumann and his co-workers have done much of the model-testing work for wall-bounded flows. Mixing length, two-equation, and Reynolds stress models have been considered. They found that the models are not as accurate as modelers would like. It is doubtful that current models can be used for engineering design without tuning for the particular flow. Schumann's group included heat transfer in many of their simulations; the models for heat transfer fared no better than those for momentum transfer. These results are especially valuable, because the experimental data for heat transfer are less complete than the comparable data for cold flows. Kuhn et al. have recently used LES to discover interesting phenomena in flow over wavy walls.

Riley and Metcalfe tested Reynolds stress models for free shear flows. Again, it was found that the quality of the models leaves something to be desired.

In summary, FTS and LES have made important contributions to turbulence modeling. Unfortunately, most of the results are negative. It is the author's opinion that these results signify that no one model, and certainly no single set of constants, can fit a large variety of turbulent flows. Models of the future will probably need to be tuned to particular zones in the flow. Zonal models will need to be joined by still other models.

III. FUTURE DIRECTIONS OF LARGE EDDY SIMULATION

A. Advances in Computer Technology

Large eddy simulation has indeed come a long way, but there is still much to be done. Future directions will be determined in large part by trends in large computer development. The supercomputer projects in the U.S. and Japan have been well publicized. Fluid dynamics computations (including meteorology and oceanography) will consume a large fraction, perhaps even a majority, of resources of these machines. LES will undoubtedly be a major beneficiary. We anticipate a major step forward in the kinds of problems that are attacked with LES in the next five years. We shall give specifics below.

VLSI technology is making it possible to build chips with ever-increasing numbers of circuits. This technology already dominates in primary computer-memory applications and is beginning to become an important factor in secondary memories. In just over ten years since the introduction of the first four-bit microprocessor, we have seen progress through eight- and sixteen-bit chips to thirty-two bit microcomputers. The current generation of chips makes it feasible to build desk-top computers that match the mainframes of ten years ago in both speed and memory. When used in clever architectures, these chips will lead to significant reductions in the cost of computation. Again, there are important consequences for LES that we shall consider below.

B. LES on Supercomputers

The present generation of large computers--the Cray-1 and the CYBER 205--are capable of approximately 100 million floating-point operations per second and have 2-8 million words of fast memory. It seems fairly certain that machines with ten times the speed and memory will become available in the next few years. Machines with yet another order of magnitude in speed (with as yet unspecified memory sizes) may become available not long after that, say in five years. As these machines will cost as much as the present large machines, access to them will remain restricted to a small number of users. They will therefore be research machines, and the tasks assigned to them in the LES area will represent the extensions of current work. Let us look at some of the possibilities.

Work on homogeneous turbulence will continue. The new machines will make it possible to routinely use grids of $128 \times 128 \times 128$ points and, in some cases, $256 \times 256 \times 256$ points. One will then be able to simulate flows at the Reynolds numbers of laboratory experiments. These will include flows in which there is a clearly defined inertial subrange. Since most applications flows have inertial subranges, we shall be able to study subgrid scale modeling under conditions which resemble those in the flows of practical importance. It will also become possible to study sheared and strained turbulence over a long enough time span to answer questions that have perplexed turbulence theorists. This development could bring an exciting advance in turbulence theory.

For free shear flows, the added capacity will allow simulation from the inception of instability to the fully developed turbulent state. One can then study the effects of various perturbations on the initial stages of free shear flows; many technological applications require this. Sound generation by turbulent flows, which can be dealt with only crudely at the present state of the art, will be open for investigation. More importantly, we shall be able to simulate spatially developing free shear flows, including some of the pressure feedback effects. Finally, it may be possible to study other flows of technological interest, such as the impingement of free shear flows on solid walls.

Combustion is an area of great and obvious importance. Application of LES to combusting flows has been slow in coming, for several reasons. Chemical reaction requires mixing at the molecular level, so the small scales must be treated accurately; present models may not be accurate enough. When this is coupled with the necessity of carrying the notoriously stiff equations for the chemical kinetics, one sees that the task is formidable. Nevertheless, the new generation of supercomputers should make it possible to begin work on the simulation of turbulent combusting flows. The potential in this area is enormous, because experimental work is also difficult.

For wall-bounded flows, the new generation of computers will bring with it the ability to fully simulate turbulent channel flow and the temporally developing boundary layer. This capability allows simulation of transition without need of a model and thereby opens up many opportunities. The simulation of fully developed channel flow will make possiible the addition of other phenomena (for example, unsteadiness) and will introduce a new way of studying and developing models for every-day applications use. As in the free shear flows, the new capability will permit the introduction of new effects.

There are further areas of potential application, but, due to the author's lack of experience in them, they will get only short mention here. One such area is meteorology. Although many LES ideas were borrowed from it, meteorology could benefit from the systematic approach to LES. In meteorology, one is forced to model far more of the physics than in engineering (due to the much higher Reynolds numbers), and experimental data are scarcer. There is therefore a need for the kind of data provided by LES. An approach in which one systematically uses data acquired at one scale to model at the next larger scale seems especially attractive. In a similar vein, applications to oceanography should be possible.

C. LES in the Field

Large computers will be cheaper and more widely available in the near future. We anticipate that engineers will begin to use LES as a tool for final design testing in approximately five years. As with any other new method, there will first be exploratory tests to determine whether the idea is sound and to get the bugs out of the method. If these are successful, there should then be a slowly increasing use of LES as an applications tool. To allow this to happen, several developments in LES itself will have to take place first.

LES may find early application to flows that are inherently three-dimensional and time-dependent, for which turbulence models must be based on ensemble averaging. If one is doing a three-dimensional, time-dependent calculation, it could as easily be a large eddy simulation as an ensemble-average one. Which whould one choose? The answer depends on which questions one is trying to answer. Consider the in-cylinder flow for an internal-combustion engine. Some questions, such as the pressure history through a cycle and, perhaps, optimum spark timing, can be answered with an ensemble-average model. Others, such as the prediction of misfire, probably require analysis of individual cycles via LES. Operationally, the difference between the two approaches lies in the length scale in the turbulence model; otherwise the same code can be used in both approaches! In meteorological flows, a similar choice arises and use of LES is indicated in weather prediction applications, but ensemble averaging might make sense for climate studies.

There are many free shear flows of technological interest, including combusting flows. For these, the future of LES is not clear at present; a number of research issues need to be resolved before we can contemplate using LES in applications. These include: the development of a method of producing inflow conditions that accurately represent a turbulent flow at the proper state of development; the development of accurate conditions for the computational outflow surface; and, finally, the development of a method of dealing with the rapid growth of length scale, i.e., a method which adapts the grid size to the local eddy size. These are areas that need to be worked on if LES is to become an applications tool.

One faces many of the same difficulties in wall-bounded flows. Spatially developing, wall-bounded flows will require methods of dealing with the in- and out-flow boundaries. Since these flows are less rapidly developing than free shear flows, the problem is one of considering a sufficiently large part of the flow while using a grid small enough to capture the important eddies. To do this at Reynolds numbers of technological interest will require eliminating the layers closest to the wall from consideration. Artificial boundary conditions of the type used by Deardorff and Schumann will have to be used. Their accuracy needs to be checked, and new models may need to be developed. We believe that this can be done by using LES at relatively low Reynolds numbers and bootstrapping to higher Reynolds numbers. Many other phenomena which occur in boundary layers should also be examined. These include: heat transfer, blowing, suction, curvature, and rotation.

These appear to be the principal directions in which LES can make important contributions in the not-too-distant future. There is much to do, and, if this program is to come to fruition, the number of people active in the field will need to increase.

IV. CONCLUSIONS

An explosion in computer capacity is coming in the next several years. Driven by the recent advances in VLSI, the technology can be expected to produce both much larger supercomputers and cheaper, but still quite large, computers capable of doing LES. This development portends significant growth of LES as a tool for turbulence research and as a top-line tool for applications. This paper has explored some lines which this work can be expected to follow.

Acknowledgments

This work has been supported in part by NASA-Ames Research Center and the Office of Naval Research. In a paper of this nature, the contributions of colleagues and students loom large. Many of them deserve mention, but, in order to avoid omitting someone, I shall mention only the one who has influenced my thinking the most, Prof. W. C. Reynolds.

References

Aupoix, B., Cousteix, J., and Liandrat, J., "Effects of Rotation on Isotropic Turbulence," Proc. 4th Symp. Turb. Shear Flows, Karlsruhe, 1983.

Bardina, J., Ferziger, J. H., and Reynolds, W. C., "Improved Turbulence Models Based on Large Eddy Simulation of Homogeneous, Incompressible Turbulent Flows," Rept. TF-19, Dept. of Mech. Engrg., Stanford University, 1983.

Baron, F., "Macrosimulation Tridimensionelle d'Ecoulement Turbulent Cisailles," thesis, Univ. of Paris VI, 1982.

Bertoglio, J., and Mathieu, J., "Study of Subgrid Models for Sheared Turbulence," Proc. 4th Symp. Turb. Shear Flows, Karlsruhe, 1983.

Cain, A. B., Reynolds, W. C., and Ferziger, J. H., "Simulation of the Transition and Early Turbulent Regions of a Free Shear Flow," Rept. TF-14, Dept. of Mech. Engrg., Stanford University, 1981.

Clark, R. A., Ferziger, J. H., and Reynolds, W. C. "Direct Simulation of Homogeneous Turbulence and Its Application to Testing Subgrid Scale Models," Rept. TF-9, Dept. of Mech. Engrg., Stanford University, 1975.

Deardorff, J. W., "Numerical Simulation of Turbulent Channel Flow at Large Reynolds Number," J. Fluid Mech., 41, 452 (1970).

Feiereisen, W. J., Reynolds, W. C., and Ferziger, J. H., "Computation of Compressible Homogeneous Turbulent Flows," Rept. TF-13, Dept. of Mech. Engrg., Stanford University, 1981.

Orszag, S. A., and Patterson, G. S., "Numerical Simulation of Turbulence," in Statistical Models and Turbulence, Springer-Verlag, 1972.

Leonard, A., "Energy Cascade in Large Eddy Simulation of Turbulent Fluid Flows," Adv. in Geophysics, 18A, 237 (1974).

McMillan, O. J., and Ferziger, J. H., "Direct Testing of Subgrid Scale Models," AIAA Journal, 17, 1340 (1979).

Moin, P., Reynolds, W. C., and Ferziger, J. H., "Large Eddy Simulation of Turbulent Channel Flow," Rept. TF-12, Dept. of Mech. Engrg., Stanford University, 1978.

Moin, P., and Kim, J. J., "Numerical Investigation of Turbulent Channel Flow," J. Fluid Mech., 118, 341 (1982).

Riley, J. J., and Metcalfe, R. W., "Simulation of Free Shear Flows," Proc. 3rd Symp. Turb. Shear Flows, Davis, 1981.

Rogallo, R. J., "Experiments in Homogeneous Turbulence," Rept. TM-81315, NASA-Ames Research Center, 1981.

Schumann, U., "Ein Verfahren zur Direkten Numerischen Simulation Strömungen in Platten- und Ringspalt-Kanalen und uber seine Anwendung zur Untersuchung von Turbulenzmodellen," dissertation, Karlsruhe, 1973.

Shirani, E., Ferziger, J. H., and Reynolds, W. C., "Simulation of Homogeneous Turbulent Flows including a Passive Scalar," Rept. TF-15, Dept. of Mech. Engrg., Stanford University, 1981.

Computing Methods in Applied Sciences and Engineering, VI
R. Glowinski and J.-L. Lions (Editors)
Elsevier Science Publishers B.V. (North-Holland)

CONVECTION OF MICROSTRUCTURES (II)

C. Bègue (AMD-BA)
T. Chacon (INRIA)
D.W. McLaughlin (U. of Arizona)
G.C. Papanicolaou (Courant Inst., NYU)
O. Pironneau (U. of Paris 13 and INRIA)

The analysis of flows with rapidly varying structures, turbulent flows, is a very complex mathematical problem that cannot be approached by direct numerical simulation. It is necessary to reduce somewhat the analysis to solving equations that do not have rapidly varying data : averaged equations. In this paper, which is a continuation of [1] presented 2 years ago, homogenization is used to average Euler and Navier-Stokes equations. An inviscid model is derived and tested on the 2-D wake behind a cylinder. A viscous model is derived on the basis of the Kolmogorov's scales.

NOTATIONS

$\langle w \rangle$: mean of w with respect to y
$\langle w \otimes w \rangle$: second order tensor with component $\langle w_i w_j \rangle$
$A:B$: $\sum_{ij} A_{ij} B_{ij}$
w : fluctuation of the mean flow
u : mean flow
q : kinetic energy of w
$\bar{a}$: Lagrangian coordinate in the flow
$C = \nabla a^T \nabla a$, $\bar{C} = \nabla \bar{a} \nabla a^T$
$v = \nabla a^T v + \langle \tilde{v} \rangle$

INTRODUCTION

In [1][2] the convection of microstructures by Euler's equations was studied :

(1) $$u_t^\varepsilon + u^\varepsilon \nabla u^\varepsilon + \nabla p^\varepsilon = 0 \qquad \nabla \cdot u^\varepsilon = 0$$

(2) $$u^\varepsilon(x,0) = U(x) + \sqrt{q_0(x)}\, W(\tfrac{x}{\varepsilon}) \;;$$

$U(x)$ is the mean velocity field initially and $W(\frac{x}{\varepsilon})$ is the fluctuation of the velocity assumed to be rapidly varying when ε is small.

The main result of the analysis in [1][2] is as follows. The mean of u^ε behaves like u which satisfies the system of equations

(3) $$u_t + u\nabla u + \nabla p = \nabla \cdot \tau, \quad \nabla \cdot u = 0$$

where the tensor τ is a function of $C = \nabla a^T \nabla a$ and q, where the Lagrangian coordinate a and the scalar field q satisfy

(4) $$a_t + u\nabla a = 0 \qquad a(x,0) = x$$

(5) $$q_t + u\nabla q = \nabla u:\tau \qquad q(x,0) = q_o(x)$$

The stress tensor τ is computed from the solution of a canonical microstructure problem in which C and q are parameters : (< > stands for the mean)

(6) $$\tilde{w}\nabla_y\tilde{w} + C\nabla_y\tilde{\pi} = 0 \ , \ \nabla_y\cdot\tilde{w} = 0 \ ; \ \tilde{w} \quad Y\text{-periodic}$$

(7) $$\frac{1}{2} < (\tilde{w} - <\tilde{w}>)\, C^{-1}(\tilde{w} - <\tilde{w}>)> = 1 \ ; \quad T_{ij}(C) = <\tilde{w}_i\tilde{w}_j>$$

(8) $$\tau = -q\nabla a^{-T}T(\nabla a^T\nabla a)\nabla a^{-1}$$

This result was obtained by an asymptotic expansion of the form

(9) $$\begin{cases} u^{\varepsilon}(x,t) = u(x,t)+(w(y,x,t)+\varepsilon u^{(1)}(y,x,t)+\varepsilon^2u^{(2)}+\ldots)\Big|_{y=\frac{a(x,t)}{\varepsilon}(1+\varepsilon\sigma(x,t))} \\ w(y,x,t) = \nabla a^{-T}(x,t)(\tilde{w}(y,(a(x,t)) - <\tilde{w}>)\sqrt{q} \end{cases}$$

It is valid so long as $\tilde{w}$ can be defined from (6)-(7), and W at t=0, in a unique manner. The main point of the result is that when the initial velocity field is made up of two fields with well separated (by ε) scales of variation (something called spectral segregation in [3]) the evolution of the large scale motion and the small scale motion can be followed for relatively long times by a system of coupled equations like (3)-(5).

While deriving (3)-(5) it was found, from the principle of conservation of global helicity $<\tilde{w}\nabla_y\times\tilde{w}>$ in (6), that σ, appearing in (9), satisfies for some function g

(10) $$\sigma_t + u\nabla\sigma = \sqrt{q}\, g(a,\nabla a).$$

Thus this resultsin a 2-equations turbulent model like the k-ε model [4] used in engineering : u computed by (3) with

(11) $$\tau = -\alpha\frac{q^2}{e}\nabla u$$

(12) $$q_t + u\nabla q - \alpha\frac{q^2}{e}u_{i,j}(u_{i,j}+u_{j,i}) - \beta q^{3/2} - \gamma\nabla\cdot(\frac{q^2}{e}\nabla q) = 0$$

(13) $$e_t + u\nabla e - \beta q\, u_{i,j}(u_{i,j}+u_{j,i}) - e^2/q - \delta\nabla\cdot(\frac{q^2}{e}\nabla e) = 0$$

The main difference between (3)-(5),(10) and (3),(11)-(13) lies in the fact that (10) is decoupled from (3) and in the fact that τ in (8) is not a diffusion tensor. In this paper we show that if the asymptotic expansion is carried further in (3) then (10) becomes couples with (3) and (for some W) τ has a diffusion part. Then we discuss the implementation of the model thus derived and we conclude by presenting some numerical results on "turbulent" flows

1. Viscous effects.

In a manner similar to [1] let us analyse the Navier-Stokes equation with large Reynolds number :

(14) $$u_t^\varepsilon + u^\varepsilon \nabla u^\varepsilon + \nabla p^\varepsilon - \nu(\varepsilon)\Delta u^\varepsilon = 0 \qquad \nabla \cdot u^\varepsilon = 0$$

If e(x,t) denotes the mean rate of energy in the fluid, then according to Kolmogorov theory [5], in a frame of reference moving with the mean flow, the oscillating part of the flow is a random process which depends only on e and ν. By dimensional analysis then, the only possible length, velocity and time scales are

(15) $$\varepsilon' = (\nu^3/e)^{1/4} \ , \ w' = (\nu e)^{1/4} \ , \ \tau' = (\nu e)^{1/2}$$

or equivalently

(16) $$\nu = (\varepsilon' e)^{4/3}/e^{2/3} \ , \ w' = (\varepsilon' e)^{1/3} \ , \ \tau' = (\varepsilon' e)^{2/3}/e$$

Thus the initial value problem relevant to turbulence modelling with (14) is

(17) $$u^\varepsilon(x,0) = U(x) + \varepsilon^{1/3} W(x, \tfrac{x}{\varepsilon}) \ , \ \nu = \mu\, \varepsilon^{4/3} \ ,$$

and one would expect an Anzats in powers of $\varepsilon^{1/3}$ begining with

$$u(x,t) + \varepsilon^{1/3} W(a(x,t)/\varepsilon, x, t).$$

Unfortunately this yields an equation for w with no non-constant solutions :

(18) $$\tilde{w}\nabla\tilde{w} + \nabla\tilde{\pi} - \mu\Delta\tilde{w} = 0.$$

So we are not able to analyze such strong viscosity effects. Instead we choose

(19) $$\nu = \mu\varepsilon^2.$$

This means that $\varepsilon^{1/3}w$ is well inside Kolmogorov's inertial range. Let us rewrite u^ε, p^ε as follows

(20) $$u^\varepsilon(x,t) = u(x,t) + \varepsilon^{1/3}w(y,x,t) + \varepsilon^{2/3}u^{(1)}(y,x,t) + \ldots + \varepsilon^{\frac{1+n}{3}}u^{(n)}(y,x,t) + \ldots$$

(21) $$p^\varepsilon(x,t) = p(x,t) + \varepsilon^{2/3}\pi(y,x,t) + \varepsilon\, p^{(1)}(y,x,t) + \ldots + \varepsilon^{\frac{n+2}{3}} p^{(n)}(y,x,t) + \ldots$$

where

(22) $$y = \varepsilon^{-1} a^\varepsilon(x,t)(1+\varepsilon\sigma(x,t))$$

(23) $$a^\varepsilon(x,t) = a(x,t) + \varepsilon^{1/3}a^{(1)}(x,t) + \varepsilon^{2/3} a^{(2)}(x,t) + \ldots$$

All functions may depend on ε also and $w, u^{(i)}, \pi, p^{(i)}$ are periodic functions or stationary random processes. The additional functions $a^{(i)}$ are introduced for convenience ; we shall use them to set to zero the means, in y, of $u^{(i+1)}, p^{(i+1)}$. The scaling factor σ really means that the size of the cell of periodicity is allowed to vary slowly. It corresponds to a time oscillation of order 1, which is also the viscous time scale of the problem (i.e. $w_{,\tau} \sim \mu\varepsilon^2 \Delta w \Rightarrow \tau = O(1)$).

Inserting the Anzats into (14) yields a set of equations for the mean flow (see the derivation in Appendix)

$$u_t + u\nabla u + \nabla p + \varepsilon^{2/3} \nabla\cdot\langle w\otimes w\rangle + \varepsilon^{4/3} \nabla\cdot\langle w\otimes u^{(2)} + u^{(2)}\otimes w\rangle = O(\varepsilon^{5/3}) \tag{24}$$

$$\nabla\cdot u = 0 \tag{25}$$

$$q_t + u\nabla q + \langle w\otimes w\rangle : \nabla u + \langle|\nabla a \nabla w|^2\rangle + \varepsilon^{1/3}A + \varepsilon^{2/3}B = O(\varepsilon) \tag{26}$$

$$\sigma_t + u\nabla\sigma + \varepsilon^{1/3}(\beta\nabla\sigma + \gamma\sigma) + \delta + \varepsilon^{1/3} = O(\varepsilon^{2/3}) \tag{27}$$

where $A, B, \beta, \gamma, \delta, \phi, w, u^{(2)}$ are complex functions of $q, \sigma, u, w, \theta^{\varepsilon}$, given in Appendix. In general (24)-(27) are coupled and (24) contains an eddy viscosity factor in the last term because $u^{(2)}$ is itself linearly dependent on u (among other things, see (A14)).

2. Convection of a periodic array of eddies.

Let us focus on the special case W odd in x/ε in (17). Then the solution of (6)-(7) remains odd when C changes with time and so does $u^{(2)}$. Therefore the system of equations is much simpler : a is given by (4), w is given by (6),(7),(9b), u is given by (24),(25), q,σ are given by (26) with A=0 and

$$B = \langle w, u_t^{(2)} + u\nabla u^{(2)} + u^{(2)}\nabla u + \nabla a^T u^{(2)} \nabla_y u^{(2)}\rangle \tag{28}$$

$$\begin{cases} \sigma'_{,t} + u\nabla\sigma' + \langle w\nabla w + \nabla\pi, \tilde{\nabla}_y \times \tilde{w}\rangle / \langle a\cdot\nabla_y w \tilde{\nabla}_y \times \tilde{w}\rangle = O(\varepsilon^{1/3}) \\ \sigma = \varepsilon^{1/3}\sigma' \end{cases} \tag{29}$$

and $u^{(2)}$ is defined by (30) with $\tilde{u}^{(2)}$ solution of (31) :

$$\begin{cases} f^{(2)} = \dfrac{1}{\sqrt{q}} \nabla a^T (w_t + u\nabla w + w\nabla u - \mu a_{\ell,j}\, a_{k,j} \dfrac{\partial^2 w}{\partial y_k \partial y_\ell}) + O(\varepsilon^{2/3}) \\ u^{(2)} = \nabla a^{-T}(\tilde{u}^{(2)} - \langle u^{(2)}\rangle) \end{cases} \tag{30}$$

$$\begin{cases} \tilde{u}^{(2)}\nabla_y \tilde{w} + \tilde{w}\,\nabla_y \tilde{u}^{(2)} + C\nabla p^{(2)} = -\mathbb{P} f^{(2)}, \ \tilde{u}^{(2)} \text{ Y-periodic} \\ \nabla_y\cdot\tilde{u}^{(2)} = 0 \ ; \ \langle\tilde{u}^{(2)} C^{-1} \tilde{w}\rangle = 0 \end{cases} \tag{31}$$

If we look for eddy viscosity terms in this system, we notice that

$$\mathbb{P}f^{(2)} = \frac{1}{\sqrt{q}} \nabla a^T (w\nabla u - \frac{1}{2} w \langle w\otimes w\rangle \nabla u) + \text{terms not explicitly dependent on } \nabla u$$

Therefore if we define $\tilde{\chi}^{ij}$ by

(32) $\tilde{\chi}^{ij}$ solution of (31) with $f_k^{(2)} = C_{ki}\, \tilde{w}_j - \frac{1}{2} \langle \tilde{w}_i \tilde{w}_j\rangle\, \tilde{w}_k$

it is easy to check that

$$\tilde{u}^{(2)} = a^{-1}_{i,m}\, a^{-1}_{j,n}\, \tilde{\chi}^m \frac{\partial u_i}{\partial x_j} + \text{terms not explicitly dependent on } \nabla u.$$

and (24) becomes ($w' = w/\sqrt{q}$ and χ^{ij} depend only on a) :

(33) $$u_t + u\nabla u + \nabla p + \nabla\cdot(q\langle w'\otimes w'\rangle + \varepsilon^{2/3}\sqrt{q}\, \frac{\partial u_i}{\partial x_j})\, {}^{ij}_{mn} = 0$$

where $\xi^{ij}_{mn} = \langle w' \chi^{mn}_j + w'_j\ \chi^{mn}_i\rangle$.

The contribution to (24) of the other terms of (30) may be negligible ; for instance the term $\sqrt{q}\, u\nabla(w')$ will yield in (24) something like $\varepsilon^{4/3}\nabla\cdot(\xi u)$ which is small compared with $u\nabla u$. In the same manner B may be neglected in (26). Thus (27) is decoupled from (24)-(26).

3. Frame invariance.

To compute $\langle w\otimes w\rangle$, ξ^{ij} and $\langle|\nabla a\nabla w|^2\rangle$ we shall use the fact that for isotropic W the final system ought not to depend upon the frame of reference in which it is written. So if $E(u) = 0$ is the mean flow equation and if R is a rotation matrix then $E(Ru) = RE(u)$. This implies that by rotation τ is changed into $R\tau R^T$. Now $\langle\tilde{w}\otimes\tilde{w}\rangle/q$ is a function of a only, so by the Rivlin-Ericksen theorem [7] there exists 3 functions of the invariants i_c of $\nabla a\nabla a^T$ such that

(34) $$\langle w\otimes w\rangle = q[\alpha(i_c)I + \beta(i_c)\nabla a\nabla a^T + \gamma(i_c)(\nabla a\nabla a^T)^2]$$

(35) $$i_c = \{\Sigma a^2_{ij}\ ,\ \sum_{i\neq j} a^2_{ik}\, a^2_{j\ell} - a_{ik}a_{i\ell}a_{jk}a_{j\ell}\}$$

It is interesting to note that the second invariant is a function of the first one for planar flows. The divergence of $q\alpha I$ is a pressure term so for planar flows τ appearing in (3) and (5) is :

(36) $$\tau = q[\beta(\Sigma a^2_{ij})I + \gamma(\Sigma a^2_{ij})\nabla a\nabla a^T]\ \nabla a\nabla a^T$$

Similarly $\langle|\nabla a\nabla w|^2\rangle$ is a function of i_c only.

4. Quasi-linearization

In [2] an attemps was made to compute the functions β and γ by solving directly the w equations (6)-(7) by a least square approach ; it turned out to be a delicate and expensive operation which is still under progress ; it is difficult to compute a continuous w (versus C) because of the bifurcations. Therefore we also studied the effect of a linear approximation of τ on the mean flow ; from (3)

(37) $$\tau = -q\,\lambda(\nabla a+\nabla a^T) + o(\|\nabla a-I\|)$$

Similarly (32), the fact that $\xi^{ij} = \xi^{ji}$ and the frame invariance yield :

(38) $$<w\otimes\xi^{ij} + \xi^{ij}\otimes w> = -\sqrt{q}\,\eta(\nabla u+\nabla u^T) + O(\|\nabla a-I\|)$$

Thus a quasi-linearized approximation of the complete system of equations is

(39) $$\begin{cases} u_t + u\nabla u + \nabla\cdot[\lambda q(\nabla a+\nabla a^T) + \eta\varepsilon^{2/3}\sqrt{q}(\nabla u+\nabla u^T)] + \nabla p = 0 & \\ \nabla\cdot u = 0 & ;\ u(x,0) = u^o(x) \\ a_t + u\nabla a = 0 & ;\ a(x,0) = x \\ q_t + u\nabla q + \lambda q(\nabla a+\nabla a^T) : \nabla u + \mu\gamma q = 0 & ;\ q(x,0) = q^o(x) \end{cases}$$

where λ,η,γ are numerical constants. The signs of λ and η are very important for stability ; they seem to depend upon W. A direct simulation of (6)-(8) gave a positive sign for λ for most W (see [2]).

5. Numerical experiments

From the numerical point of view the obvious advantages of (39), compared with (14), lie in the fact that the initial conditions are smooth and $\nu\varepsilon^2$ has been replaced by $\eta\sqrt{q}\ \varepsilon^{2/3}$.

We have tested the method on the jet problem in [2]. Here we shall investigate the wake behing a circular cylinder (fig. 1). The domain occupied by the flow being bounded, one must modify the conditions on a and set at the intake Γ_i :

(40) $$a|_{\Gamma_i} = x-ut$$

Model (39) is not able to create small structures from scratch ($q^o \equiv 0 \Rightarrow q \equiv 0$) ; thus we assume that the small structures are created by the boundary layer and mixed with the mean flow at the points of detachment.

To discretize system (39) we use a mixture of Characteristics and Finite Elements ; $\mathcal{C}_h$ and $\mathcal{C}_{2h}$ being two triangulations, one half the size of the other, we have

(41) $$\begin{cases} \frac{1}{\Delta t}\int_\Omega [u_h^{n+1}-u_h^n(x-u_h^n(x)\Delta t)]v_h\,dx + \eta\varepsilon^{2/3}\int_\Omega(\nabla u_h^{n+1}+\nabla u_h^{n+1})\nabla u_h\sqrt{q^n}\,dx \\ \qquad + \lambda\int_\Omega(\nabla a_h^n+\nabla a_h^n)\nabla v_h q_h^n\,dx = 0 \quad \forall v_h\in J_{oh} \\ u_h-u_\Gamma\in J_{oh} = \{v_{hi}\in V_h : \int_\Omega \nabla\cdot v_h p_h\,dx = 0,\ \forall p_h\in V_{2h}\ ;\ v_{h|\Gamma} = 0\} \end{cases}$$

(42) $$V_h = \{q_h \text{ continuous piecewise linear on } \mathcal{C}_h\}$$

$$(43)\qquad \begin{cases} \int_\Omega a_h^{n+1}\, v_h\, dx = \int_\Omega a_h^n(x-u_h^n(x)\Delta t) v_h\, dx \qquad \forall v_h \in V_h \\ a_h \in V_h \end{cases}$$

To compute q_h^{n+1} we set $z = \log q$, solve

$$(44)\qquad \int_\Omega q_h^{n+1} w_h\, dx = \sum_i (\exp r^{n+1}(\xi_i^{n+1})) w_h(\xi_i^{n+1}) \qquad \forall w_h \in V_h \;;\; \xi_i^{n+1} = \xi_i + u_h^n(\xi_i)\Delta t$$

and use a particule-in-cell method to compute $r^{n+1}(\xi_i)$

$$(45)\qquad r^{n+1}(\xi_i^{n+1}) = r^n(\xi_i^n) + [\lambda(\nabla a_h^n + \nabla a_h^{nT})\nabla u_h^n|_{\xi_i^n} - \mu]\Delta t\ .$$

All the particules ξ_i^n have come from the detachment points, where they escape at a constant rate with weight log q^0.

The simulation is done in 2 dimensions with 2600 nodes for u_h and about 200 time steps and 20 particules released at each time step. The test have been done with a constant viscous tensor ($\sqrt{q^n}$ constant in (49-a)).

CONCLUSION

Turbulence modelling is very difficult for non stationary flows. Homogenization provides a tool for the derivation of models with the following limitations :

- A spectral gap is needed between the small and the large structures

- Difficult mathematical hypotheses are needed for the behavior of the small structures.

Nevertheless the numerical experiments have shown that, within the stability region of the algorithm, large to small structures interactions are visible ; whether these correspond to reality will be decided by comparison with experimental data.

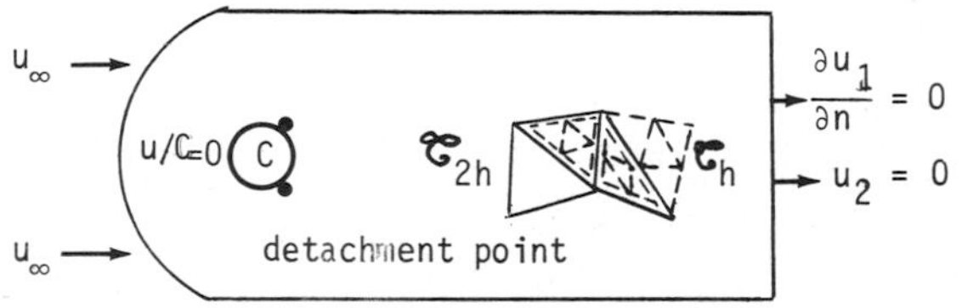

Figure 1

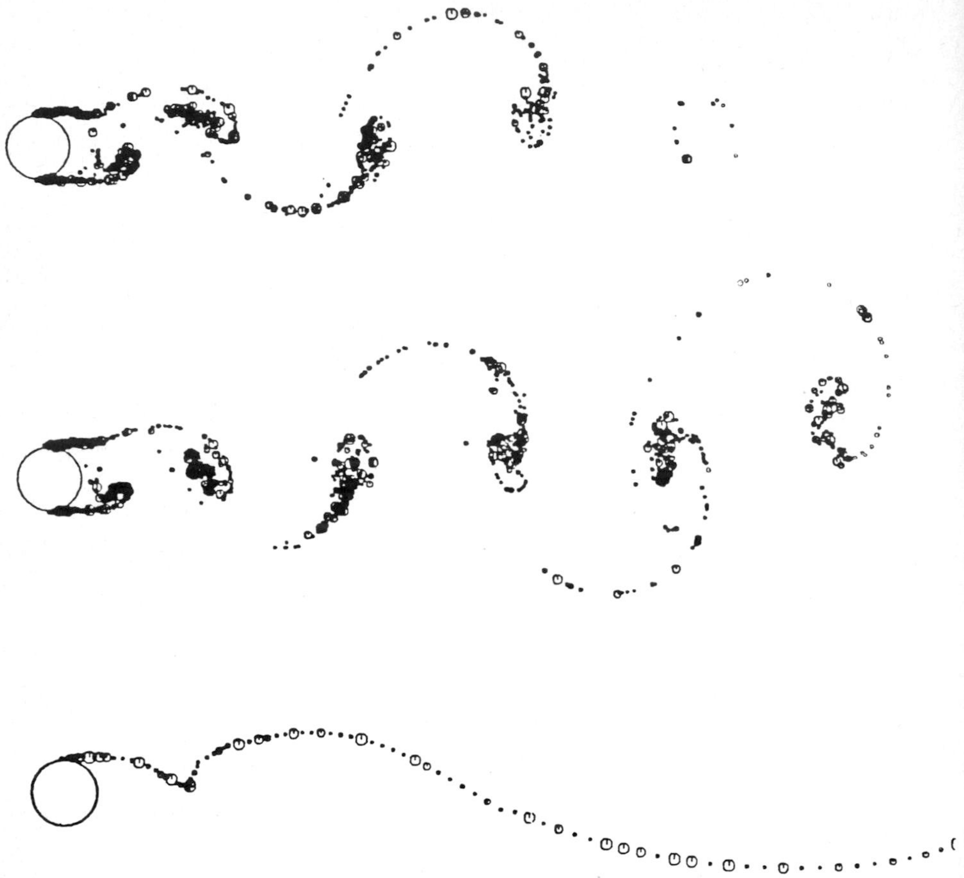

Figure 2

The top two pictures represent q at two times ; since q is discretized by point masses they are represented by small circles proportional in size to the weight of q at the center. Notice that q tends to be large in the eyes of the big eddies. The bottom picture shows the history of a point mass of q followed in time. Notice the change in weight as it travels, thereby showing the interaction between small scales and large scales.

APPENDIX

Inserting the Anzats into (14) yields (here a is replaced by θ)

$$0 = \varepsilon^{-3/2}[(\theta_t+u\nabla\theta)\cdot\nabla_y w] + \varepsilon^{-1/3}[(\theta_t+u\nabla\theta)\cdot\nabla_y u^{(1)}+(\nabla\theta^T w+\theta_t+u\nabla\theta)\cdot\nabla_y w+\nabla\theta\nabla_y\pi] + \varepsilon^0(\ldots) \tag{A1}$$

Therefore to cancel these first terms, let us define $\theta,\theta^{(1)},w,\pi$ in terms of $q = \frac{1}{2}<|w|^2>$ by

$$\theta_t + u\nabla\theta = 0 \quad \theta(x,0) = x \; ; \; C(x,t) = (\nabla\theta^\varepsilon)^T\nabla\theta^\varepsilon . \tag{A2}$$

$$\left\{\begin{array}{l} \tilde{w}\nabla_y\tilde{w} + C\nabla_y\pi = 0, \; \nabla_y\cdot\tilde{w} = 0 \\ \frac{1}{2} <(\tilde{w}- <\tilde{w}>)C^{-1}(\tilde{w}-<\tilde{w}>)> = 1 \; ; \; w' = \nabla\theta^{\varepsilon^{-T}}(\tilde{w}-<\tilde{w}>) \; ; \; w = \sqrt{q}\, w' . \end{array}\right. \tag{A3}$$

$$\theta_t^{(1)} + u\cdot\nabla\theta^{(1)} = <\tilde{w}> \sqrt{q} \; , \; \theta^{(1)}(x,0) = 0 \tag{A4}$$

Now denote by

$$L(u^{(j)}) = \sqrt{q}\, \nabla\theta^{\varepsilon^{-T}}(\tilde{w}\cdot\nabla_y(\nabla\theta^{\varepsilon^T} u^{(j)}) + \nabla\theta^{\varepsilon T}u^{(j)}+(\partial_t+u\cdot\nabla)\cdot(\theta^{(j+1)}+\sigma\theta^{(j-2)}))\nabla_y\tilde{w} + C\nabla_y p^{(j)} \tag{A5}$$

so (2.4) is

$$\left\{\begin{array}{l} 0 = L(u^{(1)}) + u_t + u\nabla u + \nabla p \\ + \varepsilon^{1/3}\{L(u^{(2)})+w_t+u\nabla w+w\nabla u+(\nabla\theta^{\varepsilon^T} u^{(1)})\cdot\nabla_y u^{(1)}-\mu\theta^\varepsilon_{\ell,j}\theta^\varepsilon_{k,j}\dfrac{\partial^2 w}{\partial y_k\partial y_\ell} \\ + [(\theta^c\sigma)_t + u\nabla(\theta^c\sigma)]\cdot\nabla_y w\} \\ + \varepsilon^{2/3}\{L(u^{(3)})+u_l^{(1)}+u\nabla u^{(1)}+u^{(1)}\nabla u +\nabla\theta^{\varepsilon^T} u^{(1)}\cdot\nabla_y u^{(2)}+\nabla\theta^{\varepsilon^T} u^{(2)}\nabla_y u^{(1)} + \nabla\pi \\ -\mu\theta^\varepsilon_{\ell,j}\theta^\varepsilon_{k,j}\dfrac{\partial^2 u^{(1)}}{\partial y_k\partial y_\ell} + [(\theta^\varepsilon\sigma)_t+u\nabla(\theta^\varepsilon\sigma)]\cdot\nabla_y u^{(1)}+w\nabla w+\nabla(\theta^\varepsilon\sigma)^T w\cdot\nabla_y w+\nabla(\theta^\varepsilon\sigma)\nabla_y\pi\} \\ + \varepsilon\{L(u^{(4)})+u_t^{(2)}+u\nabla u^{(2)}+u^{(2)}\nabla u+w\nabla u^{(1)}+u^{(1)}\nabla w+\nabla\theta^{\varepsilon^T} u^{(1)}\cdot\nabla_y u^{(3)}+\nabla\theta^{\varepsilon^T} u^{(3)}\cdot\nabla_y u^{(1)} \\ \quad +\nabla\theta^{\varepsilon^T} u^{(2)}\cdot\nabla_y u^{(2)}\cdot\nabla_y u^{(2)}+\nabla p^{(1)}-\mu\theta^\varepsilon_{\ell,j}\,\theta^\varepsilon_{k,j}\dfrac{\partial^2 u^{(2)}}{\partial y_k\partial y_\ell} + \nabla(\theta^\varepsilon\sigma)\nabla p^{(1)} \\ \qquad + \nabla(\theta^\varepsilon\sigma)^T u^{(1)}\nabla_y w +\nabla(\theta^\varepsilon\sigma)^T w\nabla_y u^{(1)}\} \\ \quad + [(\theta^\varepsilon\sigma)_t + u\nabla(\theta^\varepsilon\sigma)]\cdot\nabla_y u^{(2)} \end{array}\right. \tag{A6}$$

$$+ \varepsilon^{4/3}\{L(u^{(5)}) + u_t^{(3)} + u\nabla u^{(3)} + u^{(3)}\nabla u + w\nabla u^{(2)} + u^{(2)}\nabla w + u^{(1)}\nabla u^{(1)}$$

$$+\nabla\theta^{\varepsilon^T} u^{(2)}\cdot\nabla_y u^{(3)} + \nabla\theta^{\varepsilon^T} u^{(3)}\cdot\nabla_y u^{(2)} + \nabla\theta^{\varepsilon^T} u^{(1)}\cdot\nabla_y u^{(4)} + \nabla\theta^{\varepsilon^T} u^{(4)}\cdot\nabla_y u^{(1)} + \nabla p^{(2)}$$

$$-\mu\theta^{\varepsilon}_{\ell,j}\theta^{\varepsilon}_{k,j} \frac{\partial^2 u^{(3)}}{\partial y_k \partial y_\ell} + \nabla(\theta^{\varepsilon}\sigma)\nabla_y p^{(2)} + [(\theta^{\varepsilon}\sigma)_t + u\nabla(\theta^{\varepsilon}\sigma)]\cdot\nabla_y u^{(3)}$$

$$+ \nabla(\theta^{\varepsilon}\sigma)^T u^{(1)}\nabla_y u^{(1)} + \nabla(\theta^{\varepsilon}\sigma)^T u^{(2)}\nabla_y w + \nabla(\theta^{\varepsilon}\sigma)^T w\nabla_y u^{(2)}\} + \ldots.$$

and the divergence equation is

(A7) $$0 = \varepsilon^{-2/3}\nabla_y\cdot(\nabla\theta^T w) + \varepsilon^{-1/3}\nabla_y\cdot(\nabla\theta^T u^{(1)}) + \varepsilon^0[\nabla_y\cdot(\nabla\theta^T u^{(2)}) + \nabla\cdot u] +$$

$$+ \varepsilon^{1/3}[\nabla_y\cdot(\nabla\theta^T u^{(3)}) + \nabla\cdot w + \nabla(\theta^{\varepsilon}\sigma)^T : \nabla_y w] + \ldots$$

Thus each $u^{(j)},\theta^{(j+1)}$ will be defined by an equation of the form

(A8) $$L(u^{(j)}) = \mathbb{P}f \ , \ \tilde{\nabla}_y u^{(j)} \equiv \nabla_y\cdot(\nabla\theta^T u^{(j)}) = \mathbb{P}g \ , \ \langle u^{(j)}\rangle = 0$$

i.e.

(A9) $$u^{(j)} = \nabla\theta^{\varepsilon^{-T}}(\tilde{u}^{(j)} - \langle\tilde{u}^{(j)}\rangle)$$

(A10) $$\tilde{w}\nabla_y\tilde{u}^{(j)} + \tilde{u}^{(j)}\nabla_y\tilde{w} = \nabla\theta^{\varepsilon^T}\mathbb{P}f\ q^{-1/2} \ , \ \nabla_y\cdot\tilde{u}^{(j)} = \mathbb{P}g \ ,$$

(A11) $$\theta_t^{(j+1)} + u\nabla\theta^{(j+1)} = \langle\tilde{u}^{(j)}\rangle - (\theta^{(j-2)}\sigma)_t - u\nabla(\theta^{(j-2)}\sigma)$$

Here $\{\mathbb{P}f, \mathbb{P}g\}$ denotes the projection of $\{f,g\}$ on the range space of $\{L,\tilde{\nabla}_y\}$ which we know (see [1]) to be contained in

(A12) $$W = \{\{\hat{f},\hat{g}\}: \langle\hat{f}+\hat{g}w\rangle = 0 \ ; \ \langle\hat{f}\cdot w\rangle + \langle(p^{(j)} + \tfrac{1}{2}\tilde{w}C^{-1}\tilde{w})\hat{g}\rangle\ q = 0$$
$$\langle\hat{g}\rangle = 0 \ ; \ \langle\hat{f}(\nabla\theta^{\varepsilon^T}\nabla_y)\times w\rangle = 0\}$$

Therefore, from (A6), (A7) we find that $u^{(1)}$ is zero because it is defined by

(A13) $$L(u^{(1)}) = 0 \ , \ \tilde{\nabla}_y\cdot u^{(1)} = 0$$

Similarly $u^{(2)}$ is defined by

(A14) $$L(u^{(2)}) = \mathbb{P}\ [w_t + u\nabla w + w\nabla u - \mu\theta^{\varepsilon}_{\ell,j}\theta^{\varepsilon}_{k,j}\frac{\partial^2 w}{\partial y_k\partial y_\ell} + [(\theta^{\varepsilon}\sigma)_t + u\nabla(\theta^{\varepsilon}\sigma)]\cdot\nabla_y w] ;$$
$$\tilde{\nabla}_y\cdot u^{(2)} = 0$$

Now, by averaging (A6) and using $<L(u^j)> = -\sqrt{q}<\tilde{\nabla}_y \cdot u^{(j)} w>$ we find

(A15)
$$\begin{cases} u_t + u\nabla u + \nabla p + \varepsilon^{2/3}\nabla \cdot <w \otimes w> + \varepsilon^{4/3}\nabla \cdot <w \otimes u^{(2)} + u^{(2)} \otimes w> = 0(\varepsilon^{5/3}) \\ \nabla \cdot u = 0 \end{cases}$$

Similarly, multiply (A6) by w, average over y and use (A12). Since $q = \frac{1}{2} <w^2>$, one finds

(A16)
$$\begin{cases} q_t + u\nabla q + <w \otimes w> : \nabla u + \mu \nabla\theta^{\varepsilon^T} \nabla\theta^{\varepsilon} : <\nabla_y w_i \otimes \nabla_y w_i> \\ \qquad + \varepsilon^{1/3}\nabla \cdot <w(\pi + \frac{1}{2} w^2)> \\ \qquad + \varepsilon^{2/3} [<w, u_t^{(2)} + u\nabla u^{(2)} + u^{(2)}\nabla u + \nabla\theta^{\varepsilon^T} u^{(2)} \cdot \nabla_y u^{(2)} \\ \qquad + \mu \nabla\theta^{\varepsilon^T} \nabla\theta : <\nabla_y u_i^{(2)} \otimes \nabla_y w_i> + <((\theta^{\varepsilon}\sigma)_t + u\nabla(\theta^{\varepsilon}\sigma))\nabla_y u^{(2)}, w>] \\ \qquad = 0(\varepsilon) \end{cases}$$

And finally multiplying (A6) by $(\nabla\theta^{\varepsilon^T} \nabla_y) \times \tilde{w}$ and averaging over y yields

(A17)
$$\alpha(\sigma_t + u\nabla\sigma) + \beta\nabla\sigma + \gamma\sigma + \delta = 0(\varepsilon^{2/3})$$

with

(A18)
$$\alpha = <\theta^{\varepsilon} \cdot \nabla_y w, \tilde{\nabla}_y \times \tilde{w}> \qquad (\tilde{\nabla}_y = \nabla\theta^{\varepsilon^T} \nabla_y)$$

(A19)
$$\beta = \varepsilon^{1/3} <(\theta^{\varepsilon^T} w\nabla_y w + \theta^{\varepsilon}\nabla_y \pi)\tilde{\nabla}_y \times \tilde{w}>$$

(A20)
$$\gamma = \varepsilon^{1/3} [<\nabla\theta^{\varepsilon^T} w\nabla w + \nabla\theta^{\varepsilon}\nabla\pi, \tilde{\nabla}_y \times \tilde{w}> + <\tilde{w}><\nabla_y w, \tilde{\nabla}_y \times \tilde{w}>]$$

(A21)
$$\delta = <(w_t + u\nabla w + w\nabla u - \mu\theta^{\varepsilon}_{\ell,j}\theta^{\varepsilon}_{k,j} \frac{\partial^2 w}{\partial y_k \partial y_\ell}) \tilde{\nabla}_y \times \tilde{w} + \varepsilon^{1/3}<(w\nabla w + \nabla\pi)\tilde{\nabla}_y \times \tilde{w}>$$

REFERENCES

[1] D. McLAUGHLIN, G. PAPANICOLAOU, O. PIRONNEAU : Convection of microstructures Proc. INRIA Conf. Dec. 1981, North-Holland (R. Glowinski ed.) (to appear)

[2] D. McLAUGHLIN, G. PAPANICOLAOU, O. PIRONNEAU : Convections of microstructures and related problems (submitted to SIAM J. Appl. Math.)

[3] A. POUQUET, U. FRISCH, J.P. CHOLLET : Turbulence with spectral gap. Phys. Fluids 26(4). April 1983, p. 877-880.

[4] B.E. LAUNDER, D.B. SPALDING, Mathematical models of turbulence. Academic Press (1972).

[5] A.S. MONIN, A.M.YAGLOM, Statistical Fluid Mechanics of Turbulence, Vol. 2, p. 348 (1975) MIT Press.

[6] D. McLAUGHLIN, G. PAPANICOLAOU, O. PIRONNEAU : Simulation numérique de la turbulence par homogénéisation des structures de sous-maille. INRIA Report 188. Feb. 1983.

[7] Ph. CIARLET : Lectures on 3-D Elasticity. Tata Inst. Lecture Notes, Springer 1984.

COMBUSTION

Computing Methods in Applied Sciences and Engineering, VI
R. Glowinski and J.-L. Lions (Editors)
Elsevier Science Publishers B.V. (North-Holland)

PREDICTION OF IN-CYLINDER PROCESSES IN RECIPROCATING INTERNAL COMBUSTION ENGINES

A D Gosman

Fluids Section
Mechanical Engineering Department
Imperial College, London, UK.

Computational fluid mechanics is finding a new role as an aid to the development of fuel-efficient, low-pollution reciprocating internal combustion engines. This paper outlines the nature of the problems involved in physical and mathematical terms and reviews the efforts which have been made to devise suitable computer models of the flow, heat transfer and combustion processes within engine combustion chambers. It is shown that the task is one of considerable size and complexity and although some progress has been made, much remains to be done.

INTRODUCTION

The past two decades have seen an increasing number of applications of computational fluid dynamics (CFD) techniques outside high-technology areas such as aerospace and nuclear power where they were first extensively employed. In some instances these new excursions have been a matter of choice, while in others they have been enforced by circumstances which have imposed a faster pace on design and development than could be sustained by more traditional means.

The automobile industry falls into the latter category, for prior to the recognition in the 1960s of the adverse effects of the pollutents emitted in the engine exhaust gases, the design process was largely evolutionary and empirical: new engines were strongly based on existing designs and the changes they incorporated were arrived at by time-consuming and expensive building and testing. Such computer modelling techniques as were used were, as noted in the review of Blumberg et al (1980) largely 'zero-dimensional' in character, ie they almost completely ignored the complex spatial variations of flow, temperature and concentration within the engine combustion chambers and provided only the temporal variations of a few key properties, notably average pressure and temperature. Moreover, these methods inevitably contained numerous empirical coefficients which turned out, for reasons which will become apparent later, to be highly specific to particular engines and operating conditions.

Legislative constraints on pollutent emissions caused the manufacturers to explore more radical alternatives to existing designs (including other engine types, such as gas turbines) which brought into focus some of the weaknesses of the approach just described; but they were nonetheless usually successful in meeting the emission ceilings. However in many instances this was achieved at the expense of poorer engine performance, including increased fuel consumption. Therefore when they were later confronted by the so-called 'energy crisis', which suddenly imposed the need to substantially improve fuel economy (and gave rise to further legislation to this effect) while staying within the emission constraints, the need for more fundamentally-based design methods became even more urgent than before.

It is against this background that CFD techniques came on the scene and began to be applied to engine prediction, due to the efforts of several groups, including that of the author and his colleagues. Summaries of the historical development can be found in the publications of Butler et al (1981) and Gosman et al (1980). In what follows an outline will be provided of the nature of the problems involved in engine prediction in physical and mathematical terms: then the characteristic features of the more widely-used CFD methods will be described; and finally an evaluation will be made of what has been achieved and the remaining problems.

ENGINES AND THEIR IN-CYLINDER PROCESSES

In order to appreciate the computational task some knowledge is necessary of the main components and mode of operation of the reciprocating internal combustion engine: hence brief descriptions will be provided here.

A simple idealised form of engine is illustrated in Fig I below. Its principal features are: the cylinders, which are usually at least four in number, each containing a piston which goes up and down in reciprocating motion and in so doing drives the crankshaft, via connecting rods; the cylinder head, containing poppet valves which are cam-actuated, synchronised with the crankshaft rotation and open and close to admit or exhaust the working gases; the intake system, comprising the ducting leading from, in this example, the carburettor , in which the fuel and air are premixed; the exhaust system, comprising the ducting through which the burnt gases escape; and the ignition system, consisting of the spark plug, which ignites the fuel/air mixture at the appropriate stage and ancilliary components (not shown) which provide an electrical discharge at the required time.

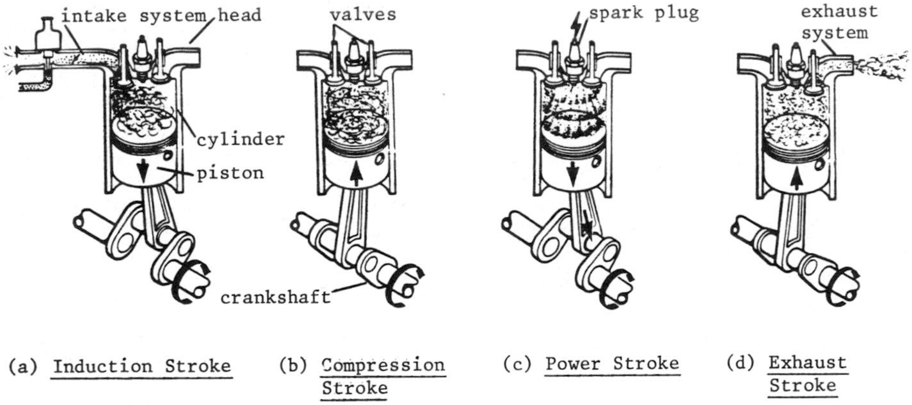

Fig 1 Illustration of idealised engine and four-stroke working cycle

Also illustrated in Fig I are the events which occur in a typical 'four-stroke' engine cycle, involving two complete revolutions of the crankshaft (usually measured as 720 degrees of crank angle (CA) rotation) during which the piston executes four traverses or 'strokes' of the length of the swept volume of the cylinder. The cycle commences with induction (diagram (a)), when the piston moves downwards from its topmost position, drawing in a fresh, ostensibly homogeneous, mixture of fuel vapour and air past the inlet valve, which has been opened by the camshaft.

There follows the compression stroke (diagram (b)), in which both valves are closed as the piston moves upwards, thus compressing the mixture, which is ignited by the spark just before (some 20–30° CA) the topmost piston position. The resulting flame propogates across the combustion chamber, defined as the volume enclosed between the piston, cylinder wall and cylinder head surfaces, so causing the pressure (and temperature) to rise and drive the piston downwards in the power stroke (diagram (c)): burning continues during the initial part of this and then ceases.

Around the stage where the piston reaches its bottom-most position the cam opens the exhaust valve and the spent gases are expelled past it by virtue of their own pressure and the subsequent upwards displacement of the piston during the exhaust stroke (diagram (d)). This completes the cycle.

Practical engines may differ from the above in two important respects: firstly, in general the shapes of the combustion chamber and inlet/exhaust passages are usually more elaborate than those shown; and secondly additional complexities may be introduced through alternative means of admitting and igniting the fuel.

Concerning the first point, Figs 2 and 3 shows representative examples of practical spark-ignition (SI) and Diesel engine configurations respectively, in which some of the elaborations can be seen: these include irregular recesses in the cylinder head (Figs 2(a) and 3(a)) or piston (Figs 2(b) and 3(b)) and inlet passages of helical form (Fig 3(b)). On the second matter, the fuel is sometimes introduced directly into the chamber in liquid form as a spray, as indicated in Fig 3.

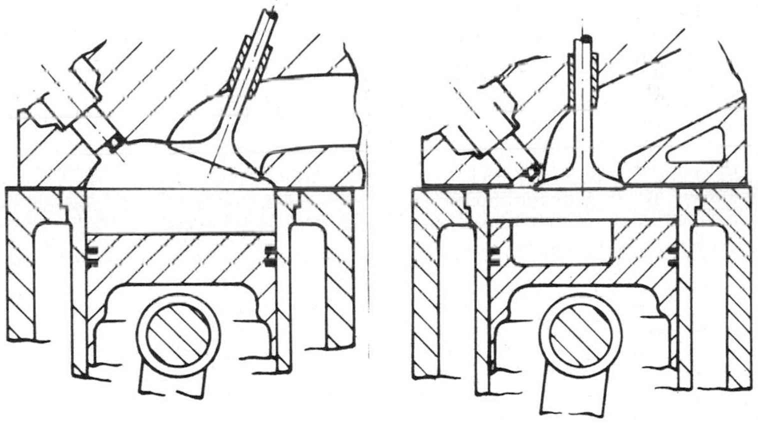

(a) Wedge recess in head (b) Bowl recess in piston

Fig 2 Illustration of some spark-ignition engine configurations

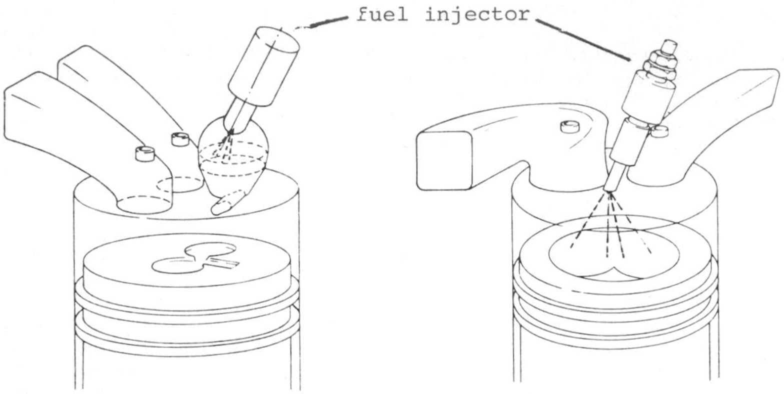

(a) Prechamber in head (b) Bowl in piston

Fig 3 Illustration of some Diesel engine configurations

This is common practice in Diesel engines, which moreover rely upon the temperature rise produced by the compression process to ignite the fuel (and are therefore termed 'compression-ignition engines', as opposed to the 'spark- ignition' type described earlier) but it is also being contemplated for certain types of 'stratified-charge' spark-ignition engine.

To what extent do these various differences matter as regards the achievement of satisfactory engine performance? This is a topic in itself, about which only a few general observations can be made here.

Perhaps the most important general characteristic of engines from the computer modelling point of view is their strong sensitivity to the fluid-dynamic processes which occur within them. This was recognised very early in their lifetime by the pioneer developers, but appeared to be somewhat forgotten by later workers, who tended to be more preoccupied with the thermodynamic and chemical reaction aspects. The re-emergence of interest in fluid flow was indeed prompted in part by the belated recognition that it is precisely the failure of the earlier 'zero dimensional' computer models to take this factor into account which severely limited their predictive capability to explore new design concepts.

The in-cylinder flow behaviour is intimately connected with the geometrical configuration of the engine and the operating conditions (ie speed, compression ratio) as is noted in, for example the review articles of Tabacynski (1979) and Gosman (1983). The geometrical influences commence during induction, when the in-cylinder flow pattern generated then is strongly influenced by the shape of the intake passage. This is why elaborate forms of passage such as that illustrated in Fig 3(b) exist: in many instances they are designed, as in this example, to impart 'swirl' (ie circumferential motion) to the incoming flow, for this type of motion tends to be the most likely to persist through to the compression stroke. This practice is widely used in Diesel engines. The motive for generating in-cylinder air motion is to enhance the fuel-air mixing and combustion processes which occur later in the compression and power strokes. Experience shows however that, swirl apart, the induction-generated motions decay very rapidly; and even swirl tends to organise itself into an unproductive solid-body structure which, due to the absence of shear, is not conducive to good mixing. Herein lies the motive for the irregular shapes which the examples of Figs 2 and 3 and most other combustion chambers exhibit: their purpose is, for the most part, to provoke a phenomenon often termed 'squish' in which, as the piston approaches the cylinder head, fluid is displaced from regions of low clearance to regions of high clearance. The motions so generated enhances combustion both directly and indirectly, by in the latter case redistributing angular momentum and so causing the swirl to depart from the solid-body form.

Clearly the strength and direction of the squish motions is determined by combustion chamber shape, so the sensitivity of engine performance to this facet is scarcely surprising. Indeed it turns out that in those engine types which rely on the squish/swirl interaction (again primarily Diesel, although there is a tendency towards this in spark-ignition engine design) there often exists an optimum inlet swirl level for a given design.

Of course the combustion process itself, and the accompanying heat and mass transfer processes must be represented in computer models. In spark ignition engines the flame is initiated at the spark plug and subsequently propogates across the chamber at a rate which depends to some extent on the chemical kinetics (ie the rate at which the fuel combines with oxygen at the molecular level) but is also strongly influenced by the 'mean' and 'turbulent' characteristics of the prevailing flow field. Of particular note is the near-proportionality between the flame speed and the 'turbulence intensity', (a characteristic of the flow field which will be defined later) and it is this, coupled with a like relation between turbulence intensity and engine speed, which is the key to the viability of the reciprocating engine. These observations clearly reinforce the strong connection between combustion and flow.

The Diesel engine differs from the above in two important ways: one is the mode of ignition which, as observed earlier, occurs spontaneously within the chamber at point(s) where the right conditions of temperature and fuel concentration are created; and the other is the manner of introduction of the fuel, in the form of a liquid spray which subsequently vaporises.

Sprays have their own characteristic features and complexities. They are generated by forcing the fuel at high pressure through a nozzle (Fig 3) where it exits at high velocity via several holes (typically 4) equally spaced circumferentially. The liquid atomises, forming, it is believed, a narrow jet-like structure containing droplets of various sizes (typically in the range 5–80μm) which may collide and coalesce, particularly in the initial part of the spray (O'Rourke and Bracco (1980)). The droplets also heat up and evaporate as the spray penetrates into the hot gases in the chamber, forming a surrounding vapour cloud in which combustion eventually takes place. They also sometimes impinge on the chamber walls if evaporation is incomplete.

As already implied, strong interactions exist between the droplet and gas fields, in which, in addition to the heat and mass interchanges, there is also momentum interchange, with the droplets being deflected by the gas motion and the gas being in turn deflected, entrained and sheared by the spray. For these reasons the Diesel engine and other types operating on the stratified charge principle are particularly sensitive to the in-cylinder flow and are therefore prime candidates for computer modelling.

The foregoing description of the in-cylinder processes if far from complete, but it should nonetheless suffice to identify the main complexities which, to summarise, are:

(a) Geometrical configurations which are of complex shape and have moving boundaries.

(b) Flow fields which are in general time-varying and three dimensional.

(c) Rapidly-propagating flame fronts which give rise to steep spatial gradients of temperature, concentration and related quantities such as density.

(d) In the case of Diesel and other fuel-injection engines, a two-phase flow structure characterised by narrow sprays containing fast-moving droplets of a range of sizes, which interact strongly with the gas phase.

(e) Close links between the mixing and combustion processes and the fine-scale turbulent components of the gas motion.

The manner in which these features have been represented in mathematical model form will now be outlined.

MATHEMATICAL MODELLING

In the absence of sprays, it is possible to write down an exact mathematical model of the in-cylinder processes; for they are governed, just as any other single- phase fluid continuum, by well-known differential conservation equations of physics, together with appropriate boundary conditions and auxiliary information. Thus the fluid motion obeys the Navier Stokes equations; and heat transfer and combustion can be described via conservation equations for energy and the masses of the participating chemical species. The dependent variables of this set are the *instantaneous* values of the three components u_i of the velocity vector in selected directions x_i, pressure p, temperature T (or a related quantity) and the concentration m_j of chemical species j, where there may be numerous j.

Direct attack on this set by numerical methods is however barely feasible and certainly not, in the present context, practical, for reasons which are well-known and connnected with the fact that the flow is turbulent. Such flows exhibit significant spatial and temporal variations on scales orders of magnitude smaller than the chamber dimensions and cycle period respectively; and hence would require very high numerical resolution, with attendant high computing overheads. Thus although numerical techniques for direct solution do exist.(notably spectral methods) their use in the present application is disqualified on grounds of cost.

Also well-known are techniques for circumventing the problem just described by avoiding the direct calculation of the small-scale motions and representing their effects indirectly, via additional equations. This is accomplished by first defining an 'averaging' or 'filtering' procedure which, when applied to u_i yields an average value U_i, instantaneous departures from which are denoted u_i', thus:

$$u_i = U_i + u_i' \qquad (1)$$

The most commonly-employed filter is time averaging, but this is clearly inappropriate for non-stationary turbulent flows of the present kind; for such an average only exists for averaging periods equal to or greater than the time for a full four-stroke cycle, and this is not of interest: what is required are temporal variations *within* the cycle.

There exist other possibilities, two of which have received attention in the engine context. The most popular of these is *ensemble averaging*, where U_i is defined as the average of a large number N of realisations at a given spatial location and given time t in the engine cycle, ie

$$U_i(t) = \lim_{N\to\infty}\left[\frac{1}{2N+1}\sum_{n=-N}^{n=N} u_i(t+n\tau)\right] \qquad (2)$$

Here τ is the period of the cycle. This approach, it should be noted, reduces to time averaging when the flow is stationary.

The second alternative is *spatial filtering* as employed in a methodology commonly referred to as 'Sub-Grid Scale' (SGS) modelling or 'Large Eddy Simulation' (LES), in which U_i now represents the fluid velocity on scales larger than some value Δ which is related to the grid spacing of a numerical method; and it follows that u_i' refers to that part of the velocity associated with scales smaller than Δ.

The second stage in the application of these averaging procedures involves substituting for u_i and other dependent variables in the Navier Stokes equations from relations like (1) and then applying the filter to the equations themselves. The result is of the general form, in Cartesian tensor notation:

$$\frac{\partial(\rho U_i)}{\partial t} + \frac{\partial}{\partial x_j}(\rho U_j U_i) = -\frac{\partial P}{\partial x_i} + \frac{\partial}{\partial x_j}(\mu S_{ij}) - \frac{\partial}{\partial x_j}(\rho u_i' u_j') + R_i \quad (3)$$

Here averaged quantities are denoted by upper-case characters, with the exception, of the fluid density ρ and viscosity μ: P stands for pressure, S_{ij} for the rate of strain tensor and R_i for 'remaining terms', including, for example, distributed sources or sinks.

The tensor $\overline{\rho u_i' u_j'}$ in equation (3) defines a set of new unknowns arising from the averaging procedure itself, which are often interpreted as effective stresses brought about by the turbulent motion. The necessity to generate additional equations connecting these stresses to known or calculable quantities (an activity which has come to be known as 'turbulence modelling') is one of several factors which have to be weighed in choosing between the averaging alternatives. The various factors have been identified and discussed by Reynolds (1980), Gosman et al (1980) and Butler et al (1981), and their findings are summarised in the following paragraphs.

The LES approach has two main attractions, these being:

(1) The ability to produce predictions on an individual cycle basis, which makes it possible, in principle, to simulate the cycle-to-cycle variations exhibited by operating engines. These are manifested in extreme cases by the phenomenon of 'misfire' which has adverse effects on performance and pollutant emissions.

(2) The restriction of the effects represented in the $\overline{\rho u_i u_j}$ terms to the small scale motions, which are believed to exhibit regularities (eg isotropy) not possessed by the larger scales, and therefore require less elaborate turbulence modelling than the latter.

There are however negative aspects as well: thus, concerning (1), LES calculations will tend to be expensive, even on an individual cycle basis, for they must always be done in three dimensions on a relatively fine grid: (Butler et al (1981) advocate a form of two-dimensional LES appropriate to axisymmetric flows, but this seems at variance with the experimental observation that turbulence is always three dimensional) so if, as will often be the case, both cycle-resolved and ensemble-average predictions are required, the cost of making many cycles of calculation could be prohibitive. As for (2), the indications are that the subgrid turbulence modelling may need to be more elaborate than was originally supposed. Finally, it is important to note that the few flows to which LES has been applied and tested thus far have been far simpler than those encountered in engines.

The main advantages of ensemble averaging are also twofold:

(1) Because it represents a generalisation of the time-averaging approach, it allows use of turbulence models developed in this context. Such models have been subjected to a much wider range of testing than LES.

(2) In situations where the mean flow is axisymmetric the calculations can be performed in two dimensions, since *all* scales of tubulence are represented in the $\rho\overline{u_i u_j}$ terms.

Here too there is a negative side, the most obvious facet being the inability to provide, by definition, predictions on an individual cycle basis. It is also claimed by some researchers that because the turbulence models must in this instance represent effects over the full range of scales, they are more difficult to devise and less likely to be accurate. Nevertheless as already noted, the balance of these arguments has thus far been in favour of ensemble averaging in most engine modeller's minds.

On the matter of the turbulence modelling of the ensemble-average $\rho\overline{u_i'u_j'}$, there also exists a range of options, but they will not be discussed here: the interested reader should consult Bradshaw (1978) and the other references just cited. The universal choice (apart from some ad-hoc approaches in early work) has been some variant of the so-called '$k-\epsilon$' model, in which k and ϵ are the ensemble- average kinetic energy of turbulence and its dissipation rate respectively. This model connects the turbulent stresses to the mean strain rate in a quasi-Newtonian manner, ie

$$\rho\overline{u_i u_j} \simeq -\mu_t S_{ij} + \frac{2}{3}\,\delta_{ij}\left(\rho k + \mu_t \frac{\partial U_i}{\partial x_j}\right) \tag{4}$$

Here μ_t is a ficticious turbulent viscosity, given by:

$$\mu_t = C_\mu \rho k^2/\epsilon \tag{5}$$

where C_μ is an empirical coefficient.

The quantities k and ϵ are obtained from their own transport equations, which run:

$$\frac{\partial}{\partial t}(\rho k) + \frac{\partial}{\partial x_j}(\rho U_j k) = \frac{\partial}{\partial x_j}\left(\frac{\mu_t}{\sigma_k}\frac{\partial k}{\partial x_j}\right) + G_{ij} - \rho\epsilon + R_k \tag{6}$$

$$\frac{\partial}{\partial t}(\rho\epsilon) + \frac{\partial}{\partial x_j}(\rho U_j \epsilon) = \frac{\partial}{\partial x_j}\left(\frac{\mu_t}{\sigma_\epsilon}\frac{\partial \epsilon}{\partial x_j}\right) + \frac{\epsilon}{k}\left(C_1 G_{ij} - C_2\rho\epsilon\right) + R_\epsilon \tag{7}$$

where: $G_{ij} = \mu_t S_{ij} \partial U_i/\partial x_j$; R_k and R_ϵ represent additional terms reflecting compressibility and other effects; and σ_k, σ_ϵ, C_1 and C_2 are further empirical coefficients. The existence of these coefficients is indicative of an important further consequence of averaging, namely the fact that an exact set of equations has been replaced, through this process, by an inexact set. This has to be borne in mind in assessing the possible sources of errors in predictions.

Heat and Mass Transfer

The application of ensemble averaging to the governing equations for these processes is straightforward in all but one respect and leads, when the resulting turbulent flux terms are modelled within the usual $k-\epsilon$ framework, to equations of the form:

$$\frac{\partial}{\partial t}(\rho F) + \frac{\partial}{\partial x_j}(\rho U_j F) = \frac{\partial}{\partial x_j}\left(\frac{\mu_t}{\sigma_f}\frac{\partial F}{\partial x_j}\right) + R_f \qquad (8)$$

where F stands for the ensemble average temperature or concentration, σ_f is the associated turbulent Prandtl or Schmidt number and R_f is an ensemble-average source/sink term, the main component of which arises from the combustion process. It is the nature of the latter which makes the evaluation of R_f not straightforward, for reasons which will now be explained.

The hydrocarbon fuels commonly employed in engines burn via a complex branched-chain mechanism involving large numbers of individual chemical species and reactions (see, eg Westbrook (1979)). Most of these reactions proceed rapidly at a rate which is a *non-linear* function (commonly called the chemical kinetic rate equation) of the *instantaneous* temperature and species concentrations. Consequences of these features are that:

(1) Simulation of the full mechanism entails solving large numbers of species concentration equations, which is costly.

(2) There are problems of 'stiffness' associated with the small time scales of the reaction process.

(3) The ensemble-averaged reaction rate R_f cannot be equated to the value given by substituting ensemble-average temperatures and concentrations into the instantaneous rate equation.

The first of these obstacles is usually by-passed by assuming an abbreviated, and therefore approximate, mechanism, involving fewer species and reactions (at the simplest level, a single irreversible reaction has been assumed). This practice is defensible on grounds of consistency of approximation with other elements of the mathematical modelling; and it also has the advantage of making the stiffness problem (2) more tractible. Obstacle (3) is however much more formidable: indeed it currently lies at the forefront of turbulent combustion research. Within the engine modelling context it has been tackled in two main ways: one has been to ignore it and evaluate R_f as indicated in (3); and the other has been to bypass it, by presuming that the reaction time scales are so much smaller than those of the turbulent mixing process that the latter becomes rate controlling (which is often the case in hydrocarbon combustion). The second approach, apart from being the more defensible of the two when mixing control prevails, also has the advantage of yielding simple, nearly-linear relations between R_f and the ensemble-mean concentrations, as shown by Bray (1978), among others.

The rate equation formulation has been used extensively in engine calculations as recorded by, for example Bracco and O'Rourke (1979) and Butler et al (1981); and the mixing model has been assessed to a limited extent by Ahmadi-Befrui et al (1981) and Gosman and Harvey (1982). Both models are able, with appropriate adjustment of empirical coefficients, to reproduce some of the trends observed in global measurements of quantities such as cylinder pressure, but neither is likely to stand up to more searching tests when more detailed data become available. Clearly therefore combustion represents another weak link in the mathematical modelling of in-cylinder processes.

Boundary Conditions

The modelling and numerical problems associated with direct calculation of wall boundary layers are well known and are discussed in the engine context by Gosman et al (1980). The usual practice is to employ simplified sub-models of these regions embodied in 'wall functions', which connect the wall stresses and heat fluxes to conditions in the interior flow. The forementioned reference describes a similar treatment of the flow through valve orifices.

Fuel Sprays

The introduction of liquid sprays brings additional uncertainties and associated modelling problems. The uncertainties, it should be noted, commence at the fundamental level, for there does not exist an 'exact' mathematical description of two-phase flows: the reasons for this are however beyond the scope of the present discussion, which will simply outline current practice.

Within the specific context of sprays, the usual practice is to treat the gas phase as a continuum, governed by appropriately-modified versions of the continuum conservation equations. Thus, if equation (8) is regarded as a prototype of this set, the modified version would read:

$$\frac{\partial}{\partial t}(\rho \theta F) + \frac{\partial}{\partial x_j}(\rho \theta U_j F) = \frac{\partial}{\partial x_j}\left(\frac{\theta \mu_t}{\sigma_f}\frac{\partial F}{\partial x_j}\right) + \theta R_f + I_f \quad (9)$$

where θ is the local ensemble-average volume fraction of the gas and I_f is an additional source/sink term associated with interactions with the liquid phase. For example when F stands for temperature, I_f represents the net heat transfer from liquid to gas.

The liquid phase is normally regarded as consisting of a polydispersed droplet field, the most 'exact' description of which is usually taken to be the *'statistical spray model'* described by Williams (1965). This has the form of an integro-differential equation for a spray probability function, which is defined in an *eight-dimensional* space whose coordinates are droplet diameter, droplet position and velocity vector components and time. The cost and complexity of obtaining solutions is a strong deterrent to its use, as can be appreciated from, for example, the work of Westbrook (1976) and Haselman and Westbrook (1978).

Most investigators have employed a droplet representation which can be regarded as a simplified derivative of the statistical spray model, sometimes termed the 'Discrete Droplet Model' (DDM). This consists of sets of Lagrangian conservation equations for the position, mass, momentum and thermal energy of a statistical sample of discrete droplets, each of which is representative of a 'parcel' of like droplets all having the same initial position, size, velocity and temperature. For example the equations for the instantaneous position $\underline{x}_d$ and velocity $\underline{u}_d$ vectors are:

$$\frac{d\underline{x}_d}{dt} = \underline{u}_d \quad (10)$$

$$\frac{d\underline{u}_d}{dt} = \frac{3C_D\rho}{4d\rho_d}|\underline{u} - \underline{u}_d|(\underline{u} - \underline{u}_d) \quad (11)$$

where d and ρ_d are the droplet diameter and density respectively, $\underline{u}$ is the instantaneous gas velocity vector and C_D is the drag coefficient between the droplet and gas. In general C_D is a non-linear function of the slip velocity ($\underline{u}-\underline{u}_d$) between these two phases. Like equations exist for droplet mass (and hence size) and temperature.

These equations can be integrated, given the initial conditions and the gas velocity field $\underline{u}$; following which the interaction terms I_f appearing in the gas phase transport equations (9) can be evaluated by summing, for a given volume element, the mass, momentum etc interchanges which occur for all droplets traversing it. The initial conditions are however unknown; for, as reviewed recently by Wu et al (1983) the atomisation process which produces the droplets is insufficiently understood to permit mathematical modelling. As a consequence, the contemporary practice is to follow the recommendation of Dukowicz (1980) and randomly sample from *presumed* probability distributions of size, velocity etc, based on the limited experimental data available.

Another source of uncertainty is the structure of, and events occurring within, the 'thick' region of the spray, so named because it is so optically dense that it has defied resolution. O'Rourke and Bracco (1980) have postulated that it consists of droplets so closely spaced that collision and coalescence occurs; and they have extended the DDM model to include representation of these processes, along with droplet proximity effects on inter-phase transfer. However, there is experimental evidence which suggests that the thick region may also contain an intact, unatomised continuous liquid core whose dimensions depend on injection and ambient conditions. No mathematical representation of this core has yet been developed.

A third major source of uncertainty is the gas velocity field for, although in early applications of the DDM this was equated with the ensemble-mean field $\underline{U}$, the fluctuations $\underline{u}'$ about this can have important effects, notably on turbulent dispersion of the droplet parcels. This matter was investigated by Dukowicz (1980) who envisaged the droplets as passing through a succession of turbulent eddies and developed a stochastic approach to represent this as follows. The integration of equations (10) and (11) is effected with $\underline{u}$ evaluated from equation (1) and $\underline{u}'$ therein obtained by randomly sampling an assumed Gaussian probability distribution, defined by the local values of $\underline{U}$ and the estimated local turbulence intensity. In this way the droplets experience perturbations from their mean trajectory, resulting in dispersion. An elaboration of this method was later developed by Gosman and Ioannides (1981) which evaluates the magnitudes and time scales of the perturbations in a more rigorous fashion, with the aid of information on these available from the k-ϵ turbulence model.

Assessment of spray models has been hindered by the lack of detailed data (especially in-cylinder) and the arbitrariness of the initial condition treatment, but within these limitations the results obtained indicate that the DDM approach is promising. This view is supported by, for example, the comparisons of O'Rourke and Bracco (1980) and Kuo and Bracco (1982) with data on spray penetration and other features, although these were not obtained at conditions wholly representative of engines.

Demonstration calculations of in-cylinder sprays have been made by Butler et al (1980), Gosman and Johns (1980) and Gosman and Harvey (1982). Selected extracts will be presented later, along with details of the numerical methods employed.

COMPUTATIONAL TECHNIQUES

Although a number of different individuals and groups have worked on techniques for solving the mathematical model equations in engine circumstances, the main centres of activity have been the Los Alamos Scientific Laboratory, Princeton University and Imperial College. Accordingly the following presentation will be structured around these institutions, although other contributors will receive mention.

Two distinct methods have eminated from the Los Alamos group, both of which were restricted in application to *two-dimensional* representations of engine combustion chambers and the processes occurring within them. However these methods, in common with all others described in this section, are all amenable in principle to straightforward extension to three dimensions. Historically however this was not undertaken until recently (with one exception, to be mentioned later) due to the need for very large, fast computers.

The first method devised at Los Alamos was based on the well-known 'ICE' technique developed there by Harlow and Amsden (1971). This is a semi-implicit, Eulerian finite-difference method employing a staggered grid arrangement of velocities and pressure, in which standard, first-order differencing is used for temporal derivatives and upwind and central differencing are employed for the convection and diffusion fluxes respectively. The latter are evaluated explicitly, with a consequent contraint on time step imposed by the usual Courant number limit, but pressure and energy are treated in a coupled implicit manner, so removing the speed-of-sound limit. This method is embodied in a computer code called 'RICE' assembled by Rivard et al (1974) which also contains provision for chemically- reacting systems; it was this code that formed the basis for later engine applications.

The mathematical modelling embodied in the original RICE was relatively primitive: it incorporated no turbulence mode or wall functions and although it could handle multi-reaction combustion processes, this was done in a purely chemical-kinetic framework, with no allowance for turbulence effects.

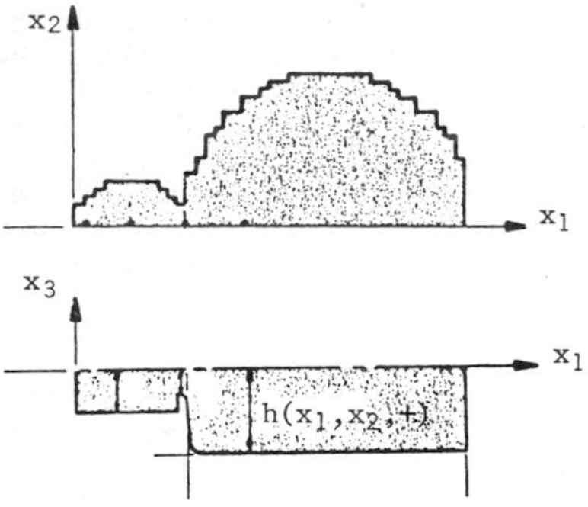

Fig 4 Grid arrangement employed in REC engine codes

RICE was further developed and applied to engine calculations by Bracco and co-workers at Princeton University, who assumed two-dimensionality in x_1-x_2 planes perpendicular to the cylinder axis and invariance in the axial (x_3) direction. The chamber geometry was approximated in the x_1-x_2 view by a castellated rectangular mesh, as illustrated in Fig 4, and effects in the x_3 direction brought about by the piston motion and recesses in the head or piston surfaces were allowed for by introducing into the governing equations a depth function $h(x_1,x_2,t)$, h being the local instantaneous distance between these two surfaces. This operation yields, when applied to the prototype equation (8), a result of the form:

$$\frac{\partial}{\partial t}(\rho h\tilde{F}) + \frac{\partial}{\partial x_i}(\rho h U_1\tilde{F}) + \frac{\partial}{\partial x_2}(\rho h U_2\tilde{F})$$

$$= \frac{\partial}{\partial x_1}\left(h\frac{\mu_t}{\sigma_f}\frac{\partial \tilde{F}}{\partial x_1}\right) + \frac{\partial}{\partial x_2}\left(h\frac{\mu_t}{\sigma_f}\frac{\partial F}{\partial x_2}\right) + h\tilde{R}_f + S_f \qquad (12)$$

where the ~ overbar denotes some (undefined) form of depth averaging and S_f represents contributions from the forementioned surfaces (eg in the energy equation S_f would contain the heat fluxes to them). (Clearly the validity of the assumption of axial invariance depends on, among other factors, h being a slowly-varying function of x_1 and x_2.) An account of the initial stages of this work is provided by Bracco and O'Rourke (1979). Additional developments which were made by the same group, which led to a series of RICE-based codes with the generic name 'REC' included:

(1) Incorporation of a version of the k-ϵ turbulence model and associated wall functions.

(2) Improvements to the numerical methodology which resulted in reduced run times.

(3) Explorations with both simplified chemical-kinetic and mixing-limited combustion models.

Further details can be found in the publications of Gupta and Syed (1979), Syed and Bracco (1979) and Grasso and Bracco (1982).

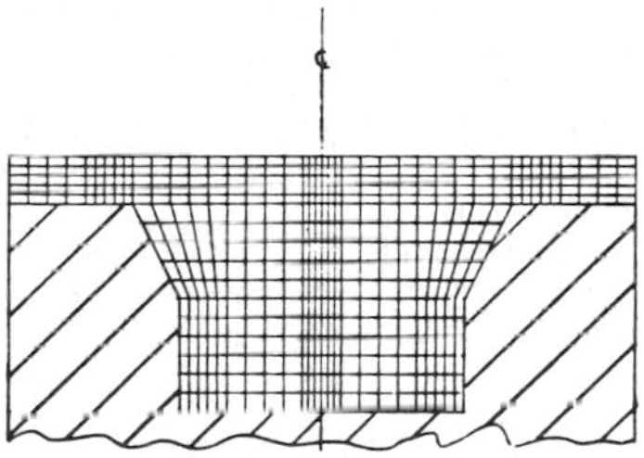

Fig 5 Grid arrangement employed in CONCHAS engine codes

The second two-dimensional method originated by the Los Alamos group is based on their well-known 'ICED-ALE' technique, originally developed by Hirt et al (1974). This is a finite-volume technique which shares many features with the ICE methodology, but has additional capabilities resulting from the introduction of a coordinate frame and computing mesh which can move in an arbitrary Eulerian/Lagrangian fashion. The form of the mesh is indicated in Fig 5: it consists of arbitrarily-shaped quadrilateral volumes, whose vertices can be moved in time in a prescribed fashion. The velocity components are located at the vertices of these volumes and all other variables are centered. The temporal and spatial differencing practices employed on this mesh are similar to those of ICE, with due account for the non-orthogonalities which may occur here.

The ICED-ALE solution algorithm operates in three phases. The first phase involves the explicit calculation of the viscous and diffusion terms in the transport equations; the second, implicit phase obtains pressure in an iterative fashion, based on the requirement of consistency between the velocity, density and pressure fields; and the final phase is a 'rezone' one in which the grid is moved, if required, to its new configuration and the convective fluxes which arise from this and the prevailing fluid motion are explicitly calculated. As with ICE, the explicit flux calculations impose the usual Courant stability condition, which however is free of the sound speed constraint due to the implicit second phase.

This methodology is clearly highly suitable for engine calculations, due to its ability to handle curved and/or moving boundaries. The first such applications were made with a version of the Los Alamos CHOLLA code, which embodies the method, but later a special-purpose engine code named CONCHAS was assembled by Butler et al (1979). This, in its original form, incorporated the same level of mathematical modelling as RICE, ie primitive turblence modelling and kinetics-based combustion representation. Further development at Los Alamos lead to an extended version called CONCHAS-SPRAY (Cloutman et al (1980)) which contains the discrete droplet spray model of Dukowicz (1980), described earlier, along with a two-dimensional LES turbulence representation, also discussed earlier. The Lagrangian droplet calculation is implicitly coupled with the gas phase solution, at the expense of an additional Courant-type time step restriction which prevents droplets from moving more than one mesh interval during this period.

The ICED-ALE based codes were also taken up and further developed by other workers, including Grasso and Bracco (1982) and El Tahry (1982) who incoporated versions of the k-ϵ turbulence model and wall functions and Traci (1981) who used an alternative turbulence model of similar form and capabilities.

Mention should also be made of a further ICE-based code called 'LDEF' which was developed at Princeton University by O'Rourke (1981) for spray calculations in constant-volume circumstances (the code was not intended for direct engine application, but rather for assessment of spray modelling): it incorporates the DDM of O'Rourke and Bracco (1980) which, it will be recalled, contains provision for collision and coalescence. Later refinements and applications of this code were made by Kuo and Bracco (1982).

Development of engine prediction methodology at Imperial College dates from 1973: a summary of the work performed up to 1979 is given by Gosman et al (1979, 1980) and developments since then have been described in other publications cited below. The finite volume employed differs from those just described in several respects, the most important of which is arguably the implicit treatment of both pressure *and* the fluxes: this avoids any stability constraint on the magnitude of the time step and allows it to be governed by accuracy considerations alone. Experience suggests that the latter requirement allows much larger time steps than the former, such that although additional effort is required to solve the implicit equations, there is an appreciable overall saving in computing time over semi-implicit methods.

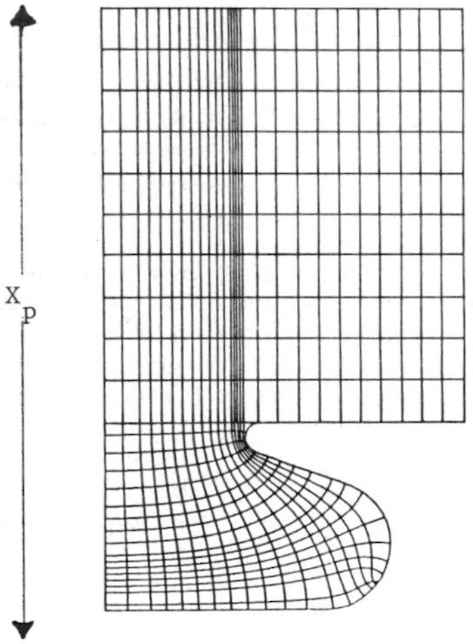

Fig 6 Grid arrangement employed in RPM method

The 'RPM' method, as it is called, employs the form of computing mesh illustrated in Fig 6, which has Eulerian/Lagrangian characteristics in the volume between the piston and the cylinder head extremities and fits recesses in either of these components with a fixed (relative to the component) general curvilinear-orthogonal arrangement. The Eulerian/Lagrangian feature, which is applied to the axial component of a cylindrical-polar frame in the region in question, is introduced via a coordinate transformation which, in its simplest form, replaces the axial coordinate x_2 by:

$$\xi \equiv x_2/X_p \qquad (13)$$

where $X_p(t)$ is the instantaneous piston displacement (Fig 6). The effect is to transform the prototype equation (8) into the form:

$$\frac{1}{X_p}\frac{\partial}{\partial t}(\rho X_p F) + \frac{1}{X_p}\frac{\partial}{\partial \xi}[\rho(U_2-U_g)F] = \frac{1}{X_p}\frac{\partial}{\partial \xi}\left(\frac{\mu_t}{\sigma_f}\frac{1}{X_p}\frac{\partial F}{\partial \xi}\right) + \text{remaining terms} \qquad (14)$$

The above equation and its curvilinear orthogonal counterpart are discretised using conservative, first-order temporal approximations and hybrid upwind/central spatial differencing.

The implicit equations are solved by an algorithm based on the well-known 'SIMPLE' procedure described by Caretto et al (1972), which obtains a provisional velocity field from the momentum equations using the prevailing pressure fields (the solutions being effected by an iterative ADI technique) and then updates these by solving a continuity-based set of pressure equations. This entire sequence, into which is also embedded the calculation of other variables such as temperature and turbulence model parameters, is repeated in an iterative fashion until convergence. A key feature of the 'RPM' varient is the use of block pressure adjustments, based on overall continuity and energy balances, which prevent the small-scale pressure variations brought about by the fluid motion from being overshadowed by the much larger changes due to the compression/expansion processes: details are given by Gosman and Watkins (1977) and more extensively by Watkins (1979). Similar procedures were later adopted for the REC and CONCHAS methods.

The mathematical modelling incorporated in the computer codes based on the RPM method included, from a very early stage, the $k-\epsilon$ turbulence model and wall functions, as recorded by Gosman et al (1980). Later publications describe extensions to include:

(1) Mixing-controlled modelling of homogenous-charge combustion (Ahmadi-Befrui et al (1981)).

(2) Discrete droplet spray modelling (Gosman and Johns (1980)).

(3) Combined kinetics/mixing modelling of Diesel combustion (Gosman and Harvey (1982)).

Concerning (2), the droplet calculation is performed semi-implicitly: the Lagrangian droplet equations are solved at the beginning of a time step and the resulting 'sources' I_f are evaluated at each cell traversed by the droplets and supplied to the gas phase calculation.

In addition to the major research programmes just described, there have been less extensive efforts at engine modelling elsewhere including the works of Griffen et al (1971) and Markatos and Mukerjee (1981) both of whom attempted three dimensional simulations. The former group employed fully explicit solution techniques and the latter used a method closely resembling RPM. Significantly, in both instances the calculations presented were performed on very coarse meshes, presumably reflecting the high cost (especially with explicit methods) of using a three dimensional grid of adequate resolution. The matter of spatial resolution, along with other facets of engine model performance, will be commented on further in the next section, where examples of applications are presented.

EXAMPLE APPLICATIONS

In this section a brief account will be given of some of the applications of the methods just described. The intention is to convey an impression of capabilities and limitations rather than make an exhaustive survey. To this end, extracts will be presented of calculations performed with one of the methods (RPM).

In view of the importance of air motion to engine performance it is of interest first to focus on studies of this particular aspect, in the absence of combustion: indeed it can be argued that accurate air motion simulation is a prerequisite for combustion prediction. The bulk of the work to date in this area has been performed at Imperial College and it is from this that the present conclusions are drawn.

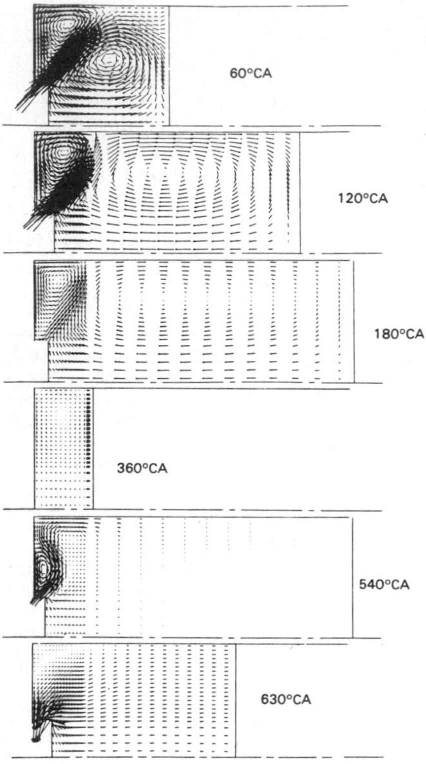

Fig 7 RPM air motion predictions for an axisymmetric engine

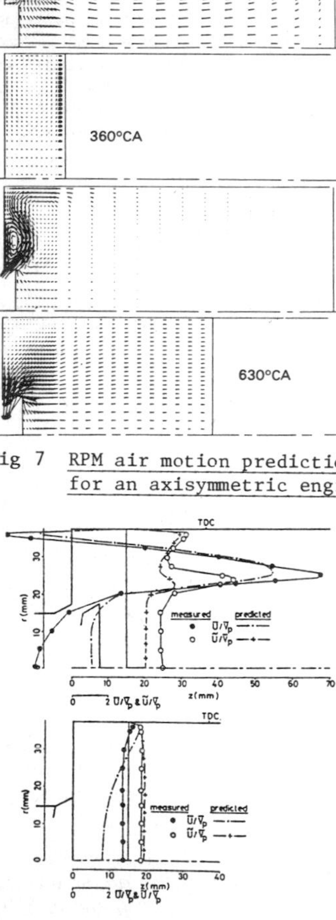

Fig 8 Comparison of RPM velocity predictions with measurement

Figure 7, taken from Gosman et al (1980) shows RPM predictions of the turbulent flow field in a model 'engine' having a planar piston and cylinder head, wtih a poppet valve centrally located in the latter. The plots depict the air motion at various stages in the engine cycle (measured in terms of degrees of crank rotation from the start of induction) throughout the four strokes. Features of interest are: the strong hollow-cone jet eminating from the valve during induction and the vortices formed on either side due to flow separation; the considerable decay of these motions during the late-induction and entire compression phases; and the regeneration of vortices near the valve as it opens during the exhaust stroke. The corresponding plots of turbulence energy (not shown) reflect the above flow behaviour, in that turbulence is selectively generated in the shear layers during induction, following which decay occurs in both the level and the inhomogeneity until the exhaust valve opens.

In a later study by Ahmadi-Befrui et al (1982) of the same type of engine the accuracy of the predictions was assessed by comparison with measurements of axial velocity (U) and turbulence intensity ($\tilde{U}$) at selected positions and times. Examples are displayed in Fig 8, close inspection of which will reveal qualitatively correct predictions, but only moderate quantitative agreement. There are several possible causes of the discrepancies, including truncation errors, uncertainties in the boundary conditions (especially at the valve entry plane) and inadequacies in the turbulence modelling. It turns out however to be difficult to assess their relative magnitudes, principally because, as noted by these authors, even with meshes as dense as 45x45, grid independence could not be achieved, presumably due to numerical diffusion and other errors arising from the (enforced) use of first-order upwind convective differencing. (The 'obvious' solution of using higher-order differencing has been explored by the present author and his colleagues, but here, as is well-known, other problems arise, such as the tendency to generate spurious extrema.) Although rigorous tests of this kind have not been performed on the other methods presented in the previous section, the fact that they also embody upwind differencing suggests they would suffer similar errors.

Despite the limitations just discussed some useful information has been derived from air motion studies, particularly on the effect of chamber geometry and swirl on the flow behaviour. Accounts of such studies are given by Gosman and Johns (1978), Gosman et al (1979) and El Tahry (1982).

Definitive assessment of combustion predictions is more difficult, due to the paucity of data, especially for truly axisymmetric circumstances, or close approximations thereto. The few comparisons which have been made are those of Ahmadi-Befrui et al (1981), who used the RPM method with mixing-controlled combustion modelling for a homogeneous-charge engine application; and Diwalker (1982) who applied CONCHAS to computations of a stratified-charge engine using kinetics-based combustion modelling. In both cases some degree of agreement with experiment was achieved after adjustment of the empirical coefficients of the combustion model.

More extensive combustion calculations have been made with the REC method, because it can in principle be applied to three-dimensional configurations, *provided* that the assumption of axial invariance holds. Such studies are described by Syed and Bracco (1979), Lee et al (1981), Basso and Rinolfi (1982) and Grasso and Bracco (1982) and encompass a range of engine types, including homogeneous-charge and stratified-charge. Although again some fitting of coefficients proved necessary, it was found that useful parametric studies could be made with this method even in circumstances where the assumptions on which it is based are clearly questionable.

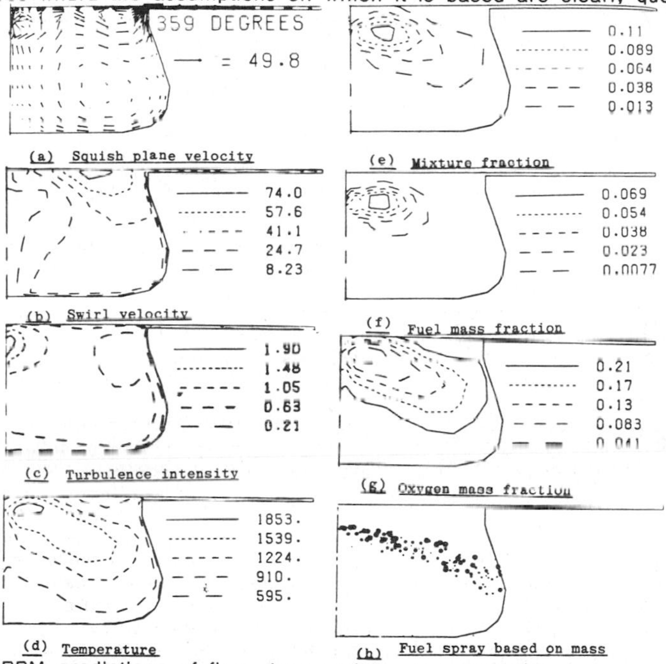

Fig 9 RPM predictions of flow, temperature, concentration and spray patterns in a Diesel engine

Direct-injection engine combustion predictions have scarcely been assessed at all, for here there is virtually no axisymmetric engine data. Demonstration calculations have been performed by Butler et al (1981), Gosman and Harvey (1982) and Harvey and Gosman (1983). The plots in Fig 9 above from the second-named study are indicative of the extent of the detailed information which these methods provide and also show the complexities of the in-cylinder processes in this type of engine. These include: strong motions generated by the combined action of squish, the spray and combustion (diagram (a)); consequent disturbance of the originally solid-body swirl structure (diagram (b)); turbulence generation in the shear layers (diagram (c)); large combustion-generated temperature and concentration gradients (diagrams (e) to (g)); and dispersal and impact of the fuel droplets on the piston bowl wall (diagram (h)).

As can be inferred from the remarks made above, one of the most urgent requirements of model development and assessment is for detailed data concerning the effects of fuel injection and combustion. Current research is directed towards this, as well as improving and extending the models themselves, towards the ultimate goal of full three-dimensional representations. Some progress in the latter area has been made by the author and his colleagues [41] with the aid of a newly-developed fully-implicit non-iterative solution algorithm.

SUMMARY AND CONCLUSIONS

The reciprocating internal combustion engine is a prime candidate for CFD techniques, but presents a formidable challenge to them. This is due to the combination of inherent geometrical complexity, including moving boundaries and numerous complex physical processes, embracing non-stationary turbulence, combustion and two-phase flow. The challenge is a dual one: to assemble adequate mathematical models of these processes and to devise suitable numerical methods for solving the model equations in engine circumstances. The first of these is arguably more the province of specialists in turbulence, combustion and two-phase flow, but in the engine field, as in many others, the CFD specialist inevitably becomes involved, if for no other reason than to provide the numerical framework for testing the models.

Mathematical models do exist for some, but not all, of the key physical processes (atomisation being the most notable exception); and when solved accurately they have been shown to yield predictions which are at least qualitatively correct and sometimes of sufficient accuracy to be useful for design purposes. They possess however many deficiencies, particularly in their treatment of non-stationary and two-phase flow phenomena.

On the computational side, techniques such as those contained in the REC, CONCHAS and RPM codes have been shown to be capable of solving the model equations in the context of two-dimensional representations of engine combustion chambers and of producing informative and useful results. However the accuracy of all existing techniques is open to question, particularly in those circumstances, often encountered in engines, when the flow is oblique to the mesh, leading to problems of numerical diffusion. Improved spatial differencing practices are undoubtedly needed, but they must be free of the spurious extrema which current so-called 'higher-order' schemes are prone to generate.

The next major step in engine modelling is the move to full three-dimensional representations and indeed, as already noted, some initial developments have already taken place. The need for three-dimensional methods is unquestionable, for without them many facets of engine behaviour will remain inaccessible to prediction, such as the in-cylinder flow patterns produced by practical induction systems. Success in this area will however require more than the straightforward extension of existing two-dimensional techniques which has been the practice thus far; for estimates of the computer resources they would need to reduce truncation errors to tolerable levels suggest they will be too costly for practical purposes.

REFERENCES

[1] Ahmadi-Befrui B, Gosman A D, Lockwood F C and Watkins A P "Multidimensional

calculation of combustion in an idealised homogeneous charge engine: a progress report. SAE Paper No 810151, (1981).

[2] Ahmadi-Befrui B, Arcoumanis C, Bicen A F, Gosman A D, Jahanbakhsh A and Whitelaw J H (1982) "Calculations and Measurements of the Flow in a Motored Model Engine and Implications for Open-Chamber, Direct-Injection Engines" in Three Dimensional Shear Flows, ed S Carmi, ASME.

[3] Blumberg P N, Lavoie G A and Tabaczynski R J (1979) "Phenomenological models for Reciprocating Internal Combustion Engines" Progress in Energy and Combustion Science, 5, pp 123-167.

[4] Bracco F V and O'Rourke P J "A Review of Initial Comparisons of Computed and Measured Two-Dimensional Unsteady Flame Fields" Prog Energy Combust Sci (1979).

[5] Bradshaw P An Introduction to Turbulence and Its Measurement Pergamon Press (1971).

[6] Bray K N C and Moss J B (1977) "A Unified Statistical Model of the Premixed Turbulent Flame" Acta Astronautica 4, pp 291-319.

[7] Butler T D, Cloutman L D, Dukowicz J K and Ramshaw J D (1979) "CONCHAS: An Arbitrary Lagrangian-Eulerian Computer Code for Multicomponent Chemically Reactive Fluid Flow at all Speeds" Los Alamos Scientific Laboratory report LA-8129-MS.

[8] Butler T D, Cloutman L D, Dukowicz J K, Ramshaw J D and Krieger R B (1980) "Toward a Comprehensive Model for Combustion in a Direct-Injection Stratified-Charge Engine" in Combustion Modelling in Reciprocating Engines, eds J N Mattavi and C A Amann, Plenum Press, New York.

[9] Butler T D, Cloutman L D, Dukowicz J K and Ramshaw J D (1981) "Multidimensional numerical simulation of reactive flow in internal combustion engines" Progress in Energy and Combustion Science, 7.

[10] Caretto L S, Gosman A D, Patankar S V and Spalding D B (1972) "Two numerical procedures for three-dimensional recirculating flows" Proc Int Conf on Numerical Methods in Fluid Dynamics, Paris.

[11] Cloutman L D, Dukowicz J K and Ramshaw J D (1980) "Numerical Simulation of Reactive Flow in Internal Combustion Engines" Proc 7th Int Conf on Numerical Methods in Fluid Dynamics, Stanford University/NASA Ames.

[12] Dukowicz J K, (1980), J Comp Phys, 35, 229.

[13] El Tahry S H (1982) "A Numerical Study on the Effects of Fluid Motion at Inlet-Valve Closure on Subsequent Fluid Motion in a Motored Engine" SAE 820035, SAE Int Congress Detroit.

[14] Gosman A D and Watkins A P (1977) "A Computer Prediction Method for Turbulent Flow and Heat Transfer in Piston/Cylinder assemblies" Proc of Symposium on Turbulent Shear Flows, Penn State University.

[15] Gosman A D, Johns R J R, Tipler W and Watkins A P (1979) "Computer Simulation of In-Cylinder Flow, Heat Transfer and Combustion: a Progress Report" Proc Int Congress on Combustion Engines, Vienna.

[16] Gosman A D, Johns R J R and Watkins A P (1980) "Development of Prediction

Methods for In-Cylinder Processes in Reciprocating Engines" in Combustion Modelling in Reciprocating Engines, eds J N Mattavi and C A Amann, Plenum Press, New York.

[17] Gosman A D and Johns R J R (1980) "Computer Analysis of Fuel-Air Mixing in Direct-Injection Engines" Proc 1980 SAE Congress, SAE 800091. Also published in SAE Transactions, 88, No 6.

[18] Gosman A D and Ioannides S I (1981) "Aspects of Computer Simulation of Liquid-Fuelled Combustors" Proc 19th AIAA Aerospace Sciences Meeting, AIAA-81-0323. Also to appear in AIAA J of Energy.

[19] Gosman A D and Harvey P S (1982) "Computer Analysis of Fuel-Air Mixing and Combustion in an Axisymmetric DI Diesel" Proc 1982 SAE Congress, SAE 820036. Also published in SAE Transactions, 89, No 9, (1981).

[20] Gosman A D (1983) "Flow Processes in Cylinders" To be published in Thermodynamics and Gas Dynamics of Internal Combustion Engines, 2, ed J H Horlock and D Winterbone, Oxford University Press.

[21] Grasso F and Bracco F V (1982) "Computed and Measured Turbulence in Axisymmetric IC Engines" Princeton University, Dept of Mechanical and Aerospace Engineering Report.

[22] Griffen M D, Diwalker R, Anderson J D Jnr and Jones E (1978) "Computational Fluid Dynamics Applied to Flows in an Internal Combustion Engine" AIAA Paper 78-57.

[23] Gupta H C and Syed S A (1979) "REC-P3 (Reciprocating Engine Combustion, Planar Geometry, Third Version): A Computer Program for Combustion in Reciprocating Engines" MAE Report No 1431, Mechanical and Aerospace Engineering Department, Princeton University.

[24] Harlow F H and Amsden A A (1979), J Comp Phys 8, 197.

[25] Haselman L C and Westbrook C K (1978) "A Theoretical Model for Two-Phase Fuel Injection in Stratified Charge Engines" SAE Paper 780318.

[26] Hirt C W, Amsden A A and Cook J L (1974) "An Arbitrary Lagrangian-Eulerian Computing Method for All Flow Speeds" J Comp Phys 14, 227.

[27] Kuo T W and Bracco F V (1982) "Computations of Drop Sizes in Pulsating Sprays and of Liquid-Core Length in Vaporizing Sprays" SAE Paper 820133, SAE Int Congress Detroit.

[28] Markatos N C and Mukerjee T (1981) "Three-Dimensional Computer Analysis of Flow and Combustion in Automotive Internal Combustion Engines" in Mathematics and Computers in Simulation XXIII 354-366.

[29] O'Rourke P J and Bracco F V (1980) "Modelling of Drop Interactions in Thick Sprays and a Comparison with Experiments" Proc of Stratified Charge Automotive Engines Conference, IMechE, London.

[30] O'Rourke P J (1981) "Collective Drop Effects in Vapourising Liquid Sprays" Princeton University, Dept of Mechanical and Aerospace Engineering, PhD Thesis 1532-T.

[31] Reynolds W C (1980) "Modelling of Fluid Motions in Engines - An Introductory Overview" Combustion Modelling in Reciprocating Engines, eds J N Mattavi

and C A Amann, Plenum Press, New York.

[32] Rivard W C, Farmer O A and Butler T D (1975) "RICE: A Computer Program for Multicomponent Chemically Reactive Flows at All Speeds" Los Alamos Scientific Laboratory report LA-5812.

[33] Syed S A and Bracco F V (1979) "Further Comparisons of Computed and Measured Divided Chamber Engine Combustion" SAE Paper 790247, SAE 1979 Congres and Exposition, Detroit.

[34] Tabaczynski R J (1979) "Turbulence and Turbulent Combustion in Spark-Ignition Engines" Progress in Energy and Combustion Sciences, Student Edition I, ed N A Chigier, Pergamon Press.

[35] Traci R M (1981) in Comparisons Between Measurement and Analysis of Fluid Motion in Internal Combustion Engine, ed P O Witze, Sandia National Laboratories Report, SAND81-8242, Livermore, CA, USA.

[36] Watkins A P (1979) "Computer Prediction of Flow and Heat Transfer in Piston/Cylinder Assemblies" PhD Thesis, University of London.

[37] Westbrook C K (1976) "Three Dimensional Numerical Modelling of Liquid Fuel Sprays" 16th Symposium (Int) on Combustion, Combustion Institute, Pittsburgh.

[38] Westbrook C K (1979) "Fuel Motion and Pollutant Formation in Stratified Charge Combustion" SAE Paper 790248.

[39] Williams F A (1965), Combustion Theory, Addison-Wesley, Reading, Mass.

[40] Wu K J, Su C C, Steinberger R L, Santavicca D A and Bracco F V (1983) "Measurements of the Spray Angle of Atomizing Jets" ASME 83-WA/FE-10, ASME Winter Meeting, Boston.

[41] Gosman A D, Tsui Y Y and Watkins A P (1983) "Calculation of 3D Air Motion in Model Engines" To be presented SAE Congress, Detroit, Feb/Mar 1984.

MULTIPHASE PROBLEMS IN OIL RESERVOIR SIMULATIONS

PROBLEMES MULTIPHASIQUES EN GISEMENTS DE PETROLE

Computing Methods in Applied Sciences and Engineering, VI
R. Glowinski and J.-L. Lions (Editors)
Elsevier Science Publishers B.V. (North-Holland)

IMPROVED ACCURACY THROUGH SUPERCONVERGENCE IN THE PRESSURE IN THE SIMULATION OF MISCIBLE DISPLACEMENT

Jim Douglas, Jr.
Department of Mathematics
University of Chicago
Chicago, Illinois, USA

A modification of a finite element procedure of Douglas, Ewing, and Wheeler for simulating miscible displacement in porous media is discussed. The pressure is approximated by a mixed finite element method and the resulting Darcy velocity is post-processed by convolution with a Bramble-Schatz kernel, with the enhanced velocity being used in the evaluation of the coefficients in the Galerkin approximation to the concentration. Optimal reflection of the superconvergent velocity approximation has been demonstrated in the error estimates, which are summarized here.

INTRODUCTION

The miscible displacement of one incompressible fluid by another in a horizontal reservoir $\Omega \subset \mathbb{R}^2$ of unit thickness is described by the system of equations

$$\phi \frac{\partial c}{\partial t} + \underset{\sim}{u}\cdot \underset{\sim}{\text{grad}}\ c - \text{div}(D(\underset{\sim}{u})\underset{\sim}{\text{grad}}\ c) = (1 - c)q^+, \tag{1.1a}$$

$$\text{div}\ \underset{\sim}{u} = q\ , \tag{1.1b}$$

for $x \in \Omega$ and $t \in [0,T] = J$, where

$$\underset{\sim}{u} = -\frac{k(x)}{\mu(c)}\ \underset{\sim}{\text{grad}}\ p, \tag{1.1c}$$

subject to an initial condition

$$c(x,0) = c_0(x), \quad x \in \Omega. \tag{1.1d}$$

In the above, c represents the concentration of one of the fluids, p the pressure, $\underset{\sim}{u}$ the Darcy velocity of the fluid, q the external flow rate and q^+ its positive part, $\phi = \phi(x)$ the porosity of the medium, $D(\underset{\sim}{u})$ a (tensor) diffusion coefficient, $k(x)$ the permeability of the medium, and $\mu(c)$ the viscosity of the fluid; see [4,5] for more details. In addition, a boundary condition must be specified; here it will be assumed that Ω is the unit square and that all functions, and spaces of functions, arising are Ω-periodic.

Let

$$a = a(c) = a(x,c) = k(x)/\mu(c),$$

$$\alpha = \alpha(c) = 1/a(c).$$

Let $\underset{\sim}{V} = H(\text{div};\Omega)$, and let $(\cdot,\cdot)$ denote the inner product in $L^2(\Omega)$ or $L^2(\Omega)^2$, as appropriate. Then a weak form of (1.1) is given by the determination of a map $\{c,\underset{\sim}{u},p\} : J \to H^1(\Omega) \times \underset{\sim}{V} \times L^2(\Omega)$ such that $c(\cdot,0) = c_0$ and, for $t \in J$,

$$(\phi \frac{\partial c}{\partial t},z) + (\underset{\sim}{u}\cdot \underset{\sim}{\text{grad}}\, c,z) + (D(\underset{\sim}{u})\underset{\sim}{\text{grad}}\, c, \underset{\sim}{\text{grad}}\, z) = ((1-c)q^+,z), \quad z \in H^1(\Omega), \tag{1.2a}$$

$$(\alpha(c)\underset{\sim}{u},\underset{\sim}{\psi}) - (\text{div}\, \underset{\sim}{\psi},p) = 0, \qquad \underset{\sim}{\psi} \in \underset{\sim}{V}, \tag{1.2b}$$

$$(\text{div}\, \underset{\sim}{u},\phi) = (q,\phi), \qquad \phi \in L^2(\Omega), \tag{1.2c}$$

$$(p,1) = 0. \tag{1.2d}$$

The numerical solution of (1.2) has been treated by Douglas, Ewing, and Wheeler [5] by approximating the concentration equation (1.2a) by a Galerkin procedure based on the space $\mathcal{M}_h = \mathcal{M}(\ell, \mathcal{T}(h_c,c))$ of C^0-piecewise polynomial functions of local degree ℓ over a quasi-regular polygonalization $\mathcal{T}(h_c,c)$ of Ω and by approximating the pressure and Darcy velocity equations (1.2b) and (1.2c) by a mixed finite element method based on a Raviart-Thomas space $\underset{\sim}{V}_h \times W_h$ of index $k \geq 0$ over a quasi-regular partition $\mathcal{T}(h_p,p)$ of Ω into triangles or rectangles. The finite element procedure of [5] is given by the finding of $\{c_h,\underset{\sim}{u},p_h\}:J \to \mathcal{M}_h \times \underset{\sim}{V}_h \times W_h$ such that $c_h(0)$ approximates c_0 to $O(h_c^{\ell+1})$ in $L^2(\Omega)$ (here and below it is assumed that the problem (1.1) has a smooth solution) and, for $t \in J$,

$$(\phi \frac{\partial c_h}{\partial t},z) + (\underset{\sim}{u}_h\cdot \underset{\sim}{\text{grad}}\, c_h,z) + (D(\underset{\sim}{u}_h)\underset{\sim}{\text{grad}}\, c_h,\underset{\sim}{\text{grad}}\, z) = ((1-c_h)q^+,z), \quad z \in \mathcal{M}_h, \tag{1.3a}$$

$$(\alpha(c_h)\underset{\sim}{u}_h,\underset{\sim}{\psi}) - (\text{div}\, \underset{\sim}{\psi},p_h) = 0, \qquad \underset{\sim}{\psi} \in \underset{\sim}{V}_h, \tag{1.3b}$$

$$(\text{div}\, \underset{\sim}{u}_h,\phi) = (q,\phi), \qquad \phi \in W_h, \tag{1.3c}$$

$$(p_h,1) = 0. \tag{1.3d}$$

Optimal order error estimates have been derived for the solution of (1.3); in particular, it has been shown [5] that (with $L^\infty(L^2)$ indicating $L^\infty(J;L^2(\Omega))$, etc.)

$$\|c - c_h\|_{L^\infty(L^2)} + \|\underset{\sim}{u} - \underset{\sim}{u}_h\|_{L^\infty(\underset{\sim}{V})} + \|p - p_h\|_{L^\infty(L^2)} \leq Q(p,c)(h_c^{\ell+1} + h_p^{k+1}), \tag{1.4a}$$

where

$$Q(p,c) \le \text{function}(\|c\|_{W^{1,\infty}(W^{\ell+1,\infty})}, \|p\|_{W^{1,\infty}(W^{2,\infty})}, \|p\|_{L^\infty(H^{k+1})}) . \quad (1.4b)$$

Recently, Douglas and Roberts [7] and Douglas and Milner [6] have derived a collection of new error estimates for mixed finite element methods for second order elliptic equations. These results, as modified to treat a periodic problem instead of a Dirichlet problem, include error bounds in Sobolev spaces of negative index and superconvergent approximation, via convolution with a Bramble-Schatz [1,2] kernel, to both the basic dependent variable (in our case, p) and the related gradient field ($\underset{\sim}{u}$). An outline of this development will be given in Section 2. It is required that $\mathcal{T}(h_p,p)$ be translation-invariant in order that the superconvergence procedure be applicable, and it is natural to take $\mathcal{T}(h_p,p)$ to be the set of squares of side length h_p related to a uniform grid over Ω.

The object of this work is to study the modification of (1.3) that occurs when u_h in (1.3a) is replaced by a superconvergent approximation to $\underset{\sim}{u}$ related to $\underset{\sim}{u}_h$. Let $K_h = K^{2k+2}_{h_p,k+2}$ be the Bramble-Schatz kernel [2] to be defined in the next section. Set (where $\{\underset{\sim}{u}_h,p_h\}$ is determined by (1.6))

$$\underset{\sim}{u}_h^* = K_h * \underset{\sim}{u}_h , \quad p_h^* = K_h * p_h . \quad (1.5)$$

Then, change the finite element procedure to become the determination of $\{c_h,\underset{\sim}{u}_h,p_h\}:J \to \mathcal{M}_h \times \underset{\sim}{V} \times W_h$ such that $\|c_h(0) - c_0\|_0 \le Q\|c_0\|_{\ell+1} h_c^{\ell+1}$ and

$$(\phi \frac{\partial c_h}{\partial t}, z) + (\underset{\sim}{u}_h^* \cdot \underset{\sim}{\text{grad}}\, c_h, z) + (D(\underset{\sim}{u}_h^*)\underset{\sim}{\text{grad}}\, c_h, \underset{\sim}{\text{grad}}\, z)$$

$$= ((1 - c_h)q^+, z), \quad z \in \mathcal{M}_h , \quad (1.6a)$$

$$(\alpha(c_h)\underset{\sim}{u}_h, \underset{\sim}{\psi}) - (\text{div}\, \underset{\sim}{\psi}, p_h) = 0, \qquad \underset{\sim}{\psi} \in \underset{\sim}{V}_h, \quad (1.6b)$$

$$(\text{div}\, \underset{\sim}{u}_h, \phi) = (\phi, \mu), \qquad \phi \in W_h, \quad (1.6c)$$

$$(p_h, 1) = 0, \quad (1.6d)$$

for $t \in J$. Note that $\mathcal{T}(h_c,c)$ has not been required to be related to a uniform grid. It can be proved that

$$\|c - c_h\|_{L^\infty(L^2)} + \|\underset{\sim}{u} - \underset{\sim}{u}_h^*\|_{L^\infty(\underset{\sim}{V})} + \|p - p_h^*\|_{L^\infty(L^2)} \le Q(p,c)\{h_c^{\ell+1} + h_p^{2k+2}\}, \quad (1.7a)$$

where now

$$Q(p,c) \le \text{function}(\|p\|_{W^{1,\infty}(W^{2,\infty}) \cap L^\infty(H^{2k+4})}, \|c\|_{W^{1,\infty}(W^{\ell+1,\infty})}). \quad (1.7b)$$

The bound (1.7a) indicates that the full possible gain in the rate of convergence with respect to h_p has been achieved. Moreover, the additional arithmetic required to replace $\underset{\sim}{u}_h$ by $\underset{\sim}{u}_h^*$ is trivial; if the partition $\mathcal{T}(h_c,c)$ is reasonably related to $\mathcal{T}(h_p,p)$, all that occurs is that, at any point at which an evaluation of $\underset{\sim}{u}_h$ is needed, a different linear combination of the parameters defining $\underset{\sim}{u}_h$ is used to evaluate $\underset{\sim}{u}_h^*$.

This paper should be regarded as a modification and continuation of [5].

2. SOME RESULTS FOR MIXED METHODS FOR ELLIPTIC PROBLEMS

Let $\mathcal{T}(h_p,p)$ be uniform, and consider the mixed method for the periodic elliptic problem associated with (1.2b-d) when the coefficient α is evaluated at $c(\cdot,t)$; i.e., for any $t \in J$, let $\{\hat{\underset{\sim}{u}}_h,\hat{p}_h\} \in \underset{\sim}{V}_h \times W_h$, where $\{\hat{\underset{\sim}{u}}_h,\hat{p}_h\}$ is periodic of period Ω, be the solution of

$$(\alpha(c)\hat{\underset{\sim}{u}}_h,\underset{\sim}{\psi}) - (\operatorname{div} \underset{\sim}{\psi},\hat{p}_h) = 0, \qquad \underset{\sim}{\psi} \in \underset{\sim}{V}_h, \tag{2.1a}$$

$$(\operatorname{div} \hat{\underset{\sim}{u}}_h,\phi) = (q,\phi), \quad \phi \in W_h, \tag{2.1b}$$

$$(\hat{p}_h,1) = 0. \tag{2.1c}$$

Let [1,2] K_h be defined as follows:

$$K_h(x) = \prod_{m=1}^{2} \Big(\prod_{i=-k}^{k} h_p^{-1} k_i' g_{k+2}(h_p^{-1} x_m - i)\Big), \tag{2.2a}$$

where

$$g_\ell(t) = (\chi_{[-1/2,1/2]} * g_{\ell-1})(t) \;, \; g_1(t) = \chi_{[-1/2,1/2]}(t) \tag{2.2b}$$

$$k'_{-i} = k'_i = \frac{1}{2} k_i \text{ for } i = 1,\dots,k, \text{ and } k'_0 = k_0 \;, \tag{2.2c}$$

$$\sum_{i=0}^{k} k_i \int_R g_k(y)(y+i)^{2n}\, dy = \delta_{0n} \;, \quad n = 0,\dots,k \;. \tag{2.2d}$$

It is known [1] that, in the periodic case considered here,

$$\|K_h * w - w\|_0 \le Q\|w\|_r h_p^r \;, \quad 0 \le r \le 2k+2, \tag{2.3a}$$

$$\|D^\nu(K_h * w)\|_s \le Q\|\partial^\nu w\|_s \;, \quad s \in \mathbb{Z}, \tag{2.3b}$$

where $D^\nu = \partial^{|\nu|}/\partial x_1^{\nu_1}\partial x_2^{\nu_2}$ and ∂^ν is the corresponding forward, divided difference with step length h_p, and

$$\|w\|_0 \le Q \sum_{|\nu|\le s} \|D^\nu w\|_{-s}, \quad 0 \le s \in Z. \tag{2.4}$$

It follows from [6] that

$$\|\partial^\nu(\underset{\sim}{u} - \hat{\underset{\sim}{u}}_h)\|_{-k-1} \le Q(c)\|p\|_{2k+4} h_p^{2k+2} \tag{2.5}$$

for $|\nu| \le k+1$. So, by (2.3), (2.4), and (2.5),

$$\|\underset{\sim}{u} - K_h * \hat{\underset{\sim}{u}}_h\|_0 \le Q(c)\|p\|_{2k+4} h_p^{2k+2}. \tag{2.6a}$$

Similarly, it follows from estimates for difference quotients for $\text{div}(\underset{\sim}{u} - \hat{\underset{\sim}{u}}_h)$ and $p - \hat{p}_h$ [6] that

$$\|\text{div}(\underset{\sim}{u} - K_h * \hat{\underset{\sim}{u}}_h)\|_0 \le Q(c)\|p\|_{2k+4} h_p^{2k+2}, \tag{2.6b}$$

$$\|p - K_h * \hat{p}_h\| \le Q(c)\|p\|_{2k+4} h_p^{2k+2}. \tag{2.6c}$$

In (2.6), $Q(c) < Q(\|c\|_{k,\infty})$.

Set $\underset{\sim}{u}_h^* = K_h * \underset{\sim}{u}_h$, as in (1.5), and $\hat{\underset{\sim}{u}}_h^* = K_h * \hat{\underset{\sim}{u}}_h$. It is of interest to note some relations between $\underset{\sim}{u}_h$, $\underset{\sim}{u}_h^*$, and $\hat{\underset{\sim}{u}}_h^*$. First, since

$$(\alpha(c_h)(\underset{\sim}{u}_h - \hat{\underset{\sim}{u}}_h), \underset{\sim}{\psi}) - (\text{div}\, \underset{\sim}{\psi}, p_h - \hat{p}_h) = ([\alpha(c) - \alpha(c_h)]\hat{\underset{\sim}{u}}_h, \underset{\sim}{\psi}), \quad \underset{\sim}{\psi} \in \underset{\sim}{V}_h, \tag{2.7a}$$

$$(\text{div}(\underset{\sim}{u}_h - \hat{\underset{\sim}{u}}_h), \phi) = 0, \quad \phi \in W_h, \tag{2.7b}$$

the known boundedness [5] of $\hat{\underset{\sim}{u}}_h$ in $L^\infty(\Omega)$ leads immediately to the bound

$$\|\underset{\sim}{u}_h - \hat{\underset{\sim}{u}}_h\|_{\underset{\sim}{V}} + \|p_h - \hat{p}_h\|_0 \le Q\|c - c_h\|_0. \tag{2.8}$$

Then, (2.3b) implies that

$$\|\underset{\sim}{u}_h^* - \underset{\sim}{u}_h^*\|_{\underset{\sim}{V}} + \|p_h^* - \hat{p}_h^*\|_0 \le Q\|c - c_h\|_0. \tag{2.9}$$

3. CONVERGENCE RESULTS

An analysis of the error in the approximate solution obtained by the procedure (1.6) can be given [3] along the same general lines as in [5]. The argument leads to the following theorem.

Theorem. Let there exist a unique solution $\{c,p\}$ of (1.1) with $\{c,p\} \in W^{1,\infty}(J;W^{\ell+1,\infty}(\Omega)) \times (W^{2,\infty}(J;W^{2,\infty}(\Omega)) \cap L^\infty(J;H^{2k+4}(\Omega))$. Let $h_p^{-1}h_c^{\ell+1} \to 0$ as $h_p + h_c \to 0$. Then, if $c_h(0)$ is determined in such a way that

$\|(c - c_h)(0)\|_0 \le Q\|c(0)\|_{\ell+1} h_c^{\ell+1}$, then there exists a unique, periodic solution $\{c_h, \underset{\sim}{u}_h, p_h\}: J \to \mathcal{M}_h \times \underset{\sim}{V}_h \times W_h$ of (1.6), and

$$\|c - c_h\|_{L^\infty(L^2)} + \|\underset{\sim}{u} - \underset{\sim}{u}_h^*\|_{L^\infty(\underset{\sim}{V})} + \|p - p_h^*\|_{L^\infty(L^2)} \le Q(p,c)(h_p^{2k+2} + h_c^{\ell+1}),$$

where Q(p,c), as described in (1.7b), is boundable in terms of the norms on the spaces listed above to which the solution {c,p} belongs.

4. REMARKS

A quite similar development can be given when the pressure is approximated by a Galerkin procedure in place of the mixed method. Let $\mathcal{N}_h$ be the periodic tensor-product of C^0-piecewise-polynomial functions of degree $k \ge 1$ over a uniform grid of length h_p in each variable. Then, for $t \in J$ determine $p_h \in \mathcal{N}_h$ such that

$$(a(c_h)\underset{\sim}{\text{grad}}\, p_h, \underset{\sim}{\text{grad}}\, w_h) = (q, w_h), \quad w_h \in \mathcal{N}_h, \tag{4.1a}$$

$$(p_h, 1) = 0. \tag{4.1b}$$

There exists [1,2] a Bramble-Schatz kernel $\underset{\sim}{M}_h = \underset{\sim}{M}(h_p, k)$ such that

$$\|\underset{\sim}{\text{grad}}\, p - \underset{\sim}{M}_h * p_h\|_0 \le Q(c)\|p\|_{2k+1} h_p^{2k}. \tag{4.2}$$

Now, set

$$\underset{\sim}{u}_h^* = -a(c_h)\underset{\sim}{M}_h * p_h, \tag{4.3}$$

and determine c_h by (1.6a). If $p_h^* = K_h * p_h$ as before, an argument similar to the one already presented leads to the estimate

$$\|c - c_h\|_{L^\infty(L^2)} + \|p - p_h^*\|_{L^\infty(L^2)} \le Q(p,c)(h_p^{2k} + h_c^{\ell+1}). \tag{4.4}$$

REFERENCES

[1] J. H. Bramble and A. H. Schatz, Estimates for spline projections, R.A.I.R.O., Analyse numérique 10 (1976), 5-37.

[2] J. H. Bramble and A. H. Schatz, Higher order local accuracy by averaging in the finite element method, Math. Comp. 31 (1977) 94-111.

[3] J. Douglas,Jr., Superconvergence in the pressure in the simulation of miscible displacement, to appear.

[4] J. Douglas, Jr., Numerical methods for the flow of miscible fluids in porous media, to appear.

[5] J. Douglas, Jr., R. E. Ewing, and M. F. Wheeler, The approximation of the pressure by a mixed method in the simulation of miscible displacement, R.A.I.R.O., Analyse numérique 17 (1983) 17-33.

[6] J. Douglas, Jr., and F. A. Milner, Interior and superconvergence estimates for mixed methods for second order elliptic equations, to appear.

[7] J. Douglas, Jr., and J. E. Roberts, Global estimates for mixed methods for second order elliptic equations, to appear.

Computing Methods in Applied Sciences and Engineering, VI
R. Glowinski and J.-L. Lions (Editors)
Elsevier Science Publishers B.V. (North-Holland)

MIXED FINITE ELEMENT METHODS FOR PETROLEUM RESERVOIR ENGINEERING PROBLEMS

Mary Fanett Wheeler

Rice University
Houston, Texas U. S. A.

and

Ruth Gonzalez

Exxon Production Research
and
Rice University
Houston, Texas U. S. A.

Mixed finite element methods are considered for computing Darcy velocities for fluid flow problems in a porous medium. It is shown that the five point block-centered finite difference method is a mixed method with special numerical quadrature formulas. Computational results with a block-centered finite difference matrix as a preconditioner for mixed method formulations using the tensor product Raviart-Thomas approximating spaces are presented.

1. INTRODUCTION

Accurate fluid velocities are essential in the numerical simulation of miscible and immiscible displacement in a porous medium. These reservoir flow problems are described by coupled systems of non-linear partial differential equations in which one of the equations known as Darcy's Law is given by

$$u = -\frac{k}{\mu} \nabla p.$$

Here u denotes the fluid velocity, k the permeability, and ∇p the gradient of pressure. This characterization of the fluid conductivity of a porous material, the permeability, was first meaningfully demonstrated by Darcy [6] in 1856.

In this paper we discuss mixed finite element methods for approximating the pair $(u;p)$ satisfying the first order system

(1.1) $\qquad u = -a\nabla p, \quad x \in \Omega,$ (Darcy's Law)

(1.2) $\qquad \nabla \cdot u = q, \quad x \in \Omega,$ (Conservation of mass)

(1.3) $\qquad u \cdot \nu = 0, \quad x \in \partial\Omega,$

where Ω is a bounded domain in $\mathbf{R}^2$ with boundary $\partial\Omega$, ν is the outward normal vector on $\partial\Omega$, and q is a sum of point sources and sinks (Dirac measures). We assume that the possibly discontinuous coefficient a satisfies $a_0 \leq a(x,y) \leq a_1$ for positive constants a_0 and a_1. Theoretical and computational results using mixed methods in the simulation of miscible and immiscible displacement can be found in [4, 7, 8, 9, 11, 15].

This paper is organized as follows. In §2, we formulate the mixed finite element method and state theoretical convergence rates for this procedure. In §3, we describe the algebraic system that arises if Ω is a rectangle and a particular choice of tensor product approximating spaces, the Raviart-Thomas spaces, are used. We also show that the five point block-centered finite difference method, one of the most frequently used procedures in the petroleum industry for computing pressure and velocities, is a mixed finite element method (first order Raviart-Thomas space) with special numerical quadrature formulas. In §4, we describe mixed method experimental results using the approximating spaces discussed in §3. Iteration counts arising from the solution of the algebraic system by preconditioned conjugate gradient procedures are presented.

2. FORMULATION OF THE MIXED FINITE ELEMENT METHOD AND THEORETICAL RESULTS

Denote by $(z,w) = \int_\Omega z \cdot w \, dx$ and $\|z\|^2 = (z,z)$, the standard L^2 inner product and norm on Ω, where $\cdot$ is multiplication (dot product) if z is a scaler (vector) function. Let $H(\text{div};\Omega)$ be the set of vector functions $v \in (L^2(\Omega))^2$ such that $\nabla \cdot v \in L^2(\Omega)$ and let

$$V = H(\text{div};\Omega) \cap \{v \,|\, v \cdot \nu = 0 \quad \text{on} \quad \partial\Omega\}.$$

Let $W = L^2(\Omega)$ and $H^\ell = H^\ell(\Omega)$, the standard Sobolev spaces.

Multiplying (1.1) by $a^{-1}v$, $v \in V$, integrating, and integrating by parts, we obtain

$$(2.1) \qquad (a^{-1}u,v) = -(\nabla p,v) = (p,\text{div } v).$$

Multiplying (1.2) by $w \in W$ and integrating the result, we note that

$$(2.2) \qquad (\text{div } u,w) = (q,w), \quad w \in W.$$

The first order system (2.1) and (2.2) is a weak form for the solution pair $(u;p)$ and motivates the definition of the mixed finite element method.

For a sequence of mesh parameters $h > 0$, we choose finite dimensional subspaces V_h and W_h with $V_h \subset V$ and $W_h \subset W$ and seek the solution pair $(U_h;P_h) \in V_h \times W_h$ satisfying

$$(2.3) \qquad (a^{-1}U_h,v) - (P_h,\text{div } v) = 0, \quad v \in V_h,$$

$$(2.4) \qquad (\text{div } U_h,w) = (q,w), \quad w \in W_h.$$

The finite dimensional subspaces V_h and W_h need certain properties in order for the convergence analysis of Brezzi [2], Falk and Osborn [13], and Raviart and Thomas [14] to hold. One of these is that $\text{div } V_h \subset W_h$. The analysis of [2, 13, 14] yields the following error estimates.

<u>Theorem 2.1</u> For smoothly distributed q,

$$\| U_h - u \| + \| \mathrm{div}(U_h - u) \| \leq C_1 \, h^{r+1}$$

and

$$\| P_h - P \| \leq C_2 \, h^{r+1},$$

where C_1 and C_2 are constants depending on the smoothness of u and p and $(r+1)$ is the best possible exponent.

If q is the sum of Dirac measures, the method can be modified by subtracting out the singularities for the leading part of the velocities at the Dirac points (point sources or sinks, the wells) and then solving for the remaining part. The procedure described below is due to Douglas Arnold.

Let $q = \sum_{j=1}^{NW} Q_j \delta_{x_j}$. Express u as the sum of a regular part u_r, and a singular part, u_s, where

$$u_s = 1/2\pi \sum_{j=1}^{NW} Q_j \nabla (\ell n |x - x^j|).$$

The regular part $u_r \in H^{1-\epsilon}$, $\epsilon > 0$, and satisfies

$$\nabla \cdot u_r = 0 \quad \text{on} \quad \Omega,$$

$$u_r \cdot \upsilon = -u_s \cdot \upsilon \quad \text{on} \quad \partial\Omega.$$

The equations (2.3) and (2.4) are modified and have the form $(U_r;P_h)$ $V_h \times W_h$, satisfying

(2.5) $\quad (1/a \, U_r, v) - (P_h, \mathrm{div}\, v) = -(1/a \, u_s, v), \quad v \in V_h,$

(2.6) $\quad (\mathrm{div}\, U_r, w) = 0, \quad w \in W_h,$

(2.7) $\quad \int_{\partial\Omega} (U_r + u_s) \cdot \upsilon \, v \cdot \upsilon \, ds = 0, \quad v \in V_h,$

with $V_h = H(\mathrm{div};\Omega)$. The approximation U_h is then defined by

$$U_h = U_r + u_s,$$

where U_r is the approximation to u_r.

For this procedure with singular q, Douglas, Ewing and one of the authors [8] extended the analysis of [2, 13, 14] to prove the following result.

<u>Theorem 2.2</u> Let q be the sum of Dirac measures. Then there exist positive constants C_3 and C_4 such that

$$\| P_h - P \| \leq C_3 h \, \ell n \, h^{-1}$$

and

$$\| U_h - u \| \leq C_4 h \, \ell n \, h^{-1}.$$

3. THE MIXED FINITE ELEMENT METHOD FOR TENSOR PRODUCT MESHES

In §2, we noted that the analyses of [2, 13, 14] assumed that div $V_h \subset W_h$. This requirement makes the following Raviart-Thomas spaces which appear strange at first glance understandable.

Let Ω be the rectangle $(0,XL) \times (0,YL)$ and let Δ_x: $0 = x_0 < x_1 < \cdots < x_{Nx} = XL$ and Δ_y: $0 = y_0 < y_1 < \cdots < y_{Ny} = YL$ be partitions of $[0,XL]$ and $[0,YL]$ respectively. For Δ a partition of $[0,L]$, defined the piecewise polynomial space

$$M_q^r(\Delta) = \{v \in C^q([0,L]): v \text{ is a polynomial of degree} \le r \text{ on each subinterval of } \Delta\},$$

where $q = -1$ refers to the discontinuous functions. Let

$$W_h^{q,r} = M_q^r(\Delta_x) \otimes M_q^r(\Delta_y),$$

$$\tilde{V}_h^{q,r} = [M_{q+1}^{r+1}(\Delta_x) \otimes M_q^r(\Delta_y)] \times [M_q^r(\Delta_x) \otimes M_{q+1}^{r+1}(\Delta_y)],$$

$$V_h^{q,r} = \tilde{V}_h^{q,r} \cap \{v: v \cdot \upsilon = 0 \quad \text{on} \quad \partial\Omega\},$$

where $h = \max_{ij} [(x_{i+1}-x_i), (y_{j+1}-y_j)]$. We remark that these spaces fulfill the necessary approximation properties for Theorem 2.1 to hold.

We now consider the linear matrix problem associated with the approximation of (2.3) and (2.4) since it is different from both the standard Galerkin or finite difference matrices. We shall assume $q = -1$ and drop the q superscript from the subspace definitions.

Let $\{v_i^x\}$ and $\{w_j^y\}$ denote bases for $M_0^{r+1}(\Delta_x) \cap H_0^1(0,XL)$ and $M_{-1}^r(\Delta_y)$ respectively. Similarly let $\{v_j^y\}$ and $\{w_i^x\}$ denote bases for $M_0^{r+1}(\Delta_y) \cap H_0^1(0,YL)$ and $M_{-1}^r(\Delta_y)$. Then bases for $(V_h^r)_x$, $(V_h^r)_y$, and W_h^r are respectively $\{v_i^x w_j^y\}$, $\{w_i^x v_j^y\}$, and $\{w_i^x w_j^y\}$. A basis for V_h^r is $((V_h^r)_x \times 0) \cup (0 \times (V_h^r)_y)$. The resulting matrix problem has the form

$$\begin{pmatrix} M_1 & 0 & -N_1 \\ 0 & M_2 & -N_2 \\ N_1^t & N_2^t & 0 \end{pmatrix} \begin{pmatrix} \alpha^1 \\ \alpha^2 \\ \beta \end{pmatrix} = \begin{pmatrix} 0 \\ 0 \\ RR \end{pmatrix}, \tag{3.1}$$

where

$$(M_1)_{i'j',ij} = (a^{-1} v_i^x w_j^y, v_{i'}^x w_{j'}^y), \tag{3.2a}$$

$$(M_2)_{i'j',ij} = (a^{-1} w_i^x v_j^y, w_{i'}^x v_{j'}^y), \tag{3.2b}$$

$$(N_1)_{i'j',ij} = (w_i^x w_j^y, (v_{i'}^x)' w_{j'}^y), \tag{3.2c}$$

$$(N_2)_{i'j',ij} = (w_i^x w_j^y, w_{i'}^x (v_{j'}^y)'), \tag{3.2d}$$

and RR contains the right-hand side of (2.4).

For $r = 0$, define

$$v_i^x = \begin{cases} (x - x_{i-1})/(x_i - x_{i-1}), & x_{i-1} \le x < x_i, \\ (x_{i+1} - x)/(x_{i+1} - x_i), & x_i \le x < x_{i+1}, \\ 0, & \text{otherwise}, \end{cases} \qquad 1 \le i \le N_x - 1, \tag{3.3}$$

and

$$w_j^y = \begin{cases} 1, & y_{j-1} < y < y_j, \\ 0, & \text{otherwise}, \end{cases} \qquad 1 \le j \le N_y - 1, \tag{3.4}$$

and similarly for v_j^y and w_i^x. Number the bases for $(V_h^0)_x$ and W_h^0 such that i changes most rapidly and for $(V_h^0)_y$ such that j changes most rapidly. The matrices M_1 and M_2 are block diagonal matrices with each block being tridiagonal. The matrix N_1 is a bidiagonal of order $(N_x-1) \times N_xN_y$ and N_2 (a matrix of order $(N_y-1) \times N_xN_y$) can be made bidiagonal by permuting rows and corresponding columns.

For $r = 1$, we choose $\{v_i^x\}_{i=1}^{2N_x-1}$ as a basis for $M_0^2(\Delta_x) \cap H_0^1(0,XL)$ as defined below. For even indices v_{2k}^x is the piecewise linear function satisfying

$$v_{2k}^x(x_k) = 1,$$
$$v_{2k}^x(x_\ell) = 0, \quad \ell \neq k,$$

and for the odd indices

$$v_{2i-1}^x(x) = \begin{cases} 4(x - x_{i-1})(x_i - x)/(x_i - x_{i-1})^2, & x \in (x_{i-1}, x_i), \\ 0, & \text{otherwise}. \end{cases}$$

We then let

$$w_{j-1}^y(y) = \begin{cases} \dfrac{y_{j-1} + \sigma_2(y_j - y_{j-1}) - y}{(\sigma_2 - \sigma_1)(y_j - y_{j-1})}, & y \in (y_{j-1}, y_j), \\ 0, & \text{otherwise}, \end{cases}$$

$$w_j^y(y) = \begin{cases} \dfrac{y - (y_{j-1} + \sigma_1(y_j - y_{j-1}))}{(\sigma_2 - \sigma_1)(y_j - y_{j-1})}, & y \in (y_{j-1}, y_j), \\ 0, & \text{otherwise} \end{cases}$$

where σ_1 and σ_2 are the two Gauss points defined on $(0,1)$. Similar definitions are used for w_i^x and v_j^y. Order $(V_h^1)_x$, $(V_h^1)_y$ and W_h^1 as before. If one uses a 3×2 Gauss quadrature rule (three Gauss points in the x direction and two in the y direction) to compute the integrals (3.2a) and 2×3 Gauss quadrature rule to compute (3.2b), then M_1 and M_2 are block diagonal matrices each block having a bandwidth of five. The matrix N_1 is of size $2N_y(2N_x - 1) \times 4N_x N_y$ while N_2 is of size $2N_x(2N_y-1) \times 4N_xN_y$. Although N_1 is not square, it has essentially a block diagonal form. N_2 can be made to have the same form by permuting the rows and corresponding columns.

The only difference in the algebraic problems arising from (2.3) - (2.4) and (2.5) - (2.7) is the right-hand side. In the removal of singularities formulation, the right-hand side has the form $(R_1, R_2, RR')^t$ where the elements of R_1 and R_2 are $(1/a\, u_s, v_i^x w_j^y)$ and $(1/a u_s, w_i^x v_j^y)$ respectively, and RR' equals RR plus the boundary integrals from (2.7).

Eliminating α_1 and α_2 from (3.1) yields a system for β given by

$$(3.5) \qquad (N^tM^{-1}N)\beta = (N_1^tM_1^{-1}N_1 + N_2^tM_2^{-1}N_2)\beta = RR.$$

For the singularities removed approach the right-hand becomes side $(RR' - N_1^tM_1^{-1}R_1 - N_2^tM_2^{-1}R_2)$.

For completeness we now wish to present a result due to Russell and one of the authors [15] that shows that the block-centered finite difference method is a mixed finite element method $(r = 0)$ with special numerical quadrature formulas. This result justifies the title of this paper and has motivated the use by the authors of the block-centered difference matrix as a preconditioner for the mixed method.

Let $x_{i+1/2} = (x_i + x_{i+1})/2$ and $y_{j+1/2} = (y_j + y_{j+1})/2$ and let $a_{n,m} = a(x_n, y_m)$. Set $\Delta x_i = x_{i+1} - x_i$, $\Delta y_j = y_{j+1} - y_j$, $\Delta x_{i+1/2} = x_{i+1/2} - x_{i-1/2}$, $\Delta y_{j+1/2} = y_{j+1/2} - y_{j-1/2}$. The block-centered finite difference approximation to $-\nabla \cdot a\nabla p = q$, in Ω, $a\nabla p \cdot \upsilon = 0$, on $\partial\Omega$, is given by

$$(3.6) \qquad \begin{aligned} &-\frac{1}{\Delta x_i}\left[a_{i+1,j+1/2}\left(\frac{p_{i+3/2,j+1/2} - p_{i+1/2,j+1/2}}{\Delta x_{i+3/2}}\right)\right. \\ &\left. - a_{i,j+1/2}\left(\frac{p_{i+1/2,j+1/2} - p_{i-1/2,j+1/2}}{\Delta x_{i+1/2}}\right)\right] \\ &- \text{similar terms in } y \text{ direction} \\ &= q_{i+1/2,j+1/2}, \quad i = 0, 1, \cdots, N_x-1, \\ &\qquad\qquad\qquad\quad\; j = 0, 1, \cdots, N_y-1, \end{aligned}$$

with the boundary reflection conditions

$$
\begin{aligned}
P_{-1/2,j+1/2} &= P_{1/2,j+1/2},\\
P_{Nx+1/2,j+1/2} &= P_{Nx-1/2,j+1/2},\\
P_{i+1/2,-1/2} &= P_{i+1/2,1/2},\\
P_{i+1/2,Ny+1/2} &= P_{i+1/2,Ny-1/2}.
\end{aligned}
\tag{3.7}
$$

Figure 1 illustrates the points used in (3.6) and (3.7).

Multiplying (3.6) by $\Delta x_i \; \Delta y_j$, the area of the cell, we obtain

$$
\begin{aligned}
&\Delta y_j((U_x)_{i+1,j+1/2} - (U_x)_{i,j+1/2})\\
&+ \Delta x_i((U_y)_{i+1/2,j+1} - (U_y)_{i+1/2,j})\\
&= \Delta y_j \; \Delta x_i q_{i+1/2,j+1/2},
\end{aligned}
\tag{3.8}
$$

where U_x and U_y are velocity components defined by

$$
(U_x)_{i+1,j+1/2} = -a_{i+1,j+1/2} \left(\frac{P_{i+3/2,j+1/2} - P_{i+1/2,j+1/2}}{\Delta x_{i+3/2}}\right), \tag{3.9a}
$$

$$
(U_y)_{i+1/2,j+1} = -a_{i+1/2,j+1} \left(\frac{P_{i+1/2,j+3/2} - P_{i+1/2,j+1/2}}{\Delta y_{j+3/2}}\right). \tag{3.9b}
$$

Let U be the unique function in V_h^0 satisfying (3.8) - (3.9), and let P be the piecewise-constant function with cell values $P_{i+1/2,j+1/2}$. We show that $(U; P) \in V_h^0 \times W_h^0$ satisfies (2.3)-(2.4) if certain numerical quadrature formulas are used. However, we first note that by reflection

$$
\begin{aligned}
(U_x)_{0,j+1/2} &= (U_x)_{Nx,j+1/2} = 0,\\
(U_y)_{i+1/2,0} &= (U_y)_{i+1/2,Ny} = 0;
\end{aligned}
$$

the boundary conditions are satisfied.

Now the left-hand side of (3.8) is equal to

$$
\Delta y_j \Delta x_i \frac{\partial}{\partial x}(U_x) + \Delta x_i \Delta y_j \frac{\partial}{\partial y}(U_y) = \int_{x_i}^{x_{i+1}} \int_{y_j}^{y_{j+1}} \nabla \cdot U \, dx,
$$

and the right-hand side is the midpoint-rule integral of q over the cell. Thus, U satisfies

$$
(\nabla \cdot U, w) = (q,w)_{Mx\,My}, \quad w \in W_h^0, \tag{3.10}
$$

where $Mx\,My$ denotes the midpoint-rule quadrature in both directions. Equation (3.9a) implies that

$$1/2(\Delta x_i + \Delta x_{i+1})\Delta y_j \frac{1}{a_{i+1,j+1/2}} (U_x)_{i+1,j+1/2}$$

$$-\Delta y_j(P_{i+1/2,j+1/2} - P_{i+3/2,j+1/2})$$

$$= 0,$$

which can be rewritten as

$$(3.11) \qquad (a^{-1}U_x, v_{i+1}^x w_{j+1}^y)_{Tx\ My} - (P, \frac{\partial}{\partial x}(v_{i+1}^x w_{j+1}^y)) = 0,$$

where v_{i+1}^x is the linear basis function (3.3) and w_{j+1}^y is the constant basis function (3.4), and Tx My denotes trapezoidal integration in the x-direction tensored with the midpoint rule in the y-direction.

Similarly,

$$(3.12) \qquad (a^{-1}U_y, w_{i+1}^x v_{j+1}^y)_{My\ Ty} - (P, \frac{\partial}{\partial y}(w_{i+1}^x v_{j+1}^y)) = 0.$$

Combining (3.10) and (3.11), we deduce that

$$(3.13) \quad (a^{-1}U_x, v_x)_{Tx\ My} + (a^{-1}U_y, v_y)_{Mx\ Ty} - (P, \nabla \cdot v) = 0, \quad v \in V_h^0.$$

By (3.10) and (3.13), the block-centered pressure and associated velocity satisfy the mixed method equations (2.3) - (2.4), if the indicated quadrature formulas are used.

We remark that one can use error estimates for quadrature formulas and the analysis of mixed methods to obtain $0(h)$ L^2-convergence for the block-centered schemes even in the case of nonuniform mesh. Moreover, this result indicates the appropriateness of harmonically averaging the coefficient a for the block-centered scheme.

As indicated in [15] another analogy can be made with block-centered finite differences and mixed methods. Both the finite difference method and the mixed methods $(q = -1)$ satisfy conservation of mass in each cell. To see the latter, recall that div $V_n^r \subset W_n^r$ and note that (2.4) implies

$$\iint_{R_{ij}} \text{div } U\, w = \iint_{R_{ij}} q\, w,$$

where w is a tensor product polynomial of degree $r - 1$ and $R_{ij} = [x_i, x_{i+1}] \times [y_j, y_{j+1}]$. Thus, if $q = 0$ on R_{ij}, then div $U = 0$ pointwise on R_{ij}.

4. COMPUTATIONAL RESULTS FOR Ω A RECTANGLE

Experimental convergence studies for selected problems with singular q have been presented in [9, 10, 12] for $r = 0$ and $r = 1$. Of particular interest was the comparison of the procedures (2.3) - (2.4) and (2.5) - (2.7); that is, the global effects of removing the singularities. In [10], numerical results for singularities removed and $r = 1$ showed first-order convergence for p and u on Ω and second-order convergence away from the singularities, as theory would indicate (Theorems 2.1 and 2.2). Without removal of the singularities, the velocity vector failed to converge on Ω. This is of course expected, because $p \in H^{1-\epsilon}(\Omega)$ and $u \in H^{0-\epsilon}(\Omega)$ for any $\epsilon > 0$. Results for $r = 0$ have been discussed in [9, 12]. We do wish to remark that if a is a rough function and highly variable, one should multiple u_s by a suitable cutoff function.

One problem that has plagued the users of mixed methods in flow problems is solving the algebraic system of equations (3.1) or (3.5). Brown [3] used an alternating-direction iteration as a preconditioner for the system (3.5) with $r = 0$ and obtained a substantial reduction in work estimates over conjugate gradient procedures.

In view of the relationship between block-centered finite differences and the mixed method $r = 0$, we now consider several test problems in which we compare the use of three preconditioners: the identity matrix (I) (conjugate gradients), the block-centered finite difference matrix, (BCM), and a block preconditioning matrix (MINV). In computing the matrix BCM, the coefficients in the finite difference equation are harmonically averaged. To avoid singularity of the BCM matrix, we normalized the pressure approximation by forcing the pressure solution to be zero at a specified node, usually the node in the upper left hand corner of the rectangle. Now the order of the matrices for block-centered finite differences and the mixed method $r = 0$ is the same, $N_x N_y$, however, for $r = 1$ the order of the mixed method is $4N_x N_y$. Following a suggestion by Axelsson [1], we used a block-centered matrix obtained by refining the mixed method grid; each cell is divided into four cells.

The block preconditioning scheme was first formulated by Concus, Golub, and Meurant [5] for standard finite difference formulations of elliptic equations and MINV is a modified incomplete block Cholesky preconditioning matrix. Here we use MINV, a preconditioner for the block-centered finite difference method, as a preconditioner for the mixed methods, $r = 0$ and $r = 1$.

The block-centered finite difference matrix A has the form

$$A = \begin{pmatrix} D_1 & A_2^t & & & \\ A_2 & D_2 & A_3^t & & \\ & \ddots & \ddots & \ddots & \\ & & A_{Ny-1} & D_{Ny-1} & A_{Ny}^t \\ & & & A_{Ny} & D_{Ny} \end{pmatrix},$$

where the D_i are tridiagonal matrices of order (N_x-1) and the A_i are diagonal. Assuming for the moment that A is positive definite, let Σ be the symmetric block diagonal matrix with the $(N_x-1) \times (N_x-1)$ blocks satisfying

$$\Sigma_1 = D_1$$

$$\Sigma_i = D_i - A_i \Sigma_{i-1}^{-1} A_i^t.$$

The block Cholesky factorization of A can be written as

$$A = (\Sigma + L)\Sigma^{-1}(\Sigma + L^t),$$

where $\Sigma + L$ is a block lower bidiagonal matrix. In [5], the matrix Σ_{i-1}^{-1} is approximated by a spare matrix Λ_{i-1}. That is, letting

$$\Delta_1 = D_1,$$

$$\Delta_i = D_i - A_i \Lambda_{i-1} A_i^t, \quad 2 \leq i \leq N_y,$$

the incomplete block preconditioning matrix for use with the conjugate gradient algorithm is

$$M = (\Delta + L)\Delta^{-1}(\Delta + L^t) = A + R,$$

where R is a block diagonal matrix

$$R = \begin{pmatrix} R_1 & & & & \\ & R_2 & & & \\ & & \cdot & & \\ & & & \cdot & \\ & & & & \cdot \\ & & & & & R_{Ny} \end{pmatrix}$$

with

$$R_1 = \Delta_1 - D_1 = 0$$

$$R_i = \Delta_i - D_i + A\Delta_{i-1}^{-1} A_i^t, \quad 2 \leq i \leq N_y.$$

In our calculations, we define

$$\Lambda_{i-1} = B(\Delta_{i-1}^{-1}, 1),$$

the band matrix consisting of the three main diagonals of Δ_{i-1}^{-1}. As shown by Concus et. al., Λ_{i-1} can be easily computed. Let Z_i be the diagonal matrix made up of the row sums of $A_i(\Delta_{i-1}^{-1} - \Lambda_{i-1})A_i^t$. Modifying Δ_i by subtracting Z_i yields a remainder with a zero row sum. The matrix MINV is M with Δ_i so modified. We remark that the block-centered finite difference matrix is positive semi-definite, not positive definite. However, the above algorithm can be implemented for this case.

We let $\Omega = (0,1) \times (0,1)$ and $q = \delta(0,0) - \delta(1,1)$. The following five choices of $a(x,y)$ were tested:

$$a(x,y) = a_1(x,y) = 1,$$

$$a(x,y) = a_2(x,y) = \begin{cases} 1, & 0 < x \leq .5, \\ 10^{-2}, & .5 < x < 1, \end{cases}$$

$$a(x,y) = a_3(x,y) = \begin{cases} 1, & 0 < x \leq .5, \\ 10^{-4}, & .5 < x < 1, \end{cases}$$

$$a(x,y) = a_4(x,y) = \begin{cases} 1, & 0 < x \leq .5, \\ 10^{-6}, & .5 < x < 1, \end{cases}$$

$$a(x,y) = a_5(x,y) = \frac{1}{1 + 100(x^2 + y^2)}.$$

In Tables 1 - 5, we present for the mixed method with $r = 0$ the number of iterations required to reduce the square of the residual error to 10^{-12} for each choice of $a(x,y)$ and each preconditioner. Uniform meshes of $h^{-1} = 10, 20, 40$ were used. Tables 6 - 10 correspond to the case $r = 1$. These results indicate that the number of iterations for the block-centered preconditioners are independent of h and a. A formal proof of this result can be derived, but it is not presented in this paper. Dashes in the tables indicate that an excess of 5000 iterations was needed for convergence.

Preliminary calculations using MINV show a dramatic decrease in the numbers of iterations over conjugate gradients and a reasonable increase over BCM. Moreover, these results appear to be independent of a. The storage requirements for MINV are substantially reduced over BCM, e.g. for $r = 0$, $2N_xN_y$ as compared with $(2N_x+1)N_xN_y$. The described calculations were run on the Cray-1S. Further investigations with these block methods is planned.

	h^{-1}	I	BCM	MINV
Standard Method	10	26	9	10
	20	69	10	13
	40	149	10	20
Singularities Removed	10	26	9	10
	20	72	10	13
	40	151	10	20

TABLE 1

Iteration Counts for the Lowest Order Mixed Method, $r = 0$, $a(x,y) = a_1(x,y)$

	h^{-1}	I	BCM	MINV
Standard Method	10	215	10	10
	20	744	10	13
	40	1850	10	20
Singularities Removed	10	201	11	11
	20	715	10	13
	40	1832	10	20

TABLE 2

Iteration Counts for the Lowest Order Mixed Method, $r = 0$, $a(x,y) = a_2(x,y)$

	h^{-1}	I	BCM	MINV
Standard Method	10	520	12	10
	20	2688	10	13
	40	--	10	21
Singularities Removed	10	422	12	11
	20	2211	10	13
	40	--	10	21

TABLE 3

Iteration Counts for the Lowest Order Mixed

Method, $r = 0$, $a(x,y) = a_3(x,y)$

	h^{-1}	I	BCM	MINV
Standard Method	10	912	14	10
	20	--	10	13
	40	--	10	22
Singularities Removed	10	636	14	11
	20	3664	10	13
	40	--	10	22

TABLE 4

Iteration Counts for the Lowest Order Mixed

Method, $r = 0$, $a(x,y) = a_4(x,y)$

	h^{-1}	I	BCM	MINV
Standard Method	10	69	10	10
	20	251	10	14
	40	831	9	23
Singularities Removed	10	77	10	11
	20	268	10	14
	40	843	10	24

TABLE 5

Iteration Counts for the Lowest Order Mixed Method, $r = 0$, $a(x,y) = a_5(x,y)$

	h^{-1}	I	BCM	MINV
Standard Method	10	82	12	16
	20	179	12	23
	40	351	12	41
Singularities Removed	10	76	12	16
	20	166	12	23
	40	345	12	41

TABLE 6

Iteration Counts for the Next Lowest Order Mixed Method, $r = 1$, $a(x,y) = a_1(x,y)$

	h^{-1}	I	BCM	MINV
Standard Method	10	880	12	16
	20	2150	12	24
	40	4523	12	42
Singularities Removed	10	791	12	16
	20	2023	12	24
	40	4414	12	42

TABLE 7

Iteration Counts for the Next Lowest Order

Method, $r = 1$, $a(x,y) = a_2(x,y)$

	h^{-1}	I	BCM	MINV
Standard Method	10	2955	12	17
	20	--	12	27
	40	--	12	44
Singularities Removed	10	2520	12	17
	20	--	12	27
	40	--	12	44

TABLE 8

Iteration Counts for the Next Lowest Order

Mixed Method, $r = 1$, $a(x,y) = a_3(x,y)$

	h^{-1}	I	BCM	MINV
Standard Method	10	--	12	17
	20	--	12	29
	40	--	12	47
Singularities Removed	10	4295	12	17
	20	--	12	29
	40	--	12	47

TABLE 9

Iteration Counts for the Next Lowest Order

Method, $r = 1$, $a(x,y) = a_4(x,y)$

	h^{-1}	I	BCM	MINV
Standard Method	10	--	12	18
	20	--	12	28
	40	--	12	47
Singularities Removed	10	285	12	18
	20	899	12	28
	40	2475	12	51

TABLE 10

Iteration Counts for the Next Lowest Order

Mixed Method, $r = 1$, $a(x,y) = a_5(x,y)$

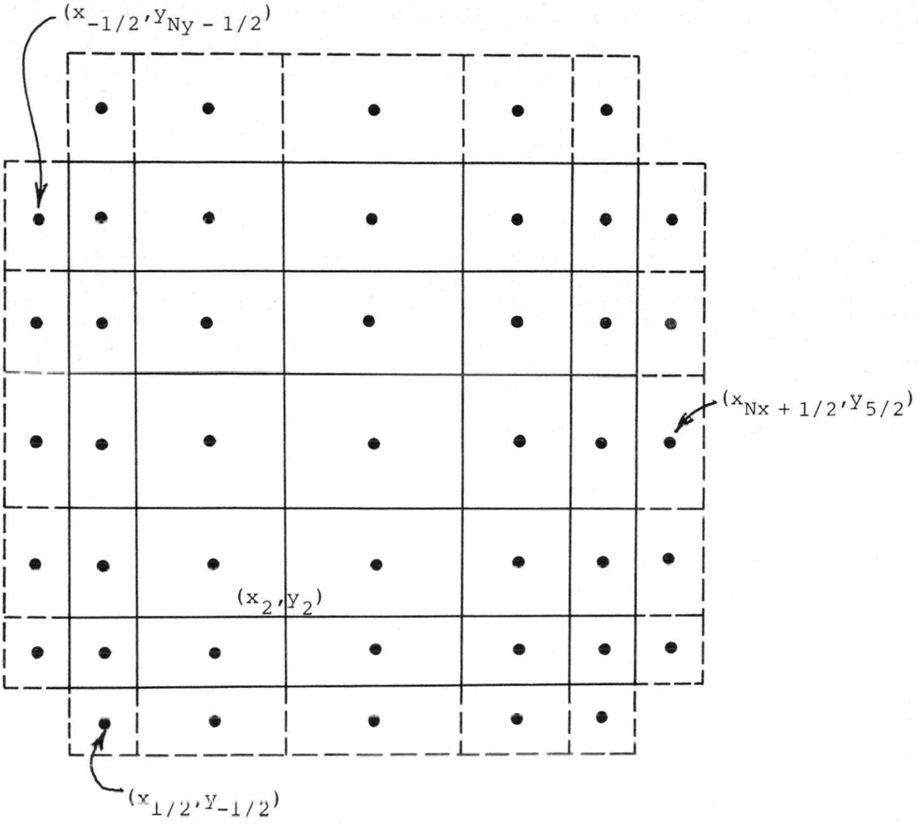

Figure 1

Block-Centered Finite Difference Grid

REFERENCES

[1] Axelsson, O. and Barker, A. V., Finite Element Solution of Boundary Value Problems (Academic Press, to appear).

[2] Brezzi, F., On the existence, uniqueness and approximation of saddle point problems arising from Lagrangian multiplier, RAIRO, Anal. Numer. 2 (1974) 129-151.

[3] Brown, D. C., Alternating-Direction Iterative Schemes for Mixed Finite Element Methods for Second Order Elliptic Problems, Ph.D. Dissertation, University of Chicago (1982).

[4] Chavent, G., Jaffre, J., Cohen, G., Dupuy, M., and Dieste, I., Simulation of two-dimensional waterflooding using mixed finite element methods, Paper 10502 presented at the SPE Sixth Symposium on Reservoir Simulation, New Orleans, La. (Feb. 1982) 147-158.

[5] Concus, P., Golub, C. H., and Meurant, G., Block preconditioning for the conjugate gradient method, to appear.

[6] Darcy, H., Les fontainer publiques de la ville de Dijon, (Dalmont, Paris 1856).

[7] Darlow, B. L., Ewing, R. E., and Wheeler, M. F., Mixed finite element methods for miscible displacement problems in porous media, Paper 10501 presented at the SPE Sixth Symposium on Reservoir Simulation, New Orleans, La. (Feb. 1982) and to appear in Soc. Pet. Eng. J.

[8] Douglas, J. Jr., Ewing, R. E., and Wheeler, M. F., The approximation of the pressure by a mixed method in the simulation of miscible displacement, RAIRO Anal. Numer. 17 (1983) 17-34.

[9] Douglas J. Jr. and Wheeler, M. F., Some recent results concerning the numerical solution of miscible displacement in porous media, in: Glowinski, R. and Lions, J. L., (eds.), Computer Methods in Applied Sciences and Engineering V (North-Holland, New York, 1982) 447-456.

[10] Ewing, R. E. and Wheeler, M. F., Computational aspects of mixed finite element methods, in: Stepleman, (ed.), Numerical Methods for Scientific Computing (North-Holland, New York, to appear).

[11] Ewing, R. E., Russell, T. R., and Wheeler, M. F., Simulation of miscible displacement using mixed methods and a modified method of characteristics, Paper 12241 to be presented at the Seventh SPE Symposium on Reservoir Simulation, San Francisco, Ca., (Nov. 1983).

[12] Ewing, R. E., Koebbi, J. V., Gonzalez, R., and Wheeler, M. F., Computing accurate velocities for fluid flow in porous media, to appear in Proceedings of TICOM Conference, Austin, Tx (Jan. 1983).

[13] Falk, R. S. and Osborn, J. E., Error estimates for mixed methods, RAIRO Anal. Numer. 14 (1980) 249-277.

[14] Raviart, P. A. and Thomas, J. M., A mixed finite element method for 2nd order elliptic problems, in: Mathematical Aspects of the Finite Element Method: lecture notes in mathematics (Springer-Verlag, Heidelberg, 1977).

[15] Russell, T. F. and Wheeler, M. F., Finite element and finite difference methods for continuous flows in porous media, in: Ewing, R. E., (ed.), Mathematics of Reservoir Simulation (SIAM Publications, Philadelphia, 1983).

Acknowledgement

The authors wish to thank Exxon Production Research for their support and assistance in this research.

Computing Methods in Applied Sciences and Engineering, VI
R. Glowinski and J.-L. Lions (Editors)
Elsevier Science Publishers B.V. (North-Holland)

SOME RECENT PROGRESS IN STEEP FRONT CALCULATIONS FOR POROUS FLOW

P. Concus, E. Kostlan, and J.A. Sethian

Lawrence Berkeley Laboratory
and
Department of Mathematics
University of California
Berkeley, California 94720
U.S.A.

A modification to the random choice method is investigated for solving the equations for multidimensional immiscible displacement in a porous medium. The principal feature of the modification is to represent multidimensional flow correctly by incorporating local properties of the flow into the one-dimensional Riemann solvers of the split random choice method. Advancing fronts are kept accurate and sharp. We perform a numerical experiment using the method to study the stabilizing effect of a small amount of physical capillary pressure on a case for which the advancing front is unstable.

INTRODUCTION

A central problem in petroleum reservoir simulation is to model the displacement of one fluid by another within a porous medium. Such problems may be characterized by the injection of a wetting fluid (e.g. water) into the reservoir at one or more points, displacing the non-wetting fluid (e.g. oil), which is being withdrawn at one or more recovery points. The nature of the front between the non-wetting fluid and the wetting fluid is of primary importance; one would like to withdraw as much oil as possible before water reaches the recovery points. A particularly troublesome phenomenon associated with the above procedure is fingering, in which small channels of water push through towards the recovery points, leaving oil behind.

In the design of numerical techniques to model petroleum reservoir flow, one is faced with the problem that the solution to the equations describing the fluid saturation within the reservoir can develop sharp fronts or discontinuities. Finite difference approximations to these equations smear out the front and may distort the nature of the interface, unless expensive mesh refinement is used. One can instead attempt to track the front between the two fluids; however, such techniques usually contain an intrinsic smoothing and thus can predict a stable interface in regimes where fingering should occur. One alternative to the above techniques, which serves as a foundation for the work presented here, is the random choice method, in which the generation of the solution at each time step is based on the exact solution of a collection of local Riemann problems.

In this paper, we apply a recently developed generalized version of the one-dimensional random choice method to two-dimensional reservoir simulation. In this improved technique, while the multi-dimensional solution is obtained through a sequence of one-dimensional problems, information about the two-dimensional solution is used to ensure that each of the one-dimensional problems correctly interprets the nature of the propagating fronts. The algorithm was first used in [5] to study the nature of fingering instability in the absence of capillary pressure. It was applied subsequently to a five-spot waterflood for the case of a stable interface, also without capillary pressure [12].

We apply the technique here to a two-dimensional five spot waterflood problem for the case of an unstable interface and permit the inclusion of capillary pressure. Our numerical results indicate that the advancing front is kept sharp, and that the method is sufficiently free of artificial numerical stabilization so that the effects of including a small amount of capillary pressure can be observed numerically. We repeat here without change much of the material presented in [12], but include in greater detail a discussion of the modifications to the random choice method.

EQUATIONS OF MOTION

We consider the equations

$$\frac{\partial s}{\partial t} + \mathbf{q}\cdot\nabla[f(s)] - \epsilon\nabla\cdot[h(s)\nabla s] = 0 \tag{1}$$

$$\nabla\cdot\mathbf{q} = 0 \tag{2}$$

$$\mathbf{q} = -\lambda(s)\nabla p \ , \tag{3}$$

which in nondimensional form (see [2]) are, respectively, the Buckley-Leverett equation, incompressibility relation, and Darcy's Law. These equations describe the flow of two immiscible, incompressible fluids through a homogeneous porous medium in the absence of gravitational effects (see [10,11]). The quantity $s = s(x,y,t)$ is the saturation (the fraction of available volume at the point (x,y) and time t filled with wetting fluid, e.g. water), $\mathbf{q} = \mathbf{q}(x,y,t)$ is the total velocity of the fluid, and $p = p(x,y,t)$ is the global pressure [3]. The quantities $f(s)$, $h(s)$, and $\lambda(s)$ are functions of the relative permeabilities and capillary pressure, which are empirically determined functions of saturation, and of the viscosities (which are assumed constant). The parameter $\epsilon \geq 0$, which we shall take to be small, measures the relative magnitude of the diffusive (capillary pressure) forces. For immiscible fluids, we take the total mobility $\lambda(s)$ to be represented by

$$\lambda(s) = s^2 + (1-s)^2/\mu \ , \tag{4}$$

where μ is the ratio of the viscosity of the non-wetting fluid to that of the wetting fluid. The corresponding fractional flow function $f(s)$ is

$$f(s) = s^2/\lambda(s) \ , \tag{5}$$

and we take $h(s) = \frac{2}{\mu}(1-s)^3 f(s)$.

Equations (2) and (3) together form an elliptic equation. For our problems of interest, (1) is hyperbolic ($\epsilon = 0$) or nearly hyperbolic ($\epsilon \ll 1$) and as such can develop discontinuities or steep fronts in s, even for arbitrarily smooth initial data. To analyze the nature of the discontinuities, we consider (1) in one space dimension $\epsilon = 0$, namely,

$$s_t + (f(s))_x = s_t + f'(s)s_x = 0 \ . \tag{6}$$

Here, we have taken the velocity as unity for simplicity.

Suppose for a given problem the fractional flow function satisfies $f_{ss} > 0$ and thus f is concave up. It can easily be shown that the characteristics for (6) are straight lines leaving the x-axis with slope

$$\frac{dx}{dt} = f'(s)\ . \tag{7}$$

Since $f_{ss} > 0$, then $f'(s)$ is an increasing function of s. Consider the initial data

$$s(x,0) = \begin{cases} s_l \ , & x<0 \\ s_r \ , & x>0 \end{cases}, \tag{8}$$

where s_l and s_r are constants. If $s_l < s_r$, then $f'(s_l) < f'(s_r)$ and information from the left is propagating slower than information from the right. The entropy condition, which requires that all characteristics reach back to the initial data, thus dictates that a rarefaction wave will form between the two (see Fig. 1). Conversely, if $s_l > s_r$, then $f'(s_l) > f'(s_r)$, and information from the left propagates faster than information from the right; the entropy condition thus dictates the formation of a shock propagating with speed $\dfrac{f(s_r) - f(s_l)}{s_r - s_l}$ (Fig. 2).

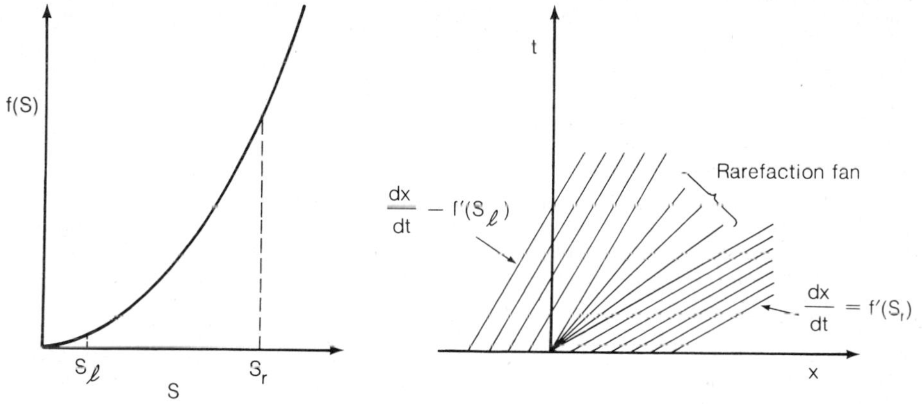

Figure 1
Riemann problem solution, convex $f(s)$, $s_l < s_r$.

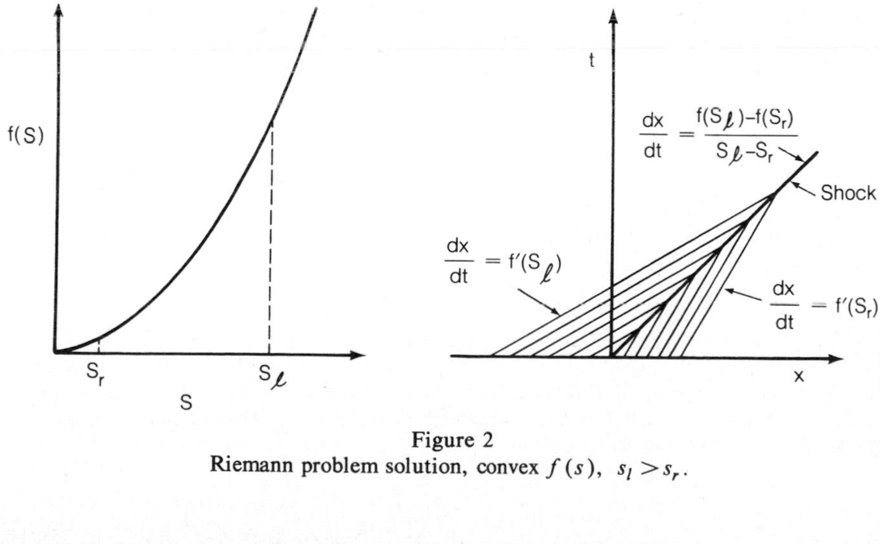

Figure 2
Riemann problem solution, convex $f(s)$, $s_l > s_r$.

In the case of more general f, one has the generalized entropy condition (Oleinik's condition E), which determines the type of admissible shocks: Given s_- and s_+, where s_- and s_+ are the limiting values to the left and the right of a discontinuity, respectively, a solution to (6) exists, is unique, and depends continuously on the initial data if the following holds

$$1) \quad \text{if } s_- < s_+ \text{ then } \quad \frac{f(s_+) - f(s_-)}{s_+ - s_-} \leq \frac{f(s_+) - f(s)}{s_+ - s}$$

$$\text{for all } s \in [s_-, s_+].$$

$$2) \quad \text{if } s_+ < s_- \text{ then } \quad \frac{f(s_+) - f(s_-)}{s_+ - s_-} \geq \frac{f(s_+) - f(s)}{s_+ - s}$$

$$\text{for all } s \in [s_+, s_-].$$

Geometrically, this condition translates into the statement that admissible shocks are those such that (1) if $s_- < s_+$, then the chord connecting $(s_-, f(s_-))$ and $(s_+, f(s_+))$ must lie below f, (2) if $s_+ < s_-$, then the chord connecting $(s_+, f(s_+))$ and $(s_-, f(s_-))$ must lie above f. In the case of a purely convex f, the above produces the situations described earlier; in particular, a shock cannot be allowed to connect the left state with the right state in Fig. 1. For non-convex f, the above condition also rules out such waves as shocks which move faster or slower than the characteristics on both of their sides.

In our problem, the fractional flow function (5) contains one inflection point (see Fig. 3). Thus the wave connecting the left and right states will be either a shock, rarefaction, or combination of the two. We determine the appropriate wave as follows. Given s_l and s_r, the left and right states respectively,

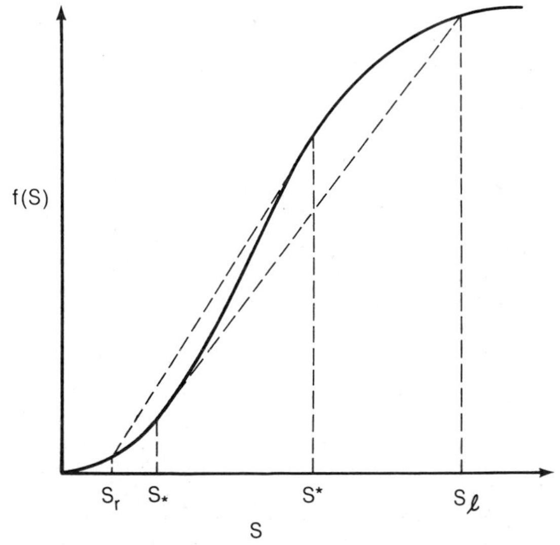

Figure 3
Fractional flow function with convex and concave hulls for compound wave solution, $s_l > s_r$.

1) If the chord connecting $(s_l, f(s_l))$ and $(s_r, f(s_r))$ does not intersect f, then the curve is convex up or down between the two points. If $f'(s_l) > f'(s_r)$, then by the generalized entropy condition a shock must connect s_l with s_r; if $f'(s_l) < f'(s_r)$, then a rarefaction connects the two states.
2) Suppose the chord connecting the left and right states intersects f. Then there exists a point s^* between s_r and s_l such that the chord from $(s_r, f(s_r))$ to $(s^*, f(s^*))$ is tangent to f at $(s^*, f(s^*))$ (the case $s_r < s_l$ is shown in Figs. 3, 4). The wave connecting s_r to s^* is a shock and the wave connecting s^* to s_l is a rarefaction.

The above choices apply to the case in which the advection speed q is unity. If q is positive, but not 1, the wave speeds are merely scaled by q. If q is negative, the roles of s_l and s_r are interchanged in the above arguments.

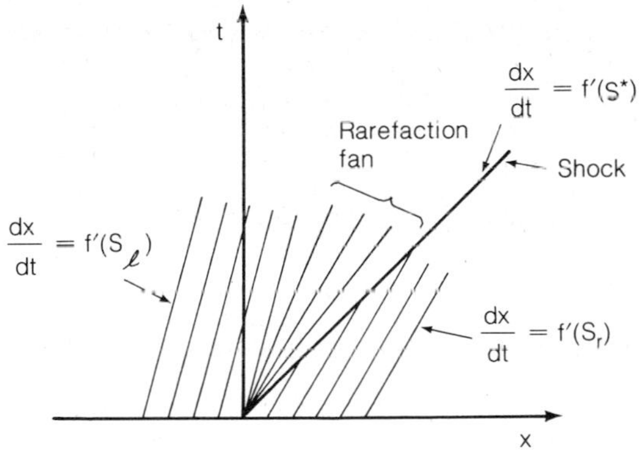

Figure 4
Riemann problem solution, compound wave, $s_l > s_r$.

NUMERICAL ALGORITHM

We consider as an example the standard diagonal geometry quarter five spot problem. That is, we assume a square domain D, centered at (½,½), with sides of unit length, and (1)-(3) holding inside the domain. At (0,0) we place a source of unit strength; at (1,1) we place a sink of unit strength. We have the initial conditions

$$s(x,y,0) = 0 \quad \text{for } (x,y) \neq (0,0) , \tag{9}$$

together with the boundary conditions

$$s(0,0,t) = 1 \tag{10}$$

$$\frac{\partial s}{\partial \nu} = \frac{\partial p}{\partial \nu} = 0 \text{ on the edges of the square ,} \tag{11}$$

where ν is the normal to the boundary. The above problem may be summarized by saying that we inject wetting fluid ($s = 1$, water) into the lower left corner with strength so that the flux is equal to unity, and withdraw non-wetting fluid ($s = 0$, oil) from the upper right corner at the same rate; at no other points can fluid enter or leave the domain.

The numerical algorithm used to solve the equations of motion is based on a generalized random choice method for the hyperbolic or nearly hyperbolic equation (1) coupled to a successive over-relaxation method for the elliptic equation (2),(3). The random choice method is a numerical technique developed in [4] and is based on a constructive existence proof introduced in [9]. We briefly describe the technique below.

Consider the equation

$$s_t + u[f(s)]_x = 0 . \tag{12}$$

Place a uniformly spaced grid of mesh length h along the x axis and let s_i^n be the value $s(ih, n\Delta t)$; thus we view s as a piecewise constant function, constant within each interval $[(i-½)h,(i+½)h]$. This produces a collection of initial value Riemann problems of the form

$$s = \begin{cases} s_l \ , & x<(i+½)h \\ s_r \ , & x>(i+½)h \end{cases} , \tag{13}$$

where $s_l = s_i^n$ and $s_r = s_{i+1}^n$. The solution is then updated in time by solving each Riemann problem exactly, with the type of wave used to connect the two states (shock, rarefaction, compound) determined as discussed above. The solution is then sampled to produce new values of s; we use a sampling based on a van der Corput sequence, see [4]. The time step Δt is chosen to be the largest possible value such that the separate Riemann problems do not interact; that is,

$$\Delta t \, |q| \max_i f'(s_i^n) \le h . \tag{14}$$

This technique was first applied to porous flow problems in [5].

A natural way to extend the above one-dimensional technique to two dimensions is to split the two-dimensional hyperbolic equation ($\epsilon = 0$)

$$s_t + \mathbf{q}\cdot\nabla[f(s)] = 0 \tag{15}$$

into two steps; first a one-dimensional problem in the x direction

$$s_t + u\frac{\partial}{\partial x}(f(s)) = 0 \tag{16}$$

is solved, followed by a one-dimensional problem in the y direction

$$s_t + v\frac{\partial}{\partial y}(f(s)) = 0 , \tag{17}$$

where $\mathbf{q} = (u,v)$. Unfortunately, a front not parallel to either the x or y axis can be misinterpreted during one of the sweeps, as may be seen by considering the following example. Suppose the interface between water and oil is lying diagonal to the grid, and the velocity vector at a point on the front is as shown in Fig. 5. It is clear that water is displacing oil, however if

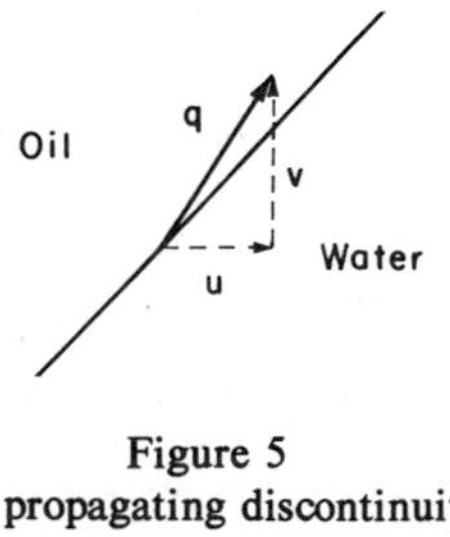

Figure 5
Obliquely propagating discontinuity front.

one looks simply at the component of the velocity vector in the x direction, oil seems to be displacing water and hence the solution to the one-dimensional problem will not accurately portray what is happening. Errors due to the above phenomenon can be significant; examples of their effect may be found in [8]. The technique we use, introduced in [5], first determines the correct orientation by analyzing the two-dimensional situation, and uses this information in each of the one-dimensional sweeps.

The details of the modified Riemann problem solutions are illustrated in Figs. 6-8. For the purely convex case, the modified (entropy violating) solutions corresponding to Figs. 1 and 2 are depicted in Figs. 6 and 7, respectively. Note that the configuration of the rarefaction shown in Fig. 7 depends on the time Δt at which the solution is to be sampled. The modified solution corresponding to the compound wave with $s_r < s_l$ for the porous flow $f(s)$ (Fig. 3) is shown in Fig. 8; the concave hull in Fig. 3, rather than the convex hull, is used in its construction. The compound wave for $s_l < s_r$ is constructed analogously. Further discussion of the algorithm can be found in [5].

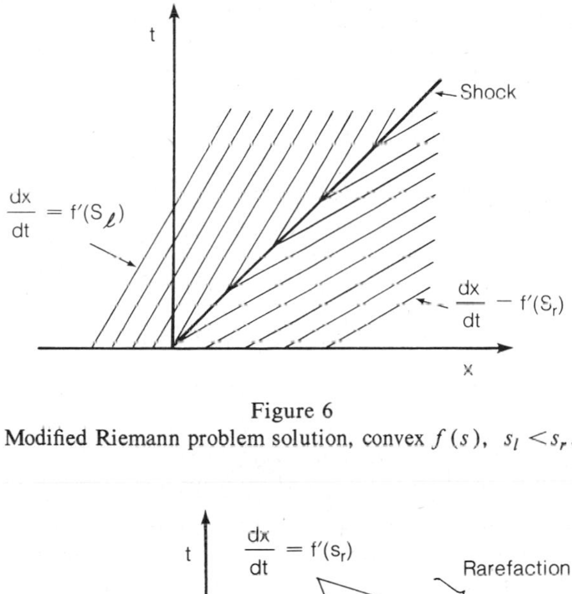

Figure 6
Modified Riemann problem solution, convex $f(s)$, $s_l < s_r$.

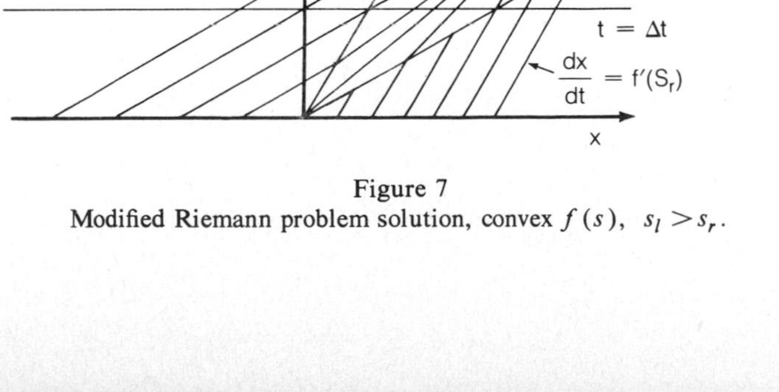

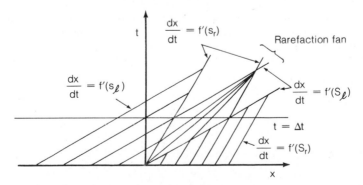

Figure 7
Modified Riemann problem solution, convex $f(s)$, $s_l > s_r$.

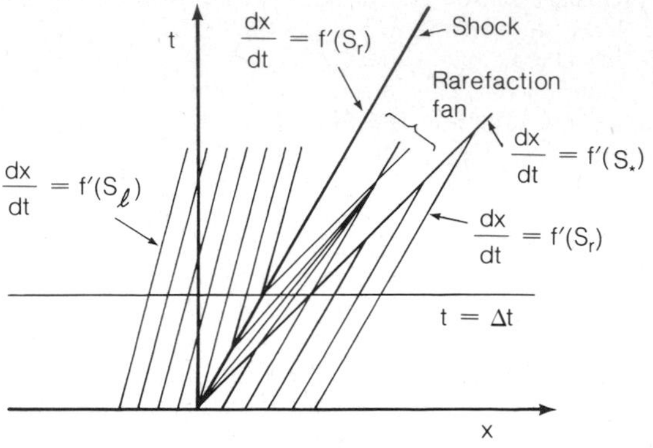

Figure 8
Modified Riemann problem solution, compound wave, $s_l > s_r$.

We now describe the algorithm for solving Equations (1)-(3). With time step Δt, let $s_{i,j}^n$ be the value of the saturation at $s(ih, jh, n\Delta t)$; here we assume that $s_{i,j}^n$ is defined on a square grid i, j of mesh length h imposed on the domain. The pressures $p_{i,j}^n$ are taken at the same grid points, and the velocities $u_{i+1/2,j}$, $v_{i,j+1/2}$ are to be evaluated at the midpoints of the sides of a cell (see Fig. 9). Assuming $s_{i,j}^n$ is known, we now describe the algorithm used to obtain $s_{i,j}^{n+1}$ from $s_{i,j}^n$.

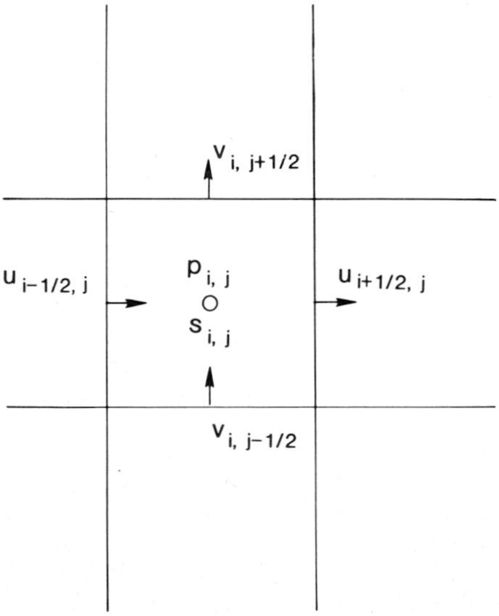

Figure 9
Staggered grid for p, s, and $\mathbf{q}$.

The first step is to calculate $p_{i,j}^n$ from $s_{i,j}^n$. Substituting (3) into (2) yields the elliptic equation for p

$$\nabla\cdot(-\lambda(s)\nabla p) = 0 . \tag{18}$$

Corresponding to differencing a typical component of (3) as

$$u_{i+½,j} \approx \frac{(\lambda(s_{i+1,j})+\lambda(s_{i,j}))\,(p_{i+1,j}-p_{i,j})}{2h} , \tag{19}$$

we approximate (18) with a centered difference scheme, which at a general interior point is

$$\begin{aligned} &-\frac{(\lambda(s_{i+1,j})+\lambda(s_{i,j}))(p_{i+1,j}-p_{i,j})}{2h^2} \\ &+\frac{(\lambda(s_{i,j})+\lambda(s_{i-1,j}))(p_{i,j}-p_{i-1,j})}{2h^2} \\ &-\frac{(\lambda(s_{i,j+1})+\lambda(s_{i,j}))(p_{i,j+1}-p_{i,j})}{2h^2} \\ &+\frac{(\lambda(s_{i,j})+\lambda(s_{i,j-1}))(p_{i,j}-p_{i,j-1})}{2h^2} \approx 0 . \end{aligned} \tag{20}$$

This scheme conserves the flux $-\lambda(s)\nabla p$ through each cell, and consequently ensures that there are no fictitious sources or sinks introduced by the finite difference approximation into the calculation.

Along the four sides, the discrete form of the boundary conditions (11) is used to modify (20). At the edges of the injection and production cells the flux is taken to be ¼, appropriately directed. From the pressures, which are obtained by solving the resulting linear system, the velocity components are calculated using (19) and its counterparts.

With u^n, v^n, p^n, and s^n known, we now solve (1) to determine s^{n+1} at the grid points. We first determine whether the two-dimensional advection velocity $\mathbf{q} = (u,v)$ causes water to displace oil or oil to displace water; this can easily be accomplished by checking the angle between $\mathbf{q}$ and ∇f. Then, we determine if the two one-dimensional sweeps for the advective term

$$s_t + u\frac{\partial}{\partial x}(f(s)) = 0 \tag{21}$$

$$s_t + v\frac{\partial}{\partial y}(f(s)) = 0 \tag{22}$$

both interpret the situation in the same way as does the two-dimensional analysis. For example, if water is displacing oil, we check to see that the situation holds for both of the one-dimensional sweeps. If so, we may simply solve (21) and then (22) using the random choice method described earlier to obtain a new value $\tilde{s}^n$ of s. Suppose, however, that one of the sweeps indicates an orientation different from the actual one described by the two-dimensional analysis, such as in Fig. 5, where, although water is clearly displacing oil, the x sweep indicates oil displacing water. In order to force the wave connecting the two states to be of the same type as that determined by the two-dimensional analysis, we switch the role of convex and concave hulls in our Riemann solution (this can be accomplished by interchanging s_l and s_r in the Riemann solver). Although the one-dimensional solution constructed may violate the generalized entropy condition (e.g. we now allow rarefaction shocks), the proper two-dimensional information is represented in each of the one-dimensional problems, and the generation of shocks and rarefaction waves due purely to grid orientation has been avoided.

To obtain the value of s^{n+1} from $\tilde{s}^n$, one finally carries out a sweep

$$s_t - \epsilon\nabla\cdot[h(s)\nabla s] = 0$$

for the contribution of the diffusive capillary-pressure term. Since ϵ is small, we take for our numerical experiments simply an explicit forward step

$$s^{n+1} = \tilde{s}^n + \epsilon\Delta t\{\nabla\cdot[h(s)\nabla s]\}_{s=\tilde{s}^n},$$

using the spatial discretization represented in (20).

RESULTS

Numerical results are presented for the standard diagonal geometry quarter five-spot test problem discussed in the previous section for waterflood of a petroleum reservoir. A 40×40 spatial grid is placed on the unit square, and the viscosity ratio for the first test problem is taken to have the value $\mu = 5$. For this value of μ, the mobility ratio M at the front is approximately 1.18, which is slightly greater than the critical value $M = 1$. Hence according to the linearized theory and the numerical results in [5], the advancing front should be unstable in the absence of capillary pressure, i.e. for $\epsilon = 0$.

Figure 10 shows the computational results for $\epsilon = 0$. Contours are drawn in saturation increments of 0.1 at three successive time values. Because the contour plotting package displaces contours that should lie on top of one another, the advancing front is shown as a band of adjacent contour lines, rather than as the sharp front produced by the numerical calculations. Observe that unstable "fingering" of the front develops spontaneously during the calculation. (The smaller undulations, which are of the order of a mesh spacing and are stable, are inherent in the random choice method and the interpolations introduced by the contour plotting program.)

Figure 11 shows the effect of including a small amount of capillary pressure ($\epsilon = 0.01$). The fingering instability of Fig. 10 is not evident in this case.

For the second test problem the viscosity ratio is taken to have the value $\mu = 10$, corresponding to which the mobility ratio at the front is $M \approx 1.40$. Figures 12, 13, and 14 show the computational results for $\epsilon = 0$, $\epsilon = 0.01$, and $\epsilon = 0.05$, respectively. Here, the results are in accord with the expectation that the front should be more unstable than for the smaller mobility ratio of Figs. 10 and 11. The effects on the fingering instability of including successively larger amounts of capillary pressure can be seen.

These results indicate that the modified random choice method holds promise as a device for studying sharp fronts in a porous medium. It appears sufficiently free of artificial numerical stabilization of an unstable front, so that the physical effects of including small amounts of capillary pressure can be investigated computationally. Computational results for the five-spot problem in the stable regime are given in [12].

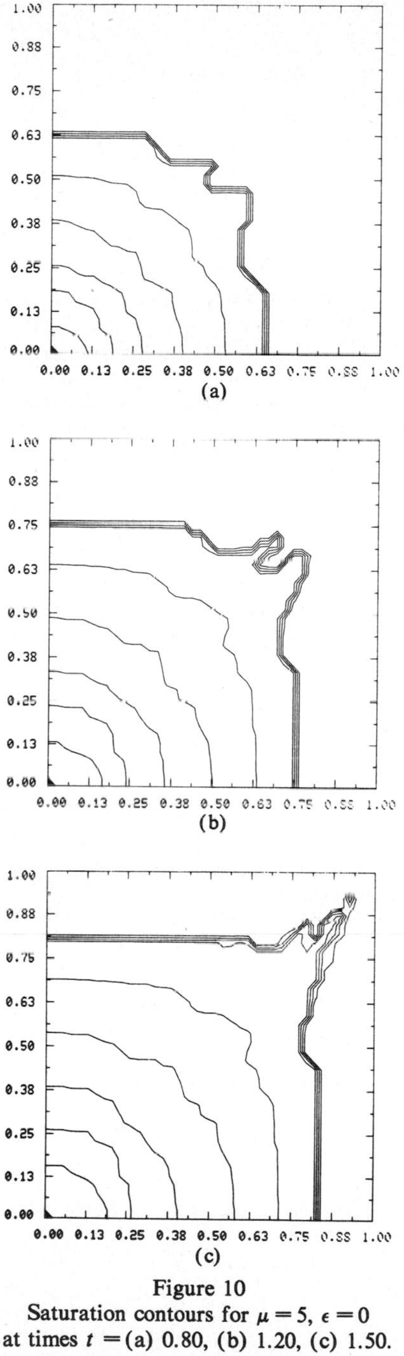

Figure 10
Saturation contours for $\mu = 5$, $\epsilon = 0$ at times $t =$ (a) 0.80, (b) 1.20, (c) 1.50.

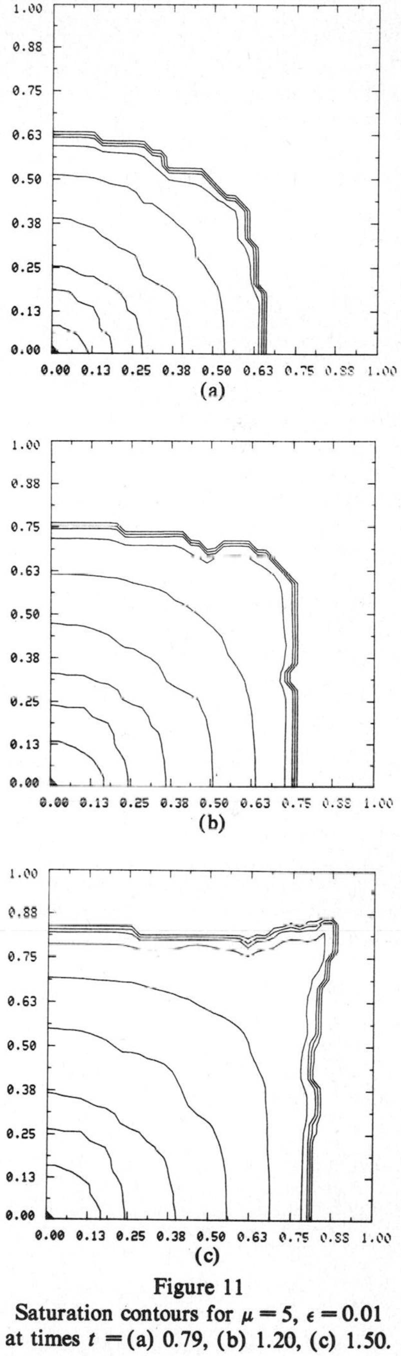

Figure 11
Saturation contours for $\mu = 5$, $\epsilon = 0.01$ at times $t =$ (a) 0.79, (b) 1.20, (c) 1.50.

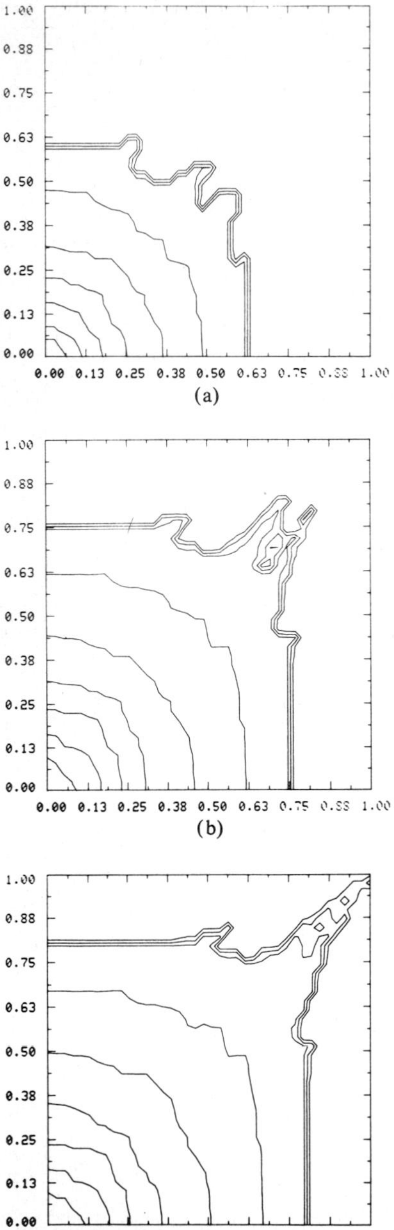

Figure 12
Saturation contours for $\mu = 10$, $\epsilon = 0$ at times $t =$ (a) 0.60, (b) 1.00, (c) 1.19.

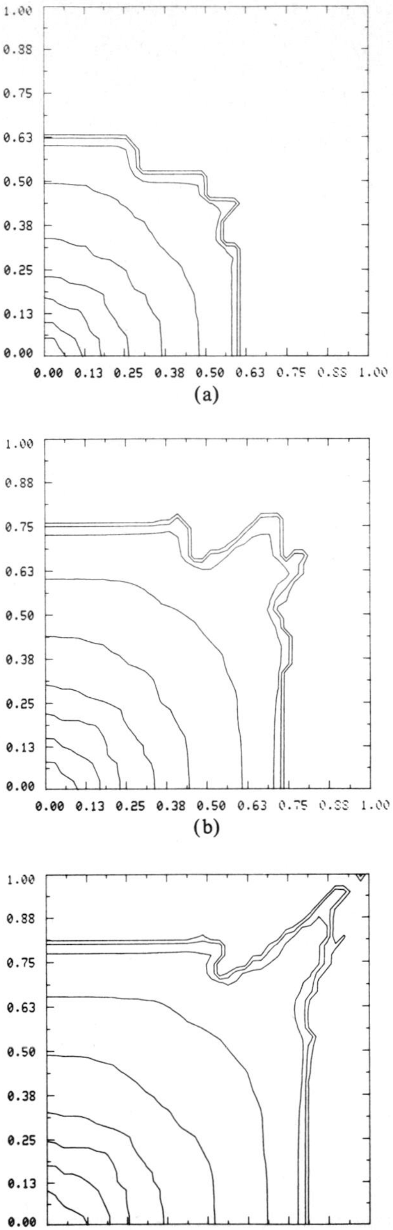

Figure 13
Saturation contours for $\mu = 10$, $\epsilon = 0.01$ at times $t =$ (a) 0.60, (b) 0.99, (c) 1.17.

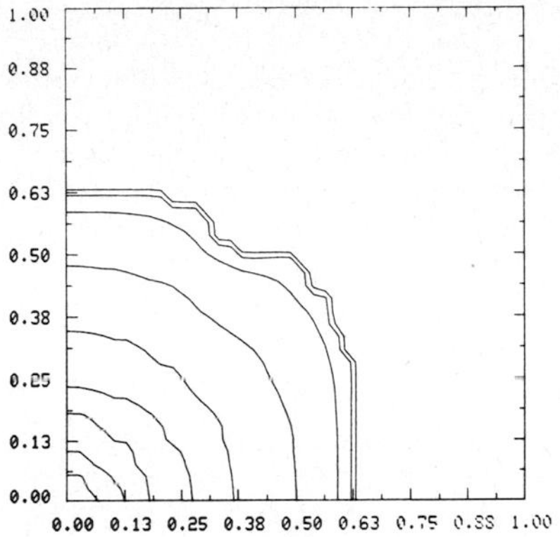

(a)

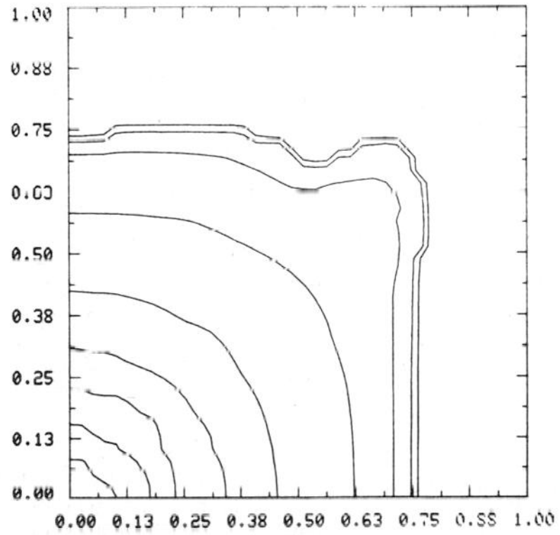

(b)

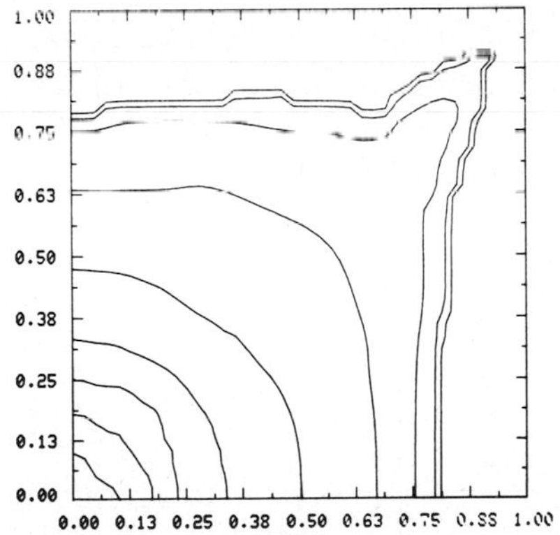

(c)

Figure 14
Saturation contours for $\mu = 10$, $\epsilon = 0.05$ at times t = (a) 0.60, (b) 0.99, (c) 1.20.

ACKNOWLEDGMENTS

We are pleased to acknowledge Alexandre Chorin for helpful conversations concerning the modified random choice method. This work was supported in part by the Director, Office of Energy Research, Office of Basic Energy Sciences, Engineering, Mathematical, and Geosciences Division of the U.S. Department of Energy under Contract DE-AC03-76SF00098, and by the International Business Machines Corporation, Palo Alto Scientific Center, Palo Alto, CA.

REFERENCES

[1] Albright, N., Concus, P., and Proskurowski, W., Numerical solution of the multidimensional Buckley-Leverett equation by a sampling method, Paper SPE 7681, Soc. Petrol. Eng. Fifth Symp. on Reservoir Simulation, Denver, CO., Feb., 1979.

[2] Anderson, C. and Concus, P., A stochastic method for modeling fluid displacement in petroleum reservoirs, in Lecture Notes in Control and Information Sci., 28 (Springer-Verlag, Berlin-Heidelberg-New York, 1980), pp. 827-841.

[3] Chavent, G., A new formulation of diphasic incompressible flows in porous media, in Lecture Notes in Math., 503 (Springer-Verlag, Berlin-Heidelberg-New York, 1976), pp. 258-270.

[4] Chorin, A.J., Random choice solution of hyperbolic systems, J. Comput. Phys., 22 (1976) 517-533.

[5] Chorin, A.J., The instability of fronts in a porous medium, Comm. Math. Phys., 91 (1983) 103-116.

[6] Colella, P., Glimm's Method for Gas Dynamics, SIAM J. Sci. Statist. Comput., 3 (1982) 76-110.

[7] Concus, P. and Proskurowski, W., Numerical solution of a nonlinear hyperbolic equation by the random choice method, J. Comput. Phys., 30 (1979) 153-166.

[8] Crandall, M. and Majda, A., The method of fractional steps for conservation laws, Numer. Math., 34 (1980) 285-314.

[9] Glimm, J., Solution in the large for nonlinear hyperbolic systems of Equations, Comm. Pure Appl. Math., 18 (1965) 697-715.

[10] Peaceman, D.W., Fundamentals of Numerical Reservoir Simulation, (Elsevier, Amsterdam-Oxford-New York, 1977).

[11] Scheidegger, A.E., The Physics of Flow in Porous Media, (University of Toronto Press, Toronto, 1974).

[12] Sethian, J.A., Chorin, A.J., and Concus, P., Numerical solution of the Buckley-Leverett equations, Paper SPE 12254, Soc. Petrol. Eng. Seventh Symp. on Reservoir Simulation, San Francisco, CA, Nov. 1983.

Computing Methods in Applied Sciences and Engineering, VI
R. Glowinski and J.-L. Lions (Editors)
Elsevier Science Publishers B.V. (North-Holland)

SIMULATION DE PUITS DANS LES MODELES DE RESERVOIRS AUX ELEMENTS FINIS

G. CHAVENT
Université Paris IX et INRIA

A method for the simulation of wells in finite element two-phase reservoir models is presented. The finite element model is linked to the well variables through a "macroelement". The macroelements are divided into sectors, in which the flow is simulated by simple one dimensional model. This approach does not require mesh refinement in the vicinity of the wells, and allows for easy implementation of dynamic opening or closing of wells.

INTRODUCTION

La simulation numérique des puits d'injection et de production dans les modèles de gisement est toujours un problème délicat. La taille d'un gisement pouvant être de quelques kilomètres, celle des mailles utilisées, tant dans les modèles différences finies qu'éléments finis, vont de quelques dizaines de mètres, pour les discrétisations les plus finies, jusqu'à quelques centaines de mètres pour les plus grossières, alors que le diamètre réel des puits est de quelques dizaines de centimètres. Le puit apparaît donc comme quasiment ponctuel vu à l'échelle du maillage. Or certaines quantités observées au cours de l'exploitation du gisement, telles les chutes de pression entre puits, les temps de percée de l'eau, dépendent de façon irrévocable du diamètre réel des puits.

Dans les modèles aux différences finies, les puits ne sont pas, en général, représentés en tant que tels : le modèle calcule des valeurs "par maille" pour la pression et la saturation, à partir desquelles on essaye d'extrapoler les valeurs aux puits. Certains auteurs (cf. Douglas [1]) ont proposé de retrancher à l'inconnue numérique les singularités aux puits, calculées a priori dans l'hypothèse d'un écoulement radial à symétrie circulaire au voisinage du puits. Cette méthode est utilisable pour la pression, et pour la saturation aux puits d'injection. Par contre, elle ne s'applique pas pour la saturation aux puits de production, où les saturations n'ont aucune raison d'avoir une symétrie circulaire de révolution.

Dans un modèle éléments finis, on pourrait être tenté de raffiner le maillage au voisinage des puits afin de représenter effectivement le bord du puit comme un bord du maillage élément fini : outre les problèmes posés par la présence de tous petits éléments en cas d'utilisation de schémas explicites, cette approche se heurte à une autre difficulté: l'ouverture et la fermeture des puits est un phénomène dynamique. En particulier, on ne connaît pas a priori, lorsqu'on commence à simuler un gisement, les emplacements des puits que l'on ouvrira en cours de simulation.

La création, à chaque ouverture de puits, d'un maillage localement raffiné, et sa suppression à chaque fermeture, est génératrice d'une complexité informatique qui, bien qu'elle ne soit pas insurmontable, reste tout de même importante.

L'idée que nous proposons ici, pour modéliser les puits dans un maillage élément fini, part de la constatation qu'au voisinage des puits l'écoulement est d'abord radial -même s'il n'a pas de symétrie de révolution.

On le représentera donc par un nombre fini d'écoulements radiaux indépendants. Ces écoulements radiaux seront reliés au reste du modèle élément fini à travers une frontière du modèle élément fini que l'on créera en retirant du maillage un ou plusieurs éléments. Cette dernière manipulation peut se faire de manière relativement simple, sans avoir a recréer entièrement le maillage. L'ensemble des écoulements radiaux liant le puit au modèle E.F. sera appelé macroélément. Chacun des écoulements radiaux pourra être calculé soit analytiquement, soit à l'aide d'un modèle numérique unidimensionel peu couteux, et pour lequel on peut se permettre de mailler jusqu'au rayon réel des puits. On pourra ainsi, incorporer directement dans le modèle les modifications des paramètres physiques des roches dans le voisinage immédiat des puits ("Skin effects").

Nous allons décrire ici le principe de la mise en oeuvre de ces macroéléments pour la représentation du puits dans un modèle de réservoir utilisant des éléments finis mixtes pour la résolution des équations en pression et des éléments linéaires discontinus avec décentrage pour la résolution des équations en saturation. Nous renvoyons à la référence [2] pour une description détaillée du modèle élément fini et des macroéléments.

II LES EQUATIONS A RESOUDRE DANS LE GISEMENT

Soit $\Omega \subset \mathbb{R}^2$, un gisement diphasique incompressible exploité par deux puits, Γ la frontière extérieure de Ω, Γ_i la frontière des puits d'injection et Γ_p celle des puits de production, cf. figure 1.

Les inconnus du problème sont alors :

$$(1)\quad \begin{cases} P = \text{pression globale du couple eau-huile (cf. Chavent [3])} \\ S = \text{saturation réduite en eau } 0 \le S \le 1 \end{cases}$$

En introduisant les inconnues auxiliaires :

$$(2)\quad \begin{cases} \vec{q} = \text{débit volumétrique total eau + huile} \\ \vec{q}_w = \text{débit volumétrique en eau} \\ \vec{r} = \text{part capillaire du débit en eau} \end{cases}$$

Les équations sont alors :

pour la pression

$$(3)\quad \nabla\vec{q} = 0$$

$$(4)\quad \vec{q} = -\psi(x)\, d(S)\nabla P + d(S)\gamma_G(S)\, \vec{q}_G$$

et pour la saturation en eau :

$$(5)\quad \phi(x)\frac{\partial S}{\partial t} + \nabla\vec{q}_w = 0$$

$$(6)\quad \vec{q}_w(x,t) = \vec{r}(x,t) + \vec{f}(x,t,S(x,t))$$

$$(7)\quad \vec{r}(x,t) = -\psi(x)\, d(S)\, \nabla\alpha(S)$$

$$(8)\quad \vec{f}(x,t,k) = b(k)\, \vec{q}(x,t) + b_G(k)\, \vec{q}_G(x)$$

où ϕ et ψ sont des paramètres liés à la porosité, la perméabilité et l'épaisseur du gisement. Les fonctions d(S) et $\gamma_G(S)$ représentent la mobilité globale et la masse volumique relative du couple eau-huile, et sont toujours strictement positives. La fonction $\alpha(S)$ est croissante et homogène à une pression capillaire ; la fonction b(S) est une fonction sans dimension croissante de 0 à 1 et représentant le flux fractionnaire de l'eau. Enfin, la fonction b_G représente la mobilité différentielle du couple eau-huile en présence de gravité, qui s'annulle pour S = 0 et S = 1.

Aux équations (3) à (8) il convient d'ajouter :

une condition initiale

(9) $S(x,0) = S_0(x)$ à t-0

et des conditions aux limites

(10) $\vec{q}\cdot\vec{\nu} = 0$, $\vec{q}_w\cdot\vec{\nu} = 0$

sur la frontière extérieure Γ, supposée fermée.

(11) $\begin{cases} \int_{\Gamma_i} \vec{q}\cdot\vec{\nu} = Q_{T_i} \\ P \text{ const sur } \Gamma_i \end{cases}$, $S = 1$ sur Γ_i

sur la frontière Γ_i du puits d'injection à débit total imposé Q_{Ti},

(12) $P = P_{bp}$ sur Γ_p , $\vec{r}\cdot\vec{\nu} = 0$ sur Γ_p

sur la frontière Γ_p du puits de production à pression de fond imposée P_{bp}.

La condition S = 1 sur Γ_i prend en compte le fait que l'on injecte de l'eau à travers Γ_i, et la condition $\vec{r}\cdot\vec{\nu} = 0$ sur Γ_p provient de ce que l'on néglige les effets capillaires sur la paroi des puits de production, par suite des vitesses importantes des fluides à cet endroit.

III PRINCIPE DE L'APPROXIMATION

On recouvre Ω par un maillage formé de triangles et de quadrilatères ne pouvant avoir en commun que des côtés entiers ou des sommets, de telle sorte que la réunion de tous ces éléments forme une bonne approximation de l'ensemble du réservoir. Ce maillage n'est pas raffiné au voisinage des puits. La seule hypothèse que l'on fait sur les positions respectives du maillage et des puits est que l'on peut, en retirant quelques éléments du maillage, créer dans le maillage des trous convexes entourant chaque puit , de façon à ce que

- le puit soit à peu près au centre du trou,
- dans le domaine couvert par le trou, l'écoulement global eau + huile est approximativement radial (ce qui n'implique pas que la saturation en eau ait une symmétrie de révolution !).

Appelons Ω_h la réunion des éléments (triangles et quadrilatères) restant et Γ_{ih} et Γ_{ph} les frontières de Ω_h entourant respectivement le puits d'injection et les puits de production (cf. figure 2).

Le principe de l'approximation va consister à résoudre les équations (3) à (9) dans Ω_h à l'aide d'un modèle élément fini, et à relier les variables du modèle E.F sur Γ_{ih} et Γ_{ph} aux variables de puits (pression de fond P_b et débit total Q_T) à l'aide d'un "macroélément" qui modélise l'écoulement entre la frontière réelle Γ_i ou Γ_p des puits et la frontière Γ_{ih} et Γ_{ph}.

Naturellement, on profitera de la nature principalement radiale des écoulements dans cette région pour simplifier le modèle du macroélément.

IV LE MODELE E.F. DANS Ω_h

Les équations en pression (3), (4), (10) sont approchées par une méthode d'éléments finis mixtes de degré le plus bas (cf. Raviart Thomas [4], dont les degrès de liberté sont :

- le débit total eau + huile Q_A à travers chacune des arêtes A du maillage, approchant $\int_A \vec{q}\cdot\vec{\nu}_A$,
- une pression P_K par élément, approchant la valeur moyenne de P sur l'élément K.
- un nombre TP_A par arête frontière de Ω_h, approchant la trace de P sur l'arête A.

Aux équations approchant (3), (4) et la partie gauche de (10), il faut ajouter des conditions aux limites sur les frontières "artificielles" Γ_{ih} et Γ_{ph}; on utilisera pour cela l'hypothèse que l'écoulement est radial entre les puits et Γ_{ih} ou Γ_{ph}, et on écrira :

$$(13)\qquad \lambda_A^n(TP_A^n - P_{bi}^n) = DIV_{K,A}\ Q_A^n \qquad \forall\ A \subset \Gamma_{ih},$$

ce qui permet de ramener la condition (11 gauche) "débit total d'injection Q_{Ti} imposé sur Γ_i" à la condition suivante sur Γ_{ih}:

$$(14)\qquad \begin{cases} \sum_{A\subset\Gamma_{ih}} Q_A = Q_{Ti} \\ TP_A^n - \dfrac{1}{\lambda_A^n} DIV_{K,A}\ Q_A^n = \text{cste} \quad \text{indépendante de } A\subset\Gamma_{ih} \end{cases}$$

De même, la condition (12 gauche) "pression de fond P_{bp} imposée sur Γ_p" se traduit sur Γ_{ph} par :

$$(15)\qquad \lambda_A^n(TP_A^n - P_{bp}) = DIV_{K,A}\ Q_A^n \qquad \forall\ A \subset \Gamma_{ph}$$

La partie "transport" $\phi \dfrac{\partial S}{\partial t} + \nabla\vec{f}$ des équations en saturation (5)(6)(8)(9), (10) est approchée par une méthode d'éléments finis discontinus linéaires sur chaque élément (cf. Lesaint-Raviart [5] et Godunov [6]), associée à un schéma explicite en temps. La partie "diffusion" de ces mêmes équations, c'est-à-dire le débit d'eau capillaire $\vec{r}$ et son équation (7), est approchée par une méthode d'élément finis mixtes du plus bas degré, (exactement comme le débit total $\vec{q}$).

Les degrés de liberté pour ces équations sont :

- les saturations $S_{K,i}$ $i = 1,2,3$ ou 4 aux sommets de chaque élément (une valeur par élément et par sommet).

- deux nombres $TS_{A,\ell}$ $\ell = 1,2$ par arête frontière de Ω_h, approchant la trace de la saturation aux points de Gauss $GA_\ell, \ell = 1,2$ de l'arête A.

- le débit d'eau capillaire R_A à travers chacune des arêtes A, approchant $\int_A \vec{r} \cdot \vec{\nu}_A$.

Aux équations approchant (5) à (9) et la partie droite de (10), il convient encore d'ajouter des conditions aux limites sur les frontières artificielles Γ_{ih} et Γ_{ph}.

Sur Γ_{ih} (frontière avec le macroélément d'injection) on imposera que les débits d'eau entrant dans le modèle E.F sont ceux fournis par le macroélément d'injection :

$$(16) \qquad -DIV_{K,A}\, QWP^n_{A,\ell} = QWPM^n_{A,\ell} \qquad \forall\, A \subset \Gamma_{ih},\ \forall\, \ell = 1,2.$$

où $QWP^n_{A,\ell}$ (resp $QWPM^n_{A,\ell}$) sont des débits d'eau partiels au $l^{ème}$-point de Gauss de l'arête A (respectivement produits par le sous-secteur A,ℓ du macroélément -cf. paragraphe V).

Sur Γ_{ph} (frontière avec le macroélément de production), on imposera que le flux d'eau capillaire est nul :

$$(17) \qquad R_A = 0 \quad \forall\, A \subset \Gamma_{ph}$$

et que le débit d'eau entrant dans chaque sous secteur du macroélément est égal au débit d'eau partiel correspondant produit par le modèle élément fini.

$$(18) \qquad QWPM^n_{A,\ell} = -DIV_{K,A}\, QWP^n_{A,\ell} \qquad \forall\, A \subset \Gamma_{ih},\ \forall\, \ell = 1,2.$$

Nous ne présenterons pas ici la mise en oeuvre de ces méthodes d'éléments finis, pour lesquelles nous renvoyons à la référence [2]. Pour des résultats partiels sur la convergence des méthodes utilisées, on pourra consulter Jaffré-Roberts [7].

V LES MACROELEMENTS

Décrivons d'abord un macroélement reliant un puits d'injection au modèle élements finis (le principe est le même dans le cas d'un puits de production).

On commence par approcher (cf. figure 3) la frontière réelle Γ_i du puits par un petit polygone convexe homothétique à la frontière Γ_{ih} du modèle E.F. On divise ensuite le milieux poreux situé entre Γ_{ih} et le polygone approchant Γ_i en autant de secteurs qu'il y a d'arêtes A dans Γ_{ih}. Comme on l'a indiqué au paragraphe IV, le débit de l'eau $QWP_{A,\ell}$, $\ell = 1, 2$ (cf. (16)) prends des valeurs différentes aux deux points de Gauss G_1 et G_2 de l'arête A. A chaque point de Gauss de A on associera donc un sous-secteur, qui sera le module élémentaire de calcul à l'intérieur du macroélément. Grâce aux hypothèses faites sur le choix de Γ_{ih}, on pourra considérer comme indépendants, les écoulements dans chaque sous-secteur; en outre, on pourra négliger les effets capillaires et de gravité à l'intérieur du macroélément.

Soit ζ l'abscisse le long d'un axe issu du puits et passant par un point de Gauss de l'arête A. Soit $SM^n_{A,\ell}(\zeta)$ le profil de saturation dans le sous-secteur A,ℓ à l'instant $t^n = n\Delta t$, et $QM^n_{A,\ell}$ le débit total eau + huile à travers le sous secteur, compté positivement dans le sens ζ. La chute de pression à travers le sous-secteur A,ℓ est obtenue, pour une répartition de saturation $SM^n_{A,\ell}$ dans le sous-secteur donnée, par :

$$(19) \qquad \lambda^n_{A,\ell}(P^n_{bi} - TP^n_A) = QM^n_{A,\ell} \qquad \forall A \subset \Gamma_{ih}, \ \forall \ell = 1,2$$

où le coefficient de Fourier $\lambda^n_{A,\ell}$ du sous-secteur est donné par

$$(20) \qquad \frac{1}{\lambda^n_{A,\ell}} = \int_{\zeta_{A,\ell \text{ puits}}}^{\zeta_{A,\ell \text{ maillage}}} \frac{d\zeta}{\zeta\, \Theta_{A\ell}\, \psi(\zeta)\, d(SM^n_{A,\ell}(\zeta))} \qquad \forall A \subset \Gamma_{ih}, \ \forall \ell = 1,2$$

où $\Theta_{A\ell}$ est un paramètre de forme du sous-secteur.

En utilisant la conservativité du débit total eau + huile à travers l'arête A, on en déduit aussitôt :

$$(21) \qquad (\lambda^n_{A,1} + \lambda^n_{A,2})(TP^n_A - P^n_{bi}) = DIV_{K,A}\, Q^n_A \qquad \forall A \subset \Gamma_{ih}$$

ce qui montre que le coefficient de Fourier λ^n_A du secteur A (cf. (13)) est :

$$(22) \qquad \lambda^n_A = \lambda^n_{A,1} + \lambda^n_{A,2} \qquad \forall A \subset \Gamma_{ih}.$$

Une fois les λ^n_A calculé par (20)(22), ils sont envoyés au programme élément finis, qui calcule alors les pressions P_K et les débits Q_A dans Ω_h, et renvoye au macroélément les débits Q_A à travers les arêtes de Γ_{ih} (et Γ_{ph})

Les débits totaux dans chaque sous secteurs sont alors :

$$(23) \qquad QM^n_{A,\ell} = - \frac{\lambda^n_{A,\ell}}{\lambda^n_A} DIV_{K,A}\, Q^n_A \qquad \forall A \subset \Gamma_{ih}, \ \forall \ell = 1,2$$

et les débits d'eau partiels entrant dans le modèle élément fini à travers les arêtes A de Γ_{ih} sont :

$$(24) \qquad QWPM^n_{A,\ell} = 2b(SM^n_{A,\ell}(\zeta_{A,\ell \text{ maillage}}))\, QM^n_{A,\ell}$$

Ces débits sont alors utilisés pour faire avancer les saturations jusqu'au pas de temps suivant, tant dans les macroéléments que dans le modèle E.F.

Dans les macroéléments, on résoud dans chaque sous-secteur, l'équation de

Buckley-Leverett suivante

$$(25)\quad \begin{cases} \zeta\, \Theta_{A\ell}(\phi)\zeta \dfrac{\partial SM_{A\ell}}{\partial t} + QM^n_{A,\ell} \dfrac{\partial}{\partial\zeta}[b(SM_{A,\ell})] = 0 \\ SM_{A,\ell} \equiv 1 \text{ pour } \zeta = z_{A,\ell}, \text{ puits} \\ SM_{A,\ell} = SM^n_{A,\ell} \text{ pour } t = t^n \end{cases}$$

où $QM^n_{A,\ell}$ a été calculé par (23).

Dans le modèle E.F., les débits d'eau partiels $QWPM^n_{A,\ell}$ calculés par (24) sont utilisés dans la condition au limite (16) sur Γ_{ih}, et la condition (17) est utilisée sur Γ_{ph}, ce qui permet de calculer les nouvelles valeurs de la saturation dans tout le domaine Ω_h.

VI RESULTATS NUMERIQUES

On montre la faisabilité de la méthode proposée en simulant un gisement carré de 600 m × 600 m exploité par 3 puits d'injection d'eau (20m³/jour chacun) et deux puits de production.

Les non-linénarité apparaissant dans les équations (3) à (8) sont représentées sur la figure 4. Il n'y a pas de pressure capillaire. Les caractéristiques de la simulation sont :

- épaisseur du gisement variant de 15 à 20 m.
- porosité 0,3.
- perméabilité absolue 100mD
- densité de l'eau : 1, viscosité de l'eau : 1 cP
- densité de l'huile : 0,75 viscosite de l'huile : 1,43 cP
- perméabilités relatives kr_1 et k_{r2} comme indiqué sur la figure 4 rapport de mobilité : 1.

On a simulé successivement le cas d'un gisement horizontal et celui d'un gisement présentant un pendage de 10% vers l'ouest.

Le maillage Ω_h est constitué de quadrilatères, et les frontières artificielles Γ_{ih} et Γ_{ph} sont des carrés de 30m de côté. Le maillage contient 95 quadrilatères, 380 inconnues en saturation et 220 inconnues de débits (c.à.d. 220 arêtes).

Les équations de Buckley-Leverett (25) dans chaque sous-secteur des macro-éléments ont étées résolues par une méthode différence finie décentrée, avec un schéma en temps explicite. On a constaté que pour obtenir des débits réguliers aux puits de production, il fallait utiliser pour les macroéléments un pas de temps bien plus petit que le maillage E.F. (on aurait pu aussi utiliser un schéma implicite dans les sous-secteurs).

On a représenté sur la figure 5 le champs de gravité $\vec{q}_{Gh}$, et le champs des débits totaux eau + huile $\vec{q}_h$ après 1000 j d'injection dans le réservoir avec pendage.

On montre enfin sur la figure 6 les répartitions de saturation dans Ω_h calculées avec le modèle élément fini, dans le cas d'un gisement horizontal (au centre) et penchant vers l'ouest (à droite). A titre de comparaison, on a fait figurer à gauche les résultats obtenus avec un modèle classique aux différences finies sur un maillage 24×24.

REMERCIEMENT

Cette étude a été faite dans le cadre d'une collaboration entre l'INRIA, l'IFP et la SNEA(P), que nous remercions pour l'autorisation de publier les résultats numériques.

REFERENCES

[1] Douglas, J,Jr., Ewing, R.E., Wheeler, M.F., *The approximation of the pressure by a mixed method in the simulation of miscible displacement* R.A.I.R.O. Anal. Num. 17(1983) 17-33.

[2] Chavent, G., Cohen, G., Jaffré, J., *Discontinuous upwinding and mixed finite elements for two phase flows in reservoir simulation*, à paraître dans Computer methods in applied mechanics and engineering.

[3] Chavent, G., *A new formulation of diphasic incompressible flow in porous media* in Applications of methods of functional analysis to problems in mechanics, Marseille (1975), Lecture Notes in Mathematics 503, (Springer Berlin, 1976).

[4] Raviart, P.A., Thomas, J.M., *A mixed finite element method for second order elliptic problems* in Mathematical aspects of the finite element method, Rome (1975), Lecture Notes in Mathematics 606, (Springer, Berlin, 1977).

[5] Lesaint, P., Raviart, P.A., *On a finite element method for solving the neutron transport equation* in Carl de Boor. (eds.), Mathematical aspects of finite elements in partial differential equations, (Academic Press, New York 1974).

[6] Godunov, S.K., *Finite difference method for numerical computation of discontinuous solution of the equations of fluid dynamics* Math. Sb. 47 (1959) 271.

[7] Jaffre, J., Roberts, J.E., *Upstream weighting and mixed finite elements in the simulation of miscible displacement* , Rapport INRIA (1983).

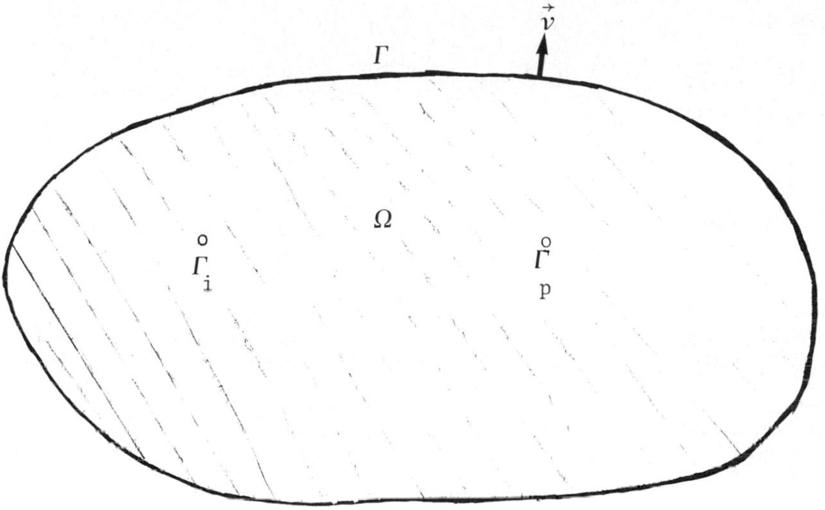

FIGURE 1

LE GISEMENT Ω ET LES DEUX PUITS D'INJECTION (Γ_i) ET DE PRODUCTION (Γ_p)

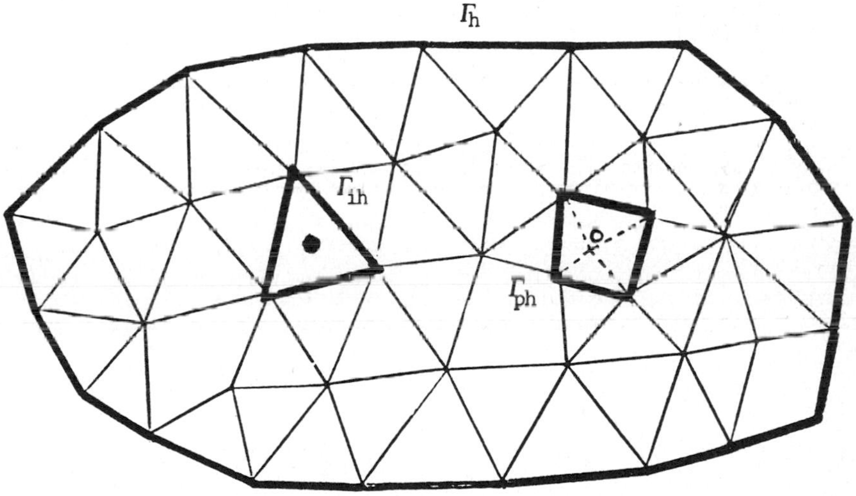

● PUITS D'INJECTION.

○ PUITS DE PRODUCTION.

FIGURE 2

LE MAILLAGE ELÉMENTS FINIS

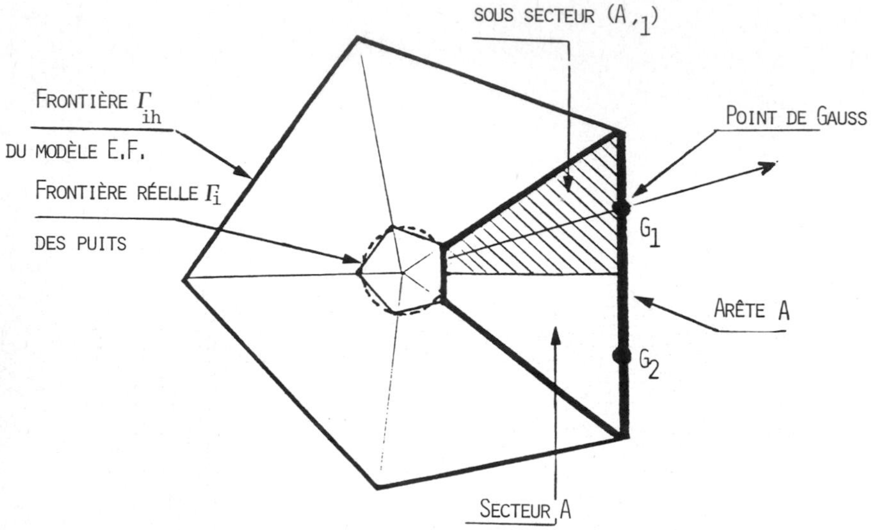

FIGURE 3
GÉOMÉTRIE D'UN MACROÉLÉMENT

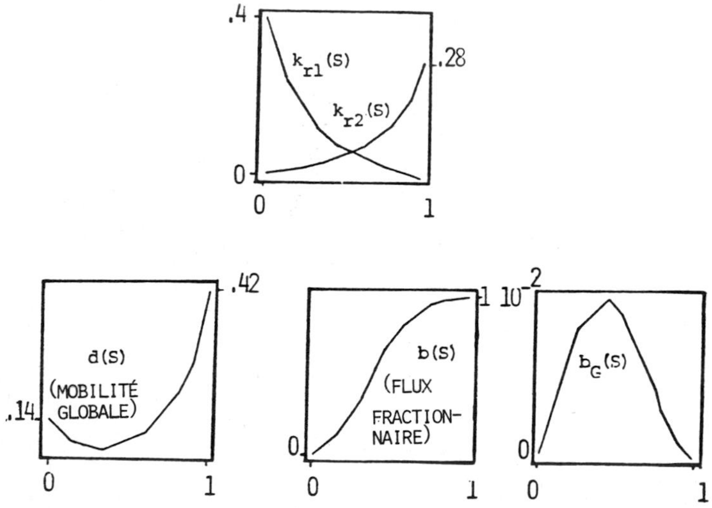

FIGURE 4 : LES NON-LINÉARITÉS DES EQUATIONS D'ETAT

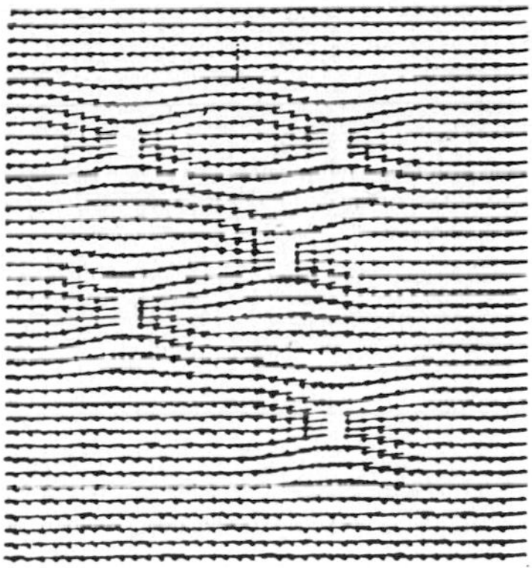

CHAMPS DE GRAVITÉ

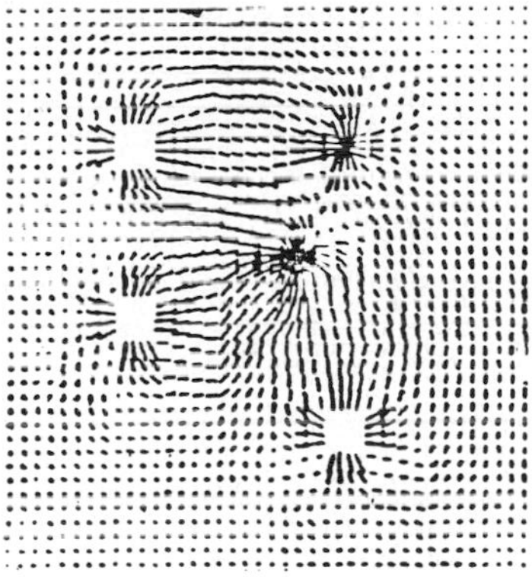

CHAMPS DES DÉBITS TOTAUX
EAUX + HUILE APRÈS 1000 JOURS
D'INJECTION.

FIGURE 5

FIGURE 6 LES RÉPARTITIONS CALCULÉES DE SATURATION

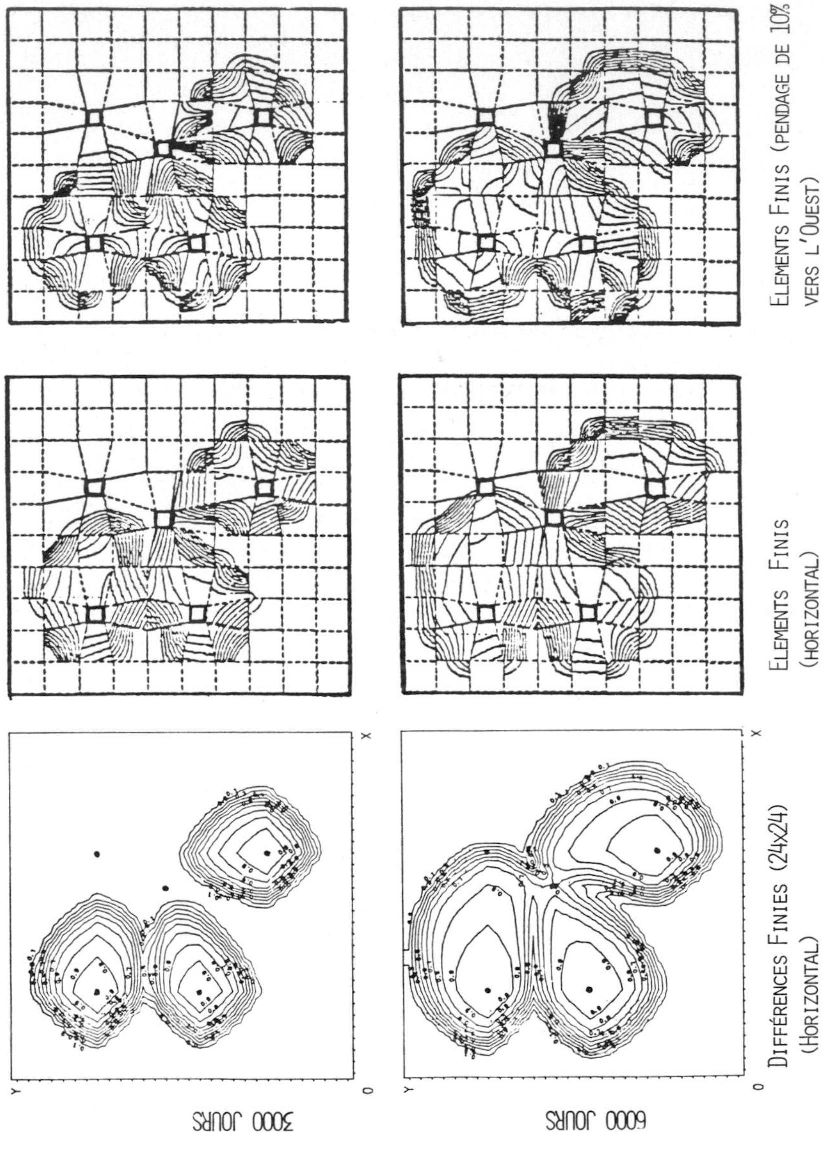

SEMI-CONDUCTORS

SEMI-CONDUCTEURS

Computing Methods in Applied Sciences and Engineering, VI
R. Glowinski and J.-L. Lions (Editors)
Elsevier Science Publishers B.V. (North-Holland)

DIFFUSION D'IMPURETES AVEC OXYDATION

A. GERODOLLE*, A. MARROCCO**, S. MARTIN*

* CNET-CNS, BP 98, 38243 Meylan Cédex

** INRIA, BP 105, 78153 Le Chesnay Cédex

INTRODUCTION

Nous présentons dans cet article une méthode numérique permettant de simuler la diffusion d'impuretés dans du silicium, ce dernier subissant en plus une oxydation par une partie de sa surface (LOCOS process). L'étude et les résultats présentés ici sont la suite de précédents travaux exposés dans [1], [2], et dans lesquels on étudiait des algorithmes numériques permettant de résoudre le problème non linéaire lié à la diffusion d'une impureté dans du silicium et puis le couplage entre plusieurs impuretés ainsi que le phénomène de ségrégation (discontinuité des concentrations) se produisant à l'interface Silicium-Oxyde de Silicium (S_i-$S_i O_2$). L'oxydation, (phénomène très important intervenant lors de la fabrication de circuits intégrés), n'avait pas encore été considérée, du point de vue numérique, par les auteurs. L'oxydation modifie la géométrie du silicium (dans lequel se produit aussi la diffusion) conduisant donc à des problèmes délicats du point de vue numérique. Les processus d'oxydation et de diffusion sont liés mais si l'on tient compte du fait que la cinétique de la ségrégation est plus rapide que celle de l'oxydation, on peut en quelque sorte (et pour une première approche), découpler les deux phénomènes.

Plusieurs approches dans la modélisation de ce phénomène sont possibles et la plus utilisée [3], [4], [5], est celle qui consiste à ramener le domaine physique étudié (le silicium), à un domaine fixé. Ce procédé nécessite de connaître explicitement et analytiquement la loi de croissance de l'oxyde ; cette loi, dont l'expression varie selon les auteurs, est ajustée expérimentalement et se présente en général sous la forme $y = f(x,t)$ (famille de fonctions dépendant du temps t). On effectue donc un changement de coordonnées ramenant le domaine variable à un domaine fixé. Les équations de diffusion sont évidemment changées et il apparait un terme de transport.

Nous avons choisi dans cette étude de travailler dans le plan physique, c'est à dire avec un domaine variable (ce domaine variable, sera, évidemment le silicium car c'est la partie du transistor qui nous intéresse). L'oxydation et la ségrégation sont modélisées par un flux à l'interface S_i-S_i O_2 pouvant être entrant ou sortant selon l'impureté considérée. Dans cette modélisation, on n'a plus besoin de connaître explicitement la loi de croissance de l'oxyde ; ce dont on a besoin, ce sont les déplacements que subissent les points de la frontière du Silicium dans un laps de temps donné (le pas de discrétisation en temps). Ces déplacements peuvent provenir, par exemple, d'un module extérieur, simulant la croissance de l'oxyde à l'aide de la diffusion de l'oxygène dans l'oxyde [6], [7].

Pour les essais numériques présentés ici, (premier test), nous nous sommes donné la loi de croissance d'oxyde du type de celle déjà utilisée dans [3], [4], [5] et nous calculons des déplacements à partir de cette loi. L'approximation numérique est faite par la méthode des éléments finis en espace et par un schéma implicite en temps. Les problèmes non linéaires aux différents pas de temps, sont résolus par des algorithmes déjà exposés dans [1], [2], et les techniques utilisées sont essentiellement itératives.

MODELE PHYSIQUE

Les lois de diffusion des dopants (tels que l'Arsenic, le Bore, le phosphore) sont celles déjà décrites dans [1], [2] ; ces lois tiennent compte de l'action du champ électrique interne, de la concentration de l'impureté, ainsi que du clustering.

Si $\vec{J}_i(C_j)$ est le flux de l'impureté i dans le domaine Ω, la loi de conservation pour l'impureté i s'écrit, [pas de source à l'intérieur du domaine].

$$\frac{\partial C_i}{\partial t} = - \operatorname{div} \vec{J}_i(C_j) \tag{1}$$

et la condition aux limites à l'interface S_i-S_i O_2 tenant compte de l'avancée de l'interface ainsi que de la ségrégation (découlant aussi d'une loi de conservation, peut se formuler ainsi :

$$\vec{J}_i.\vec{n} = (1-\alpha h_i)\, C_{S_i,i}\, \vec{V}_{OX}.\vec{n} \tag{2}$$

où $\vec{n}$ est la normale à l'interface S_i-S_i O_2 (Γ),

$\vec{V}_{OX}$ est la vitesse d'avancée de l'interface,

α est le coefficient de dilatation entre oxyde et silicium.

$C_{S_i,i}$ est la concentration en impuretés i dans le S_i (Silicium),

h_i est le coefficient de ségrégation ($C_{OX,i} = h_i\, C_{S_i,i}$).

Sur les autres parties de la frontière, on peut avoir des conditions plus classiques de type Neumann homogène (ou Dirichlet homogène ou non).

La valeur numérique des coefficients retenue pour les simulations est : α = 1/0.44, h_i = 3.3 pour le Bore et h_i = 0.1 pour l'Arsenic ou le phosphore.

La fonction decrivant l'avancée de l'oxyde dans le silicium est modélisée par :

$$U(x,t) = U_0 + \left[\frac{U(t) - U_0}{2}\right] \operatorname{erfc}\left(\frac{x-a + \delta[U(t)-U_0]}{\sqrt{2}\, R_0\, U(t)}\right) \tag{3}$$

où R_0 est le rapport entre la vitesse latérale et la vitesse verticale d'avancée de la frontière, δ est un paramètre qui comme R_0 dépend de la technologie utilisée.
a est l'abscisse du bord du masque de nitrure (barrière pour l'oxygène),
U_0 est l'épaisseur initiale d'oxyde,
U(t) est la fonction donnant la loi unidimensionnelle [8] de croissance de l'oxyde (loi de type parabolique),
et erfc est la fonction erreur complémentaire donnée par :

(4) $$\mathrm{erfc}(x) = \frac{2}{\sqrt{\pi}} \int_x^\infty e^{-t^2}\, dt.$$

Rappelons toutefois que ce qui interviendra dans le module de simulation, ce sont les déplacements nodaux liés au passage du temps de t à $t+\Delta t$, et que ces déplacements peuvent avoir été calculés par un autre module.

DEFORMATION DU MAILLAGE

Durant l'oxydation, le silicium (Ω) change de forme. Etant donné que l'on veut aussi résoudre un problème de diffusion dans Ω à l'aide des éléments finis, il faut disposer avant tout d'un maillage sur Ω. (plus précisément, une triangulation car on utilisera des éléments finis P1). Différentes solutions sont envisageables à ce niveau ; nous avons retenu celle qui consiste à déformer une triangulation existante, en gardant la même topologie pour les éléments.

Etant donné une triangulation sur Ω, on se donne donc les déplacements pour les noeuds appartenant à la frontière de Ω, et l'on veut calculer les déplacements à imposer aux noeuds internes de façon à obtenir un nouveau maillage qui soit topologiquement correct (pas de chevauchement ou de retournement). Un tel problème pourrait être formulé et résolu comme un problème d'élasticité classique 2D par exemple (en formulation déplacements sur le maillage existant), mais il ne faut pas oublier que cela doit être répété à chaque pas de temps du processus de diffusion, et donc risque de devenir prohibitif du point de vue temps calcul.

On peut aussi penser à extrapoler pour un maillage de type éléments finis une méthode itérative donnée par A.A. AMSDEN et C.W. HIRT (dans Journal of Computational Physics 11, 348-359, 1973), mais cela ne semble facile à utiliser que lorsque les domaines considérés sont convexes, ce qui n'est malheureusement pas le cas pour nous.

Nous avons donc élaboré une méthode suffisamment robuste, du moins pour les cas qui nous intéressent, et qui s'apparente peut-être à une formulation de type éléments frontière pour un problème d'élasticité. (Voir figure 1 - maillage initial et figure 2 maillage obtenu après 40 étapes ; correspondant à 1 heure d'oxydation).

Soit $\mathcal{K}$ l'ensemble des noeuds frontières et soit $\vec{D}_K$, $K \in \mathcal{K}$ le déplacement imposé (connu) au noeud K, W_K est un poids que l'on affecte au déplacement de ce noeud K, par exemple, W_K est la somme des segments frontières entourant le noeud K.

Le déplacement d'un noeud intérieur i est alors calculé par :

(5) $$\vec{D}_i = \frac{1}{\alpha_i} \sum_{K \in \mathcal{K}} W_K\, \alpha_{K_i}\, \vec{D}_K$$

où α_{K_i} est en quelque sorte un facteur d'influence.

(6) $$\alpha_{K_i} = \frac{1}{d_{K_i}{}^{\beta}}$$

d_{K_i} étant la distance du noeud i au noeud K,

β étant un paramètre qui d'après les essais numériques semble lié à la "dureté" du maillage ($\beta \geq 2$, pour les essais numériques nous avons pris $\beta = 4$),

α_i est un facteur de normalisation et vaut

(7) $$\alpha_i = \sum_{K \in \mathcal{K}} \alpha_{K_i}\, W_K.$$

Procédant ainsi, (différentes variantes sont possibles), nous obtenons une méthode robuste de déformation de maillages (supportant même des déformations assez brutales) (voir figure 3) et relativement peu coûteuse en temps calcul. De plus, si les déplacements imposés sur le bord varient régulièrement, il est possible de ne pas considérer tous les points frontière dans (5) mais par exemple 1 sur 2 (pour des résultats comparables, mais avec un temps calcul divisé par 2!).

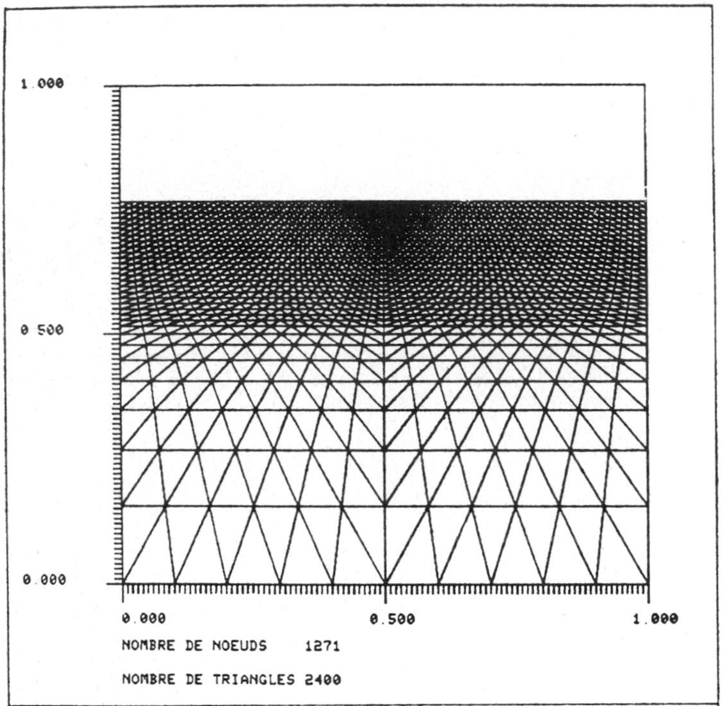

Figure 1 : Maillage initial

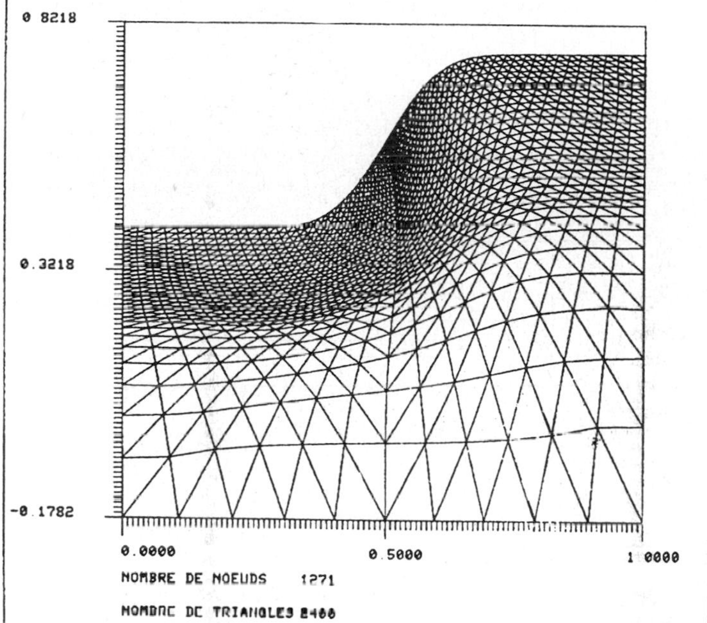

Figure 2 : Maillage obtenu après 40 pas de temps (1 heure)
Selon la loi donnée en (3)

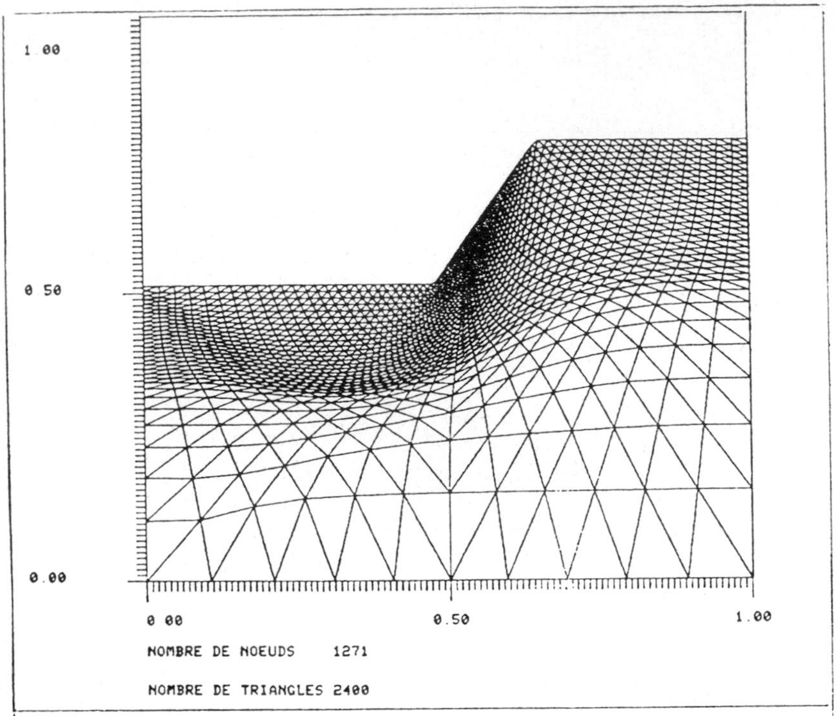

Figure 3 : Déformation du maillage obtenue en une seule étape

DIFFUSION AVEC DOMAINE VARIABLE

En ce qui concerne le problème de diffusion, nous reprenons la même méthodologie que celle exposée dans nos précédents articles [1], [2]. Le schéma en temps utilisé ici est le schéma totalement implicite, et l'approximation en espace se fait toujours par la méthode des éléments finis P1.

Reste à prendre en compte le fait que le domaine varie au cours du temps et à "bien interpréter" du point de vue discret la condition de flux (2).

La formulation variationnelle de notre problème de diffusion sera écrite sur le domaine Ω^{n+1} (c.a.d. Ω au temps n+1). Un élément de l'espace discret approximant l'espace fonctionnel dans lequel on cherche les concentrations peut s'écrire :

$$C = \sum_i C_i w_i \tag{8}$$

avec $\{w_i\}$ base de l'espace, et C_i valeurs nodales.

La dérivée en temps dans (1) approchée selon le schéma implicite s'exprime par

$$\frac{\partial C}{\partial t} \sim \frac{C^{n+1}-C^n}{\tau} = \frac{1}{\tau} \sum_i \left[C_i^{n+1}\, w_i^{n+1} - C_i^n\, w_i^n\right] \tag{9}$$

que l'on remplacera dans notre schéma par :

$$(10)\qquad \sum_i \left(\frac{C_i^{n+1} - C_i^n}{\tau}\right) w_i^{n+1} + \sum_i C_i^n \left(\frac{w_i^{n+1} - w_i^n}{\tau}\right)$$

Le premier terme dans (10) correspond au terme classique d'approximation pour le schéma implicite, le second correspond à la correction à approter du fait du domaine variable.

L'interprétation discrète de la condition de flux (2), conduirait à une intégration sur le domaine compris entre Γ^n et Γ^{n+1} (Γ^n = frontière de Ω à l'instant n) ; celle-ci est approchée par une intégrale calculée sur Γ^n ou Γ^{n+1}. Finalement la formulation discrète que nous avons retenue pour notre problème est la suivante (passage de n à n+1).

Connaissant C^n sur Ω^n, on calcule C^{n+1} sur Ω^{n+1} en résolvant le problème non linéaire $\forall$ i,

$$(11)\qquad \int_{\Omega^{n+1}} C^{n+1}.w_i^{n+1}\,dx + \tau\int_{\Omega^{n+1}} D(C^{n+1})\nabla C^{n+1}.\nabla w_i^{n+1}\,dx +$$
$$+ \int_{\Gamma_{n+1}} \frac{\alpha h}{2}(-\vec{depl}).\vec{n}\, C^{n+1}.w_i^{n+1}\,dx = + \int_{\Omega^{n+1}} \tilde{C}^n.w_i^{n+1}\,dx -$$
$$- \int_{\Omega^{n+1}} (\tilde{C}^n - C^n) w_i^{n+1}\,dx - \int_{\Gamma_n} \vec{depl}.\vec{n}\left[\frac{C^n_{\Gamma_{n+1}} + C^n}{2}\right] w_i^n\,d\Gamma +$$
$$+ \int_{\Gamma_n} \frac{\alpha h}{2}\vec{depl}.\vec{n}\, C^n.w_i^n\,d\Gamma.$$

Dans (11), certains termes du second membre (termes communs) peuvent être regroupés, de même que les intégrales sur Γ_n peuvent être exprimées par des intégrales sur Γ_{n+1}, on a aussi

$$(12)\qquad \tilde{C}^n = \sum_i C_i^n\, w_i^{n+1}$$

$$\text{et } \vec{depl} = \vec{V}_{ox} \times \tau.$$

A l'aide de (11), nous avons simulé la diffusion avec oxydation en partant d'une condition initiale correspondant à une double implantation ionique: tout d'abord, une implantation de Bore, sur toute la surface avec une dose 10^{14} et une énergie de 100 Kev, puis une implantation ionique d'arsenic, à travers un masque avec une dose de 5.10^{15} et une énergie de 80 Kev.

La diffusion et l'oxydation (en atmosphère humide) s'est poursuivie pendant 1 heure à 1000 degrés C. Voir figures 4, 5, 6 donnant les isovaleurs des concentrations de Bore, d'Arsenic, à l'instant final, ainsi que la jonction réalisée (figure 6).

Nous remercions F. HECHT, chercheur à l'INRIA pour les fructueuses discussions que nous avons eu avec lui pour la formulation du problème de diffusion en domaine variable. Une version plus détaillée de cet article sera publiée dans les proceedings du colloque.

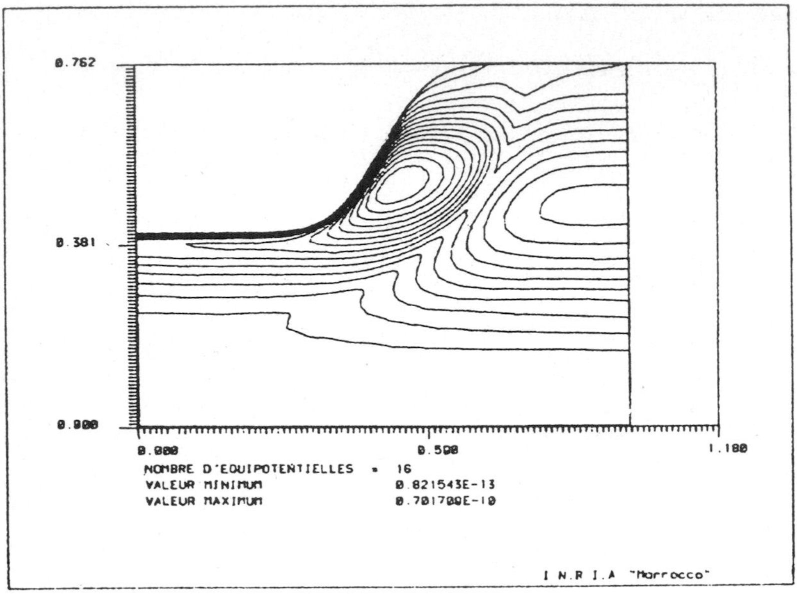

Figure 4 : Isovaleurs de la concentration en Bore

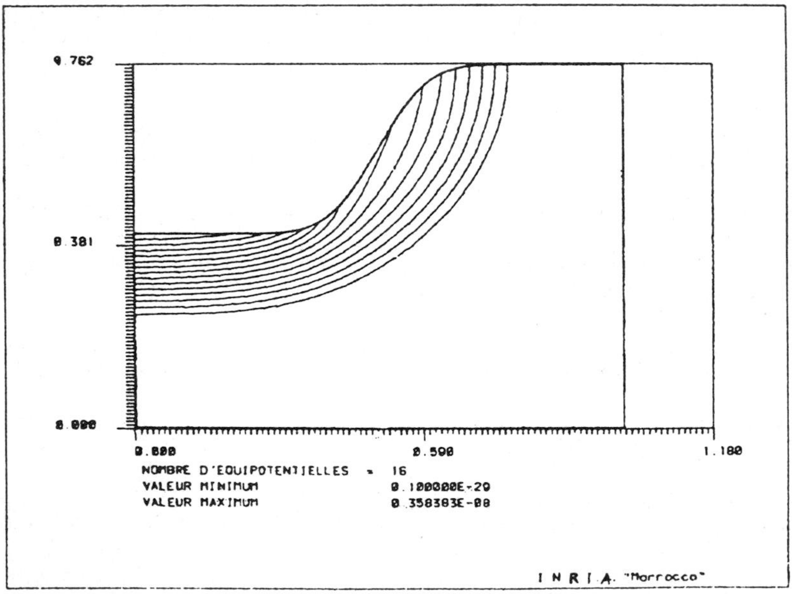

Figure 5 : Isovaleurs de la concentration en Arsenic

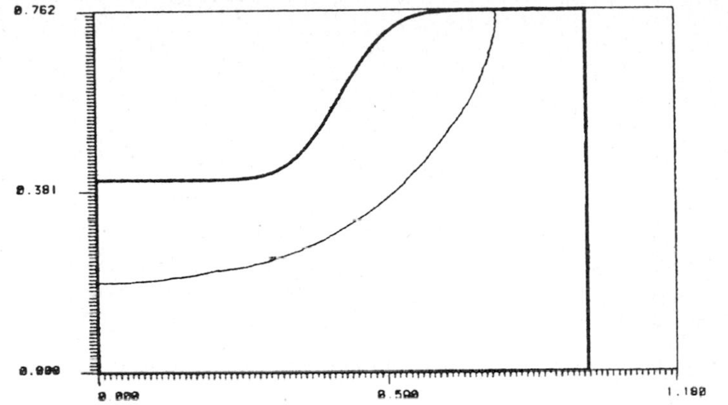

Figure 6 : Jonction réalisée

BIBLIOGRAPHIE

[1] E. CAQUOT, E. FERNANDEZ-CARA, A. MARROCCO. Résolution numérique d'un problème non linéaire lié à la diffusion d'impuretés dans du silicium. COMPUTING METHODS IN APPLIED SCIENCES AND ENGINEERING, V pp 599-618., North Holland 1982. Proceedings of the fifth International Symposium a computing method in Applied Sciences and Engineering Versailles, dec. 14-18, 1981.

[2] E. CAQUOT, E. FERNANDEZ-CARA, A. MARROCCO. 2-D simulation of coupled diffusion and segregation of impurities in V.L.S.I. process using finite elements. Joint SIAM-IEEE Conference on numerical simulation of V.L.S.I. devices, Boston nov 1982.

[3] B.R. PENUMALLI. Lateral oxydation and redistribution of dopants 2nd Int. Conference on numerical Analysis of semiconductor device and IC'S (NASECODE II). Dublin, Ireland, June 17-19, 1981.

[4] R. TIELERT. Numerical simulation of impurity redistribution near mask edges. NATO advanced study institute on Process and device simulation for MOS-VLSI Circuits SOGESTA - Urbino July 12-23, 1982.

[5] M. KUMP, R.W. DUTTON, Two dimensional process simulation-SUPRA, NATO advanced study Institute on Process and device simulation for MOS-VLSI Circuits SOGESTA - Urbino, July 12-23, 1982.

[6] E. CAQUOT, Two dimensional simulation of local oxidation in V.L.S.I. processes JOINT. SIAM-IEEE Conference on numerical simulation of V.L.S.I. devices, Boston nov 1982.

[7] A. PONCET, Finite element simulation of local oxidation of silicion. Presented at the summer course on V.L.S.I. process and device modeling ; Heverlee, Belgium, June 7-10, 1983.

[8] B.E. DEAL, A.S. GROVE. General relationship for the thermal oxidation of silicon . J. Appl. Phys., Vol 36, pp. 3770-3778, (1965).

Computing Methods in Applied Sciences and Engineering, VI
R. Glowinski and J.-L. Lions (Editors)
Elsevier Science Publishers B.V. (North-Holland)

FINITE-ELEMENT SOLUTION OF THE SEMICONDUCTOR TRANSPORT EQUATIONS

B. M. Grossman, E. M. Buturla and P. E. Cottrell

IBM General Technology Division
Essex Junction, Vt., USA

Numerical simulation plays an increasingly important role in the design of semiconductor devices. The operating behavior of a semiconductor device is described by three nonlinear partial-differential equations — the semiconductor transport equations. These equations are solved by the finite-element method in one, two or three dimensions, for transient or steady-state conditions. This paper presents a numerical solution technique for the transport equations and a comparison of modeled results with experimental data.

I. INTRODUCTION

Silicon chips that contain over one-half million electrical circuits have been developed in laboratories in the United States [1]. The devices and interconnections in such Very-Large-Scale-Integrated (VLSI) circuits have linear dimensions that approach the wavelength of visible light. The technology that led up to these developments has progressed rapidly. Initially, semiconductor devices were structurally and operationally one-dimensional and were adequately described by analytic models. As semiconductor technology progressed into the integrated circuit era, the density of circuits increased and devices were made microscopically small. In state-of-the-art semiconductor devices, physical features are approximately equal in size in all three directions. Two- and three-dimensional physical effects, that were unimportant in older technologies, are now dominant. An analytic solution of the equations that determine the operating behavior of a semiconductor device in more than one dimension is not possible for practical problems. As a result, multi-dimensional numerical models have been developed.

The development of a two- or three-dimensional numerical model for semiconductor devices is a formidable task due to the complexity of the problem and the high degree of nonlinearity in the governing equations. In this work, we describe the numerical algorithm in the FInite ELement Device AnalYsis (FIELDAY) program [2]. FIELDAY is a general purpose program that is used extensively within IBM for designing semiconductor devices. The program solves the semiconductor transport equations in one, two or three dimensions for steady-state or transient operating conditions, by the finite-element method. The continuum model and system of equations solved by FIELDAY are described in section II; the discrete model and numerical method of solution are described in detail in section III. Three applications of the program are presented in section IV and the important features of this work are summarized in section V.

II. SEMICONDUCTOR TRANSPORT EQUATIONS

The flow of electrons and holes in a semiconductor are described by Poisson's equation (1), and the equations of electron (2) and hole (3) current continuity.

$$\nabla^2\psi = -\frac{q}{\varepsilon}(p - n + N_D - N_A + N_Q) \quad (1)$$

$$\nabla \cdot \mathbf{J}_n = q(\frac{\partial n}{\partial t} + R_n) \quad (2)$$

$$\nabla \cdot \mathbf{J}_p = -q(\frac{\partial p}{\partial t} + R_p) \quad (3)$$

$$\mathbf{J}_n = q[D_n \nabla n - \mu_n n \nabla(\psi + \Delta V_c)] \quad (4)$$

$$\mathbf{J}_p = -q[D_p \nabla p + \mu_p p \nabla(\psi - \Delta V_v)] \quad (5)$$

These coupled, nonlinear partial differential equations form the semiconductor transport equations [3,4]. The three unknown quantities are the the space-charge potential (ψ) and the electron (n) and hole (p) mobile charge densities at each instant of time. The variables that appear in Eqs. (1) – (5) are listed in Table I.

TABLE I.

D_n, D_p	Diffusion coefficient for electrons and holes
ε	Permittivity
$\mathbf{J}_n, \mathbf{J}_p$	Current density for electron and hole carriers
K	Boltzmann's constant
μ_n, μ_p	Electron and hole mobilities
n	Electron concentration
N_A	Acceptor concentration
N_D	Donor concentration
N_Q	Concentration of fixed charge particles
p	Hole concentration
ψ	Space – charge potential
q	Electronic charge
R_n, R_p	Electron and hole recombination rates
T	Operating temperature
$\Delta V_c, \Delta V_v$	Changes in the conduction and valence band edges

In the FIELDAY model, the electron and hole mobilities are functions of electric-field strength $|\Delta\psi|$ and impurity density [5], and the diffusion coefficients are related to the mobilities by the Einstein relationship. The recombination-generation mechanisms include carrier generation due to avalanche multiplication [6] and photo-generation, as well as

Auger and Schockley-Read-Hall recombination [7–9]. These recombination-generation mechanisms couple the two current continuity equations and introduce strong nonlinearities, particularly for the case of avalanche multiplication.

A. BOUNDARY CONDITIONS

To solve Eqs. (1) – (3), boundary conditions for the space-charge potential and the electron and hole densities at contacts and along the entire edge of the semiconductor device model are required. At a contact, the potential is either a fixed value or determined by an equation of constraint when current sources or simple circuit elements are specified [10]. At ohmic contacts, the mobile charge densities are set equal to their thermal equilibrium values with space-charge neutrality assumed. At Schottky contacts, carrier concentrations are either set to fixed values or modulated by a thermionic recombination velocity [11]. Along the noncontact boundaries of the device, the normal components of the electric field, and the electron and hole current densities are all equal to zero.

III. NUMERICAL METHOD

Solving equations (1) – (3) in two or three dimensions by analytic techniques, *i.e.*, integration and application of boundary conditions is not possible for practical problems. Instead, a numerical approach is necessary. The continuum problem is approximated by a discrete model that divides the two- or three-dimensional space domain of the device into a finite number of triangular or prismatic elements. The unknown variables (ψ, n and p) are evaluated at the points (nodes) that form the vertices of these elements. In the FIELDAY program, Poisson's equation is transformed into a discrete representation by the finite-element method and the current continuity equations are transformed by a hybrid finite-element finite-difference technique [12]. For time dependent problems, the change of carrier densities with time in the current continuity equations are approximated by Euler's method. The discrete equations are solved by one of two numerical algorithms. The decoupled or Gummel's algorithm solves the discrete Poison's and current continuity equations sequentially until a self-consistent solution is obtained [13]. The second method, referred to as the coupled algorithm, solves the complete set of discrete equations simultaneously using Newton's method [14,15]. Either approach results in large, sparse matrix equations which are solved by a direct solution technique.

A. DISCRETE MODEL

i) Poisson's Equation

The functional for Poisson's equation is given by

$$\int_V \frac{\varepsilon}{2}(\nabla\psi)^2 dv - \int_V q\psi(p - n + N_D - N_A + N_Q)dv \tag{6}$$

where V is the volume of the domain. The functional is the energy stored in the static electric field and assemblage of charge in the device. At a stable operating point, which physically occurs when a solution to equations (1) – (3) is attained, the energy in the system is a minimum and the first variation of (6) with respect to ψ is zero. In each element, ψ is approximated by a linear interpolating function. Substituting the interpolating function into (6), taking the first variation, integrating over the space of a single element and setting the resulting expression equal to zero, yields the following matrix equation:

$$[\mathbf{A}]\ \{\psi\} = [\mathbf{B}]\ \{p - n + N_{\mathrm{D}} - N_{\mathrm{A}} + N_{\mathrm{Q}}\} \tag{7}$$

Matrix [**A**] is a symmetric matrix and its terms are a function of element geometry and permittivity. The components of column vectors $\{\psi\}$ and $\{p - n + N_{\mathrm{D}} - N_{\mathrm{A}} + N_{\mathrm{Q}}\}$ are the mid-band potentials and charged particle densities at each of the nodes of the element, respectively. Matrix [**B**] is a diagonal matrix which assigns a weighted area or volume to each node of the finite element. [**A**] and [**B**] are third order matrices for two-dimensional triangular elements or sixth order matrices for three-dimensional prismatic elements.

The choice of nodal weighing schemes, or the form of the volume distribution matrix [**B**], does not receive much attention in most applications of the finite-element method. However, in semiconductor problems, it is common for the charge density to vary by several orders of magnitude over microscopic distances. In the discrete model, charge density is converted to charge at a node by a volume weight in matrix [**B**]. Therefore, the volume associated with a node is extremely important. The traditional lumped method, common in structural analysis and heat transfer problems, assigns equal portions of volume to each node of the element. The consistent method assumes the same order of approximation for charge density over the element as for the potential [16]. Either of these methods results in anomalous oscillations of potential with respect to position in semiconductor problems [17].

To avoid these anomalies, FIELDAY uses a unique approach. Within an element, the volume closest to a node is associated with that node. For a two-dimensional triangular element, this area is defined by the perpendicular bisectors of the sides of the element [12], as shown in Fig. 1. For a prismatic element [18,19], this construction is extended in the third dimension, as shown in Fig. 2. The volume associated with each node is the appropriate portion of the triangular face times one-half the length of the prism. Under severely nonlinear operating conditions, anomalous results in a device simulation may still appear when *obtuse* elements with large length-to-height aspect ratios (*e.g.*, greater than 50:1) are placed in regions where large amounts of current flow. Avoiding obtuse elements of this type is only a minor restriction for most semiconductor device applications.

ii) Current Continuity Equations
The current continuity equations are solved with a hybrid finite-element finite-difference scheme. The finite-element division of space is used along with a finite-difference technique to approximate current flow between the network of nodes. Using Gauss' theorem, equation (2) is transformed into

$$\int_{\mathrm{S}} \mathbf{J}_n \cdot d\mathbf{s} = \int_{\mathrm{V}} q\left(\frac{\partial n}{\partial t} + R_n\right) dv\ , \tag{8}$$

where the surface S encloses the volume V. In the discrete model, current density is assumed constant along the edges of an element. The current flowing between adjacent nodes is the product of the current density along the edge and a flux cross-section. In acute and right triangular elements, the flux cross-sections are the lines that bound the area associated with each node. In acute or right prismatic elements the flux cross-sections are the planes that bound the nodal volume weight described previously. This

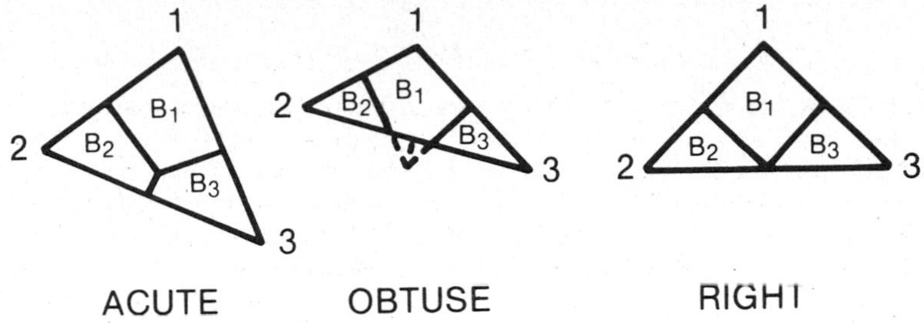

Figure 1. Area distribution for acute, obtuse and right triangular finite elements.

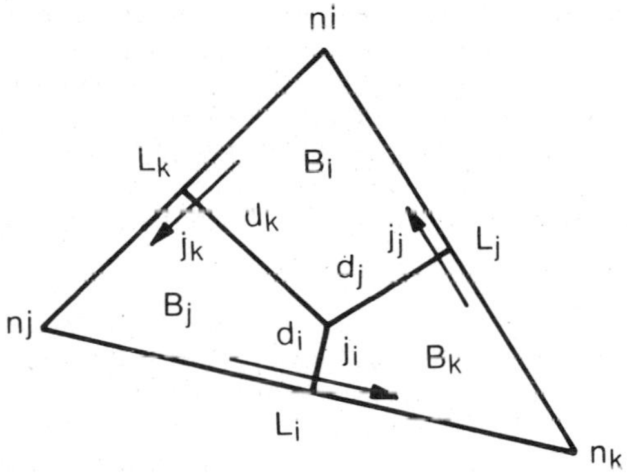

Figure 3. Current continuity in triangular finite element.

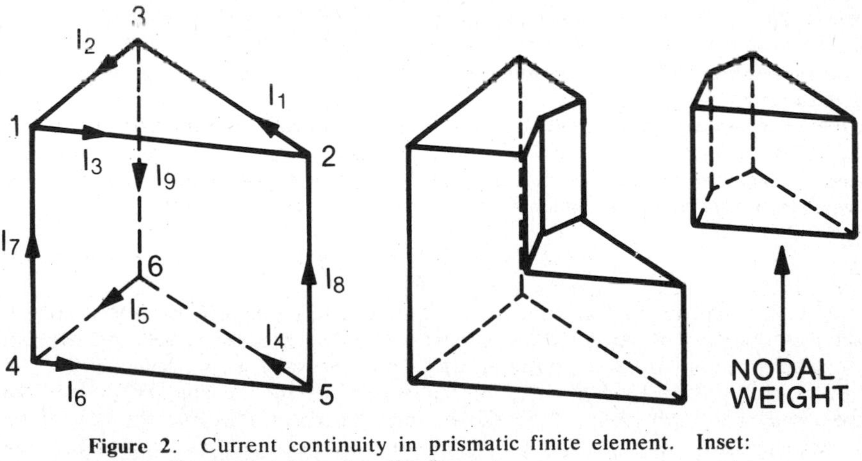

Figure 2. Current continuity in prismatic finite element. Inset: volume distribution for prismatic element.

assures that the flux cross-sections that surround a node, and the volume weight associated with a node, are consistent with the distribution scheme for Poisson's equation.

The discrete approximations for the electron and hole currents flowing from node i to node j along side k, as shown in Fig. 3, are [20,21]

$$I_{n,k} = qD_n \frac{d_k}{l_k}[n_j Z(\Delta_k) - n_i Z(-\Delta_k)] \text{ and} \tag{9}$$

$$I_{p,k} = -qD_p \frac{d_k}{l_k}[p_j Z(-\Delta_k) - p_i Z(\Delta_k)] , \tag{10}$$

where d_k and l_k are the flux cross-section and length of side k, respectively. The Bernoulli function is defined by

$$Z(\Delta_k) = \frac{\Delta_k}{e^{\Delta_k} - 1}$$

where $\Delta_k = \beta(\psi_j - \psi_i)$ and $\beta = q/KT$. Evaluation of the Bernoulli function about the point $\Delta_k=0$ requires special attention. Equations (9) and (10) are derived from continuum expressions (4) and (5) for electron and hole current densities, respectively.

The discrete approximation for electron current (9) is derived as follows. Using the Einstein relationship, $D_n=\mu_n/\beta$, (5) is rewritten as

$$J_n = qD_n \, exp\,(\beta\psi)\, \nabla[n\, exp\,(-\beta\psi)] . \tag{11}$$

Dividing (11) by $qD_n\, exp\,(\beta\psi)$ and integrating from node i to node j, gives

$$n_j\, exp\,(-\beta\psi_j) - n_i\, exp\,(-\beta\psi_i) = \frac{J_n}{qD_n} \int_i^j exp\,(-\beta\psi) d\xi , \tag{12}$$

where it is assumed that J_n and D_n are constant along the path of integration. Assuming that ψ varies linearly between node i and node j, the right-hand-side of (12) can be rewritten as

$$\frac{J_n l_k \beta}{q\, D_n D_k} \int_{\psi_i}^{\psi_j} exp\,(-\beta\psi) d\psi = \frac{J_n l_k}{q\, D_n \Delta_k}[exp\,(-\beta\psi_i) - exp\,(-\beta\psi_j)] . \tag{13}$$

An expression for the discrete electron current density J_n is obtained by setting the right-hand side of (13) equal to the left-hand side of (12). Multiplying J_n by d_k yields Eq. (9) above. Equation (10) for holes is derived by following a similar procedure.

In a single element, discrete equations for the unknown carrier densities are obtained at each node. Electron current continuity at node i is satisfied by the equation

$$\sum_{\substack{k=\text{edges that} \\ \text{intersect at } i}} I_k = -B_i\left(\frac{\partial n_i}{\partial t} + R_i\right) \tag{14}$$

When (9) is substituted into (14) and the partial derivative of electron density with respect to time is approximated by a first-order finite difference, the following elemental matrix equation is formed:

$$\left([\mathbf{C}] + [\mathbf{B}]\frac{1}{\Delta t}\right)\{n(t + \Delta t)\} = [\mathbf{B}]\left\{\frac{n(t)}{\Delta t} - R_n\right\} , \tag{15}$$

where [**C**] and R_n are evaluated at time t. The components of column vector $\{n(\Delta t + t)\}$ are the unknown electron densities at the nodes of the element, at the next time step. For two-dimensional problems,

$$[\mathbf{C}] = \begin{bmatrix} C_2Z(\Delta_2) + C_3Z(-\Delta_3) & -C_3Z(\Delta_3) & -C_2Z(-\Delta_2) \\ -C_3Z(-\Delta_3) & C_3Z(\Delta_3) + C_1Z(-\Delta_1) & -C_1Z(\Delta_1) \\ -C_2Z(\Delta_2) & -C_1Z(-\Delta_1) & C_1Z(\Delta_1) + C_2Z(-\Delta_2) \end{bmatrix} \tag{16}$$

where $C_k = qD_nd_k/l_k$. For three-dimensional problems, [**C**] is given by

$$\begin{bmatrix} C_2Z(\Delta_2)+C_3Z(-\Delta_3)+C_7(\Delta_7), & C_3Z(\Delta_3), & C_2Z(\Delta_2), & C_7Z(\Delta_7), & 0, & 0 \\ -C_3Z(-\Delta_3), & C_1Z(-\Delta_1)+C_3Z(\Delta_3)+C_8Z(\Delta_8), & -C_1Z(\Delta_1), & 0, & -C_8Z(-\Delta_8), & 0 \\ -C_2Z(\Delta_2), & -C_1Z(-\Delta_1), & C_1Z(\Delta_1)+C_2Z(-\Delta_2)+C_9Z(-\Delta_9), & 0, & 0, & -C_9Z(\Delta_9) \\ -C_7Z(\Delta_7), & 0, & 0, & C_5Z(\Delta_5)+C_6Z(-\Delta_6)+C_7Z(-\Delta_7), & -C_6Z(\Delta_6), & -C_5Z(-\Delta_5) \\ 0, & C_8Z(\Delta_8), & 0, & -C_6Z(-\Delta_6), & C_4Z(-\Delta_4)+C_6Z(\Delta_6)+C_8Z(-\Delta_8), & -C_4Z(\Delta_4) \\ 0, & 0, & -C_9Z(-\Delta_9), & -C_5Z(\Delta_5), & -C_4Z(-\Delta_4), & C_4Z(\Delta_4)+C_5Z(-\Delta_5)+C_9Z(\Delta_9) \end{bmatrix} \tag{17}$$

A similar development is followed for the hole current continuity equations. The complete set of discrete equations are

$$\{F_1\} = [\mathbf{A}]\{\psi\} - [\mathbf{B}]\{p-n + N_D-N_A + N_Q\} , \tag{18}$$

$$\{F_2\} = \left([\mathbf{C}] + [\mathbf{B}]\frac{1}{\Delta t}\right)\{n(t + \Delta t)\} - [\mathbf{B}]\left\{\frac{n(t)}{\Delta t} - R_n\right\} \quad \textit{and} \tag{19}$$

$$\{F_3\} = \left([\mathbf{D}] + [\mathbf{B}]\frac{1}{\Delta t}\right)\{p(t + \Delta t)\} - [\mathbf{B}]\left\{\frac{p(t)}{\Delta t} - R_p\right\} . \tag{20}$$

The hole current continuity matrix [**D**] that appears in Eq. (20) has the same form as [**C**] with Δ_k replaced by $-\Delta_k$ and D_n replaced by D_p.

B. LINEARIZATION

Two different linearization techniques have been implemented to solve Eqs. (18)–(20). The first technique decouples Eq. (18) from (19) and (20), linearizes (18) and solves the three equations serially. The second method solves all three equations simultaneously. Both approaches begin with an estimate of the solution. The initial guess may consist of the results of a previously found solution. Or, if previous results are not available, the potential of each donor or acceptor region is set equal to the potential of the contact associated with that region and the mobile carrier densities are set equal to their space-charge neutral, thermal-equilibrium values.

Potential, distance and charged particle density are transformed to dimensionless quantities in both solution methods. Voltage, length and particle density are normalized by β, intrinsic Debye length and the intrinsic concentration of the semiconductor, respectively.

i) Decoupled Algorithm

The decoupled (Gummel's) algorithm first solves Eq. (18) for ψ by Newton's method. Then, Eqs. (19) and (20) are solved for n and p. These new values of n and p are substituted into (18) and the cycle begins again.

Expanding the residual $\{F_1\}$ in Taylor series about ψ at the kth Poisson-loop iteration, setting $\{F_1(\psi^{k+1})\}$ equal to zero and neglecting higher order terms, gives the elemental matrix equation

$$-\{F_1(\psi^k)\} = ([\mathbf{A}] + [\mathbf{B}]\ \{p^k + n^k\})\ \{\Delta\psi\}, \tag{21}$$

where the increment $\Delta\psi^{k+1} \equiv (\psi^{k+1} - \psi^k)$ and ψ^k is the dimensionless space-charge potential. With the assumption of Boltzmann statistics, the dimensionless mobile charge densities at the kth iteration are given by

$$\{n^k\} = \{\ exp(\psi^k - \phi_n)\}\quad and \tag{22}$$

$$\{p^k\} = \{\ exp(\phi_p - \psi^k)\}\ , \tag{23}$$

where ϕ_n and ϕ_p are the dimensionless electron and hole quasi-Fermi potentials, respectively. The exponentials in (22) and (23) are not explicitly evaluated since overflow may occur. Instead, ϕ_n and ϕ_p are assumed constant during Poisson-loop iterations, and the mobile charge densities at the k+1 iteration are given by

$$\{n^{k+1}\} = \{n^k\ exp(\Delta\psi^{k+1})\} \tag{24}$$

$$\{p^{k+1}\} = \{p^k\ exp(-\Delta\psi^{k+1})\} \tag{25}$$

Although the quasi-Fermi potentials are never explicitly evaluated, they are implicitly updated when new values for n and p are obtained from a solution of (19) and (20). This procedure is continued until a consistent set of ψ, n and p are obtained.

The decoupled solution method is attractive from a developmental point of view since the Poisson and current continuity components of the algorithm can be programmed and tested separately. Also, the three global N×N systems of linear equations that result, where N is the number of free nodes in the model, may be solved independently. However, the computation time required for the convergence of a high-current, highly nonlinear problem may be exceedingly long. Convergence is linear at best. For the simulation of devices operating at high current, the coupled solution algorithm presents a useful alternative.

ii) Coupled Algorithm

The coupled algorithm solves Eqs. (18) – (20) for ψ, n and p, simultaneously. Applying Newton's method to (18) – (20) yields the elemental system of equations

$$\begin{Bmatrix} -F_1 \\ -F_2 \\ -F_3 \end{Bmatrix} = \begin{bmatrix} \frac{\partial F_1}{\partial \psi} & \frac{\partial F_1}{\partial n} & \frac{\partial F_1}{\partial p} \\ \frac{\partial F_2}{\partial \psi} & \frac{\partial F_2}{\partial n} & \frac{\partial F_2}{\partial p} \\ \frac{\partial F_3}{\partial \psi} & \frac{\partial F_3}{\partial n} & \frac{\partial F_3}{\partial p} \end{bmatrix} \begin{Bmatrix} \Delta\psi \\ \Delta n \\ \Delta p \end{Bmatrix} \tag{26}$$

In this expression $\Delta\psi$, Δn and Δp are vectors that represent the changes in potential, and electron and hole densities with each iteration, at the nodes of the finite element. Similarly, the terms in the Jacobian matrix in (26) symbolically represent submatrices of the form

$$\frac{\partial F_i}{\partial \xi} \equiv \frac{\partial(F_{i1}, F_{i2}, \bullet \bullet \bullet, F_{ij})}{\partial(\xi_1, \xi_2, \bullet \bullet \bullet, \xi_j)}, \tag{27}$$

where i=(1,2,3) and j is the number of nodes in the element. The Jacobian matrix in (26) is ninth order for triangular elements and eighteenth order for six-node prismatic elements. The system of equations in (26) are iterated until the residuals F_1, F_2 and F_3 are acceptably small.

The coupled algorithm is attractive because convergence is quadratic. However, it is far more complex and difficult to implement. Also, the coupled algorithm solves a 3N×3N global matrix that requires far more computer memory than the three N×N matrices that are solved separately in Gummel's algorithm. At low current levels, Gummel's algorithm generally reaches a solution more quickly because less computation is required to assemble the matrices. However, at high current levels, Gummel's algorithm may not converge and the coupled approach is necessary.

iii) Accelerated Convergence

In both linearization schemes, the elemental matrices are loaded into large global matrices which must be solved repeatedly. The global matrices usually are ill conditioned. So, a judicious choice of solution method is important. To assure convergence, The FIELDAY program uses a direct solution technique [22].

Most of the computational effort in a simulation is expended in assembling and factoring the global matrix during Newton iterations. In the FIELDAY program, the global Jacobian matrix is evaluated and factored only every m^{th} iteration [23], where m is adjusted as the solution proceeds. For most problems, this technique results in a 30 to 50% savings in computation time. The greatest savings occur in large problems where more time is spent solving the matrix equations.

The optimal value for m may be determined by the following method. Consider a general Newton procedure P(m) in which each evaluation of the Jacobian matrix, or "full" step, is followed by m "quick" steps. The Jacobian matrix is not recomputed during the quick steps. As the process converges asymptotically, the residual Γ at the i+1 full step is $\Gamma_{i+1} \leq h\Gamma_i^{m+2}$, where h is a constant [24]. Suppose one full step is c times more expensive than one quick step. Then, the cost of each "super" step in P(m) is proportional to $c+m$. Since P(m) has $m+2$ order convergence, the number of super steps required for a given level of accuracy is proportional to $1/\log(m+2)$. Therefore, the total cost of solution is proportional to

$$\frac{c+m}{\log\,(m+2)} \,. \tag{28}$$

The optimal value of m minimizes the expression in (28) above. Notice that as c increases, it becomes more economical to perform many quick steps.

In practice, an initial guess may not be close enough to a solution for P(m) to converge asymptotically. Safeguards that monitor convergence have been implemented. If divergent behavior is observed, the quick steps are replaced by full steps.

IV. APPLICATIONS

In this section, three applications of the FIELDAY model are presented. First, the threshold voltages for Insulated-Gate Field-Effect Transistors IGFETs of different channel length are modeled and compared with experimental data. In the second example, the avalanche-breakdown current-voltage characteristics of an IGFET are simulated and also compared with measured data. The final example proposes a design for controlling punchthrough current in ultrasmall FETs.

Example 1
As the channel length of an IGFET is reduced, the surface potential near the source is modulated by the drain bias and the amount of drain-to-source current, at a given gate voltage, is increased. The effect of short channels on IGFET threshold voltage is illustrated by Fig. 4. Devices with channel lengths of 10.0, 2.0, 1.3, 1.0, and 0.7 μm were simulated on a two-dimensional mesh of 1620 nodes. Measurements were made on wide devices with channel lengths that ranged between 10.0 and 0.7 μm. The simulated and measured reductions in threshold voltage that appear in Fig. 4 are relative to the threshold voltage of the 10μm IGFET.

Example 2
At sufficiently high electric-field strength, charge carriers generated by impact ionization contribute to the operating currents of a semiconductor device and may lead to avalanche breakdown. Modeled and experimental data [25] for the avalanche breakdown of an

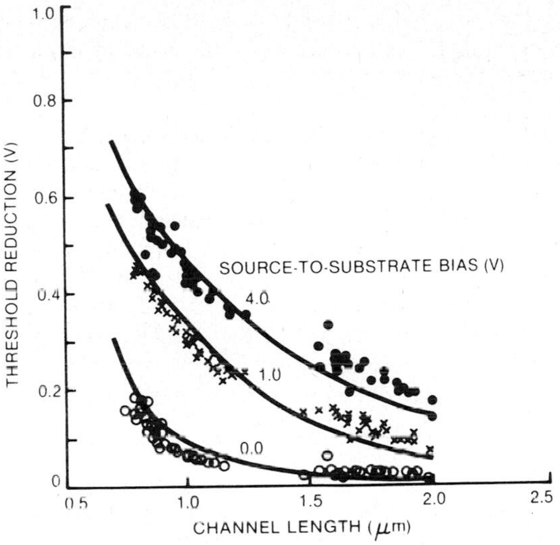

Figure 4. Measured (•, x, o) and simulated (—) reduction in threshold voltage with channel length at source-to-substrate biases of 0.0, 1.0 and 4.0V, and a drain bias of 4.0V. The simulated threshold voltages had been shifted by −0.56V to account for the work function and effective oxide-charge differences between the actual devices and the model [2].

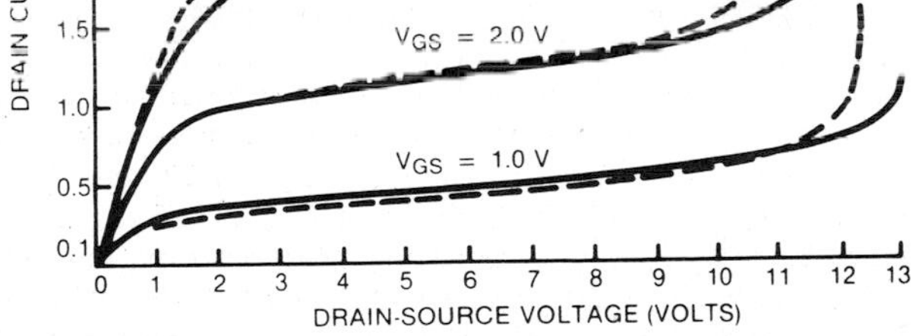

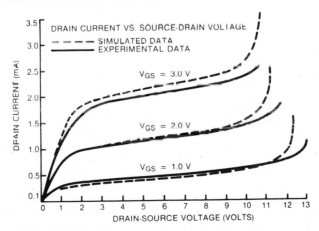

Figure 5: Measured (- - -) and simulated (—) drain current *vs.* drain-to-source voltage at gate-to-source biases V_{GS} of 1.0, 2.0 and 3.0V, and a source-to-substrate bias of −2.0V. The channel length and width are 1.9μm and 30μm, respectively. Significant avalanche-generation of charged carriers occurs at drain-to-source voltages above 10V [25].

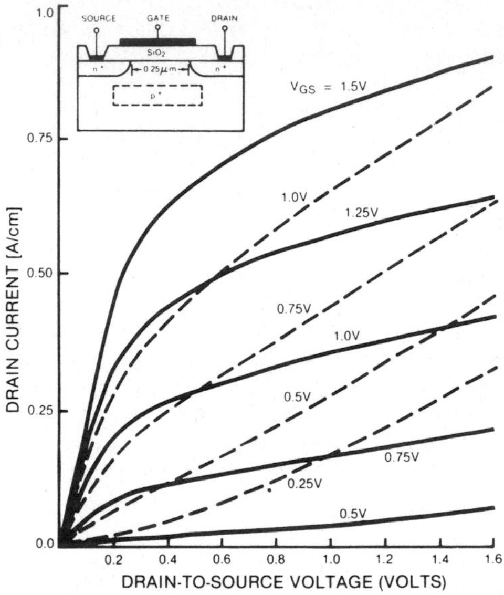

Figure 6. Simulated drain-to-source current-voltage characteristics for $0.25\mu m$ n-channel IGFETs with (—) and without (- - -) buried p^+ layer. Inset: cross-section of modeled IGFET. Oxide thickness = 200Å. Gaussian distributed p^+ implant located $0.47\mu m$ below Si surface.

n-channel IGFET appear in Fig. 5. The channel length and channel width of the device characterized were 1.9 and 30.0 μm, respectively. A two-dimensional model of 435 nodes and 792 elements was used in this simulation.

Example 3
Researchers have fabricated IGFETs that were designed to operate in the punchthrough mode [26,27]. Devices of this type have lower vertical electric fields at their semiconductor surfaces. As a result, they may show improved performance with respect to the injection of hot carriers into the gate insulator and the degradation of the mobility of charge carriers in the channel. However, difficulty in keeping punchthrough devices in a normally-off state for a practical range of drain-to-source voltage, at low gate bias, limit their usefulness as driver transistors. This application of the model shows that punchthrough current in an ultrasmall n-channel IGFET (L=0.25μm) may be controlled by placing a heavily doped p^+ layer underneath the conducting channel [28]. The heavily doped layer is placed deep enough so that the conducting channel at the surface remains punched through. Modeled source-to-drain current-voltage characteristics for IGFETs with and without the heavily doped layer appear in Fig. 6.

V. SUMMARY

A numerical model that simulates the operating behavior of semiconductor devices has been described. The model solves the the semiconductor transport equations in two or three dimensions, for steady-state or transient conditions. The transport equations consist of Poisson's equation and the equations of electron and hole current continuity. Poisson's equation is transformed into a discrete analogue by the (Ritz) finite-element method and the current continuity equations are discretized by a hybrid finite-element finite-difference technique. In these equations, the finite-element volume-distribution scheme required special attention since the the fixed and mobile charge densities in a semiconductor device frequently vary several orders in magnitude over microscopic distances. The discrete equations are solved by either a coupled or decoupled algorithm. The decoupled algorithm is more efficient for weakly nonlinear problems. The coupled algorithm requires significantly more computer memory, but is absolutely necessary for strongly nonlinear, ill-conditioned problems.

Over the past several years, the model has been applied to a wide variety of bipolar and field-effect transistors [2,10,12,18,19]. In this work, three applications of the model were presented. In two of the applications, simulated results were compared with measured data. The close correlation between the simulated and experimental data illustrates the accuracy of the model, the integrity of its computational methods and the credibility of its results.

REFERENCES

[1] H. Kalter, P. Coppens, W. Ellis, P. Heudorfer, T. Leasure, C. Miller, Q. Nguyen, R. Papritz, C. Patton, M. Poplowski and W. Van Der Hoven, "An Experimental 120ns one-half meagbit dynamic RAM with plate push cell," presented at the 1983 Symposium on VLSI Technology, Maui, HA, Sept 1983.

[2] E. M. Buturla, P. E. Cottrell, B. M. Grossman and K. A Salsburg, "Finite-element analysis of semiconductor devices: the FIELDAY program," *IBM J. Res. Develop.* **25**, 218–231 (1981).

[3] W. Van Roosbroeck, "Theory of flow of electrons and holes in germanium and other semiconductors," *Bell Syst. Tech. J.* **29**, 560–562 (1950).

[4] R. J. Van Overstraeten, H. J. DeMan and R. P. Mertens, "Transport equations in heavy doped silicon," *IEEE Trans. Electron Devices* **ED–20**, 290–298 (1973).

[5] D. M. Caughey and R. E. Thomas, "Carrier mobilities in silicon empirically related to doping and field," *Proc. of the IEEE* **55**, 2192–2193 (1967).

[6] C. R. Crowell and S. M. Sze, "Temperature dependence of avalanche multiplication in semiconductors," *Appl. Phys. Lett.* **9**, 242–244 (1966).

[7] J. Dziewior and W. Schmid, "Auger coefficients for highly doped and excited silicon," *Appl. Phys. Lett.* **31**, 346–348 (1977).

[8] R. N. Hall, "Electron-hole recombination in germanium," *Phys. Rev.* **87**, 387 (1952).

[9] W. Shockley and W. T. Read, "Statistics of the recombination of holes and electrons," *Phys. Rev.* **87**, 835–842 (1952).

[10] B. M. Grossman and M. J. Hargrove, "Numerical solution of the semiconductor transport equations with current boundary conditions," *Trans. Electron Devices* **ED–30**, 1092–1096 (1983).

[11] C. R. Crowell and S. M. Sze, "Current transport in metal-semiconductor barriers," *Solid-State Electron.* **9**, 1035–1048 (1966).

[12] P. E. Cottrell and E. M. Buturla, "Two-dimensional static and transient simulation of mobile carrier transport in a semiconductor device," in: B. T. Browne and J. J. H. Miller (eds.), *Numerical Analysis of Semiconductor Devices, Proceedings of the NASECODE I Conference* (Boole Press, Dublin, Ireland, 1979), pp. 31–64.

[13] H. K. Gummel, "A self-consistent iterative scheme for one-dimensional steady state transistor calculations," *IEEE Trans. Electron Devices* **ED–11**, 445–465 (1964).

[14] G. D. Hachtel, M. Mack and R. R. O'Brien, "Semiconductor device analysis via finite elements," *Proceedings of the Eighth Asilomar Conference on Circuits and Systems*, Pacific Grove, CA, 1974.

[15] E. M. Buturla and P. E. Cottrell, "Simulation of semiconductor transport using coupled and decoupled solution techniques," *Solid-State Electron.* **23**, 331–334 (1980).

[16] J. S. Archer, "Consistent matrix formulations for structural analysis using influence coefficient techniques," presented at the First American Institute of Aeronautics and Astronautics Annual Meeting, June 1964, paper No. 64-488.

[17] P. E. Cottrell and E. M. Buturla, "Application of the finite element method to semiconductor transport," presented at the Asilomar Conference on Circuits and Systems, Pacific Grove, CA, 1976.

[18] E. M. Buturla, P. E. Cottrell, B. M. Grossman, M. B. Lawlor, C. T. McMullen and K. A. Salsburg, "Three dimensional simulation of semiconductor devices," *IEEE International Solid State Circuits Conference Digest of Technical Papers*, San Francisco, Feb. 1980, pp. 76–77.

[19] E. M. Buturla, P. E. Cottrell, B. M. Grossman, C. T. McMullen and K. A. Salsburg, "Three-dimensional transient finite element analysis of the semiconductor transport equations," in: B. T. Browne and J. J. H. Miller (eds.), *Numerical Analysis of Semiconductor Devices, Proceedings of the NASECODE II Conference* (Boole Press, Dublin, Ireland, 1981), pp. 160–165.

[20] D. L. Scharfetter and H. K. Gummel, "Large-signal analysis of a silicon read diode oscillator," *IEEE Trans. Electron Devices* **ED–16**, 64–77 (1969).

[21] J. Slotboom, "Mathematical 2-dimensional model of semiconductor devices," *Electron. Lett.* **7**, 47–50 (1971).

[22] International Business Machines Program Product, *Subroutine Library–Mathematics User's Guide*, Order No. SH12–5300–1.

[23] Ortega and Reinbolt, *Iterative Solutions of Nonlinear Equations in Several Variables* (Academic Press, New York, 1970), p. 187.

[24] *Ibid.*, p. 316.

[25] Experimental data obtained by William W. Walker, IBM General Technology Division, Essex Junction, VT.

[26] P. Richman, "Modulation of space-charge limited current flow in insulated-gate field-effect tetrodes," *Trans. Electron Devices* **ED–16**, 759–766 (1969).

[27] T. Nakamura, M. Yamamoto, H. Ishikawa and M. Shinoda, "Submicron channel MOSFET's logic under punchthrough," *IEEE J. Solid-State Circuits* **SC–13**, 572–577 (1978).

[28] L. M. Dang and M. Konaka, "A two-dimensional analysis of triode-like characteristics of short-channel MOSFET's," *IEEE Trans. Electron Devices* **ED–27**, 1533–1539 (1980).

Computing Methods in Applied Sciences and Engineering, VI
R. Glowinski and J.-L. Lions (Editors)
Elsevier Science Publishers B.V. (North-Holland)

SOME RECENT RESULTS AND OPEN QUESTIONS IN NUMERICAL SIMULATION OF SEMICONDUCTOR DEVICES

Michael S. Mock

Department of Applied Mathematics
The Hebrew University
Jerusalem, Israel

The present state of the art in numerical models of semiconductor devices is surveyed, with emphasis on the discretization of the stationary continuity equations and methods for solving the discrete stationary problem. These are the problems which are likely to cause the most trouble in the development of useful codes for this problem in three space dimensions, as are becoming of increased interest in applications.

A new result in this area is announced, on a proof of convergence of the "Scharfetter-Gummel" method in an arbitrary number of dimensions.

1. Introduction

The small size and relative complexity of modern semiconductor devices and integrated circuits have made the classical methods of analysis of these devices increasingly unsatisfactory. In the design of such devices, increased dependence is being placed on the results of numerical simulation. The mathematical model on which most of these simulations are based is essentially that proposed by von Roosbroeck in 1950 [9], although a similar system had been previously studied in the context of the flow of ions in a solution. Let $\Omega \subset R^n$, $n = 1,2,$ or 3, be a bounded open region with piecewise smooth boundary $\partial\Omega$. In a system of units in which q, the magnitude of the electron charge, kT/q, the Boltzmann voltage, and n_i, the intrinsic carrier density, are set equal to unity, our model may be written in the form

(1.1) $$\nabla \cdot (\varepsilon \nabla \psi) = u - v - N,$$

(1.2) $$\frac{\partial u}{\partial t} = \nabla \cdot j_u - R,$$

(1.3) $$\frac{\partial v}{\partial t} = - \nabla \cdot j_v - R,$$

(1.4) $$j_u = \mu_u (\nabla u - u\nabla\psi - u \nabla \log n_{i,e}),$$

(1.5) $j_v = -\mu_v(\nabla v + v\nabla\psi - v\nabla \log n_{i,e})$.

In (1.1-1.5), the dependent variables are u, the local density of free (conduction band) electrons, v, the local density of valence band holes (which act like free, positively charged particles), and ψ, the electrostatic potential; these are functions of $x \in \Omega$, $t \geq 0$. The electron and hole current densities are denoted respectively by j_u, j_v. This system is supplemented, of course, by expressions for the various other quantities appearing: the dielectric constant ε, the net doping function N, and the effective intrinsic carrier density $n_{i,e}$ are typically given functions of x; the carrier mobilities μ_u, μ_v functions of x, the electric field $-\nabla\psi$, and possibly also the carrier densities u,v; and the net recombination R a function of u,v, and possibly also the electric field. However, the details of these expressions are not important for the present discussion and will be left unspecified. Similarly, the details of the boundary conditions need not be elaborated here; it suffices to say the Dirichlet conditions for u,v,ψ, are prescribed on part of the boundary $\partial\Omega$, corresponding to the contacts to the device, and homogeneous Neumann boundary conditions generally applied on the remainder of the boundary.

A change of variable is often made which puts the stationary continuity equations (1.2, 1.3) in symmetric form. Introducing the quasi-Fermi potentials ϕ_u, ϕ_v and their exponentials η, ρ by the relations

(1.6) $u = n_{i,e} \exp(\psi - \phi_u) = n_{i,e} \exp(\psi)\,\eta$,

(1.7) $v = n_{i,e} \exp(\phi_v - \psi) = n_{i,e} \exp(-\psi)\rho$,

the expressions for the current densities may be put in the form

(1.8) $j_u = -\mu_u\, u\, \nabla\phi_u = \mu_u\, n_{i,e} \exp(\psi)\nabla\eta$,

(1.9) $j_v = -\mu_v\, v\, \nabla\phi_v = -\mu_v\, n_{i,e} \exp(-\psi)\nabla\rho$.

2. Discretization of the stationary problem

The stationary problem associated with the system (1.1-1.5), obtained by setting the time derivative terms in (1.2, 1.3) equal to zero, has been the more important from the point of view of application of the results. For the various parameters in their regions of physical interest, this problem is uniformly elliptic, but rather strongly nonlinear. Based on the existence of estimates in L_p for u,v, the existence of reasonably smooth stationary solutions is readily proved [4], chapter 2. Thus the questions immediately confronting numerical analysts are how to discretize the system (1.1-1.5), and how to solve the discrete nonlinear system which results from such a discretization. Although both of these questions have been intensively studied since the mid-sixties, in neither case can the situation be called satisfactory.

The trouble in discretizing this system is that many of the variables which appear, including u,v,n,ρ, $\exp(\pm\psi)$, and often also the quasi-Fermi potentials ϕ_u,ϕ_v, vary so rapidly that their relative variations cannot be reasonably resolved on a mesh of practical size. Thus although finite-difference or finite-element methods can readily be applied to this problem, the associated asymptotic error estimates are not realistic, and the results obtained can be quite disappointing. This has often been confirmed by experiment.

The Poisson equation (1.1) is clearly not the major source of trouble. Both standard second order finite-difference methods and finite-element methods with piecewise linear basis functions have routinely been successfully applied to this equation. The finite-element treatment is theoretically preferable, at least in higher dimensions, because of less required regularity of the electrostatic potential function ψ. At the discrete level, there are two differences between the finite-difference and finite-element formulations of the Poisson equation: one is in the discrete form of the derivative term, and one is the difference between the use of point-values of the carrier densities and doping function and the use of appropriate moments of these functions. As we shall see below, it is the latter difference which is most important.

The stationary continuity equations are not as easily discretized, because of the large relative variation of the dependent variables. Nevertheless, it was shown in [8] that in one dimension, in the absence of recombination, assuming constant carrier mobilities and piecewise linear electrostatic potential and $\log n_{i,e}$ functions that the continuity equations can be solved exactly. In this case the electron continuity equation may be written

$$(2.1)\qquad \frac{d}{dx}\left(e^{\psi'}\frac{d\eta}{dx}\right) = 0\ ,\quad \psi' = \psi + \log n_{i,e}$$

and the discrete form, on a uniform mesh of size h, is

$$(2.2)\qquad \frac{1}{h}\left[f(\psi'_{i+1},\psi'_i)\left(\frac{\eta_{i+1}-\eta_i}{h}\right) - f(\psi'_i,\psi'_{i-1})\left(\frac{\eta_i-\eta_{i-1}}{h}\right)\right] = 0,$$

$$i = 2,3,\ldots,M-1$$

with

$$(2.3)\qquad f(y,z) = \frac{e^{-y}-e^{-z}}{z-y}\ ,\quad y \neq z\ ;\quad f(y,y) = e^{y}.$$

In (2.2), η_i,ψ'_i are the approximate values of η,ψ', respectively, at the point $x_i = (i-1)h$. Boundary values are applied for $i = 1$ and $i = M$. Of course, an entirely analogous equation is used for the holes.

In practice, the carrier mobilities μ_u,μ_v and the effective intrinsic carrier density $n_{i,e}$ are expected to be relatively slowly varying functions, and the effect of the recombination term is often not large. Thus the generalization of (2.2) to the cases where the mobility is variable and recombination is included is easily made.

From (2.2), we obtain point-values of the variable η, or of the electron density via $u_i = \exp(\psi'_i)\eta_i$. However, in this case the values of η and ψ are also known between the mesh points, so that appropriate moments of the carrier densities can also be computed. For example, the moment of the electron density with the "tent function" of unit area centered at x_i, denoted by Λ_i, is given by

$$(2.4) \qquad \langle u, \Lambda_i \rangle = u_i + (u_{i+1} - u_i)\, g(\psi'_{i+1} - \psi'_i) + (u_{i-1} - u_i)\, g(\psi'_{i-1} - \psi'_i)$$

with $\langle , \rangle$ denoting the real L_2 inner product Ω over and $g(\cdot)$ given by

$$(2.5) \qquad g(z) = \frac{e^z - 1 - z - \frac{1}{2} z^2}{z^2 (e^z - 1)}, \quad z \neq 0, \; g(0) = 1/6.$$

A complete discussion of this method is given in [4], Chapter 3. This discussion concludes with an experiment designed to test the importance of computing carrier density moments for use in the Poisson equation. In this experiment, the results of three methods are compared, as applied to a forward biased p-n junction with specially computed doping profile such that the exact solution is known in closed form for comparison purposes. In each method, the continuity equations are discretized by the method (2.2). In one method, point-values of the carrier densities are used in the Poisson equation. In the second method, a three-point, "Simpson's rule" average is used for the carrier densities, as would be obtained from (2.4) if ψ' were constant. In the third method, the carrier density moments are computed according to (2.4), and the corresponding expression for holes. (Throughout this discussion and in general, the corresponding expressions for holes can be obtained from the transformation $u \leftrightarrow v$, $\psi \leftrightarrow -\psi$, applied to the expressions for electrons.) In this test problem, virtually identical results were obtained with the first two methods, while the third method was substantially superior, with some of the errors a factor of ten smaller than were obtained with the other two methods. We conjecture that the importance of using consistent carrier density moments is likely to increase further in higher dimensions, because of increased growth of the higher derivatives of the electrostatic potential. Unfortunately, to date no such experiments have been made.

The interpretation of the Scharfetter-Gummel scheme (2.2) as given above is obviously valid only in one dimension. Nevertheless, the obvious generalization of this method to higher dimensions has been used with considerable, perhaps remarkable, sucess. A finite-difference analysis of this method in two dimensions, c.f. [4], leads to an overly pessimistic result: the method can be expected to give accurate results only if the flow of current is essentially in one of the grid directions in regions where large changes occur in the values of electrostatic potential between neighboring mesh points. Thus until fairly recently, an explanation for the apparent success of this method in higher dimensions was an open question.

A much more satisfactory analysis of this method is now available [7]. In two dimensions (we describe here only the case of zero recombination for simplicity), the Scharfetter-Gummel method may be though of as part of the discrete system obtained from a projection method discretization of the first order system

$$\nabla\eta + \frac{\exp(-\psi')}{\mu_u} \nabla \times \theta = 0 . \tag{2.6}$$

In the discretization of (2.6), the stream function θ is approximated by a piecewise constant function, with appropriate divided differences replacing the indicated derivatives. The accuracy of the method is seen to depend essentially only on the accuracy of the approximation of the stream function, in a relatively weak topology. As the physical current density, and hence the stream function, are expected to vary much less rapidly than say, the carrier density, this is a satisfactory explanation of the accuracy of the method. The accuracy of the method is first or second order in the mesh size, depending on how the finite elements are chosen and in particular how the boundary is treated.

We wish to announce here the existence of a proof of the accuracy of the Scharfetter-Gummel method in arbitrary dimensions. The argument is a generalization of that given in [7], and depends on the weak approximation of the current density function from a finite-dimensional space. It is not clear whether accuracy of more than first order in the mesh size can be established by this analysis, but at least some ambiguity in the application of the method, as occurs for example with the use of tetrahedral elements, is removed.

Thus given an approximation to the electrostatic potential function, the

Scharfetter-Gummel method is a reasonable procedure for obtaining point-values of the carrier densities. The necessity of identifying the values obtained as point-values, as opposed to local averages, is very clear in the above mentioned analysis, and we have seen that in a solution of the full system (1.1-1.5) it is moments of the carrier densities which are needed and not point-values.

How to reasonably accomplish this is a major open question at the present time. Below we present three possible approaches, all of which are in need of testing and possible further development.

We first note from (1.8, 1.9) that in order to compute the carrier density moments, it suffices to obtain the respective current densities. Let γ be a given test function; to compute $\langle u,\gamma\rangle$ we first find a function ξ such that

(2.7) $\nabla\cdot\xi = \gamma e^{\psi'}\ ,\quad x\in\Omega\ ,\quad \nu\cdot\xi = 0\ ,\quad x\in\partial\Omega_{ins},$

where ν is the outward normal vector and $\partial\Omega_{ins}$ is that part of the boundary on which Neumann boundary conditions are applied.

(2.8) $$\langle u,\gamma\rangle = \langle\eta,\nabla\cdot\xi\rangle = \int_{\partial\Omega}\eta\nu\cdot\xi\, dS - \langle\nabla\eta,\xi\rangle = \int_{\partial\Omega}\eta\nu\cdot\xi\, dS - \langle\frac{e^{-\psi'}}{\mu_u} j_u,\xi\rangle\ .$$

We note that given γ, (2.7) does not determine ξ uniquely; appropriate methods for choosing these functions are discussed in [4], chapter 3.

There are two fairly obvious methods for determing approximations to the current density functions. One method, which is particularly attractive when recombination can be ignored (extensions of this exist when recombination is important, c.f. [4], [7]) is to discretize the stream function equation

(2.9) $$\nabla\times\left(\frac{\exp(-\psi')}{\mu u}\nabla\times\theta\right) = 0$$

which follows immediately from (2.6). In two dimensions this leads to a scalar, second order elliptic system, which is an excellent candidate for discretization by the finite-element method, as the dependent variable θ is relatively slowly varying, whereas the coefficient $\exp(-\psi')$ is very rapidly varying.

In three dimensions, we can choose a finite-dimensional space of vector-valued, divergence-free functions, denoted by X, and obtain our approximation θ_h by Galerkin's method: we find θ_h in X, satisfying the essential boundary conditions for θ (c.f. [4], chapter 2) and such that

$$(2.10) \qquad < \frac{\exp(-\psi')}{\mu_u} \nabla\times\theta_h, \nabla\times\alpha > \; = 0$$

for all $\alpha \in X$ satisfying the homogeneous form of the essential boundary conditions for θ. An argument analogous to the uniqueness proof for the stream function in three dimensions as given in [4] then shows that the magnitude of the error $\theta - \theta_h$ will then be dependent simply on the approximation properties of the space X.

A second method, computationally much simpler but probably somewhat less accurate than that of determining the stream functions, is to note that the Scharfetter-Gummel procedure, in addition to giving point-values of the carrier densities, gives approximate values of the current density. Indeed, this is the basic error estimate obtained in [7]. The effect of recombination is also automatically included in this procedure. These discrete values of the current density can obviously be extended by interpolation, and the resulting current density function used in (2.8). Such an interpolation is justified, in principle, on grounds of the relative smoothness of the current density.

A third approach is to approximate the quasi-Fermi potentials ϕ_u, ϕ_v from a finite-dimensional space, as well as the electrostatic potential ψ. Then an approximation to the carrier densities is available everywhere, so the desired moments are easily computed. There are, however, two potentially serious difficulties with this method: one is the accuracy with which the quasi-Fermi potentials can be so approximated, particularly in higher dimensions; the second is related to the method for solving the discrete system so obtained, as discussed in the following section.

3. Solution of Discrete Systems

Discretization of the stationary problem for the system (1.1-1.5) leads to a large discrete nonlinear system. In practice, several hundred or several thousand solutions of such systems, corresponding to variations in the various parameters, are required in a serious study of the behavior of a given device.

Thus reasonably fast solution algorithms are needed, but to date there has been relatively little progress in this area. Below we shall briefly describe the methods which are presently in use, and offer an explanation of the difficulty of finding better methods.

Despite of the obvious inefficiency in more than one space dimension, Newton's method with some form of parametrization is often used to solve the full, coupled, discrete stationary system. This is particularly true for the so-called finite-element codes, in which the stencil corresponding to any one of the three basic equations would be irregular in any event.

For many or most problems, a more efficient method is one pioneered by H. Gummel [1]. This is also an iterative method: given an approximation to the electrostatic potential distribution, denoted by ψ_m, the corresponding carrier density distributions u_m, v_m are found from

$$(3.1) \qquad u_m = n_{i,e}\, e^{\psi_m} \eta_m, \quad v_m = n_{i,e}\, e^{-\psi_m} \rho_m$$

$$(3.2) \qquad \nabla \cdot (\mu_u n_{i,e}\, e^{\psi_m} \nabla \eta_m) = R(u_{m-1}, v_{m-1}) + (u_m - u_{m-1}) \frac{\partial R}{\partial u} (u_{m-1}, v_{m-1})$$

$$(3.3) \qquad \nabla \cdot (\mu_v n_{i,e}\, e^{-\psi_m} \nabla \rho_m) = R(u_{m-1}, v_{m-1}) + (v_m - v_{m-1}) \frac{\partial R}{\partial v} (u_{m-1}, v_{m-1})$$

and the boundary conditions for the carrier densities. Then the "improved" approximation to the electrostatic potential ψ_{m+1} is obtain from

$$(3.4) \qquad \nabla \cdot (\varepsilon \nabla \psi_{m+1}) = (u_m + v_m)(\psi_{m+1} - \psi_m) = u_m - v_m - N$$

and the boundary conditions for the electrostatic potential.

The separate, sequential solution of the three basic equations in Gummel's method results in a substantial improvement in computational efficiency with respect to Newton's method. More itations are often required with Gummel's method than with Newton's, for the simple reason that the asymptotic rate of convergence of Gummel's method is only linear. As against that, in practice Gummel's method generally converges with a much poorer "initial guess" than would be required for Newton's method, thus reducing the need for parametrized sequences of solutions. An additional advantage of Gummel's method arises when additional variables are introduced, for example when the continuity equations are formulated and discretized in terms of the stream functions.

A partial explanation for the success of Gummel's method is given by some results concerning monotone operators [11]. We consider the mapping of ψ_m into $u_m - v_m - N$, determined by the solution of (3.1 - 3.3). Under conditions close to equilibrium and ignoring recombination, this mapping is provably monotone in the sense [4] that

$$\text{(3.5)} \qquad \langle \psi_\alpha - \psi_\beta \, , \; u_\alpha - v_\beta - u_\beta + v_\beta \rangle \geq 0$$

where ψ_α, ψ_β are two approximations to the electrostatic potential, each satisfying the appropriate boundary conditions, and $u_\alpha, v_\alpha, u_\beta, v_\beta$ are the corresponding carrier densities. (The equilibrium situation corresponds to each of the quasi-Fermi potentials ϕ_u, ϕ_v constant throughout the device, so that the carrier densities are simply exponentials of the electrostatic potential.)

Let Y be a subset of $C(\Omega)$ containing a function ψ_α, corresponding to a solution of the stationary problem, and such that (3.5) holds for all ψ_β within Y. It follows [11] that the solution function ψ_α is unique within Y, and that, with some assumptions about the precise shape of Y, an iteration of "contracting average" type will converge to the solution for any initial approximation within Y. Gummel's method may be regarded as a method of this type; thus we have an explanation of the apparent global convergence property. Nevertheless, two questions remain open in this area: the region Y is apparently very much larger than the proof in [4] indicates; although some physical arguments can be given for this, a satisfactory justification is not known. Secondly, there is the choice of the "stabilizing factor" $(u_m + v_m)$ in (3.4); convergence is expected for this factor sufficiently large, but the expected rate of convergence deteriorates as this factor is increased more than necessary. This choice is much more successful than might be expected; again a satisfactory mathematical justification is unavailable. The equation (3.4) was obtained in [1] by considering the linearization of the Poisson equation, in the form

$$\nabla \cdot (\varepsilon \nabla \psi) = n_{i,e}\, \eta\, e^{\psi} - n_{i,e}\, \rho\, e^{-\psi} - N, \tag{3.6}$$

and ignoring the variations in η and ρ. Indeed, under equilibrium conditions with η and ρ simply constants, Gummel's method and Newton's method for this problem coincide.

To see that there is something special about Gummel's method, that it's convergence is not simply a question of "weak coupling" between the equations, it is helpful to consider the following alternative type of iteration scheme: given approximations ψ_m, u_m, v_m, we first find "improved" approximations to the quasi-Fermi potentials $\phi_{u,m+1}, \phi_{v,m+1}$ from relations of the form

$$\nabla \cdot (\mu_u\, u_m\, \nabla \phi_{u,m+1}) = -R, \tag{3.7}$$

$$\nabla \cdot (\mu_v\, v_m\, \nabla \phi_{v,m+1}) = R, \tag{3.8}$$

then find ψ_{m+1} from (3.4) or some suitable modification, and finally find u_{m+1}, v_{m+1} from

$$u_{m+1} = n_{i,e} \exp(\psi_{m+1} - \phi_{u,m+1}), \quad v_{m+1} = n_{i,e} \exp(\phi_{v,m+1} - \psi_{m+1}). \tag{3.9}$$

Numerous experiments of this type have been done. Although the boundedness of the iterates is readily obtained, even local convergence fails except under the most special conditions.

This is also one reason why the quasi-Fermi potentials ϕ_u, ϕ_v are not often approximated from a finite-dimensional space. If the discretization were accomplished in such a manner, then the solution of (3.2) or (3.3) for η_m, ρ_m, respectively, would still require the solution of a nonlinear system within each Gummel iteration.

There are, of course, examples of problems for which Gummel's method is completely unsatisfactory. A recently encountered example is that of problems with "floating" bulk neutral regions, i.e. regions in which $|N|$ is very large, which are not connected to an ohmic contact, so that the majority carrier quasi-Fermi potential is determined by the flow of current into and out of the region. The solution of such a problem in two dimensions, in the context of a model of a silicon-on-saphire (SOS) device, is reported in [10], having been accomplished by integrating the time-dependent equations towards infinity. Such a problem in one dimension was also solved in [5], by using the result of several hundred Gummel iterations as the initial approximation for a Newton's method computation. Obviously both of these methods are of quite limited applicability, in view of the computation inefficiency.

We close this section with a discussion of why good iterative methods for this problem have been so hard to find. A proof of convergence in the large for an iteration scheme implies uniqueness in the large of the solution. Indeed, often in proving uniqueness, as opposed to existence, of the solution one obtains a suggestion of a method for computing the solution. Thus we could reasonably expect uniqueness results for this problem to precede the finding of iterative methods with good convergence properties. Indeed, for bias conditions sufficiently close to equilibrium (both quasi-Fermi potentials very nearly constant) uniqueness results follow immediately from the monotone condition [3.5] [4], and this provides an explanation for the convergence of Gummel's method.

Unfortunately, except under conditions close to equilibrium, very little is known about the uniqueness of the stationary solution for the system (1.1-1.5), and what is known is mostly negative results. In principle, the region Ω can contain not just one device, but an entire integrated circuit, with any number of stable stationary states. Thus one cannot expect any uniqueness result without a major assumption to simplify the geometry of the problem considered. In this respect, the assumption of a problem in only one dimension is very useful mathematically, as the current densities are then constant except for the effects of recombination, and it is clear on physical grounds that the multiple, stationary states of an integrated circuit are intimately related to the spatial variations of the flow of current.

Avalanche generation in the recombination expression is known to cause multiple solutions, so we must simplify the expression for R; for simplicity we shall ignore recombination entirely. Similarly, the dependence of the carrier mobilities on electric field can cause multiple solutions, so we shall assume constant mobilities for simplicity. For additional simplicity, we shall take ε constant and $n_{i,e}$ to be equal to unity as well. Thus we are led to consider the nonlinear two-point boundary value problem

$$(3.10) \qquad \varepsilon \frac{d^2\psi}{dx^2} = \eta e^{\psi} - \rho e^{-\psi} - N,$$

$$(3.11) \qquad \frac{d}{dx}\left(e^{\psi} \frac{d\eta}{dx}\right) = 0,$$

$$(3.12) \qquad \frac{d}{dx}\left(e^{-\psi} \frac{d\rho}{dx}\right) = 0 , \quad 0 < x < L;$$

$$(3.13) \qquad \eta(0) = \rho(0) = 1,\ \eta(L),\ \rho(L),\ \psi(0),\ \psi(L),\ N(x),\quad 0 < x < L \text{ specified},$$

with the specified values of $\eta(L)$ and $\rho(L)$ strictly positive. For this problem, it is proved in [4], chapter 2, that the solution is unique, provided that either $|\log \eta(L)|$ or $|\log \rho(L)|$ does not exceed log 2, so that the device can be described as "almost unipolar", i.e. current flow is predominantly of one type.

The proof of this theorem is not especially difficult. Since the solution of the corresponding equilibrium problem, with $\eta(L) = \rho(L) = 1$, is obviously unique, it follows that if multiple, regular solutions exist then there must be a critical solution for some smaller values of $|\log \eta(L)|$, $|\log \rho(L)|$. Some work is then done to show that this is impossible. We note in passing that from this proof, it follows that this problem can reliably be solved by Newton's method, letting $|\log \eta(L)|$ and $|\log \rho(L)|$ take on a sequence of values beginning at zero. Our primary concern, however, is whether this final assumption is necessary, for it is obviously seriously restrictive. The answer, qualitatively, is yes; in [5], an example of a problem of the type (3.10-3.13) is constructed for which multiple solutions exist for $\log \rho(L) = -\log \eta(L)$ above a relatively modest value in the context of physical devices. Thus we are not optimistic about positive uniqueness results for this problem.

4. The Time-Dependent Problem

There is now an increasing interest in numerical computations for the time-dependent problem (1.1-1.5), resulting from an interest in simulating specific, transient phenomena: the effects of parasitic devices in integrated circuits, the effects of perturbations such as those caused by ionizing radiation, etc. Such computations (in one space dimension) have been performed since the late sixties. It is well known that the system (1.1-1.5) can be numerically integrated by using pure, backward time differences in (1.2, 1.3) and solving a coupled, nonlinear discrete system for the updated variables ψ, u, v, simultaneously at each time step. Of course, in more than one space dimension, this is an inefficient procedure, for the same reason that Newton's method is an inefficient method for the stationary problem.

At first glance, it appears that the system (1.1-1.5) is easily separated for purposes of numerical integration, for example by solving (1.2, 1.4) for the updated electron density (using the latest available values for the electrostatic potential and the hole density functions), then solving (1.3, 1.5) for the updated hole density, and finally solving (1.1) for the updated electrostatic potential. Each step in this process clearly corresponds to solving a well-posed problem, and each step separately can easily be made stable, for example by using backward time differences in (1.2, 1.3). However, the coupled system suffers from a serious stiffness problem; all such procedures are unstable unless the time step Δt satisfies (c.f. [4], chapter 5)

$$\Delta t \leq 2\varepsilon/(\mu_u u + \mu_v v), \tag{4.1}$$

which in general is a prohibitively severe restriction.

Nevertheless, a number of methods are known [4] which permit the separate solution of the various equations while maintaining unconditional stability. Perhaps not surprisingly, all of these methods are based on modifications of the Poisson equation (1.1), as used for the computations. At present, the most promising such method is as follows: differentiating (1.1) with respect to time and using (1.2-1.5) to replace the time derivative terms, we obtain an evolution equation for the electrostatic potential.

$$(4.2)\qquad \nabla\cdot\left(\varepsilon\nabla\frac{\partial\psi}{\partial t}\right) + \nabla\cdot((\mu_u u+\mu_v v)\nabla\psi) - \nabla\cdot(\mu_u \nabla u-\mu_v \nabla v)$$

$$+ \nabla\cdot((\mu_u u-\mu_v v)\nabla \log n_{i,e}) = 0.$$

There is, of course, a physical significance to this equation; it is an expression of conservation of total current, including displacement current. It is not difficult to show that if the functions ψ,u,v, satisfy (1.2-1.5, 4.2) and if the Poisson equation (1.1) is satisfied at the initial time point, then it will remain satisfied. Furthermore, (4.2) is a better equation for discretization, from the point of view of stability, than is (1.1). Indeed, the procedure of solving (1.2, 1.4) for the updated electron density, then solving (1.3, 1.5) for the updated hole density (using backward time differences in both cases), and then finding the updated electrostatic potential ψ_m from

$$(4.3)\qquad \nabla\cdot\left(\varepsilon\,\frac{\nabla\psi_m-\nabla\psi_{m-1}}{\Delta t}\right) + \nabla\cdot((\mu_u u_m+\mu_v v_m)\nabla\psi_m) - \nabla\cdot(\mu_u\nabla u_m-\mu_v\nabla v_m)$$

$$+ \nabla\cdot((\mu_u u_m - \mu_v v_m)\ \nabla\log n_{i,e}) = 0,$$

is stable, independently of the value of the time step [2,4]. (In (4.3), subscripts $m,m-1$ denote time level; the functions u_m and v_m are available for use in (4.3), having been determined in the first two steps.)

In addition to being stable, this procedure is efficiently applied to circuits containing several devices, coupled together with conventional circuit elements [2,3,4]. Some refinements of this basic method, aimed at reducing the accumulated error in the Poisson equation, have been incorporated into an operational computer code as described in [6].

5. References

1. H. K. Gummel, A self-consistent iterative scheme for one-dimensional steady state transistor calculations, IEEE Trans. Elec. Dev. ED-11 (1964), 455-465.
2. M. S. Mock, A time-dependent numerical model of the insulated-gate field-effect transistor, Solid State Electronics 24 (1981), 959-966.
3. M. S. Mock, Time-dependent simulation of coupled devices, in: B.T. Browne and J.J.H. Miller, eds., Numerical analysis of semiconductor devices and intedrated circuits (Boole Press, Dublin, 1981).
4. M. S. Mock, Analysis of mathematical models of semiconductor devices (Boole Press, Dublin, 1983).
5. M. S. Mock, An example of nonuniqueness of stationary solutions in semiconductor device models, Compel 1 (1982), 165-174.
6. M. S. Mock, Time-dependent simulation of integrated circuits, in: B.T. Browne and J.J.H. Miller, eds., Proceedings of the Nasecode III conference (Boole Press, Dublin, 1983).
7. M. S. Mock, Analysis of a discretization algorithm for stationary continuity equations in semiconductor device models, Compel 2 (1983)
8. D. L. Scharfetter and H.K. Gummel, Large signal analysis of a silicon Read diode oscillator, IEEE Trans. Elec. Dev. ED-16 (1969), 64-77.
9. W. von Roosbroeck, Theory of flow of electrons and holes in germanium and other semiconductors, Bell Sys. Tech. J. 29 (1950), 560-607.
10. K. Yamaguchi, in: B. T. Browne and J.J.H. Miller, eds., Proceedings of the Nasecode III conference (Boole Press, Dublin, 1983).
11. E. H. Zarntonello, Solving functional equations by contractive averaging, Math. Research Center Technical Report No. 160, Madison, Wisc., (June 1960).